Organic Reactions

ADVISORY BOARD

John E. Baldwin	Andrew S. Kende
Peter Beak	Steven V. Ley
Virgil Boekelheide	James A. Marshall
George A. Boswell, Jr.	Blaine C. McKusick
Engelbert Ciganek	Jerrold Meinwald
Donald J. Cram	Leo A. Paquette
David Y. Curtin	Gary H. Posner
Samuel Danishefsky	Hans J. Reich
Heinz W. Gschwend	Charles Sih
Stephen Hanessian	Barry M. Trost
Ralph F. Hirschmann	Milán Uskokovic
Herbert O. House	James D. White
Robert C. Kelly	

FORMER MEMBERS OF THE BOARD NOW DECEASED

Roger Adams	Louis F. Fieser
Homer Adkins	John R. Johnson
Werner E. Bachmann	Willy Leimgruber
A. H. Blatt	Frank C. McGrew
Theodore L. Cairns	Carl Niemann
Arthur C. Cope	Harold R. Snyder
William G. Dauben	Boris Weinstein

Organic Reactions

VOLUME 57

EDITORIAL BOARD

LARRY E. OVERMAN, *Editor-in-Chief*

ANDRÉ CHARETTE STUART W. MCCOMBIE
DENNIS CURRAN T. V. RAJANBABU
SCOTT E. DENMARK JAMES H. RIGBY
VITTORIO FARINA WILLIAM R. ROUSH
LOUIS HEGEDUS AMOS B. SMITH, III
LAURA KIESSLING PETER WIPF
MICHAEL J. MARTINELLI

ROBERT BITTMAN, *Secretary*
Queens College of The City University of New York, Flushing, New York

JEFFERY B. PRESS, *Secretary*
Brewster, New York

EDITORIAL COORDINATORS

ROBERT M. JOYCE
Sun City Center, Florida

LINDA S. PRESS
Brewster, New York

ASSOCIATE EDITORS

EVAN G. ANTOULINAKIS
HUW M. L. DAVIES
ROBERT M. MORIARTY
OM PRAKASH
ALBERT M. VAN LEUSEN
DAAN VAN LEUSEN

JOHN WILEY & SONS, INC.

New York • Chichester • Weinheim • Brisbane • Singapore • Toronto

This text is printed on acid-free paper.

Published by John Wiley & Sons, Inc.

Copyright © 2001 by Organic Reactions, Inc. All rights reserved.

Published simultaneously in Canada.

No part of this publication may be reproduced, stored in a retrieval system or transmitted in any form or by any means, electronic, mechanical, photocopying, recording, scanning or otherwise, except as permitted under Sections 107 or 108 of the 1976 United States Copyright Act, without either the prior written permission of the Publisher, or authorization through payment of the appropriate per-copy fee to the Copyright Clearance Center, 222 Rosewood Drive, Danvers, MA 01923, (978)750-8400, fax (978)750-4744. Requests to the Publisher for permission should be addressed to the Permissions Department, John Wiley & Sons, Inc., 605 Third Avenue, New York, NY 10158-0012, (212)850-6011, fax (212)850-6008, E-Mail: PERMREQ@WILEY.COM.

For ordering and customer service, call 1-800-CALL-WILEY.

Library of Congress Catalog Card Number 42-20265

ISBN 0-471-43511-2

Printed in the United States of America

10 9 8 7 6 5 4 3 2 1

PREFACE TO THE SERIES

In the course of nearly every program of research in organic chemistry the investigator finds it necessary to use several of the better-known synthetic reactions. To discover the optimum conditions for the application of even the most familiar one to a compound not previously subjected to the reaction often requires an extensive search of the literature; even then a series of experiments may be necessary. When the results of the investigation are published, the synthesis, which may have required months of work, is usually described without comment. The background of knowledge and experience gained in the literature search and experimentation is thus lost to those who subsequently have occasion to apply the general method. The student of preparative organic chemistry faces similar difficulties. The textbooks and laboratory manuals furnish numerous examples of the application of various syntheses, but only rarely do they convey an accurate conception of the scope and usefulness of the processes.

For many years American organic chemists have discussed these problems. The plan of compiling critical discussions of the more important reactions thus was evolved. The volumes of *Organic Reactions* are collections of chapters each devoted to a single reaction, or a definite phase of a reaction, of wide applicability. The authors have had experience with the processes surveyed. The subjects are presented from the preparative viewpoint, and particular attention is given to limitations, interfering influences, effects of structure, and the selection of experimental techniques. Each chapter includes several detailed procedures illustrating the significant modifications of the method. Most of these procedures have been found satisfactory by the author or one of the editors, but unlike those in *Organic Syntheses* they have not been subjected to careful testing in two or more laboratories.

Each chapter contains tables that include all the examples of the reaction under consideration that the author has been able to find. It is inevitable, however, that in the search of the literature some examples will be missed, especially when the reaction is used as one step in an extended synthesis. Nevertheless, the investigator will be able to use the tables and their accompanying bibliographies in place of most or all of the literature search so often required.

Because of the systematic arrangement of the material in the chapters and the entries in the tables, users of the books will be able to find information desired by reference to the table of contents of the appropriate chapter. In the interest of economy the entries in the indices have been kept to a minimum, and, in particular, the compounds listed in the tables are not repeated in the indices.

The success of this publication, which will appear periodically, depends upon the cooperation of organic chemists and their willingness to devote time and effort to the preparation of the chapters. They have manifested their interest already by the almost unanimous acceptance of invitations to contribute to the work. The editors will welcome their continued interest and their suggestions for improvements in *Organic Reactions*.

Chemists who are considering the preparation of a manuscript for submission to *Organic Reactions* are urged to write either secretary before they begin work.

CONTENTS

CHAPTER	PAGE

1. INTERMOLECULAR METAL-CATALYZED CARBENOID CYCLOPROPANATIONS
    Huw M. L. Davies and Evan G. Antoulinakis 1

2. OXIDATION OF PHENOLIC COMPOUNDS WITH ORGANOHYPERVALENT IODINE REAGENTS
    Robert M. Moriarty and Om Prakash 327

3. SYNTHETIC USES OF TOSYLMETHYL ISOCYANIDE (TosMIC)
    Daan van Leusen and Albert M. van Leusen 417

CUMULATIVE CHAPTER TITLES BY VOLUME 667

AUTHOR INDEX, VOLUMES 1–57 681

CHAPTER AND TOPIC INDEX, VOLUMES 1–57 685

Organic Reactions

CHAPTER 1

INTERMOLECULAR METAL-CATALYZED CARBENOID CYCLOPROPANATIONS

Huw M. L. Davies and Evan G. Antoulinakis

Department of Chemistry, State University of New York at Buffalo, Buffalo, New York 14260-3000

CONTENTS

	Page
Acknowledgments	2
Introduction	3
Mechanism	4
Scope and Limitations	7
Cyclopropanation of Alkenes	7
Carbenoids Derived from Diazoacetates and Related Systems	8
Diastereoselectivity	8
Asymmetric Induction Using Chiral Auxiliaries	9
Asymmetric Induction Using Chiral Catalysts	9
Carbenoids Derived from Diazoacetoacetates, Diazomalonates, and Related Systems	12
Asymmetric Induction	13
Carbenoids Derived from Vinyldiazoacetates, Phenyldiazoacetates, and Related Systems	13
Diastereoselectivity	14
Asymmetric Induction Using Chiral Auxiliaries	15
Asymmetric Induction Using Chiral Catalysts	15
Cyclopropanation of Dienes	16
Reactions of Carbenoids with Furans	19
Reactions of Carbenoids with Pyrroles	21
Synthetic Utility	23
Comparison with other Methods	29
Experimental Conditions	32
Synthesis of Diazo Compounds	32

Organic Reactions, Vol. 57, Edited by Larry E. Overman et al.
ISBN 0-471-43511-2 © 2001 Organic Reactions, Inc. Published by John Wiley & Sons, Inc.

General Reaction Conditions 35
Safety Considerations 35
EXPERIMENTAL PROCEDURES 36
  General Procedures for the Preparation of Carbenoid Precursors 36
    N-Diazoacetylmorpholine [Use of Succinimidyl Diazoacetate] 36
    (E)-1-Diazo-4-phenyl-3-buten-2-one [Detrifluoroacetylative Diazo Group
      Transfer Using 4-Dodecylbenzenesulfonyl Azide] 37
    Methyl (E)-2-Diazo-4-phenyl-3-butenoate [Diazo Transfer Reaction with
      p-Acetamidobenzenesulfonyl Azide] 38
  General Procedures for Carbenoid Cyclopropanation Reactions 38
    Ethyl (1S,2S)-2-Phenylcyclopropanecarboxylate and Ethyl (1S,2R)-2-
      Phenylcyclopropanecarboxylate [Asymmetric Cyclopropanation of Styrene
      with Ethyl Diazoacetate in the Presence of Copper Bisoxazoline Catalyst 23] . . 38
    Ethyl 2-Phenylcyclopropanecarboxylate [Reaction of Ethyl Diazoacetate with
      Styrene in the Presence of $Rh_2(OAc)_4$] 39
    Ethyl (1S,2S)-2-Phenylcyclopropanecarboxylate and Ethyl (1S,2R)-2-
      Phenylcyclopropanecarboxylate [Asymmetric Cyclopropanation of Styrene
      with Ethyl Diazoacetate in the Presence of $Rh_2(5S\text{-MEPY})_4$ (**25**)] . . . 39
    Ethyl (1R,2R)-2-Phenylcyclopropanecarboxylate and Ethyl (1R,2S)-
      2-Phenylcyclopropanecarboxylate [Asymmetric Cyclopropanation
      of Styrene with Ethyl Diazoacetate in the Presence of
      trans-$RuCl_2$(Pybox-*ip*)(ethylene) (**28**)] 40
    Ethyl 2,5-Dioxabicyclo[4.1.0]heptanecarboxylate [Reaction of Ethyl Diazoacetate
      with p-Dioxene in the Presence of Copper-bronze] 40
    (3aS^*, 8aR^*)-2,2-Dimethyl-2,3,3a,6,7,8a-hexahydro-5H-1,8-
      dioxacyclopent[a]indene-3,4-dione [Reaction of 2-Diazo-1,3-cyclohexanedione
      with 2,2-Dimethyl-3(2H)-furanone in the Presence of $Rh_2(OAc)_4$] . . 41
    (1S,2S)-Methyl 2β-Phenyl-1β-[2-(Z)-styryl]cyclopropane-1α-carboxylate
      [Asymmetric Cyclopropanation of Styrene with Methyl (E)-2-Diazo-4-
      phenyl-3-butenoate in the Presence of $Rh_2(S\text{-DOSP})_4$ (**46**)] 41
TABULAR SURVEY 41
  Chart 1. Non-Chiral Catalyst Complexes 44
  Chart 2. Chiral Copper C_2-Symmetric Catalyst Complexes. 47
  Chart 3. Chiral Copper Non-C_2-Symmetric Catalyst Complexes 55
  Chart 4. Chiral Rhodium Catalysts. 62
  Chart 5. Other Chiral Metal Complexes. 65
  Table I. Diastereoselective Cyclopropanation of Alkenes Using Diazoacetates
    and Related Systems 68
  Table II. Diastereoselective Cyclopropanation of Alkenes Using Diazoacetoacetates,
    Diazomalonates, and Related Systems 157
  Table III. Diastereoselective Cyclopropanation of Alkenes Using Vinyldiazoacetates
    and Phenyldiazoacetates 173
  Table IV. Asymmetric Cyclopropanation of Alkenes Using Chiral Auxiliaries. . . 179
  Table V. Asymmetric Cyclopropanation of Alkenes Using Chiral Copper Catalysts . 184
    A. C_2-Symmetric Complexes 184
    B. Non-C_2-Symmetric Complexes 209
  Table VI. Asymmetric Cyclopropanation of Alkenes Using Chiral Rhodium Catalysts . 223
  Table VII. Asymmetric Cyclopropanation of Alkenes Using Miscellaneous Chiral
    Catalysts. 244
  Table VIII. Reactions of Carbenoids with Dienes 255
  Table IX. Reactions of Carbenoids with Furans 288
  Table X. Reactions of Carbenoids with Pyrroles 301
REFERENCES 313

ACKNOWLEDGMENTS

We are indebted to our colleagues in the Davies Group (Dr. Rebecca Calvo, Jason De Meese, Dr. Monica Grazini, Dr. Timothy Gregg, Dr. Tore Hansen, Dr. L. Mark Hodges, Melinda Hodges, Dr. Darrin Hopper, Michael Levitt, Dr. Tadamichi Nagashima, Stephen Panaro, Dr. Pingda Ren, Leatte Rusiniak, Douglas Stafford, Dr. Robert Townsend, Chandrasekar Venkataramani, and Yasuno Yokota) at the State University of New York at Buffalo for their assistance in assembling the references, and their assistance in reviewing the manuscript. We also thank Angela Davies and Diane Witczak for their assistance in assembling the references and their patience with us through the preparation of this manuscript, as well as Catherine Antoulinakis for her assistance in organizing the references and record keeping. We gratefully acknowledge the guidance and assistance of the editorial staff of *Organic Reactions,* in particular Dr. Linda Press and Professor Amos B. Smith III, during the preparation of this chapter.

Warning: Great care should be taken in handling diazo compounds because they are potentially toxic and can have explosive properties. Therefore, diazo compounds should be handled carefully and all reactions should be carried out in a well-vented fume hood behind a blast shield.

INTRODUCTION

The metal-catalyzed decomposition of diazo compounds in the presence of alkenes is a well-established reaction. Since the original *Organic Reactions* review on the reaction of ethyl diazoacetate with alkenes and aromatic compounds in 1970,[1] several new developments have revolutionized this area of chemistry. Most notably, major advances have been made in catalyst design such that highly chemoselective, diastereoselective and enantioselective carbenoid transformations can now be achieved. Furthermore, it has been recognized that a wide array of carbenoid structures can be utilized in this chemistry, leading to a broad range of synthetic applications.

This chapter comprises coverage of the metal-catalyzed intermolecular cyclopropanations of diazo compounds containing at least one adjacent electron-withdrawing group. The coverage of diazoacetate chemistry will be limited to material since 1970 because the previous *Organic Reactions* review[1] covers the earlier literature. The alkene component is limited to alkenes, dienes, furans, and pyrroles because these are the systems that have resulted in the greatest developments since the 1970 review. Metal-carbenoid intermediates derived from diazo compounds undergo a variety of useful reactions, including cyclopropanation, insertion, and ylide formation. In recent years several excellent reviews have appeared on various aspects of this chemistry.[2-28] Three recent reviews have focused on asymmetric intermolecular cyclopropanations.[3,29,30] Several books and reviews on carbenoid chemistry have major sections on intermolecular cyclopropanations. Because of the historical central prominence of carbenoids derived from diazoacetates, most reviews have tended to focus on this class of carbenoids. In this chapter, a comparison is presented of the chemical differences that exist among the major classes of carbenoids that contain adjacent electron-withdrawing groups. The extensive nature of the topic precludes coverage of related reactions such as the metal-catalyzed decomposition of diazoalkanes,[31] phenyldiazoalkanes,[32] or vinyldiazoalkanes[33] that lack adjacent electron-withdrawing functionality. Other cyclopropanation reactions such as the Simmons-Smith reaction,[34-36] photochemical or thermal decomposition of diazo compounds in the presence of alkenes,[1] and cyclopropanation using stoichiometric metal carbenes[37,38] are not covered.

MECHANISM

Despite the great synthetic utility of diazo compounds in cyclopropanations, definitive mechanistic studies on the metal-catalyzed cyclopropanations are lacking.[39-42] Reasonable mechanistic models have been rationalized on the basis of product distribution, and especially the stereochemical outcome of various carbenoid reactions.[8,9,13] Because of the rapid catalytic turnovers of these reactions, structural information about the intermediates is difficult to obtain. Recently, some significant advances have been made that will likely have a major impact on the mechanistic understanding of these transformations. Stable ruthenium-carbenoid complexes have been characterized by X-ray crystallography.[43-45] As these systems are also capable of inducing catalytic carbenoid transformations, the X-ray crystallographic data lead to definitive information about the key metal carbenoid intermediate in catalytic reactions. Additionally, some major advances have been made in the use of molecular modeling to probe the structure of carbenoid intermediates,[46-48] and as more advanced computational methods are used, very useful information regarding the validity of various carbenoid transition state models should be forthcoming.

It is generally agreed that the reaction proceeds as shown in Eq. 1.[49] Interaction of diazo compound **1** with the metal forms diazonium complex **2**, which then

$$N_2=\underset{R^2}{\overset{R^1}{\diagdown}} \xrightarrow{M} \underset{R^2}{\overset{\overset{+}{N_2}\;R^1}{M-\diagdown}} \xrightarrow{-N_2} M=\underset{R^2}{\overset{R^1}{\diagdown}} \xrightarrow[-M]{\overset{R^3\;\;R^5}{\underset{R^4\;\;\;R^6}{\diagup\!\!\!=\!\!\!\diagdown}}} \underset{R^4\;\;\;\;R^6}{\overset{R^1\;R^2}{R^3\triangle R^5}}$$

1   **2**   **3**   **4**   **5**

(Eq. 1)

extrudes nitrogen to form carbenoid intermediate **3**. Reaction of alkene **4** with **3** forms cyclopropane **5** and regenerates the active catalyst. For most catalysts, attack of the alkene on the carbenoid is generally considered to occur without coordination of the alkene to metal, although there is evidence that prior coordination does occur for palladium-based catalysts and copper(I) triflate.[37,50] An alternative view that the reaction involves metallocyclobutane intermediates analogous to the chemistry of Fischer carbenes has been presented in a few publications,[51-54] although supporting evidence is limited.

If the cyclopropanation does occur without prior coordination of the alkene, the trajectory of the approach of the alkene to the metal-carbene complex becomes a critical element for control of regiochemistry and stereochemistry. The most established model for alkene approach to the metal-carbenoid complex was proposed by Doyle to explain the stereochemistry in cyclopropanations by diazoacetates. In the latest version of the Doyle model (**6**),[8,9,32,49] the alkene is con-

6
Doyle's Model

7
Kodadek's Model

8
Davies' Model

sidered to approach the carbene close to end-on, and cyclopropanation occurs in a nonsynchronous mode with charge build-up in the transition state. Interaction between the ester of the carbenoid and the partial positive charge of the alkene in the transition state is a major cause of the stereocontrol. The metal surface is considered to be a "wall" and the stereochemistry is controlled by ensuring that the bulky functionality of the alkene points away from the metal surface. A similar model was used to explain the diastereoselectivity observed in cyclopropanations with nitrodiazomethane.[55]

A second model, **7**, was developed by Kodadek to explain unusual Z-stereoselectivity of bulky rhodium porphyrin catalysts.[56,57] The alkene is considered to approach the carbene in a side-on approach with no charge build-up in the transition state. Evidence to support the lack of charge build-up was seen in the lack of alkene substituent effects on the relative rates of cyclopropanations, and the absence of deuterium isotope effects. To complete the cyclopropanation, rotation of the alkene, either inward or outward, is required. The direction of this rotation is dependent on the steric effects of the catalyst and the carbenoid substituents.

A third model has been proposed by Davies to explain the extremely high diastereoselectivity that occurs in cyclopropanation reactions of vinyldiazoacetates and aryldiazoacetates.[6,58] This model, **8**, is a hybrid of models **6** and **7**. The alkene approaches the vinylcarbenoid close to side-on in a nonsynchronous mode. The approach occurs on the side of the electron-withdrawing group with the bulky functionality of the alkene pointing away from the face of the rhodium complex. A trans alkene is unreactive because it is unable to avoid having a substituent pointing directly toward the rhodium surface. As the reaction proceeds, the alkene would need to rotate outward to form the cyclopropane ring, where R would end up on the same side as the vinyl group, leading to the observed stereochemistry. The fact that the diastereocontrol is enhanced by electron-donating groups on the alkene[59] while trans alkenes are unreactive[58] is good evidence to support this side-on nonsynchronous model. The subtle differences between these models may reflect that the actual interaction between the carbenoid and alkene is variable, and is dependent upon the types of carbenoid and catalyst employed.

In certain instances the product isolated in the metal-catalyzed reaction of diazo compounds with unsaturated systems is not the simple cyclopropanated product. These types of reactions are especially prevalent when the carbenoid contains two electron-withdrawing groups and the alkene or aromatic system has electron-donating substituents.[60-64] Examples of these types of reactions are shown in Eqs. 2–4.[65-68]

When the cyclopropane is not the isolated product there is often ambiguity about whether these products are derived directly from the carbenoid on reaction with the alkene component or from rearrangement of the initially formed cyclopropane **9** (Eq. 5).[49] The zwitterionic intermediate **10** could in principle be formed from either source. Furthermore, depending on the reaction, the zwitterionic intermediate may still be associated with the catalyst, which would enable the catalyst to affect asymmetric induction and product distribution. As cyclopropanation and formation of products derived from zwitterionic intermediates are often competing processes,[60-64] both reactions are covered in this review.

In recent years, with the advent of numerous excellent methods for asymmetric induction in various carbenoid transformations, several transition state models have been developed to explain how asymmetric induction is obtained.[6,14,49,69,70] A discussion of the various predictive models that have been proposed for asymmetric induction will not be presented in this chapter. Many of these models require further verification before it would be appropriate to discuss them in a review of this type.

SCOPE AND LIMITATIONS

Cyclopropanation of Alkenes

The most widely exploited reaction of carbenoids is the cyclopropanation of alkenes. In this discussion, the carbenoids will be divided into three major groups (**11–13**), because the chemistry is very much dependent upon the car-

$$M=\overset{H}{\underset{EWG}{\bigg\langle}} \qquad M=\overset{EWG}{\underset{EWG}{\bigg\langle}} \qquad M=\overset{EWG}{\underset{R^1}{\bigg\langle}}$$

EWG = CO_2R, COR,   EWG = CO_2R,   EWG = CO_2R or COR
NO_2, $PO(OR)_2$ or SO_2R COR or NO_2 R^1 = vinyl or phenyl

      11   **12**   **13**

benoid structure. The first group will be carbenoids containing a single electron-withdrawing group. This will include carbenoids derived from diazoacetates and related systems.[8,9,15,20,71] The diazoacetate system is by far the most extensively studied system. In recent years, the development of chiral catalysts to achieve asymmetric cyclopropanations with the diazoacetate system has been an immensely popular field of research.[3,6,14,30,49,69] The second group will be carbenoids containing two electron-withdrawing groups. This includes carbenoids derived from diazoacetoacetates, diazomalonates,[61] and diazodiketones. Also, carbenoids derived from diazopyruvates[72] are included with this group. Often, the products from this group of carbenoid precursors are not cyclopropanes but isomeric structures derived from zwitterionic intermediates.[60-62] The third group will be carbenoids containing both an electron-withdrawing group and an electron-donating group such as vinyl or phenyl.[4-7] This includes carbenoids derived from vinyldiazoacetates, phenyldiazoacetates, and related systems. Carbenoids containing a combination of donor and acceptor groups exhibit unique stereoselectivity in their chemistry[59,73,74] that warrants their consideration as a separate class.

Carbenoids Derived from Diazoacetates and Related Systems

The decomposition of alkyl diazoacetates in the presence of alkenes is an excellent method for the synthesis of cyclopropanes (Eq. 6).[1,8,9,15,20,49,71] The reaction is extremely general, with electron-rich, electron-neutral, and even slightly

$$R^1\text{—CH=CH}_2 + N_2\text{=CH—CO}_2R \xrightarrow{\text{catalyst}} R^1\text{—}\triangle\text{—CO}_2R \qquad (Eq.\ 6)$$

electron-deficient alkenes subject to cyclopropanation. Even though monosubstituted alkenes are the most common substrates, a range of substitution patterns, from monosubstituted to tetrasubstituted, can be tolerated. The traditional catalysts for these transformations were copper based,[1] but in recent years dirhodium complexes have become the catalysts of choice for most reactions.[75] Not only do these catalysts permit very mild reaction conditions, but also fine-tuning of the reaction is possible by appropriate choice of dirhodium ligands. The most commonly used catalyst is dirhodium tetraacetate.[76,77] Highly electron-deficient ligands such as trifluoroacetate lead to more electrophilic carbenoids while electron-donating ligands such as acetamide lead to less reactive and more selective carbenoids.[78] Complexes of a number of other metals, such as palladium,[79,80] ruthenium,[51,81–84] cobalt,[85–89] osmium,[90,91] and iron[92] have also been used as catalysts in these diazo decomposition reactions. One of the most common side-reactions in the chemistry of this class of carbenoids is the formation of carbene dimers,[1,12,93] although in general this problem can be alleviated by using very slow addition of the diazo compound to the reaction mixture.[94]

Similar cyclopropanations have been carried out with a range of other diazo compounds containing a single electron-withdrawing group. These include keto,[95] nitro,[96] sulfonyl,[97] and phosphonyl[98] groups. In many instances the reported yields are relatively low, but several of the studies were carried out using copper catalysis, and it would be reasonable to expect that greatly improved yields would be achieved in many instances if the dirhodium tetracarboxylates were used as catalysts.

Diastereoselectivity. One of the major limitations of cyclopropanations by diazoacetates is that, in general, the cyclopropanations are not particularly stereoselective.[49,71] Standard reactions of styrene with ethyl diazoacetates **14a** and **14b** yield the diastereomeric products **15** and **16** (Eq. 7), their E to Z ratios dependent upon the size of the ester moiety. The use of ethyl diazoacetate (**14a**) results in a diastereoselectivity of less than 2:1 favoring the E isomer (**15a**) over the Z isomer (**16a**). Considerable enhancement in the diastereoselectivity can be achieved by using bulky ester derivatives, such as that derived from 3,5-bis(tert-butyl)-4-hydroxytoluene (**14b**).[78] In combination with rhodium acetamide $Rh_2(acam)_4$ as catalyst, the diastereoselectivity of the reaction of styrene with **14b** is improved to 98:2 $E:Z$ (**15b:16b**). The diastereoselectivity of diazoacetate cyclopropanations is only moderately influenced by most catalysts.[8,9,15] If the cat-

$$\text{Ph}\diagup\!\!=\ +\ N_2\!\!=\!\!\diagup\text{COR}\ \xrightarrow{\text{catalyst}}\ \text{Ph}\triangleright\text{COR}\ +\ \text{Ph}\triangleright\text{COR}$$

14 → **15** + **16** (Eq. 7)

	R	catalyst	15:16
a	OEt	Rh$_2$(OAc)$_4$	62:38
a	OEt	Rh$_2$(acam)$_4$	60:40
b	BHT	Rh$_2$(OAc)$_4$	84:16
b	BHT	Rh$_2$(acam)$_4$	98:2

BHT = 2,6-(t-Bu)$_2$-4-MeC$_6$H$_2$O

alysts are extremely bulky, however, such as dirhodium tetra(2,6-disubstituted benzoates)[99] or sterically crowded rhodium porphyrins,[100–102] the diastereoselectivity can be altered leading to a slight preference for the Z isomer. A major improvement in diastereoselectivity favoring the formation of E cyclopropanes has been discovered recently for reactions catalyzed by ruthenium complexes.[103–107]

Asymmetric Induction Using Chiral Auxiliaries. Several attempts have been made to develop an asymmetric cyclopropanation using chiral auxiliaries although these have not been very successful.[106,108–111] Cyclopropanation with l-menthyl diazoacetate results in high yields but the asymmetric induction is very poor, yielding <2% enantiomeric excess (ee).[112] Attempts to enhance the asymmetric induction by using auxiliaries containing a carbonyl group that may interact with the carbenoid have also been carried out.[110] This approach has not been effective with the diazoacetate system. Reaction of **17** with styrene results in the formation of cyclopropanes **18** and **19** with low enantioselectivity (Eq. 8).[110]

17 →[1. Rh$_2$(OAc)$_4$, PhCH=CH$_2$; 2. NaOEt/EtOH] (25%) E:Z = 2:1 → **18** 14% ee + **19** 13% ee   (Eq. 8)

Asymmetric Induction Using Chiral Catalysis. The development of chiral catalysis for asymmetric carbenoid cyclopropanations is of considerable mechanistic and synthetic significance. The observation by Nozaki and co-workers[113,114] of low enantioselectivity in the reaction of ethyl diazoacetate with styrene catalyzed by copper catalyst **20** (Eq. 9) is a landmark study because it demonstrates that the carbene is associated with the copper catalyst during the cyclopropanation step.

The first report of a highly enantioselective cyclopropanation was described by Aratani using the copper catalyst **21** (Eq. 10).[112,115–117] In these studies, the highest asymmetric induction was obtained using menthyl esters. Even though the menthyl ester with an achiral catalyst results in low asymmetric induction, its reaction with **21** results in considerable improvement in both enantioselectivity

$$Ph\diagup\!\!=\; + \;N_2\!\!=\!\!\diagdown CO_2Et \xrightarrow[(72\%)\ E:Z = 2.3:1]{20} Ph\triangleleft CO_2Et \; + \; Ph\triangleleft\!\!\cdot CO_2Et$$
6% ee   6% ee

(Eq. 9)

$$Ph\diagup\!\!=\; + \;N_2\!\!=\!\!\diagdown COR \xrightarrow{\text{catalyst}} Ph\triangleleft COR \; + \; Ph\triangleleft\!\!\cdot COR$$

(Eq. 10)

R	catalyst	E:Z	% ee (E)	% ee (Z)
O-menthyl-l	21	2.6:1	90	59
O-menthyl-l	22	4.6:1	97	95
BHT	23	15.7:1	99	—

Ar = 2-OctO-5-t-BuC$_6$H$_4$
Oct = octanoate

21   **22**   **23**

and diastereoselectivity compared to ethyl diazoacetate. The improved stereoselectivity is primarily due to the increased size of the ester, since the double stereoselection that is observed in this system is rather moderate. The next major breakthrough in this area was the discovery by Pfaltz that C$_2$ symmetric semicorrin copper complexes **22** are superb catalysts for asymmetric induction.[69,118–123] Since then, several groups have developed various C$_2$ symmetric copper catalysts for asymmetric induction. The catalyst that has enjoyed the most general use to date is the bisoxazoline catalyst **23**, developed by Evans and Masamune.[124–126]

As rhodium catalysis tends to result in much cleaner reactions than copper systems, several groups have explored the use of various chiral rhodium complexes for asymmetric cyclopropanation.[3,29,30,49] Dirhodium(II) tetracarboxylates have been generally unsuccessful in asymmetric intermolecular cyclopropanations by diazoacetates,[127,128] but recently very promising results were achieved with the biphenylcarboxylate catalyst Rh$_2$(S-BDME)$_4$ (**24**).[129] In contrast, various rhodium(II) amide catalysts have been used in intermolecular cyclopropanations with high asymmetric induction. The Rh$_2$(S-MEPY)$_4$ catalyst (**25**), developed by

Doyle, was the first catalyst that was shown to result in reasonably high enantioselectivity in intermolecular cyclopropanation, but since then other catalysts such as **26** and **27** have been found that result in even higher enantioselectivity. In all of these systems, a bulky diazoacetate results in higher enantioselectivity than ethyl diazoacetate.

24: $Rh_2(S\text{-BDME})_4$

25: $Rh_2(5S\text{-MEPY})_4$

26

27: $Rh_2(S\text{-PTPI})_4$

The recent development of the C_2 symmetric ruthenium catalyst **28** for asymmetric cyclopropanation offers one major advantage over both rhodium and copper catalysts because high enantioselectivity and diastereoselectivity can be obtained with this system without resorting to using extremely bulky ester derivatives.[103,104,106,107] Illustrative examples are shown in Eq. 11.

$$Ph\text{-CH=CH}_2 + N_2\text{=CH-CO}_2\text{menthyl-}l \xrightarrow[\text{(83\%) }E\text{:}Z = 97\text{:}3]{\textbf{28}} \quad \text{(Eq. 11)}$$

Ph—△—CO₂ menthyl-*l*   +   Ph—△—CO₂ menthyl-*l*

96% ee   80% ee

Carbenoids Derived from Diazoacetoacetates, Diazomalonates, and Related Systems

Metal-carbenoid intermediates flanked by two electron-withdrawing groups are highly electrophilic systems. Furthermore, because of their ability to stabilize a negative charge at the carbenoid carbon, products derived from zwitterionic intermediates are commonly observed, especially when the reaction is carried out with electron-rich alkenes.[65,67,72,130–136] Similar chemistry occurs with carbenoids derived from diazopyruvates.[72,137] Reaction of **29** with cyclohexene results in the formation of the cyclopropane **30**, whereas reaction with dihydropyran forms the fused dihydrofuran **31** (Eq. 12).

catalyst	X	30 (%)	31 (%)
CuSO$_4$	CH$_2$	(76)	(0)
Cu(acac)$_2$	O	(0)	(57)

(Eq. 12)

The carbenoid derived from 2-diazo-1,3-cyclohexanedione (**32**) is highly susceptible to [3 + 2] annulation chemistry.[67,136,138,139] An illustrative example is the reaction of **32** with dihydrofuranone **33** that results in the formation of the tricyclic system **34** (Eq. 13).[136]

(Eq. 13)

A second side-reaction that is commonly seen with this class of carbenoid intermediates is C-H insertion.[61,140] Reaction of dimethyl diazomalonate with cyclohexene generates a mixture of the cyclopropane **35** and the C-H insertion product **36** (Eq. 14). In more functionalized cyclohexenes, the C-H insertion can become the dominant reaction pathway. The C-H insertion product could be derived either by a direct C-H insertion mechanism or through proton exchange

(Eq. 14)

from a zwitterionic intermediate that could be directly formed on reaction of the carbenoid with the alkene.

Asymmetric Induction. The standard asymmetric protocols that have been developed for the diazoacetate system have not been generally used with the other carbenoid systems. However, two reports described highly enantioselective transformations with the 2-diazo-1,3-cyclohexanedione system. A communication with limited details described highly asymmetric cyclopropanations with the copper catalyst **37** leading to the spirocyclopropane **38** in 98% ee (Eq. 15).[141] A recent report described highly enantioselective [3 + 2] annulations using the rhodium prolinate catalyst **39** leading to the tricyclic system **40** in 96% ee (Eq. 16).[142] Both of these reactions represent most unusual cases of asymmetric

(Eq. 15)

(Eq. 16)

induction because as the faces of the carbenoid are not enantiotopic, the enantioselectivity must be due to alkene face selectivity. This indicates that the trajectory of alkene approach in the 2-diazo-1,3-cyclohexanedione system must be very demanding, although no detailed model has been proposed to explain the enantioselectivity observed in this system.

Carbenoids Derived from Vinyldiazoacetates, Phenyldiazoacetates, and Related Systems

In recent years, it has become clear that carbenoids derived from vinyldiazoacetates have a very different reactivity profile to carbenoids derived from diazoacetates.[4,6,7] Intermolecular cyclopropanations will only occur with monosubstituted alkenes, 1,1-disubstituted alkenes, and *cis*-1,2-disubstituted alkenes.[58] Furthermore, many of these reactions are highly diastereoselective, again differing from the typical results obtained with the diazoacetate system.

Diastereoselectivity. Cyclopropanations with vinyldiazoacetate **41** are generally highly diastereoselective.[59,143] The highest diastereoselectivity is obtained with electron-rich alkenes, such as styrene and vinyl ethers, and with vinylcarbenoids lacking an electron-withdrawing group on the vinyl portion (Eq. 17).[59,143] In many of these reactions, the second diastereomer cannot be observed in the NMR spectrum of the crude reaction mixtures.

(Eq. 17)

R	(%)	E:Z
Ph	(94)	>95:5
OEt	(80)	>95:5

Spectacular chemoselectivity and diastereoselectivity were demonstrated during a short synthesis of the ether analog of acetomycin (Eq. 18).[144] Decomposition of the vinyldiazoacetate **42** in the presence of an *E/Z* mixture of ethyl 1-propenyl ether results in the formation of **43**, containing three stereogenic centers, as a

(Eq. 18)

single diastereomer. Only the Z vinyl ether is capable of reacting with the carbenoid, and the high diastereoselectivity typical of the vinyldiazoacetate system is obtained.

Similar highly diastereoselective cyclopropanations have been reported for the phenyldiazoacetate system, as illustrated in Eq. 19.[73,74,145,146] A comparison study of a range of carbenoid systems concludes that these highly diastereoselective

(74%) *E:Z* = 94:6

(Eq. 19)

cyclopropanations occur only in the case of carbenoids that are flanked by both an electron-withdrawing group and an electron-releasing group such as vinyl or phenyl.[73]

Asymmetric Induction Using Chiral Auxiliaries. In contrast to the results in the diazoacetate system, chiral auxiliaries have been found to be very effective in asymmetric vinyldiazoacetate cyclopropanations.[147,148] Inexpensive α-hydroxy esters are excellent chiral auxiliaries for these transformations. The highest asymmetric induction is obtained using (R)-pantolactone as the chiral auxiliary, as illustrated for the reaction of the vinyldiazoacetate **44** with styrene, which results in the formation of vinylcyclopropane **45** in 97% diastereomeric excess (de) (Eq. 20).[148] An even less expensive, effective chiral auxiliary is methyl (S)-lactate, although in general the asymmetric induction observed is lower than that achieved using (R)-pantolactone.[148]

$$\text{Ph} + \text{N}_2\text{=vinyldiazoacetate} \xrightarrow[0°]{Rh_2(OAc)_4} \textbf{45 (84\%) 97\% de} \qquad (Eq.\ 20)$$

44

45 (84%) 97% de

Asymmetric Induction Using Chiral Catalysts. Although rhodium(II) prolinates are poor chiral catalysts for diazoacetate cyclopropanations, they are extremely effective for asymmetric cyclopropanations with vinyldiazoacetates.[5,58,128] The highest degrees of asymmetric induction are obtained when the reactions are carried out in nonpolar solvents at low temperatures. Consequently, the most effective catalyst to date is $Rh_2(S\text{-DOSP})_4$ (**46**) which is soluble in pentane even at $-78°$. $Rh_2(S\text{-DOSP})_4$-catalyzed decomposition of the vinyldiazoacetate **41** in the presence of styrene at $-78°$ results in the formation of cyclopropane **47** in 98% ee (Eq. 21).[58]

$$\text{Ph} + \text{N}_2\text{=C(CO}_2\text{Me)CH=CHPh} \xrightarrow[\text{pentane, }-78°]{\textbf{46}:\ Rh_2(S\text{-DOSP})_4} \textbf{47} \qquad (Eq.\ 21)$$

41

47 (68%) 98% ee

Cyclopropanation of Dienes

The reaction of diazo compounds with dienes is a useful synthetic process because the resulting vinylcyclopropanes can be used in further synthetic transformations.[149] The reaction with dienes offers an interesting test of regioselectivity wherein the two double bonds compete for the active carbenoid intermediate. Several systematic studies on the reaction of ethyl diazoacetate with 1-substituted dienes and 2-substituted dienes demonstrate that reasonable selectivity between the double bonds can be achieved in appropriately substituted dienes.[150,151] In 1-substituted dienes, a chloro substituent gives the highest regiochemistry (**48:49**, Eq. 22) favoring the sterically less crowded double bond,[151] whereas in 2-substituted dienes, a methoxy substituent gives the highest regioselectivity (**50:51**, Eq. 23).[151] As is typical of diazoacetate cyclopropanations, low diastereoselectivity is obtained in these reactions.

R	(%)	**48:49**
Me	(61)	88:12
Ph	(84)	98:2
OMe	(90)	89:11
Cl	(73)	>99:1

(Eq. 22)

R	(%)	**50:51**
Me	(99)	62:38
Ph	(91)	69:31
OMe	(88)	>99:1
Cl	(76)	79:21

(Eq. 23)

The reaction of dimethyl diazomalonate with dienes can lead either to cyclopropanation (Eq. 24)[152] or to [4 + 1] cycloaddition (Eq. 25).[53] The outcome of

(Eq. 24)

[Structure diagram showing reaction of TBDMSO-diene with N₂=C(CO₂Me)₂ catalyzed by Rh₂(OAc)₄ giving product 53 (35%)]

(Eq. 25)

this chemistry is dependent on the electronic structure of the diene. Unpolarized dienes tend to form cyclopropanation products (**52**), whereas electron-rich dienes tend to form the [4 + 1] annulation products (**53**). The [4 + 1] annulation products may be derived from zwitterionic intermediates, although it has been postulated that a metallocyclohexene intermediate may be involved in these reactions.[53]

The reaction of vinylcarbenoids with dienes leads to the formation of *cis*-divinylcyclopropanes.[4,7] Cyclopropanations with vinyldiazoacetates are highly diastereoselective, and in most instances no evidence for the formation of the *trans*-divinylcyclopropanes is observed. As *cis*-divinylcyclopropanes undergo a Cope rearrangement under mild conditions,[153,154] the combined cyclopropanation/Cope rearrangement is a very direct method for the synthesis of cycloheptadienes of defined stereochemistry. Most effective in this regard is the reaction of vinyldiazoacetates with dienes, because in most instances the cyclopropanation is highly diastereoselective and the resulting *cis*-divinylcyclopropanes rearrange to cycloheptadienes under the reaction conditions.[4,7] An illustrative example is the reaction of vinyldiazoacetate **54** with cyclopentadiene that results in the formation of the *endo*-bicyclooctadiene **56** in 98% yield.[155,156] The stereoselective nature of this reaction is attributable to the requirement for a boat-like transition state for the rearrangement of the *cis*-divinylcyclopropane **55**.

[Reaction scheme showing cyclopentadiene + vinyldiazoacetate 54 with Rh₂(OAc)₄ giving intermediate 55 then product 56 (98%)]

(Eq. 26)

A further notable feature of vinyldiazoacetate reactions is that the carbenoid is very sensitive to both steric and electronic effects, which leads to the possibility of highly regioselective cyclopropanations with unsymmetrical dienes.[156,157] An example of this phenomenon is seen in the key annulation step in the synthesis of tremulenolide A.[158] Reaction of **57** with the (*E*/*Z*)-diene **58** results in the formation of a single [3 + 4] annulation product **60** in 49% yield. Even though the diene is not electronically biased, carbenoids derived from vinyldiazoacetates do not undergo intermolecular cyclopropanation with trans alkenes. Consequently, the cis double bond in **58** is selectively cyclopropanated leading to *cis*-divinylcyclopropane **59**, which rearranges through a boat-like transition state to **60**.

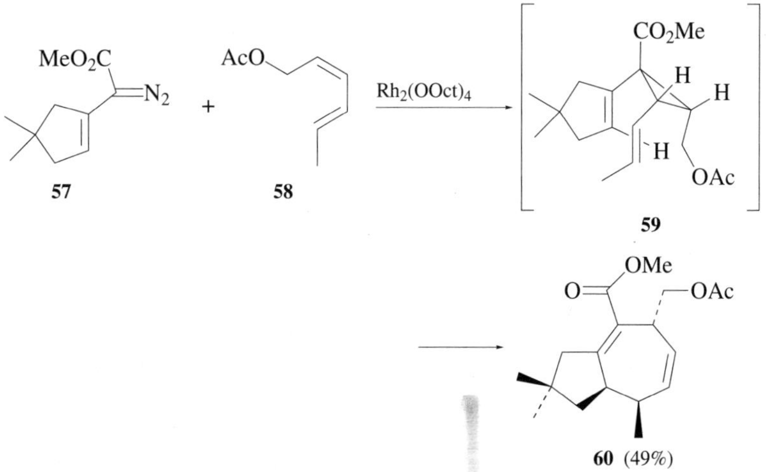

(Eq. 27)

An unusual side-reaction occurs with vinylcarbenoids derived from **61** that lack a substituent on the vinyl terminus, as seen in Eq. 28.[159] In addition to the [3 + 4] annulation product **63**, the bicyclo[2.2.1] system **65** is formed. This system is considered to be formed by reaction of the vinylcarbenoid at the vinylogous

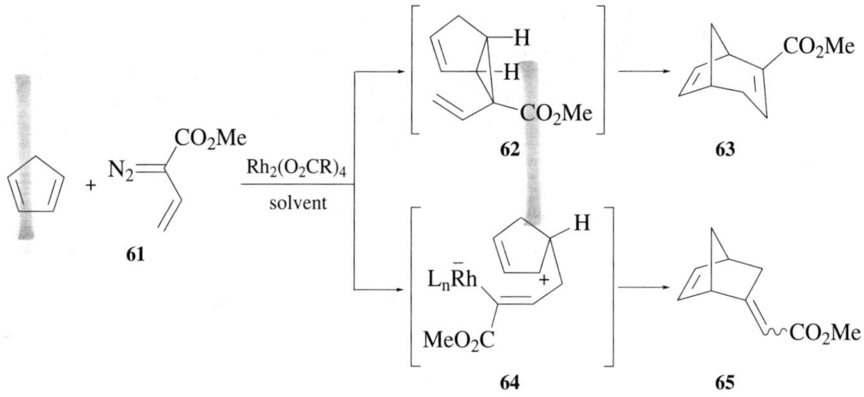

(Eq. 28)

catalyst	solvent	**63:65**
$Rh_2(OAc)_4$	CH_2Cl_2	67:33
$Rh_2(TFA)_4$	CH_2Cl_2	32:68
$Rh_2(OPiv)_4$	hexane	98:2

position instead of at the carbenoid center leading to zwitterionic intermediate **64** instead of divinylcyclopropane **62**. Use of a combination of relatively electron-rich catalysts, such as dirhodium tetrapivalate, and nonpolar solvents can circumvent the reactivity at the vinylogous position.

Both the chiral auxiliary and catalyst approaches have been effectively used for asymmetric cyclopropanation of dienes. A commercially important transformation is the reaction of menthyl diazoacetate with **66** to form the vinylcyclopropanes **67**, important intermediates in the synthesis of pyrethroids (Eq. 29). The reaction of chiral vinyldiazoacetates with dienes can occur with high asymmetric induction as illustrated in Eqs. 30[148] and 31.[157] Reaction of **68** with diene

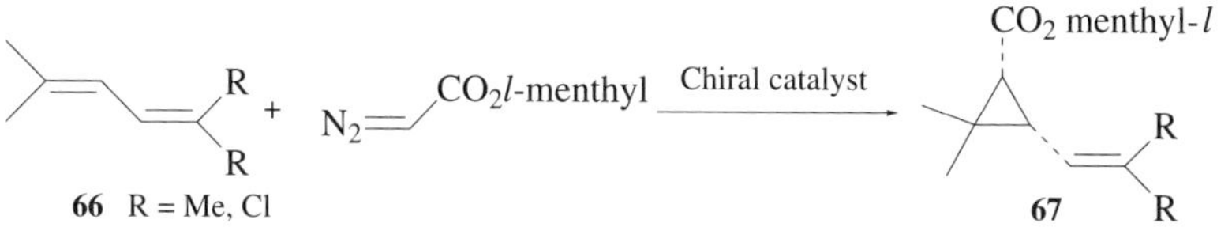

66 R = Me, Cl **67**

(Eq. 29)

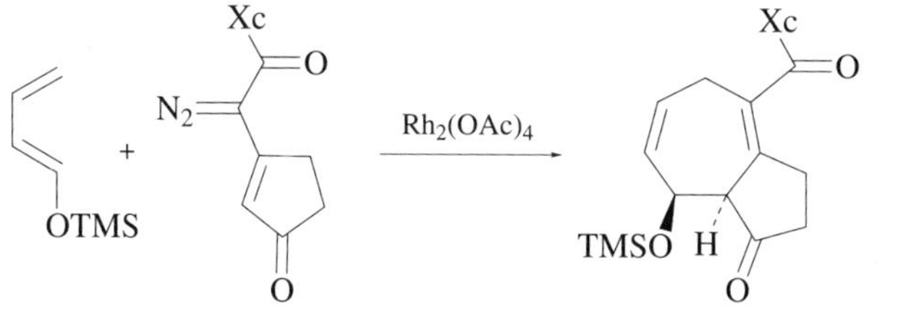

68 **69** Xc = (*R*)-pantolactone **70** (80%) >90% de

(Eq. 30)

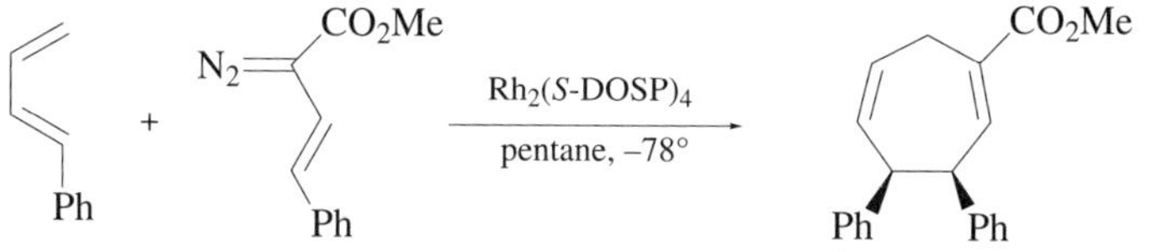

41 **71** (83%) 98% ee

(Eq. 31)

69 results in the formation of the bicyclic system **70** in greater than 90% de. Rh_2(*S*-DOSP)-catalyzed decomposition of vinyldiazoacetate **41** in the presence of *trans*-1-phenyl-1,3-butadiene results in the formation of the *cis*-diphenylcycloheptadiene **71** in 98% ee.

Reactions of Carbenoids with Furans

The reaction of carbenoids with furans usually leads to the unraveling of the heterocycle resulting in the formation of differentially functionalized dienes in good yield.[54,160–171] A re-examination of the reaction between ethyl diazoacetate and furan revealed the formation of four products, furanocyclopropane **72**, two isomeric dienes **73** and **74**, and a trace of alkylated product **75** in 66% overall

[Eq. 32 scheme: furan + N₂=CH-CO₂Et with Rh₂(OAc)₄ catalyst gives 72 (34%, furanocyclopropane with CO₂Et), 73 (20%, OHC-CH=CH-CH=CH-CO₂Et), 74 (10%, OHC-CH=CH-CH=CH-CO₂Et isomer), and 75 (2%, 2,3-dihydrofuran with =CH-CO₂Et)]

yield (Eq. 32).[172,173] The furanocyclopropane **72** is unstable, however, and on prolonged standing or on treatment with iodine rearranges cleanly to the Z,E diene **73**. Furans with electron-withdrawing groups such as esters[172,173] and unsaturated esters[172,173] can participate in this chemistry, but competing cyclopropanation of the vinyl group occurs with vinylfurans.[174]

The reaction of 2-diazo-1,3-cyclohexanedione (**32**) with furans results in the formation of [3 + 2] annulation products such as **76** (Eq. 33).[67] The regiochem-

[Eq. 33 scheme: furan + 32 (2-diazo-1,3-cyclohexanedione) with Rh₂(OAc)₄ gives 76 (90%), a fused bicyclic product]

istry of these reactions is most puzzling because they would involve the intermediacy of the less stabilized of the two possible zwitterionic intermediates. A mechanism has been proposed in which the reaction proceeds by an initial cyclopropanation, which is then followed by selective ring opening to the zwitterionic intermediate.[67] Alternatively, it has been suggested that the zwitterionic intermediate is directly formed, and the regiochemistry is controlled by steric factors that govern the approach of the furan to the carbenoid.[175] Limited attempts have been made to achieve asymmetric induction in this reaction using a dirhodium tetra(binaphthylphosphate) catalyst.[176]

Reactions of vinylcarbenoids with furans offer another level of complexity because the furanocyclopropanes in these cases would be divinylcyclopropanes capable of Cope rearrangement in addition to electrocyclic ring opening to trienes.[166,167] As illustrated in Eq. 34, the product distribution is very dependent on furan structure. With 2,5-disubstituted furans, bicyclo[3.2.1] systems **77** are formed exclusively, but with furan or 2-substituted furans, trienes **78** are also produced. In these reactions, the trienes are considered to be derived from zwitterionic intermediates arising from attack of the carbenoid at the α-position of the furan, whereas the furanocyclopropanes cleanly rearrange to **77**.

Asymmetric [3 + 4] annulations between vinylcarbenoids and furans are readily achieved using chiral auxiliaries, as demonstrated in the asymmetric syn-

[Eq. 34 scheme]

R^1	R^2	77 (%)	78 (%)
H	H	(62)	(26)
Me	H	(8)	(74)
OMe	H	(0)	(92)
Me	Me	(70)	(0)

(Eq. 34)

thesis of 8-oxabicyclo[3.2.1]octan-3-ones.[177] These oxabicyclic systems are very versatile intermediates for organic synthesis but typically they have been prepared in racemic form by the [3 + 4] annulation between allyl cations and furans.[178] An illustrative example is the reaction of siloxyvinyldiazoacetate **79** with furan, which generates the 8-oxabicyclo[3.2.1]octadiene **80** in 82% yield and 94% de (Eq. 35).[177]

79 Xc = (R)-pantolactone   **80** (82%) 94% de

(Eq. 35)

Reactions of Carbenoids with Pyrroles

The reaction of carbenoids with pyrroles commonly leads to either substitution or cyclopropanation products, depending on the functionality on nitrogen.[179] The reaction with N-alkylated pyrroles leads exclusively to substitution products. In view of the pharmaceutical importance of certain pyrrolylacetates, the reaction of alkyl diazoacetates with pyrrole has been extensively studied.[66,180,181] Both the 2- and 3-alkylated products, **81** and **82**, can be formed; the ratio is dependent on

[Eq. 36 scheme]

R	catalyst	81 + 82 (%)	81:82
Me	Cu(SO$_4$)$_2$	(58)	66:34
Me	Cu(acac)$_2$	(36)	91:9
t-Bu	Cu(acac)$_2$	(34)	<2:98

(Eq. 36)

the catalysts and the size of the N-alkyl and ester groups. Some illustrative examples of the trends that are observed are shown in Eq. 36.[180] This has been interpreted as evidence that transient cyclopropane intermediates are not involved in this reaction because if this were the case the catalyst should not have influenced the isomer distribution. Instead, the reaction is believed to proceed by dipolar intermediates, whereby product control is determined by the position of electrophilic attack by the carbenoid.

The reaction of ethyl diazoacetate with N-acylated pyrroles in the presence of cuprous bromide generates the 2-azabicyclo[3.1.0]hex-3-ene system **83** and some of the diadduct **84** (Eq. 37).[182] Higher yields of **83** can be obtained by using $Rh_2(OAc)_4$.[183] Heating of **83** in the presence of cuprous bromide causes rearrangement of **83** to the 2-pyrrolylacetate **85**, which is considered to arise from a zwitterionic intermediate. In contrast, on flash vacuum pyrolysis **83** is transformed to dihydropyridine **86**. A plausible mechanism for the formation of **86** is the rearrangement of **83** to an acyclic dieneimine analogous to the furan ring opening, which then undergoes a 6π electrocyclization to **86**.

In a manner analogous to furan cycloadditions as illustrated in Eq. 33, the reaction of diazodimedone with N-BOC-pyrrole results in an intriguing [3 + 2] annulation product **88** (Eq. 38).[67] The regiochemistry would require the formation of zwitterionic intermediate **87** where the carbenoid is attached to the 3-position of the pyrrole. It has been suggested that the preferential formation of **87** rather than the more stable 2-substituted zwitterionic structure is due to either the regiochemistry of ring opening of the pyrrolocyclopropane[67] or the steric demands for approach of the carbenoid to the pyrrole.[175]

The reaction between vinylcarbenoids and pyrroles is a general method for the stereoselective construction of tropanes (Eq. 39).[184,185] Side-reactions due to vinyl terminus reactivity occur with vinylcarbenoids lacking functionality at the vinyl terminus, but this reactivity can generally be avoided by using nonpolar sol-

(Eq. 38)

(Eq. 39)

vents.[185] Asymmetric reactions using a chiral catalyst such as $Rh_2(S\text{-}TBSP)_4$ are not particularly effective in this case because the pyrrole is too electron-rich and leads to products derived from zwitterionic intermediates in addition to tropanes.[175] The asymmetric synthesis of tropanes, however, can be achieved by reaction of S-lactate derivatives like **89** with N-BOC-pyrrole, leading to tropanes such as **90** with respectable yields and diastereoselectivity.[175,186] The sense of asymmetric induction using α-hydroxy esters as chiral auxiliaries is opposite on going from furans[177] to pyrroles.[175] This has been explained as a change in facial selectivity on the heterocycle during non-synchronous cyclopropanation with predominant initial bond formation on furan occurring at the 2-position, and on N-BOC-pyrrole at the 3-position.

SYNTHETIC UTILITY

The stereoselective synthesis of cyclopropanes is a transformation of considerable importance. The cyclopropane unit is present in a number of natural and commercial products.[187,188] Furthermore, a number of stereoselective ring opening and ring expansion reactions of cyclopropanes have been developed into versatile synthetic methods.[149,153,154,189–193] As can be seen from this chapter, the metal-catalyzed decomposition of diazo compounds in the presence of alkenes is an extremely general method for the synthesis of cyclopropanes. With the development of a range of excellent chiral catalysts, the majority of these cyclopropanations can, in principle, be achieved with high asymmetric induction. The diastereoselectivity in these cyclopropanations has been difficult to control but major improvements have now been made since the understanding of the effect of carbenoid structure[59,78] and catalysts[78,99,106] on diastereoselectivity has improved.

The early commercial interest in devising catalysts for asymmetric cyclopropanation was directed primarily toward the asymmetric synthesis of pyrethroids,[194] such as chrysanthemic acid **91** and permethrinic acid **92**. Similarly, the asymmetric cyclopropanation of 2-methylpropene was developed as a direct route to **93**, which is the cyclopropane constituent of the antibiotic cilastatin.[117,126] Various conformationally constrained cyclopropane amino acids[187,188] such as **94–96**[58,195,196] have also been prepared by intermolecular cyclopropanations.

Because of the strain associated with the cyclopropane system, a variety of ring-opening and ring-expansion reactions can occur. These reactions have been extensively reviewed.[149,153,154,189–193] Only a few illustrative examples of the most useful synthetic processes will be described here.

Nucleophile-induced ring-opening of cyclopropanes generally proceeds with inversion of configuration.[190] Only very powerful soft nucleophiles will react with monoactivated cyclopropanes but a much wider range of nucleophiles can be used if two electron-withdrawing groups are present on the cyclopropane.[197,198] Ring opening of cyclopropanes by nitrogen nucleophiles has been used for the synthesis of alkaloids.[199] An elegant application of cuprate-induced ring opening of cyclopropanes was demonstrated by Corey for the asymmetric synthesis of the antidepressant (+)-sertraline (**99**).[200] Cuprate-induced ring opening of enantiomerically pure **97** generates the diaryl derivative **98**, which is readily converted to (+)-sertraline by a series of standard reactions.

(Eq. 40)

Vinyl ethers are very efficiently cyclopropanated and the resulting donor/acceptor-substituted cyclopropanes (**100**) have been widely used in organic synthesis.[149,190] Ring opening can be achieved under very mild conditions resulting in a versatile approach to 1,4-difunctionalized compounds. Illustrative general examples are shown in Eqs. 41–43.[62] On treatment with mild acid, ring opening

$$\text{TMSO}\diagup\!\!\!\triangle\!\!\!\diagdown\text{COX} \quad \underset{X = R'}{\xrightarrow{H^+}} \quad \underset{\textbf{101}}{R\!\!-\!\!\text{CO}\!\!-\!\!\text{CH}_2\text{CH}_2\!\!-\!\!\text{CO}\!\!-\!\!R'} \qquad (\text{Eq. 41})$$

$$\xrightarrow[X = OR']{\text{PhSeCl}} \quad \underset{\textbf{102}}{R\!\!-\!\!\text{CO}\!\!-\!\!\text{CH}_2\!\!-\!\!\text{CH(SePh)}\!\!-\!\!\text{CO}_2R'} \qquad (\text{Eq. 42})$$

$$\xrightarrow[X = OMe]{\text{TMSI/(TMS)}_2\text{NH}} \quad \underset{\textbf{103}}{R\!\!-\!\!\text{C(OTMS)}\!\!=\!\!\text{CH}\!\!-\!\!\text{CH}_2\!\!-\!\!\text{CO}_2\text{Me}} \qquad (\text{Eq. 43})$$

R, R' = various alkyl

of **100** occurs to form 1,4-dicarbonyl compounds **101**, which can be used for the synthesis of cyclopentenones or furans. Alternatively, treatment of **100** with phenylselenyl chloride generates selenylated products **102**,[201] while treatment with a catalytic amount of trimethylsilyl iodide and bis(trimethylsilyl)amine results in the formation of silylated products **103**.[202]

The enolate of the cyclopropane **104** is readily formed, and can undergo aldol reactions or alkylations.[53,149,203–208] The resulting aldol products can undergo ring opening to a variety of products such as β,γ-unsaturated ketones, furanones, tetrahydrofurans, dihydrofurans, and macrocycles.[206] Illustrative examples of the synthetic potential of this chemistry are shown in Eqs. 44[209] and 45.[203] Alkylation of the enolate from **104**, followed by treatment with fluoride, generates the ring-opened product **105**, which on standing undergoes a smooth transformation

(Eq. 44)

104 → **105** → **106** (86%)

Reagents: 1. LDA, (bromo-diene); 2. NEt$_3$·HF; then 20°, 4 days

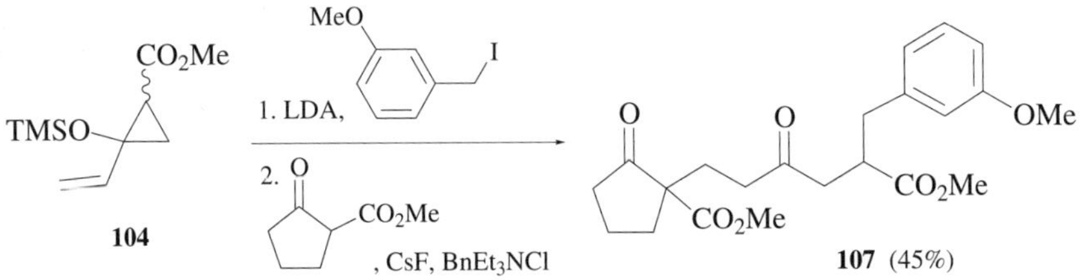

(Eq. 45)

to the Diels-Alder cycloadduct **106**.[209] Alternatively, alkylation of the vinylcyclopropane **104** followed by Michael addition-induced ring opening generates **107**, which can be converted to a mixture of steroid products in 56% combined yield.[203]

An elegant strategy for the synthesis of fused cyclopentanoids comprises the reaction between donor/acceptor substituted cyclopropanes and phosphonium salt **108** (Eq. 46).[210,211] Reaction of **108** with cyclopropane **109** generates bicyclic system **110**. Further conversion of **110** to **111** enables the annulation sequence to be repeated to form triquinane derivatives such as **112**.

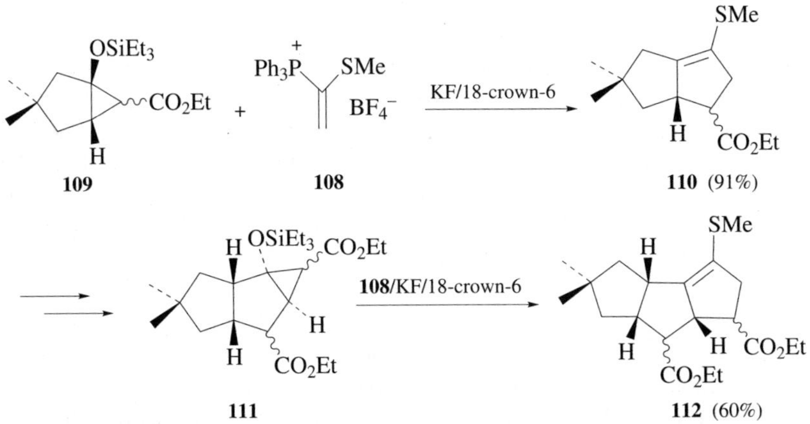

(Eq. 46)

The vinylcyclopropane/cyclopentene rearrangement has been extensively used in organic synthesis.[189] Traditionally, flash vacuum pyrolysis conditions are required to achieve the rearrangement,[189] but when the vinylcyclopropane contains donor and acceptor substituents, mild Lewis acid conditions can be used.[212,213] The combination of an intramolecular cyclopropanation of a diene followed by rearrangement to a cyclopentene has been used in the total synthesis of several natural products. An example of the use of a product derived from an intermolecular reaction is shown in Eq. 47.[213] Diethylaluminum chloride catalyzed rearrangement of **113** to cyclopentene **114** occurs with full control of relative stereochemistry.

The Cope rearrangement of *cis*-divinylcyclopropanes is thermally allowed and is a powerful method for the stereoselective synthesis of cycloheptadienes.[7,153,154] Divinylcyclopropanes can be prepared from the reaction of ketocarbenoids with dienes followed by methylenation of the keto group. An example of this approach is the formal synthesis of (±)-quadrone shown in Eq. 48.[214] Decomposition of

ethyl diazoacetate in the presence of the bicyclic compound **115** results in selective cyclopropanation to form cyclopropane **116**. Further modification of **116** generates the *trans*-divinylcyclopropane **117**, which on thermolysis followed by desilylation produces the tricyclic system **118**. Conversion of **118** to the ketone **119** completes the formal synthesis of (±)-quadrone.

A major advance in the use of carbenoids to prepare *cis*-divinylcyclopropanes **120** is the highly diastereoselective reaction between vinyldiazoacetates and dienes. Compounds such as **120** generally rearrange at or below room temperature to the cycloheptadienes **121** with full control of relative stereochemistry (Eq. 49).[4,7] Furthermore, as chiral auxiliaries and chiral catalysts are very effective for asymmetric vinylcarbenoid cyclopropanations, the enantioselective synthesis of cycloheptadienes is readily achieved.[157] This reaction is applicable to a wide range of dienes including furans[177] and *N*-acylated pyrroles.[175]

R^1 = EWG;
R^2-R^{10} = various groups including
H, alkyl, aryl, Cl, Osilyl

121 (47-87%) 73-98% ee

(Eq. 49)

Another useful rearrangement of 2-vinylcyclopropanecarboxylates is the homodienyl rearrangement that leads to the formation of β,γ-unsaturated esters of defined alkene geometry.[193] The illustrative example shown in Eq. 50 leads to the vinylsilane **122**.

122 (98%)

(Eq. 50)

In many instances, the reaction between a carbenoid and an alkene does not lead to the isolation of cyclopropanes. This is especially prevalent when two electron-withdrawing groups are attached to the carbenoid and the alkene is electron-rich. A very useful reaction is the [3 + 2] annulation that occurs on reaction of 2-diazo-1,3-cyclohexanedione with vinyl ethers. This chemistry has been elegantly developed for the construction of natural products, such as (−)-pseudosemiglabrin (**123**)[136] and aflatoxin B$_2$ (**124**).[215]

The furan fragmentation that can occur on reaction of carbenoids with dienes is a useful method for the synthesis of *Z,E* dienes. Numerous applications of this

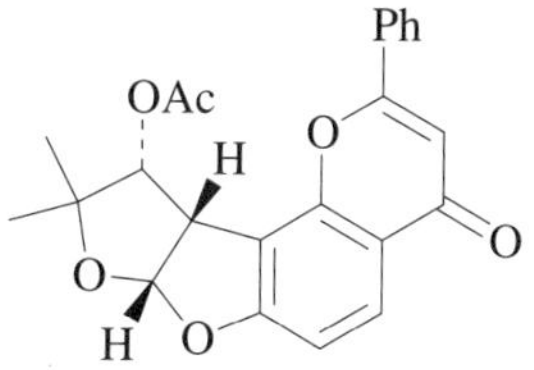

123 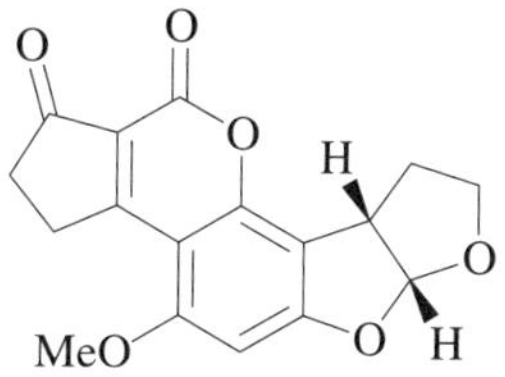 **124**

chemistry to the synthesis of leukotrienes have been reported[163,164,168,169] as illustrated for the preparation of 12-hydroxyeicosatetraenoic acid (**125**).[169] Reaction of the carbenoid precursor **126** with furan in the presence of rhodium(II) acetate generates a furanocyclopropane which on standing rearranges to predominantly the Z,E isomer **127**. Using fairly standard reaction conditions, **127** is readily converted to **125**.

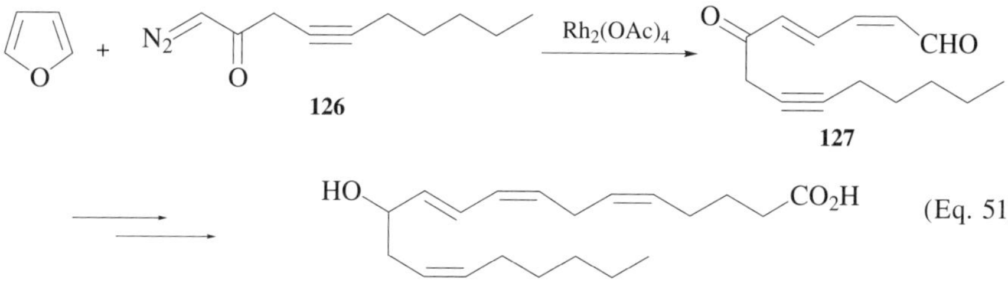

$$\text{furan} + \mathbf{126} \xrightarrow{Rh_2(OAc)_4} \mathbf{127} \longrightarrow \mathbf{125} \quad \text{(Eq. 51)}$$

The reaction of pyrroles with carbenoids results in the formation of alkylation products unless the pyrrole nitrogen is N-acylated.[1,179] This direct method for the preparation of pyrrole acetates has been used to prepare compounds of pharmaceutical interest such as **128**[180] and **129**.[216,217]

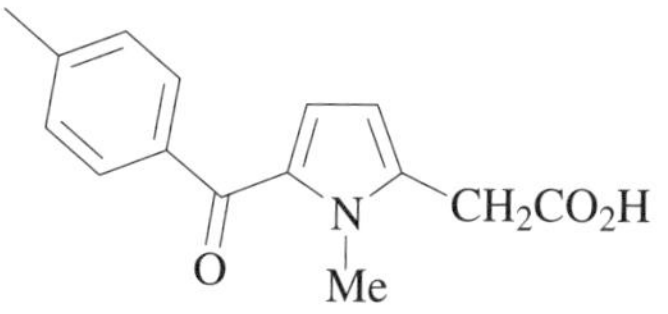

128 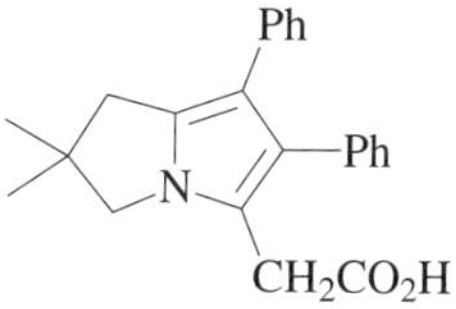 **129**

COMPARISON WITH OTHER METHODS

Because of the vastness of the topic, the discussion of intermolecular carbenoid cyclopropanations has been limited to metal-catalyzed reactions of diazo compounds containing at least one electron-withdrawing group. Another source of carbenoids that has been used extensively are iodonium ylides.[218–222] It has been suggested that this carbenoid source is safer than diazo compounds, but overall it has not been demonstrated that they offer significant advantages over diazo compounds.[221,222]

Metal-catalyzed decomposition of diazomethane in the presence of alkenes is a general synthetic process for the formation of cyclopropanes.[31] An alternative and more stable carbenoid precursor that can be used is silyldiazomethane.[223] Palladium acetate is the favored catalyst for these reactions[31] and it is likely that these reactions do not involve discrete metal carbenoid species.[37,50] As cyclopropanation of alkenes containing electron-withdrawing groups is possible,[31] products similar to those from the reaction of diazoacetates and unpolarized alkenes can be formed. Chiral auxiliaries on the alkene have proved to be effective for asymmetric induction as illustrated in Eq. 52.[224,225] The results for chiral catalysis, however, have tended to be disappointing,[226] but reasonable levels of asymmetric induction are obtained using the semicorrin copper complex **22** (Eq. 53).[123]

Other precursors to carbenoid species are phenyldiazomethanes,[32,227] vinyldiazomethanes,[33,228] and alkynyldiazomethanes.[33] Illustrative examples of cyclopropanations with these reagents are shown in Eqs. 54–56. These carbenoid precursors have not been extensively used in organic synthesis because they are unstable and difficult to handle.

R^1, R^2, R^3 = H, various alkyl, aryl, alkoxy

R = alkyl

R = alkyl

The use of carbenoid intermediates in intramolecular cyclopropanations is a very useful transformation because fused cyclopropanes are readily formed, and unlike the intermolecular version, full diastereocontrol is assured.[2,12,20,229] Illustrative examples are shown in Eqs. 57[230] and 58.[231] Further manipulation of the

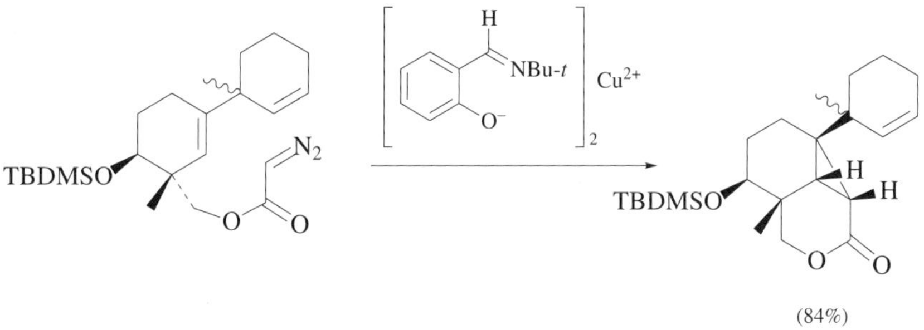

(Eq. 57)

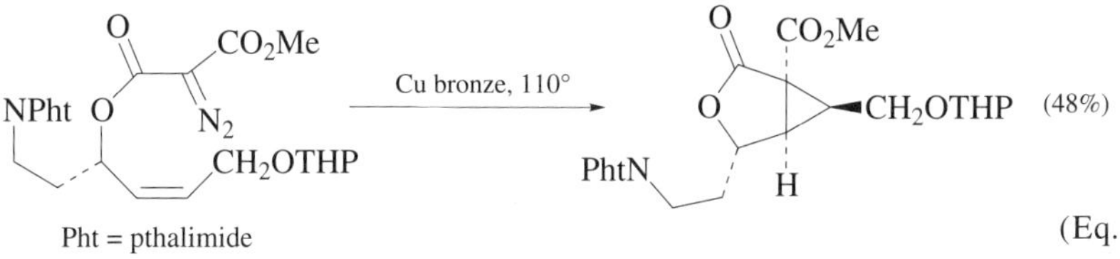

(Eq. 58)

fused cyclopropanes has led to the synthesis of numerous natural products.[232] The asymmetric synthesis of fused cyclopropanes can be effectively achieved using a number of catalysts. The most effective to date appear to be the Doyle $Rh_2(S\text{-MEPY})_4$ catalyst **25** and related rhodium amide catalysts,[233–237] and the Evans C_2-symmetric copper bisoxazoline catalyst **23**.[126,238]

An alternative method for the cyclopropanation of alkenes is the Simmons-Smith reaction.[36] In this venerable reaction diiodomethane in the presence of zinc reagents is the carbenoid source. Improvements have been made by using samarium iodide instead of zinc reagents.[35] The asymmetric version of the reaction has now been achieved either by using chiral auxiliaries or chiral catalysis, as illustrated in Eq. 59[239–242] and Eq. 60.[243,244] Although the Simmons-Smith reaction is

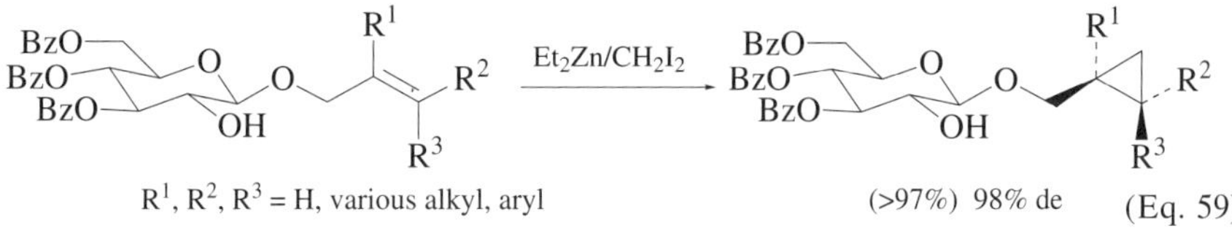

(Eq. 59)

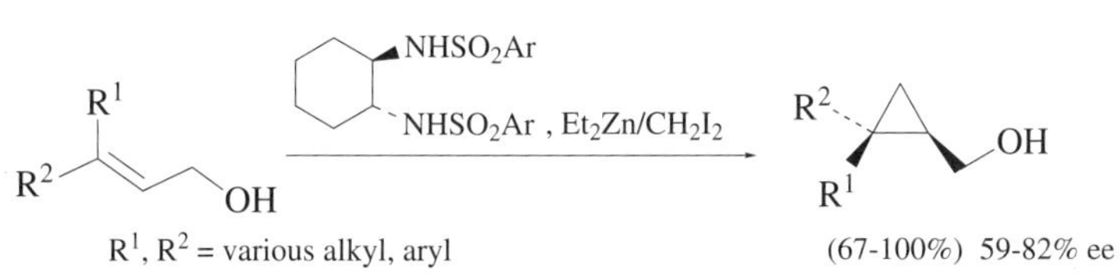

(Eq. 60)

a powerful reaction for methylene transfer to an alkene, unlike the metal-catalyzed decomposition of diazo compounds, is not applicable to a broad range of carbenoid functionality.

Cyclopropanes may also be formed from the reaction of alkenes with Fisher carbene complexes[245-249] (Eq. 61) and other stoichiometric organometallic

$$MeO_2C-CH=CH_2 + (CO)_5Cr=C(OMe)(CH=CHPh) \longrightarrow \text{cyclopropane}(65\%) + \text{cyclopropane}(16\%)$$

(Eq. 61)

sources of carbenes.[37] Even though a range of diverse structures can be obtained using these organometallic complexes, the reactions are not particularly attractive because a stoichiometric quantity of the carbene complex is required.

A number of other methods are also available for the stereoselective synthesis of cyclopropanes.[24] Some of the most general are the following: 1,3-dipolar cycloaddition of diazoalkanes to various alkenes followed by extrusion of nitrogen;[250,251] 1,4-addition of ylides to electron-deficient alkenes;[252] reaction of 1-phenylseleno-2-silylethenes with electron-deficient alkenes;[253-255] and palladium-catalyzed tandem alkylation and cyclization of stabilized anions to 1,4-dichloro-2-butene.[256]

EXPERIMENTAL CONSIDERATIONS

Synthesis of Diazo Compounds

In order to be able to exploit metal-catalyzed cyclopropanation, practical methods are required for the synthesis of the diazo compounds. Excellent summaries of the methods available for the synthesis of diazo compounds have been published.[12,22] Only a general overview is presented here. The most common methods for preparing diazo compounds containing at least one electron-withdrawing group are the following:

1. Reaction of activated carboxylic acid derivatives with diazoalkanes.

The reaction of diazomethane with acid chlorides or anhydrides is a well-established method for the synthesis of diazo ketones (Eq. 62).[257-260] The reaction

$$\text{Cl-C(O)-R} \xrightarrow{CH_2N_2} N_2=CH-C(O)-R \quad \text{(Eq. 62)}$$

can be carried out with a very broad range of substrates and can be extended to functionalized diazoalkanes.[12] The major drawback with this synthetic method is the inherent danger of handling diazoalkanes.[259,261] *Diazomethane needs to be handled extremely carefully because it is prone to explode even on contact with ground-glass joints.*[259,261]

2. Diazo transfer reactions using arylsulfonyl azides.

The diazo transfer reaction is a very practical method for the synthesis of diazo compounds (Eq. 63).[12,22] A wide range of diazo transfer agents have been devel-

$$\underset{EWG^2}{\overset{EWG^1}{\diagdown}} \xrightarrow[\text{base}]{ArSO_2N_3} N_2 {=}\!\!\underset{EWG^2}{\overset{EWG^1}{\diagup}} \qquad (Eq.\ 63)$$

oped.[262] Traditionally, toluenesulfonyl azide (**130**) has been used as the diazo transfer agent, but owing to its potential explosive properties,[263,264] safer reagents have been developed. A detailed overview of the safety issues regarding arylsulfonyl azides is available.[262] The two most practical reagents are *p*-dodecylbenzenesufonyl azide (**131**)[265–267] and *p*-acetamidobenzenesulfonyl azide (*p*-ABSA, **132**).[268,269] The first is ideally suited for the synthesis of crystalline diazo com-

130   **131**   **132**: *p*-ABSA

pounds because the sulfonamide byproduct is a highly soluble liquid that can be removed by trituration of the product. *p*-ABSA is commercially available and is ideal for noncrystalline diazo compounds because the sulfonamide byproduct is highly crystalline and can be removed by trituration, leaving the diazo compound in solution. Various bases can be used in the reaction but the most commonly used are triethylamine and DBU.[12]

The diazo transfer reaction is ideally suited for the synthesis of compounds containing two electron-withdrawing groups, although phenyldiazoacetates and vinyldiazoacetates are readily formed from the diazo transfer reaction if DBU is used as base.[156,268] Convenient methods have been developed for the preparation of diazo compounds with a single electron-withdrawing group, by following the diazo transfer reaction with deformylation (Eq. 64).[270] Acetyl,[78] benzoyl,[271,272] and trifluoroacetyl[110,273] groups have also been commonly used as the second electron-withdrawing group that would be lost in a deacylation step.

$$\underset{CHO}{\overset{CO_2R}{\diagdown}} \xrightarrow[\text{base}]{ArSO_2N_3} N_2{=}\!\!\diagup\!\!{CO_2R} \qquad (Eq.\ 64)$$

R = various alkyl, aryl

3. **Thermolysis of tosylhydrazones.**

A well-established method for the synthesis of diazo compouds is the thermolysis of tosylhydrazones.[274] A particularly important application of this method is for the synthesis of alkyl diazoacetates by reaction of alcohols with tosylimidoglycolate **133** (Eq. 65).[275]

$$\text{TosNHN}=\underset{\textbf{133}}{\overset{\text{COCl}}{\diagdown}} \xrightarrow[\text{NEt}_3]{\text{ROH}} \text{N}_2=\overset{\text{CO}_2\text{R}}{\diagdown} \quad \text{(Eq. 65)}$$

R = various alkyl, aryl

4. **Functionalization of a diazocarbonyl derivative.**

In recent years it has become apparent that useful reactions can be carried out on diazo compounds without destruction of the diazo functionality. The chemistry of the succinamide **134**, which can be converted to a range of diazoacetate derivatives, is of notable practical application (Eq. 66).[233,276] The diazomalonyl chloride **135** is another useful reagent for the synthesis of elaborate diazomalonate derivatives (Eq. 67).[277] Diazoacetoacetates may be silylated or reduced and then dehydrated to form vinyldiazoacetates (Eqs. 68[144,278] and 69).[279] The enolate of diazoacetate is also a useful reagent capable of undergoing aldol and alkylation reactions.[280]

$$\text{N}_2=\overset{\text{O}}{\diagdown}\text{O-N(succinimide)} \xrightarrow{\text{Nu}^-} \text{N}_2=\overset{\text{O}}{\diagdown}\text{Nu} \quad \text{(Eq. 66)}$$

134

$$\text{N}_2=\underset{\textbf{135}}{\overset{\text{CO}_2\text{R}}{\underset{\text{COCl}}{\diagdown}}} \xrightarrow{\text{Nu}^-} \text{N}_2=\overset{\text{CO}_2\text{R}}{\underset{\text{CONu}}{\diagdown}} \quad \text{(Eq. 67)}$$

R = various alkyl, aryl

$$\text{N}_2=\overset{\text{CO}_2\text{R}}{\diagdown}\overset{\text{O}}{\underset{\text{R'}}{\diagdown}} \xrightarrow{\text{TBSOTf/NEt}_3} \text{N}_2=\overset{\text{CO}_2\text{R}}{\diagdown}\overset{}{\underset{\text{TBSO} \quad \text{R'}}{\diagdown}} \quad \text{(Eq. 68)}$$

R = various alkyl
R' = H, alkyl

$$N_2 = \overset{CO_2R}{\underset{\underset{R'}{O}}{\big|}} \xrightarrow[\text{2. POCl}_3]{\text{1. NaBH}_4} N_2 = \overset{CO_2R}{\underset{R'}{\big|}}$$ (Eq. 69)

R = various alkyl
R' = H, alkyl, alkoxy, thioalkoxy

A variety of miscellaneous methods are also available for the synthesis of diazo compounds containing at least one electron-withdrawing group. These include diazotization of amines,[281,282] decomposition of tosylhydrazones,[283] and reaction of oximes with chloramine.[284]

EXPERIMENTAL CONDITIONS

General Reaction Conditions

General reaction conditions for intermolecular carbenoid cyclopropanations are very practical.[75] Most of the catalysts are stable compounds or can be generated in situ from stable precursors. Rhodium(II) carboxylates are indefinitely stable in the open atmosphere. Copper(I) catalysts are generally formed in situ by reaction of the ligand with a copper salt such as copper(I) triflate. Many of the catalysts are introduced into the reaction as copper(II) complexes, but copper(II) is reduced to copper(I) under the reaction conditions, and it is most likely that copper(I) salts are the active catalysts in these reactions.

The diazo compound is usually added slowly to a stirred solution of the catalyst and the alkene. Usually, an excess of alkene is used in order to minimize side-reactions such as carbene dimerization, which can be very prevalent in the diazoacetate system. It has been shown, however, that good yields of cyclopropanes can be obtained with just one equivalent of alkene as long as syringe pump techniques are used for slow addition of the diazo compound.[94] Typically, the reaction is carried out under an inert atmosphere and anhydrous conditions. The standard solvent for many of the carbenoid reactions is dichloromethane but many other solvents, such as benzene, fluorobenzene, hexane, pentane, and diethyl ether have been used. It has become increasingly recognized that solvents can have a major effect on these reactions. Products derived from zwitterionic intermediates can be minimized by using nonpolar solvents.[63,64,185] Also, the enantioselectivity displayed by certain catalysts can be very solvent dependent.[58,74]

Safety Considerations

Great care should be taken in handling diazo compounds because they are potentially toxic and can have explosive properties. Even though diazo compounds containing electron-withdrawing functionalities are much more stable than their diazoalkane counterparts, detailed studies on the potential dangers of many of the diazo compounds described in this review are not available. Therefore, diazo compounds should be handled carefully and all reactions should be carried out in a well-vented fume hood, behind a blast shield.

There have been reports in the literature of unpredicted explosions occurring when using toluenesulfonyl azide[263,264] and methanesulfonyl azide.[263,264] It is thus advisable that the more stable arylsulfonyl azides such as p-dodecylbenzenesulfonyl azide[265–267] and p-acetamidobenzenesulfonyl azide (p-ABSA)[268,269] be used instead of toluenesulfonyl azide or methanesulfonyl azide to carry out the diazo transfer reactions.

EXPERIMENTAL PROCEDURES

General Procedures for the Preparation of Carbenoid Precursors

The following experiments are typically carried out behind a blast shield in a one- or three-necked flask equipped with a magnetic stirring bar (or a mechanical stirrer for large scale reactions) and either a dropping funnel or rubber septum. All reactions are performed under an inert atmosphere (nitrogen or argon). Addition of diazo compounds to the reaction mixture is accomplished via dropping funnel, cannula, or syringe pump.

N-Diazoacetylmorpholine [Use of Succinimidyl Diazoacetate].[233,276] Glyoxylic acid chloride (p-toluenesulfonyl)hydrazone[285] (103.3 g, 0.500 mol) in CH_2Cl_2 (1.00 L) was added over 2 hours to a mechanically stirred suspension of N-hydroxysuccinimide (63.25 g, 0.550 mol) and Na_2CO_3 (79.5 g, 0.750 mol) in dry CH_2Cl_2 (0.75 L) maintained at 0°. The resulting mixture was stirred for an additional hour and then warmed to room temperature, where it was maintained for 3 hours, after which time it was filtered though a sand plug and then Celite®. The filtrate was concentrated under reduced pressure to provide crude succinimidyl diazoacetate as a brown-yellow solid. Recrystallization from CH_2Cl_2/hexanes gave a light yellow crystalline product (43.0 g, 0.235 mol, 47% yield),[233] mp 113.5–115.0° (lit.[276] mp 119–120°); IR ($CHCl_3$) 2100 cm^{-1}; ^{1}H NMR ($CDCl_3$) δ 2.85 (s, 4 H), 5.13 (br s, 1 H); ^{13}C NMR ($CDCl_3$) δ 25.4, 45.0, 162, 169.3; MS (m/z): 184 (M + 1, 21); Anal. Calcd. for $C_6H_5N_3O_4$: C, 39.03; H, 2.73; N, 22.95; O, 34.97. Found: C, 39.59; H, 2.8; N, 22.98; O, 35.1.

Morpholine (1.74 g, 20 mmol) in THF (5 mL) was treated with succinimidyl diazoacetate (1.83 g, 10 mmol) at room temperature for 1 hour. The solvent was evaporated in vacuo, and the product was purified by chromatography on silica gel to give an amber yellow oil (1.51 g, 97%); IR (CHCl$_3$) 2100 cm^{-1}; ^{1}H NMR (CDCl$_3$) δ 3.39 (m, 4 H), 3.66 (m, 4 H), 4.98 (s, 1 H); ^{13}C NMR (CDCl$_3$) δ 43.9, 66.4, 130.8, 164.89; MS (m/z): 155 (M$^+$, 70).

Ph—CH=CH—C(O)—CH$_3$
1. LiHMDS, THF, CF$_3$CO$_2$CH$_2$CF$_3$
2. 4-dodecylbenzenesulfonyl azide,
Et$_3$N, H$_2$O, CH$_3$CN
→ Ph—CH=CH—C(O)—CH=N$_2$

(E)-1-Diazo-4-phenyl-3-buten-2-one [Detrifluoroacetylative Diazo Group Transfer Using 4-Dodecylbenzenesulfonyl Azide].[273] A 500-mL, three-necked, round-bottomed flask was equipped with a mechanical stirrer, nitrogen inlet adapter, and 150-mL pressure-equalizing dropping funnel fitted with a rubber septum. The flask was charged with dry THF (70 mL) and hexamethyldisilazane (15.9 mL, 0.075 mol), and then cooled in an ice-water bath while a 2.50 M solution of n-butyllithium (28.8 mL, 0.072 mol) in hexane was added dropwise over 5–10 minutes. After 10 minutes, the resulting solution was cooled to −78° in a dry ice-acetone bath, and a solution of of trans-4-phenyl-3-buten-2-one (10.0 g, 0.068 mol) in dry THF (70 mL) was added dropwise over 25 minutes. The dropping funnel was washed with THF (2 × 5 mL) and then replaced with a rubber septum. The yellow reaction mixture was stirred for 30 minutes at −78°, and then 2,2,2-trifluoroethyl trifluoroacetate (10.1 mL, 0.075 mol) was added rapidly in one portion via syringe. After 10 minutes, the reaction mixture was poured into a 1-L separatory funnel containing diethyl ether (100 mL) and 5% aqueous HCl (200 mL). The aqueous layer was separated and extracted with diethyl ether (50 mL). The combined organic layers were washed with saturated NaCl solution (200 mL), dried over anhydrous Na$_2$SO$_4$, filtered, and concentrated under reduced pressure using a rotary evaporator to afford a yellow oil (18.61 g). This yellow oil was immediately dissolved in acetonitrile (70 mL) and transferred to a 500-mL, one-necked flask equipped with a magnetic stirring bar and a 150-mL pressure equalizing dropping funnel fitted with a nitrogen inlet adapter. Water (1.2 mL, 0.069 mol), triethylamine (14.3 mL, 0.103 mol), and a solution of 4-docecylbenzenesulfonyl azide[267] (35.74 g, 0.103 mol) in acetonitrile (10 mL) were then added sequentially via the dropping funnel. The resulting yellow solution was stirred at room temperature for 6.5 hours and then poured into a 1-L separatory funnel containing diethyl ether (100 mL) and aqueous 5% NaOH (200 mL). The organic layer was separated, washed successively with 5% aqueous NaOH (3 × 200 mL), water (4 × 200 mL), and saturated NaCl (200 mL), dried over anhydrous Na$_2$SO$_4$, filtered, and concentrated at reduced pressure to yield the crude reaction product as a light brown oil (23.17 g). The crude reaction product was purified by column chromatography (silica gel, 5–10% diethyl ether/hexane) to furnish 9.80 g (83% yield) of (E)-1-diazo-4-phenyl-3-buten-2-one (mp 68–69°) as a bright yellow solid. IR (CCl$_4$) 3150–3000, 2090, 1645, 1600, 1445, 1360, 1180, 1140, 1095, 1070, 970, 690 cm^{-1}; ^{1}H NMR (CDCl$_3$) δ 5.54 (s, 1 H), 6.60 (d, J = 15.8 Hz, 1 H), 7.30–7.34 (m, 3 H), 7.46–7.49 (m, 2 H), 7.57 (d, J = 15.8 Hz, 1 H); ^{13}C NMR (CDCl$_3$) δ 55.8, 123.5, 127.8, 128.5,

129.9, 134.0, 140.1, 184.0; Anal. Calcd for $C_{10}H_8N_2O$: C, 69.76; H, 4.68; N, 16.27. Found: C, 69.65; H, 4.84, N, 16.32.

Ph−CH=CH−CH₂−CO₂Me →(p-ABSA, DBU / CH₃CN, 0°)→ Ph−CH=CH−C(=N₂)−CO₂Me

Methyl (E)-2-Diazo-4-phenyl-3-butenoate [Diazo Transfer Reaction with p-Acetamidobenzenesulfonyl Azide].[156] DBU (2.39 g, 15.7 mmol) was added to a stirred solution of methyl 4-phenylbutenoate[286] (2.05 g, 11.6 mmol) and p-acetamidobenzenesulfonyl azide[268,269] (2.8 g, 11.8 mmol) in acetonitrile (75 mL) at 0°. After the mixture was stirred for 4 hours, saturated aqueous NH₄Cl solution was added, and the mixture was extracted twice with CH_2Cl_2. The organic layer was then dried ($MgSO_4$), and the solvent was evaporated under reduced pressure. The residue was triturated with ether/pentane (50:50), filtered, and the solvent was evaporated under reduced pressure. Further purification of the product by chromatography on silica gel with ether/petroleum ether (1:4) as the eluent gave methyl (E)-2-diazo-4-phenyl-3-butenoate (2.08 g, 89% yield) as a red solid; IR (neat) 3010, 2940, 2040, 1695, 1620, 1590, 1440 cm^{-1}; ^{1}H NMR (CDCl₃) δ 3.81 (s, 3 H), 6.15 (d, J = 16.2 Hz, 1 H), 6.44 (d, J = 16.2 Hz, 1 H), 7.34–7.15 (m, 5 H).

Ph−CH=CH₂ + N₂=CH−CO₂Et →(23/CuOTf, CHCl₃, rt)→ Ph−△−CO₂Et (1S,2S) + Ph−△−CO₂Et (1S,2R)

General Procedures for Carbenoid Cyclopropanation Reactions

Ethyl (1S,2S)-2-Phenylcyclopropanecarboxylate and Ethyl (1S,2R)-2-Phenylcyclopropanecarboxylate [Asymmetric Cyclopropanation of Styrene with Ethyl Diazoacetate in the Presence of Copper Bisoxazoline Catalyst 23].[126] To a suspension of CuOTf (0.0068 g, 0.027 mmol) was added a solution of **23** (8.1 mg, 0.028 mmol) in CHCl₃ (9 mL). After 1 hour, the mixture was passed through a filter cannula comprised of a needle with the hub packed with glass wool. To this was added styrene (1.6 mL, 14 mmol), and then a solution of ethyl diazoacetate (0.285 mL, 2.71 mmol) in CHCl₃ (10 mL) was added dropwise over 1.5 hours. After 14 hours, the mixture was concentrated in vacuo to a green oil. This was purified by flash chromatography (3 × 30 cm silica, 10:90 EtOAc/hexane as eluent). The products were isolated as a clear oil (0.399 g, 77%, trans:cis = 73:27). The isomers obtained were separated by medium pressure liquid chromatography (Michel-Miller column, 3:97–10:90 EtOAc/hexane as eluent) and identified. Ethyl (1S,2S)-2-phenylcyclopropanecarboxylate: mp 37.7–39.1°; IR (CCl₄) 3090, 3070, 3040, 2990, 2910, 2880, 1730, 1610, 1500, 1480, 1460, 1440, 1410, 1390, 1370, 1340, 1325, 1305, 1290, 1265, 1220, 1180, 1120, 1100, 1080, 1045, 1020, 950, 935, 850 cm^{-1}; ^{1}H NMR (CDCl₃) δ 7.1–7.4 (m, 5 H), 4.17 (q, J = 7.1 Hz, 2 H), 2.52 (ddd, J = 4.1, 6.4, 9.2 Hz, 1 H), 1.90 (ddd,

J = 4.2, 5.3, 8.4 Hz, 1 H), 1.60 (m, J = 4.4, 5.1, 9.4 Hz, 1 H), 1.31 (ddd, J = 4.6, 6.5, 8.4 Hz, 1 H), 1.28 (t, J = 7.1 Hz, 3 H); [α]$_D$ + 296° (c 0.88, CHCl$_3$); 99% ee. Ethyl (1S,2R)-2-phenylcyclopropanecarboxylate: IR (thin film) 3090, 3060, 3030, 2980, 2940, 2910, 2880, 1730, 1605, 1495, 1480, 1450, 1440, 1400, 1380, 1355, 1270, 1220, 1180, 1155, 1095, 1080, 1030, 955, 865, 850, 790, 750, 720, 695 cm^{-1}; ^{1}H NMR (CDCl$_3$) δ 7.1–7.3 (m, 5 H), 3.87, (q, J = 7.1 Hz, 2 H), 2.58 (m, 1 H), 2.08 (ddd, J = 5.6, 7.8, 9.3 Hz, 1 H), 1.71 (m, J = 5.3, 7.5 Hz, 1 H), 1.34 (ddd, J = 5.1, 7.9, 8.6 Hz, 1 H), 0.97 (t, J = 7.1 Hz, 3 H); [α]$_D$ + 18.6° (c 1.01, CHCl$_3$); 97% ee.

Ethyl 2-Phenylcyclopropanecarboxylate [Reaction of Ethyl Diazoacetate with Styrene in the Presence of Rh$_2$(OAc)$_4$].[78] Ethyl diazoacetate (57 mg, 0.50 mmol) dissolved in anhydrous CH$_2$Cl$_2$ (3.0 mL) was added at a controlled rate over a 5-hour period to a stirred mixture of styrene (520 mg, 5.0 mmol) and Rh$_2$(OAc)$_4$ (2.2 mg, 0.005 mmol) in CH$_2$Cl$_2$ (3.0 mL) at room temperature. Two hours after addition was complete the reaction solution was passed through a short alumina column with CH$_2$Cl$_2$ as the eluent to remove the catalyst. Solvent and excess styrene were distilled under reduced pressure to give ethyl trans- and cis-2-phenylcyclopropanecarboxylate (88 mg, 93%, 62:38).

Ethyl (1S,2S)-2-Phenylcyclopropanecarboxylate and Ethyl (1S,2R)-2-Phenylcyclopropanecarboxylate [Asymmetric Cyclopropanation of Styrene with Ethyl Diazoacetate in the Presence of Rh$_2$(5S-MEPY)$_4$ (25)].[108,287] To a light blue solution of styrene (1.04 g, 10.0 mmol) and Rh$_2$(5S-MEPY)$_4$ (25) (7.7 mg, 0.010 mmol) in anhydrous CH$_2$Cl$_2$ (20 mL) under N$_2$ was added ethyl diazoacetate (0.114 g, 1.00 mmol) in CH$_2$Cl$_2$ (10 mL) through a syringe pump at a rate of 0.5 mL/hour. After addition was complete, the mixture was filtered through a 1-cm plug of silica gel to separate the catalyst, and the plug was eluted with CH$_2$Cl$_2$ (30 mL). The excess styrene was removed and the residue was purified by bulb-to-bulb distillation at 60–80°/0.05 Torr to give the products as a clear oil (0.112 g, 59%). GC analysis was performed prior to and following distillation without noticeable change in isomer ratios (trans:cis = 56:44). Diastereomer ratios were obtained using a capillary methyl silicone column, and enantiomer separation was performed by GC, using a Chiraldex-γ-cyclodextrin-TFA column (58% ee trans, 33% ee cis).

Ethyl (1R,2R)-2-Phenylcyclopropanecarboxylate and Ethyl (1R,2S)-2-Phenylcyclopropanecarboxylate [Asymmetric Cyclopropanation of Styrene with Ethyl Diazoacetate in the Presence of trans-RuCl$_2$(Pybox-ip)(ethylene) (28)].[106] To a solution of trans-RuCl$_2$(Pybox-ip)(ethylene) (28) (30 mg, 0.06 mmol) and styrene (1.7 mL, 15 mmol) in CH$_2$Cl$_2$ (2.0 mL) was added a dichloromethane solution of ethyl diazoacetate (3.0 mmol, ~1 M) through a microsyringe controlled by a mechanical feeder (~4 μL/drop, ~0.4 mL/hour) for 8 hours at 20–25° under an argon atmosphere. After stirring for an additional 10 hours, the reaction mixture was concentrated under reduced pressure. The residual oil was subjected to silica gel column chromatography with hexane-ether as eluent to give an oily mixture (416 mg, 73% yield) of methyl trans-2-phenylcyclopropane-1-carboxylate and methyl cis-2-phenylcyclopropane-1-carboxylate (trans:cis = 91:9). Enantiomeric purity was measured by GLPC (Astec, Chiraldex B-DA, 30 m × 0.25 mm) (89% ee trans, 79% ee cis).

Ethyl 2,5-Dioxabicyclo[4.1.0]heptanecarboxylate [Reaction of Ethyl Diazoacetate with p-Dioxene in the Presence of Copper-bronze].[95] Ethyl diazoacetate (2.85 g, 25 mmol) was added dropwise to a stirred suspension of copper-bronze (300 mg) in p-dioxene (4.5 g, 50 mmol), kept at 80° under nitrogen, and thereafter the stirring and heating was continued for 0.5 hour. After removal of the excess p-dioxene by vacuum distillation the mixture was filtered through Celite. The catalyst was washed with ether and the combined filtrates were evaporated. Distillation of the residue yielded 3.45 g (80% yield) of a colorless liquid, ethyl 2,5-dioxabicyclo[4.1.0]heptanecarboxylate: bp 57–59° (0.1 Torr); IR (C=O) 1718 cm^{-1}; ^{1}H NMR δ 1.20 (t, J = 7 Hz, Me, 3 H), 2.12 (t, J = 3 Hz, COCH, 1 H), 3.63 (s, 4 H, 2 OCH$_2$), 3.89 (d, J = 3 Hz, 2 OCH, 2 H), 4.07 (q, J = 7 Hz, OCH$_2$ of OEt, 2 H); ^{13}C NMR δ 13.9 (Me), 24.4 (C-7), 56.8 (C-1, C-6), 60.0 (OCH$_2$), 62.5 (C-3, C-4), 170.2 (C=O).

(3a*S**,8a*R**)-2,2-Dimethyl-2,3,3a,6,7,8a-hexahydro-5*H*-1,8-dioxacyclopent[*a*]indene-3,4-dione [Reaction of 2-Diazo-1,3-cyclohexanedione with 2,2-Dimethyl-3(2*H*)-furanone in the Presence of Rh$_2$(OAc)$_4$].[136] To a solution of Rh$_2$(OAc)$_4$ (0.192 g, 0.43 mmol) and 2,2-dimethyl-3(2*H*)-furanone (4.871 g, 43.44 mmol) in fluorobenzene (50 mL) was added a solution of 2-diazo-1,3-cyclohexanedione (3.0 g, 21.7 mmol) in fluorobenzene (4 mL) at room temperature. The reaction mixture was stirred for 10 hours. Evaporation and filtration through Celite˙ with 25% ethyl acetate in hexane afforded a viscous oil which upon treatment with 20% ether in hexane gave 2.703 g (56%) of a gummy solid: ^{1}H NMR (300 MHz, CDCl$_3$), δ 6.52 (d, J = 6.6 Hz, 1 H), 3.95 (d, J = 6.6 Hz, 1 H), 2.54–2.01 (m, 6 H), 1.32 (s, 6 H); IR (KBr) 2956, 1740, 1628, 1397, 1330, 1186, 1132, 1100, 915 cm^{-1}; EIMS (m/z, rel. intensity): 222, (M$^+$, 100), 204 (14), 179 (22), 176 (13), 152 (11), 136 (10), 124 (4), 108 (4); HRMS m/z (M$^+$) for C$_{12}$H$_{14}$O$_4$ calcd 222.0892, found 222.0894.

$$Ph\diagup\diagdown + \underset{Ph}{\overset{N_2}{\diagup\diagdown}}CO_2Me \xrightarrow{Rh_2(S\text{-DOSP})_4\ (46)}_{pentane,\ -78°} Ph\diagdown\triangle\diagup\diagdown Ph$$
$$CO_2Me$$
$$(1S, 2S)$$

(1*S*,2*S*)-Methyl 2β-Phenyl-1β-(2-(*Z*)-styryl)cyclopropane-1α-carboxylate [Asymmetric Cyclopropanation of Styrene with Methyl (*E*)-2-Diazo-4-phenyl-3-butenoate in the Presence of Rh$_2$(*S*-DOSP)$_4$ (46)].[58] A mixture of styrene (44.2 g, 424 mmol) and Rh$_2$(*S*-DOSP)$_4$ (46) (1.58 g, 0.85 mmol) in pentane (350 mL) was stirred at $-78°$ under an argon atmosphere. To this solution was added methyl (*E*)-2-diazo-4-phenyl-3-butenoate (17.2 g, 84.8 mmol) in pentane (0.12 M) over 30 minutes, and the reaction mixture was then stirred at $-78°$ for 24 hours. The mixture was then concentrated in vacuo, and the residue was purified on silica using ether/petroleum ether (0:100 to 10:90) as the eluent to give (1*S*,2*S*)-methyl 2β-phenyl-1β-[2-(*Z*)-styryl]cyclopropane-1α-carboxylate (16.05 g, 68%) as a white solid (mp 57–60°; 98% ee); IR (CHCl$_3$) 3110, 3090, 3060, 2980, 2950, 2880, 1735 cm^{-1}; ^{1}H NMR[148] (CDCl$_3$) δ 1.85 (dd, J = 7.3, 5.1 Hz, 1 H), 2.05 (dd, J = 9.1, 5.1 Hz, 1 H), 3.04 (dd, J = 9.1, 7.3 Hz, 1 H), 3.77 (s, 3 H), 6.15 (d, J = 15.9, 1 H), 6.37, (d, J = 15.9 Hz, 1 H), 7.12–7.28 (m, 10 H); ^{13}C NMR (CDCl$_3$) δ 18.5, 33.2, 34.9, 52.3, 124.0, 126.1, 126.7, 127.2, 127.9, 128.3, 129.0, 133.0, 135.4, 137.0, 174.1; [α]$^{25}_D$ -166° (c 1.1, CHCl$_3$); Anal. Calcd. for C$_{19}$H$_{18}$O$_2$: C, 81.99; H, 6.52. Found: C, 81.74; H, 6.53.

TABULAR SURVEY

Tables I-III cover achiral cyclopropanations of alkenes. They are divided according to the three major types of carbenoid systems: those with a single electron-withdrawing group, those with two electron-withdrawing groups, and those with both an electron-withdrawing and an electron-donating group. Charts appearing before the Tables section are used to show structures of more complicated catalysts and assign them numbers for use in the Tables section. Chart 1 depicts non-chiral catalysts.

Tables IV-VII cover the asymmetric cyclopropanation of alkenes. The three major types of carbenoids are described together in these tables. In Table IV the use of chiral auxiliaries is described, while in Tables V-VII, the use of chiral catalysis is described. Because of the vast range of chiral catalysts that have been used for asymmetric cyclopropanation, Tables V-VII are divided according to the metal used, copper, rhodium, and miscellaneous metals. Charts 2–5 depict chiral catalysts.

Tables VII-X cover both achiral and asymmetric reactions of carbenoids with dienes, furans and pyrroles. The three major types of carbenoids are described together in these tables.

Entries in the Tables are ordered by increasing carbon count (boldface type) of the carbenoid precursor. Where the carbenoid precursor structure contains one or more "R" groups, the carbon count for the precursor with the substituents defined is also given (plain type). For a particular carbon count of carbenoid precursor, the alkene substrates are also ordered by increasing carbon count, followed by increasing hydrogen count. Protecting groups are omitted from the carbon count. Unspecified yields are indicated by (—), while missing or unspecified trans/cis ratios, enantiomeric excesses, and absolute stereochemistries are indicated by —. The term "ee" is used to describe the asymmetric induction of the cyclopropanation. In order to avoid confusion with the formation of cis and trans cyclopropane isomers, the term "ee" is used even when chiral auxiliaries exist on the carbenoid and the resulting cyclopropanes, prior to removal of the auxiliary, are actually diastereomers.

Note that some "R" groups, especially in Table IV, are designated by the compound from which the ester in the carbenoid precursor is derived, such as pantolactone and methyl lactate. This convention allows a clearer representation of the precursor than the alternative chemical name.

The tables contain all examples that could be found in the literature from 1970 through June 1999.

ABBREVIATIONS

abs	absolute
p-ABSA	*p*-acetamidobenzenesulfonyl azide
Ac	acetyl
acac	acetylacetonate
aq	aqueous
BHT	2,6-bis-*tert*-butyl-4-methylphenolyl
Bn	benzyl
BOC	*tert*-butoxycarbonyl
BNP	binaphthyl phosphate
BSI	*N*-*tert*-butyl salicylimide
Bz	benzoyl
CBZ	benzyloxycarbonyl
COD	cyclooctadiene
Cp	cyclopentadiene
Cy	cyclohexyl
DBU	1,5-diazabicyclo[5.4.0]undec-5-ene

DCE	1,2-dichloroethane
DCM	dicyclohexylmethyl
DDQ	2,3-dichloro-5,6-dicyano-1,4-benzoquinone
de	diastereomeric excess
Diphos	1,2-bis(diphenylphosphino)ethane
DME	1,2-dimethoxyethane
DMF	N,N-dimethylformamide
DPMS	diphenylmethylsilyl
Dppe	1,2-bis(diphenylphosphino)ethane
EDA	ethyl diazoacetate
ee	enantiomeric excess
ether	diethyl ether
c-hex	cyclohexyl
Hex	Hexane
LDA	lithium diisopropyl amide
LiHMDS	lithium hexamethyldisilylamide
MOM	methoxymethyl
Naph	naphthyl
Oct	octanoate
OEP	octaethylporphyrin
PE	polyethylene
Pfb	perfluorobutyrate
Pip	piperidine
Piv	pivalate
PMB	p-methoxybenzyl
PNB	p-nitrobenzyl
POM	pivaloyloxymethyl
Pr	propyl
pyr	pyridine
rt	room temperature
sat.	saturated
SC	supercritical
TBAF	tetrabutylammonium fluoride
TBDMS	*tert*-butyldimethylsilyl
TCE	1,1,2,2-tetrachloroethane
TES	triethylsilyl
Tf	trifluoromethanesulfonyl
TFA	trifluoroacetic acid
THF	tetrahydrofuran
THP	2-tetrahydropyranyl
TMP	tetramesitylporphyrin
TMS	trimethylsilyl
TPA	triphenyl acetate
TpFPP	tetra(hexafluorophenyl)porphyrin
Ts	p-toluenesulfonyl
p-TsOH	p-toluenesulfonic acid
TTP	tetra(p-tolyl)porphyrin

CHART 1. NON-CHIRAL CATALYST COMPLEXES

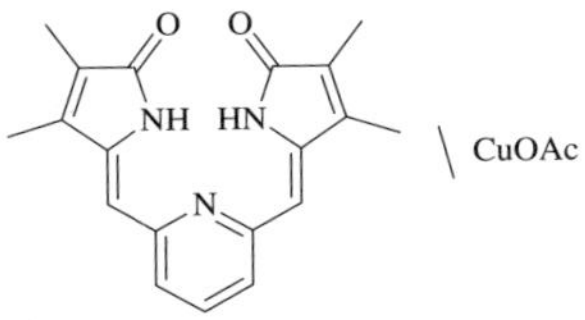

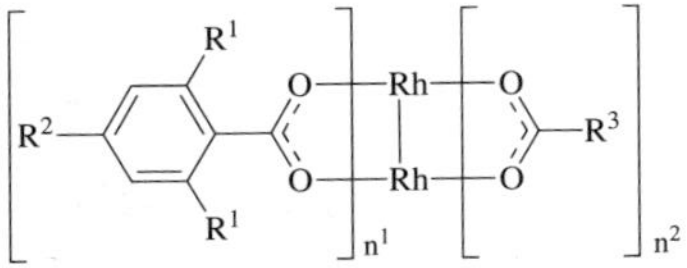

	R^1	R^2	R^3	n^1	n^2
7a:	Ph	Ph	Me	3	1
7b:	Ph	Ph	Me	4	0
7c:	4-MeC$_6$H$_4$	4-MeC$_6$H$_4$	Me	3	1
7d:	4-MeC$_6$H$_4$	4-MeC$_6$H$_4$	Me	4	0
7e:	4-biphenyl	Ph	Me	3	1
7f:	4-biphenyl	Ph	Me	4	0
7g:	4-t-BuC$_6$H$_4$	Ph	Me	2	2
7h:	4-t-BuC$_6$H$_4$	Ph	Me	3	1
7i:	4-MeC$_6$H$_4$	4-MeC$_6$H$_4$	t-Bu	3	1
7j:	4-MeC$_6$H$_4$	4-MeC$_6$H$_4$	CF$_3$	3	1

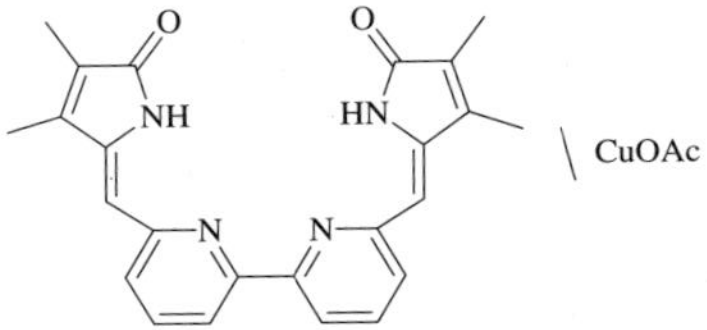

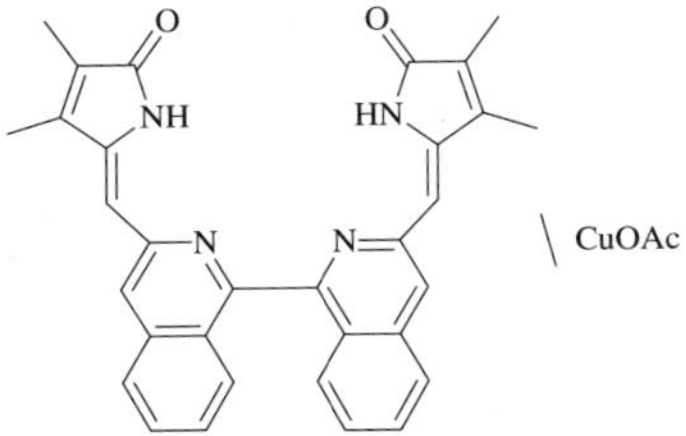

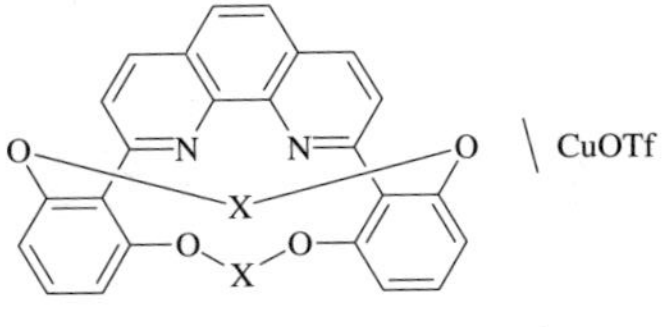

4: X = —CH$_2$(CH$_2$OCH$_2$)$_2$CH$_2$—
5: X = —CH$_2$(CH$_2$OCH$_2$)$_3$CH$_2$—

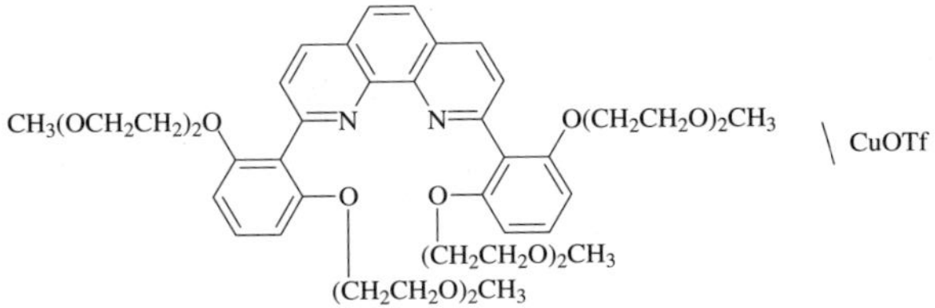

CHART 1. NON-CHIRAL CATALYST COMPLEXES (*Continued*)

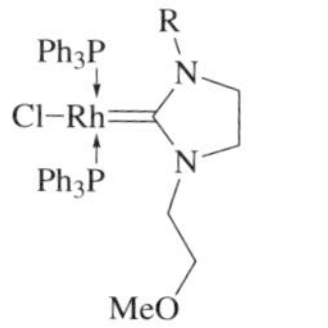

	R
8a:	Me
8b:	CH$_2$CH$_2$OMe

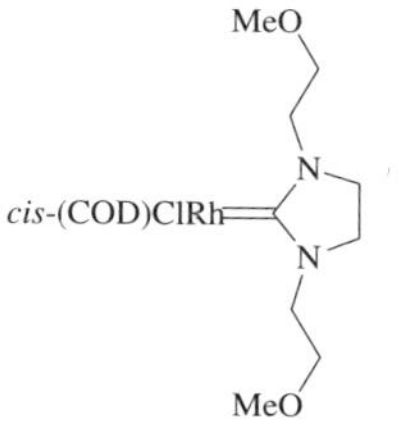

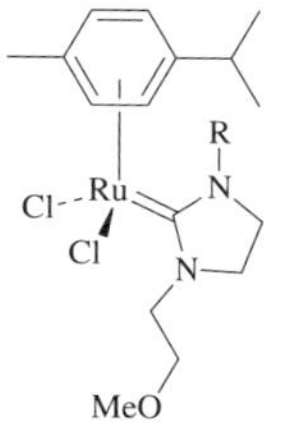

	R
10a:	Me
10b:	CH$_2$CH$_2$OMe

	R
11a:	Me
11b:	CH$_2$CH$_2$OMe

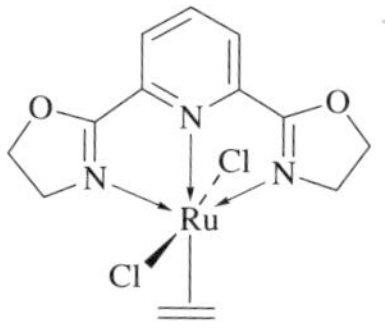

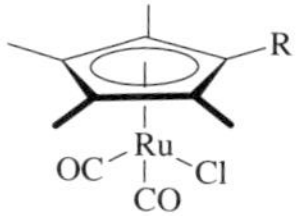

13a: R = Me
13b: R = CH(OH)Ph
13c: R = C(O)Ph
13d: R = (1,3-dioxolan-2-yl)
13e: R = CH=CHC(O)Me
13f: R = CH=CHC(O)(2-thienyl)

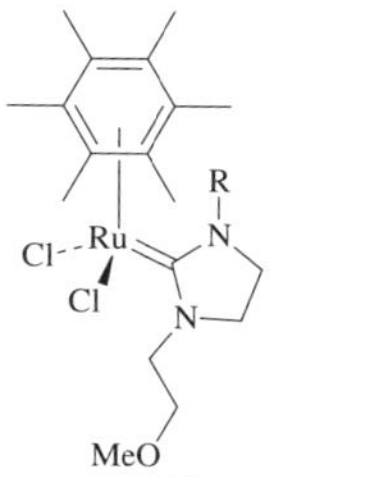

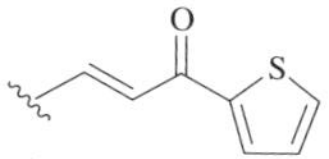

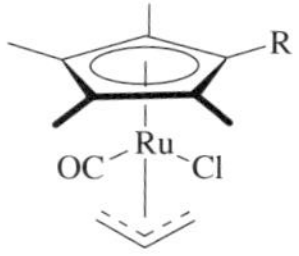

14a: R = Me
14b: R = CH(OH)Ph
14c: R = C(O)Ph
14d: R = (1,3-dioxolan-2-yl)
14e: R = (4,5-bis(CO$_2$Et)-1,3-dioxolan-2-yl)
14f: R = CH=CHC(O)Ph

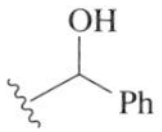

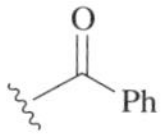

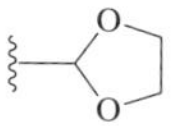

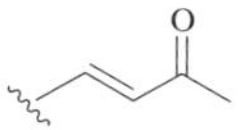

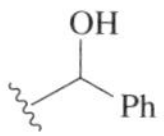

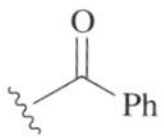

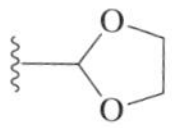

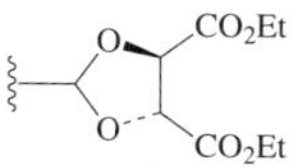

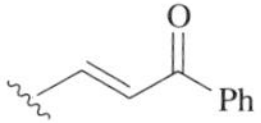

CHART 1. NON-CHIRAL CATALYST COMPLEXES (*Continued*)

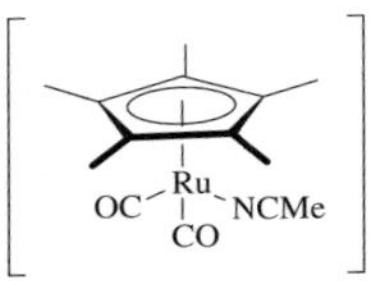

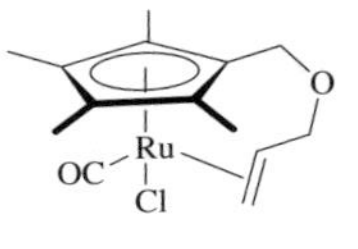

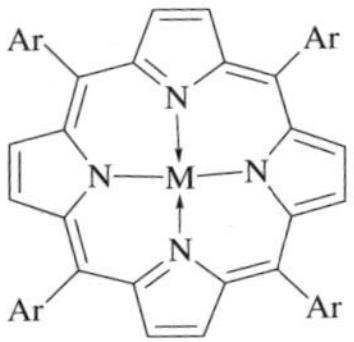

	Ar	M
25a:	4-MeC₆H₄	Ru, TTP
25b:	4-MeC₆H₄	Os, TTP
25c:	4-MeC₆H₄	Rh, TTP
25d:	mesityl	Ru, TMP
25e:	mesityl	Os, TMP
25f:	mesityl	Rh, TMP
25g:	C₆F₅	Fe, TpFPP

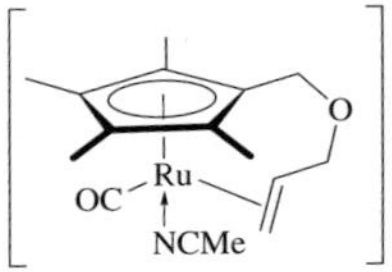

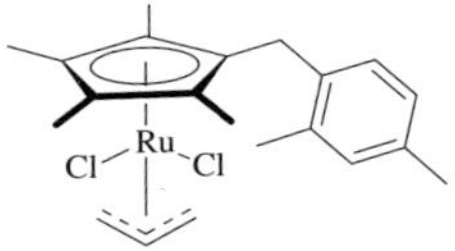

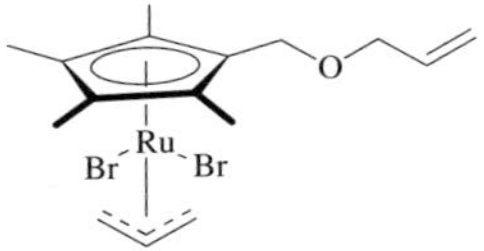

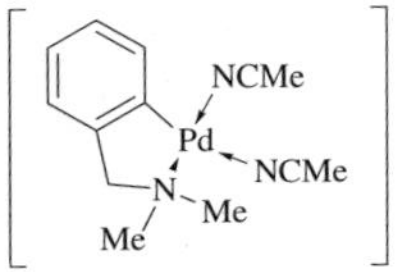

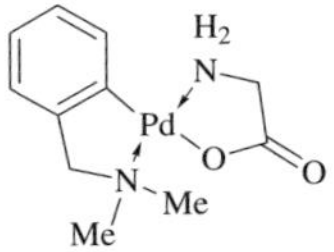

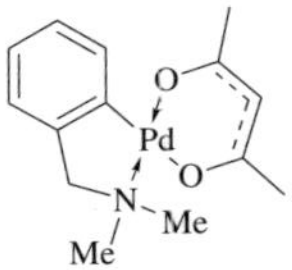

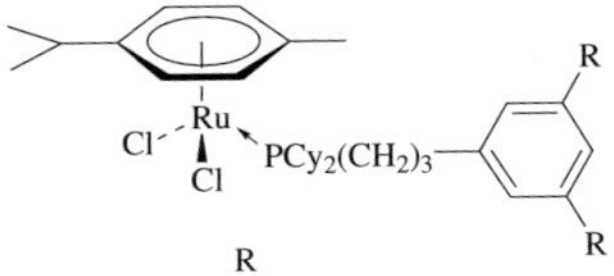

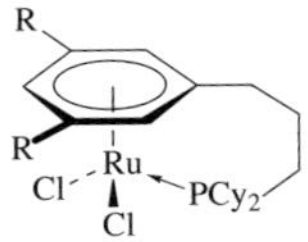

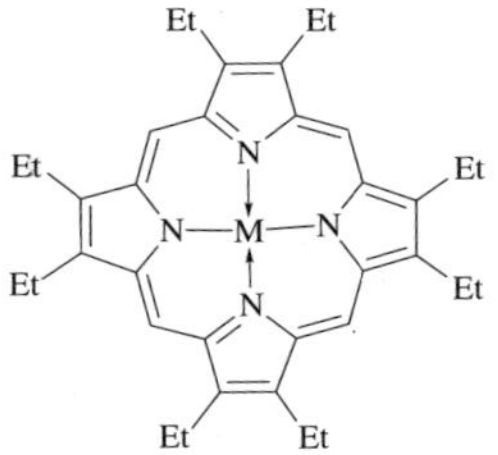

	R
23a:	H
23b:	Me

	R
24a:	H
24b:	Me

	M
26a:	Ru, OEP
26b:	Os, OEP

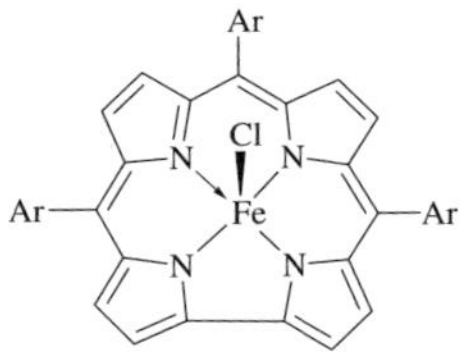

27: Ar = C₆F₅

46

CHART 2. CHIRAL COPPER C$_2$-SYMMETRIC CATALYST COMPLEXES

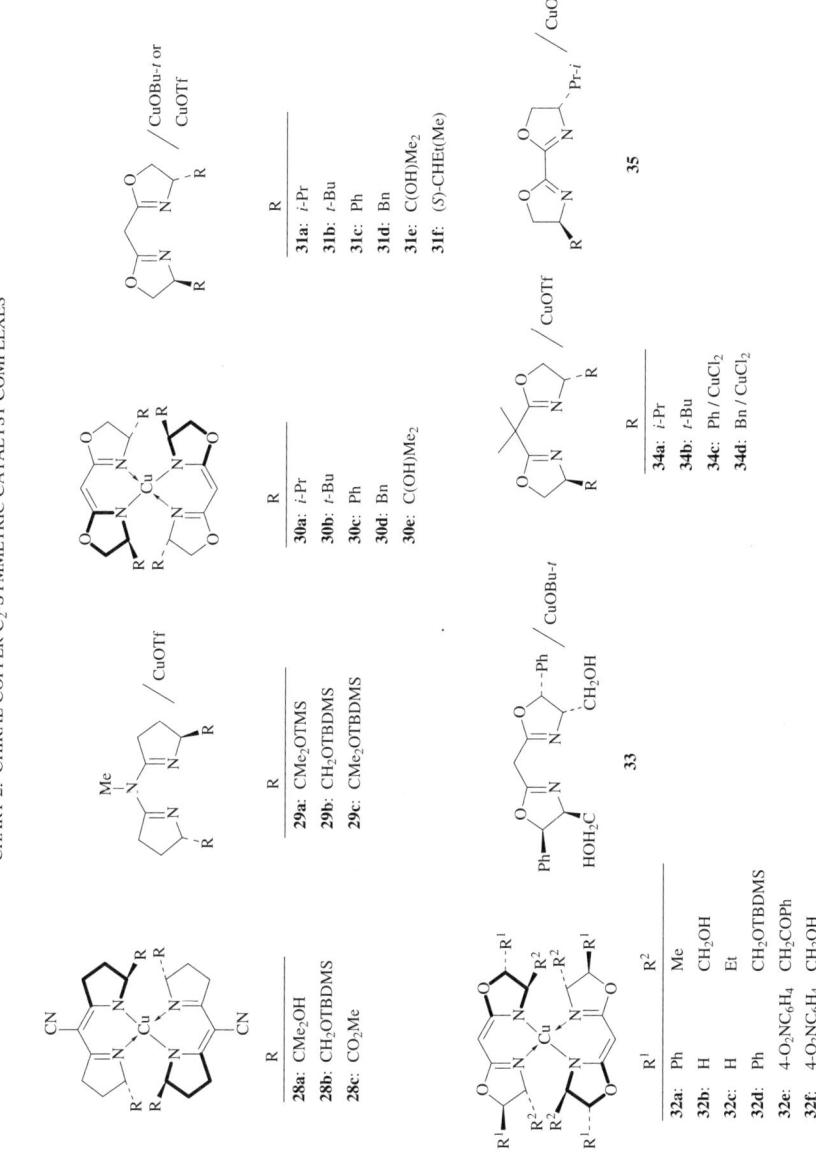

CHART 2. CHIRAL COPPER C_2-SYMMETRIC CATALYST COMPLEXES (*Continued*)

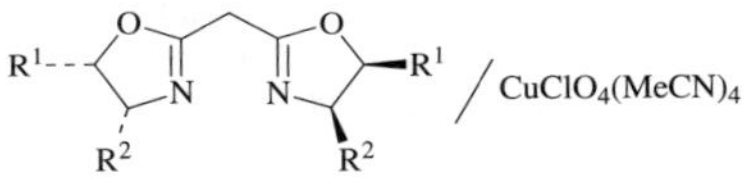

	R^1	R^2
36a:	Ph	Ph
36b:	Ph	Me
36c:	*t*-Bu	Et

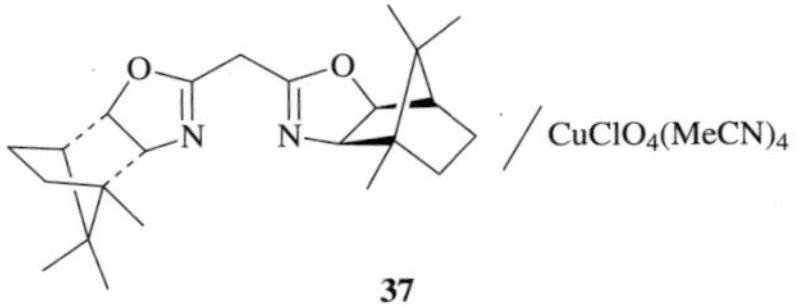

37

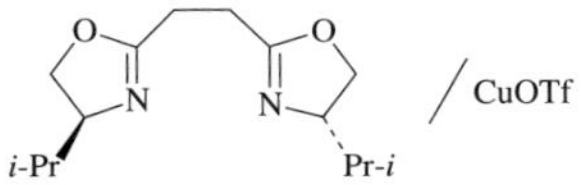

38

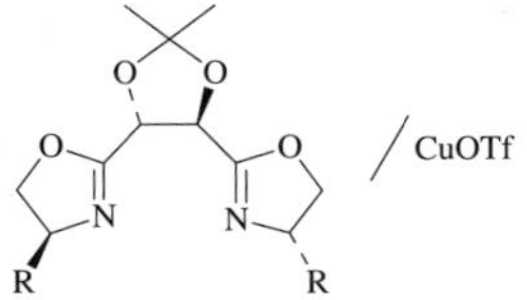

	R
39a:	*i*-Pr
39b:	*t*-Bu
39c:	Ph
39d:	Bn
39e:	*i*-Bu

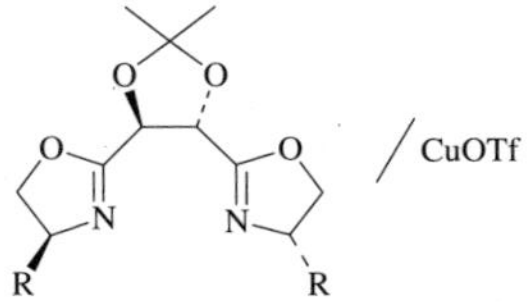

	R
40a:	*i*-Pr
40b:	*t*-Bu
40c:	Ph
40d:	Bn

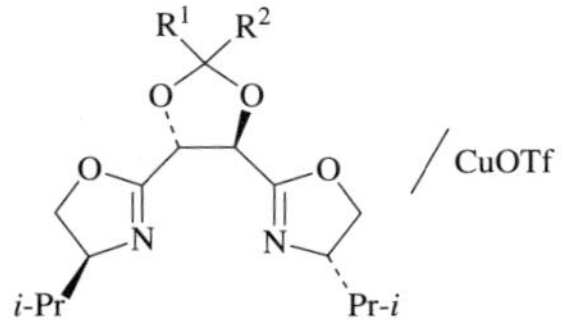

	R^1	R^2
41a:	H	H
41b:	Et	Et
41c:	*n*-Bu	*n*-Bu
41d:	Me	H
41e:	Me	Ph
41f:	Ph	H

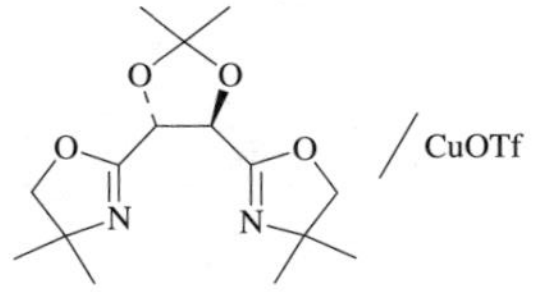

42

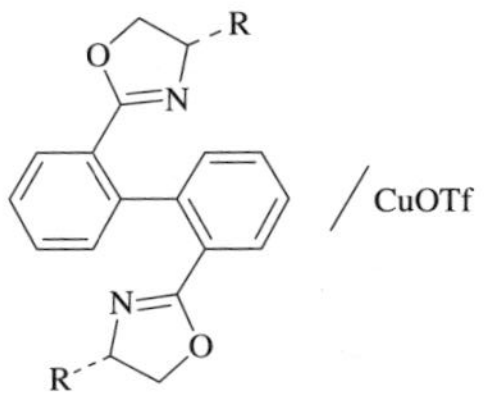

	R
43a:	*i*-Pr
43b:	*t*-Bu

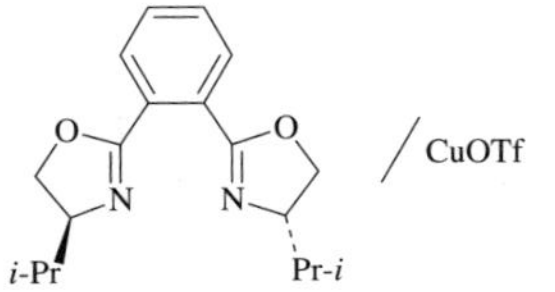

44

CHART 2. CHIRAL COPPER C_2-SYMMETRIC CATALYST COMPLEXES (*Continued*)

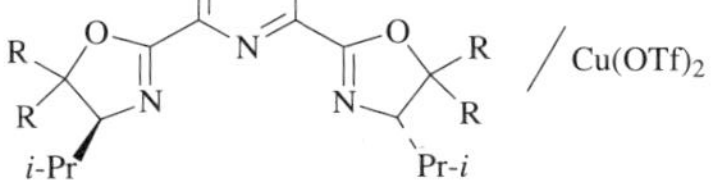

	R
45a:	Et
45b:	Ph
45c:	H

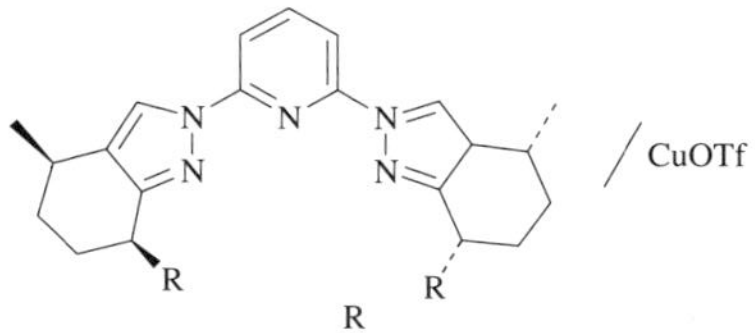

	R
46a:	*i*-Pr
46b:	*t*-Bu

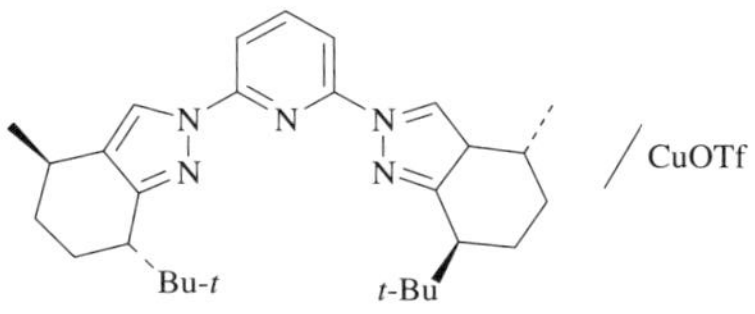

47

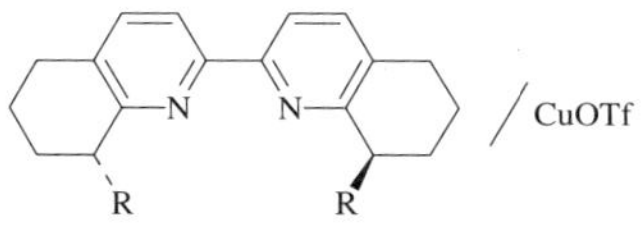

	R
48a:	Me
48b:	*i*-Pr
48c:	$CMe_2(OMe)$
48d:	TMS
48e:	TES
48f:	$CMe_2(OTBDMS)$

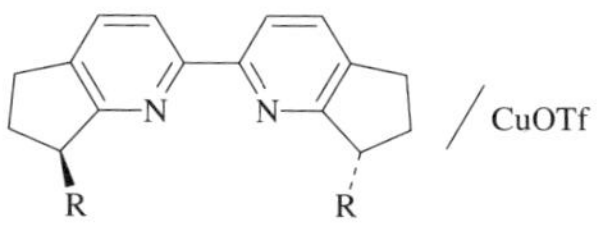

	R
49a:	Me
49b:	$CMe_2(OTBDMS)$

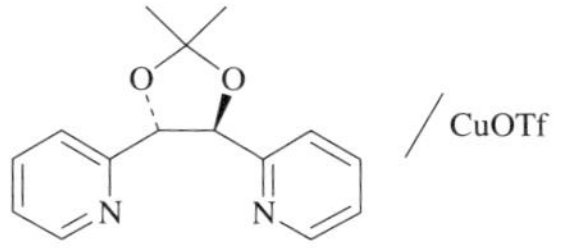

50

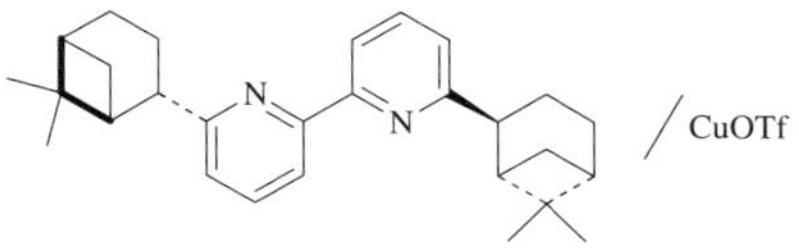

51

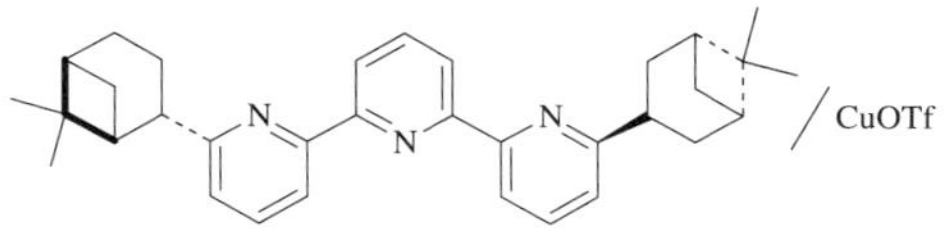

52

CHART 2. CHIRAL COPPER C_2-SYMMETRIC CATALYST COMPLEXES (*Continued*)

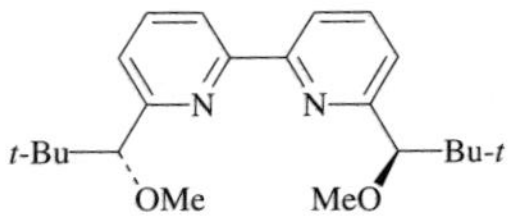

53a: /CuOTf

53b: /$CuCl_2$

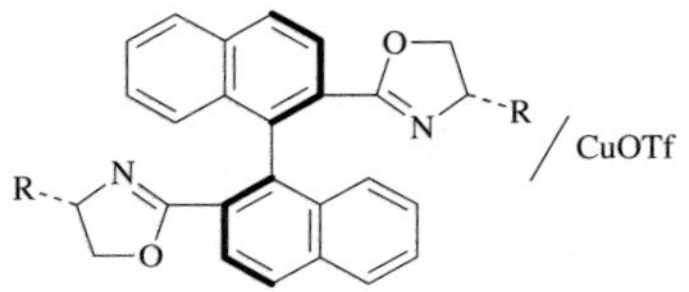

54a:
54b:
54c:
54d:

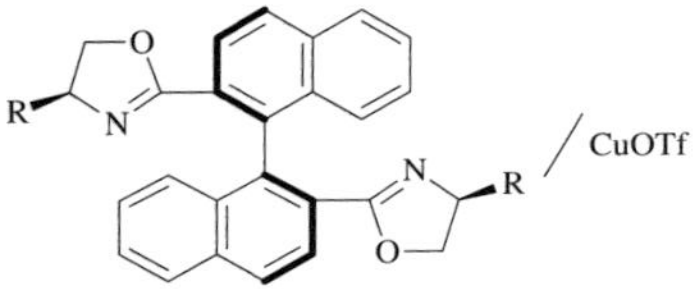

55a:
55b:

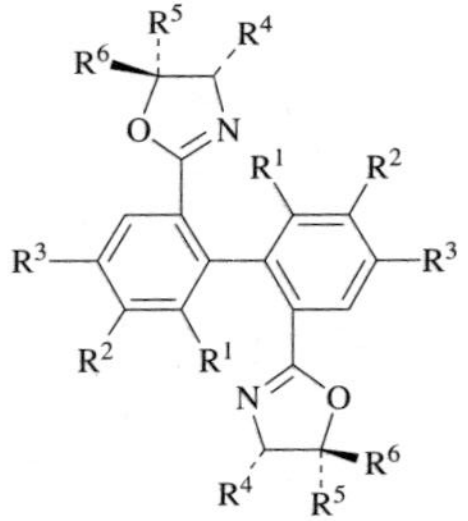

56a:
56b:
56c:
56d:
56e:
56f:
56g:
56h:
56i:
56j:
56k:

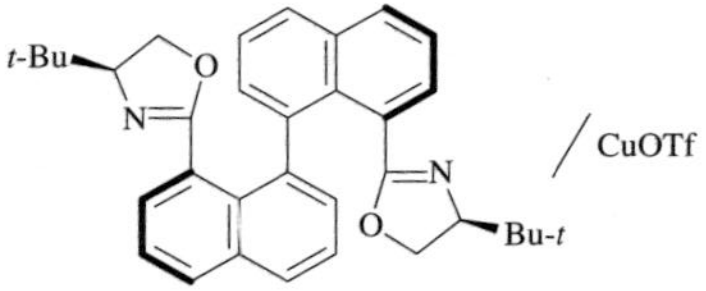

57

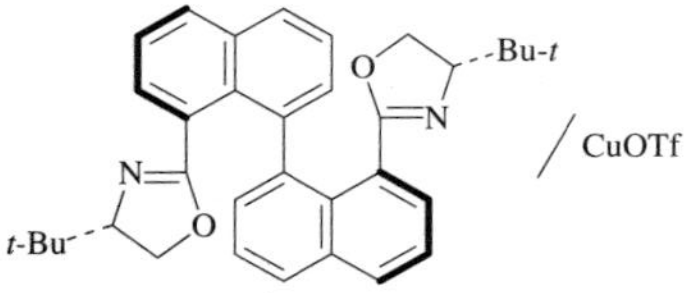

58

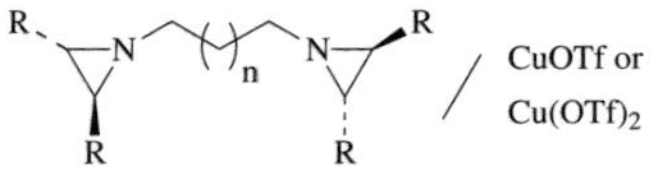

	n
59a:	0
59b:	3
59c:	0
59c:	0

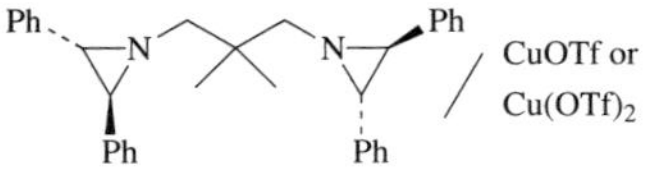

60

CHART 2. CHIRAL COPPER C$_2$-SYMMETRIC CATALYST COMPLEXES (*Continued*)

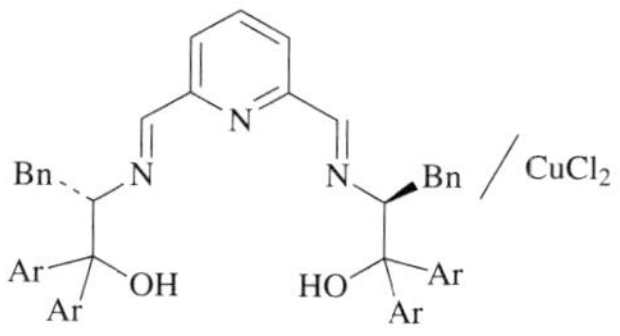

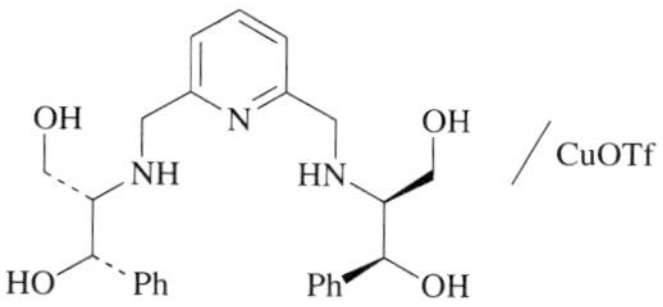

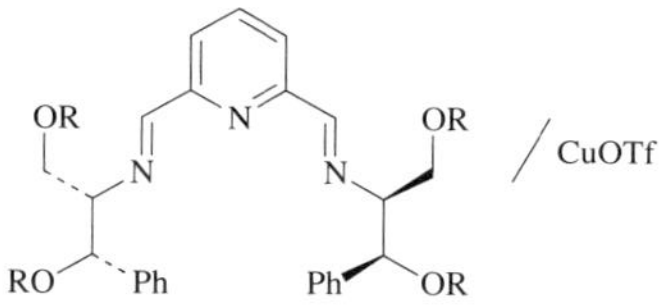

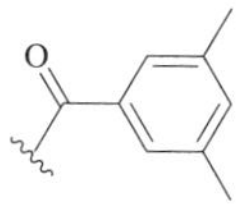

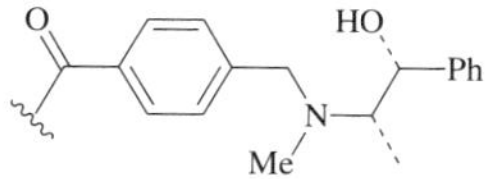

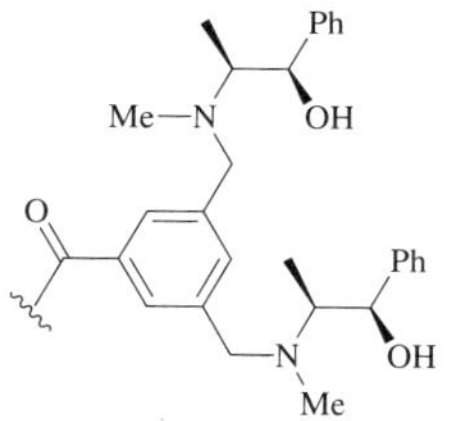

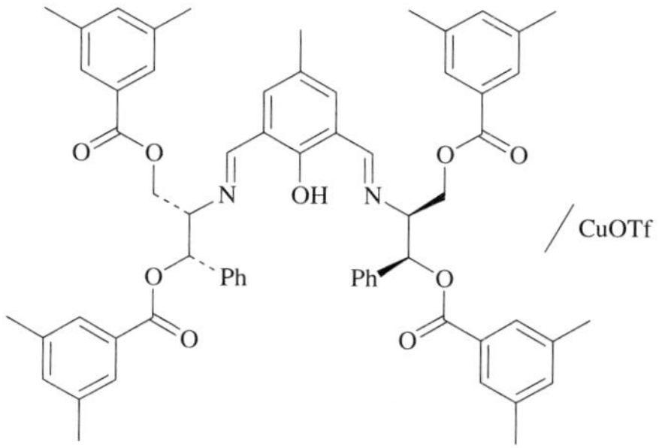

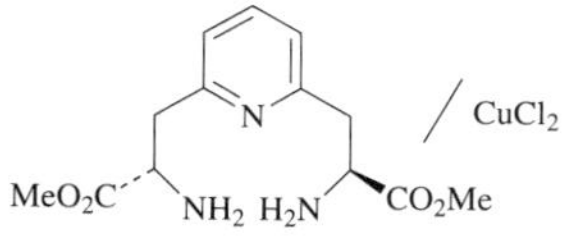

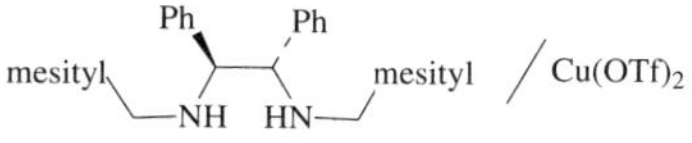

CHART 2. CHIRAL COPPER C_2-SYMMETRIC CATALYST COMPLEXES (*Continued*)

	R
67a:	i-Pr
67b:	(S)-CHPh(Me)
67c:	(R)-CHPh(Me)
67d:	Ph
67e:	(R)-CH(CH$_2$OH)(Et)
67f:	(S)-CH(CH$_2$OH)(Et)
67g:	D-norephedrinyl
67h:	L-norephedrinyl
67i:	(S)-CH(CPh$_2$OH)(Pr-i)

	R
68a:	D-norephedrin
68b:	L-norephedrin

70: Ar = 2-(n-C$_8$H$_{17}$O)-5-t-BuC$_6$H$_3$

71: Ar = 2-(n-C$_8$H$_{17}$O)-5-t-BuC$_6$H$_3$

69

	R^1	R^2	R^3
72a:	Cl	Cl	H
72b:	Me	Me	H
72c:	Cl	Cl	Cl
72d:	Br	Br	Br
72e:	Me	Me	Me
72f:	i-Pr	i-Pr	i-Pr
72g:	OH	H	H

CHART 2. CHIRAL COPPER C$_2$-SYMMETRIC CATALYST COMPLEXES (*Continued*)

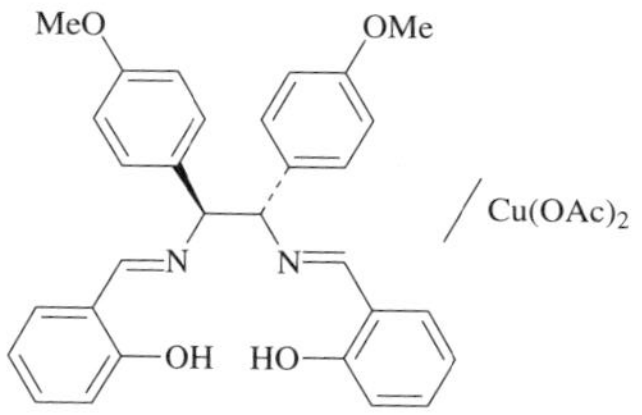

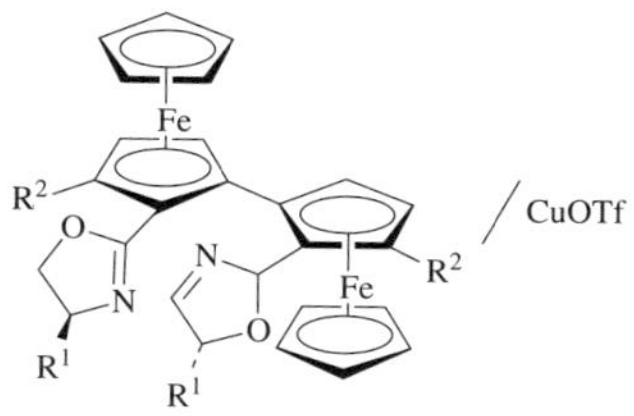

	R^1	R^2
74a:	*i*-Pr	H
74b:	*t*-Bu	H
74c:	*t*-Bu	TMS
74d:	*t*-Bu	TES

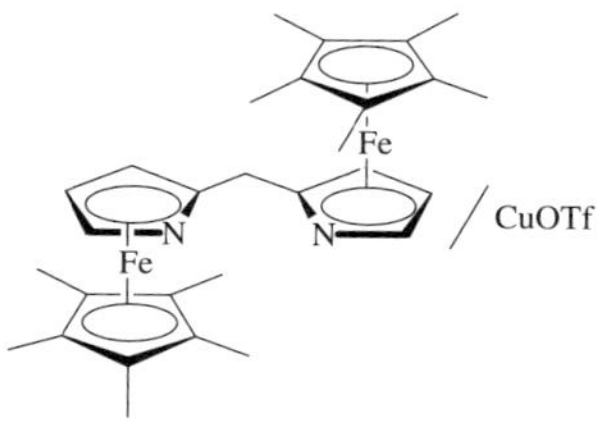

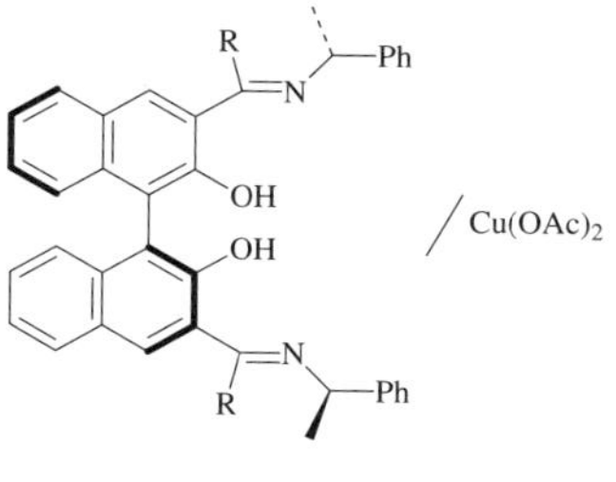

76a:
76b:
76c:

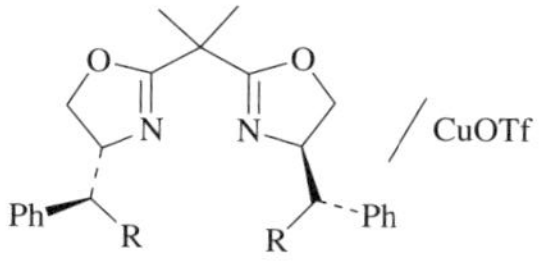

77a:
77b:
77c:
77d:
77e:
77f:

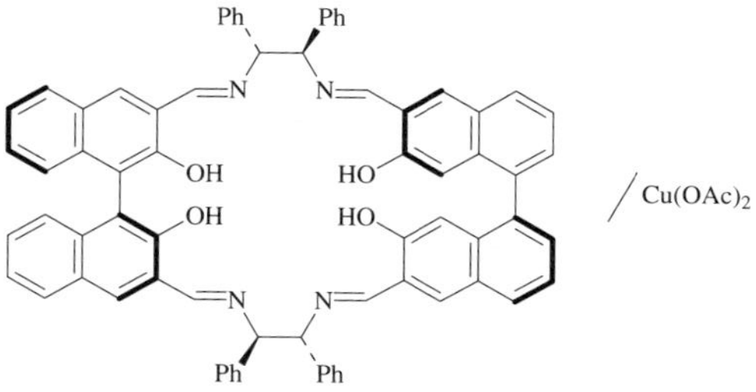

CHART 2. CHIRAL COPPER C_2-SYMMETRIC CATALYST COMPLEXES (*Continued*)

79a:
79b:

80

81

82

83

84

85

86

87

88

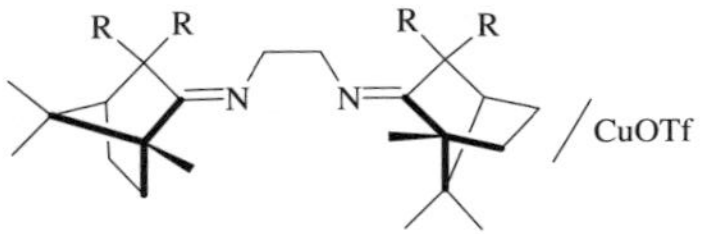

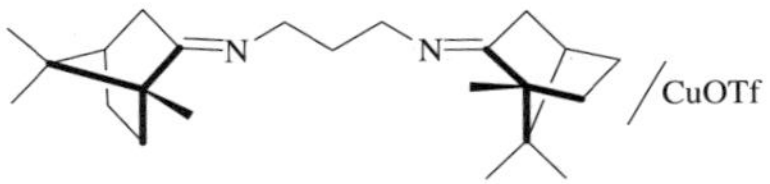

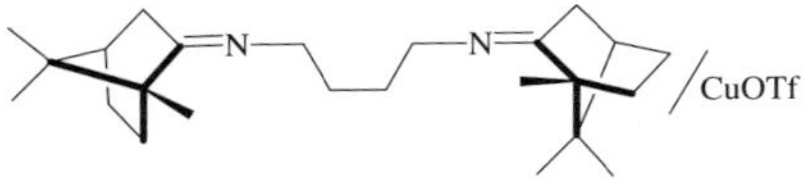

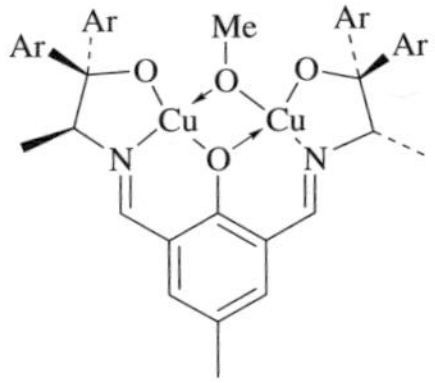

	Ar
89a:	Ph
89b:	2-(n-$C_8H_{17}O$)-5-t-BuC_6H_3
89c:	2-(i-PrO)-5-t-BuC_6H_3
89d:	2,5-(t-BuO)-5-t-BuC_6H_2
89e:	2,5-BnO-5-t-BuC_6H_2

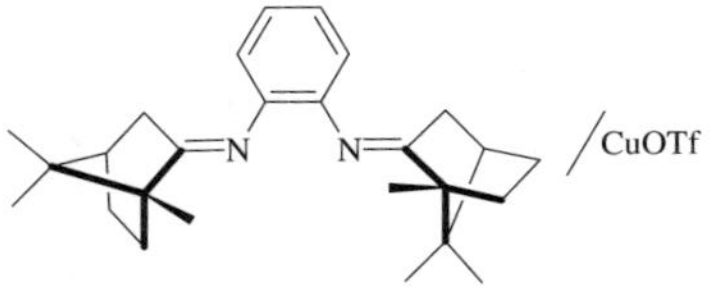

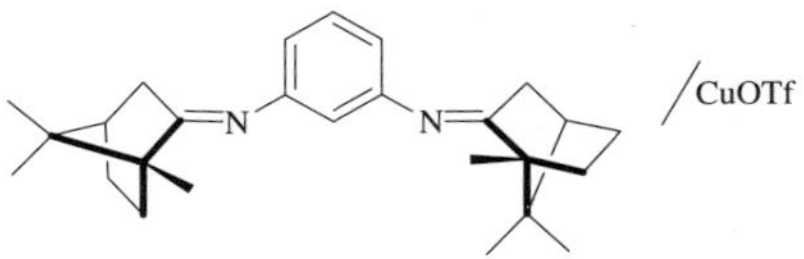

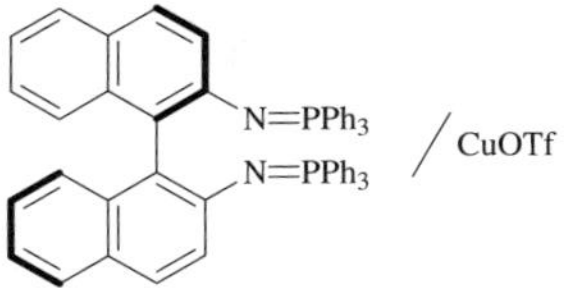

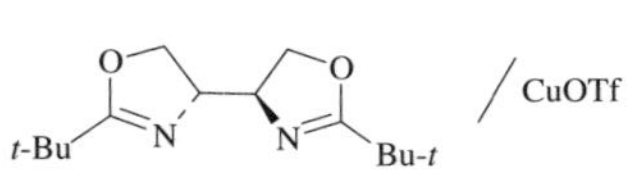

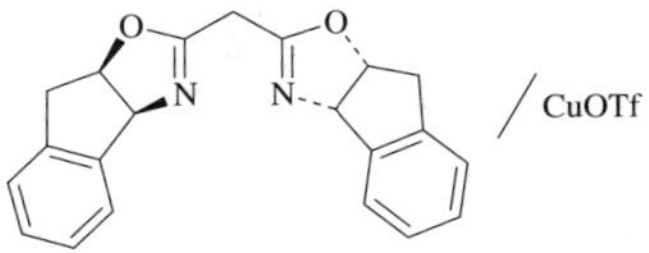

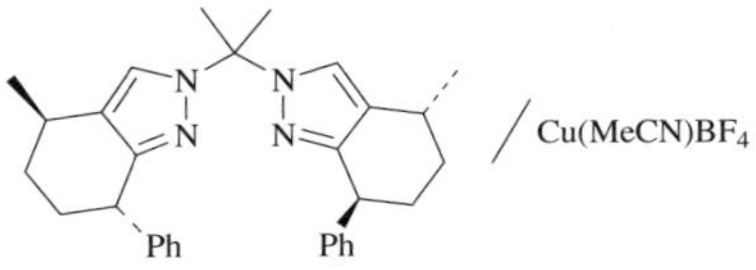

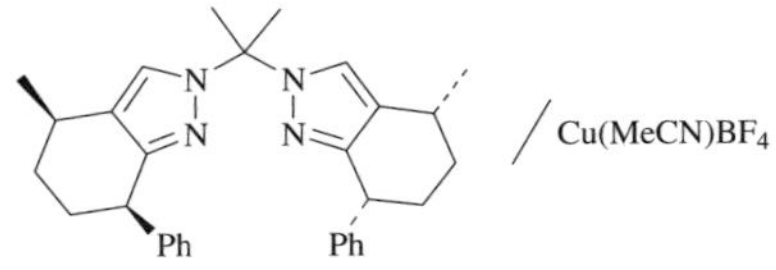

CHART 3. CHIRAL COPPER NON-C_2-SYMMETRIC CATALYST COMPLEXES

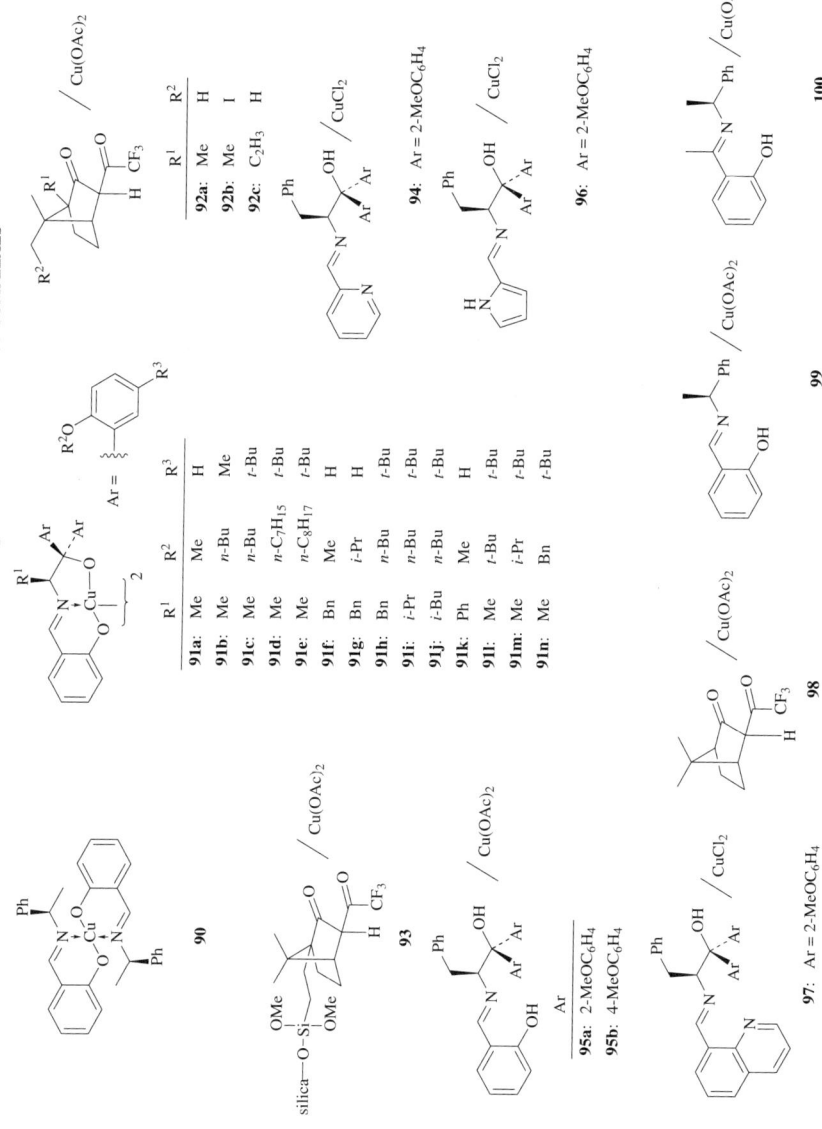

CHART 3. CHIRAL COPPER NON-C_2-SYMMETRIC CATALYST COMPLEXES (*Continued*)

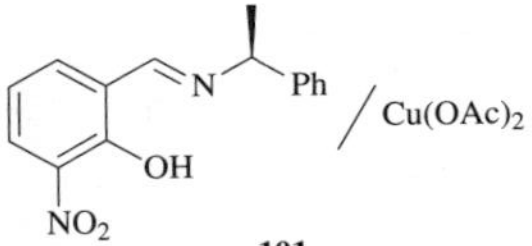

101

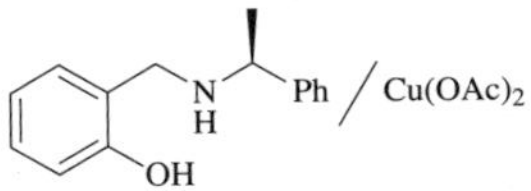

102

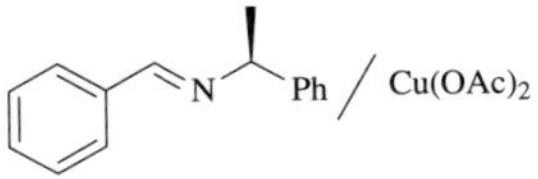

103

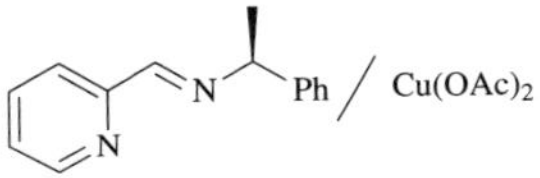

104

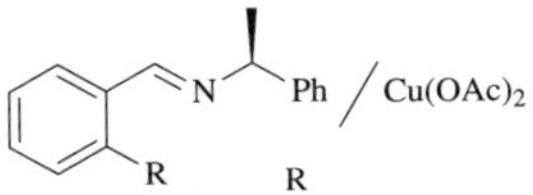

	R
105a:	PPh_2
105b:	NH_2

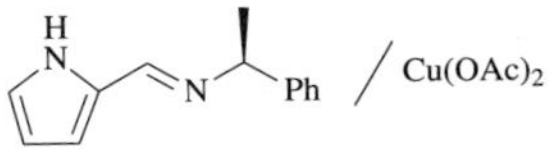

106

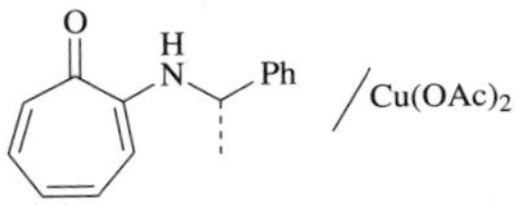

107

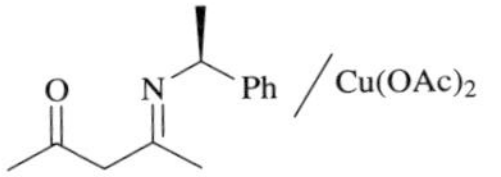

108

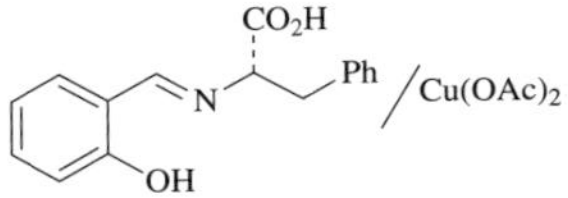

109

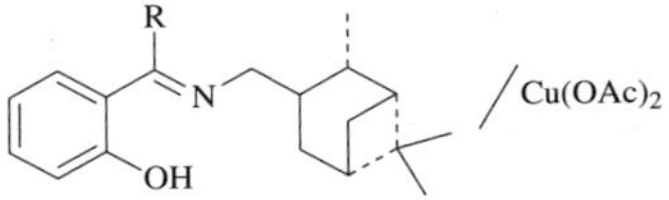

	R
110a:	H
110b:	Me

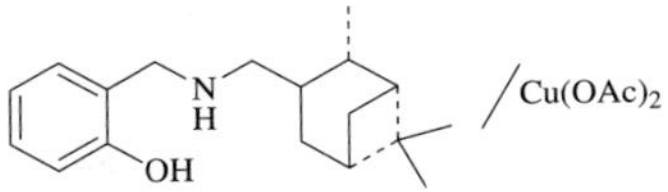

111

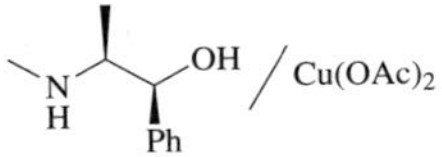

112

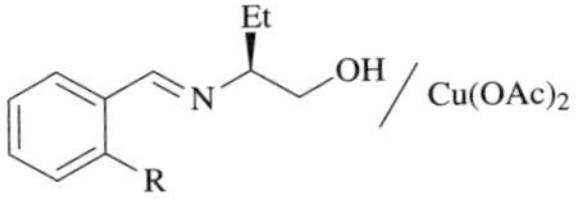

	R
113a:	H
113b:	OH

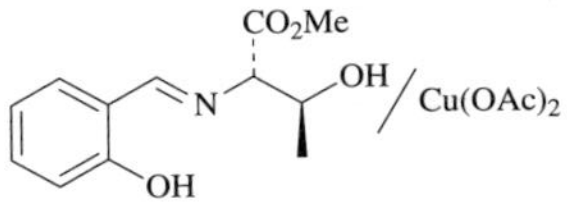

114

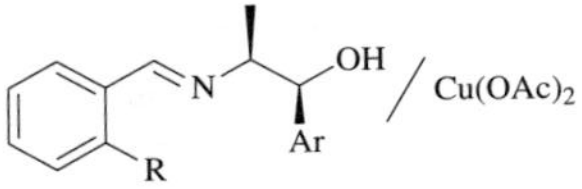

	R	Ar
115a:	H	Ph
115b:	OH	Ph
115c:	OH	2-(n-$C_8H_{17}O$)-3-t-BuC_6H_3

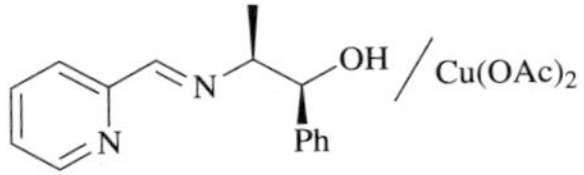

116

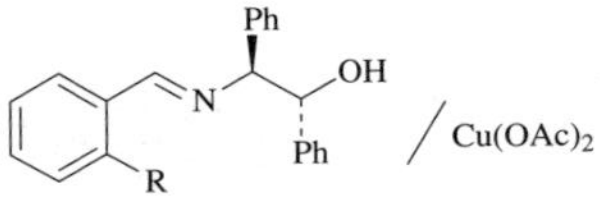

	R
117a:	H
117b:	OH

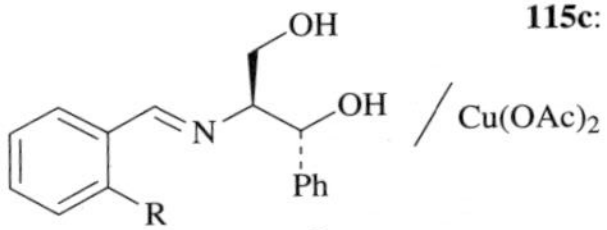

	R
118a:	H
118b:	OH

CHART 3. CHIRAL COPPER NON-C_2-SYMMETRIC CATALYST COMPLEXES (Continued)

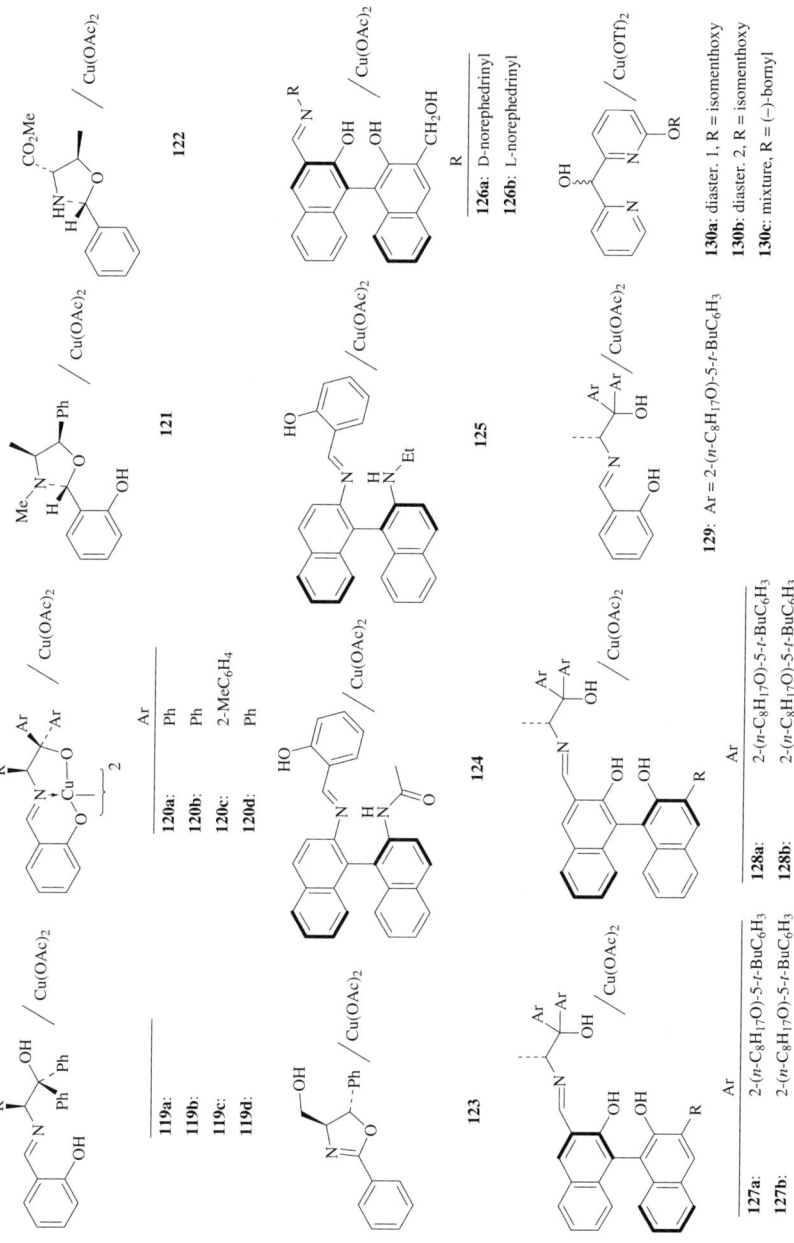

CHART 3. CHIRAL COPPER NON-C_2-SYMMETRIC CATALYST COMPLEXES (*Continued*)

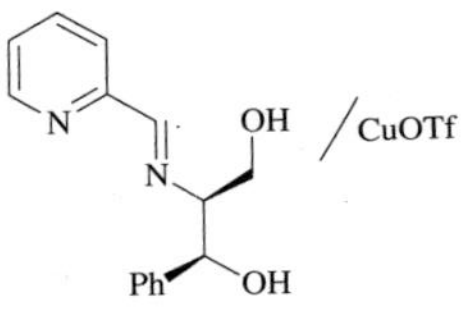

131

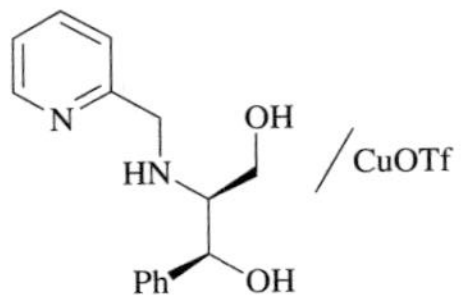

132

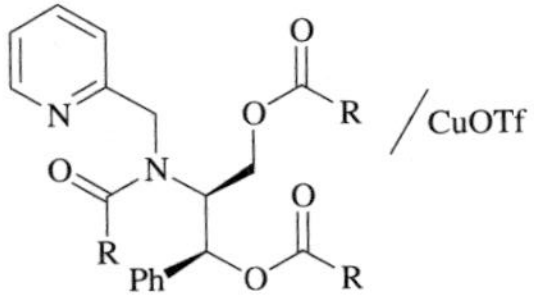

R	
133a:	Me
133b:	Ph

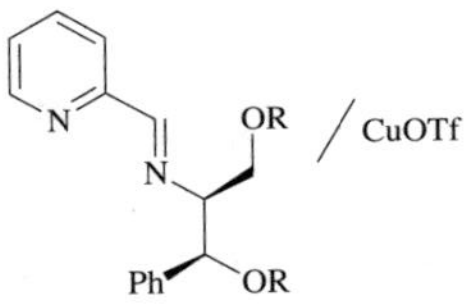

134a: R = 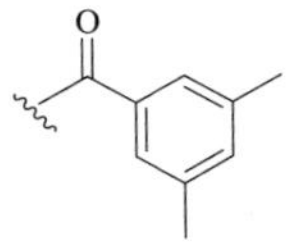

134b: R = 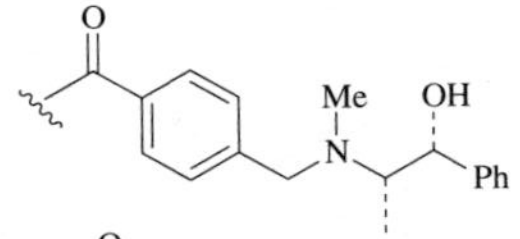

134c: R = 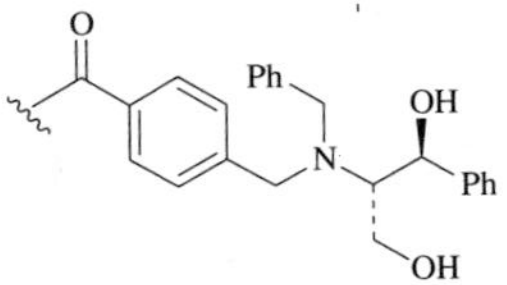

134d: R = 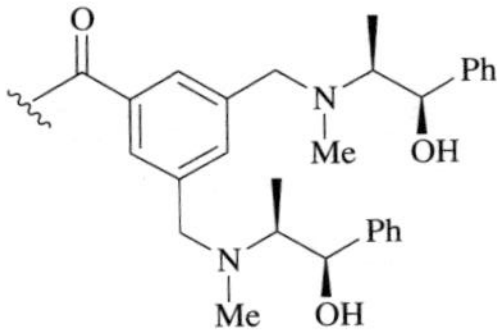

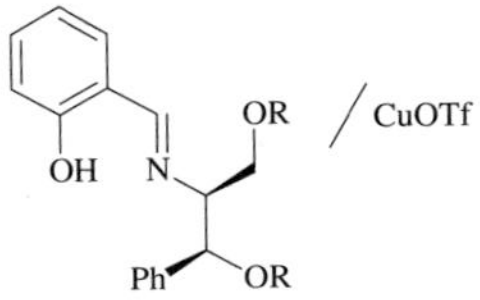

135a: R = 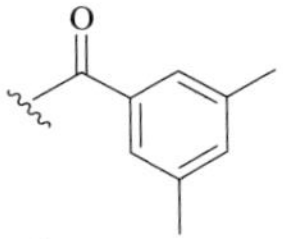

135b: R = 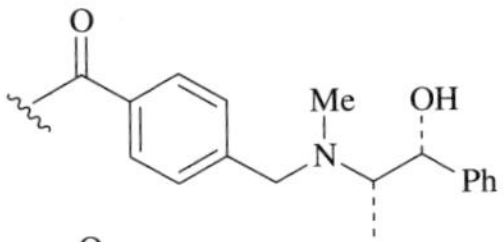

135c: R = 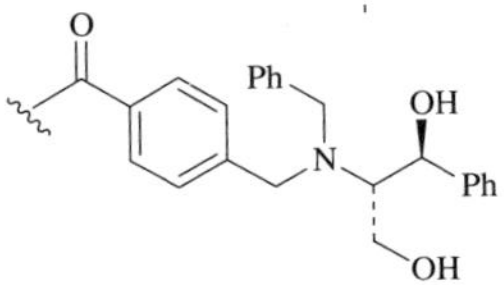

135d: R = 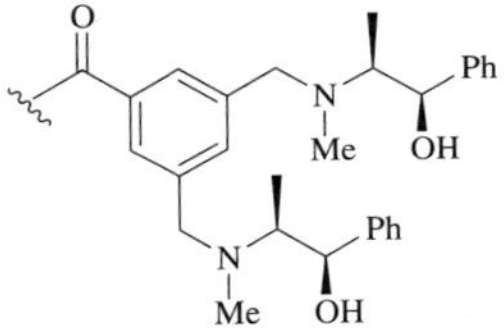

CHART 3. CHIRAL COPPER NON-C_2-SYMMETRIC CATALYST COMPLEXES (*Continued*)

	R^1	R^2	R^3	R^4	R^5
141a:	H	H	Ph	H	CH_2OH
141b:	H	H	H	CPh_2OH	H
141c:	H	Me	H	CPh_2OH	H
141d:	H	Ph	Ph	H	$CH_2C_6H_4OH$-4
141e:	NO_2	H	Ph	H	CH_2OH
141f:	NO_2	H	H	CPh_2OH	H
141g:	NO_2	Me	H	CPh_2OH	H
141h:	NO_2	Ph	Ph	H	CPh_2OH

	R^1	R^2
143a:	Ph	4-MeC_6H_4
143b:	Ph	1-naphthyl
143c:	Ph	2-naphthyl
143d:	Ph	$2,4,6$-$Me_3C_6H_2$
143e:	Ph	CF_3
143f:	Ph	4-t-BuC_6H_4
143g:	t-Bu	4-MeC_6H_4

	R
142a:	H
142b:	Me

CHART 3. CHIRAL COPPER NON-C_2-SYMMETRIC CATALYST COMPLEXES (*Continued*)

	R^1	R^2	R^3	R^4
148a:	Me	H	H	H
148b:	Et	H	H	H
148c:	H	i-Pr	H	H
148d:	H	CH(Me)Et	H	H
148e:	H	CH$_2$CHMe$_2$	H	H
148f:	H	t-Bu	H	H
148g:	Ph	H	H	H
148h:	H	Bn	H	H
148i:	H	Me	Ph	H
148j:	H	CH$_2$OH	H	Ph
148k:	H	CPh$_2$OH	H	H
148l:	H	CH$_2$CPh$_2$OH	Ph	Ph

	R
147a:	Me
147b:	Bn
147c:	i-Pr
147d:	t-Bu
147e:	Ph

	R^1	R^2
149a:	H	H
149b:	H	Cl
149c:	H	Me
149d:	H	Ph
149e:	Cl	H
149f:	Br	H
149g:	Me	H

150a: R =

150b: R =

150c: R =

151a: R =

151b: R =

151c: R =

CHART 3. CHIRAL COPPER NON-C$_2$-SYMMETRIC CATALYST COMPLEXES

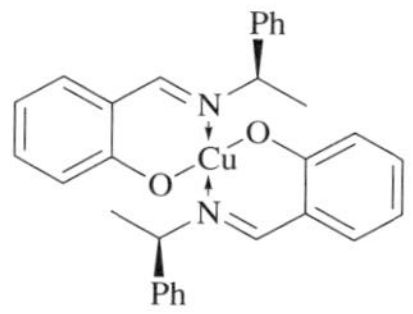

90

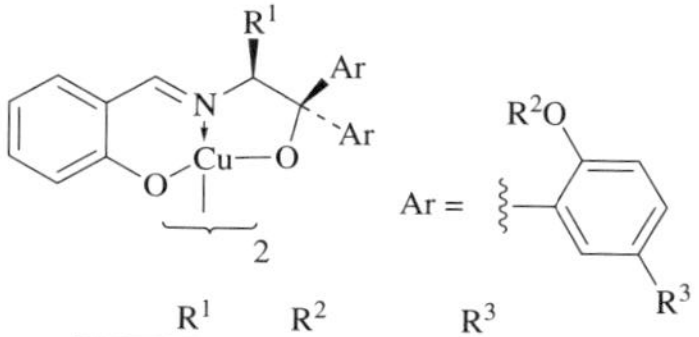

91

Ar =

	R^1	R^2	R^3
91a:	Me	Me	H
91b:	Me	n-Bu	Me
91c:	Me	n-Bu	t-Bu
91d:	Me	n-C$_7$H$_{15}$	t-Bu
91e:	Me	n-C$_8$H$_{17}$	t-Bu
91f:	Bn	Me	H
91g:	Bn	i-Pr	t-Bu
91h:	Bn	n-Bu	t-Bu
91i:	i-Pr	n-Bu	t-Bu
91j:	i-Bu	n-Bu	t-Bu
91k:	Ph	Me	H
91l:	Me	t-Bu	t-Bu
91m:	Me	i-Pr	t-Bu
91n:	Me	Bn	t-Bu

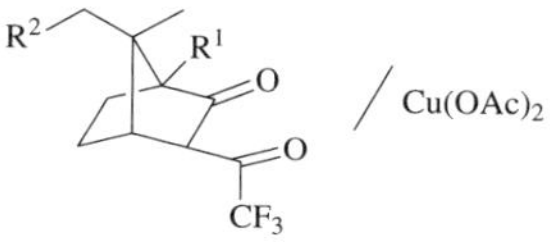

	R^1	R^2
92a:	Me	H
92b:	Me	I
92c:	C$_2$H$_3$	H

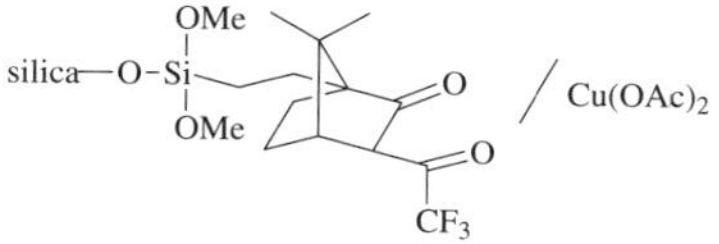

93

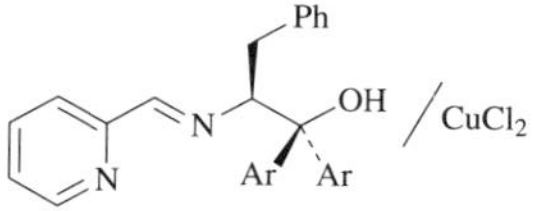

94: Ar = 2-MeOC$_6$H$_4$

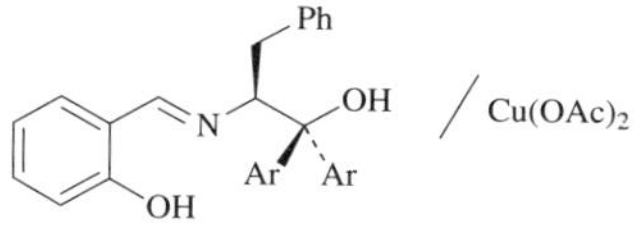

95a: 2-MeOC$_6$H$_4$
95b: 4-MeOC$_6$H$_4$

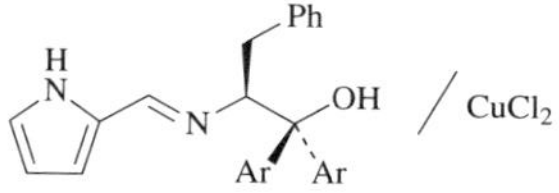

96: Ar = 2-MeOC$_6$H$_4$

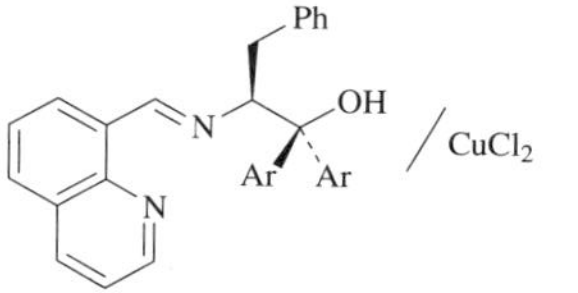

97: Ar = 2-MeOC$_6$H$_4$

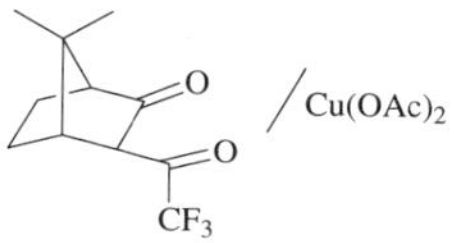

98

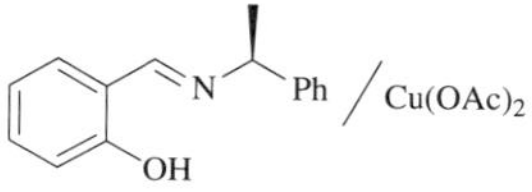

99

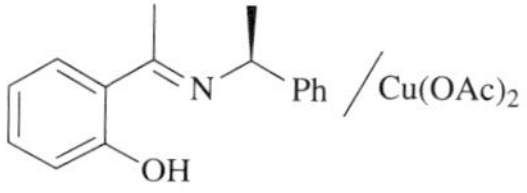

100

CHART 4. CHIRAL RHODIUM CATALYSTS

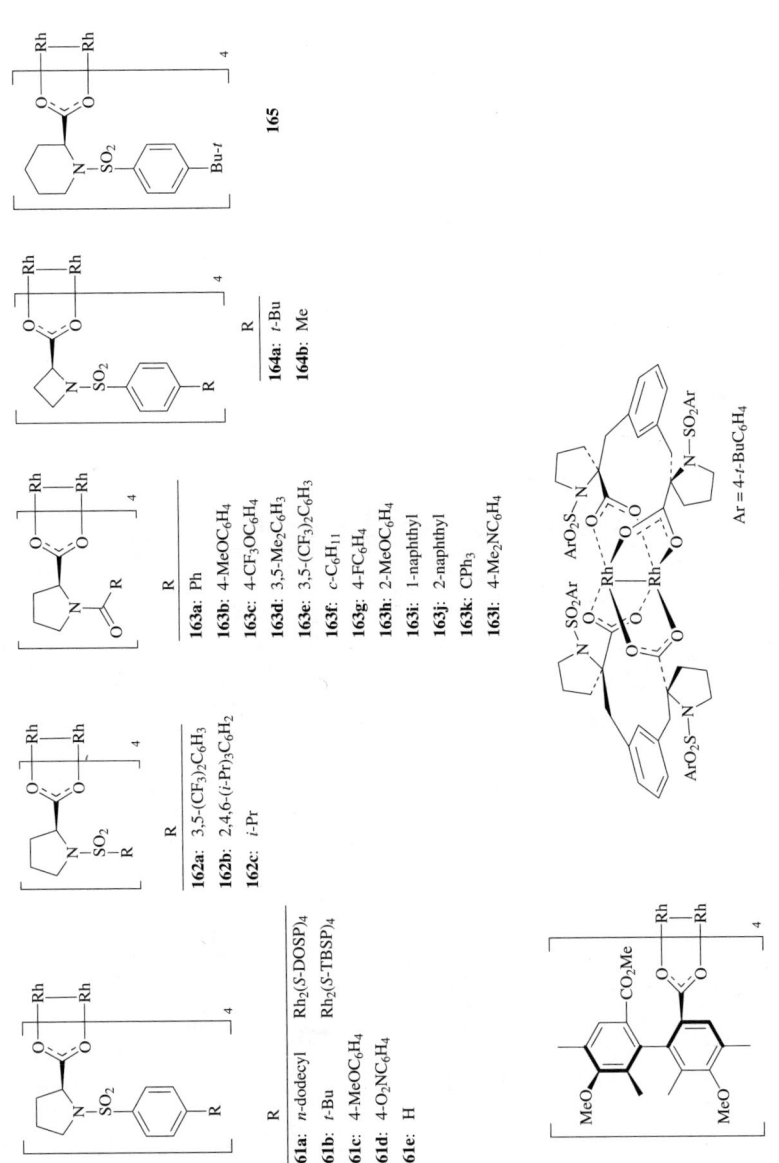

CHART 4. CHIRAL RHODIUM CATALYSTS (*Continued*)

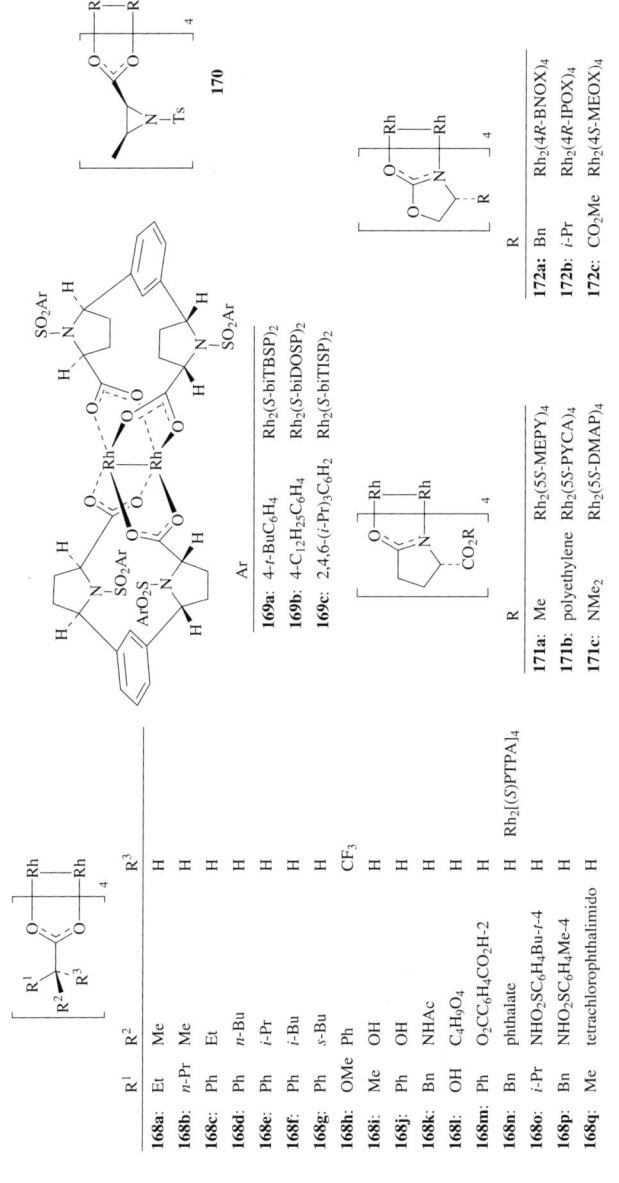

CHART 4. CHIRAL RHODIUM CATALYSTS (*Continued*)

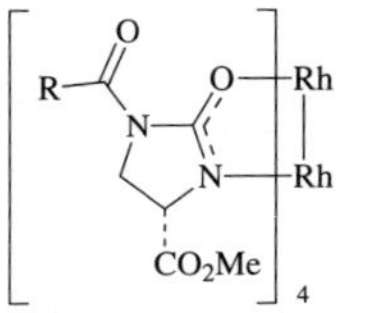

	R	
174a:	Ph	Rh$_2$(4S-MBOIM)$_4$
174b:	4-*t*-BuC$_6$H$_4$	Rh$_2$(4S-TBOIM)$_4$

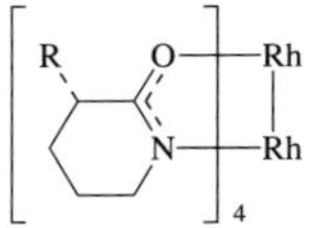

175: R =

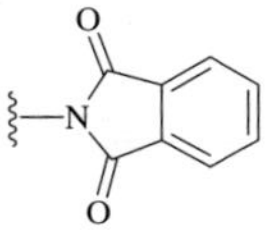

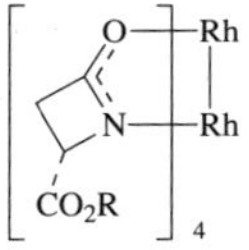

| 176a: | Rh$_2$(4S-BNAZ)$_4$ |
| 176b: | Rh$_2$(4S-IBAZ)$_4$ |

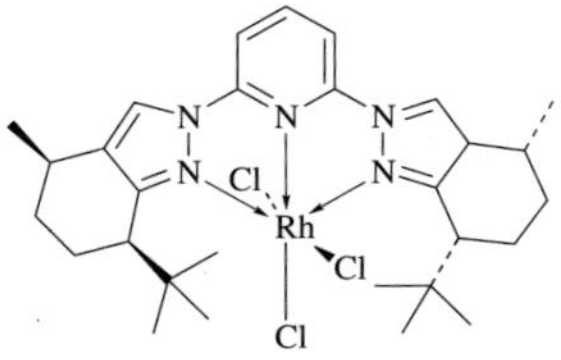

177

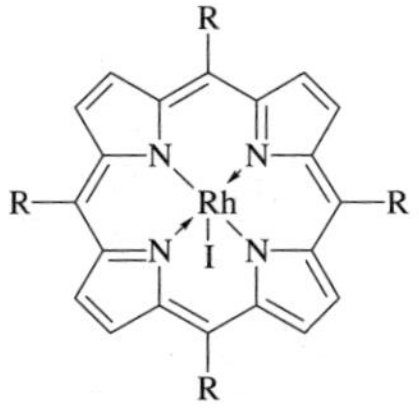

178a: R =

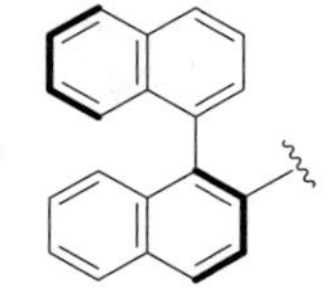

178b: R =

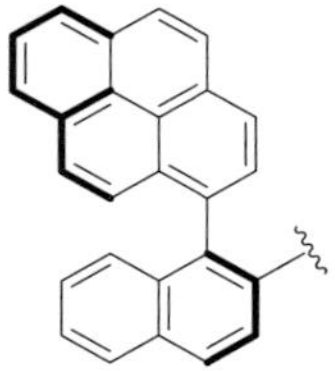

CHART 5. OTHER CHIRAL METAL CATALYSTS

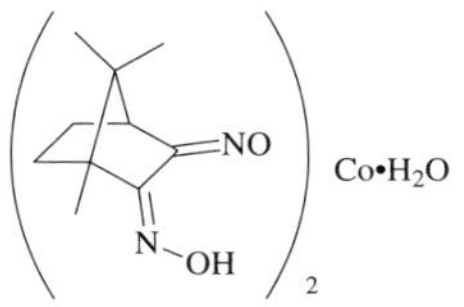

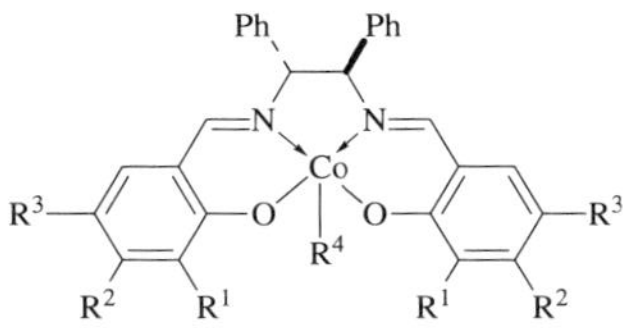

	R¹	R²	R³	R⁴
181a	t-Bu	H	H	I
181b	t-Bu	H	H	Br
181c	t-Bu	H	t-Bu	I
181d	t-Bu	H	t-Bu	Br
181e	Me	H	H	I
181f	H	t-Bu	H	I
181g	H	t-Bu	H	Br
181h	H	H	t-Bu	I
181i	H	H	t-Bu	Br
181j	H	H	H	I
181k	H	H	H	Br
181l	H	H	OMe	Br

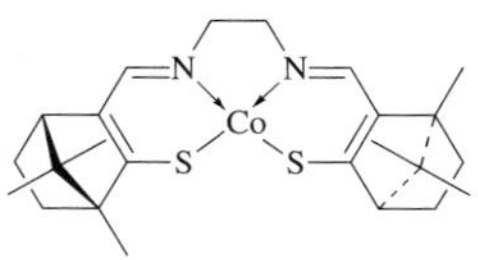

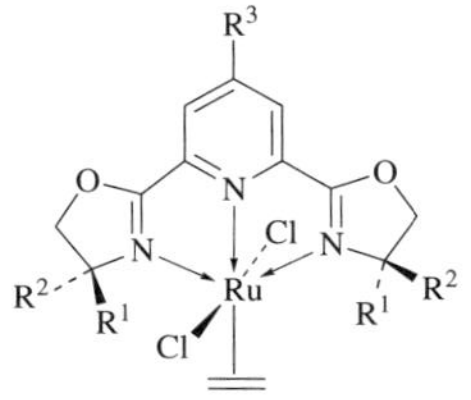

180

183

184

179

	R¹	R²	R³
182a	H	Et	H
182b	s-Bu	H	H
182c	Bz	H	H
182d	H	Ph	H
182e	i-Pr	H	NMe₂
182f	i-Pr	H	OMe
182g	i-Pr	H	H
182h	i-Pr	H	Cl
182i	i-Pr	H	CO₂Me

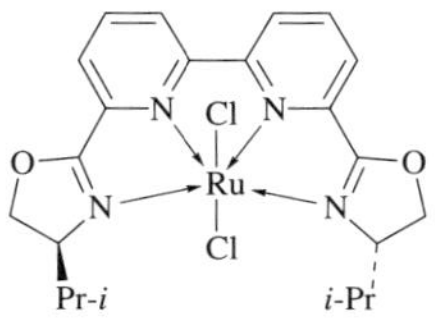

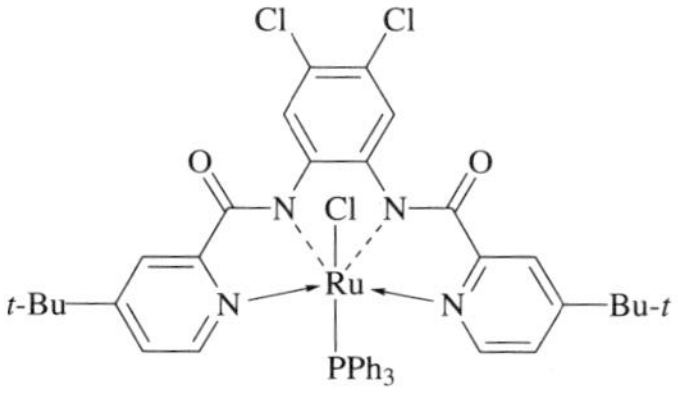

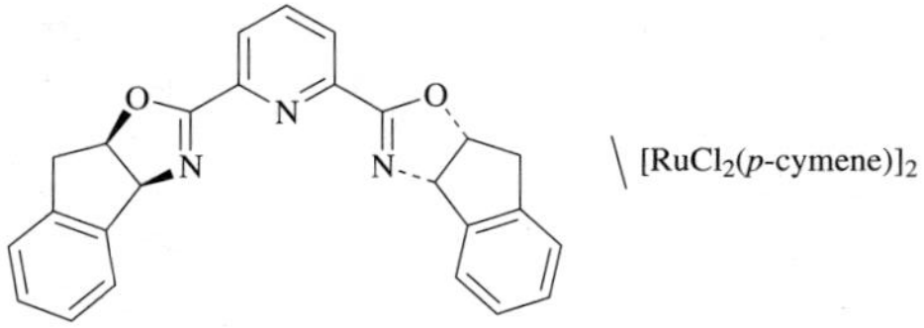

CHART 5. OTHER CHIRAL METAL CATALYSTS (*Continued*)

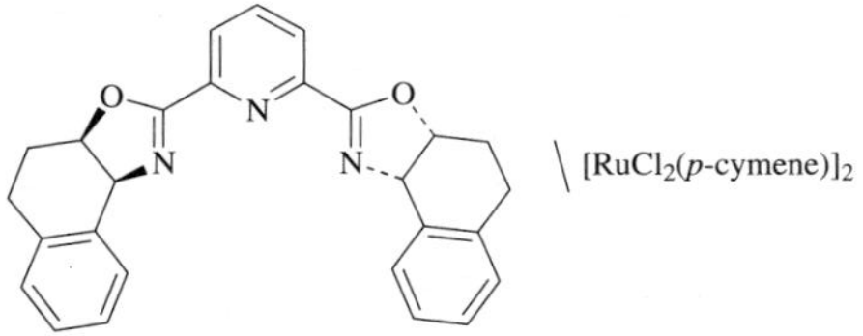

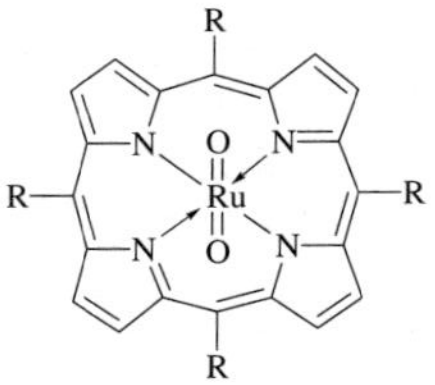

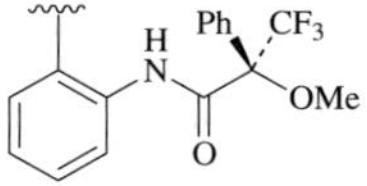

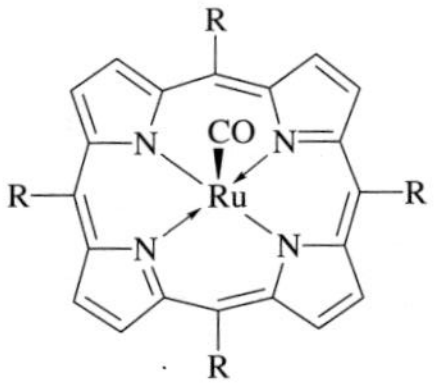

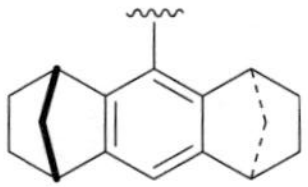

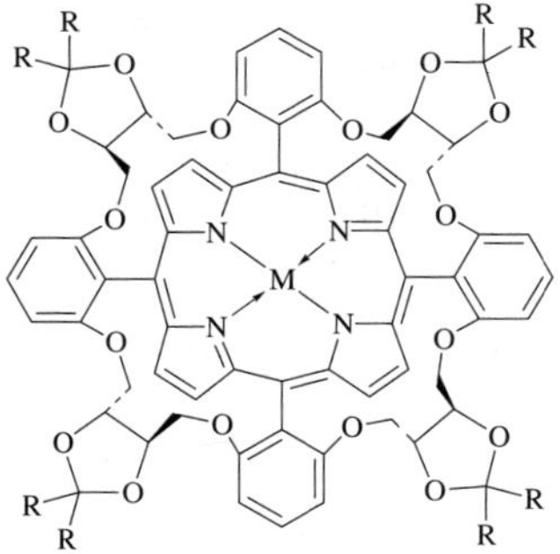

	R	M
189a:	Me	Ru
189b:	Me	Rh
189c:	Me	Fe
189d:	Ph	Ru

CHART 5. OTHER CHIRAL METAL CATALYSTS (*Continued*)

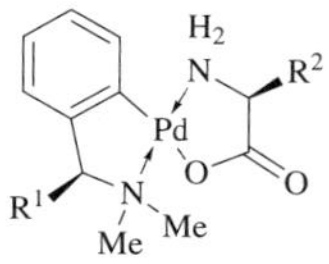

	R^1	R^2
190a:	H	Me
190b:	H	Bz
190c:	H	*i*-Pr
190d:	Me	H
190e:	Me	Me
190f:	Me	Bz
190g:	Me	*i*-Pr

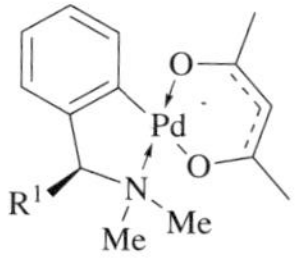

191

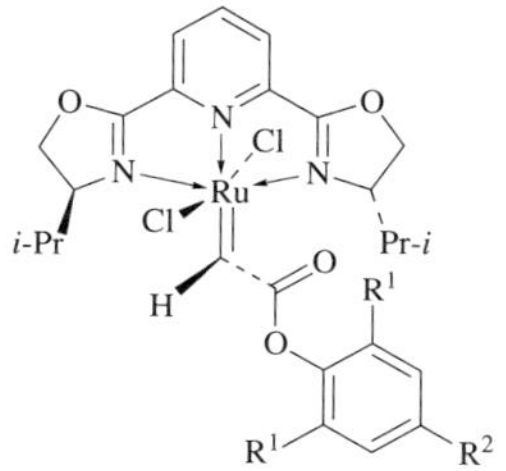

192a: R^1 = Me, R^2 = Me
192b: R^1 = *i*-Pr, R^2 = H

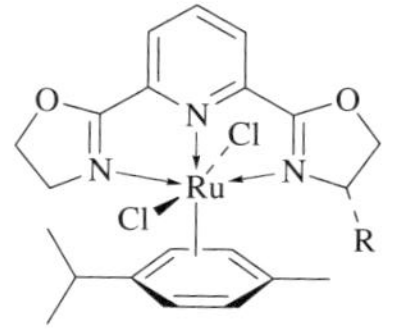

193a: R = *i*-Pr
193b: R = *t*-Bu

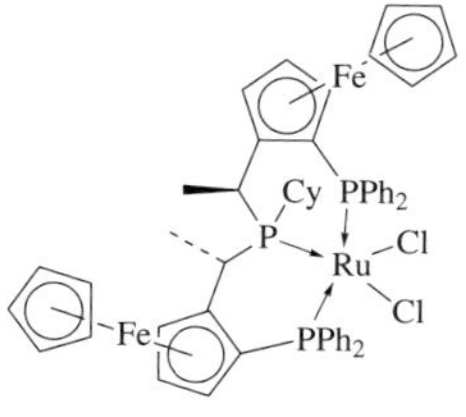

194

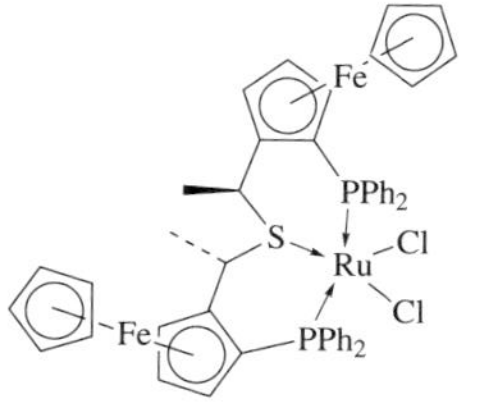

195

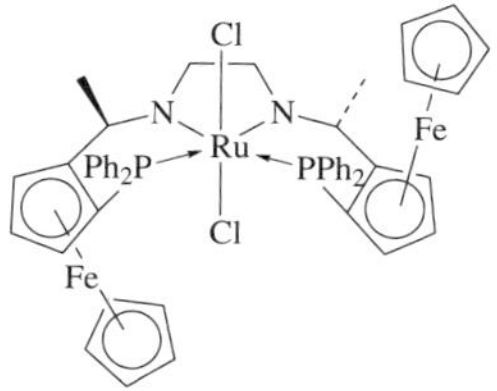

196

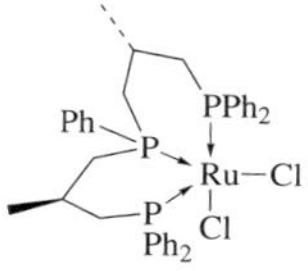

197

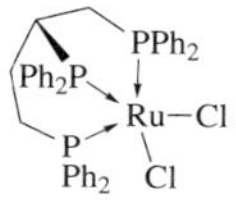

198

TABLE 1. DIASTEREOSELECTIVE CYCLOPROPANATION OF ALKENES USING DIAZOACETATES AND RELATED SYSTEMS

(See Chart 1 for catalyst structures.)

Carbenoid Precursor	Substrate	Conditions	Product(s) and Yield(s) (%)	Refs.
C₁ $N_2\text{=}\!\!\!\diagup\!\!\!-NO_2$	isobutylene	Rh₂(OAc)₄, Et₂O	cyclopropane-NO₂ (4)	288
	cyclohexene	Rh₂(OAc)₄, Et₂O	bicyclic-NO₂ (14) *exo:endo* = 67:33	288
	2,3-dimethyl-2-butene	Rh₂(OAc)₄, Et₂O	cyclopropane-NO₂ (15)	288
	benzodioxepine	Rh₂(OAc)₄	spirocycle-CN (64)	289
C₂ $N_2\text{=}\!\!\!\diagup\!\!\!-CN$ $N_2\text{=}\!\!\!\diagup\!\!\!-CO_2R$, R = Et	Cl₂C=CCl₂	Rh₂(OPiv)₄, neat, 60°	Cl₄-cyclopropane-CO₂Et (0)	290
C₄	Cl₂C=CHCl	Rh₂(OPiv)₄, neat, 60°	Cl₃-cyclopropane-CO₂Et (0)	290
	cis-ClCH=CHCl	Rh₂(OPiv)₄, CH₂Cl₂, rt	Cl₂-cyclopropane-CO₂Et (32)	290
		Cu, DCE, 80°	(1)	290

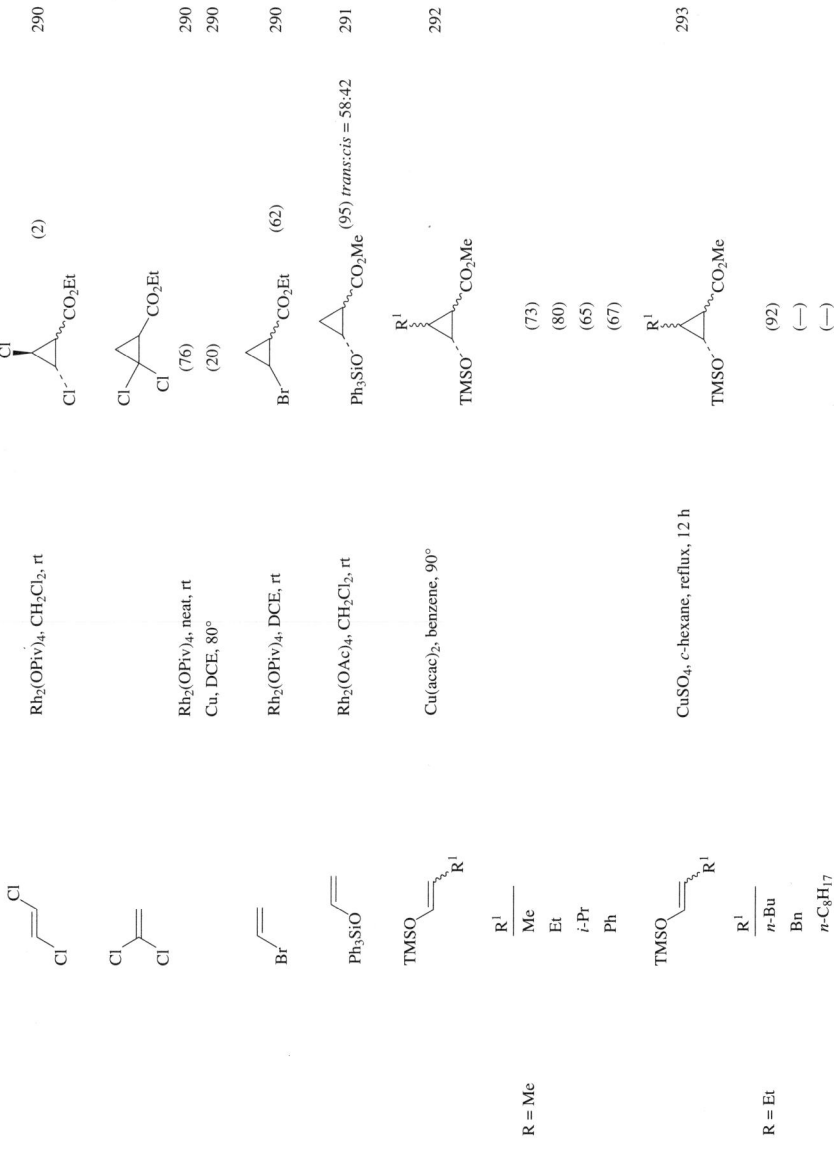

TABLE I. DIASTEREOSELECTIVE CYCLOPROPANATION OF ALKENES USING DIAZOACETATES AND RELATED SYSTEMS (Continued)

Carbenoid Precursor	Substrate	Conditions	Product(s) and Yield(s) (%)	Refs.
C_3 $R = Me$	R^1Me_2SiO R^2 R^3 R^4	$Cu(acac)_2$, reflux	R^1Me_2SiO R^3 R^4 R^2 CO_2Me	294

R^1	R^2	R^3	R^4	Solvent		trans:cis
Me	H	H	H	benzene	(58)	72:28
Me	H	H	H	ethyl acetate	(45)	72:28
Me	Me	H	H	benzene	(65)	65:35
Me	Me	H	H	ethyl acetate	(70)	65:35
Me	n-C_5H_{11}	H	H	benzene	(81)	61:39
Me	i-Pr	H	H	benzene	(68)	58:42
Me	i-Pr	H	H	ethyl acetate	(80)	58:42
Me	t-Bu	H	H	benzene	(87)	52:48
Me	t-Bu	H	H	ethyl acetate	(85)	52:48
t-Bu	t-Bu	H	H	benzene	(69)	58:42
t-Bu	t-Bu	H	H	ethyl acetate	(80)	58:42
Me	vinyl	H	H	benzene, 100°	(60)	64:36
Me	vinyl	H	H	ethyl acetate	(39)	64:36
Me	Ph	H	H	benzene	(75)	52:48
Me	Ph	H	H	ethyl acetate	(78)	52:48
t-Bu	Ph	H	H	benzene	(78)	60:40
t-Bu	Ph	H	H	ethyl acetate	(77)	60:40
Me	Ph	H	Me	benzene	(71)	55:45
Me	Ph	H	Me	ethyl acetate	(71)	55:45
Me	Ph	Me	Me	benzene	(72)	54:46
Me	Ph	Me	Me	ethyl acetate	(80)	54:46
Me	Me	Me	Me	benzene	(86)	68:32
Me	Me	Me	Me	ethyl acetate	(77)	68:32
Me	H	Me	Me	benzene	(79)	75:25
Me	H	Me	Me	ethyl acetate	(67)	75:25

295

R^1O–$C(R^2)$=$C(R^3)$–OR^1 + (R = Et diazo) $\xrightarrow{\text{Cu(acac)}_2,\text{ benzene, reflux}}$ cyclopropane with R^1O, R^1O, R^2, R^3, CO_2Et

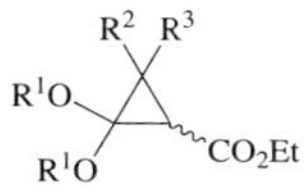

R^1	R^2	R^3		solvent	yield	ratio
t-Bu	H	Me	Me	benzene	(77)	77:23
Me	H	H	Me	benzene	(73)	78:22
Me	H	H	Me	ethyl acetate	(70)	78:22

C$_4$ R = Et

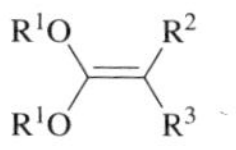

R^1	R^2	R^3	
Me	H	H	(—)
Et	H	H	(—)
Me	Me	H	(—)
Et	Me	H	(—)
Me	Et	H	(75)
Me	Ph	H	(35)
Me	Me	Me	(—)

296

TMSO–C(R^1)=CH–C(=O)R^2 $\xrightarrow{\text{1. CuSO}_4,\text{ cyclohexane, reflux}}_{\text{2. H}^+}$ R^1C(=O)–CH(CO$_2$Et)–CH$_2$–C(=O)R^2

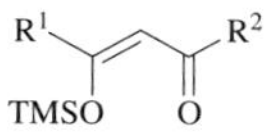

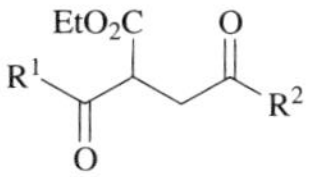

R = Et

R^1	R^2	
Ph	OEt	(73)
Me	OEt	(63)
n-Bu	OEt	(69)
n-C$_6$H$_{13}$	OEt	(58)
i-Bu	OEt	(63)
CH$_2$Bn	OEt	(75)
i-Bu	i-Bu	(35)
n-C$_5$H$_{11}$	n-C$_5$H$_{11}$	(41)
Me	t-Bu	(52)

TABLE 1. DIASTEREOSELECTIVE CYCLOPROPANATION OF ALKENES USING DIAZOACETATES AND RELATED SYSTEMS (*Continued*)

Carbenoid Precursor	Substrate	Conditions	Product(s) and Yield(s) (%)	Refs.
R = Et	enaminone with R¹, R², R³ (OEt/Me/H combinations)	Cu(acac)₂, CH₂Cl₂, 60°, 3 d	pyrrole product (—)	297
R = Et	vinylene carbonate	Cu(OTf)₂, benzene, 0°	bicyclic carbonate-cyclopropane (33)	95
	isocyanide NC-CH=CH₂	Mo(CO)₆, neat, rt, 72 h	*trans:cis* 58:42 (46)	298
		Mo(CO)₆, neat, 65°, 7 h	52:48 (46)	298, 299
		Mo₂(OAc)₄, neat, 65°	52:48 (42)	299
C₃ R = Me	(TMS)₂N-allyl derivatives	Rh₂(OAc)₄, CH₂Cl₂, rt	*trans:cis* 66:34 (67); 60:40 (68); — (26)	300

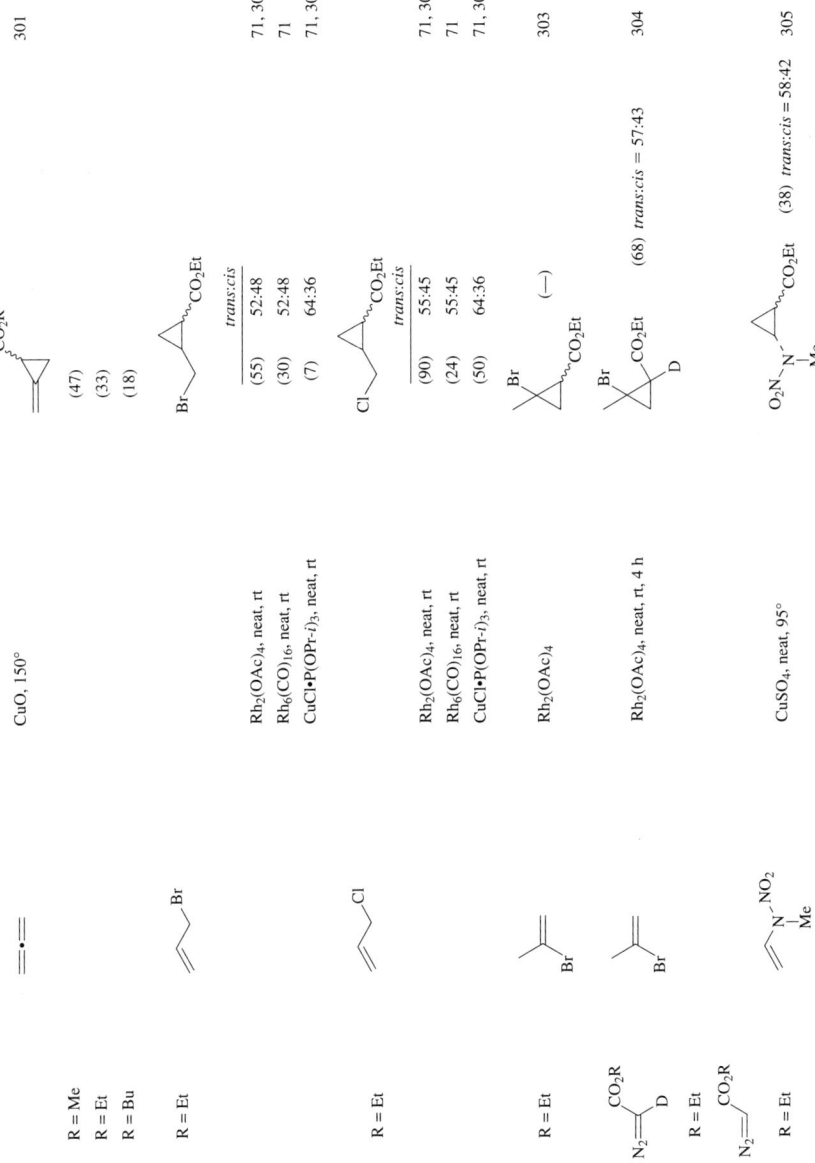

TABLE 1. DIASTEREOSELECTIVE CYCLOPROPANATION OF ALKENES USING DIAZOACETATES AND RELATED SYSTEMS (*Continued*)

	Carbenoid Precursor	Substrate	Conditions	Product(s) and Yield(s) (%)	Refs.
C_3	R = Me	(TMS)$_2$N⁀⁀⁀	Cu(acac)$_2$, DCE, 80°	(TMS)$_2$N⁀⁀⁀△⁀⁀⁀CO$_2$Me (77) *trans:cis* = 85:15	300
C_4	R = Et	NC⁀⁀⁀		NC⁀⁀⁀△⁀⁀⁀CO$_2$Et	
				trans:cis	
			Mo(CO)$_6$, neat, rt, 72 h	(77) 41:59	298
			Mo(CO)$_6$, neat, 65°, 3 h	(79) 41:59	298, 299
			Mo$_2$(OAc)$_4$, neat, 65°	(72) 44:56	299
			Mo(CO)$_6$, benzene, 65°	(80) —	299
			Mo(CO)$_4$(PPh$_3$)$_2$, benzene, 65°	(81) —	299
			Mo(CO)$_4$(pip)$_2$, benzene, 65°	(0) —	299
			Mo(CO)$_4$(PPh$_3$)(pip), benzene, 65°	(63) —	299
			Mo(N$_2$)$_2$(dppe)$_2$, benzene, 65°	(15) —	299
	R = Et	▢	CuOTf, cyclohexane, 0°, 8 h	⟨bicyclic⟩ CO$_2$Et (64) *exo:endo* = 80:20	306
C_3	R = Me	⟨2,5-dihydrofuran⟩	Cu(acac)$_2$, 100°, 3 h	⟨bicyclic⟩ CO$_2$Me (51)	307
			[Ru$_2$(CO)$_4$(μ-OAc)$_2$]$_n$, CH$_2$Cl$_2$, rt	(62) *trans:cis* = >97:3	308
	R = Me	⟨2,3-dihydrofuran⟩	Cu(acac)$_2$, 100°, 3 h	⟨bicyclic⟩ CO$_2$Me (26)	307

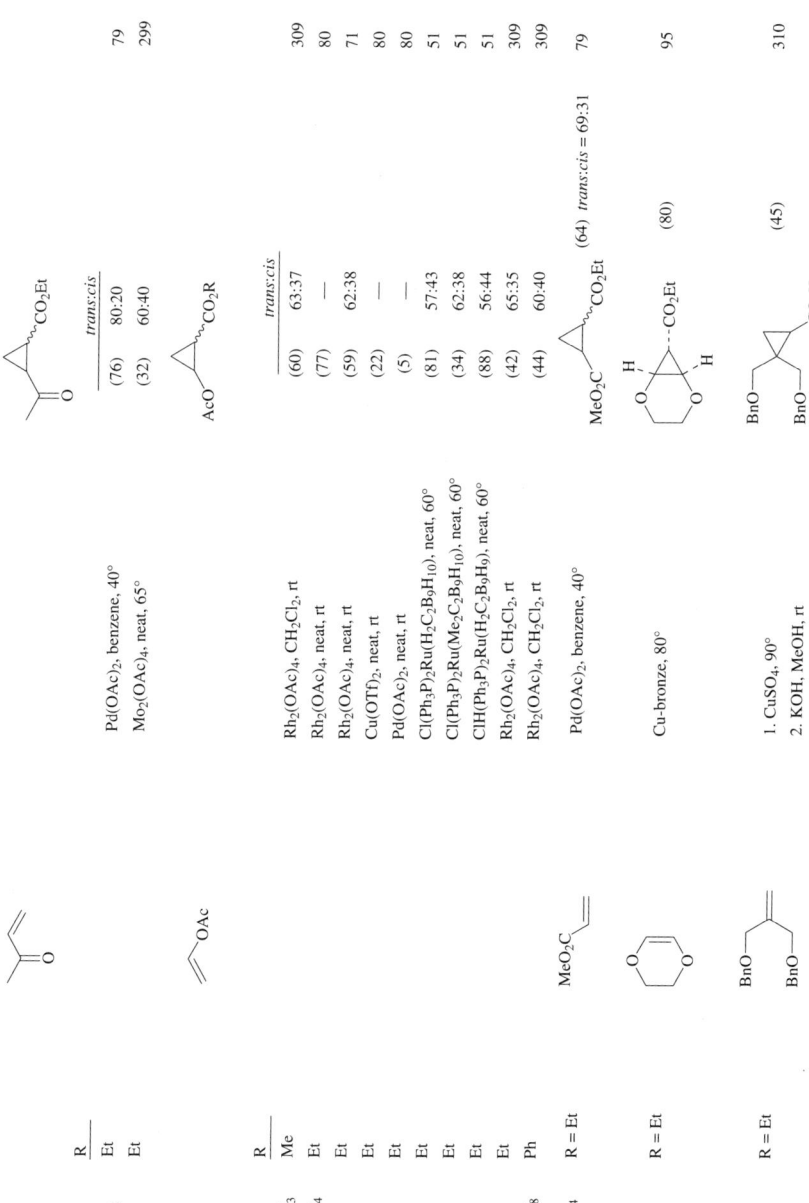

TABLE 1. DIASTEREOSELECTIVE CYCLOPROPANATION OF ALKENES USING DIAZOACETATES AND RELATED SYSTEMS (*Continued*)

Carbenoid Precursor	Substrate	Conditions	Product(s) and Yield(s) (%)	Refs.
R = Et	TMSO⟩=⟨OTMS	Cu, cyclohexane, 80°	TMSO–△–OTMS, CO₂Et (60)	311
C₃ R = Me	TMSO⟩=⟨	Cu(acac)₂, benzene, 90° Cu(acac)₂, EtOAc, 100°	TMSO–△–CO₂Me (66) (81) *trans:cis* = 75:25	312 313
C₄ R = Et	⟩=⟨–Br	Rh₂(OAc)₄, neat, rt	CO₂Et–△–Br (25) *trans:cis* = 69:31	71
R = Et	TBDMSO–⟩–NHBOC	Pd(OAc)₂, Et₂O, rt	TBDMSO–△–NHBOC, CO₂Et (88)	314
			CO₂R–△–	
R / Me	⟩=⟨	Rh₂(OAc)₄, neat, rt	(54)	80
C₃ Me		Pd(OAc)₂, neat, rt	(24)	80
C₄ Et		Cu(OTf)₂, Et₂O, reflux, 6 h	(33)	315
Et		CuOTf, 0°	(>50)	316
			CO₂Me–△–	
R / Me	⟩=	Pd(OAc)₂, neat, rt	(21)	80
C₃ Me				
C₄ Et		CuOTf, 0°	(>50)	316

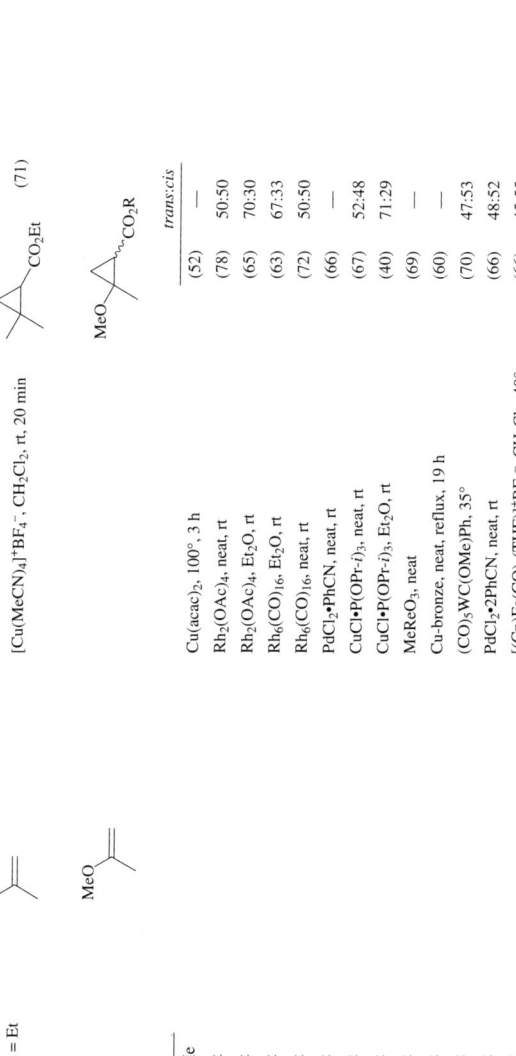

R = Et

[Cu(MeCN)₄]⁺BF₄⁻, CH₂Cl₂, rt, 20 min   (71)   317

	R			trans:cis	
C₃	Me				
C₄	Et	Cu(acac)₂, 100°, 3 h	(52)	—	307
	Et	Rh₂(OAc)₄, neat, rt	(78)	50:50	71
	Et	Rh₂(OAc)₄, Et₂O, rt	(65)	70:30	94
	Et	Rh₆(CO)₁₆, Et₂O, rt	(63)	67:33	94
	Et	Rh₆(CO)₁₆, neat, rt	(72)	50:50	71, 318
	Et	PdCl₂•PhCN, neat, rt	(66)	—	318
	Et	CuCl•P(OPr-i)₃, neat, rt	(67)	52:48	71, 358
	Et	CuCl•P(OPr-i)₃, Et₂O, rt	(40)	71:29	94
	Et	MeReO₃, neat	(69)	—	319
	Et	Cu-bronze, neat, reflux, 19 h	(60)	—	320
	Et	(CO)₅WC(OMe)Ph, 35°	(70)	47:53	321
	Et	PdCl₂•2PhCN, neat, rt	(66)	48:52	71
	Et	[(Cp)Fe(CO)₂(THF)]⁺BF₄⁻, CH₂Cl₂, 40°	(66)	45:55	322

TABLE 1. DIASTEREOSELECTIVE CYCLOPROPANATION OF ALKENES USING DIAZOACETATES AND RELATED SYSTEMS (Continued)

Carbenoid Precursor	Substrate	Conditions	Product(s) and Yield(s) (%)		Refs.
R				trans:cis	
C$_3$ Me		[Ru$_2$(CO)$_4$(μ-OAc)$_2$]$_n$, CH$_2$Cl$_2$, rt	(89)	82:18	308
Me		Ru$_2$(CO)$_4$(μ-OAc)$_2$(MeCN)$_2$, CH$_2$Cl$_2$, rt	(54)	92:8	308
Me		Ru$_2$(CO)$_4$(μ-OAc)$_2$(MeCN)$_2$, neat, 36°	(83)	84:16	308
C$_4$ Et		CuCl•P(OPh)$_3$, neat, rt	(64)	69:31	71
Et		Cu(OTf)$_2$, neat, 0°	(55)	71:29	71
Et		Cu(acac)$_2$, neat, rt	(15)	62:38	71
Et		Rh$_2$(OAc)$_4$, Et$_2$O, rt	(75)	62:38	94
Et		Rh$_2$(OAc)$_4$, rt	(60-93)	62:38	323
Et		Rh$_2$(O$_2$CC$_3$F$_7$-n)$_4$, rt	(60-93)	57:43	323
Et		Rh$_2$(NHCOCH$_3$)$_4$, rt	(60-93)	74:26	323
Et		Rh$_2$(NHCOCH$_3$)$_4$, CH$_2$Cl$_2$, rt, 5 h	(72)	trans	78
Et		Rh$_2$(OAc)$_4$, CH$_2$Cl$_2$, rt, 5 h	(62)	trans	78
Et		Rh$_2$(O$_2$CC$_3$H$_7$)$_4$, CH$_2$Cl$_2$, rt, 5 h	(60)	trans	78
Et		Rh$_2$(NHCOCF$_3$)$_4$, CH$_2$Cl$_2$, rt, 5 h	(62)	trans	78
Et		Rh$_2$(OCSCH$_3$)$_4$, CH$_2$Cl$_2$, rt, 5 h	(60)	trans	78
Et		Rh$_2$(O$_2$CCF$_3$)$_4$, CH$_2$Cl$_2$, rt, 5 h	(59)	trans	78
Et		Rh$_2$(O$_2$CC$_3$F$_7$-n)$_4$, CH$_2$Cl$_2$, rt, 5 h	(57)	trans	78
Et		Rh$_2$(acac)$_4$, CH$_2$Cl$_2$, rt, 5 h	(72)	trans	78
Et		Rh$_2$(OAc)$_4$, neat, rt	(88)	63:37	71
Et		Rh$_6$(CO)$_{16}$ neat, rt	(62)	63:37	71
Et		CuCl•P(OPr-i)$_3$, neat, rt	(61)	66:34	71
Et		PdCl$_2$•2PhCN, neat, rt	(43)	60:40	71

	R	Conditions	(%)	Ratio	Ref.
	Et	$Rh_2(OAc)_4$, neat, rt	(85)	—	80
	Et	$Cu(OTf)_2$, neat, rt	(0)	—	80
	Et	$Pd(OAc)_2$, neat, rt	(42)	—	80
	Et	$Rh_6(CO)_{16}$, neat, rt	(62)	—	318
	Et	$PdCl_2 \bullet PhCN$, neat, rt	(43)	—	318
	Et	$CuCl \bullet P(OPr\text{-}i)_3$, neat, rt	(61)	—	318
	Et	Iron meso-tetra(pentafluorophenyl) porphyrin chloride, CH_2Cl_2, rt, 4 h	(—)	77:23	57
	Et	Iron meso-tetramesitylporphyrin chloride/cobaltocene, CH_2Cl_2, rt, 4 h	(—)	80:20	57
	Et	Iron octaethylporphyrin chloride/cobaltocene, CH_2Cl_2, rt, 4 h	(—)	82:18	57
	Et	$(CO)_5WC(OMe)Ph$, 35°	(38)	58:42	321
	Et	$[(Cp)Fe(CO)_2(THF)]^+BF_4^-$, CH_2Cl_2, 40°	(66)	55:45	322
C_{10}	2,3,4-Me$_3$-3-pentyl	$Rh_2(OAc)_4$, rt	(62-99)	70:30	323
	2,3,4-Me$_3$-3-pentyl	$Rh_2(O_2CC_3F_7\text{-}n)_4$, rt	(62-99)	60:40	323
	2,3,4-Me$_3$-3-pentyl	$Rh_2(NHCOCH_3)_4$, rt	(62-99)	81:19	323
	2,3,4-Me$_3$-3-pentyl	$Rh_2(OAc)_4$, CH_2Cl_2, rt, 5 h	(67)	trans	78
	2,3,4-Me$_3$-3-pentyl	$Rh_2(acac)_4$, CH_2Cl_2, rt, 5 h	(80)	trans	78
C_{12}	d-menthyl	$Rh_2(OAc)_4$, CH_2Cl_2, rt	(88)	63:37	359
	3-i-Pr-2-Me-3-heptyl	$Rh_2(OAc)_4$, CH_2Cl_2, rt, 5 h	(70)	trans	78
C_{13}	2,6-(t-Bu)$_2$-4-MeC$_6$H$_2$	$Rh_2(OAc)_4$, CH_2Cl_2, rt, 5 h	(71)	trans	78
C_{17}	2,6-(t-Bu)$_2$-4-MeC$_6$H$_2$	$Rh_2(acac)_4$, CH_2Cl_2, rt, 5 h	(85)	trans	78

C_4 R = Et $Rh_2(OAc)_4$, CH_2Cl_2, 1 h   (52)   324

TABLE 1. DIASTEREOSELECTIVE CYCLOPROPANATION OF ALKENES USING DIAZOACETATES AND RELATED SYSTEMS (*Continued*)

	Carbenoid Precursor	Substrate	Conditions	Product(s) and Yield(s) (%)		Refs.
C_3	R = Me	(OTMS-cyclopentene)	Cu(acac)$_2$, benzene, 100°	(72) *trans:cis* = 54:46		294, 312
C_4	R = Et	MeO—OMe	Cu-bronze, 80°	(83)		95
		(cyclopentene)			*exo:endo*	
	R					
	Et		Rh$_2$(O$_2$CH)$_4$, 60°	(—)	62:38	325
	Et		Rh$_2$(OAc)$_4$, 60°	(—)	71:29	325
	Et		Rh$_2$(O$_2$CBu-n)$_4$, 60°	(—)	71:29	325
	Et		Rh$_2$(OAc)$_4$, neat, rt	(95)	—	80
	Et		Rh$_2$(OAc)$_4$, neat, rt	(96)	68:32	71
	Et		Cu(OTf)$_2$, neat, rt	(60)	—	80
	Et		Pd(OAc)$_2$, neat, rt	(60)	—	80
	Et		Cl(Ph$_3$P)$_2$Ru(H$_2$C$_2$B$_9$H$_{10}$), neat, 60°	(49)	51:49	51
	Et		Cl(Ph$_3$P)$_2$Ru(Me$_2$C$_2$B$_9$H$_{10}$), neat, 60°	(26)	51:49	51
	Et		ClH(Ph$_3$P)$_2$Ru(H$_2$C$_2$B$_9$H$_9$), neat, 60°	(53)	50:50	51
	Et		Rh$_2$(OAc)$_4$, CH$_2$Cl$_2$, rt, 5 h	(72)	*exo*	78
	Et		CuOTf, DCE, rt	(61)	77:23	326
	Et		**4**, DCE, rt	(64)	94:6	326
	Et		**5**, DCE, rt	(66)	97:3	326
	Et		**6**, DCE, rt	(51)	98:2	326
C_{13}	3-i-Pr-2-Me-3-heptyl		Rh$_2$(OAc)$_4$, CH$_2$Cl$_2$, rt, 5 h	(97)	*exo*	78
C_{17}	2,6-(t-Bu)$_2$-4-MeC$_6$H$_2$		Rh$_2$(OAc)$_4$, CH$_2$Cl$_2$, rt, 5 h	(69)	*exo*	78

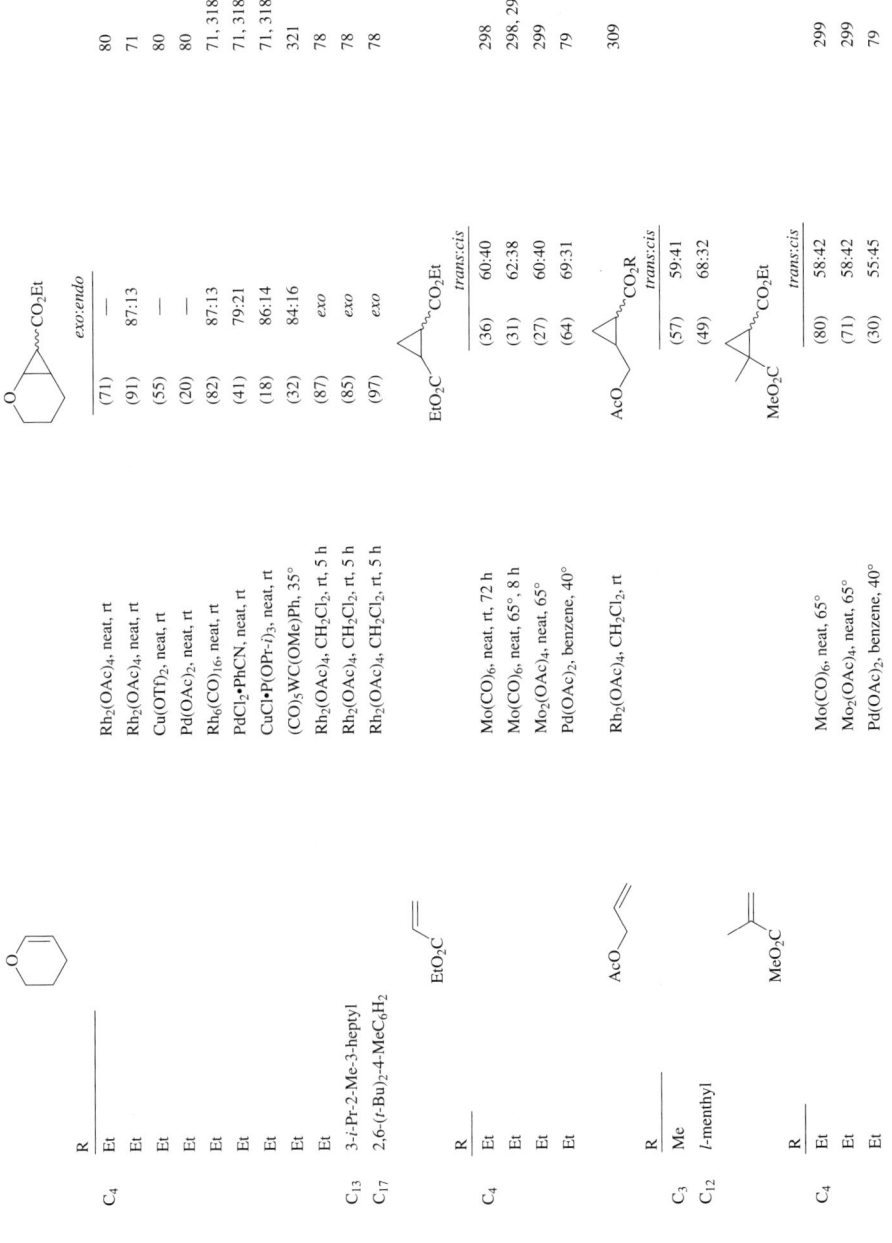

TABLE 1. DIASTEREOSELECTIVE CYCLOPROPANATION OF ALKENES USING DIAZOACETATES AND RELATED SYSTEMS (*Continued*)

Carbenoid Precursor	Substrate	Conditions	Product(s) and Yield(s) (%)	Refs.
R = Et	(1,3-dioxepine)	CuSO$_4$, 75°	(bicyclic CO$_2$Et product) (—)	327
R = Et	CH$_2$=C(Br)(i-Pr)	CuSO$_4$, neat, 90°, 10 h	(Br, i-Pr cyclopropane-CO$_2$Et) (28)	328
R = Et	(Me)$_2$C=C(Br)(Me)	CuSO$_4$, neat, 90°, 10 h	(Br cyclopropane-CO$_2$Et) (24)	328
R = Et	CH$_2$=CH–i-Pr	Rh$_2$(OAc)$_4$, neat, rt	(i-Pr cyclopropane-CO$_2$Et) (58) *trans:cis* = 67:33	71
R = Et	(Me)$_2$C=CHMe	Rh$_2$(OAc)$_4$, neat, rt [(Cp)Fe(CO)$_2$(THF)]$^+$BF$_4^-$, CH$_2$Cl$_2$, 40°	(cyclopropane-CO$_2$Et) (25) *trans:cis* = 69:31 (0)	71 322

	R	Conditions	Product		Refs.
			![cyclopropane-CO2R]		
	Et	Rh$_2$(OAc)$_4$, neat, rt	(70)	*trans*	80
	Et	Cu(OTf)$_2$, neat, rt	(30)	—	80
	Et	Pd(OAc)$_2$, neat, rt	(5)	—	80
	Et	Rh$_2$(OAc)$_4$, CH$_2$Cl$_2$, rt, 5 h	(60)	*trans*	78
	Et	Rh$_2$(acac)$_4$, CH$_2$Cl$_2$, rt, 5 h	(58)	*trans*	78
C$_{10}$	2,3,4-Me$_3$-3-pentyl	Rh$_2$(OAc)$_4$, CH$_2$Cl$_2$, rt, 5 h	(64)	*trans*	78
	2,3,4-Me$_3$-3-pentyl	Rh$_2$(acac)$_4$, CH$_2$Cl$_2$, rt, 5 h	(66)	*trans*	78
C$_{13}$	3-i-Pr-2-Me-3-heptyl	Rh$_2$(OAc)$_4$, CH$_2$Cl$_2$, rt, 5 h	(65)	*trans*	78
	3-i-Pr-2-Me-3-heptyl	Rh$_2$(acac)$_4$, CH$_2$Cl$_2$, rt, 5 h	(65)	*trans*	78
C$_{17}$	2,6-(t-Bu)$_2$-4-MeC$_6$H$_2$	Rh$_2$(OAc)$_4$, CH$_2$Cl$_2$, rt, 5 h	(96)	*trans*	78

propenyl ether substrate:

	R	Conditions	Yield	*trans:cis*	Refs.
C$_4$	Et	Rh$_2$(OAc)$_4$, rt	(60-93)	63:37	323
	Et	Rh$_2$(O$_2$CC$_3$F$_7$)$_4$, rt	(60-93)	57:43	323
	Et	Rh$_2$(NHCOCH$_3$)$_4$, rt	(60-93)	74:26	323
C$_{10}$	2,3,4-Me$_3$-3-pentyl	Rh$_2$(OAc)$_4$, rt	(62-99)	68:32	323
	2,3,4-Me$_3$-3-pentyl	Rh$_2$(O$_2$CC$_3$F$_7$)$_4$, rt	(62-99)	58:42	323
	2,3,4-Me$_3$-3-pentyl	Rh$_2$(NHCOCH$_3$)$_4$, rt	(62-99)	79:21	323

	R	Conditions	Product	Refs.
C$_3$	R = Me	Cu(acac)$_2$, 100°, 3 h	(83)	307

	R	Conditions	Product	Refs.
C$_3$	Me	Cu(acac)$_2$, 100°, 3 h	(29)	307
C$_4$	Et	Cu-bronze, methylcyclohexane, reflux, 4 h	(—) *trans:cis* = 70:30	329

TABLE 1. DIASTEREOSELECTIVE CYCLOPROPANATION OF ALKENES USING DIAZOACETATES AND RELATED SYSTEMS (*Continued*)

Carbenoid Precursor	Substrate	Conditions	Product(s) and Yield(s) (%)	Refs.
R = Et	5-Cl-furanyl-CH=CH$_2$	CuSO$_4$, 65°	cyclopropane-furanyl(Cl)-CO$_2$Et (36)	330
R = Et	5-Br-furanyl-CH=CH$_2$	CuSO$_4$, 65°	cyclopropane-furanyl(Br)-CO$_2$Et (22)	330
R = Et	5-O$_2$N-furanyl-CH=CH$_2$	CuSO$_4$, 65°	cyclopropane-furanyl(O$_2$N)-CO$_2$Et (19)	330

Substrate: glycal with OR^1, R^{2}O, OR^3 substituents; Product: fused cyclopropane bearing CO$_2$Et on sugar with OR^1, RO^2, OR^3.

R	R^1	R^2	R^3	Conditions	Yield (%)	Refs.
Et	Bn	Bn	Bn	Rh$_2$(OAc)$_4$, CH$_2$Cl$_2$, 1 h	(59)	324
Et	TBDMS	TBDMS	TBDMS	Rh$_2$(OAc)$_4$, CH$_2$Cl$_2$, 1 h	(85)	324
Et	Bz	Bz	Bz	Rh$_2$(OAc)$_4$, CH$_2$Cl$_2$, 1 h	(75)	324
Et	TBDMS	Ac	Ac	Rh$_2$(OAc)$_4$, CH$_2$Cl$_2$, 14 h	(89)	331
Et	TBDMS	TBDMS	TBDMS	Rh$_2$(OAc)$_4$, CH$_2$Cl$_2$, 14 h	(81)	331
Et	TIPS	TIPS	TIPS	Rh$_2$(OAc)$_4$, CH$_2$Cl$_2$, 14 h	(61)	331
Et	Bn	Bn	Bn	Rh$_2$(OAc)$_4$, CH$_2$Cl$_2$, 14 h	(35)	331
Et	Ac	Ac	Ac	Rh$_2$(OAc)$_4$, CH$_2$Cl$_2$, 14 h	(62)	331
Et	Ac	Ac	Ac	Cu(acac)$_2$, CH$_2$Cl$_2$, 14 h	(4)	331
Et	Ac	Ac	Ac	PdCl$_2$, CH$_2$Cl$_2$, 14 h	(0)	331
Et	Ac	Ac	Ac	Pd(OAc)$_2$, CH$_2$Cl$_2$, 14 h	(0)	331
Et	TBDMS	H	H	Rh$_2$(OAc)$_4$, CH$_2$Cl$_2$, 14 h	(0)	331
Et	TBDMS	TBDMS	TBDMS	Cu, MeOBu-*t*, reflux	(92)	332
Et	Bn	Bn	Bn	Cu, MeOBu-*t*, reflux	(34)	332

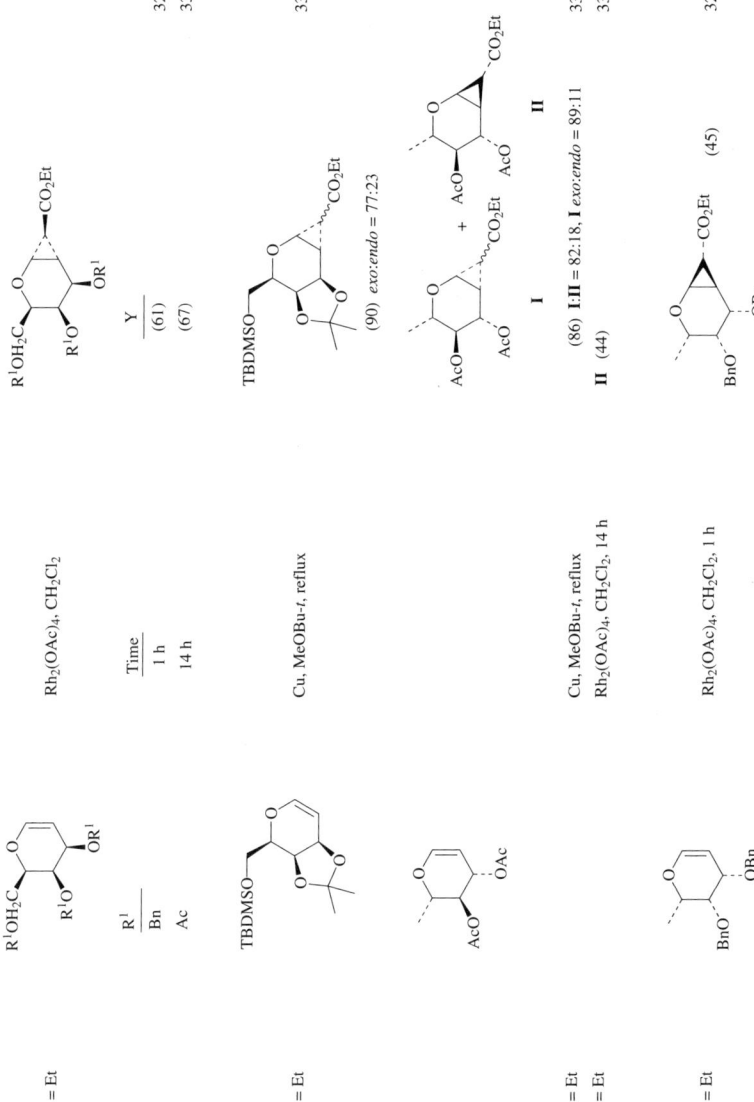

TABLE 1. DIASTEREOSELECTIVE CYCLOPROPANATION OF ALKENES USING DIAZOACETATES AND RELATED SYSTEMS (*Continued*)

Carbenoid Precursor	Substrate	Conditions	Product(s) and Yield(s) (%)	Refs.
R = Et	(cyclohexadiene)	Rh$_2$(OAc)$_4$, neat, rt, 10 h	(80) *trans:cis* —	77
Et		Cu(OTf)$_2$, neat, rt, 10 h	(60) —	77
Et		Pd(OAc)$_2$, neat, rt, 10 h	(37) —	77
Et		Cl(Ph$_3$P)$_2$Ru(H$_2$C$_2$B$_9$H$_{10}$), neat, 60°	(61) 29:71	51
Et		Cl(Ph$_3$P)$_2$Ru(Me$_2$C$_2$B$_9$H$_{10}$), neat, 60°	(34) 31:69	51
Et		ClH(Ph$_3$P)$_2$Ru(H$_2$C$_2$B$_9$H$_9$), neat, 60°	(63) 29:71	51
R = Et	(bicyclopropylidene)	Rh$_2$(OAc)$_4$, CH$_2$Cl$_2$, 0°, 12 h	(77)	333
C$_3$ R = Me	(1,2-bis-OTMS cyclohexene)	Cu(acac)$_2$, solvent, reflux	*trans:cis* benzene (63) >95:5; ethyl acetate (66) >95:5	294, 312; 294
C$_4$ R = Et	(OTMS cyclopentene with CH$_2$OTBDMS)	Cu(acac)$_2$, benzene, reflux	(75)	334

TABLE 1. DIASTEREOSELECTIVE CYCLOPROPANATION OF ALKENES USING DIAZOACETATES AND RELATED SYSTEMS (Continued)

Carbenoid Precursor	Substrate	Conditions	Product(s) and Yield(s) (%)		Refs.
				exo:endo	
R					
C_3 Me		$Cu(acac)_2$, 100°, 3 h	(62)	—	307
Me		$Rh_2(OAc)_4$, neat, rt	(80)	—	80
Me		$Rh_2(OAc)_4$, neat, rt, 10 h	(84)	—	77
Me		$Rh_2(O_2CC_3H_7\text{-}n)_4$, neat, rt, 10 h	(85)	—	77
Me		$Cu(OTf)_2$, neat, rt, 10 h	(54)	—	77
Me		$[Ru_2(CO)_4(\mu\text{-}OAc)_2]_n$, CH_2Cl_2, rt	(68)	79:21	308
Me		$Pd(OAc)_2$, neat, rt	(15)	—	80
C_4 Et		$MeReO_3$, neat	(71)	—	319
Et		$Rh_2(OAc)_4$, rt	(60-93)	79:21	323
Et		$Rh_2(O_2CC_3F_7\text{-}n)_4$, rt	(60-93)	67:33	323
Et		$Rh_2(NHCOCH_3)_4$, rt	(60-93)	91:9	323
Et		Iodorhodium(III)-meso-tetraphenylporphyrin, neat, 60°	(62)	54:46	102
Et		Iodorhodium(III)-meso-tetraphenylporphyrin, $CDCl_3$, 55°	(76)	—	335
Et		Iodorhodium(III)-meso-tetraphenylporphyrin, DCE, 60°	(62)	57:43	101
Et		Iodorhodium(III)-meso-tetraphenylporphyrin, DCE, 60°	(83)	46:54	101
Et		Iodorhodium(III)-meso-cyanooctaethyl-porphyrin, DCE, 60°	(38)	60:40	101
Et		Iodorhodium(III)-meso-tetra-o-tolyl-porphyrin, DCE, 60°	(53)	55:45	101
Et		$[OsCl_2(p\text{-cymene})]_2$, neat, 80°	(14)	61:39	91
Et		Zeolite NaCuX-38, 83°	(—)	—	336

Et	Rh$_2$(O$_2$CH)$_4$, 60°	(95)	69:31	325
Et	Rh$_2$(OAc)$_4$, rt	(83)	79:21	325
Et	Rh$_2$(O$_2$CPr-n)$_4$, rt	(93)	87:13	325
Et	Rh$_2$(O$_2$CC$_{17}$H$_{35}$-n)$_4$, rt	(97)	77:23	325
Et	Rh$_2$(O$_2$CPh)$_4$, 60°	(96)	72:28	325
Et	Rh$_2$(O$_2$CPr-n)$_4$, 60°	(95)	70:30	325
Et	Rh$_2$[O$_2$C(CH$_2$)$_2$Ph]$_4$, rt	(97)	75:25	325
Et	Rh$_2$(O$_2$CCH$_2$CPh$_3$)$_4$, rt	(96)	62:38	325
Et	Rh$_2$[O$_2$CC$_6$H$_4$OH-2]$_4$, 60°	(91)	65:35	325
Et	Rh$_2$[O$_2$CC$_6$H$_4$Bz-2]$_4$, rt	(96)	54:46	325
Et	Rh$_2$[O$_2$C-2,4-Cl$_2$C$_6$H(NO$_2$)$_2$-3,5]$_4$, rt	(30)	48:52	325
Et	Rh$_2$(O$_2$CC$_6$H$_4$CF$_3$-2)$_4$, rt	(98)	64:36	325
Et	Rh$_2$(O$_2$CC$_6$H$_4$CF$_3$-3)$_4$, rt	(98)	83:17	325
Et	Rh$_2$(O$_2$CC$_6$H$_2$Me$_3$-2,4,6)$_4$, 60°	(91)	67:33	325
Et	Rh$_2$(O$_2$CC$_2$H$_4$COMe)$_4$, rt	(98)	76:24	325
Et	Rh$_2$(O$_2$CCOPh)$_4$, 60°	(92)	70:30	325
Et	Rh$_2$(O$_2$CCF$_3$)$_4$, rt	(91)	69:31	325
Et	Rh$_2$(OAc)$_4$, Et$_2$O, rt	(80)	82:18	94
Et	Rh$_6$(CO)$_{16}$, Et$_2$O, rt	(43)	75:25	94
Et	Rh$_6$(CO)$_{16}$, neat, rt	(88)	—	318
Et	Rh$_2$(OAc)$_4$, neat, rt	(88)	—	80
Et	Cu(OTf)$_2$, neat, rt	(54)	—	80
Et	Pd(OAc)$_2$, neat, rt	(21)	—	80
Et	PdCl$_2$•PhCN, neat, rt	(31)	—	318
Et	CuCl•P(OPr-i)$_3$, neat, rt	(28)	—	318
Et	CuCl•P(OPr-i)$_3$, Et$_2$O, rt	(40)	93:7	94
Et	CuCl•PMe$_3$, neat, reflux	(77)	—	337

TABLE 1. DIASTEREOSELECTIVE CYCLOPROPANATION OF ALKENES USING DIAZOACETATES AND RELATED SYSTEMS (Continued)

Carbenoid Precursor	Substrate	Conditions	Product(s) and Yield(s) (%)		Refs.
Et		$RuCl_2(PPh_3)_3$, neat, 60°, 4 h	(4)	73:27	52, 338
Et		$RuCl_2(PPh_3)_3/C_6H_6$, neat, 60°, 4 h	(5)	77:23	52
Et		$RuH_3[Si(OEt)_3](PPh_3)_2$, neat, 60°, 4 h	(3)	75:25	52
Et		$Ru[Si(OEt)_3]_2(PPh_3)_2$, neat, 60°, 4 h	(4)	72:28	52
Et		$(CO)_5WC(OMe)Ph$, 35°	(33)	78:22	321
Et		$Rh_2(OAc)_4$, neat, rt	(90)	79:21	71
Et		$Rh_6(OAc)_{16}$, neat, rt	(88)	80:20	71
Et		$CuCl•P(OPr-i)_3$, neat, rt	(28)	87:13	71
Et		$PdCl_2•2PhCN$, neat, rt	(31)	69:31	71
Et		$CuCl•P(OPh)_3$, neat, rt	(25)	88:12	71
Et		$Cu(OTf)_2$, neat, rt	(80)	87:13	71
Et		Cu-bronze, neat, rt	(23)	88:12	71
Et		$Cu(acac)_2$, neat, 60°	(18)	87:13	71
Et		Cu-bronze-K10 montmorillonite, CH_2Cl_2, rt	(—)	62:38	339
Et		Cu-bronze-bentonite, CH_2Cl_2, rt	(—)	78:22	339
Et		$[(Cp)Fe(CO)_2(THF)]^+BF_4^-$, CH_2Cl_2, 40°	(0)	—	322
Et		$OsCl_2(PPh_3)_3$, neat, 60°, 4 h	(1)	—	338
Et		$[Cu(MeCN)_4]^+BF_4^-$, CH_2Cl_2, rt, 10 min	(30)	85:15	317
Et		$Pd(OAc)_2$, neat, rt, 10 h	(21)	—	77
Et		$Cl(PH_3P)_2Ru(H_2C_2B_9H_{10})$, neat, 60°	(75)	77:23	51
Et		$Cl(PH_3P)_2Ru(Me_2C_2B_9H_{10})$, neat, 60°	(32)	69:31	51
Et		$ClH(PH_3P)_2Ru(H_2C_2B_9H_9)$, neat, 60°	(79)	78:22	51
Et		$Rh_2(NHCOCH_3)_4$, CH_2Cl_2, rt, 5 h	(89)	exo	78
Et		$Rh_2(OAc)_4$, CH_2Cl_2, rt, 5 h	(79)	exo	78
Et		$Rh_2(O_2CC_3H_7\text{-}n)_4$, CH_2Cl_2, rt, 5 h	(77)	exo	78
Et		$Rh_2(NHCOCF_3)_4$, CH_2Cl_2, rt, 5 h	(74)	exo	78
Et		$Rh_2(OCSCH_3)_4$, CH_2Cl_2, rt, 5 h	(75)	exo	78

R	Conditions	(Yield)	Ratio	Ref.
Et	Rh₂(O₂CCF₃)₄, CH₂Cl₂, rt, 5 h	(67)	exo	78
Et	Rh₂(O₂CC₃F₇-n)₄, CH₂Cl₂, rt, 5 h	(66)	exo	78
Et	Iodorhodium(III)-meso-tetra-(2,4,6-Me₃C₆H₂)porphyrin, neat, 60°	(92)	45:55	99
Et	Rh₂(OPiv)₄, neat, 60°	(98)	72:28	99
Et	Rh₂(O₂CCPh₃)₄, neat, 60°	(96)	74:26	99
Et	Rh₂(O₂C-dehydroabietyl)₄, neat, 60°	(96)	74:26	99
Et	Rh₂(O₂C-adamantyl-1)₄, neat, 60°	(97)	68:32	99
Et	Rh₂(O₂C-anthracenyl-9)₄, neat, 60°	(93)	64:36	99
Et	Rh₂(O₂C-mesityl)₄, neat, 60°	(97)	59:41	99
Et	Rh₂(acac)₄, CH₂Cl₂, rt, 5 h	(89)	exo	78
Et	CuOTf, DCE, rt	(37)	84:16	326
Et	**4**, DCE, rt	(61)	97:3	326
Et	**5**, DCE, rt	(49)	98:2	326
Et	**6**, DCE, rt	(31)	99:1	326
Et	**7a**, neat, 60°	(78)	33:67	99
Et	**7b**, neat, 60°	(0)	—	99
Et	**7c**, neat, 60°	(80)	28:72	99
Et	**7d**, neat, 60°	(0)	—	99
Et	**7e**, neat, 60°	(68)	26:74	99
Et	**7f**, neat, 60°	(0)	—	99
Et	**7g**, neat, 60°	(—)	70:30	99
Et	**7h**, neat, 60°	(71)	27:73	99
Et	**7i**, neat, 60°	(77)	42:58	99
Et	**7j**, neat, 60°	(52)	28:72	99
Et	**23b**, neat, 60°, 4 h	(22)	74:26	340
Et	**24b**, neat, 60°, 4 h	(5)	78:22	340
CH₂CH₂Cl	CuSO₄, neat, 90°	(80)	—	341
CH₂CH₂Br	Rh₂(OAc)₄, 60°	(65)	—	342
n-Bu	Rh₂(OAc)₄, neat, rt, 10 h	(80)	—	77
C₆ n-Bu	Cu(OTf)₂, neat, rt, 10 h	(64)	—	77

TABLE 1. DIASTEREOSELECTIVE CYCLOPROPANATION OF ALKENES USING DIAZOACETATES AND RELATED SYSTEMS (Continued)

Carbenoid Precursor	Substrate	Conditions	Product(s) and Yield(s) (%)		Refs.
	t-Bu	Rh$_2$(OAc)$_4$, neat, rt	(89)	—	80
	t-Bu	Pd(OAc)$_2$, neat, rt	(19)	—	80
C$_{10}$	2,3,4-Me$_3$-3-pentyl	Rh$_2$(OAc)$_4$, rt	(62-99)	90:10	323
	2,3,4-Me$_3$-3-pentyl	Rh$_2$(O$_2$CC$_3$F$_7$-n)$_4$, rt	(62-99)	80:20	323
	2,3,4-Me$_3$-3-pentyl	Rh$_2$(NHCOCH$_3$)$_4$, rt	(62-99)	92:8	323
	2,3,4-Me$_3$-3-pentyl	Rh$_2$(OAc)$_4$, CH$_2$Cl$_2$, rt, 5 h	(90)	exo	78
	2,3,4-Me$_3$-3-pentyl	Rh$_2$(acac)$_4$, CH$_2$Cl$_2$, rt, 5 h	(92)	exo	78
	2,3,4-Me$_3$-3-pentyl	Rh$_2$(acac)$_4$, CH$_2$Cl$_2$, rt, 5 h	(96)	exo	78
C$_{13}$	3-i-Pr-2-Me-3-heptyl	Rh$_2$(OAc)$_4$, CH$_2$Cl$_2$, rt, 5 h	(92)	exo	78
	3-i-Pr-2-Me-3-heptyl	Rh$_2$(OAc)$_4$, CH$_2$Cl$_2$, rt, 5 h	(91)	exo	78
C$_{17}$	2,6-(t-Bu)$_2$-4-MeC$_6$H$_2$	Rh$_2$(OAc)$_4$, CH$_2$Cl$_2$, rt, 5 h	(92)	exo	78
	2,6-(t-Bu)$_2$-4-MeC$_6$H$_2$	Rh$_2$(acac)$_4$, CH$_2$Cl$_2$, rt, 5 h			
C$_4$	R = Et	CuSO$_4$ and Cu-bronze, octane, reflux, 4 h	(—)		343
	R = Et	CuSO$_4$ and Cu-bronze, octane, reflux, 4 h	(—) I:II = 75:25		343

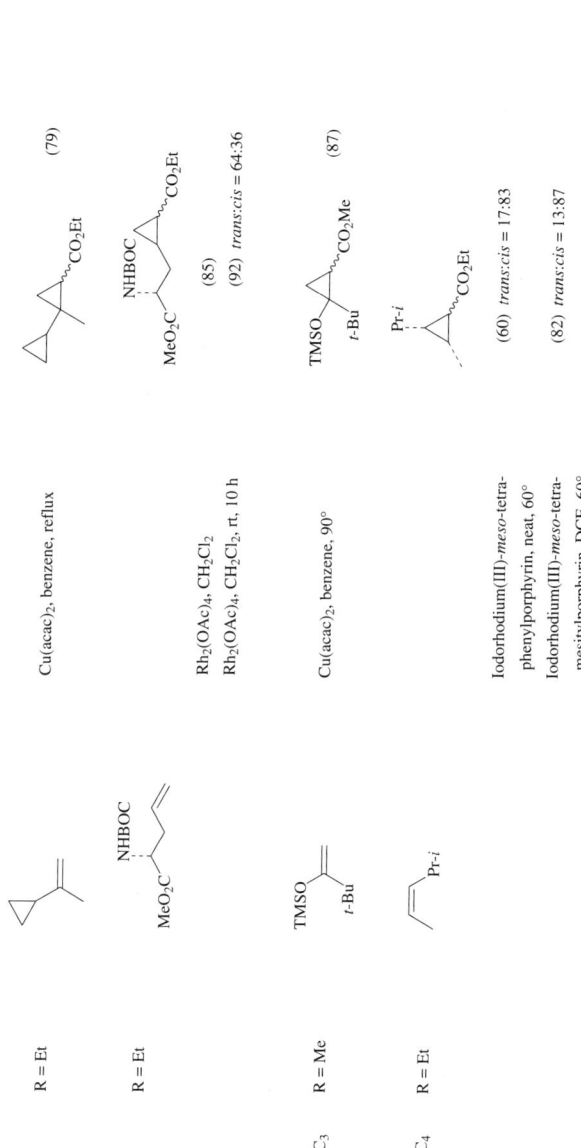

TABLE I. DIASTEREOSELECTIVE CYCLOPROPANATION OF ALKENES USING DIAZOACETATES AND RELATED SYSTEMS (*Continued*)

Carbenoid Precursor	Substrate	Conditions	Product(s) and Yield(s) (%)		Refs.
R	n-C$_4$H$_9$		n-C$_4$H$_9$△CO$_2$R		
				trans:cis	
C$_3$ Me		Rh$_2$(OAc)$_4$, neat, rt	(86)	—	80
Me		Cu(OTf)$_2$, neat, rt	(36)	—	80
Me		Pd(OAc)$_2$, neat, rt	(30)	—	80
Me		[Ru$_2$(CO)$_4$(μ-OAc)$_2$]$_n$, CH$_2$Cl$_2$, rt	(67)	67:33	308
C$_4$ Et		Rh$_2$(O$_2$CH)$_4$, 40°	(95)	54:46	325
Et		Rh$_2$(OAc)$_4$, rt	(86)	60:40	325
Et		Rh$_2$(OAc)$_4$, neat, rt	(95)	60:40	71
Et		Rh$_2$(O$_2$CPr-*n*)$_4$, rt	(96)	57:43	325
Et		Rh$_2$(O$_2$CPr-*n*)$_4$, 40°	(98)	57:43	325
Et		Rh$_2$(O$_2$CC$_{17}$H$_{35}$-*n*)$_4$, rt	(96)	60:40	325
Et		Rh$_2$(O$_2$CPh)$_4$, 40°	(95)	57:43	325
Et		Rh$_2$(O$_2$C(CH$_2$)$_2$Ph)$_4$, rt	(95)	56:44	325
Et		Rh$_2$(O$_2$CCH$_2$CPh$_3$)$_4$, rt	(98)	49:51	325
Et		Rh$_2$[O$_2$CC$_6$H$_4$OH-2]$_4$, 40°	(93)	52:48	325
Et		Rh$_2$[O$_2$CC$_6$H$_4$Bz-2]$_4$, rt	(97)	45:55	325
Et		Rh$_2$[O$_2$C-2,4-Cl$_2$C$_6$H(NO$_2$)$_2$-3,5]$_4$, 40°	(55)	50:50	325
Et		Rh$_2$[O$_2$CC$_6$H$_4$CF$_3$-2]$_4$, rt	(100)	52:48	325
Et		Rh$_2$[O$_2$CC$_6$H$_4$CF$_3$-3]$_4$, rt	(100)	57:43	325
Et		Rh$_2$[O$_2$CC$_6$H$_2$Me$_3$-2,4,6]$_4$, 40°	(95)	55:45	325
Et		Rh$_2$(O$_2$C$_2$H$_4$COMe)$_4$, rt	(93)	55:45	325
Et		Rh$_2$(O$_2$CCOPh)$_4$, 40°	(87)	57:43	325
Et		Rh$_2$(O$_2$CCF$_3$)$_4$, rt	(69)	58:42	325
Et		CuSO$_4$, 100°	(50)	—	347
Et		[OsCl$_2$(*p*-cymene)]$_2$, neat, 60°	(1)	—	91
Et		OsCl$_2$(PPh$_3$)$_3$, neat, 50°, 4 h	(1)	—	338

Et	Rh$_2$(OAc)$_4$, rt	(60-93)	58:42	323
Et	Rh$_2$(O$_2$CC$_3$F$_7$)$_4$, rt	(60-93)	55:45	323
Et	Rh$_2$(NHCOCH$_3$)$_4$, rt	(60-93)	63:37	323
Et	Iodorhodium(III)-*meso*-tetraphenylporphyrin, DCE, 60°	(68)	58:42	101
Et	Iodorhodium(III)-*meso*-tetramesitylporphyrin, DCE, 60°	(85)	53:47	101
Et	Rh$_2$(OAc)$_4$, CH$_2$Cl$_2$, rt, 5 h	(60)	*trans*	78
Et	Rh$_2$(acac)$_4$, CH$_2$Cl$_2$, rt, 5 h	(63)	*trans*	78
Et	Iodorhodium(III)-*meso*-tetra(2,4,6-Me$_3$C$_6$H$_2$)porphyrin, neat, 60°	(85)	53:47	99
Et	Hydridotris(3,5-Me$_2$-1-pyrazolyl)borateCu(C$_2$H$_4$), DCE, rt	(63)	50:50	348
Et	Cl(Ph$_3$P)$_2$Ru(H$_2$C$_2$B$_9$H$_{10}$), neat, 60°	(87)	58:42	51
Et	Cl(Ph$_3$P)$_2$Ru(Me$_2$C$_2$B$_9$H$_{10}$), neat, 60°	(23)	64:36	51
Et	ClH(Ph$_3$P)$_2$Ru(H$_2$C$_2$B$_9$H$_9$), neat, 60°	(92)	58:42	51
Et	RuCl$_2$(PPh$_3$)$_3$, neat, 50°, 4 h	(6)	60:40	52, 338
Et	RuCl$_2$(PPh$_3$)$_3$/C$_6$H$_6$, neat, 50°, 4 h	(9)	62:38	52
Et	RuH$_3$[Si(OEt)$_3$](PPh$_3$)$_2$, neat, 50°, 4 h	(6)	68:32	52
Et	Ru[Si(OEt)$_3$]$_2$(PPh$_3$)$_2$, neat, 50°, 4 h	(6)	65:35	52
Et	7a, neat, 60°	(59)	37:63	99
Et	7c, neat, 60°	(83)	34:66	99
Et	7e, neat, 60°	(89)	33:67	99
CH$_2$CN	Rh$_2$(OAc)$_4$, CH$_2$Cl$_2$, rt	(85)	—	349
CH$_2$CH$_2$Br	Rh$_2$(OAc)$_4$, 60°	(55)	—	342
C$_{10}$ 2,3,4-Me$_3$-3-pentyl	Rh$_2$(OAc)$_4$, rt	(62-99)	71:29	323
2,3,4-Me$_3$-3-pentyl	Rh$_2$(O$_2$CC$_3$F$_7$-n)$_4$, rt	(62-99)	58:42	323
2,3,4-Me$_3$-3-pentyl	Rh$_2$(NHCOCH$_3$)$_4$, rt	(62-99)	78:22	323
2,3,4-Me$_3$-3-pentyl	Rh$_2$(OAc)$_4$, CH$_2$Cl$_2$, rt, 5 h	(71)	*trans*	78
2,3,4-Me$_3$-3-pentyl	Rh$_2$(NHCOCH$_3$)$_4$, CH$_2$Cl$_2$, rt, 5 h	(81)	*trans*	78

TABLE 1. DIASTEREOSELECTIVE CYCLOPROPANATION OF ALKENES USING DIAZOACETATES AND RELATED SYSTEMS (*Continued*)

Carbenoid Precursor	Substrate	Conditions	Product(s) and Yield(s) (%)		Refs.
C_{12} *d*-menthyl		$Rh_2(OAc)_4$, neat, rt	(95)	60:40	309
C_{15} 3-*i*-Pr-2-Me-3-heptyl		$Rh_2(OAc)_4$, CH_2Cl_2, rt, 5 h	(74)	*trans*	78
3-*i*-Pr-2-Me-3-heptyl		$Rh_2(NHCOCH_3)_4$, CH_2Cl_2, rt, 5 h	(83)	*trans*	78
C_{17} 2,6-(*t*-Bu)$_2$-4-MeC$_6$H$_2$		$Rh_2(OAc)_4$, CH_2Cl_2, rt, 5 h	(75)	*trans*	78
2,6-(*t*-Bu)$_2$-4-MeC$_6$H$_2$		$Rh_2(NHCOCH_3)_4$, CH_2Cl_2, rt, 5 h	(93)	*trans*	78

Substrate: Et–CH=CH–Et (cis)

Product:

$$\text{Et} \overset{CO_2R}{\underset{\text{Et}}{\triangle}}$$

	R					
C_3	Me		$Rh_2(OAc)_4$, neat, rt	(56)		80, 77
	Me		$Rh_2(OPiv)_4$, neat, rt, 10 h	(68)		77
C_4	Et		$Cu(OTf)_2$, neat, rt	(15)		80, 77
	Et		$Pd(OAc)_2$, neat, rt	(15)		80, 77
	Et		$MeReO_3$, neat	(63)		319
C_6	*n*-Bu		$Rh_2(OAc)_4$, neat, rt	(98)		80, 77
	n-Bu		$Rh_2(OPiv)_4$, neat, rt, 10 h	(100)		77

t-Bu-CH=CH2 → t-Bu-[cyclopropane]-CO2R

	R			trans:cis	
C4	Et	Rh2(OAc)4, rt	(60-93)	74:26	323
	Et	Rh2(O2CC3F7)4, rt	(60-93)	60:40	323
	Et	Rh2(NHCOCH3)4, rt	(60-93)	83:17	323
	Et	Rh2(OAc)4, neat, rt	(87)	81:19	71
	Et	Rh6(CO)16, neat, rt	(42)	82:18	71
	Et	CuCl•P(OPr-i)3, neat, rt	(23)	88:12	71
	Et	Cu(OTf)2, neat, rt	(54)	85:15	71
	Et	Cu/bronze, neat, rt	(5)	89:11	71
	Et	Cu(acac)2, neat, 60°	(20)	91:9	71
	Et	PdCl2•2PhCN, neat, rt	(34)	71:29	71
	Et	Rh2(OAc)4, CH2Cl2, rt, 5 h	(75)	trans	78
	Et	Rh2(NHCOCH3)4, CH2Cl2, rt, 5 h	(83)	trans	78
C10	2,3,4-Me3-3-pentyl	Rh2(OAc)4, rt	(62-99)	86:14	323
	2,3,4-Me3-3-pentyl	Rh2(O2CC3F7)4, rt	(62-99)	64:36	323
	2,3,4-Me3-3-pentyl	Rh2(NHCOCH3)4, rt	(62-99)	89:11	323
	2,3,4-Me3-3-pentyl	Rh2(OAc)4, CH2Cl2, rt, 5 h	(84)	trans	78
	2,3,4-Me3-3-pentyl	Rh2(NHCOCH3)4, CH2Cl2, rt, 5 h	(89)	trans	78
C12	d-menthyl	Rh2(OAc)4, CH2Cl2, rt	(87)	81:19	309
C13	3-i-Pr-2-Me-3-heptyl	Rh2(OAc)4, CH2Cl2, rt, 5 h	(89)	trans	78
C17	2,6-(t-Bu)2-4-MeC6H2	Rh2(OAc)4, CH2Cl2, rt, 5 h	(93)	trans	78
	2,6-(t-Bu)2-4-MeC6H2	Rh2(NHCOCH3)4, CH2Cl2, rt, 5 h	(99)	trans	78

TABLE 1. DIASTEREOSELECTIVE CYCLOPROPANATION OF ALKENES USING DIAZOACETATES AND RELATED SYSTEMS (*Continued*)

	Carbenoid Precursor	Substrate	Conditions	Product(s) and Yield(s) (%)	Refs.
C_4	R = Et		Iron *meso*-tetra-(pentafluorophenyl)-porphyrin chloride, CH_2Cl_2, rt, 2 h	CO$_2$Et (—)	57
				CO$_2$Et	
		R			
C_3		Me	$Rh_2(OAc)_4$, neat, rt, 10 h	(70)	77
		Me	$Rh_2(OPiv)_4$, neat, rt, 10 h	(75)	77
		Me	$Cu(OTf)_2$, neat, rt, 10 h	(30)	77
		Me	$Pd(OAc)_2$, neat, rt, 10 h	(<8)	77
		Me	$[Ru_2(CO)_4(\mu\text{-}OAc)_2]_n$, CH_2Cl_2, rt	(47)	308
		Me	$Ru_2(CO)_4(\mu\text{-}OAc)_2(MeCN)_2$, CH_2Cl_2, 60°	(12)	308
C_4		Et	$MeReO_3$, neat	(57)	319
		CH_2CN	$Rh_2(OAc)_4$, CH_2Cl_2, rt	(80)	349
		$(CH_2)_2Br$	$Rh_2(OAc)_4$, 60°	(65)	342
C_6		$(CH_2)_2NMe_2$	$Cu/CuSO_4$, neat, 110°	(40)	350
C_8		$(CH_2)_2NEt_2$	$Cu/CuSO_4$, neat, 110°	(40)	350
C_4	R = Et		Cu-bronze, neat, reflux, 19 h	CO$_2$Et (73)	320

					trans:cis	
	R = Et	(alkene: butyl vinyl ether)	Rh$_2$(OAc)$_4$, neat, rt	(84)	63:37	71
			Rh$_6$(CO)$_{16}$, neat, rt	(69)	64:36	71
			CuCl•P(OPr-i)$_3$, neat, rt	(51)	67:33	71
			PdCl$_2$•2PhCN, neat, rt	(34)	62:38	71
			Rh$_6$(CO)$_{16}$, neat, rt	(69)	—	318
			PdCl$_2$•PhCN, neat, rt	(34)	—	318
			CuCl•P(OPr-i)$_3$, neat, rt	(51)	—	318
			Cl(Ph$_3$P)$_2$Ru(H$_2$C$_2$B$_9$H$_{10}$), neat, 60°	(93)	64:36	51
			Cl(Ph$_3$P)$_2$Ru(Me$_2$C$_2$B$_9$H$_{10}$), neat, 60°	(46)	65:35	51
			ClH(Ph$_3$P)$_2$Ru(H$_2$C$_2$B$_9$H$_9$), neat, 60°	(91)	62:38	51
			(7PPh$_2$-8-H-7,8-C$_2$B$_9$H$_{10}$) Rh(PPh$_3$)$_2$, neat, 4 h, 80°	(88)	59:41	351
	R = Et	(alkene: MeO/i-Pr substituted)	Cu, 95°, 1 h	(85)		352
	R = Et	(alkene: (MeO)$_2$C=CMe$_2$)	Cu-bronze, methylcyclohexane, reflux, 4 h	(—)		329

	R				
C$_3$	Me	(norbornene)	CuSO$_4$, DME, 110°	(—)	353
C$_4$	Et		Rh$_2$(OAc)$_4$, neat, rt	(88)	80
	Et		Cu(OTf)$_2$, neat, rt	(47)	80
	Et		Pd(OAc)$_2$, neat, rt	(95)	80
	Et		Zeolite NaCuX-57	(75)	336

TABLE 1. DIASTEREOSELECTIVE CYCLOPROPANATION OF ALKENES USING DIAZOACETATES AND RELATED SYSTEMS (*Continued*)

Carbenoid Precursor	Substrate	Conditions	Product(s) and Yield(s) (%)	Refs.
R = Et	(norbornene)	(π-allyl)PdCl, neat, 0°, 5 h, 2 eq EDA	EtO₂C—[I] CO₂Et + EtO₂C—[II]—CO₂Et + EtO₂C—[III]—CO₂Et I:II:III = 1.6:2:1	354

R				exo:endo	
Et	(norbornene)	Cu(OTf)₂	CO₂R (norbornane)	(—)	30:70	355
Et		Rh(II), rt, 12 h		(95)	—	356
Et		Iodorhodium(III)-*meso*-tetraphenylporphyrin, neat, 60°		(71)	35:65	102
Et		Rh₂(OAc)₄, rt		(60-93)	67:33	323
Et		Rh₂(O₂CC₃F₇)₄, rt		(60-93)	60:40	323
Et		Rh₂(NHCOCH₃)₄, rt		(60-93)	78:22	323
Et		Rh₂(OAc)₄, neat, rt		(95)	—	80
Et		Cu(OTf)₂, neat, rt		(95)	—	80
Et		Pd(OAc)₂, neat, rt		(87)	—	80
Et		Iodorhodium(III)-*meso*-tetraphenylporphyrin, DCE, 60°		(71)	—	101
Et		Iodorhodium(III)-*meso*-tetramesitylporphyrin, DCE, 60°		(76)	—	101
Et		Iodorhodium(III)-*meso*-cyanooctaethylporphyrin, DCE, 60°		(67)	—	101
Et		Iodorhodium(II)-*meso*-tetra-*o*-tolylporphyrin, DCE, 60°		(74)	—	101

	R	Conditions	(Yield)	Ratio	Ref.
	Et	Zeolite NaCuX-57	(78)	—	336
	Et	(7PPh$_2$-8-H-7,8-C$_2$B$_9$H$_{10}$)Rh(PPh$_3$)$_2$, neat, 4 h, 80°	(9)	59:41	351
	Et	RuCl(p-cymene)(TsNC$_6$H$_4$NH$_2$-4), neat, 80°	(0)	—	357
	Et	Rh$_2$(OAc)$_4$, CH$_2$Cl$_2$, rt, 5 h	(67)	exo	78
	Et	Rh$_2$(acac)$_4$, CH$_2$Cl$_2$, rt, 5 h	(82)	exo	78
	Et	Iodorhodium(III)-meso-tetra-(2,4,6-Me$_3$C$_6$H$_2$)porphyrin, neat, 60°	(76)	32:68	99
	Et	7a, neat, 60°	(39)	31:69	99
	Et	7c, neat, 60°	(9)	45:55	99
	Et	23b, neat, 60°, 4 h	(0)	—	340
	Et	24b, neat, 60°, 4 h	(0)	—	340
C$_{10}$	2,3,4-Me$_3$-3-pentyl	Rh$_2$(OAc)$_4$, rt	(62-99)	76:24	323
	2,3,4-Me$_3$-3-pentyl	Rh$_2$(O$_2$CC$_3$F$_7$)$_4$, rt	(62-99)	83:17	323
	2,3,4-Me$_3$-3-pentyl	Rh$_2$(NHCOCH$_3$)$_4$, rt	(62-99)	67:33	323
	2,3,4-Me$_3$-3-pentyl	Rh$_2$(OAc)$_4$, CH$_2$Cl$_2$, rt, 5 h	(76)	exo	78
	2,3,4-Me$_3$-3-pentyl	Rh$_2$(NHCOCH$_3$)$_4$, CH$_2$Cl$_2$, rt, 5 h	(83)	exo	78
C$_{13}$	3-i-Pr-2-Me-3-heptyl	Rh$_2$(NHCOCH$_3$)$_4$, CH$_2$Cl$_2$, rt, 5 h	(86)	exo	78
	3-i-Pr-2-Me-3-heptyl	Rh$_2$(OAc)$_4$, CH$_2$Cl$_2$, rt, 5 h	(82)	exo	78
C$_{17}$	2,6-(t-Bu)$_2$-4-MeC$_6$H$_2$	Rh$_2$(OAc)$_4$, CH$_2$Cl$_2$, rt, 5 h	(95)	exo	78
	2,6-(t-Bu)$_2$-4-MeC$_6$H$_2$	Rh$_2$(NHCOCH$_3$)$_4$, CH$_2$Cl$_2$, rt, 5 h	(92)	exo	78

TABLE 1. DIASTEREOSELECTIVE CYCLOPROPANATION OF ALKENES USING DIAZOACETATES AND RELATED SYSTEMS (*Continued*)

Carbenoid Precursor	Substrate	Conditions	Product(s) and Yield(s) (%)	Refs.
C₃ R = Me	(methylenecyclopropane with CO₂Me)	CuSO₄, 90°	(cyclopropane product with two CO₂Me) (60)	358
R = Me	(ethylidene with CO₂Me)	CuSO₄, 90°	(cyclopropane with two CO₂Me) (—)	358
R = Me	(ethylidene with CO₂Me)	CuSO₄, 90°	(cyclopropane with two CO₂Me) (—)	358
R = Me	(OTMS cyclohexylidene)	Cu(acac)₂, benzene, heat Temp 90° 100°	(TMSO cyclopropane with CO₂Me) (58) (75) *trans:cis* = 74:26	312 294
R = Me	(OTMS cycloheptene)	Cu(acac)₂, solvent, reflux Solvent benzene ethyl acetate	(OTMS bicyclic with CO₂Me) (76) *trans:cis* = 53:47 (77) *trans:cis* = 53:47	294, 312 194

Substrate	Conditions	Product(s) and Yield(s) (%)	Refs.
R = Me (methylcyclohexenyl-OTMS)	Cu(acac)$_2$, ethyl acetate, reflux	(74) OTMS/CO$_2$Me cyclopropane	294
R = Me (methylcyclohexenyl-OTMS)	Cu(acac)$_2$, ethyl acetate, reflux	(62) trans:cis = 60:40	294
R = Et (N-CO$_2$Me dihydropyridine)	Cu/bronze, neat, 135°	(95) exo:endo = 67:33	359

C$_4$ cycloheptene substrate:

R	Conditions	Product (%)	exo:endo	Refs.
Et	Rh$_2$(O$_2$CH)$_4$, 60°	(—)	66:34	325
Et	Rh$_2$(OAc)$_4$, 60°	(—)	69:31	325
Et	Rh$_2$(O$_2$CBu-n)$_4$, 60°	(—)	69:31	325
Et	Rh$_2$(OAc)$_4$, neat, rt	(75)	—	80
Et	Cu(OTf)$_2$, neat, rt	(30)	—	80
Et	Pd(OAc)$_2$, neat, rt	(40)	—	80
Et	Cl(Ph$_3$P)$_2$Ru(H$_2$C$_2$B$_9$H$_{10}$), neat, 60°	(89)	67:33	51
Et	Cl(Ph$_3$P)$_2$Ru(Me$_2$C$_2$B$_9$H$_{10}$), neat, 60°	(26)	63:37	51
Et	ClH(Ph$_3$P)$_2$Ru(H$_2$C$_2$B$_9$H$_9$), neat, 60°	(88)	67:33	51

| R = Et (methylcyclohexene) | CuCl•PMe$_3$, neat, reflux | (40) CO$_2$Et cyclopropane | 337 |
| R = Et | Rh$_2$(OAc)$_4$, neat, rt | (80) trans:cis = 73:27 | 71 |

TABLE 1. DIASTEREOSELECTIVE CYCLOPROPANATION OF ALKENES USING DIAZOACETATES AND RELATED SYSTEMS (*Continued*)

Carbenoid Precursor	Substrate	Conditions	Product(s) and Yield(s) (%)	Refs.
R				
	(cyclohexenyl-OMe)		(cyclopropane product with OMe, CO$_2$Et) *trans:cis*	
Et		Rh$_2$(OAc)$_4$, neat, rt	(78) 71:29	71
Et		Rh$_2$(OAc)$_4$, Et$_2$O, rt	(80) 71:29	94
Et		MeReO$_3$, neat	(74) —	319
Et		Rh$_6$(CO)$_{16}$, neat, rt	(59) 72:28	71, 318
Et		PdCl$_2$•PhCN, neat, rt	(39) 60:40	71, 318
Et		CuCl•P(OPr-i)$_3$, neat, rt	(54) 81:19	71, 318
Et		CuCl•P(OPh)$_3$, neat rt	(35) 82:18	71
Et		Cu(acac)$_2$, neat, 60°	(44) 83:17	71
C$_3$ R = Me	(2,2-dimethyl-4-methyl dihydrofuran)	Cu(acac)$_2$	(cyclopropane, CO$_2$Me) (54)	360
C$_4$ R = Et	(2-ethoxy dihydropyran)	Rh$_6$(CO)$_{16}$, neat, rt	(EtO-bicyclic, CO$_2$Et) (70)	318
		PdCl$_2$•PhCN, neat, rt	(33)	
		CuCl•P(OPr-i)$_3$, neat, rt	(42)	
R = Et	(MeO-methyl dihydropyran)	Cu(F$_3$CCOCHCOCF$_3$)$_2$, benzene, reflux, 3 h	(MeO-bicyclic, CO$_2$Et) (57)	361

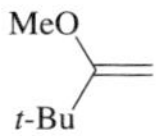

	Alkene	Conditions	Product(s) and Yield(s) (%)	Refs.

R				
Et	MeO—C(=CH₂)—t-Bu	Cu, 100°, 1 h	t-Bu(MeO)C—CH—CO₂Et (79)	352
Et		Rh₂(OAc)₄, neat, rt	(70) *trans:cis* 42:58	71
Et		Rh₆(CO)₁₆, neat, rt	(83) 42:58	71
Et		CuCl•P(OPr-*i*)₃, neat, rt	(7) 55:45	71
Et		PdCl₂•2PhCN, neat, rt	(28) 38:62	71

C₃

| R = Me | (CH₃)₂C=CH—t-Bu | 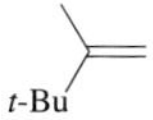 Cu(acac)₂, 100°, 3 h 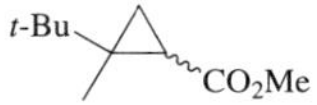 | t-Bu—cyclopropane(Me)₂—CO₂Me (79) | 307 |

C₄

R = Et	(EtO)₂OP—CH₂CH₂—CH=CH₂	CuSO₄, cyclohexane, reflux	  (EtO)₂OP—CH₂—cyclopropane—CO₂Et (30) *trans:cis* = 75:25	345
R = Et	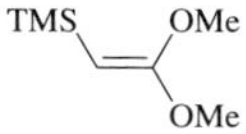 TMS—CH=C(OMe)(OMe)	Cu(acac)₂, benzene, reflux	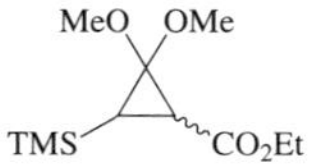 MeO(OMe)—cyclopropane(TMS)—CO₂Et (72)	362
R = Et	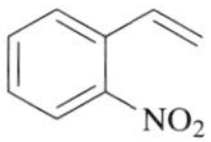 2-NO₂-C₆H₄-CH=CH₂	Pd(OAc)₂, neat, rt 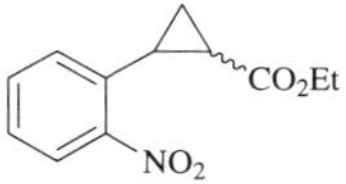	2-NO₂-C₆H₄—cyclopropane—CO₂Et (73)	80

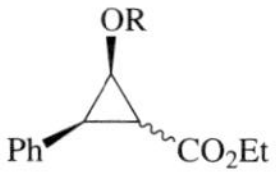

R¹	Ph—CH=CH—OR¹		Ph—cyclopropane(OR)—CO₂Et *trans:cis*	
Me		CuSO₄, benzene, 75°	(23) —	363
Me		Rh₂(OAc)₄, neat, rt	(60) 67:33	71
t-Bu		CuSO₄, benzene, 75°	(51) —	363

105

TABLE 1. DIASTEREOSELECTIVE CYCLOPROPANATION OF ALKENES USING DIAZOACETATES AND RELATED SYSTEMS (Continued)

Carbenoid Precursor	Substrate	Conditions	Product(s) and Yield(s) (%)	Refs.
	Ph⟵		Ph▱CO$_2$R trans:cis	
R				
C$_3$ Me		Cu(acac)$_2$, 100°, 3 h	(52) —	307
Me		[Ru$_2$(CO)$_4$(μ-OAc)$_2$]$_n$, CH$_2$Cl$_2$, rt	(95) 62:38	308
Me		Ru$_2$(CO)$_4$(μ-OAc)$_2$(MeCN)$_2$, CH$_2$Cl$_2$, rt	(58) 70:30	308
Me		Ru$_2$(CO)$_4$(μ-OAc)$_2$(MeCN)$_2$, neat, 60°	(94) 60:40	308
Me		**13a**, neat, 60°, 4 h	(66) 65:35	364
Me		**14a**, neat, 60°, 4 h	(75) 71:29	364
Me		**14c**, neat, 60°, 4 h	(57) 83:17	364
Me		**14f**, neat, 60°, 4 h	(53) 86:14	364
C$_4$ Et		Rh$_2$(O$_2$CH)$_4$, 60°	(98) 51:49	325
Et		Rh$_2$(OAc)$_4$, rt	(91) 60:40	325
Et		Rh$_2$(O$_2$CPr-n)$_4$, rt	(99) 62:38	325
Et		Rh$_2$(O$_2$CC$_{17}$H$_{35}$-n)$_4$, rt	(98) 60:40	325
Et		Rh$_2$(O$_2$CC$_6$H$_5$)$_4$, 60°	(97) 55:45	325
Et		Rh$_2$(O$_2$CPr-n)$_4$, 60°	(100) 51:49	325
Et		Rh$_2$[O$_2$C(CH$_2$)$_2$C$_6$H$_5$]$_4$, rt	(100) 54:46	325
Et		Rh$_2$[O$_2$CCH$_2$C(C$_6$H$_5$)$_3$]$_4$, rt	(100) 45:55	325
Et		Rh$_2$[O$_2$CC$_6$H$_4$OH-2]$_4$, 60°	(91) 46:54	325
Et		Rh$_2$[O$_2$CC$_6$H$_4$COPh-2]$_4$, rt	(98) 46:54	325
Et		Rh$_2$[O$_2$C-2,4-Cl$_2$C$_6$H(NO$_2$)$_2$-3,5]$_4$, rt	(65) 48:52	325
Et		Rh$_2$[O$_2$CC$_6$H$_4$CF$_3$-2)$_4$, rt	(86) 51:49	325
Et		Rh$_2$[O$_2$CC$_6$H$_4$CF$_3$-3)$_4$, rt	(87) 55:45	325
Et		Rh$_2$[O$_2$CC$_6$H$_2$Me$_3$-2,4,6)$_4$, 60°	(98) 50:50	325
Et		Rh$_2$[O$_2$CC$_2$H$_4$COMe]$_4$, rt	(98) 60:40	325
Et		Rh$_2$[O$_2$CCOC$_6$H$_5$]$_4$, 60°	(89) 51:49	325
Et		Rh$_2$(O$_2$CCF$_3$)$_4$, rt	(99) 53:47	325

	Catalyst/Conditions	Yield	Ratio	Ref.
Et	Rh$_2$(OAc)$_4$, rt	(92)	60:40	80
Et	Rh$_2$(O$_2$CBu-t)$_4$, rt	(60)	60:40	80
Et	Rh$_2$(O$_2$CC$_6$H$_{13}$-n)$_4$, rt	(95)	57:43	80
Et	Rh$_2$(O$_2$CCF$_3$)$_4$, rt	(66)	47:53	80
Et	RhCl$_3$•3H$_2$O, rt	(7)	—	80
Et	RhCl(PPh$_3$)$_3$	(12)	—	80
Et	Mo$_2$(OAc)$_4$, rt	(5)	—	80
Et	Ru$_2$(OAc)$_4$Cl, rt	(38)	64:36	80
Et	PdCl$_2$, rt	(70)	—	80
Et	Pd(OAc)$_2$, rt	(98)	67:33	80
Et	PdCl$_2$•2PhCN, rt	(65)	70:30	80
Et	Pd(PPh$_3$)$_4$, rt	(57)	69:31	80
Et	Pd/C, rt	(0)	—	80
Et	Cu(acac)$_2$, rt	(65)	68:32	80
Et	Cu(OTf)$_2$, rt	(80)	—	80
Et	CuCl•P(OPh)$_3$, neat, rt	(84)	71:29	71
Et	Cu(OTf)$_2$, neat, 60°	(97)	66:34	71
Et	Cu-bronze, neat, 60°	(53)	66:34	71
Et	Cu(acac)$_2$, neat, 60°	(71)	72:28	71
Et	Hydridotris(3,5-dimethyl-1-pyrazolyl)-borateCu(C$_2$H$_4$), DCE, rt	(75)	43:57	348, 365
Et	(7-PPh$_2$-8-H-7,8-C$_2$B$_9$H$_{10}$)Rh(PPh$_3$)$_2$, neat, 4 h, 100°	(91)	53:47	351
Et	(7-PPh$_2$-8-Me-7,8-C$_2$B$_9$H$_{10}$)Rh(PPh$_3$)$_2$, neat, 4 h, 100°	(92)	52:48	351
Et	(7-PPh$_2$-8-H-7,8-C$_2$B$_9$H$_{10}$)Rh(1,5-cyclooctadiene), neat, 4 h, 100°	(91)	54:46	351
Et	RuCl$_2$(PPh$_3$)$_3$, neat, 20°, 4 h	(37)	67:33	52
Et	RuCl$_2$(PPh$_3$)$_3$, neat, 30°, 4 h	(43)	66:34	338
Et	RuCl$_2$(PPh$_3$)$_3$, neat, 40°, 4 h	(55)	61:39	52
Et	RuCl$_2$(PPh$_3$)$_3$, neat, 60°, 4 h	(93)	56:44	52, 338
Et	RuCl$_2$(PPh$_3$)$_3$/C$_6$H$_6$, neat, 20°, 4 h	(38)	68:32	52

TABLE 1. DIASTEREOSELECTIVE CYCLOPROPANATION OF ALKENES USING DIAZOACETATES AND RELATED SYSTEMS (*Continued*)

Carbenoid Precursor	Substrate	Conditions	Product(s) and Yield(s) (%)		Refs.
Et		$RuCl_2(PPh_3)_3/C_6H_6$, neat, 40°, 4 h	(53)	62:38	52
Et		$RuCl_2(PPh_3)_3/C_6H_6$, neat, 60°, 4 h	(89)	56:44	52
Et		$RuH_3Si(OEt)_3[(PPh_3)_2$, neat, 20°, 4 h	(33)	71:29	52
Et		$RuH_3Si(OEt)_3[(PPh_3)_2$, neat, 40°, 4 h	(48)	71:29	52
Et		$RuH_3Si(OEt)_3[(PPh_3)_2$, neat, 60°, 4 h	(85)	70:30	52
Et		$Ru[Si(OEt)_3]_2(PPh_3)_2$, neat, 20°, 4 h	(41)	65:35	52
Et		$Ru[Si(OEt)_3]_2(PPh_3)_2$, neat, 40°, 4 h	(54)	62:38	52
Et		$Ru[Si(OEt)_3]_2(PPh_3)_2$, neat, 60°, 4 h	(91)	60:40	52
Et		$RuCl_2(AsPh_3)_3$, neat, 60°, 4 h	(92)	58:42	52
Et		$RuCl_2(SbPh_3)_3$, neat, 60°, 4 h	(94)	70:30	52
Et		(5,10,15,20-tetraphenyl)porphyrinato)Ru-(diethoxycarbonyl)carbene MeOH, CH_2Cl_2, rt	(85)	93:7	366
Et		(5,10,15,20-tetraphenyl)porphyrinato)Ru-(diethoxycarbonyl)carbene MeOH, toluene, reflux	(66)	—	366
Et		$(CO)_5WC(OMe)Ph$, 35°	(41)	62:38	321
Et		$Pd(OAc)_2$, benzene, 40°	(76)	67:33	79
Et		$[(Cp)Fe(CO)_2(THF)]^+BF_4^-$, CH_2Cl_2, 40°	(68)	15:85	322
Et		$OsCl_2(PPh_3)_3$, neat, 30°, 4 h	(18)	71:29	338
Et		$OsCl_2(PPh_3)_3$, neat, 60°, 4 h	(52)	69:31	338
Et		$Rh_6(CO)_{16}$ neat, rt	(87)	—	318
Et		$[Rh(PE-CO_2)_2]_2$, toluene, 100°	(96)	64:36	367
Et		$PdCl_2 \cdot PhCN$, neat, rt	(52)	—	318
Et		$CuCl \cdot P(OP-i)_3$, neat, rt	(80)	—	318
Et		$Rh_2(OAc)_4$, neat, rt	(93)	62:38	71
Et		$Rh_6(CO)_{16}$ neat, rt	(86)	63:37	71
Et		$CuCl \cdot P(OP-i)_3$, neat, rt	(88)	74:26	71
Et		$PdCl_2 \cdot 2PhCN$, neat, rt	(52)	62:38	71
Et		$[Cu(MeCN)_4]^+BF_4^-$, CH_2Cl_2, rt, 10 min	(60)	62:38	317
Et		$[Cu(MeCN)_2(PPh_3)_2]^+BF_4^-$, CH_2Cl_2, rt, 2 h	(55)	71:29	317
Et		$[Cu(MeCN)(PPh_3)_3]^+BF_4^-$, CH_2Cl_2, rt, 16 h	(27)	70:30	317

Et	Cu(BF$_4$)(PCy$_3$)$_2$, CH$_2$Cl$_2$, rt, 4.5 h	(29)	80:20	317
Et	Rh$_2$(NHCOCH$_3$)$_4$, CH$_2$Cl$_2$, rt, 5 h	(68)	trans	78
Et	Rh$_2$(OAc)$_4$, CH$_2$Cl$_2$, rt, 5 h	(62)	trans	78
Et	Rh$_2$(O$_2$CC$_3$H$_{7-n}$)$_4$, CH$_2$Cl$_2$, rt, 5 h	(60)	trans	78
Et	Rh$_2$(NHCOCF$_3$)$_4$, CH$_2$Cl$_2$, rt, 5 h	(59)	trans	78
Et	Rh$_2$(OCSCH$_3$)$_4$, CH$_2$Cl$_2$, rt, 5 h	(62)	trans	78
Et	Rh$_2$(O$_2$CCF$_3$)$_4$, CH$_2$Cl$_2$, rt, 5 h	(56)	trans	78
Et	Rh$_2$(O$_2$CC$_3$F$_{7-n}$)$_4$, CH$_2$Cl$_2$, rt, 5 h	(52)	trans	78
Et	Rh$_2$(NHCOCH$_3$)$_4$, CH$_2$Cl$_2$, rt, 5 h	(68)	trans	78
Et	Pt(PPh$_3$)$_4$, neat, 60°	(76)	61:39	368
Et	Pt(PPh$_3$)$_4$, neat, 100°	(95)	61:39	368
Et	PtCl$_2$, neat, 60°	(48)	62:38	368
Et	PtCl$_2$, neat, 100°	(94)	61:39	368
Et	[PtCl$_2$(cyclohexene)]$_2$, neat, 60°	(49)	61:39	368
Et	[PtCl$_2$(cyclohexene)]$_2$, neat, 100°	(94)	61:39	368
Et	cis-PtCl$_2$(benzonitrile)$_2$, neat, 60°	(47)	61:39	368
Et	cis-PtCl$_2$(benzonitrile)$_2$, neat, 100°	(92)	61:39	368
Et	cis-PtCl$_2$(pyr)$_2$, neat, 60°	(34)	62:38	368
Et	cis-PtCl$_2$(pyr)$_2$, neat, 100°	(88)	61:39	368
Et	cis-PtCl$_2$(PPh$_3$)$_2$, neat, 60°	(75)	62:38	368
Et	cis-PtCl$_2$(PPh$_3$)$_2$, neat, 100°	(92)	62:38	368
Et	trans-PtCl$_2$(PPh$_3$)$_2$, neat, 60°	(69)	61:39	368
Et	trans-PtCl$_2$(PPh$_3$)$_2$, neat, 100°	(91)	62:38	368
Et	PtBr$_2$, neat, 60°	(43)	61:39	368
Et	PtBr$_2$, neat, 100°	(89)	61:39	368
Et	PtI$_2$, neat, 60°	(58)	61:39	368
Et	PtI$_2$, neat, 100°	(66)	61:39	368
Et	PtI$_2$(NH$_3$)$_2$, neat, 60°	(67)	62:38	368
Et	PtI$_2$(NH$_3$)$_2$, neat, 100°	(91)	61:39	368

TABLE 1. DIASTEREOSELECTIVE CYCLOPROPANATION OF ALKENES USING DIAZOACETATES AND RELATED SYSTEMS (*Continued*)

Carbenoid Precursor	Substrate	Conditions	Product(s) and Yield(s) (%)		Refs.
Et		PtCl$_4$, neat, 60°	(79)	58:42	368
Et		PtCl$_4$, neat, 100°	(85)	54:46	368
Et		PtBr$_4$, neat, 60°	(54)	62:38	368
Et		PtBr$_4$, neat, 100°	(92)	61:39	368
Et		Pt(C$_2$H$_4$)(PPh$_3$)$_2$, CH$_2$Cl$_2$, rt, 24 h	(—)	—	369
Et		Cl(PPh$_3$)$_2$Ru(H$_2$C$_2$B$_9$H$_{10}$), neat, 60°	(86)	59:41	51
Et		Cl(PPh$_3$)$_2$Ru(Me$_2$C$_2$B$_9$H$_{10}$), neat, 60°	(81)	55:45	51
Et		ClH(PPh$_3$)$_2$Ru(H$_2$C$_2$B$_9$H$_7$), neat, 60°	(93)	58:42	51
Et		[RuH(7-PPh$_2$-8-H-C$_2$B$_9$H$_{10}$)(PPh$_3$)$_2$], neat, 60°	(74)	61:39	81
Et		[RuH(7-PPh$_2$-8-H-C$_2$B$_9$H$_{10}$)(PPh$_3$)$_2$], neat, 100°	(97)	61:39	81
Et		[RuH(7-PPh$_2$-8-Me-C$_2$B$_9$H$_{10}$)(PPh$_3$)$_2$], neat, 60°	(78)	62:38	81
Et		[RuH(7-PPh$_2$-8-Me-C$_2$B$_9$H$_{10}$)(PPh$_3$)$_2$], neat, 100°	(96)	61:39	81
Et		RuCl(*p*-cymene)[TsN(CH$_2$)$_2$NH$_2$], neat, 60°	(78)	64:36	357
Et		RuCl(*p*-cymene)[TsN(CH$_2$)$_2$NH$_2$], neat, 100°	(89)	63:37	357
Et		RuCl(*p*-cymene)[TsN(CH$_2$)$_2$NH$_2$], neat, 60°	(94)	63:37	357
Et		RuCl(*p*-cymene)[TsN(CH$_2$)$_2$NH$_2$], neat, 100°	(81)	60:40	357
Et		RuCl(*p*-cymene)[TsN(CH$_2$)$_2$NH$_2$], neat, 60°	(80)	62:38	357
Et		RuCl(*p*-cymene)[TsN(CH$_2$)$_2$NH$_2$], neat, 100°	(90)	62:38	357
Et		RuCl(*p*-cymene)[TsN-2-H$_2$NC$_6$H$_4$], neat, 60°	(75)	62:38	357
Et		RuCl(*p*-cymene)[TsN-2-H$_2$NC$_6$H$_4$], neat, 100°	(87)	61:39	357
Et		RuCl(*p*-cymene)(2-TsNpyridine), neat, 60°	(65)	62:38	357
Et		RuCl(*p*-cymene)(2-TsNpyridine), neat, 100°	(85)	65:35	357
Et		RuCl(*p*-cymene)[2-(TsNCH$_2$)pyridine], neat, 60°	(77)	65:35	357
Et		RuCl(*p*-cymene)[2-(TsNCH$_2$)pyridine], neat, 100°	(86)	65:35	357
Et		Cu/bronze-K10 montmorillonite, CH$_2$Cl$_2$, rt	(—)	44:56	339
Et		Cu/bronze-bentonite, CH$_2$Cl$_2$, rt	(—)	64:36	339
Et		Cu/bronze-laponite, CH$_2$Cl$_2$, rt	(—)	58:42	339

Et	Iron meso-tetra-p-tolylporphyrin, CH$_2$Cl$_2$, rt, 1 h	(—)	90:10	57
Et	Iron meso-tetra-p-tolylporphyrin chloride, CH$_2$Cl$_2$, 40°, 10 h	(—)	85:15	57
Et	Iron meso-tetra-p-tolylporphyrin chloride/cobaltocene, CH$_2$Cl$_2$, rt, 2 h	(—)	90:10	57
Et	Iron meso-tetra-p-MeOphenylporphyrin chloride/cobaltocene, CH$_2$Cl$_2$, rt, 3 h	(—)	90:10	57
Et	Iron meso-tetra(pentafluorophenyl)porphyrin chloride, CH$_2$Cl$_2$, rt, 6 h	(—)	86:14	57
Et	Iron meso-tetramesitylporphyrin chloride/cobaltocene, CH$_2$Cl$_2$, rt, 2 h	(—)	93:7	57
Et	Iron octaethylporphyrin chloride/cobaltocene, CH$_2$Cl$_2$, rt, 4 h	(—)	91:9	57
Et	Osmium meso-tetra-p-tolylporphyrin-(CO)(pyr), toluene, rt, 2 h	(44)	90:10	90
Et	Osmium meso-tetra-p-tolylporphyrin dimer, toluene, rt, 2 h	(79)	91:9	90
Et	Ru(meso-tetraphenylporphyrin)(CO), neat, rt, 4 h	(—)	93:7	84
Et	Ru(meso-tetramesitylporphyrin)(CO), neat, rt, 4 h	(—)	89:11	84
Et	Ru(meso-tetramesitylporphyrin)(O)$_2$, neat, rt, 4 h	(—)	88:12	84
Et	[Ru(meso-tetraphenylporphyrin)(CO)(EtOH)], CH$_2$Cl$_2$, rt, 8 h	(45)	90:10	370
Et	[Ru(meso-tetramesitylporphyrin)(CO)(EtOH)], CH$_2$Cl$_2$, rt, 8 h	(68)	89:11	370
Et	[Ru(octaethylporphyrin)(CO)(EtOH)], CH$_2$Cl$_2$, rt, 8 h	(25)	92:8	370

TABLE 1. DIASTEREOSELECTIVE CYCLOPROPANATION OF ALKENES USING DIAZOACETATES AND RELATED SYSTEMS (*Continued*)

Carbenoid Precursor	Substrate	Conditions	Product(s) and Yield(s) (%)		Refs.
Et		[Ru(tetrapropylporphyrin)(CO)(EtOH)], CH$_2$Cl$_2$, rt, 8 h	(39)	92:8	370
Et		CuOTf, DCE, rt	(70)	59:41	326
Et		**1**, DCE, 60°, 4 h	(79)	73:27	371
Et		**2**, DCE, 60°, 4 h	(89)	73:27	371
Et		**3**, DCE, 60°, 4 h	(78)	72:28	371
Et		**4**, DCE, rt	(83)	86:14	326
Et		**5**, DCE, rt	(70)	85:15	326
Et		**6**, DCE, rt	(66)	82:18	326
Et		**8a**, neat, 80°, 4 h	(71)	75:25	372
Et		**8b**, neat, 80°, 4 h	(69)	75:25	372
Et		**9**, neat, 80°, 4 h	(91)	76:24	372
Et		**10a**, neat, 80°, 4 h	(26)	76:24	372
Et		**10b**, neat, 60°, 4 h	(38)	82:18	372
Et		**10b**, neat, 80°, 4 h	(52)	82:18	372
Et		**11a**, neat, 80°, 4 h	(44)	76:24	372
Et		**11b**, neat, 60°, 4 h	(34)	67:33	372
Et		**11b**, neat, 60°, 13 h	(44)	73:27	372
Et		**11b**, neat, 80°, 4 h	(54)	75:25	372
Et		**11b**, neat, 80°, 8 h	(56)	75:25	372
Et		**11b**, neat, 100°, 2 h	(58)	78:22	372
Et		**11b**, neat, 100°, 4 h	(59)	77:23	372
Et		**12**, CH$_2$Cl$_2$, 60°	(54)	89:11	106
Et		**13a**, neat, 60°, 4 h	(70)	72:28	364
Et		**13a**, neat, 100°, 4 h	(72)	82:18	364
Et		**13b**, neat, 60°, 4 h	(62)	66:34	364
Et		**13b**, neat, 100°, 4 h	(—)	—	364
Et		**13c**, neat, 60°, 4 h	(75)	65:35	364
Et		**13c**, neat, 100°, 4 h	(96)	71:29	364

13d, neat, 60°, 4 h	Et	(73)	64:36	364
13d, neat, 100°, 4 h	Et	(94)	76:24	364
13e, neat, 60°, 4 h	Et	(57)	66:34	364
13e, neat, 100°, 4 h	Et	(90)	79:21	364
13f, neat, 60°, 4 h	Et	(71)	69:31	364
13f, neat, 100°, 4 h	Et	(84)	79:21	364
14a, neat, 60°, 4 h	Et	(69)	72:28	364
14a, neat, 100°, 4 h	Et	(57)	83:17	364
14b, neat, 60°, 4 h	Et	(57)	83:17	364
14b, neat, 100°, 4 h	Et	(72)	75:25	364
14c, neat, 60°, 4 h	Et	(60)	84:16	364
14c, neat, 100°, 4 h	Et	(84)	71:29	364
14d, neat, 60°, 4 h	Et	(57)	84:16	364
14d, neat, 100°, 4 h	Et	(71)	82:18	364
14e, neat, 60°, 4 h	Et	(63)	88:12	364
14e, neat, 100°, 4 h	Et	(66)	85:15	364
14f, neat, 60°, 4 h	Et	(56)	87:13	364
14f, neat, 100°, 4 h	Et	(73)	76:24	364
15, neat, 60°, 4 h	Et	(93)	62:38	364
15, neat, 100°, 4 h	Et	(93)	61:39	364
16, neat, 60°, 4 h	Et	(78)	81:19	364
16, neat, 100°, 4 h	Et	(77)	75:25	364
17, neat, 60°, 4 h	Et	(52)	61:39	364
17, neat, 100°, 4 h	Et	(91)	61:39	364
18, neat, 60°, 4 h	Et	(83)	68:32	364
18, neat, 100°, 4 h	Et	(60)	79:21	364
19, neat, 60°, 4 h	Et	(42)	66:34	364
19, neat, 100°, 4 h	Et	(68)	78:22	364
20, CH_2Cl_2, rt, 24 h	Et	(60)	64:36	364
21, CH_2Cl_2, rt, 24 h	Et	(52)	65:35	364
22, CH_2Cl_2, rt, 24 h	Et	(55)	65:35	364

TABLE 1. DIASTEREOSELECTIVE CYCLOPROPANATION OF ALKENES USING DIAZOACETATES AND RELATED SYSTEMS (*Continued*)

Carbenoid Precursor	Substrate	Conditions	Product(s) and Yield(s) (%)		Refs.
Et		23a, neat, rt, 4 h	(23)	61:39	340
Et		23a, neat, 40°, 4 h	(51)	59:41	340
Et		23a, neat, 60°, 4 h	(71)	57:43	340
Et		23b, neat, rt, 4 h	(25)	62:38	340
Et		23b, neat, 40°, 4 h	(50)	57:43	340
Et		23b, neat, 60°, 4 h	(71)	55:45	340
Et		24a, neat, rt, 4 h	(19)	63:37	340
Et		24a, neat, 40°, 4 h	(45)	63:37	340
Et		24a, neat, 60°, 4 h	(68)	62:38	340
Et		24b, neat, rt, 4 h	(12)	63:37	340
Et		24b, neat, 40°, 4 h	(43)	62:38	340
Et		24b, neat, 60°, 4 h	(66)	62:38	340
Et		25g, CH$_2$Cl$_2$, rt, 3 h	(47)	85:15	373
Et		27, CH$_2$Cl$_2$, rt, 3 h	(66)	70:30	373
Et		MeReO$_3$, neat	(81)	—	319
Et		Iodorhodium(III)*meso*-tetraphenylporphyrin, neat, 60°	(71)	47:53	102
Et		Rh$_2$(OAc)$_4$, rt	(95)	62:38	323
Et		Rh$_2$(O$_2$CC$_3$F$_7$-n)$_4$, rt	(81)	52:48	323
Et		Rh$_2$(NHCOCH$_3$)$_4$, rt	(89)	68:32	323
Et		Iodorhodium(III)*meso*-tetraphenylporphyrin, DCE, 60°	(71)	53:47	101
Et		Iodorhodium(III)*meso*-tetramesitylporphyrin, DCE, 60°	(78)	51:49	101
Et		Iodorhodium(III)*meso*-tetra(*p*-tolyl)porphyrin, 0°	(68)	—	374
Et		[OsCl$_2$(*p*-cymene)]$_2$, neat, 60°	(59)	64:36	91
Et		[OsCl$_2$(*p*-cymene)]$_2$, neat, 80°	(78)	63:37	91

	NMe$_2$	Rh$_2$(OAc)$_4$, rt	(74)	69:31	323
	NMe$_2$	Rh$_2$(O$_2$CC$_3$F$_7$)$_4$, rt	(67)	60:40	323
	NMe$_2$	Rh$_2$(NHCOCH$_3$)$_4$, rt	(70)	71:29	323
C$_6$	t-Bu	Rh$_2$(OAc)$_4$, CH$_2$Cl$_2$, rt	(68)	66:34	309
	t-Bu	**13a**, neat, 60°, 4 h	(66)	72:28	364
	t-Bu	**14a**, neat, 60°, 4 h	(65)	86:14	364
	t-Bu	**14c**, neat, 60°, 4 h	(60)	94:6	364
	t-Bu	**14f**, neat, 60°, 4 h	(58)	95:5	364
C$_8$	N(Pr-i)$_2$	Rh$_2$(OAc)$_4$, rt	(53)	98:2	323
	N(Pr-i)$_2$	Rh$_2$(O$_2$CC$_3$F$_7$)$_4$, rt	(51)	92:8	323
	N(Pr-i)$_2$	Rh$_2$(NHCOCH$_3$)$_4$, rt	(47)	99:1	323
C$_{10}$	CMe(Pr-i)$_2$	Rh$_2$(OAc)$_4$, rt	(95)	71:29	323
	CMe(Pr-i)$_2$	Rh$_2$(O$_2$CC$_3$F$_7$)$_4$, rt	(83)	52:48	323
	CMe(Pr-i)$_2$	Rh$_2$(NHCOCH$_3$)$_4$, rt	(87)	82:18	323
	2,3,4-Me$_3$-3-pentyl	Rh$_2$(OAc)$_4$, CH$_2$Cl$_2$, rt, 5 h	(71)	trans	78
	2,3,4-Me$_3$-3-pentyl	Rh$_2$(NHCOCH$_3$)$_4$, CH$_2$Cl$_2$, rt, 5 h	(87)	trans	78
C$_{13}$	3-i-Pr-2-Me-3-heptyl	Rh$_2$(OAc)$_4$, CH$_2$Cl$_2$, rt, 5 h	(73)	trans	78
	3-i-Pr-2-Me-3-heptyl	Rh$_2$(NHCOCH$_3$)$_4$, CH$_2$Cl$_2$, rt, 5 h	(93)	trans	78
C$_{17}$	2,6-(t-Bu)$_2$-4-MePh	Rh$_2$(OAc)$_4$, CH$_2$Cl$_2$, rt, 5 h	(84)	trans	78
	2,6-(t-Bu)$_2$-4-MePh	Rh$_2$(NHCOCH$_3$)$_4$, CH$_2$Cl$_2$, rt, 5 h	(98)	trans	78

TABLE 1. DIASTEREOSELECTIVE CYCLOPROPANATION OF ALKENES USING DIAZOACETATES AND RELATED SYSTEMS (*Continued*)

Carbenoid Precursor	Substrate	Conditions	Product(s) and Yield(s) (%)		Refs.
R	R^1			trans:cis	
Me	$OCHF_2$	$Cu/CuSO_4$, TCE, 100°	(42)	50:50	375
Me	OAc	$Cu/CuSO_4$, TCE, 100°	(60)	50:50	375
Et	Cl	$Rh_2(O_2CH)_4$, 60°	(—)	55:45	325
Et	Cl	$Rh_2(OAc)_4$, 60°	(—)	61:39	325
Et	Cl	$Rh_2(O_2CPr\text{-}n)_4$, 60°	(—)	62:38	325
Et	Cl	$Pd(OAc)_2$	(—)	—	376
Et	Cl	$[OsCl_2(p\text{-cymene})]_2$, neat, 60°	(57)	65:35	91
Et	Cl	$[OsCl_2(p\text{-cymene})]_2$, neat, 80°	(75)	66:34	91
Et	Cl	$Cl(Ph_3P)_2Ru(H_2C_2B_9H_{10})$, neat, 60°	(91)	55:45	51
Et	Cl	$Cl(Ph_3P)_2Ru(Me_2C_2B_9H_{10})$, neat, 60°	(73)	56:44	51
Et	Cl	$ClH(Ph_3P)_2Ru(H_2C_2B_9H_7)$, neat, 60°	(96)	56:44	51
Et	Cl	$Pd(OAc)_2$, neat, rt	(86)	—	80
Et	Cl	$RuCl_2(PPh_3)_3$, neat, 60°, 4 h	(89)	64:36	52, 338
Et	Cl	$RuCl_2(PPh_3)_3/C_6H_6$, neat, 60°, 4 h	(87)	71:29	52
Et	Cl	$RuH_3[Si(OEt)_3](PPh_3)_2$, neat, 60°, 4 h	(81)	71:29	52
Et	Cl	$Ru[Si(OEt)_3]_2(PPh_3)_2$, neat, 60°, 4 h	(83)	71:29	52
Et	Cl	$OsCl_2(PPh_3)_3$, neat, 60°, 4 h	(47)	70:30	338
Et	Cl	$(7\text{-}PPh_2\text{-}8\text{-}H\text{-}7,8\text{-}C_2B_9H_{10})Rh(PPh_3)_2$, neat, 4 h, 100°	(89)	66:34	351
Et	Cl	$(7\text{-}PPh_2\text{-}8\text{-}Me\text{-}7,8\text{-}C_2B_9H_{10})Rh(PPh_3)_2$, neat, 4 h, 100°	(92)	63:37	351
Et	Cl	$(7\text{-}PPh_2\text{-}8\text{-}H\text{-}7,8\text{-}C_2B_9H_{10})\text{-}Rh(1,5\text{-cyclooctadiene})$, neat, 4 h, 100°	(87)	66:34	351

Et	Cl	[RuH(7-PPh$_2$-8-H-C$_2$B$_9$H$_{10}$)(PPh$_3$)$_2$], neat, 100°	(94)	67:33	81
Et	Cl	[RuH(7-PPh$_2$-8-Me-C$_2$B$_9$H$_{10}$)(PPh$_3$)$_2$], neat, 100°	(93)	68:32	81
Et	Cl	RuCl(p-cymene)(TsNC$_6$H$_4$NH$_2$-2), neat, 80°	(85)	71:29	357
Et	Cl	Hydridotris(3,5-dimethyl-1-pyrazolyl)borateCu(C$_2$H$_4$), DCE, rt	(—)	58:42	365
Et	Cl	[Ru(*meso*-tetraphenylporphyrin)(CO)], neat, rt, 4 h	(—)	93:7	84
Et	Cl	[Ru(*meso*-tetramesitylporphyrin)(CO)], neat, rt, 4 h	(—)	89:11	84
Et	Cl	[Ru(*meso*-tetraphenylporphyrin)(CO)(EtOH)], CH$_2$Cl$_2$, rt, 8 h	(44)	93:7	370
Et	Cl	[Ru(*meso*-tetramesitylporphyrin)(CO)(EtOH)], CH$_2$Cl$_2$, rt, 8 h	(53)	91:9	370
Et	Cl	[Ru(octaethylporphyrin)(CO)(EtOH)], CH$_2$Cl$_2$, rt, 8 h	(23)	92:8	370
Et	Cl	[Ru(tetrapropylporphyrin)(CO)(EtOH)], CH$_2$Cl$_2$, rt, 8 h	(21)	93:7	370
Et	Cl	**13a**, neat, 60°, 4 h	(75)	76:24	364
Et	Cl	**13d**, neat, 60°, 4 h	(71)	69:31	364
Et	Cl	**14a**, neat, 60°, 4 h	(76)	82:18	364
Et	Cl	**14d**, neat, 60°, 4 h	(63)	87:13	364
Et	Cl	**14f**, neat, 60°, 4 h	(60)	87:13	364
Et	Cl	**23b**, neat, 60°, 4 h	(72)	69:31	340
Et	Cl	**24b**, neat, 60°, 4 h	(67)	70:30	340
Et	F	(7-PPh$_2$-8-H-7,8-C$_2$B$_9$H$_{10}$)Rh(PPh$_3$)$_2$, neat, 4 h, 100°	(93)	62:38	351
Et	F	Pd(OAc)$_2$	(—)	—	376
Et	Br	Pd(OAc)$_2$	(—)	—	376

TABLE 1. DIASTEREOSELECTIVE CYCLOPROPANATION OF ALKENES USING DIAZOACETATES AND RELATED SYSTEMS (*Continued*)

Carbenoid Precursor	Substrate	Conditions	Product(s) and Yield(s) (%)		Refs.
Et	Br	(7-PPh$_2$-8-H-7,8-C$_2$B$_9$H$_{10}$)Rh(PPh$_3$)$_2$, neat, 4 h, 100°	(90)	68:32	351
Et	Br	RuCl$_2$(PPh$_3$)$_3$, neat, 60°, 4 h	(85)	68:32	52
Et	Br	RuCl$_2$(PPh$_3$)$_3$/C$_6$H$_6$, neat, 60°, 4 h	(84)	61:39	52
Et	Br	RuH$_3$[Si(OEt)$_3$](PPh$_3$)$_2$, neat, 60°, 4 h	(80)	68:32	52
Et	Br	Ru[Si(OEt)$_3$]$_2$(PPh$_3$)$_2$, neat, 60°, 4 h	(80)	62:38	52
Et	NO$_2$	Hydridotris(3,5-dimethyl-1-pyrazolyl)borateCu(C$_2$H$_4$), DCE, rt	(—)	—	365
Et	NO$_2$	Pd(OAc)$_2$	(—)	—	52
Et	CN	Pd(OAc)$_2$	(—)	—	52
Et	CF$_3$	Hydridotris(3,5-dimethyl-1-pyrazolyl)borateCu(C$_2$H$_4$), DCE, rt	(—)	50:50	365
Et	Me	Rh$_2$(O$_2$CH)$_4$, 60°	(—)	55:45	325
Et	Me	Rh$_2$(OAc)$_4$, 60°	(—)	60:40	325
Et	Me	Rh$_2$(O$_2$CPr-n)$_4$, 60°	(—)	60:40	325
Et	Me	Pd(OAc)$_2$	(—)	—	376
Et	Me	[OsCl$_2$(p-cymene)]$_2$, neat, 60°	(66)	67:33	91
Et	Me	[OsCl$_2$(p-cymene)]$_2$, neat, 80°	(85)	61:39	91
Et	Me	Cl(PPh$_3$)P$_2$Ru(H$_2$C$_2$B$_9$H$_{10}$), neat, 60°	(89)	60:40	51
Et	Me	Cl(PPh$_3$)P$_2$Ru(Me$_2$C$_2$B$_9$H$_{10}$), neat, 60°	(77)	60:40	51
Et	Me	ClH(PPh$_3$)P$_2$Ru(H$_2$C$_2$B$_9$H$_9$), neat, 60°	(93)	63:37	51
Et	Me	Pd(OAc)$_2$, neat, rt	(81)	—	80
Et	Me	[(Cp)Fe(CO)$_2$(THF)]$^+$BF$_4^-$, CH$_2$Cl$_2$, 40°	(66)	40:60	322
Et	Me	RuCl$_2$(PPh$_3$)$_3$, neat, 60°, 4 h	(91)	65:35	52, 338
Et	Me	RuCl$_2$(PPh$_3$)$_3$/C$_6$H$_6$, neat, 60°, 4 h	(91)	63:37	52
Et	Me	RuH$_3$[Si(OEt)$_3$](PPh$_3$)$_2$, neat, 60°, 4 h	(83)	67:33	52
Et	Me	Ru[Si(OEt)$_3$]$_2$(PPh$_3$)$_2$, neat, 60°, 4 h	(90)	65:35	52
Et	Me	OsCl$_2$(PPh$_3$)$_3$, neat, 60°, 4 h	(61)	64:36	338
Et	Me	(7-PPh$_2$-8-H-7,8-C$_2$B$_9$H$_{10}$)Rh(PPh$_3$)$_2$, neat, 4 h, 100°	(90)	60:40	351

Et	Me	(7-PPh$_2$-8-Me-7,8-C$_2$B$_9$H$_{10}$)Rh(PPh$_3$)$_2$, neat, 4 h, 100°	(93)	58:42	351
Et	Me	(7-PPh$_2$-8-H-7,8-C$_2$B$_9$H$_{10}$) Rh(1,5-cyclooctadiene), neat, 4 h, 100°	(94)	59:41	351
Et	Me	[RuH(7-PPh$_2$-8-H-C$_2$B$_9$H$_{10}$)(PPh$_3$)$_2$], neat, 100°	(96)	66:34	81
Et	Me	[RuH(7-PPh$_2$-8-Me-C$_2$B$_9$H$_{10}$)(PPh$_3$)$_2$], neat, 100°	(96)	65:35	81
Et	Me	RuCl(p-cymene)(TsNC$_6$H$_4$NH$_2$-2), neat, 80°	(93)	65:35	357
Et	Me	[Ru(*meso*-tetraphenylporphyrin)(CO)(EtOH)], CH$_2$Cl$_2$, rt, 8 h	(58)	92:8	370
Et	Me	[Ru(*meso*-tetramesitylporphyrin)(CO)(EtOH)], CH$_2$Cl$_2$, rt, 8 h	(66)	88:12	370
Et	Me	[Ru(octaethylporphyrin)(CO)(EtOH)], CH$_2$Cl$_2$, rt, 8 h	(36)	91:9	370
Et	Me	[Ru(tetrapropylporphyrin)(CO)(EtOH)], CH$_2$Cl$_2$, rt, 8 h	(52)	91:9	370
Et	Me	13a, neat, 60°, 4 h	(73)	75:25	364
Et	Me	13d, neat, 60°, 4 h	(67)	69:31	364
Et	Me	14a, neat, 60°, 4 h	(65)	77:23	364
Et	Me	14d, neat, 60°, 4 h	(59)	86:14	364
Et	Me	14f, neat, 60°, 4 h	(52)	86:14	364
Et	Me	23b, neat, 60°, 4 h	(74)	63:37	340
Et	Me	24b, neat, 60°, 4 h	(70)	69:31	340
Et	OMe	Pd(OAc)$_2$	(—)	—	376
Et	OMe	[OsCl$_2$(p-cymene)]$_2$, neat, 60°	(61)	65:35	91
Et	OMe	[OsCl$_2$(p-cymene)]$_2$, neat, 80°	(84)	62:38	91
Et	OMe	Pd(OAc)$_2$, neat, rt	(79)	—	80
Et	OMe	Iron *meso*-tetra-p-tolylporphyrin chloride, CH$_2$Cl$_2$, 40°, 5 h	(—)	85:15	57

TABLE 1. DIASTEREOSELECTIVE CYCLOPROPANATION OF ALKENES USING DIAZOACETATES AND RELATED SYSTEMS (*Continued*)

Carbenoid Precursor	Substrate	Conditions	Product(s) and Yield(s) (%)		Refs.
Et	OMe	Iron *meso*-tetra(pentafluorophenyl)-porphyrin chloride, CH_2Cl_2, rt, 7 h	(—)	85:15	57
Et	OMe	Iron *meso*-tetramesitylporphyrin-chloride/cobaltocene, CH_2Cl_2, rt, 3 h	(—)	92:8	57
Et	OMe	$RuCl_2(PPh_3)_3$, neat, 60°, 4 h	(92)	61:39	52, 338
Et	OMe	$RuCl_2(PPh_3)_3/C_6H_6$, neat, 60°, 4 h	(87)	59:41	52
Et	OMe	$RuH_3[Si(OEt)_3](PPh_3)_2$, neat, 60°, 4 h	(87)	62:38	52
Et	OMe	$Ru[Si(OEt)_3]_2(PPh_3)_2$, neat, 60°, 4 h	(90)	61:39	52
Et	OMe	$OsCl_2(PPh_3)_3$, neat, 60°, 4 h	(59)	62:38	338
Et	OMe	(7-PPh$_2$-8-H-7,8-C$_2$B$_9$H$_{10}$)Rh(PPh$_3$)$_2$, neat, 4 h, 100°	(87)	62:38	351
Et	OMe	[RuH(7-PPh$_2$-8-H-C$_2$B$_9$H$_{10}$)(PPh$_3$)$_2$], neat, 100°	(90)	62:38	81
Et	OMe	[RuH(7-PPh$_2$-8-Me-C$_2$B$_9$H$_{10}$)(PPh$_3$)$_2$], neat, 100°	(89)	64:36	81
Et	OMe	RuCl(*p*-cymene)(TsNC$_6$H$_4$NH$_2$-2), neat, 80°	(87)	70:30	357
Et	OMe	Hydridotris(3,5-Me$_2$-1-pyrazolyl)-borateCu(C$_2$H$_4$), DCE, rt	(—)	58:42	365
Et	OMe	[Ru(*meso*-tetraphenylporphyrin)-(CO)(EtOH)], CH_2Cl_2, rt, 8 h	(71)	89:11	370
Et	OMe	[Ru(*meso*-tetramesitylporphyrin)-(CO)(EtOH)], CH_2Cl_2, rt, 8 h	(81)	88:12	370
Et	OMe	[Ru(octaethylporphyrin)(CO)(EtOH)], CH_2Cl_2, rt, 8 h	(41)	85:15	370
Et	OMe	[Ru(tetrapropylporphyrin)(CO)(EtOH)], CH_2Cl_2, rt, 8 h	(44)	88:12	370
Et	OMe	**23b**, neat, 60°, 4 h	(72)	68:32	340
Et	OMe	**24b**, neat, 60°, 4 h	(68)	69:31	340

Et	Et	Pd(OAc)$_2$	(—)	—	376
Et	OEt	Pd(OAc)$_2$	(—)	—	376
Et	CO$_2$Et	Pd(OAc)$_2$	(—)	—	376
Et	NMe$_2$	Pd(OAc)$_2$, neat, rt	(0)	—	80
Et	i-Pr	Pd(OAc)$_2$	(—)	—	376
Et	t-Bu	Rh$_2$(O$_2$CH)$_4$, 60°	(—)	53:47	325
Et	t-Bu	Rh$_2$(OAc)$_4$, 60°	(—)	57:43	325
Et	t-Bu	Rh$_2$(O$_2$CPr-n)$_4$, 60°	(—)	58:42	325
Et	t-Bu	Pd(OAc)$_2$	(—)	—	376
Et	t-Bu	[OsCl$_2$(p-cymene)]$_2$, neat, 60°	(57)	62:38	91
Et	t-Bu	[OsCl$_2$(p-cymene)]$_2$, neat, 80°	(79)	60:40	91
Et	t-Bu	Cl(Ph$_3$P)$_2$Ru(H$_2$-C$_2$B$_9$H$_{10}$), neat, 60°	(88)	58:42	51
Et	t-Bu	Cl(Ph$_3$P)$_2$Ru(Me$_2$-C$_2$B$_9$H$_{10}$), neat, 60°	(77)	60:40	51
Et	t-Bu	ClH(Ph$_3$P)$_2$Ru(H$_2$-C$_2$B$_9$H$_9$), neat, 60°	(89)	57:43	51
Et	t-Bu	RuCl$_2$(PPh$_3$)$_3$, neat, 60° ,4 h	(90)	58:42	52, 338
Et	t-Bu	RuCl$_2$(PPh$_3$)$_3$/C$_6$H$_6$, neat, 60°, 4 h	(87)	64:36	52
Et	t-Bu	RuH$_3$[Si(OEt)$_3$](PPh$_3$)$_2$, neat, 60°, 4 h	(83)	64:36	52
Et	t-Bu	Ru[Si(OEt)$_3$]$_2$(PPh$_3$)$_2$, neat, 60°, 4 h	(85)	63:37	52
Et	t-Bu	OsCl$_2$(PPh$_3$)$_3$, neat, 60°, 4 h	(49)	57:43	338
Et	t-Bu	(7-PPh$_2$-8-H-7,8-C$_2$B$_9$H$_{10}$)Rh(PPh$_3$)$_2$, neat, 4 h, 100°	(93)	65:35	351
Et	t-Bu	[RuH(7-PPh$_2$-8-H-C$_2$B$_9$H$_{10}$)(PPh$_3$)$_2$], neat, 100°	(93)	67:33	81
Et	t-Bu	[RuH(7-PPh$_2$-8-Me-C$_2$B$_9$H$_{10}$)(PPh$_3$)$_2$], neat, 100°	(91)	68:32	81
Et	t-Bu	RuCl(p-cymene)(TsNC$_6$H$_4$NH$_2$-2), neat, 80°	(88)	67:33	357
Et	t-Bu	**23b**, neat, 60°, 4 h	(67)	65:35	340
Et	t-Bu	**24b**, neat, 60°, 4 h	(65)	70:30	340

TABLE 1. DIASTEREOSELECTIVE CYCLOPROPANATION OF ALKENES USING DIAZOACETATES AND RELATED SYSTEMS (*Continued*)

Carbenoid Precursor	Substrate	Conditions	Product(s) and Yield(s) (%)		Refs.
R = Et	PhO⁀⫽	Cu, xylene, 120°	PhO⁀⟨▲⟩CO₂Et		
				trans:cis	
		Cu, xylene, 120°	(42)	—	377
		(CO)₅WC(OMe)Ph, 35°	(28)	55:45	321
		Rh₂(OAc)₄, neat, rt	(92)	58:42	71
R = Et	(furan-dioxolane with =CH₂)	CuSO₄, heptane, 70°	(furan-dioxolane spiro cyclopropane CO₂Et) (55)		378
R = Et	(glucose tetraacetate vinyl ether)	Cu-bronze, benzene, reflux, 24 h	(glucose tetraacetate cyclopropyl CO₂Et) (50)		379
R = Et	(bicyclic cyclopentene)	Rh₂(OAc)₄	(tricyclic cyclopropane CO₂Et) (24)		380
R = Et	(bicyclic cyclopentene OTBDMS)	Rh₂(OAc)₄	(tricyclic cyclopropane OTBDMS CO₂Et) (78)		380, 381

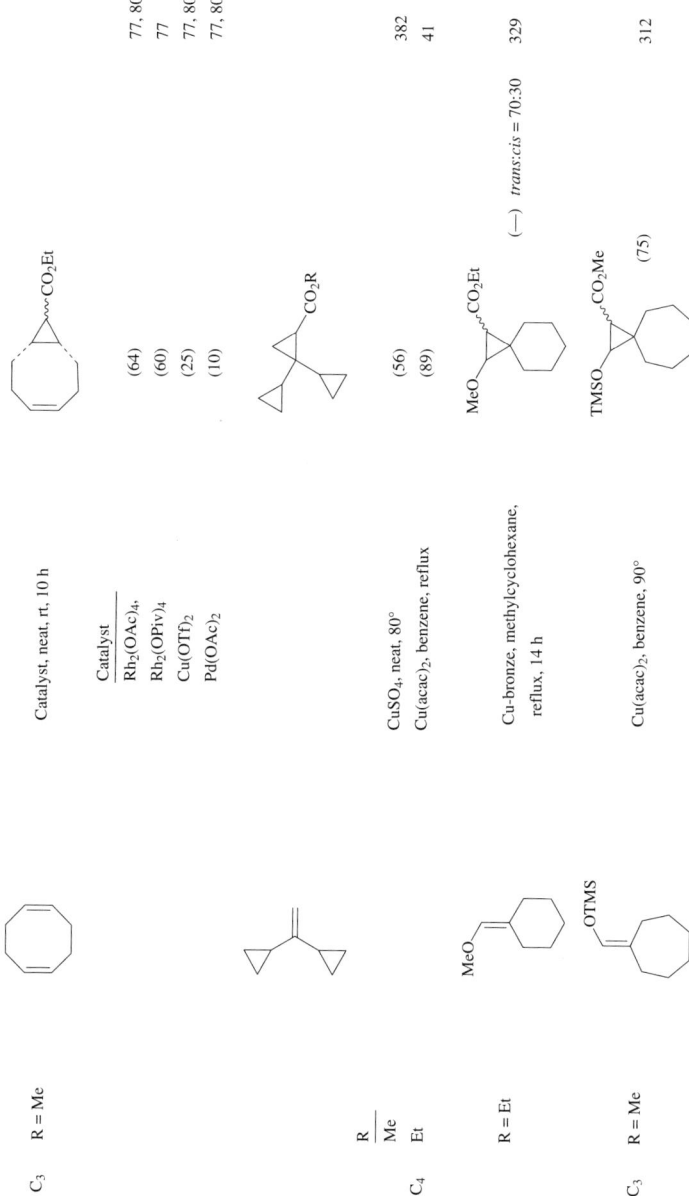

TABLE 1. DIASTEREOSELECTIVE CYCLOPROPANATION OF ALKENES USING DIAZOACETATES AND RELATED SYSTEMS (*Continued*)

Carbenoid Precursor	Substrate	Conditions	Product(s) and Yield(s) (%)		Refs.
				exo:endo	
C_4 R = Et	(cyclooctene)	$Rh_2(O_2CH)_4$, 60°	(—)	58:42	325
		$Rh_2(OAc)_4$, 60°	(—)	56:44	325
		$Rh_2(OAc)_4$, neat, rt	(95)	—	80
		$Rh_2(O_2CPr\text{-}n)_4$, 60°	(—)	57:43	325
		$Cu(OTf)_2$, neat, rt	(28)	—	80
		$Pd(OAc)_2$, neat, rt	(20)	—	80
		$[Rh(PE\text{-}CO_2)_2]_2$, toluene, 100°	(80)	78:22	367
		$CuSO_4$, neat, rt to 135°	(—)	33:67	383
		$[OsCl_2(p\text{-cymene})]_2$, neat, 60°	(13)	41:59	91
		$[OsCl_2(p\text{-cymene})]_2$, neat, 80°	(26)	46:54	91
		$OsCl_2(PPh_3)_3$, neat, 60°, 4 h	(1)	53:47	338
		$Cl(PPh_3)_2Ru(H_2C_2B_9H_{10})$, neat, 60°	(91)	63:37	51
		$Cl(PPh_3)_2Ru(Me_2C_2B_9H_{10})$, neat, 60°	(16)	47:53	51
		$ClH(PPh_3)_2Ru(H_2C_2B_9H_9)$, neat, 60°	(87)	51:49	51
		$RuCl_2(PPh_3)_3$, neat, 60°, 4 h	(5)	61:39	52, 338
		$RuCl_2(PPh_3)_3/C_6H_6$, neat, 60°, 4 h	(6)	58:42	52
		$RuH_3[Si(OEt)_3](PPh_3)_2$, neat, 60°, 4 h	(4)	65:35	52
		$Ru[Si(OEt)_3]_2(PPh_3)_2$, neat, 60°, 4 h	(4)	65:35	52
		$(7\text{-}PPh_2\text{-}8\text{-}H\text{-}7,8\text{-}C_2B_9H_{10})Rh(PPh_3)_2$, neat, 4 h, 100°	(86)	59:41	351
		$(7\text{-}PPh_2\text{-}8\text{-}Me\text{-}7,8\text{-}C_2B_9H_{10})Rh(PPh_3)_2$, neat, 4 h, 100°	(85)	60:40	351
		$(7\text{-}PPh_2\text{-}8\text{-}H\text{-}7,8\text{-}C_2B_9H_{10})\text{-}Rh(1,5\text{-cyclooctadiene})$, neat, 4 h, 100°	(88)	58:42	351

Catalyst			
Pt(PPh$_3$)$_4$, neat, 100°	(42)	73:27	368
PtCl$_2$, neat, 100°	(39)	68:32	368
[PtCl$_2$(cyclohexene)]$_2$, neat, 100°	(54)	67:33	368
cis-PtCl$_2$(benzonitrile)$_2$, neat, 100°	(54)	66:34	368
cis-PtCl$_2$(pyr)$_2$, neat, 100°	(22)	83:17	368
cis-PtCl$_2$(PPh$_3$)$_2$, neat, 100°	(29)	67:33	368
trans-PtCl$_2$(PPh$_3$)$_2$, neat, 100°	(33)	65:35	368
PtBr$_2$, neat, 100°	(29)	72:28	368
PtI$_2$, neat, 100°	(54)	66:34	368
PtI$_2$(NH$_3$)$_2$, neat, 100°	(31)	74:26	368
PtCl$_4$, neat, 100°	(71)	64:36	368
PtBr$_4$, neat, 100°	(42)	74:26	368
[RuH(7-PPh$_2$-8-H-C$_2$B$_9$H$_{10}$)(PPh$_3$)$_2$], neat, 100°	(51)	54:46	81
[RuH(7-PPh$_2$-8-Me-C$_2$B$_9$H$_{10}$)(PPh$_3$)$_2$], neat, 100°	(65)	48:52	81

R = Et

Catalyst, neat	Temp		exo:endo	357
RuCl(p-cymene)[TsN(CH$_2$)$_2$NH$_2$]	60°	(34)	58:42	
RuCl(p-cymene)[TsN(CH$_2$)$_2$NH$_2$]	100°	(66)	60:40	
RuCl(p-cymene)[TsN(CH$_2$)$_2$NH$_2$]	60°	(10)	61:39	
RuCl(p-cymene)[TsN(CH$_2$)$_2$NH$_2$]	100°	(89)	53:47	
RuCl(p-cymene)[TsN(CH$_2$)$_2$NH$_2$]	60°	(12)	65:35	
RuCl(p-cymene)[TsN(CH$_2$)$_2$NH$_2$]	100°	(58)	65:35	
RuCl(p-cymene)[TsN-o-NH$_2$Ph]	60°	(52)	51:49	
RuCl(p-cymene)[TsN-o-NH$_2$Ph]	100°	(83)	61:39	
RuCl(p-cymene)[2-TsNpyridine]	60°	(20)	59:41	
RuCl(p-cymene)[2-TsNpyridine]	100°	(88)	59:41	
RuCl(p-cymene)[2-(TsNCH$_2$)pyridine]	60°	(45)	73:27	
RuCl(p-cymene)[2-(TsNCH$_2$)pyridine]	100°	(82)	56:44	

TABLE 1. DIASTEREOSELECTIVE CYCLOPROPANATION OF ALKENES USING DIAZOACETATES AND RELATED SYSTEMS (Continued)

Carbenoid Precursor	Substrate	Conditions	Product(s) and Yield(s) (%)	Refs.
R = Et	(cyclooctene)		(cyclopropane-CO$_2$Et fused to cycloheptane) *exo:endo*	
		Hydridotris(3,5-dimethyl-1-pyrazolyl)-borateCu(C$_2$H$_4$), DCE, rt	(78) 99:1	348
		Ru$_2$(OAc)$_4$(MeOH)$_2$, neat, 60°	(12) 68:32	384
		Ru$_2$(O$_2$CCF$_3$)$_4$, neat, 60°	(99) 38:62	384
		RuCl$_2$(PPh$_3$)$_3$, neat, 60°	(82) 62:38	384
		[(C$_6$H$_6$)RuCl$_2$]$_2$, neat, 60°	(42) 62:38	384
		[(p-cymene)RuCl$_2$]$_2$, neat, 60°	(44) 60:40	384
		(p-cymene)RuCl$_2$•PPh$_3$, neat, 60°	(43) 67:33	384
		(p-cymene)RuCl$_2$•(p-cymene)$_3$, neat, 60°	(0) —	384
		23a, neat, 60°, 4 h	(47) 59:41	340
		23b, neat, 60°, 4 h	(51) 65:35	340
		24a, neat, 60°, 4 h	(19) 60:40	340
		24b, neat, 60°, 4 h	(8) 65:35	340
R = Et	(1,2-dimethylcyclohexene)	CuCl•PMe$_3$, neat, reflux	(cyclopropane-CO$_2$R on cyclohexane) (51)	337
R = Et	(MeO)(MeO)C=cyclopentane	Cu(acac)$_2$, benzene, 85°	(spiro cyclopropane with MeO, MeO, CO$_2$Et) (48)	295
R = Et	n-C$_6$H$_{13}$-CH=CH$_2$	Catalyst, neat, 60°, 4 h	n-C$_6$H$_{13}$-cyclopropane-CO$_2$Et *trans:cis*	340
		Catalyst		
		23b	(46) 56:44	
		24b	(9) 60:40	

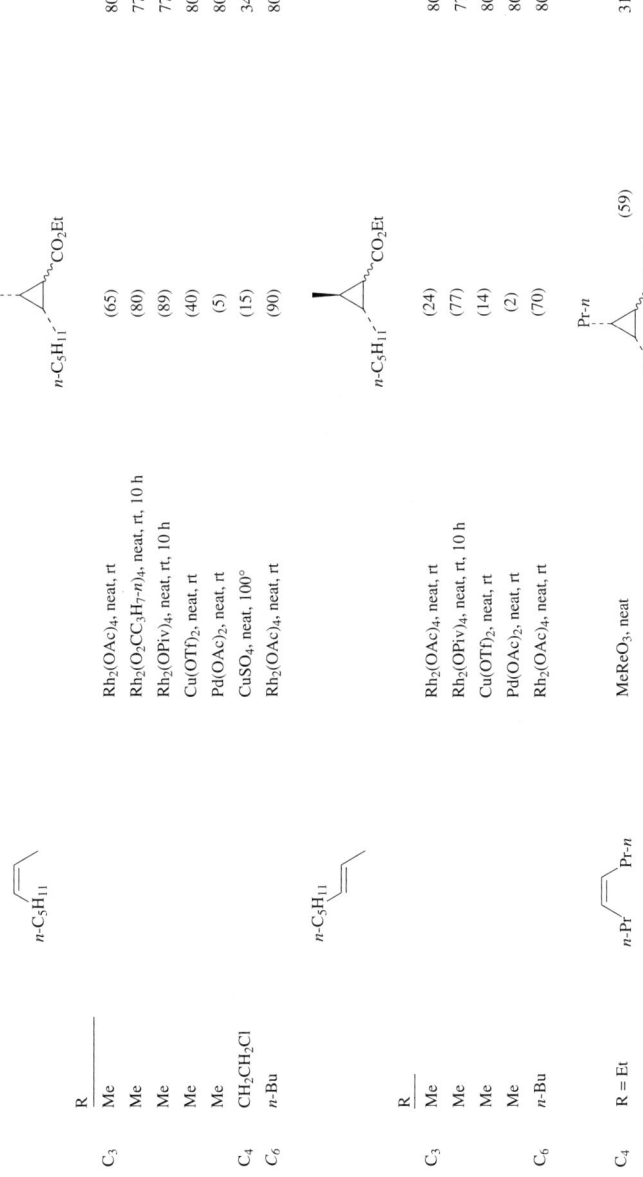

	R			
C_3	Me	Rh$_2$(OAc)$_4$, neat, rt	(65)	80, 77
	Me	Rh$_2$(O$_2$CC$_3$H$_{7}$-n)$_4$, neat, rt, 10 h	(80)	77
	Me	Rh$_2$(OPiv)$_4$, neat, rt, 10 h	(89)	77
	Me	Cu(OTf)$_2$, neat, rt	(40)	80, 77
	Me	Pd(OAc)$_2$, neat, rt	(5)	80, 77
C_4	CH$_2$CH$_2$Cl	CuSO$_4$, neat, 100°	(15)	341
C_6	n-Bu	Rh$_2$(OAc)$_4$, neat, rt	(90)	80, 77
C_3	Me	Rh$_2$(OAc)$_4$, neat, rt	(24)	80, 77
	Me	Rh$_2$(OPiv)$_4$, neat, rt, 10 h	(77)	77
	Me	Cu(OTf)$_2$, neat, rt	(14)	80, 77
	Me	Pd(OAc)$_2$, neat, rt	(2)	80, 77
C_6	n-Bu	Rh$_2$(OAc)$_4$, neat, rt	(70)	80, 77
C_4	R = Et	MeReO$_3$, neat	(59)	319

TABLE 1. DIASTEREOSELECTIVE CYCLOPROPANATION OF ALKENES USING DIAZOACETATES AND RELATED SYSTEMS (*Continued*)

	Carbenoid Precursor	Substrate	Conditions	Product(s) and Yield(s) (%)	Refs.
		n-Pr⁀⁀⁀Pr-n	Catalyst, neat, rt, 10 h	n-Pr⁀△⁀Pr-n, CO₂R	77
	R				
C₃	Me			Catalyst	
				Rh₂(OAc)₄ (7)	
	Me			Rh₂(O₂CC₃H₇-n)₄ (54)	
	Me			Rh₂(OPiv)₄ (60)	
	Me			Cu(OTf)₂ (8)	
	Me			Pd(OAc)₂ (12)	
C₄	Et			Cu(OTf)₂ (23)	
C₆	n-Bu			Rh₂(OAc)₄ (70)	
	n-Bu			Rh₂(OPiv)₄ (85)	
C₄	R = Et	TMS⁀⁀⁀OTBDMS	Cu-bronze, CuSO₄, 118°	EtO₂C⁀△⁀TMS, OTBDMS (59)	385
	R				
C₃	Me	O=C(Ph)O-CH=CH₂	Rh₂(OAc)₄, CH₂Cl₂, rt	PhC(O)O-△-CO₂R *trans:cis* (54) 62:38	309
C₆	t-Bu			(49) 67:33	
C₄	R = Et	TMSO⁀⁀⁀Ph	CuSO₄, cyclohexane, reflux, 12 h	Ph, TMSO-△-CO₂Et (—)	293

	R	Conditions	(yield)	E:Z	Ref.
C_3	Me	$Cu(acac)_2$, 100°, 3 h	(67)	—	307
	Me	$[Ru_2(CO)_4(\mu\text{-}OAc)_2]_n$, CH_2Cl_2, rt	(91)	40:60	308
C_4	Et	$Rh_2(O_2CH)_4$, 60°	(—)	49:51	325
	Et	$Rh_2(OAc)_4$, 60°	(—)	50:50	325
	Et	$Rh_2(O_2CPr\text{-}n)_4$, 60°	(—)	50:50	325
	Et	$[Rh(PE\text{-}CO_2)_2]_2$, toluene, 100°	(96)	62:38	367
	Et	$MeReO_3$, neat	(60)	60:40	319
	Et	$[OsCl_2(p\text{-}cymene)]_2$, neat, 60°	(62)	40:60	91
	Et	$[OsCl_2(p\text{-}cymene)]_2$, neat, 80°	(79)	41:59	91
	Et	$Cl(Ph_3P)_2Ru(H_2C_2B_9H_{10})$, neat, 60°	(81)	50:50	51
	Et	$Cl(Ph_3P)_2Ru(Me_2C_2B_9H_{10})$, neat, 60°	(73)	42:58	51
	Et	$ClH(Ph_3P)_2Ru(H_2C_2B_9H_7)$, neat, 60°	(87)	50:50	51
	Et	$Pd(OAc)_2$, neat, rt	(42)	—	80
	Et	Iron meso-tetra-p-tolylporphyrin chloride, CH_2Cl_2, rt, 1 h	(—)	81:19	57
	Et	Iron meso-tetra-p-tolylporphyrin chloride, CH_2Cl_2, 40°, 18 h	(—)	77:23	57
	Et	Iron meso-tetra(pentafluorophenyl)porphyrin chloride, CH_2Cl_2, rt, 3 h	(—)	52:48	57
	Et	Iron meso-tetramesitylporphyrin chloride/cobaltocene, CH_2Cl_2, rt, 8 h	(—)	75:25	57
	Et	Iron octaethylporphyrin chloride/cobaltocene, CH_2Cl_2, rt, 4 h	(—)	79:21	57
	Et	Osmium meso-tetra-p-tolylporphyrin(CO)(pyr), toluene, rt, 2 h	(39)	74:26	90
	Et	$[(Cp)Fe(CO)_2(THF)]^+BF_4^-$, CH_2Cl_2, 40°	(60)	40:60	322
	Et	$(7\text{-}PPh_2\text{-}8\text{-}H\text{-}7,8\text{-}C_2B_9H_{10})Rh(PPh_3)_2$, neat, 4 h, 100°	(94)	53:47	351

TABLE 1. DIASTEREOSELECTIVE CYCLOPROPANATION OF ALKENES USING DIAZOACETATES AND RELATED SYSTEMS (*Continued*)

	Carbenoid Precursor	Substrate	Conditions	Product(s) and Yield(s) (%)		Refs.
	Et		RuCl$_2$(PPh$_3$)$_3$, neat, 60°, 4 h	(94)	53:47	338
	Et		OsCl$_2$(PPh$_3$)$_3$, neat, 60°, 4 h	(55)	51:49	338
	Et		[RuH(7-PPh$_2$-8-H-C$_2$B$_9$H$_{10}$)(PPh$_3$)$_2$], neat, 100°	(98)	51:49	81
	Et		[RuH(7-PPh$_2$-8-Me-C$_2$B$_9$H$_{10}$)(PPh$_3$)$_2$], neat, 100°	(97)	50:50	81
	Et		RuCl(p-cymene)(TsNC$_6$H$_4$NH$_2$-2), neat, 80°	(92)	51:49	357
	Et		Cu/bronze-K10 montmorillonite, CH$_2$Cl$_2$, rt	(—)	44:56	339
	Et		Cu/bronze-bentonite, CH$_2$Cl$_2$, rt	(—)	55:45	339
	Et		[Ru(*meso*-tetraphenylporphyrin)(CO)], neat, rt, 4 h	(—)	76:24	84
	Et		[Ru(*meso*-tetramesitylporphyrin)(CO)], neat, rt, 4 h	(—)	62:38	84
	Et		[Ru(*meso*-tetramesitylporphyrin)(O)$_2$], neat, rt, 4 h	(—)	60:40	84
	Et		[Ru(*meso*-tetraphenylporphyrin)(CO)(EtOH)], CH$_2$Cl$_2$, rt, 8 h	(50)	74:26	370
	Et		[Ru(*meso*-tetramesitylporphyrin)(CO)(EtOH)], CH$_2$Cl$_2$, rt, 8 h	(68)	63:37	370
	Et		[Ru(octaethylporphyrin)(CO)(EtOH)], CH$_2$Cl$_2$, rt, 8 h	(36)	67:33	370
	Et		[Ru(tetrapropylporphyrin) (CO)(EtOH)], CH$_2$Cl$_2$, rt, 8 h	(23)	73:27	370
	Et		RuCl$_2$(PPh$_3$)$_3$, neat, 60°, 4 h	(94)	53:47	52
	Et		RuCl$_2$(PPh$_3$)$_3$/C$_6$H$_6$, neat, 60°, 4 h	(90)	53:47	52
	Et		RuH$_3$[Si(OEt)$_3$](PPh$_3$)$_2$, neat, 60°, 4 h	(86)	53:47	52
	Et		Ru[Si(OEt)$_3$]$_2$(PPh$_3$)$_2$, neat, 60°, 4 h	(92)	51:49	52
	Et	23b,	neat, 60°, 4 h	(82)	58:42	340
	Et	24b,	neat, 60°, 4 h	(78)	65:35	340
C$_{12}$	l-menthyl		Rh$_2$(OAc)$_4$, CH$_2$Cl$_2$, rt	(80)	53:47	309
	d-menthyl		Rh$_2$(OAc)$_4$, CH$_2$Cl$_2$, rt	(—)	—	309

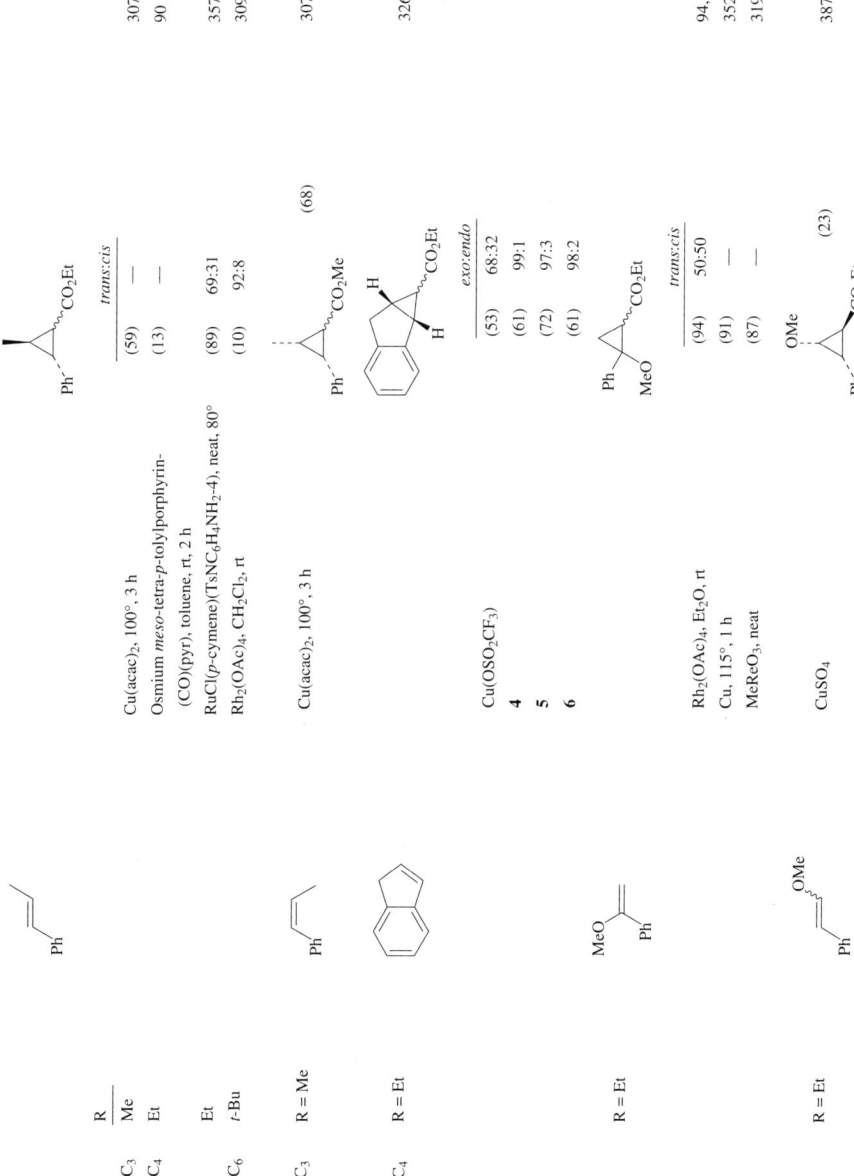

	R				trans:cis	
C₃	Me		Cu(acac)₂, 100°, 3 h	(59)	—	307
C₄	Et		Osmium *meso*-tetra-*p*-tolylporphyrin-(CO)(pyr), toluene, rt, 2 h	(13)	—	90
C₆	Et		RuCl(*p*-cymene)(TsNC₆H₄NH₂-4), neat, 80°	(89)	69:31	357
	t-Bu		Rh₂(OAc)₄, CH₂Cl₂, rt	(10)	92:8	309
C₃	R = Me		Cu(acac)₂, 100°, 3 h	(68)		307

				exo:endo	
C₄	R = Et	Cu(OSO₂CF₃)			326
		4	(53)	68:32	
		5	(61)	99:1	
		6	(72)	97:3	
			(61)	98:2	

				trans:cis	
R = Et		Rh₂(OAc)₄, Et₂O, rt	(94)	50:50	94, 71
		Cu, 115°, 1 h	(91)	—	352
		MeReO₃, neat	(87)	—	319

| R = Et | CuSO₄ | (23) | | | 387 |

TABLE 1. DIASTEREOSELECTIVE CYCLOPROPANATION OF ALKENES USING DIAZOACETATES AND RELATED SYSTEMS (Continued)

Carbenoid Precursor	Substrate	Conditions	Product(s) and Yield(s) (%)	Refs.
R = Et	indane-R^1	CuSO$_4$, neat, 110°, 3 h	EtO$_2$C~~~⟨⟩~~~R^1 + [bicyclic II with CO$_2$Et and R^1] **I:II** R^1 = H: (70) 71:29 R^1 = Me: (75) 80:20	388
R = Et	[methylcyclohexene-CO$_2$Me]	Cu-bronze, heat	cyclopropane-CO$_2$Et / CO$_2$Me (—) exo:endo = 2:1	380
R = Et	[dioxolane-spiro-cycloheptene]	Cu, 110°, 30 min	[spiro dioxolane cyclopropane-CO$_2$Et] (52)	389
R = Et	[OMe-cyclohexene with methyl]	Cu-bronze, methylcyclohexane, reflux, 7 h	MeO-cyclohexyl-cyclopropane-CO$_2$Et (—) mixture of stereoisomers = 70:30	329
R = Et	C$_7$H$_{15-n}$ alkene	[Rh(PE-CO$_2$)$_2$]$_2$, toluene, 100°	RO$_2$C~~cyclopropane~~C$_7$H$_{15-n}$ (79) trans:cis = 63:37	367
R = Me	TMSO / R^1 ; TMS–NBOC R^1 = Ph R^1 = cyclohexyl	1. Cu(acac)$_2$, benzene, 80° 2. Bu$_4$NF, CH$_2$Cl$_2$	R^1–C(=O)–CH$_2$–CO$_2$Me with NBOC (37) (50)	390

C$_3$

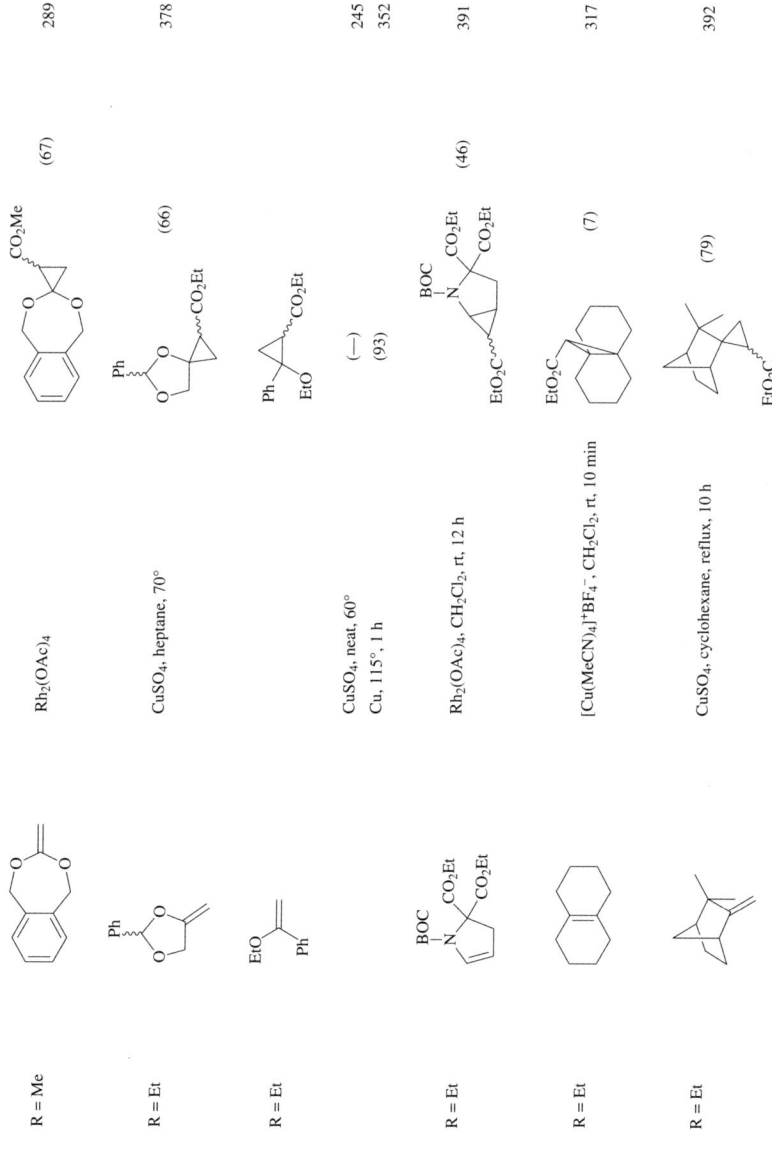

TABLE 1. DIASTEREOSELECTIVE CYCLOPROPANATION OF ALKENES USING DIAZOACETATES AND RELATED SYSTEMS (*Continued*)

	Carbenoid Precursor	Substrate	Conditions	Product(s) and Yield(s) (%)	Refs.
	R = Et		Cu(acac)$_2$, benzene, 90°	(—)	393
C$_3$	R = Me		Cu(acac)$_2$, ethyl acetate, 100°	(74)	294
C$_{12}$	R = d,l-menthyl		Cu(acac)$_2$, cyclohexane, 40°	(88)	394
C$_4$	R = Et		CuSO$_4$, 100°	trans:cis (59) —	347
			Osmium *meso*-tetra-*p*-tolylporphyrin-(CO)(pyr), toluene, rt, 2 h	(32) 81:19	90
			23b, neat, 60°, 4 h	(42) 58:42	340
			24b, neat, 60°, 4 h	(8) 64:36	340
	R = Et		Rh$_2$(OAc)$_4$, Et$_2$O	(—)	395

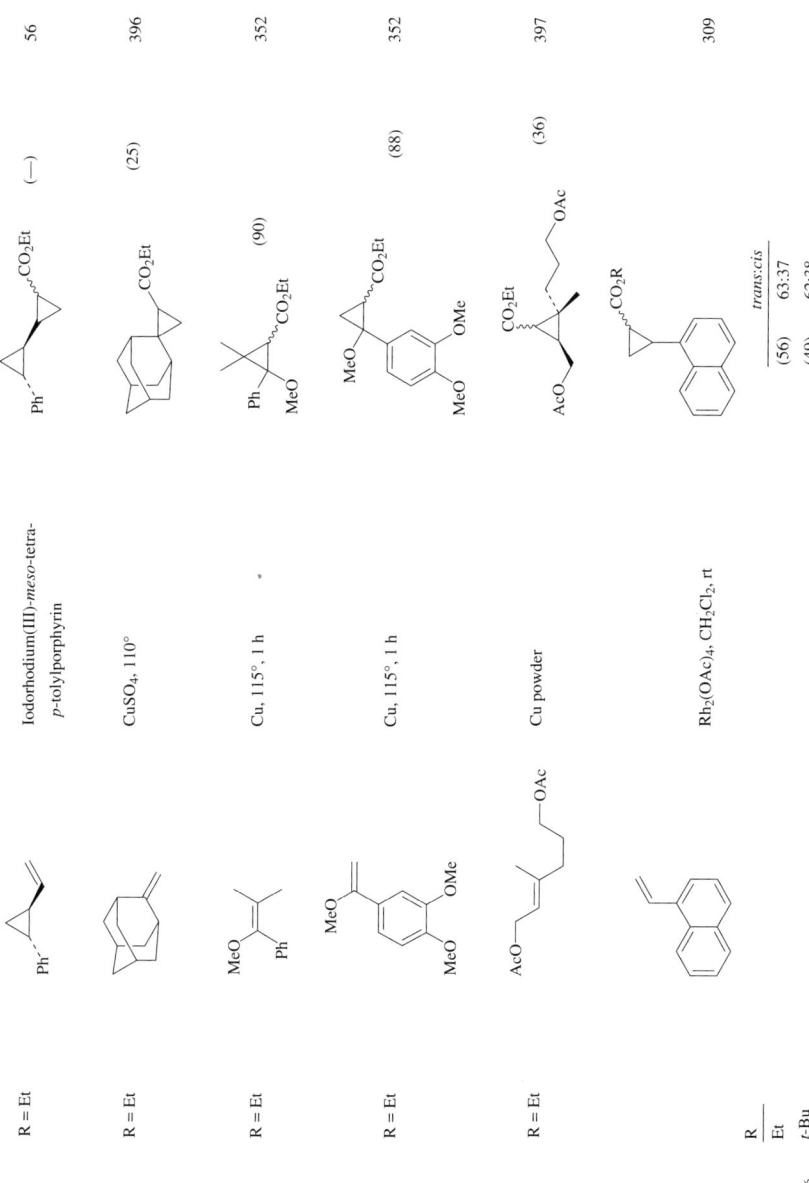

TABLE 1. DIASTEREOSELECTIVE CYCLOPROPANATION OF ALKENES USING DIAZOACETATES AND RELATED SYSTEMS (*Continued*)

	Carbenoid Precursor	Substrate	Conditions	Product(s) and Yield(s) (%)		Refs.
C_4	R = Et	(sugar with OBn groups)	$Rh_2(OAc)_4$, CH_2Cl_2, 1 h	(pyran product) (43)		324
	R = Et	(adamantyl vinyl)	$CuSO_4$, 80°	(adamantyl cyclopropane CO_2Et) (59)	*trans:cis* = >95:5	398
C_3	R = Me	(Pr-i, CO$_2$Me alkene)	$CuSO_4$, neat, 0°	(cyclopropane with Pr-i, CO_2Me, CO_2Me) (76)		399
C_4	R = Et	(OTMS cyclopentene CO_2Bu-t)	$CuSO_4$, benzene	(spiro OTMS, CO_2Et, CO_2Bu-t) (65)		210
	R = Et	(MeO-CH=cyclohexyl-Bu-t)	$CuCl\cdot P(OEt)_3$, hexane, rt, 1.5 h	(EtO_2C, MeO, Bu-t) (30)		400
	R = Et	($C_{10}H_{21}$-n allyl)		EtO_2C—(cyclopropane)—$C_{10}H_{21}$-n	*trans:cis*	
			$CuSO_4$, 100°	(63)	—	347
			23b, neat, 60°, 4 h	(44)	58:42	340
			24b, neat, 60°, 4 h	(12)	67:33	340
			(7-PPh$_2$-8-H-7,8-$C_2B_9H_{10}$)Rh(PPh$_3$)$_2$, neat, 4 h, 100°	(73)	60:40	351
			[RuH(7-PPh$_2$-8-H-$C_2B_9H_{10}$)(PPh$_3$)$_2$], neat, 100°	(59)	58:42	81
			[RuH(7-PPh$_2$-8-Me-$C_2B_9H_{10}$)(PPh$_3$)$_2$], neat, 100°	(61)	58:42	81

Reactant	Conditions	Products (%)	Ref.

R = Et, Rh₂(OAc)₄ → (70) 401 + (20) 402

R = Et, CuSO₄, neat, 90°, 2 h → (50) 214

R = Et, Rh₂(OAc)₄ → (90)

R = Et, CuSO₄, benzene, 90° → (70) exo:endo = 80:20, 403

TABLE 1. DIASTEREOSELECTIVE CYCLOPROPANATION OF ALKENES USING DIAZOACETATES AND RELATED SYSTEMS (*Continued*)

Carbenoid Precursor	Substrate	Conditions	Product(s) and Yield(s) (%)	Refs.
R = Et	(geranyl acetone-type alkene)	$CuSO_4$, neat, 90°, 2 h	cyclopropane with CO_2Et and dimethyl groups, pendant chain with ketone (50)	404
R = Et	phenanthrene	Iodorhodium(III)-*meso*-tetraphenylporphyrin, neat, 60°	phenanthrene-fused cyclopropane with CO_2Et (51)	102, 101
R = Et	$Ph_2C=CH_2$ (1,1-diphenylethylene)	1. $CuSO_4$, benzene 2. Saponification	Ph, Ph-cyclopropane-CO_2H (55)	405
C_{12} R = *l*-menthyl	same	$Rh_2(OAc)_4$, CH_2Cl_2, rt	Ph, Ph-cyclopropane-CO_2menthyl-*l* (63)	309
C_4 R = Et	*cis*-stilbene	1. $CuSO_4$, benzene 2. Saponification	Ph, Ph-cyclopropane-CO_2H (81)	405
R = Et	*trans*-stilbene	Hydridotris(3,5-dimethyl-1-pyrazolyl)borateCu(C_2H_4), DCE, rt	Ph, Ph-cyclopropane-CO_2Et (0)	348
R = Et	*trans*-stilbene	1. $CuSO_4$, benzene 2. Saponification	Ph, Ph-cyclopropane-CO_2H (77)	405

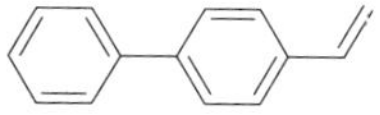	Rh$_2$(OAc)$_4$, CH$_2$Cl$_2$, rt			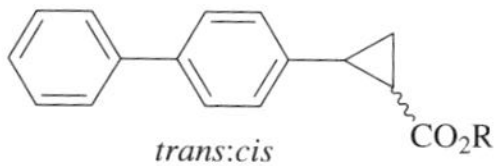 309

C$_6$

R	trans:cis	
Et	(55)	60:40
t-Bu	(51)	67:33

C$_4$ R = Et

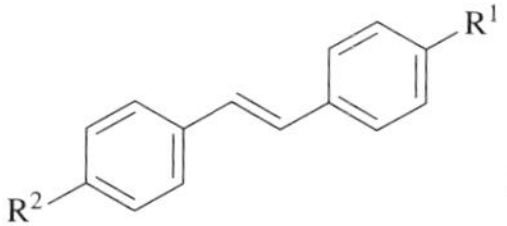

1. CuSO$_4$, benzene, 75°, 4 h
2. NaOH/MeOH/H$_2$O, 50°

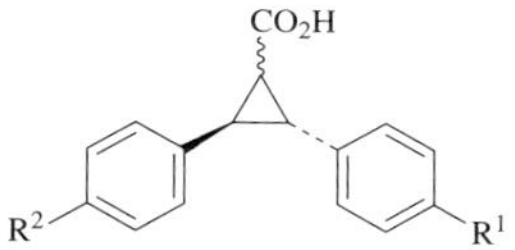

406

R^1	R^2	
H	H	(21)
Me	Me	(71)
OMe	OMe	(50)
Cl	Cl	(50)
H	Me	(30)
H	Cl	(14)
H	CF$_3$	(8)
Me	Cl	(50)
Me	CF$_3$	(6)
CF$_3$	CF$_3$	(<1)
Br	Me	(38)

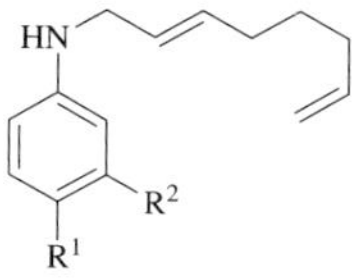

Cu bis[N-(R,S)-α-PhEt-salicylaldiminate], benzene, 80°

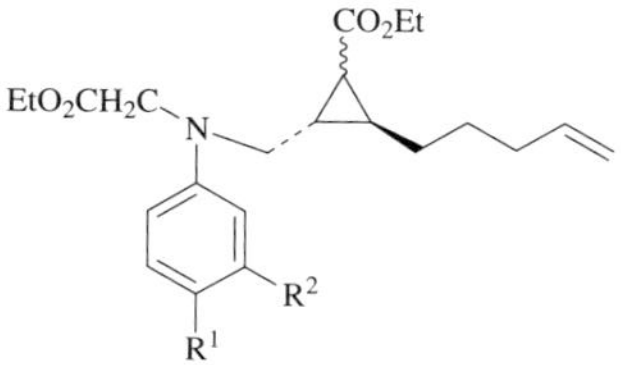

407

R = Et

R^1	R^2	
H	H	(68)
Me	H	(67)
H	Me	(65)

TABLE 1. DIASTEREOSELECTIVE CYCLOPROPANATION OF ALKENES USING DIAZOACETATES AND RELATED SYSTEMS (*Continued*)

Carbenoid Precursor	Substrate	Conditions	Product(s) and Yield(s) (%)	Refs.
R = Et	(indane-CO₂Bu-*t*)	Cu(acac)₂, THF, 65°, 9 h	(*t*-BuO₂C-indane-CO₂Et) (45)	408
R = Et	C₁₂H₂₅-*n* alkene	CuSO₄, 100°	EtO₂C-cyclopropane-C₁₂H₂₅-*n* (65)	347
R = Et	(quinone-SCH₂CO₂H tricyclic)	Rh₂(OAc)₄	(product with CO₂Et, SCH₂CO₂H) (75)	401
R = Et	Ac-N(Ar)-CH₂CH=CH-... with R¹ = H, Me	Cu bis[*N*-(*R*,*S*)-α-PhEt-salicylaldiminate], benzene, 80°	(cyclopropane product with CO₂Et) (60), (14)	407
R = Et	Ph,Ph-dioxolane-methylene	CuSO₄	(Ph,Ph-dioxolane-cyclopropane-CO₂Et) (—)	409
R = Et	(spiroketal dihydronaphthofuran methylene)	Rh₂(OAc)₄, Et₂O/THF	(spiro cyclopropane-CO₂Et product) (63) *E*:*Z* = 50:50	410

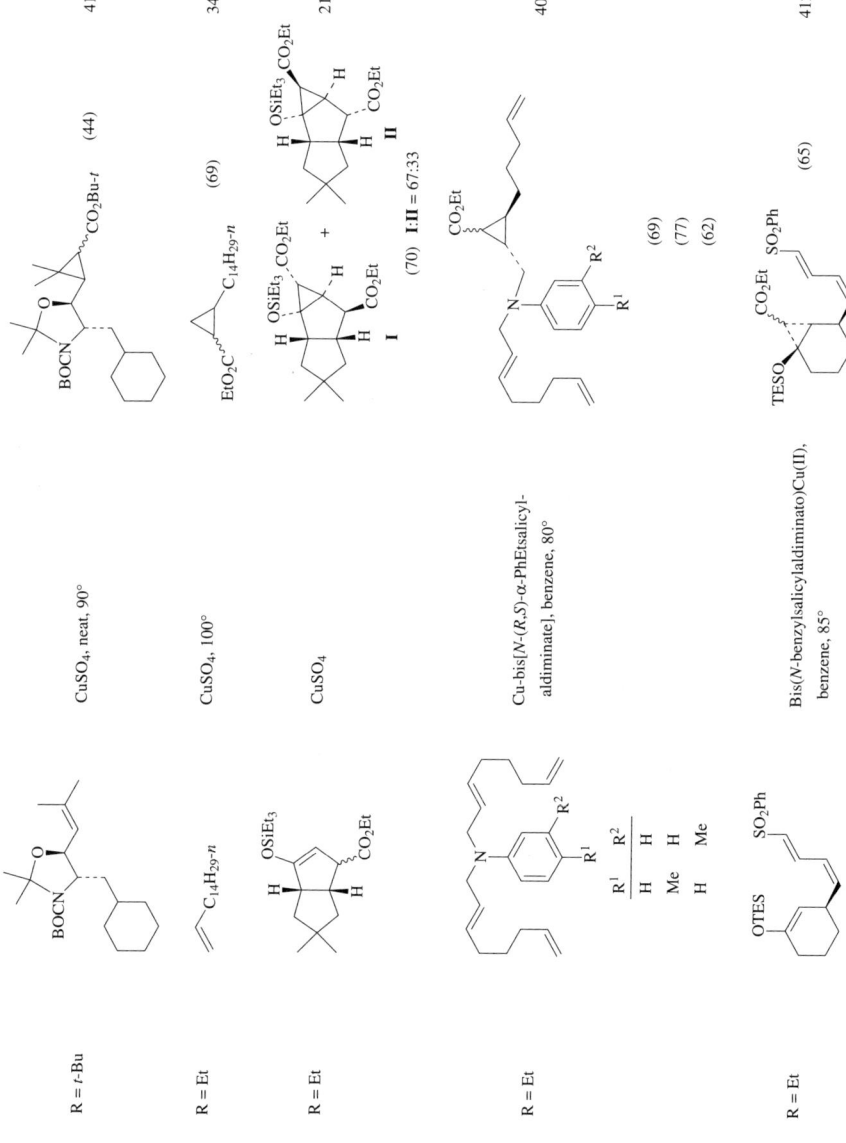

TABLE 1. DIASTEREOSELECTIVE CYCLOPROPANATION OF ALKENES USING DIAZOACETATES AND RELATED SYSTEMS (*Continued*)

Carbenoid Precursor	Substrate	Conditions	Product(s) and Yield(s) (%)	Refs.
R = Et	(steroid with OAc groups and butenyl side chain)	Rh$_2$(OAc)$_4$, CH$_2$Cl$_2$, rt, 18 h	(cyclopropanated steroid with CO$_2$Et) (92)	413
R = Et	(steroid with OMe)	CuI·P(OMe)$_3$, neat, rt	(cyclopropanated steroid with CO$_2$Et) (74)	414
C$_3$				
(N$_2$CHC(O)Me)	(1,4-dioxene)	Cu-bronze, 80°	(bicyclic with Ac) (54)	95
	(1-OTMS-cyclohexene)	Cu(acac)$_2$, 60°	(OTMS cyclohexenyl product) (40)	415
	(cyclohexylidene with OR) R: Me / Ac	Cu-bronze, neat, 90°, 5 h / Cu-bronze, cyclohexane, reflux, 10 h	(spirocyclopropane with OR, Ac) (—) / (—)	416 / 416

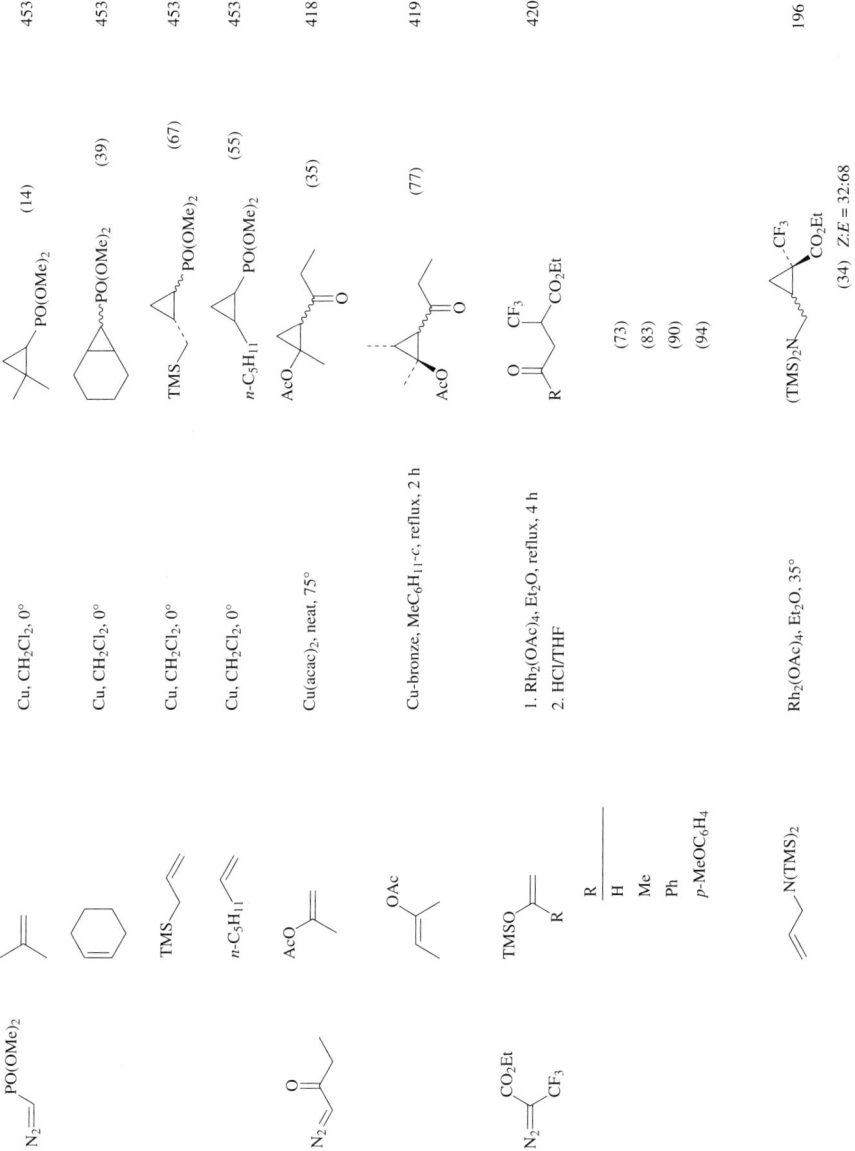

TABLE 1. DIASTEREOSELECTIVE CYCLOPROPANATION OF ALKENES USING DIAZOACETATES AND RELATED SYSTEMS (*Continued*)

Carbenoid Precursor	Substrate	Conditions	Product(s) and Yield(s) (%)	Refs.
N$_2$=C(CH$_3$)CO$_2$Et	steroid with OTMS, TMSO, H OTMS	1. Rh$_2$(OAc)$_4$, Et$_2$O, reflux, 4 h 2. HCl/THF	steroid product with CF$_3$, CO$_2$Et, O, HO, H, OH (77)	420
	cyclopentene	Cu(acac)$_2$, neat, reflux	I (CO$_2$Et bicyclic) + II (CO$_2$Et bicyclic) (—) **I:II = 6:1**	418
	cyclohexene	Cu(acac)$_2$, neat, reflux	I (CO$_2$Et bicyclic) + II (CO$_2$Et bicyclic) (—) **I:II = 7:1**	418
	n-BuO–CH=CH$_2$	Cu-bronze	n-BuO–cyclopropane–CO$_2$Et (66)	68
N$_2$=CH–PO(OEt)$_2$	cyclopentene	CuOTf, CH$_2$Cl$_2$, 5°	bicyclic–PO(OEt)$_2$ (62)	98, 421
	3,4-dihydro-2H-pyran	CuOTf, CH$_2$Cl$_2$, 5°	O–bicyclic–PO(OEt)$_2$ (71)	98, 421

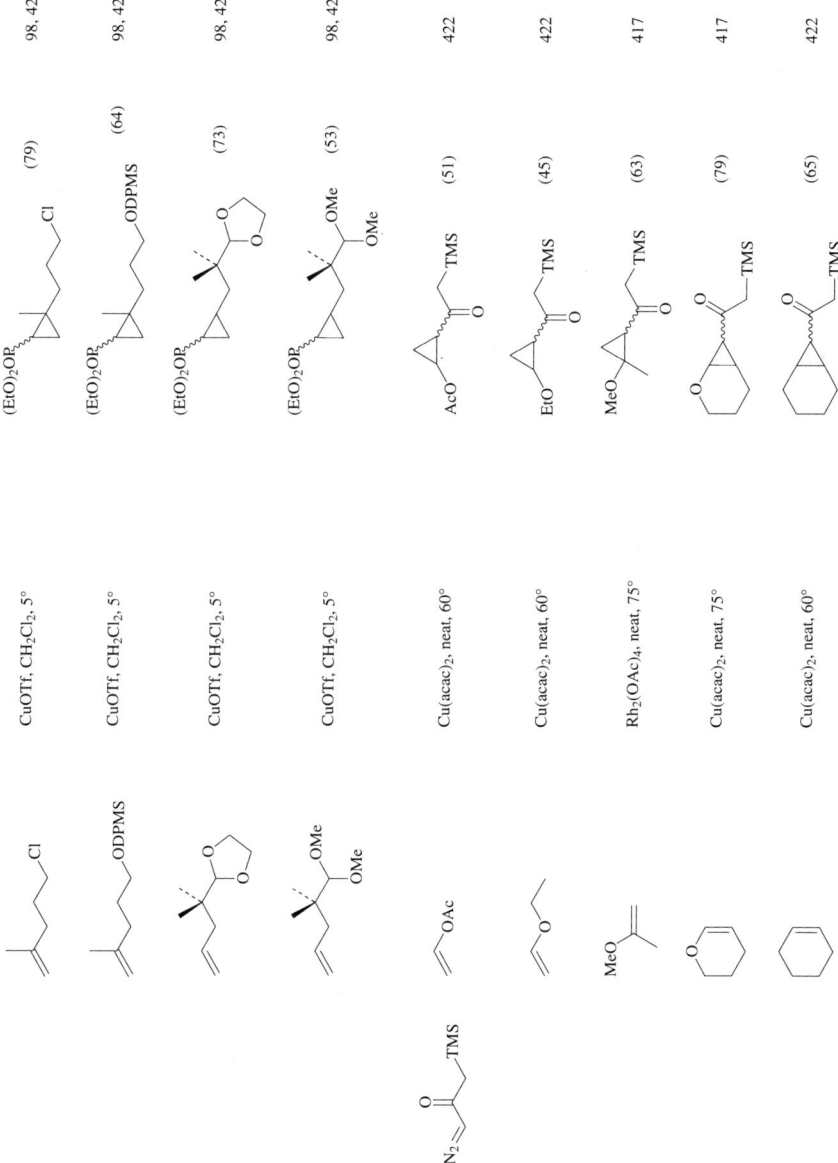

TABLE 1. DIASTEREOSELECTIVE CYCLOPROPANATION OF ALKENES USING DIAZOACETATES AND RELATED SYSTEMS (*Continued*)

Carbenoid Precursor	Substrate	Conditions	Product(s) and Yield(s) (%)	Refs.
	n-C$_3$H$_7$	Cu(acac)$_2$, neat, 60°	n-C$_3$H$_7$ —TMS (57)	422
	n-C$_4$H$_9$–O	Cu(acac)$_2$, neat, 75°	n-C$_4$H$_9$–O —TMS (64)	417
	(norbornene)	Cu(acac)$_2$, benzene, 60°	—TMS (60)	422
	(cycloheptene)	Cu(acac)$_2$, neat, 60°	—TMS (53)	422
	styrene	Cu(acac)$_2$, neat, 60°	Ph —TMS (55)	422
	Ph–O–CH=CH$_2$	Cu(acac)$_2$, Et$_2$O, 75°	Ph–O —TMS (46)	417
	(cyclooctene)	Cu(acac)$_2$, neat, 60°	—TMS (50)	422
	n-C$_6$H$_{13}$	Cu(acac)$_2$, neat, 60°	n-C$_6$H$_{13}$ —TMS (49)	422

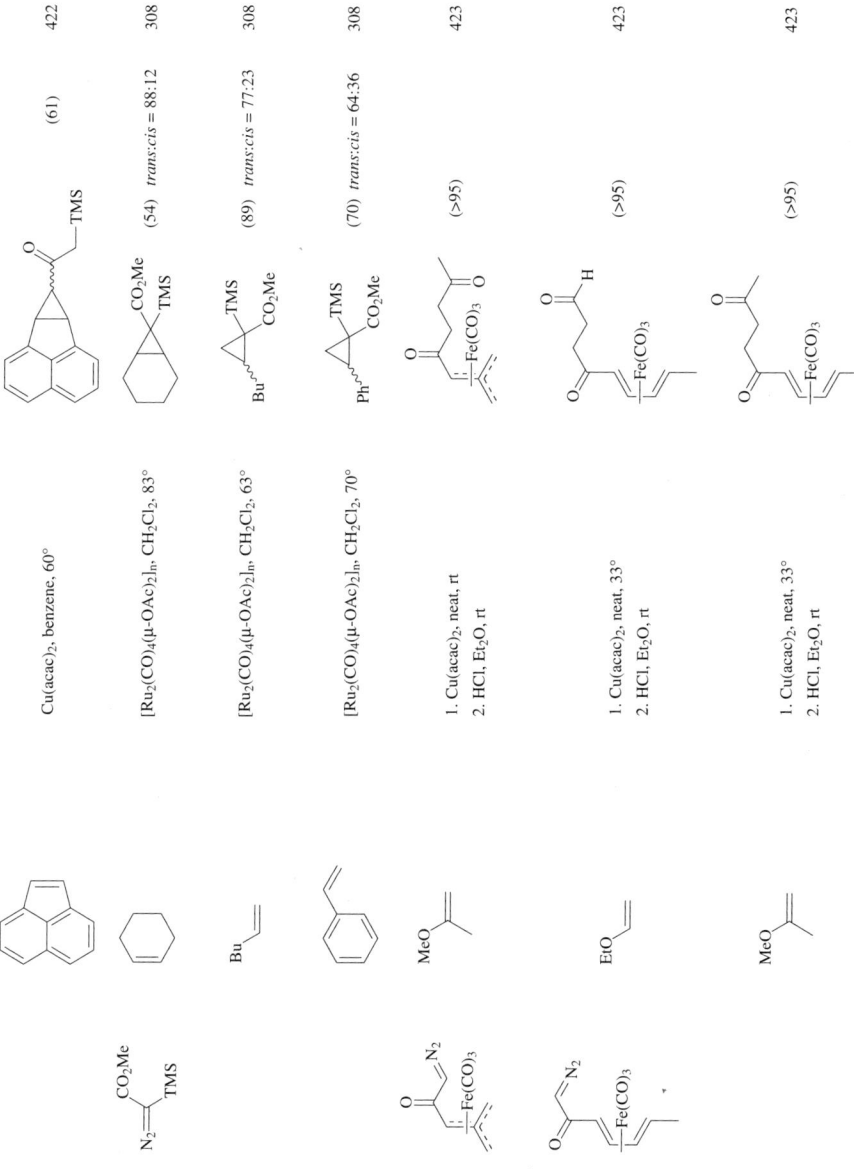

TABLE 1. DIASTEREOSELECTIVE CYCLOPROPANATION OF ALKENES USING DIAZOACETATES AND RELATED SYSTEMS (*Continued*)

Carbenoid Precursor	Substrate	Conditions	Product(s) and Yield(s) (%)	Refs.
C_8 N_2=CH–C(O)–Ph	(CH$_3$)$_2$C=C(CH$_3$)$_2$	Cu(acac)$_2$, neat, 60°	(75)	423
	NC, R / C=CH$_2$ with R	Mo(CO)$_6$, neat	R / Temp / Time H / rt / 70 h (63) *trans:cis* 69:31 H / 65° / 7 h (77) 69:31 Me / rt / 72 h (19) 55:45 Me / 65° / — (58) 55:45	298 298, 299 298 298, 299
	CH$_3$–C(O)–CH=CH$_2$	Mo(CO)$_6$, neat, 65°	(52) *trans:cis* = 71:29	299
	EtO$_2$C–CH=CH$_2$	Mo(CO)$_6$, neat	Temp / Time rt / 70 h (48) *trans:cis* 57:43 65° / 8 h (72) 63:37	298 298, 299
	MeO$_2$C–C(CH$_3$)=CH$_2$	Mo(CO)$_6$, neat, 65°	(85) *trans:cis* = 66:34	299

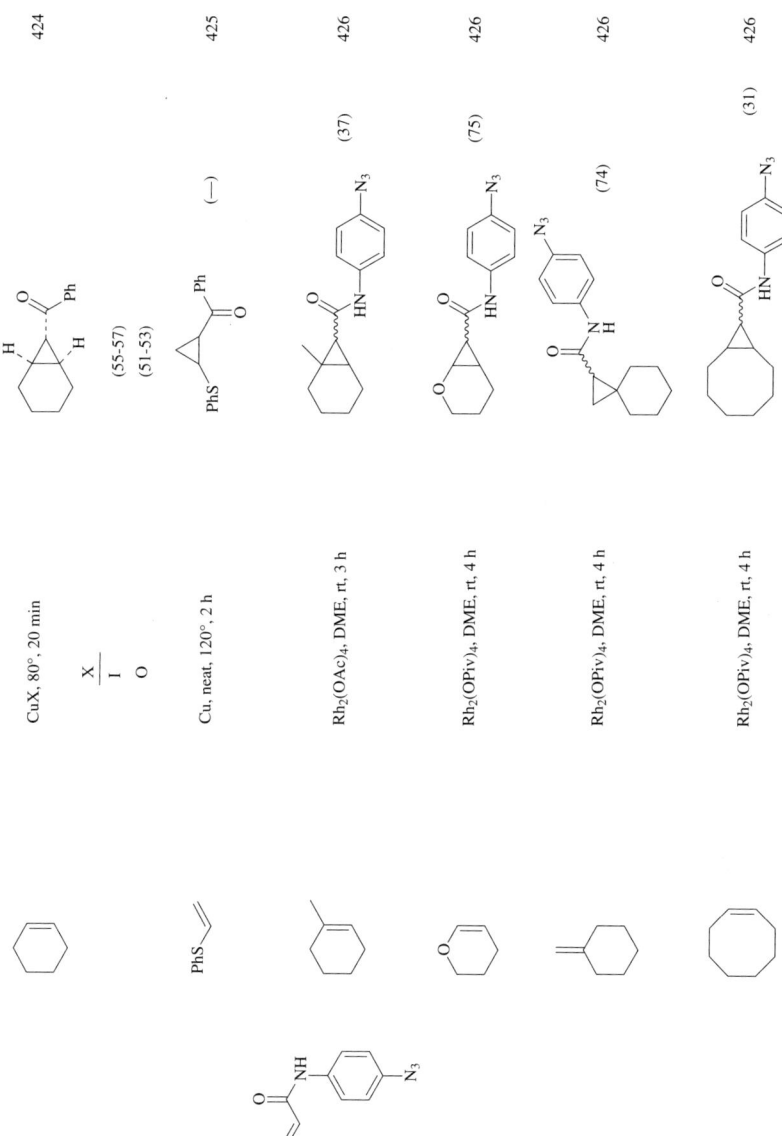

TABLE 1. DIASTEREOSELECTIVE CYCLOPROPANATION OF ALKENES USING DIAZOACETATES AND RELATED SYSTEMS (*Continued*)

Carbenoid Precursor	Substrate	Conditions	Product(s) and Yield(s) (%)	Refs.
		Rh$_2$(OPiv)$_4$, DME, rt, 5 h Rh$_2$(OPiv)$_4$, DME, 80°, 4 h Pd(OAc)$_2$, DME, rt, 4 h	(22) (21) (0)	426
		Rh$_2$(OPiv)$_4$, DME, rt, 4 h Rh$_6$(CO)$_{16}$, DME, rt, 4 h	(6) (2)	426
		Rh$_2$(OPiv)$_4$, DME, rt, 4 h	(30)	426
		Rh$_2$(OAc)$_4$, DME, rt, 1 h	(31)	426
		Rh$_2$(OPiv)$_4$, DME, rt, 4 h	(83)	426

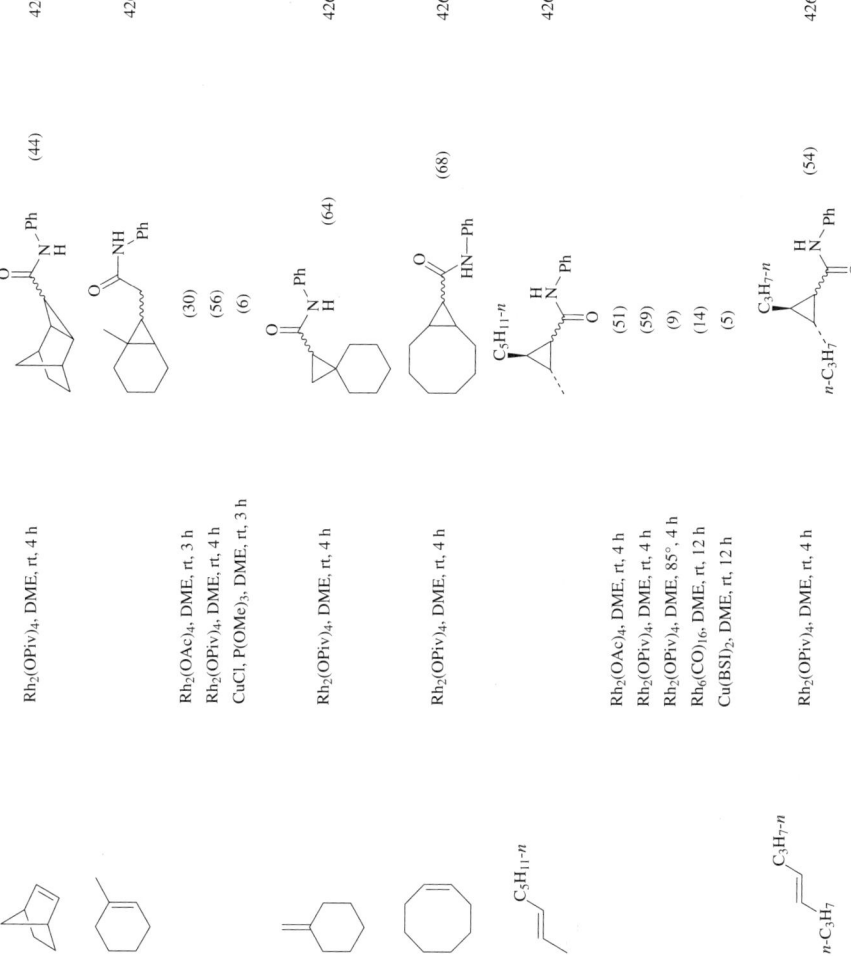

TABLE 1. DIASTEREOSELECTIVE CYCLOPROPANATION OF ALKENES USING DIAZOACETATES AND RELATED SYSTEMS (*Continued*)

Carbenoid Precursor	Substrate	Conditions	Product(s) and Yield(s) (%)	Refs.
		Rh₂(OPiv)₄, DME, rt, 2.5 h	(13)	426
		Rh₂(OPiv)₄, DME, rt, 4 h	(18)	426
		Rh₂(OAc)₄, DME, rt, 4 h Rh₂(OPiv)₄, DME, rt, 4 h	(42) (55)	426
		Pd(OAc)₂, neat, reflux, 1.5 h	(63)	427
		Rh₂(OAc)₄ Pd(OAc)₂	(71) (37)	428, 429
		(Bu₂S)₂CuI, 5–10°	(56)	424

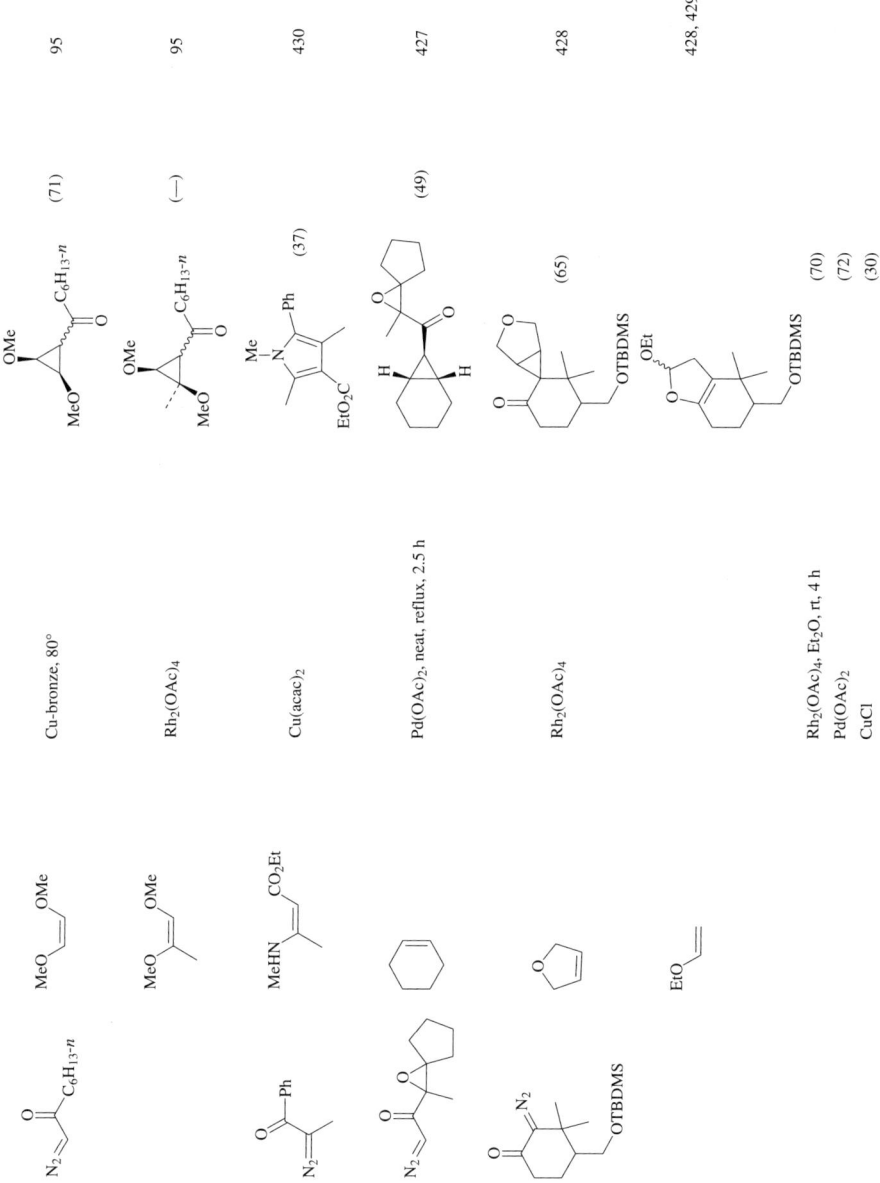

TABLE 1. DIASTEREOSELECTIVE CYCLOPROPANATION OF ALKENES USING DIAZOACETATES AND RELATED SYSTEMS (*Continued*)

Carbenoid Precursor	Substrate	Conditions	Product(s) and Yield(s) (%)	Refs.
C10				
(diazoketone with OTBDMS)	EtO-allyl	Rh₂(OAc)₄ Pd(OAc)₂ CuCl	(75) (72) (35)	428, 429
(α-diazo aryl ketone, OAc)	cyclohexene	Rh₂(OAc)₄, benzene	(50)	431
(α-diazo ketone with Ph-epoxide)	cyclohexene	Pd(OAc)₂, neat, reflux, 0.5 h	(33)	427
C11				
(α-diazo ester CO₂Me)	MeO-CH=CH-OMe	Cu-bronze, 80°	(76)	95
C12				
(α-diazo ketone with Me,Ph-epoxide)	cyclohexene	Pd(OAc)₂, neat, reflux, 5.5 h	(39)	427

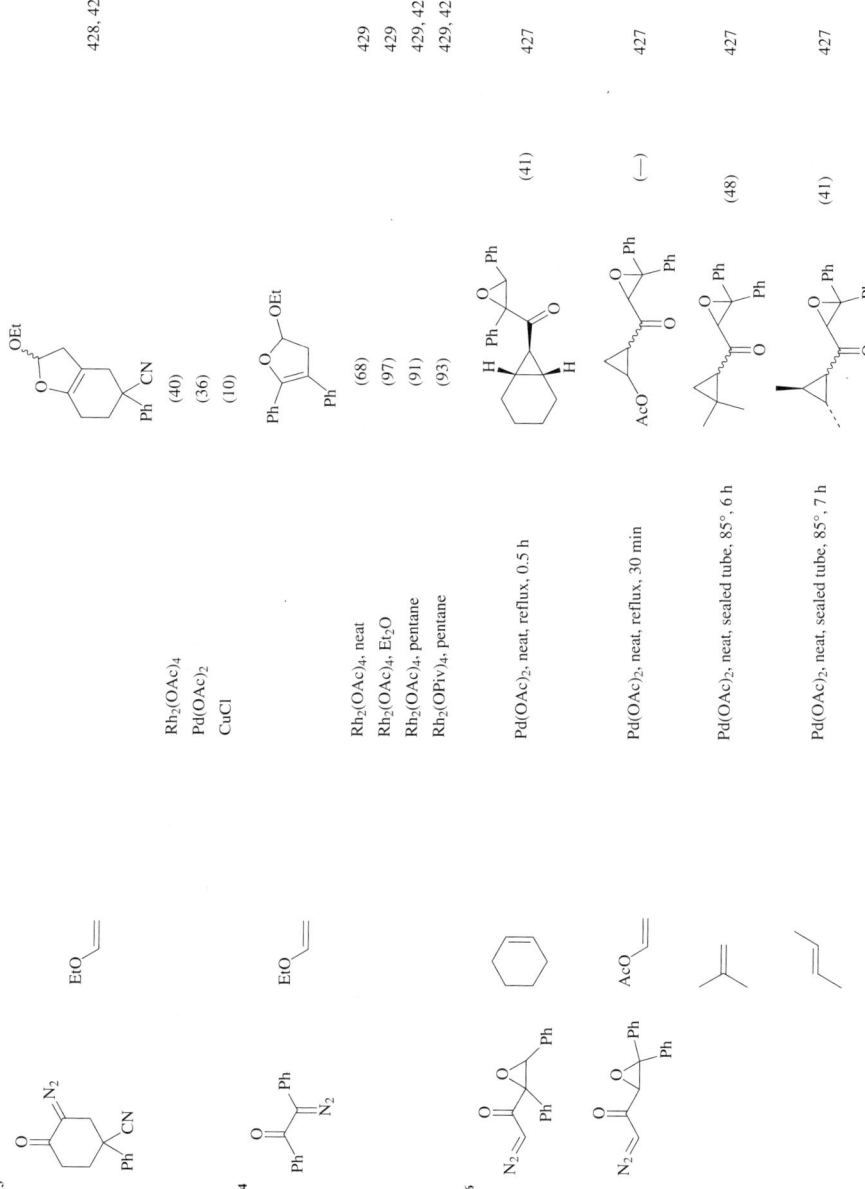

TABLE I. DIASTEREOSELECTIVE CYCLOPROPANATION OF ALKENES USING DIAZOACETATES AND RELATED SYSTEMS (*Continued*)

Carbenoid Precursor	Substrate	Conditions	Product(s) and Yield(s) (%)	Refs.
	(allyl/propene)	Pd(OAc)₂, neat, sealed tube, 85°, 7 h	(38) *trans:cis* = 67:33	427
	cyclopentene	Pd(OAc)₂, neat, reflux, 6 h	(25) *exo:endo* = 83:17	427
	cyclohexene	Pd(OAc)₂, neat, reflux, 5 h	(54)	427
	2,3-dimethyl-2-butene	Pd(OAc)₂, neat, reflux, 6 h	(—)	427
	styrene (Ph)	Pd(OAc)₂, CH₂Cl₂, rt, 10 days	(16)	427
C₃₀ (diazo cyclobutanone with Ph substituents)	cyclohexene	Cu, heat, 20 min	(60)	432

TABLE II. DIASTEREOSELECTIVE CYCLOPROPANATION OF ALKENES USING DIAZOACETOACETATES, DIAZOMALONATES, AND RELATED SYSTEMS

Carbenoid Precursor	Substrate	Conditions	Product(s) and Yield(s) (%)	Refs.

C_2

$\underset{CN}{\overset{NO_2}{N_2{=}{<}}}$

Substrate:
$\underset{R^1 \quad R^2}{\overset{R^3 \quad R^4}{>={<}}}$

R^1	R^2	R^3	R^4
Me	H	Me	H
Me	Me	H	H
Me	H	H	Me
—(CH$_2$)$_4$—		H	H
Me	Me	Me	H
Ph	H	H	H

Conditions: Rh$_2$(OAc)$_4$, CH$_2$Cl$_2$

Products:
$\underset{R^1 \quad R^2}{\overset{O_2N \quad CN}{\underset{R^3 \quad\quad R^4}{\triangle}}}$ **I** + $\underset{R^1 \quad R^2}{\overset{NC \quad NO_2}{\underset{R^3 \quad\quad R^4}{\triangle}}}$ **II**

	I+II	I:II
	(50)	—
	(40)	94:6
	(30)	—
	(40)	>95:5
	(35)	—
	(55)	67:33

Refs: 433

$\underset{CF_3}{\overset{NO_2}{N_2{=}{<}}}$

Substrate:
$\underset{R^1 \quad R^2}{\overset{R^3 \quad R^4}{>={<}}}$

R^1	R^2	R^3	R^4
Me	H	Me	H
Me	Me	Me	Me
Ph	H	H	H

Conditions: Rh$_2$(OAc)$_4$, CH$_2$Cl$_2$

Products:
$\underset{R^1 \quad R^2}{\overset{O_2N \quad CF_3}{\underset{R^3 \quad\quad R^4}{\triangle}}}$ **I** + $\underset{R^1 \quad R^2}{\overset{F_3C \quad NO_2}{\underset{R^3 \quad\quad R^4}{\triangle}}}$ **II**

	I+II	I:II
	(6)	—
	(0)	—
	(30)	50:50

Refs: 433

TABLE II. DIASTEREOSELECTIVE CYCLOPROPANATION OF ALKENES USING DIAZOACETOACETATES, DIAZOMALONATES, AND RELATED SYSTEMS (*Continued*)

Carbenoid Precursor	Substrate		Conditions	Product(s) and Yield(s) (%)	Refs.
$N_2\begin{array}{c}CO_2R\\CO_2R\end{array}$	$\begin{array}{c}R^1\\R^2\end{array}\!\!\!=$			$R^1\!\!\!\!\begin{array}{c}\\\triangle\end{array}\!\!\!\!\begin{array}{c}CO_2R\\CO_2R\end{array}$ with R^2	

C$_3$

R	R^1	R^2			
C$_5$ Me	H	COMe	Cu(acac)$_2$	(10)	434
Me	H	4-ClC$_6$H$_4$	Cu(F$_6$acac)$_2$	(45)	435
Me	H	4-(CN)C$_6$H$_4$	Cu(F$_6$acac)$_2$	(33)	435
Me	H	Ph	Cu(F$_6$acac)$_2$	(57)	435
Me	H	CH$_2$CH$_2$PO(OEt)$_2$	CuSO$_4$	(17)	345
Me	H	4-MeC$_6$H$_4$	Cu(F$_6$acac)$_2$	(60)	435
Me	H	4-MeOC$_6$H$_4$	Cu(F$_6$acac)$_2$	(66)	435
Et	Br	CH$_2$OTBDMS	Rh$_2$(OAc)$_4$	(75)	438
C$_7$ t-Bu	Me	Me	Cu powder	(—)	65
C$_{11}$ t-Bu	H	4-MeOC$_6$H$_4$	CuI•P(OMe)$_3$	(87)	437
C$_{15}$ c-C$_6$H$_{11}$	H	Ph	CuOTf	(76)	436

C$_5$ R = Me	[isopropenylcyclopropane]	Cu(acac)$_2$	[cyclopropane with CO$_2$Me, CO$_2$Me, Me, cyclopropyl substituents] (84)	41, 344
C$_5$ R = Me	[dicyclopropylmethylene]	Cu(acac)$_2$	[cyclopropane with CO$_2$Me, CO$_2$Me, two cyclopropyl substituents] (95)	41, 344

Rh₂(OAc)₄	(74) (83) (80)	439
CuCl	(12) (75-80)	439
Rh₂(OAc)₄, 20° CuCl, 80°	(54) (2) (28) (35)	439
Cu(acac)₂, 80 °C	(68) (87)	434

TABLE II. DIASTEREOSELECTIVE CYCLOPROPANATION OF ALKENES USING DIAZOACETOACETATES, DIAZOMALONATES, AND RELATED SYSTEMS (*Continued*)

Carbenoid Precursor	Substrate	Conditions	Product(s) and Yield(s) (%)	Refs.
R = Me	(vinyl dioxolane)	Cu(acac)$_2$, 80 °C	cyclopropane I + 8-membered II + dioxane III; I (3), II (4), III (30) for R^1=Me; (3),(10),(32) for R^1=Ph	434
R = Me	(allyl dioxolane)	Cu-bronze	(38)	440
R = Me	(butenyl dioxolane)	Cu-bronze	(70)	440
R = Me	(pentenyl dioxolane)	1. Cu-bronze 2. MeOH/H$_2$O/HCl	(49)	440
R = Me	TMS-allene, R^1 = H, Me, n-Bu	Rh$_2$(OAc)$_4$	(33), (79), (80)	441

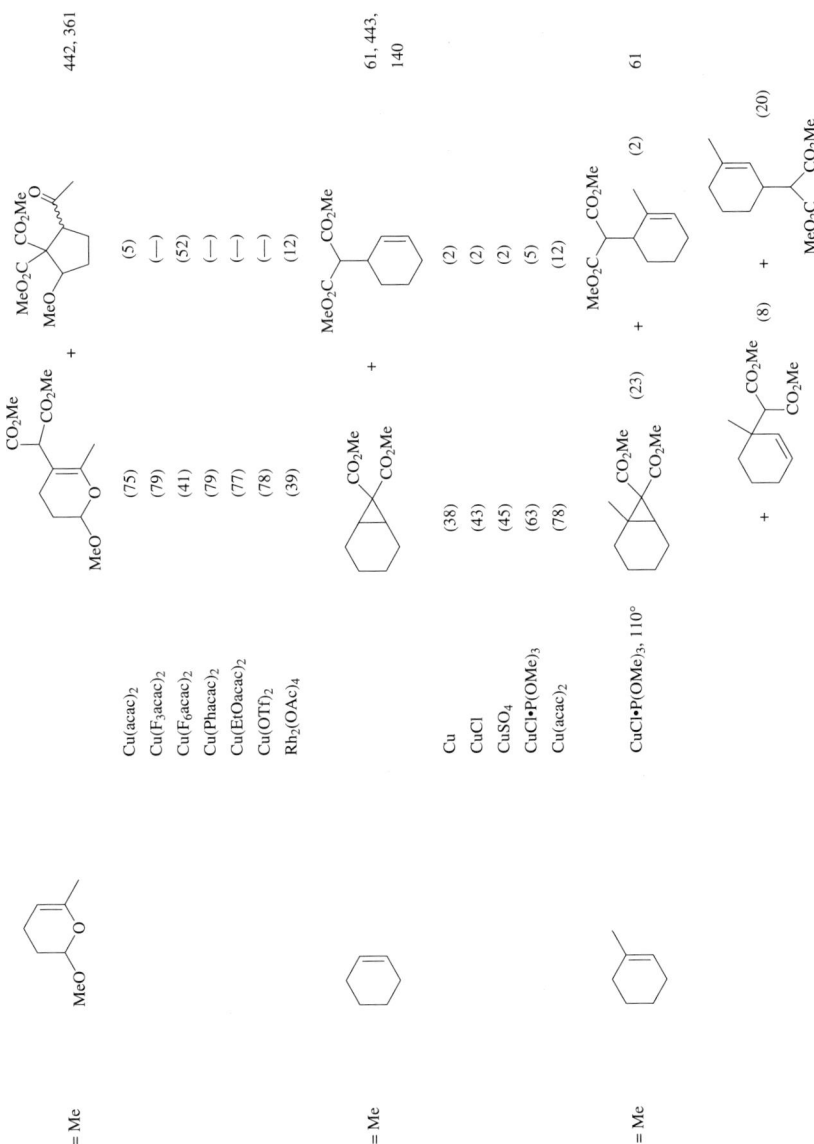

TABLE II. DIASTEREOSELECTIVE CYCLOPROPANATION OF ALKENES USING DIAZOACETOACETATES, DIAZOMALONATES, AND RELATED SYSTEMS (*Continued*)

Carbenoid Precursor	Substrate	Conditions	Product(s) and Yield(s) (%)	Refs.
R = Me	1,2-dimethylcyclohexene	CuCl•P(OMe)$_3$	MeO$_2$C-CH(CO$_2$Me)-cyclohexenyl (74) + isomer (13)	61
R = Me	cyclohexene	CuI•P(OMe)$_3$	bicyclic di(CO$_2$Me) (73)	61, 140
R = Me	propenyl-Bu-n	CuI•P(OMe)$_3$, 98°	MeO$_2$C-C(CO$_2$Me)-cyclopropane-Bu-n (76)	443
R = Me	propenyl-Bu-n (cis)	CuI•P(OMe)$_3$, 98°	MeO$_2$C-C(CO$_2$Me)-cyclopropane-Bu-n (96)	443
R = Me	cycloheptene	CuI•P(OMe)$_3$	bicyclic di(CO$_2$Me) (80)	61, 140
R = Me	1-methoxycyclohexene	CuI•P(OMe)$_3$	MeO$_2$C-C(CO$_2$Me)-OMe-cyclohexene (56)	68
R = Et	PhCH=CH$_2$	Cu powder, 100°	EtO$_2$C-C(CO$_2$Et)-cyclopropane-Ph (35)	444
C$_5$ R = Me	3-methyl-1-methoxycyclohexene	CuI•P(OMe)$_3$	OMe-cyclohexene-CH(CO$_2$Me)$_2$ (39)	68

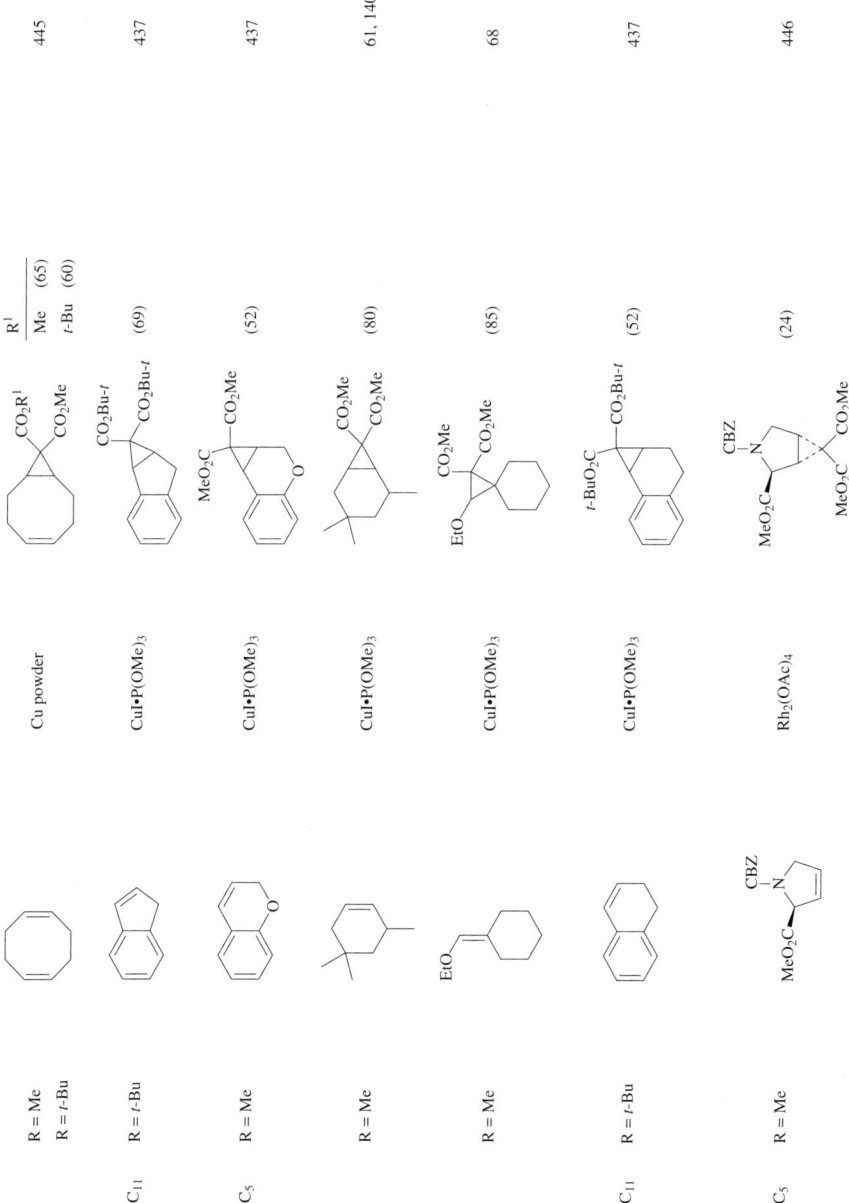

TABLE II. DIASTEREOSELECTIVE CYCLOPROPANATION OF ALKENES USING DIAZOACETOACETATES, DIAZOMALONATES, AND RELATED SYSTEMS (*Continued*)

Carbenoid Precursor	Substrate	Conditions	Product(s) and Yield(s) (%)	Refs.
R = Me		Cu-bronze	(64)	447
		CuSO$_4$	I + II + III (33-45) I:II:III = 66:29:5	448
C$_4$ R = Me		CuSO$_4$	(30-37)	448
C$_5$ R = Et		Rh$_2$(OAc)$_4$, CH$_2$Cl$_2$		72

R^1	R^2	R^3	R^4	
Me	OMe	H	H	(58)
Me	H	OMe	H	(90)
Me	H	H	OMe	(82)
Et	H	Me	Me	(79)
Bu	H	H	H	(62)
Et	OEt	Me	H	(37)
Me	Ph	H	H	(64)

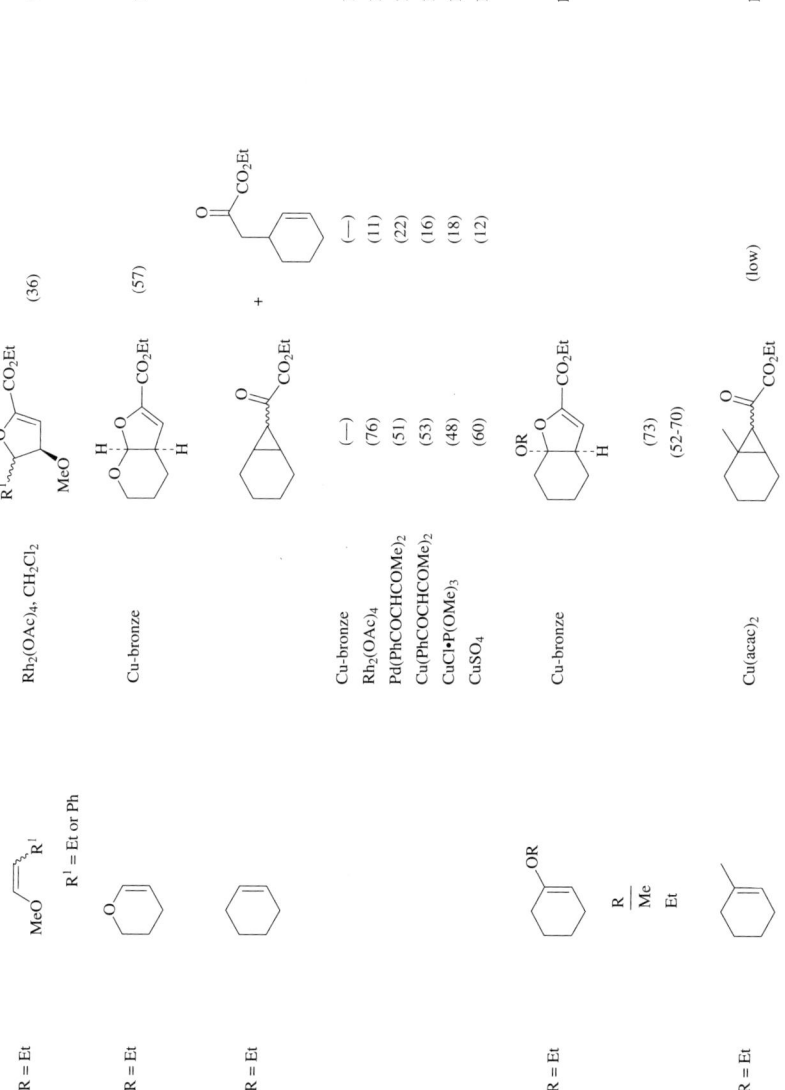

TABLE II. DIASTEREOSELECTIVE CYCLOPROPANATION OF ALKENES USING DIAZOACETOACETATES, DIAZOMALONATES, AND RELATED SYSTEMS (*Continued*)

Carbenoid Precursor	Substrate	Conditions	Product(s) and Yield(s) (%)	Refs.
R = Et	4-R-C₆H₄-CH=CH₂	Cu(acac)₂	I (cis) + II (trans) cyclopropanes; R=H: (49) I+II, 75:25 I:II; R=Cl: (52) I+II, 75:25 I:II	137
R = Et	1-Cl-1-Ph-CH=CH₂	Cu(acac)₂	Cl,Ph-cyclopropane-CO₂Et (low)	137
R = Et	MeO-CH=cyclohexylidene	Cu(acac)₂	MeO-spiro furan CO₂Me (50)	72
C₄ CH₃C(O)C(N₂)C(O)R; R = Me (C₅), OEt (C₆)	3,4-dihydro-2H-pyran	Cu(F₆acac)₂	bicyclic furan-COR; R=Me (72), R=OEt (67)	65
R = Me (C₅), OEt (C₆)	n-BuO-CH=CH₂	Cu(F₆acac)₂	n-BuO-dihydrofuran-COR; R=Me (76), R=OEt (75)	65, 68

166

			I	II	III	IV	
C$_5$	R=OMe, Me	Cu(F$_6$acac)$_2$	(16) (4)	(4) (—)	(6) (12)	(36) (—)	65
C$_6$	R = OEt						65
C$_5$	R=OMe, Me	Cu(F$_6$acac)$_2$ / Cu(acac)$_2$	(25) (27)				65, 68
	R = OMe	Cu(F$_6$acac)$_2$	(21)				65
C$_6$	R = OEt	Cu(F$_6$acac)$_2$	(74)				65

TABLE II. DIASTEREOSELECTIVE CYCLOPROPANATION OF ALKENES USING DIAZOACETOACETATES, DIAZOMALONATES, AND RELATED SYSTEMS (*Continued*)

Carbenoid Precursor	Substrate	Conditions	Product(s) and Yield(s) (%)		Refs.

Carbenoid Precursor:

N_2=C(NO$_2$)(CO$_2$Et)

R^1	R^2	R^3	R^4			I+II	I:II	
Me	H	Me	H	Rh$_2$(OAc)$_4$, CH$_2$Cl$_2$		(75)	—	55, 433
Me	Me	H	H			(65)	80:20	55, 433
Me	H	H	Me			(0)	—	55, 433
OAc	H	H	H			(55)	75:25	55
i-Pr	H	H	H			(22)	57:43	55
—(CH$_2$)$_4$—		H	H			(35)	86:14	55, 433
Me	Me	Me	Me			(0)	—	55, 433
n-Bu	H	H	H			(35)	52:48	55
t-Bu	H	H	H			(3)	44:56	55
Ph	H	H	H			(75)	89:11	55, 433
Ph	H	Me	H			(20)	71:29	55
Ph	H	H	Me			(20)	67:33	55

Products I and II:

I: cyclopropane with O$_2$N and CO$_2$Et on C1, R^3/R^1 and R^4/R^2 on other carbons

II: cyclopropane with EtO$_2$C and NO$_2$ on C1 (opposite stereochemistry)

C$_5$:

N_2=C(H)(CO$_2$Et) (α-diazo ester with CHO)

Substrate: AcO–CH=CH$_2$; Conditions: Rh$_2$(OAc)$_4$

Products: 2-AcO-4-(CO$_2$Et)-2,5-dihydrofuran (13) + cyclopropane with AcO, CO$_2$Et, CHO (17) ; Ref. 132

Substrate: 3,4-dihydro-2H-pyran ; Conditions: Rh$_2$(OAc)$_4$

Products: bicyclic pyranofuran with CO$_2$Et (75) + dioxole with CO$_2$Et and vinyl ether (8) ; Ref. 132

			I+II	I:II
			(80)	—
			(50)	80:20
			(30)	80:20
			(0)	—
			(83)	67:33

TABLE II. DIASTEREOSELECTIVE CYCLOPROPANATION OF ALKENES USING DIAZOACETOACETATES, DIAZOMALONATES, AND RELATED SYSTEMS (*Continued*)

Carbenoid Precursor	Substrate	Conditions	Product(s) and Yield(s) (%)	Refs.

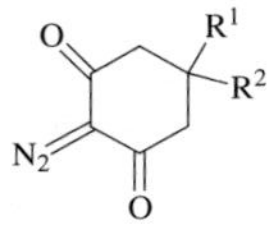

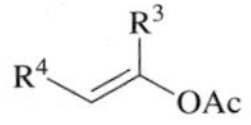

C_6

R^1	R^2		R^3	R^4			
H	H		H	H	1. $Rh_2(OAc)_4$	(69)	133, 134
H	H		Me	H	2. *p*-TsOH, toluene	(62)	134
Me	H		H	H		(59)	134
Me	H		H	Me (*E* + *Z*)		(41)	134
Me	Me		H	H		(71)	134
Me	Me		Me	H		(64)	134

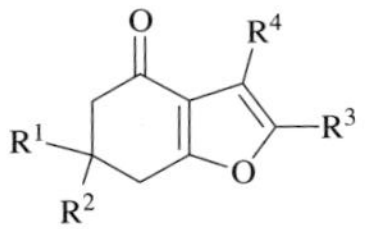

C_8

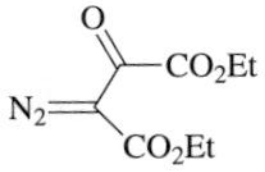

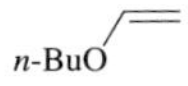

n-BuO-CH=CH$_2$; $Rh_2(OAc)_4$; (95) ; 132

C_8

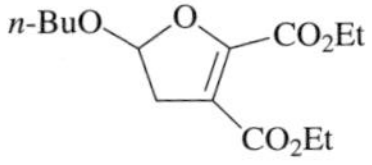

$N_2C(NO_2)(COPh)$

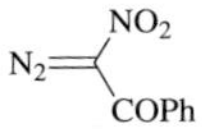

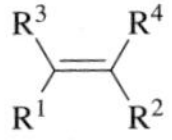

R^1	R^2	R^3	R^4		I+II	I:II	
Me	H	Me	H	$Rh_2(OAc)_4$, CH_2Cl_2	(58)	—	433
Me	Me	H	H		(45)	9:91	
—$(CH_2)_4$—	H	H			(20)	<5:95	
Me	Me	Me	Me		(0)	—	
Ph	H	H	H		(75)	14:86	

C₈						

TABLE II. DIASTEREOSELECTIVE CYCLOPROPANATION OF ALKENES USING DIAZOACETOACETATES, DIAZOMALONATES, AND RELATED SYSTEMS (*Continued*)

Carbenoid Precursor	Substrate	Conditions	Product(s) and Yield(s) (%)	Refs.
	norbornene	Rh$_2$(OAc)$_4$	(16)	450
	methylenecyclohexane	Rh$_2$(OAc)$_4$	(46)	450
	4-R-styrene	Rh$_2$(OAc)$_4$	R = Cl (80-82); Br (52-53); NO$_2$ (46-48); H (63-72); Me (54-59); OMe (0)	450

TABLE III. DIASTEREOSELECTIVE CYCLOPROPANATION OF ALKENES USING VINYL AND PHENYL DIAZOACETATES

Carbenoid Precursor	Substrate		Conditions	Product(s) and Yield(s) (%)		Refs.
	R^1	R^2		cyclopropane (I-type)	cyclopentene (II-type)	
C_4 N_2=C(R)(CO$_2$R), R=						
C_5 R=Me	n-Bu	H	Rh$_2$(OOct)$_4$, pentane	(91)	(0)	212
	Et	H	Rh$_2$(OOct)$_4$, pentane	(87)	(0)	212
	Et	Me	Rh$_2$(OOct)$_4$, pentane	(76)	(0)	451, 143
	n-Bu	H	Rh$_2$(OAc)$_4$, CH$_2$Cl$_2$	(37)	(24)	143
	n-Bu	H	Rh$_2$(OPiv)$_4$, CH$_2$Cl$_2$	(0)	(32)	452, 143
C_8 R=t-Bu	n-Bu	H	Rh$_2$(OPiv)$_4$, pentane	(98)	(0)	451, 143
	n-Bu	H	Rh$_2$(OAc)$_4$, CH$_2$Cl$_2$	(33)	(22)	143
	n-Bu	H	Rh$_2$(OPiv)$_4$, CH$_2$Cl$_2$	(0)	(21)	144
	Et	Me	Rh$_2$(OPiv)$_4$, pentane	(37)	(24)	451, 143
C_{19} R=2,6-(t-Bu)$_2$-4-MeC$_6$H$_2$	n-Bu	H	Rh$_2$(OAc)$_4$, CH$_2$Cl$_2$	(0)	(90)	143
	Et	H	Rh$_2$(OAc)$_4$, CH$_2$Cl$_2$	(0)	(78)	143
	Et	Me	Rh$_2$(OAc)$_4$, CH$_2$Cl$_2$	(0)	(75)	452

Carbenoid Precursor	Substrate	Conditions	Product(s) and Yield(s) (%)	Refs.
C_5 R = Me	2,3-dihydrofuran	1. Rh$_2$(OOct)$_4$, pentane; 2. Et$_2$AlCl, CH$_2$Cl$_2$	**I** (66)	143
C_{19} R = 2,6-(t-Bu)$_2$-4-MeC$_6$H$_2$	2,3-dihydrofuran	Rh$_2$(OAc)$_4$, CH$_2$Cl$_2$	**II** (42)	132

TABLE III. DIASTEREOSELECTIVE CYCLOPROPANATION OF ALKENES USING VINYL AND PHENYL DIAZOACETOACETATES (*Continued*)

Carbenoid Precursor	Substrate	Conditions	Product(s) and Yield(s) (%)	Refs.
C$_8$ R = *t*-Bu	(2,3-dihydrofuran)	Rh$_2$(OOct)$_4$, pentane	(74)	452
C$_{19}$ R = 2,6-(*t*-Bu)$_2$-4-MeC$_6$H$_2$	MeO–CH=CH–Ph (*E*)	Rh$_2$(OAc)$_4$, CH$_2$Cl$_2$	(85)	451
	MeO–CH=CH–Ph (*Z*)	Rh$_2$(OAc)$_4$, CH$_2$Cl$_2$	(44) + (25)	451
	BuO–CH=CH$_2$	Rh$_2$(OAc)$_4$, CH$_2$Cl$_2$	cyclopropane (15)/(0)/(0) + cyclopentene (58)/(0)/(45)	451

C$_4$ N$_2$=C(CO$_2$R^1)–C(R^2)=CH$_2$

	R^1	R^2
C$_6$	Me	Me
C$_9$	Me	*t*-Bu
C$_{11}$	Me	Ph

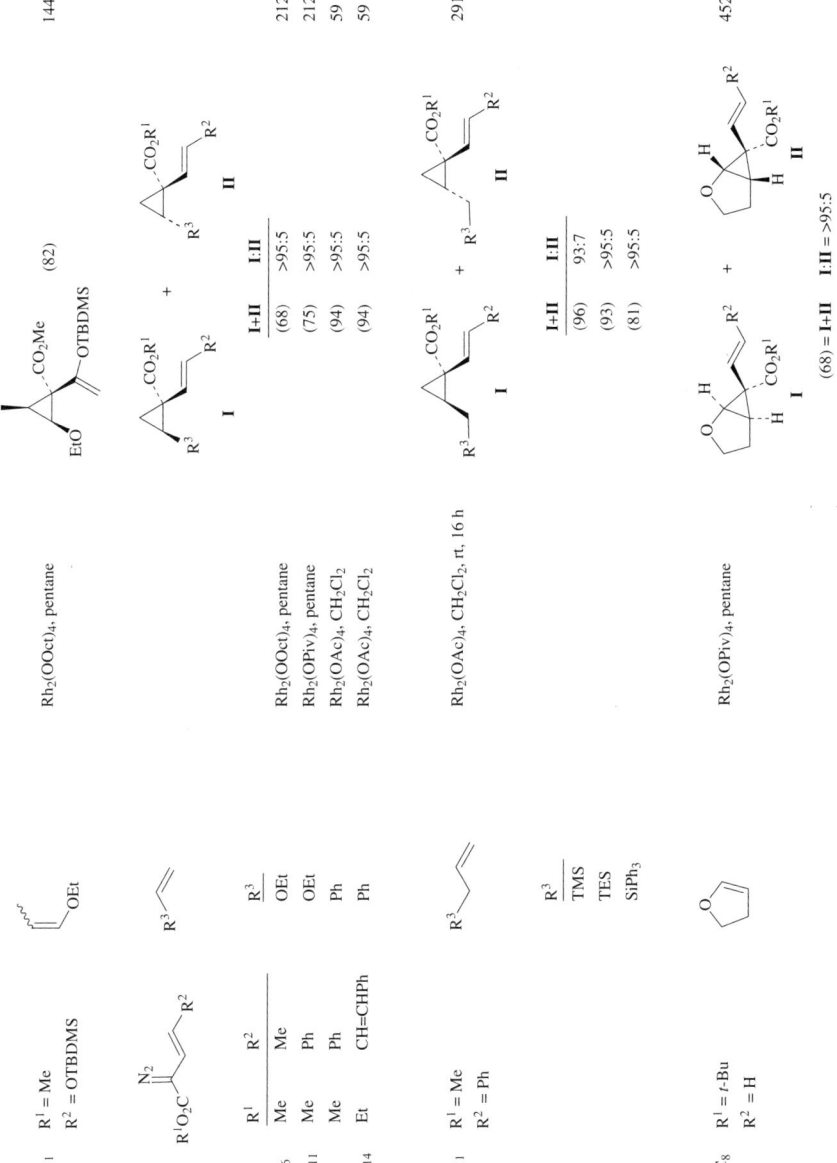

TABLE III. DIASTEREOSELECTIVE CYCLOPROPANATION OF ALKENES USING VINYL AND PHENYL DIAZOACETATES (*Continued*)

Carbenoid Precursor	Substrate	Conditions	Product(s) and Yield(s) (%)	Refs.
R¹ R² Me H (C₅) Me Me (C₆) Me Ph (C₁₁)	(dihydropyran)	Rh₂(OOct)₄, pentane Rh₂(OOct)₄, pentane Rh₂(OPiv)₄, pentane	I + II, I:II (56) >95:5 (73) >95:5 (66) >95:5	212
R¹ = Me, R² = Ph	(Z)-4-MeO-cinnamyl methyl ether	Rh₂(OOct)₄, pentane	(39)	212
R¹ R² R³ Me H Me (C₆) Me H Ph (C₁₁) t-Bu OTBDMS H (C₁₄)	EtO-vinyl ether	Rh₂(OPiv)₄, pentane Rh₂(OPiv)₄, pentane Rh₂(OOct)₄, pentane	(83) (80) (82)	212 212 144

C$_9$

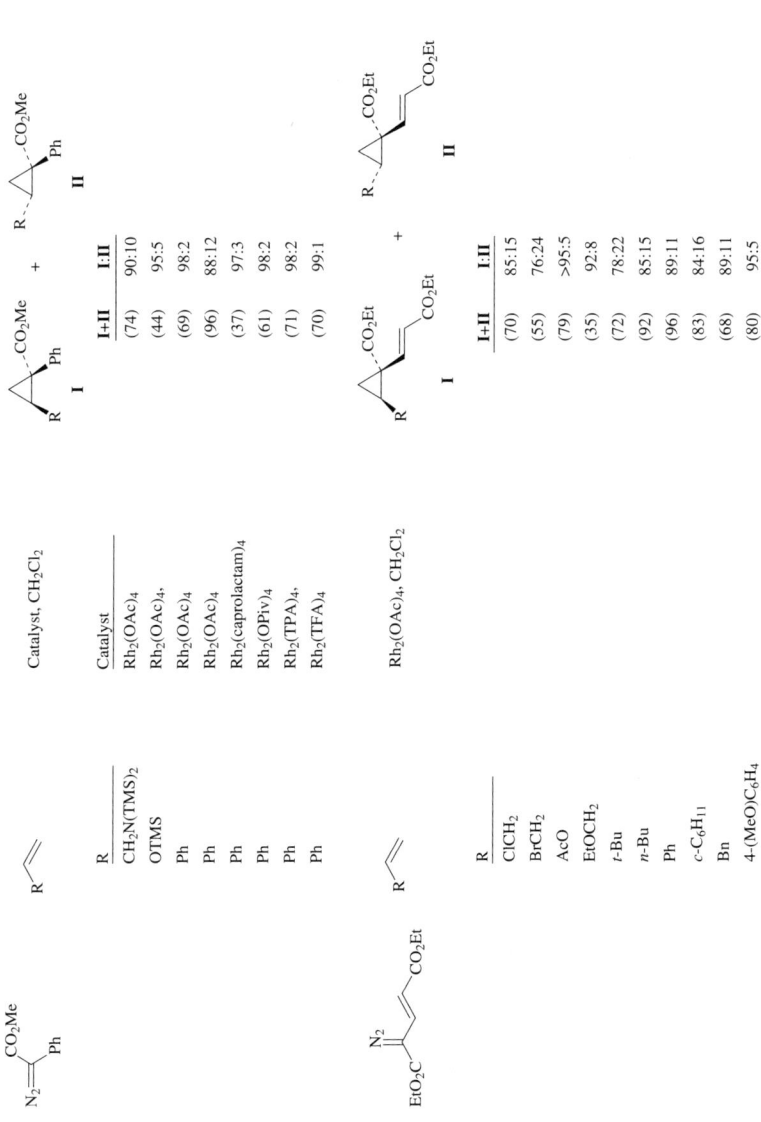

R	I+II	I:II	
CH$_2$N(TMS)$_2$	(74)	90:10	196
OTMS	(44)	95:5	292
Ph	(69)	98:2	146
Ph	(96)	88:12	145
Ph	(37)	97:3	74
Ph	(61)	98:2	146
Ph	(71)	98:2	146
Ph	(70)	99:1	146

R	I+II	I:II	
ClCH$_2$	(70)	85:15	59
BrCH$_2$	(55)	76:24	
AcO	(79)	>95:5	
EtOCH$_2$	(35)	92:8	
t-Bu	(72)	78:22	
n-Bu	(92)	85:15	
Ph	(96)	89:11	
c-C$_6$H$_{11}$	(83)	84:16	
Bn	(68)	89:11	
4-(MeO)C$_6$H$_4$	(80)	95:5	

TABLE III. DIASTEREOSELECTIVE CYCLOPROPANATION OF ALKENES USING VINYL AND PHENYL DIAZOACETATES (*Continued*)

Carbenoid Precursor	Substrate	Conditions	Product(s) and Yield(s) (%)	Refs.
N_2=C(P(OMe)$_2$)(Ph)	R—CH=CH$_2$ (R = n-C$_5$H$_{11}$, CH$_2$TMS)	Cu powder	R-cyclopropane-C(P(OMe)$_2$)(Ph) (73), (76)	453
	cyclohexene	Cu powder	bicyclic-C(P(OMe)$_2$)(Ph) (95)	453
C$_{11}$				
N_2=C(P(OEt)$_2$)(Ph)	cyclohexene	Cu powder	bicyclic H,H-C(P(OEt)$_2$)(Ph) (54) + cyclohexyl-CH(Ph)(P(OEt)$_2$) (4) + bi(cyclohexenyl) (—)	454

TABLE IV. ASYMMETRIC CYCLOPROPANATION OF ALKENES USING CHIRAL AUXILIARIES

	Carbenoid Precursor	Substrate	Conditions	Product(s) and Yield(s) (%)							Refs.

C_2: $N_2\text{=}\overset{CO_2R}{}$

Substrate: $Ph\text{—}CH\text{=}CH_2$

Products I and II: $Ph\cdots\triangle\text{-}CO_2R$ (I) + $Ph\cdots\triangle\text{-}CO_2R$ (II)

					ee (%)		abs stereochem		
	R			(I+II)	I:II	I	II	I	II	Refs.
C_8	(R)-pantolactone		$Rh_2(OAc)_4$, DCE, reflux	(72)	76:24	28	30	—	—	109
C_{12}	d-menthyl		$Rh_2(OAc)_4$, CH_2Cl_2, reflux, 4 h	(79)	68:32	5	13	—	—	109, 108
	l-menthyl		$Rh_2(OAc)_4$, CH_2Cl_2, reflux, 4 h	(82)	68:32	5	13	1R,2R	1R,2S	109, 108
	d-menthyl		Ru, DCE, 60°, 18 h	(72)	94:6	39	5	1S,2S	1R,2S	106
	l-menthyl		Ru, DCE, 60°, 18 h	(83)	97:3	39	17	1R,2R	1S,2R	106
	l-menthyl		CuCl, neat, 50°	(88)	68:32	—	—	—	—	111
	(−)-bornyl		CuCl, neat, 50°	(90)	74:26	—	—	—	—	111
	(+)-bornyl		CuCl, neat, 50°	(—)	74:26	—	—	—	—	111
C_{14}	(+)-trans-2-Ph-c-C_6H_{10}OH		$Rh_2(OAc)_4$, CH_2Cl_2, reflux	(71)	75:25	31	45	—	—	109

C_4: $N_2\text{=}\overset{CO_2R}{\underset{}{}}\text{—OTBDMS}$

Substrate: $R^1\text{—CH=CH—}OR^2$

Products I and II (cyclopropanes with OR^2, OTBDMS, CO_2R, R^1 substituents)

$R = (R)$-pantolactone

R^1	R^2
H	Bu
Me	Et
—($CH_2)_2$—	
Ph	Me
4-ClC_6H_4	Me

Conditions: $Rh_2(OOct)_4$, hexane, reflux

		ee (%)		abs stereochem	
I+II	I:II	I	II	I	II
(76)	95:5	71	—	—	—
(76)	95:5	86	—	—	—
(94)	95:5	68	—	—	—
(79)	95:5	71	—	—	—
(73)	95:5	80	—	—	—

Refs. 456

TABLE IV. ASYMMETRIC CYCLOPROPANATION OF ALKENES USING CHIRAL AUXILIARIES (Continued)

Carbenoid Precursor	Substrate	Conditions	Product(s) and Yield(s) (%)	Refs.
C₈ (isopropyl oxazolidinone diazoacetyl)	Ph–CH=CH₂	1. Rh₂(OAc)₄, CH₂Cl₂, reflux 2. EtONa, EtOH	Ph⋯CO₂Et (I) + Ph⋯CO₂Et (II) (20–24) **I:II** = 64:36 **I** ee = 13% (1R,2R), **II** ee = —	110
C₁₂ (camphorsultam derivative)	Ph–CH=CH₂	Catalyst (Chart 1), CH₂Cl₂, rt	Ph⋯CO₂X_c (I) + Ph⋯CO₂X_c (II)	455

Catalyst	Time	I+II	I:II	de (%) I	de (%) II	abs stereochem I	abs stereochem II
25a	5 h	(60)	94:6	57	99	1S,2S	1S,2R
25b	72 h	(30)	79:21	35	65	1S,2S	1S,2R
25c	72 h	(5)	41:59	28	74	1S,2S	1S,2R
25d	7 d	(50)	74:26	80	50	1R,2R	1R,2S
25e	8 d	(5)	69:31	54	29	1R,2R	1R,2S
25f	24 h	(11)	33:67	82	30	1R,2R	1R,2S
25g	24 h	(33)	73:27	63	65	1S,2S	1S,2R
25g	—	(10)	72:28	67	63	1S,2S	1S,2R
26a	5 h	(60)	96:4	54	84	1S,2S	1S,2R
26b	72 h	(10)	76:24	47	77	1S,2S	1S,2R
27	—	(41)	47:53	25	66	1S,2S	1S,2R

C₁₃ (phenyl oxazolidinone diazoacetyl)	Ph–CH=CH₂	1. Rh₂(OAc)₄, CH₂Cl₂, 22° 2. EtONa, EtOH, 0°	Ph⋯CO₂Et (I) + Ph⋯CO₂Et (II) (35–40) **I:II** = 64:36 **I** ee = 14% (1R,2R), **II** ee = 13% (1R,2S)	110

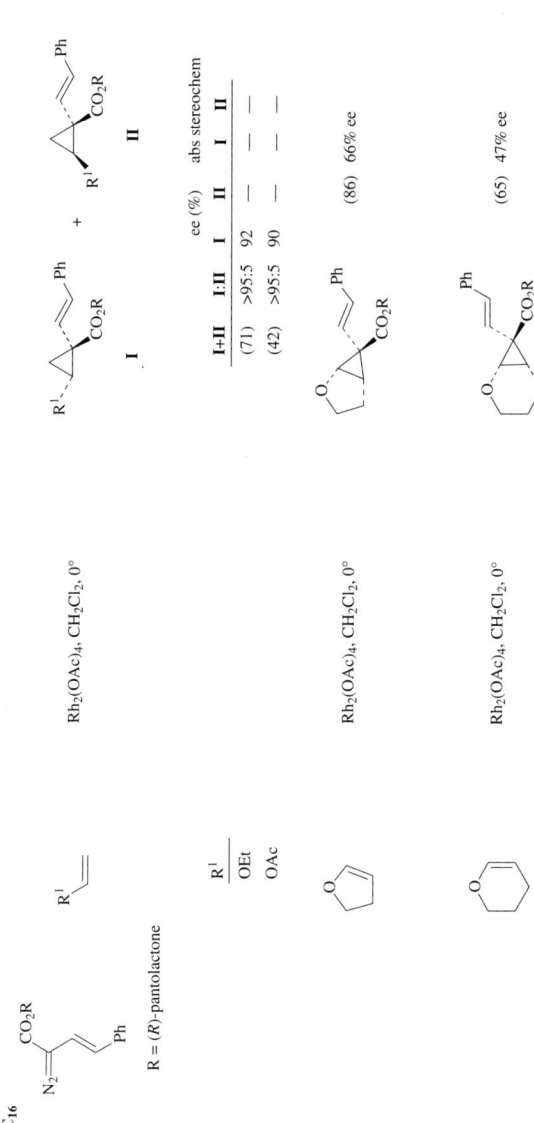

TABLE IV. ASYMMETRIC CYCLOPROPANATION OF ALKENES USING CHIRAL AUXILIARIES (Continued)

	Carbenoid Precursor	Substrate	Conditions	Product(s) and Yield(s) (%)						Refs.
			Catalyst, CH_2Cl_2, reflux	I + II		ee (%)		abs stereochem		
				(I+II)	I:II	I	II	I	II	
C_{12}	d-menthyl		$Rh_2(OAc)_4$	(81)	>95:5	3	—	—	—	147
C_{14}	(±)-5-OH-butyrolactone		$Rh_2(OAc)_4$	(90)	>95:5	42	—	—	—	147, 148
	(S)-methyl lactate		$Rh_2(OAc)_4$	(83)	>95:5	67	—	1S,2S	—	147, 148
	(S)-methyl lactate		$Rh_2(OPiv)_4$	(83)	>95:5	67	—	1S,2S	—	147
C_{15}	(±)-3-OH-methyl butanoate		$Rh_2(OAc)_4$	(77)	>95:5	12	—	—	—	148
C_{16}	(R)-pantolactone		$Rh_2(OAc)_4$	(91)	>95:5	89	—	1R,2R	—	147, 148
	(R)-pantolactone		$Rh_2(O_2CF_3)_4$	(95)	>95:5	78	—	1R,2R	—	147, 148
	(R)-pantolactone		$Rh_2[O(NH)CCH_3]_4$	(37)	>95:5	76	—	1R,2R	—	147, 148
	(R)-pantolactone		$Rh_2(OPiv)_4$	(95)	>95:5	69	—	1R,2R	—	147, 148
	(R)-pantolactone		$Rh_2(OOct)_4$	(84)	>95:5	87	—	1R,2R	—	147, 148
	(R)-pantolactone		$Rh_2(OOct)_4$,[a]	(84)	>95:5	97	—	1R,2R	—	147, 148
	(R)-pantolactone		$Rh_2(O_2CH)_4$	(42)	>95:5	89	—	1R,2R	—	148
	(R)-pantolactone		$Rh_2[(S)-O_2CCH(OH)(Ph)]_4$	(89)	>95:5	17	—	1R,2R	—	147, 148
	(R)-pantolactone		$Rh_2[(R)-O_2CCH(OH)(Ph)]_4$	(95)	>95:5	81	—	1R,2R	—	147, 148
	(S)-2-OH-3-Me-methyl butanoate		$Rh_2(OAc)_4$	(82)	>95:5	78	—	1S,2S	—	148
C_{17}	(S)-2-OH-3,3-Me_2-methyl butanoate		$Rh_2(OAc)_4$	(81)	>95:5	79	—	1S,2S	—	148
	(±)-methyl mandelate		$Rh_2(OAc)_4$	(71)	>95:5	59	—	—	—	147, 148
C_{19}	(±)-methyl mandelate		$Rh_2(OPiv)_4$	(81)	>95:5	60	—	—	—	147

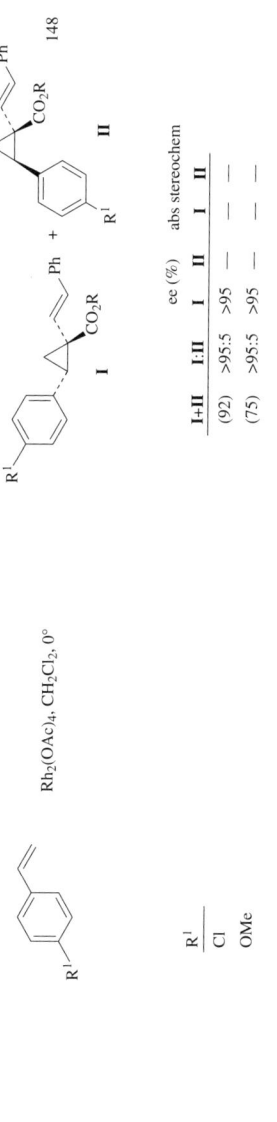

R^1	I+II	I:II	ee (%) I	ee (%) II	abs stereochem I	abs stereochem II
Cl	(92)	>95:5	>95	—	—	—
OMe	(75)	>95:5	>95	—	—	—

[a] This reaction was carried out at 0°.

TABLE V. ASYMMETRIC CYCLOPROPANATION OF ALKENES USING CHIRAL COPPER CATALYSTS

Carbenoid Precursor	Substrate	Conditions	Product(s) and Yield(s) (%)	Refs.

A. C_2-Symmetric Complexes (See Chart 2 for Catalyst Structures)

C_3

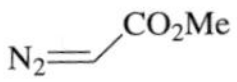

$N_2\!=\!\!\!\diagup\!\!\!CO_2Me$

TMSO—=

TMSO⋯▷⋯CO_2Me (I) + TMSO—▷—CO_2Me (II)

			I+II	I:II	ee (%) I	ee (%) II	abs stereochem I	abs stereochem II	
		28a, benzene, 80°	(43)	75:25	33	12	—	—	457
		34b, DCE, rt, 1 h	(55)	66:34	73	76	—	—	458
		36a•CuOTf, DCE, rt, 3 h	(50)	>97:3	11	—	—	—	458

TBDMSO—=

TBDMSO⋯▷⋯CO_2Me (I) + TBDMSO—▷—CO_2Me (II)

		34b, DCE, rt, 1 h	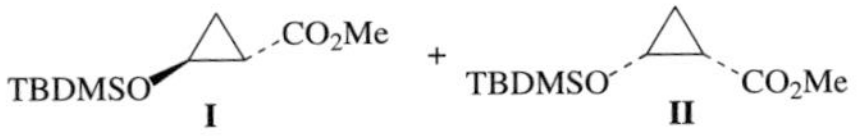 (66) I:II = 68:32 I ee = 75%, II ee = 80%		458

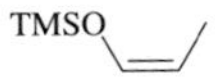

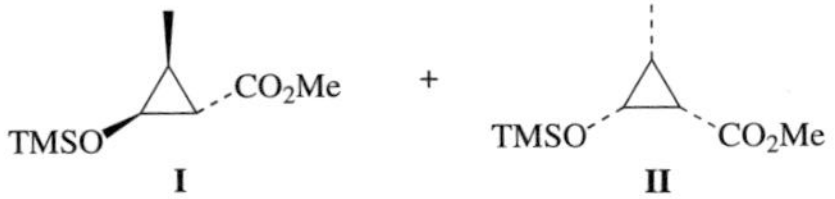

TMSO—

TMSO⋯▷⋯CO_2Me (I) + TMSO—▷—CO_2Me (II)

			I+II	I:II	ee (%) I	ee (%) II	abs stereochem I	abs stereochem II	
		28a, benzene, 70°	(40)	85:15	40	15	1S	1R	457
		34b, DCE, rt, 1 h	(39)	>97:3	49	—	1R	—	458
		36a•CuOTf, DCE, rt, 3 h	(64)	>97:3	28	—	1S	—	458

TMSO–cyclopentene + :CHCO₂Me → I + II

	I+II	I:II	ee (%) I	ee (%) II	abs stereochem I	abs stereochem II	
28a, benzene, 80°	(54)	52:48	43	15	1R	1R	457
28a, CHCl₃, rt	(54)	19:81	11	79	1S,5R,6S	1R,5S,6S	459
29a, CHCl₃, rt	(90)	32:68	43	85	1S,5R,6S	1R,5S,6S	459
34b, CHCl₃, rt	(56)	27:73	87	92	1R,5S,6R	1S,5R,6R	459
85, CHCl₃, rt	(53)	30:70	6	24	1S,5R,6S	1R,5S,6S	459
86, CHCl₃, rt	(35)	59:41	63	41	1R,5S,6R	1S,5R,6R	459

TMSO–cyclohexene + :CHCO₂Me → I + II

	I+II	I:II	ee (%) I	ee (%) II	abs stereochem I	abs stereochem II	
29a, DCE, 82°	(47)	39:61	—	69	1S,6R,7S	1R,6S,7S	459
29a, DCE, rt	(40)	24:76	—	76	1S,6R,7S	1R,6S,7S	
34b, CHCl₃, rt	(7)	73:27	—	11	1R,6S,7R	1S,6R,7R	
34b, toluene, reflux	(13)	22:78	—	18	1R,6S,7R	1S,6R,7R	

TABLE V. ASYMMETRIC CYCLOPROPANATION OF ALKENES USING CHIRAL COPPER CATALYSTS (*Continued*)

Carbenoid Precursor	Substrate	Conditions	Product(s) and Yield(s) (%)							Refs.
						I		**II**		
						ee (%)		abs stereochem		
			I+II	I:II	I	II	I	II		
TMSO / Ph	TMSO / Ph	**28a**, benzene, 80°	(53)	55:45	48	40	—	—		457
		28a, DCE, rt	(74)	72:28	80	63	—	—		459
		29a, DCE, rt	(70)	64:36	64	68	—	—		459
		34b, CHCl₃, rt	(82)	52:48	89	96	—	—		459
		34b, DCE, rt, 1 h	(69)	59:41	56	77	—	—	458, 460
		36a·CuOTf, DCE, rt, 3 h	(0)	—	—	—	—	—		458
						I		**II**		
						ee (%)		abs stereochem		
			I+II	I:II	I	II	I	II		
TMSO / Ph		**34b**, DCE, rt, 1 h	(59)	87:13	22	30	1R	1R	458, 460
		36a·CuOTf, DCE, rt, 3 h	(41)	63:37	23	25	1S	1R	458
TMSO / Ph		**34b**, DCE, rt, 1 h	(0)							458
TMSO		**34b**, DCE, rt, 1 h	(42) I:II = 15:85 I ee = —, II ee = 26%							458

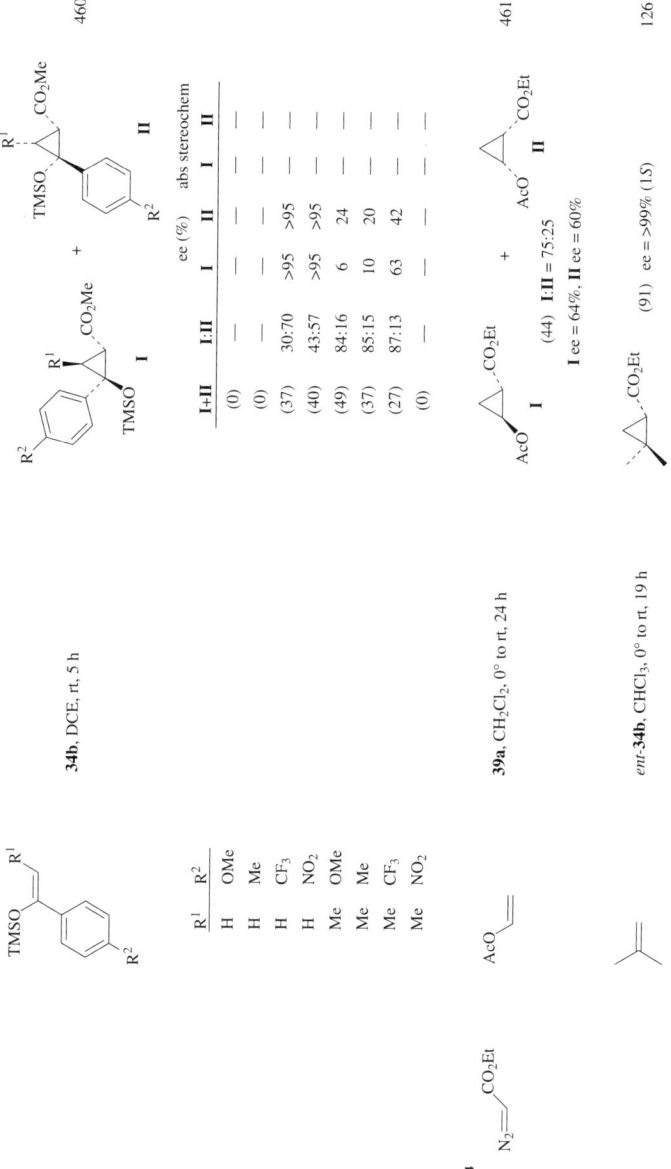

TABLE V. ASYMMETRIC CYCLOPROPANATION OF ALKENES USING CHIRAL COPPER CATALYSTS (*Continued*)

Carbenoid Precursor	Substrate	Conditions	Product(s) and Yield(s) (%)	Refs.
	TMSO-cyclopentene	Catalyst, CHCl$_3$, rt	TMSO-cyclopropane-CO$_2$Et (**I**) + TMSO-cyclopropane-CO$_2$Et (**II**) |		ee (%)		abs stereochem	
	I+II	**I:II**	**I**	**II**	**I**	**II**
29a	(70)	27:73	56	56	1S,5R,6S	1R,5S,6S
34b	(46)	25:75	40	85	1R,5S,6R	1S,5R,6R
85	(60)	35:65	0	0	—	—		459
	TMSO-cyclohexene	**29a**, DCE, rt	TMSO-bicyclic-CO$_2$Et (**I**) + TMSO-bicyclic-CO$_2$Et (**II**) (25) **I:II** = 33:67 **I** ee = —%, **II** ee = 72%	459
	1-hexene	**53b**, AgOTf, CHCl$_3$, rt, 20 h	n-C$_4$H$_9$-cyclopropane-CO$_2$Et (**I**) + n-C$_4$H$_9$-cyclopropane-CO$_2$Et (**II**) (42) **I:II** = 73:27 **I** ee = 76%, **II** ee = 83%	462
	Cl$_2$(CF$_3$)C-CH$_2$-CH=C(CH$_3$)$_2$	**61**, toluene, 80°	Cl$_2$(CF$_3$)C-CH$_2$-cyclopropane(CH$_3$)$_2$-CO$_2$Et (**I**) + Cl$_2$(CF$_3$)C-CH$_2$-cyclopropane(CH$_3$)$_2$-CO$_2$Et (**II**) (—) **I:II** = 40:60 **I** ee = 0%, **II** ee = 26%	463

C$_4$ N_2=CHCO$_2$Et   Ph-CH=CH$_2$

Ph△CO$_2$Et (I) + Ph△``CO$_2$Et (II)

	(I+II)	I:II	ee (%) I	ee (%) II	abs stereochem I	abs stereochem II	
28a, DCE, rt, 16 h	(65)	78:22	85	68	1S,2S	1S,2R	118, 69
28b, DCE, rt, 5 h	(95)	63:37	24	14	1S,2S	1S,2R	464
28b, DCE, 40°, 16 h	(60–80)	75:25	59	45	1S,2S	1S,2R	69, 118
28c, DCE, 60°, 16 h	(60–80)	74:26	23	19	1S,2S	1S,2R	69, 118
29a, CHCl$_3$, rt, 18 h	(80)	75:25	94	68	1S,2S	1S,2R	120
29b, CHCl$_3$, rt, 18 h	(40)	75:25	66	43	1S,2S	1S,2R	120
29c, CHCl$_3$, rt, 18 h	(45)	77:23	95	90	1S,2S	1S,2R	120
30a, PhNHNH$_2$, DCE, rt, 24 h	(72)	71:29	46	31	1R,2R	1S,2R	124
30b, PhNHNH$_2$, DCE, rt, 24 h	(80)	75:25	90	77	1R,2R	1S,2R	124
30c, PhNHNH$_2$, DCE, rt, 24 h	(81)	70:30	60	52	1R,2R	1S,2R	124
30d, PhNHNH$_2$, DCE, rt, 24 h	(76)	71:29	36	15	1R,2R	1S,2R	124
31a, CH$_2$Cl$_2$, rt, 19 h	(—)	64:36	64	48	1R,2R	1R,2S	126
31b, DCE, rt, 5 h	(60)	62:38	69	61	1R,2R	1R,2S	464
ent-**31b**, CHCl$_3$, rt, 19 h	(—)	77:23	98	93	1S,2S	1S,2R	126
31c, DCE, rt, 24 h	(81)	70:30	60	52	1R,2R	1R,2S	465
32a, PhNHNH$_2$, DCE, rt, 24 h	(78)	71:29	28	30	1S,2S	1R,2S	124
32b, PhNHNH$_2$, DCE, rt, 24 h	(88)	75:25	48	36	1S,2S	1R,2S	124
32c, PhNHNH$_2$, DCE, rt, 24 h	(78)	72:28	19	31	1S,2S	1R,2S	124
34a, CH$_2$Cl$_2$, 0° to rt, 19 h	(—)	69:31	49	45	1R,2R	1R,2S	126
ent-**34b**, CH$_2$Cl$_2$, rt, 19 h	(77)	73:27	99	97	1S,2S	1S,2R	126
34b, laponite, CH$_2$Cl$_2$, rt, 19 h	(39)	64:36	69	64	1R,2R	1R,2S	466
ent-**34c**, DCE, rt, 5 h	(90)	70:30	40	42	1S,2S	1S,2R	464
34c, laponite, CH$_2$Cl$_2$, 16 h	(44)	47:53	46	3	1R,2R	1R,2S	466, 467
34d, EtNO$_2$, 23 h	(—)	64:36	45	40	1R,2R	1R,2S	467
34d, laponite, CH$_2$Cl$_2$, rt, 40 h	(39)	55:45	31	26	1R,2R	1R,2S	466, 467
34d, bentonite, CH$_2$Cl$_2$, rt, 23 h	(40)	58:42	34	21	1R,2R	1R,2S	466, 467

TABLE V. ASYMMETRIC CYCLOPROPANATION OF ALKENES USING CHIRAL COPPER CATALYSTS (*Continued*)

Carbenoid Precursor	Substrate	Conditions	Product(s) and Yield(s) (%)				Refs.
	34d, K10 montmorillonite, CH$_2$Cl$_2$, rt, 120 h	(11)	52:48	30	18	1R,2R 1R,2S	466
	35, CHCl$_3$, rt, 19 h	(—)	66:34	3	8	1R,2R 1R,2S	126
	38, CH$_2$Cl$_2$, 0° to rt, 24 h	(49)	64:36	59	30	—	461, 468
	39a, CH$_2$Cl$_2$, 0° to rt, 24 h	(76)	70:30	84	65	1R,2R 1R,2S	461, 468
	39a, CHCl$_3$, 0° to rt, 24 h	(54)	67:33	73	—	1R,2R	461
	39a, neat, 0° to rt, 24 h	(48)	74:26	63	—	1R,2R	461
	39a, THF, 0° to rt, 24 h	(34)	80:20	37	—	1R,2R	461
	39a, toluene, 0° to rt, 24 h	(36)	78:22	59	—	1R,2R	461
	39a, pentane, 0° to rt, 24 h	(—)	—	—	—	—	461
	39b, CH$_2$Cl$_2$, 0° to rt, 24 h	(78)	64:36	2	—	1R,2R	461, 468
	39c, —	(64)	80:20	38	21	1R,2R 1R,2S	469
	*ent-***39c**, —	(69)	79:21	49	34	1S,2S 1S,2R	469
	39d, CH$_2$Cl$_2$, 0° to rt, 24 h	(77)	68:32	50	39	1R,2R 1R,2S	461, 468
	39e, CH$_2$Cl$_2$, 0° to rt, 24 h	(80)	73:27	77	70	1R,2R 1R,2S	461
	40a, CH$_2$Cl$_2$, 0° to rt, 24 h	(77)	67:33	68	73	1R,2R 1R,2S	461
	40a, CH$_2$Cl$_2$, rt, 19 h	(64)	72:28	74	77	1R,2R 1R,2S	470
	40b, CH$_2$Cl$_2$, 0° to rt, 24 h	(85)	70:30	84	85	1R,2R 1R,2S	461
	40b, CH$_2$Cl$_2$, rt, 19 h	(56)	75:25	49	38	1R,2R 1R,2S	470
	41a, CH$_2$Cl$_2$, 0° to rt, 24 h	(53)	70:30	50	—	1R,2R	461
	41b, CH$_2$Cl$_2$, 0° to rt, 24 h	(71)	73:27	74	—	1R,2R	461
	41c, CH$_2$Cl$_2$, 0° to rt, 24 h	(60)	54:46	25	—	1R,2R	461
	41d, CH$_2$Cl$_2$, 0° to rt, 24 h	(59)	67:33	51	—	1R,2R	461
	41e, CH$_2$Cl$_2$, 0° to rt, 24 h	(50)	72:28	69	—	1R,2R	461
	41f, CH$_2$Cl$_2$, 0° to rt, 24 h	(26)	74:26	66	—	1R,2R	461
	42, CH$_2$Cl$_2$, 0° to rt, 24 h	(58)	58:42	18	—	1S,2S	461
	43a, CH$_2$Cl$_2$, rt, 24 h	(72)	74:26	49	59	1R,2R 1R,2S	471
	43b, CH$_2$Cl$_2$, rt, 24 h	(69)	68:32	74	84	1R,2R 1R,2S	471
	44, CH$_2$Cl$_2$, 0° to rt, 24 h	(12)	64:36	8	—	—	461, 468

Entry						
45a, CHCl$_3$, rt, 20 h	(86)	64:36	26	0	1R,2R	1R,2S	472
45b, CHCl$_3$, rt, 20 h	(93)	66:34	28	8	1R,2R	1R,2S	472
46a, CHCl$_3$, rt, 6 h	(73)	63:37	40	41	1R,2R	1R,2S	473
46a, CHCl$_3$, rt, 6 h	(58)	61:39	20	16	1R,2R	1R,2S	473
46b, CHCl$_3$, rt, 6 h	(82)	66:34	15	12	1R,2R	1R,2S	473
46b, CHCl$_3$, 0° to rt, 6 h	(63)	65:35	24	22	1R,2R	1R,2S	473
47, CHCl$_3$, rt, 6 h	(49)	64:36	48	50	1S,2S	1S,2R	473
47, CHCl$_3$, 0° to rt, 6 h	(77)	64:36	67	66	1S,2S	1S,2R	473
50, CH$_2$Cl$_2$, rt, 2 h	(50)	68:32	4	1	1R,2R	1S,2R	474
51, CH$_2$Cl$_2$, rt, 24 h	(5)	61:39	10	32	1R,2R	1S,2R	474
52, CH$_2$Cl$_2$, rt, 2 h	(60)	78:22	4	5	1S,2S	1S,2R	474
53a, CHCl$_3$, rt, 5 h	(50)	80:20	89	74	1R,2R	1R,2S	462
53b, AgOTf, CHCl$_3$, rt, 20 h	(48)	79:21	88	72	1R,2R	1R,2S	462
53b, AgOTf, CHCl$_3$, 0°, 20 h	(52)	80:20	91	82	1R,2R	1R,2S	462
54a, CHCl$_3$, rt, 5 h	(58)	60:40	62	61	1R,2R	1R,2S	475
54b, CHCl$_3$, rt, 5 h	(59)	59:41	87	86	1R,2R	1R,2S	475
54c, CHCl$_3$, rt, 5 h	(43)	67:33	55	57	1R,2R	1R,2S	475
54d, CHCl$_3$, rt, 5 h	(59)	68:32	16	8	1R,2R	1R,2S	475
55a, CHCl$_3$, rt, 5 h	(38)	64:36	4	4	1S,2S	1S,2R	475
55b, CHCl$_3$, rt, 5 h	(52)	70:30	3	3	1S,2S	1S,2R	475
P-ent-56a, CHCl$_3$, rt, 4 h	(—)	69:31	14	18	—	—	476
M-56b, CHCl$_3$, rt, 4 h	(—)	70:30	55	59	1R,2R	1R,2S	476
P-56b, CHCl$_3$, rt, 4 h	(—)	73:27	0	0	—	—	476
M-56c, CHCl$_3$, rt, 4 h	(—)	78:22	49	53	—	—	476
P-56c, CHCl$_3$, rt, 4 h	(—)	80:20	0	0	—	—	476
M-56d, CHCl$_3$, rt, 4 h	(—)	65:35	33	34	—	—	476
P-56d, CHCl$_3$, rt, 4 h	(—)	68:32	8	8	—	—	476
M-56e, CHCl$_3$, rt, 4 h	(—)	67:33	34	13	—	—	476
M-56f, CHCl$_3$, rt, 4 h	(—)	65:35	48	70	—	—	476
P-56g, CHCl$_3$, rt, 4 h	(—)	70:30	66	70	—	—	476

TABLE V. ASYMMETRIC CYCLOPROPANATION OF ALKENES USING CHIRAL COPPER CATALYSTS (*Continued*)

Carbenoid Precursor	Substrate	Conditions	Product(s) and Yield(s) (%)					Refs.
		P-**56i**, CHCl$_3$, rt, 4 h	(—)	67:33	88	89	—	—	476
		P-**56j**, CHCl$_3$, rt, 4 h	(—)	72:28	44	49	—	—	476
		M-**56j**, CHCl$_3$, rt, 4 h	(—)	90:10	0	10	—	—	476
		M-**56k**, CHCl$_3$, rt, 4 h	(—)	85:15	46	49	—	—	476
		57, CH$_2$Cl$_2$, rt	(21)	50:50	0	0	—	—	477
		58, CH$_2$Cl$_2$, rt	(76)	66:34	67	90	1*R*,2*R*	1*R*,2*S*	477
		59a, CHCl$_3$, 0° to rt, 24 h	(82)	75:25	60	52	1*S*,2*S*	1*S*,2*R*	478, 479, 480
		59b, CHCl$_3$, 0° to rt, 24 h	(43)	66:33	8	5	1*S*,2*S*	1*S*,2*R*	478
		59c, CHCl$_3$, 0° to rt, 24 h	(60)	75:25	58	—	1*S*,2*S*	—	480
		59d, CHCl$_3$, 0° to rt, 24 h	(73)	75:25	90	—	1*S*,2*S*	—	480
		60, CHCl$_3$, 0° to rt, 24 h	(60)	66:33	19	20	1*S*,2*S*	1*S*,2*R*	478
		62, —	(—)	—	<2	—	—	—	481
		63a, —	(—)	—	11	—	—	—	481
		63b, —	(—)	—	0	—	—	—	481
		63c, —	(—)	—	0	—	—	—	481
		64, —	(—)	—	11	—	—	—	481
		65, —	(—)	67:33	20	20	—	—	482
		66, PhNHNH$_2$, DCE, rt	(88)	74:26	86	58	1*S*,2*S*	1*S*,2*R*	483
		67a, —	(60)	67:33	12	2	1*R*,2*R*	1*R*,2*S*	484
		67b, —	(55)	74:26	9	24	1*S*,2*S*	1*S*,2*R*	484
		67c, —	(50)	73:27	8	5	1*R*,2*R*	1*S*,2*R*	484
		67d, —	(40)	65:35	2	10	1*R*,2*R*	1*R*,2*S*	484
		67e, —	(25)	68:32	3	9	1*S*,2*S*	1*S*,2*R*	484
		67f, —	(60)	71:29	8	40	1*R*,2*R*	1*S*,2*R*	484
		67g, —	(70)	78:22	20	36	1*R*,2*R*	1*S*,2*R*	484
		67h, —	(70)	75:25	12	25	1*S*,2*S*	1*S*,2*R*	484

Entry	(%)	Ratio			Config	Config	Ref
68a, —	(60)	71:29	4	2	1R,2R	1S,2R	484
68b, —	(60)	74:26	4	5	1S,2S	1R,2S	484
69, neat, 50°, 2 h	(8)	65:35	2	8	1R,2R	1R,2S	485
70, neat, 50°, 2 h	(55)	70:30	18	58	1R,2R	1R,2S	485
71, neat, 50°, 2 h	(60)	70:30	66	70	1R,2R	1R,2S	485
72a, CH$_2$Cl$_2$, rt, 22 h	(75)	79:21	34	47	—	—	486
73, neat, 80°	(—)	—	—	—	—	—	513
74a, DCE, 45°, 18 h	(47)	66:34	1.7	27	1S,2S	1R,2S	487
74b, DCE, rt, 18 h	(49)	67:33	40	73	1R,2R	1R,2S	487
74c, DCE, rt, 18 h	(58)	66:34	70	83	1R,2R	1R,2S	487
74d, DCE, rt, 18 h	(52)	61:39	62	74	1R,2R	1R,2S	487
75, CH$_2$Cl$_2$, rt	I (39)	76:24	73	44	1S,2S	1S,2R	488
76a, —	(—)	—	—	—	—	—	489
76b, —	(17)	—	5	—	1R,2R	—	489
76c, —	(—)	—	—	—	—	—	489
77a, DCE, rt, 23 h	(70)	75:25	69	66	1R,2R	1R,2S	490
77b, DCE, rt, 23 h	(69)	71:29	36	28	1R,2R	1R,2S	490
77c, DCE, rt, 23 h	(91)	65:35	57	51	1R,2R	1R,2S	490
77d, DCE, rt, 23 h	(85)	66:34	64	60	1R,2R	1R,2S	490
77e, DCE, rt, 23 h	(85)	70:30	18	17	1R,2R	1R,2S	490
77f, DCE, rt, 23 h	(83)	62:38	75	85	1R,2R	1R,2S	490
78, —	(—)	—	37	—	—	—	489
79a, DCE, rt, 23 h	(71)	60:40	32	30	1R,2R	1R,2S	491, 464
79b, DCE, rt, 23 h	(78)	58:42	15	15	1R,2R	1R,2S	491
80, DCE, rt, 23 h	(76)	61:39	7	5	1S,2S	1R,2S	491
81, DCE, rt, 23 h	(71)	57:43	2	9	1R,2R	1R,2S	491
82, DCE, rt, 23 h	(72)	62:38	8	14	1R,2R	1R,2S	491
83, DCE, rt, 23 h	(62)	62:38	2	4	1R,2R	1R,2S	491
84, DCE, −20°, 3 h	(—)	50:50	74	90	1R,2R	1S,2R	492
87, CH$_2$Cl$_2$, 12 h	(40)	51:49	12	10	1S,2S	1R,2S	493
88, CH$_2$Cl$_2$, 12 h	(42)	44:56	6	14	1R,2R	1S,2R	493
89a, CH$_3$CN, PhNHNH$_2$, 50°	(41)	87:13	67	58	—	—	494

193

TABLE V. ASYMMETRIC CYCLOPROPANATION OF ALKENES USING CHIRAL COPPER CATALYSTS (*Continued*)

Carbenoid Precursor	Substrate	Conditions	Product(s) and Yield(s) (%)			Refs.
		89a, CH_2Cl_2, $PhNHNH_2$, 50°	(37)	90:10	38 35	494
		89a, DCE, $PhNHNH_2$, 50°	(37)	84:16	48 28	494
		89a, toluene, $PhNHNH_2$, 50°	(35)	88:12	54 50	494
		89b, CH_3CN, $PhNHNH_2$, 50°	(65)	71:29	81 92	494
		89c, CH_3CN, $PhNHNH_2$, 50°	(55)	73:27	86 92	494
		89c, CH_2Cl_2, $PhNHNH_2$, 50°	(—)	—	77 —	494
		89c, THF, $PhNHNH_2$, 50°	(—)	—	64 —	494
		89c, DCE, $PhNHNH_2$, 50°	(—)	—	83 —	494
		89c, DMF, $PhNHNH_2$, 50°	(—)	—	85 —	494
		89d, CH_3CN, $PhNHNH_2$, 50°	(64)	73:27	87 93	494
		89e, CH_3CN, $PhNHNH_2$, 50°	(59)	75:25	87 93	494

$$\text{R-C}_6\text{H}_4\text{-cyclopropyl-CO}_2\text{Et (I)} \quad + \quad \text{R-C}_6\text{H}_4\text{-cyclopropyl-CO}_2\text{Et (II)}$$

					ee (%)		abs stereochem		
R			I+II	I:II	I	II	I	II	
Cl		**53a**, $CHCl_3$, rt, 20 h	(65)	83:17	87	83	—	—	462
Cl		**53b**, AgOTf, $CHCl_3$, rt, 20 h	(61)	84:16	89	84	—	—	462
Me		**53a**, $CHCl_3$, rt, 20 h	(72)	79:21	88	83	—	—	462
Me		**53b**, AgOTf, $CHCl_3$, rt, 20 h	(63)	80:20	88	83	—	—	462
OMe		**53a**, $CHCl_3$, rt, 20 h	(78)	75:25	77	63	—	—	462
OMe		**53b**, AgOTf, $CHCl_3$, rt, 20 h	(60)	75:25	77	63	—	—	462
t-Bu		**28b**, DCE, rt, 5 h	(76)	65:35	42	28	1S,2S	1S,2R	464
t-Bu		**31b**, DCE, rt, 5 h	(87)	66:34	80	66	1R,2R	1R,2S	464
t-Bu		**79a**, DCE, rt, 5 h	(71)	60:40	32	30	1R,2R	1R,2S	464

Alkene	Conditions	Product(s) and ratio	Ref.

Structures and data (reading rows):

- TMSO/Ph alkene; **29a**, DCE, rt → Ph/TMSO cyclopropane-CO₂Et (I) + TMSO/Ph cyclopropane-CO₂Et (II); (81) I:II = 68:32, I ee = 77%, II ee = 77%; 459
- Same alkene; **34b**, CHCl₃, rt → (77) I:II = 56:44, I ee = 90%, II ee = 95%; —

- n-C₅H₁₁ alkene (trans); **48d**, CH₂Cl₂, rt, 2 h → n-C₅H₁₁ cyclopropane-CO₂Et (I) + n-C₅H₁₁ cyclopropane-CO₂Et (II); (64) I:II = 48:52, I ee = 72%, II ee = 75%; 495

- C₃H₇-n alkene; **48d**, CH₂Cl₂, rt, 2 h → n-C₃H₇ cyclopropane-CO₂Et; (59) ee = 56% (2R,3R); 495

- Ph/Me isobutenyl type (Ph, =CH₂ with Me); **66**, PhNHNH₂, DCE, rt → Ph/Me cyclopropane-CO₂Et (I) + Ph/Me cyclopropane-CO₂Et (II); (98) I:II = 58:42, I ee = 84% (1S), II ee = 74% (1S); 483

- cis-Ph alkene; **53b**, AgOTf, CHCl₃, rt, 20 h → Ph cyclopropane-CO₂Et (I) + Ph cyclopropane-CO₂Et (II); (60) I:II = 96:4, I ee = 47%, II ee = 32%; 462

TABLE V. ASYMMETRIC CYCLOPROPANATION OF ALKENES USING CHIRAL COPPER CATALYSTS (*Continued*)

Carbenoid Precursor	Substrate	Conditions	Product(s) and Yield(s) (%)	Refs.

Substrate: Ph-CH=CH- (styrene)

Products: **I** (trans Ph/CO$_2$Et cyclopropane) + **II** (cis Ph/CO$_2$Et cyclopropane)

Conditions	I+II	I:II	ee (%) I	ee (%) II	abs stereochem I	abs stereochem II	Refs.
48d, CH$_2$Cl$_2$, rt, 2 h	(71)	26:74	25	90	1S,2S,3S	1S,2R,3R	495
53b, AgOTf, CHCl$_3$, rt, 20 h	(57)	43:57	49	42	—	—	462
66, PhNHNH$_2$, DCE, rt	(49)	51:49	52	—	1S	—	483

Substrate: 4-MeO-C$_6$H$_4$-CH=CH-

Products: **I** + **II** (aryl = 4-MeOC$_6$H$_4$ cyclopropane carboxylates)

Conditions	I+II	I:II	ee (%) I	ee (%) II	abs stereochem I	abs stereochem II	Refs.
48d, CH$_2$Cl$_2$, rt, 2 h	(48)	39:61	32	91	1S,2S,3S	1S,2R,3R	495
66, PhNHNH$_2$, DCE, rt	(37)	61:39	—	62	—	—	483

Substrate: Ph$_2$C=CH$_2$

Product: Ph$_2$-cyclopropane-CO$_2$Et

Conditions	Yield (%)	ee (%)	abs stereochem	Refs.
31d, PhNHNH$_2$, CH$_2$Cl$_2$, rt	(48)	48	1R	496
ent-**32e**, PhNHNH$_2$, CH$_2$Cl$_2$, rt	(56)	47	1R	496
ent-**32f**, PhNHNH$_2$, CH$_2$Cl$_2$, rt	(54)	23	1S	496
ent-**34b**, CHCl$_3$	(70)	>99	1S	126
53b, AgOTf, CHCl$_3$, rt, 20 h	(56)	83	—	462
59a, CHCl$_3$, 0° to rt, 24 h	(74)	63	—	478, 479

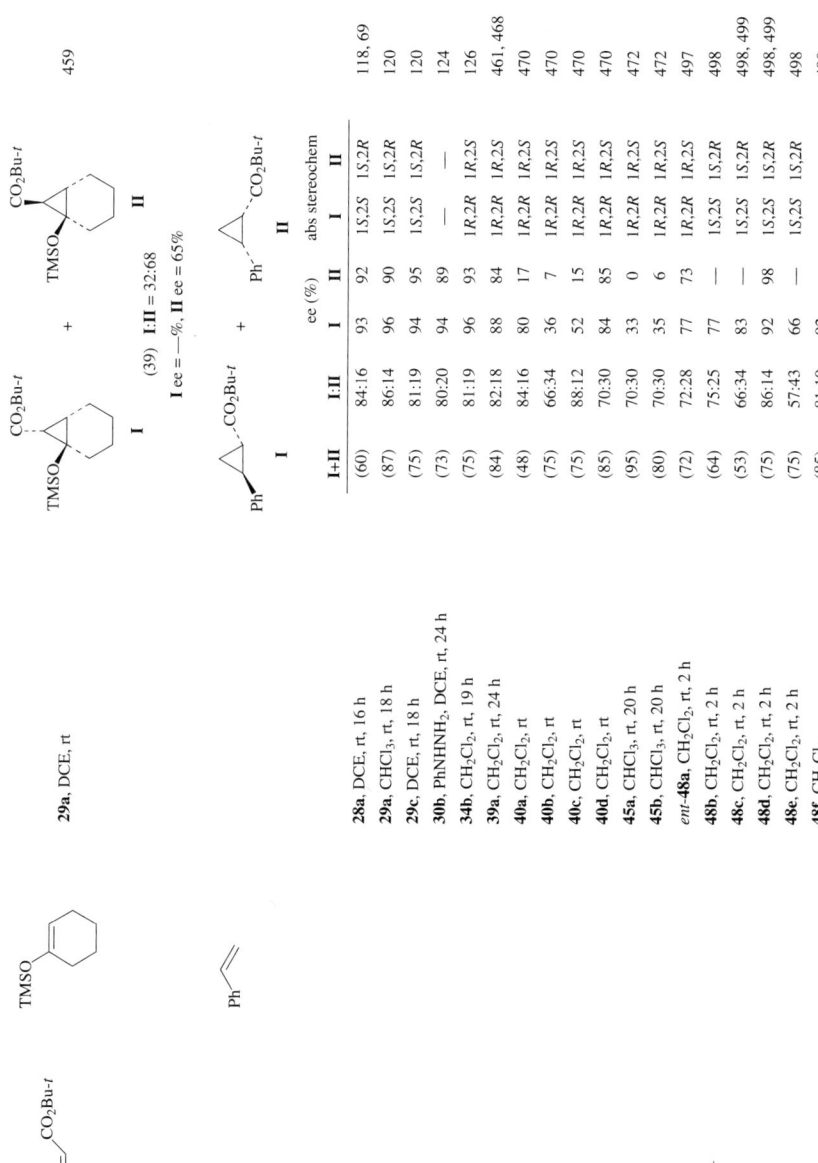

	I+II	I:II	ee (%) I	ee (%) II	abs stereochem I	abs stereochem II	
28a, DCE, rt, 16 h	(60)	84:16	93	92	1S,2S	1S,2R	118, 69
29a, CHCl$_3$, rt, 18 h	(87)	86:14	96	90	1S,2S	1S,2R	120
29c, DCE, rt, 18 h	(75)	81:19	94	95	1S,2S	1S,2R	120
30b, PhNHNH$_2$, DCE, rt, 24 h	(73)	80:20	94	89	—	—	124
34b, CH$_2$Cl$_2$, rt, 19 h	(75)	81:19	96	93	1R,2R	1R,2S	126
39a, CH$_2$Cl$_2$, rt, 24 h	(84)	82:18	88	84	1R,2R	1R,2S	461, 468
40a, CH$_2$Cl$_2$, rt	(48)	84:16	80	17	1R,2R	1R,2S	470
40b, CH$_2$Cl$_2$, rt	(75)	66:34	36	7	1R,2R	1R,2S	470
40c, CH$_2$Cl$_2$, rt	(75)	88:12	52	15	1R,2R	1R,2S	470
40d, CH$_2$Cl$_2$, rt	(85)	70:30	84	85	1R,2R	1R,2S	470
45a, CHCl$_3$, rt, 20 h	(95)	70:30	33	0	1R,2R	1R,2S	472
45b, CHCl$_3$, rt, 20 h	(80)	70:30	35	6	1R,2R	1R,2S	472
ent-48a, CH$_2$Cl$_2$, rt, 2 h	(72)	72:28	77	73	1R,2R	1R,2S	497
48b, CH$_2$Cl$_2$, rt, 2 h	(64)	75:25	77	—	1S,2S	1S,2R	498
48c, CH$_2$Cl$_2$, rt, 2 h	(53)	66:34	83	—	1S,2S	1S,2R	498, 499
48d, CH$_2$Cl$_2$, rt, 2 h	(75)	86:14	92	98	1S,2S	1S,2R	498, 499
48e, CH$_2$Cl$_2$, rt, 2 h	(75)	57:43	66	—	1S,2S	1S,2R	498
48f, CH$_2$Cl$_2$	(85)	81:19	92	—	—	—	499

TABLE V. ASYMMETRIC CYCLOPROPANATION OF ALKENES USING CHIRAL COPPER CATALYSTS (Continued)

Carbenoid Precursor	Substrate	Conditions	Product(s) and Yield(s) (%)					Refs.
	R—⟨=⟩ (R = Cl, OMe)	**49a**, CH_2Cl_2, rt, 2 h	(81)	69:31	72	67	1R,2R 1R,2S	497
		ent-**49b**, CH_2Cl_2	(76)	81:19	94	—	— —	499
		53a, $CHCl_3$, rt, 20 h	(75)	90:10	92	71	1R,2R 1R,2S	462
		54a, $CHCl_3$, rt, 5 h	(48)	72:28	72	71	1R,2R 1R,2S	475
		54b, $CHCl_3$, rt, 5 h	(67)	72:28	89	92	1R,2R 1R,2S	475
		66, $PhNHNH_2$, DCE, rt	(55)	83:17	72	—	1S,2S —	483
		72a, CH_2Cl_2, rt, 22 h	(77)	81:19	50	28	— —	486
		74b, DCE, rt, 18 h	(31)	73:27	37	62	1R,2R 1R,2S	487
		74c, DCE, rt, 18 h	(36)	74:26	73	90	1R,2R 1R,2S	487
		75, CH_2Cl_2, rt	**I** (63)	86:14	87	82	— —	488

R—⟨△⟩-CO₂Bu-t (**I**) + R—⟨△⟩···CO₂Bu-t (**II**)

R		
Cl		
OMe		

		ee (%)		abs stereochem	
	I:II	I	II	I	II
(72)	82:18	95	98	—	—
(73)	90:10	83	>99	—	—

TMSO—C(=)—Ph   **29a**, DCE, rt   Ph⟨△⟩···CO₂Bu-t / TMSO (**I**)   +   TMSO⟨△⟩···CO₂Bu-t / Ph (**II**)   459

(82) **I:II** = 65:35
I ee = 70%, **II** ee = 60%

n-C_6H_{13}—⟨=⟩   **48d**, CH_2Cl_2, rt, 2 h n-C_6H_{13}⟨△⟩···CO₂Bu-t (**I**) + n-C_6H_{13}⟨△⟩···CO₂Bu-t (**II**)   498

(65) **I:II** = 85:15
I ee = 91%, **II** ee = —

TABLE V. ASYMMETRIC CYCLOPROPANATION OF ALKENES USING CHIRAL COPPER CATALYSTS (Continued)

Carbenoid Precursor	Substrate	Conditions	Product(s) and Yield(s) (%)	Refs.
	TMSO–C(Ph)=CHMe	**48b**, DCE, rt, 1 h	Ph(TMSO)–cyclopropane–CO₂Bu-t (I) + TMSO–Ph–cyclopropane–CO₂Bu-t (II) (30) **I:II** = 79:21; **I** ee = 19%, **II** ee = 51%	458
	CH₂=C(Ph)₂	**39a**, CH₂Cl₂, 0°, 21 h	Ph₂–cyclopropane–CO₂Bu-t	461, 468
		53a, CHCl₃, rt, 20 h	ee (%) abs stereochem (90) 92 — (56) 87 —	462
C₉				
N₂=CH–CO₂Me (with Ph)	CH₂=CHPh	**34b**, CHCl₃	Ph–cyclopropane–CO₂Me (I) + Ph–cyclopropane–CO₂Me (II) (54) **I:II** = 99:1; **I** ee = 8%	74
N₂=CH–CO₂CH(Pr-i)₂	CH₂=CHPh	**66**, DCE, PhNHNH₂, rt	Ph–cyclopropane–CO₂CH(Pr-i)₂ (I) + Ph–cyclopropane–CO₂CH(Pr-i)₂ (II) (77) **I:II** = 89:11; **I** ee = 86% (1S,2S)	483
C₁₀				
N₂=CH–CO₂Ar (Ar = 2,6-Me₂C₆H₃)	CH₂=CHPh	**34b**, CHCl₃, rt	Ph–cyclopropane–CO₂Ar (I) + Ph–cyclopropane–CO₂Ar (II) (68) **I:II** = 86:14 Ar = 2,6-Me₂C₆H₃ **I** ee = 97% (1R,2R), **II** ee = 96% (1R,2S)	126

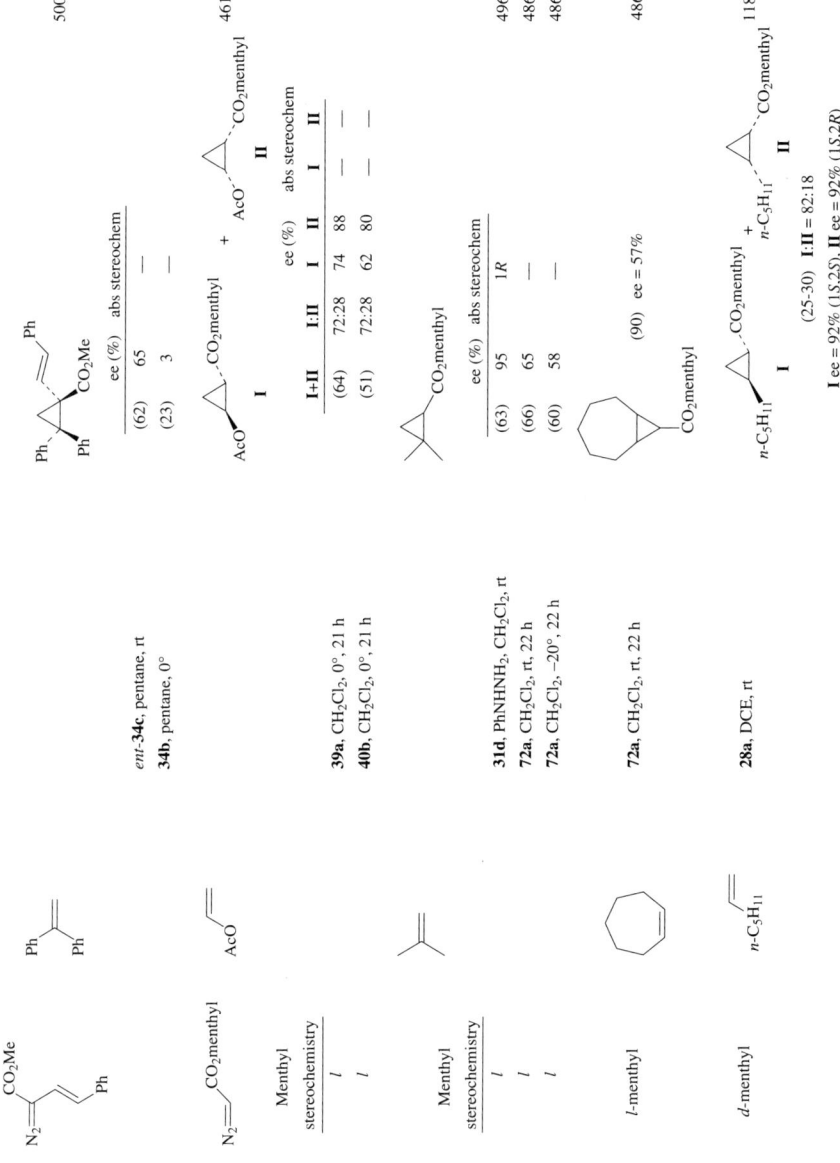

TABLE V. ASYMMETRIC CYCLOPROPANATION OF ALKENES USING CHIRAL COPPER CATALYSTS (Continued)

Carbenoid Precursor	Substrate	Conditions	Product(s) and Yield(s) (%)	Refs.
l-menthyl	t-Bu⁀	**36a**, CH_2Cl_2, 0°, 14 h	(75) **I:II** = 88:12 **I** ee = 95%, **II** ee = 80% **I** t-Bu-△-CO₂menthyl + **II** t-Bu-△-CO₂menthyl	125

Menthyl stereochemistry		Conditions	I+II	I:II	ee (%) I	ee (%) II	abs stereochem I	abs stereochem II	Refs.
	Ph⁀								
							Ph-△-CO₂menthyl **I** + Ph-△-CO₂menthyl **II**		
d		**28a**, DCE, rt, 16 h	(60-70)	82:18	97	95	1S,2S	1S,2R	118, 69
l		**28a**, DCE, rt, 16 h	(65-75)	85:15	91	90	1S,2S	1S,2R	118, 69
d		**29a**, DCE, rt	(89)	84:16	98	99	1S,2S	1S,2R	120
d		**29c**, DCE, rt	(75)	84:16	98	99	1S,2S	1S,2R	120
d		**30b**, PhNHNH₂, DCE, rt	(71)	84:16	98	80	1S,2S	—	124
d		**30b**, PhNHNH₂, DCE, rt	(77)	86:14	98	96	1S,2S	—	124
l		**31a**, DCE, rt	(60-80)	84:16	13	5	1R,2R	1R,2S	121
l		**31b**, DCE, rt	(60-80)	86:14	19	5	1R,2R	1R,2S	121
l		**31b**, PhNHNH₂, DCE, rt	(60-80)	87:13	96	97	1R,2R	1R,2S	121
l		**31c**, DCE, rt	(60-80)	83:17	55	50	1R,2R	1R,2S	121
l		**31d**, DCE, rt	(60-80)	86:14	19	9	1R,2R	1R,2S	121
d		ent-**31e**, DCE, rt	(60-80)	83:17	90	90	1S,2S	1S,2R	121
d		ent-**31e**, PhNHNH₂, DCE, rt	(60-80)	83:17	78	71	1S,2S	1S,2R	121
l		**31f**, DCE, rt	(60-80)	85:15	9	9	1R,2R	1R,2S	121
l		ent-**33**, DCE, rt	(60-80)	82:18	34	56	1R,2R	1R,2S	121
l		**39a**, CH_2Cl_2, 0°, 24 h	(86)	85:15	89	89	1R,2R	1R,2S	461, 468
l		**43a**, CH_2Cl_2, rt, 24 h	(60)	79:21	70	87	1R,2R	1R,2S	471
l		**43b**, CH_2Cl_2, rt, 24 h	(60)	81:19	84	92	1R,2R	1R,2S	471

202

	Conditions	(%)	ratio	%ee	%ee			Ref.
d	54a, CHCl$_3$, rt, 5 h	(51)	84:16	75	52	1R,2R	1R,2S	475
l	54a, CHCl$_3$, rt, 5 h	(50)	68:32	87	91	1R,2R	1R,2S	475
d	54b, CHCl$_3$, rt, 5 h	(68)	86:14	90	87	1R,2R	1R,2S	475
l	54b, CHCl$_3$, rt, 5 h	(60)	68:32	95	97	1R,2R	1R,2S	475
d	66, PhNHNH$_2$, DCE, rt	(55)	90:10	88	—	1S,2S	1S,2R	483
l	66, PhNHNH$_2$, DCE, rt	(86)	91:9	94	—	1S,2S	1S,2R	483
d	66, PhNHNH$_2$, DCE, 0°	(50)	93:7	96	66	1S,2S	1S,2R	483
l	ent-70, neat, 50°, 3 h	(68)	89:11	38	11	1S,2S	1S,2R	485
l	ent-70, neat, 50°, 3 h	(67)	82:18	76	79	1R,2R	1R,2S	485
l	72a, CH$_2$Cl$_2$, rt, 22 h	(81)	83:17	57	43	—	—	486
l	72a, CH$_2$Cl$_2$, 0°, 22 h	(92)	85:15	55	63	—	—	486
l	ent-72a, CH$_2$Cl$_2$, rt, 22 h	(96)	82:18	57	85	1R,2R	1R,2S	486
l	72b, CH$_2$Cl$_2$, rt, 22 h	(85)	75:25	39	6	—	—	486
l	72c, CH$_2$Cl$_2$, rt, 22 h	(95)	81:19	50	41	—	—	486
l	72d, CH$_2$Cl$_2$, rt, 22 h	(84)	69:31	15	6	—	—	486
l	72e, CH$_2$Cl$_2$, rt, 22 h	(87)	73:27	36	9	—	—	486
l	72f, CH$_2$Cl$_2$, rt, 22 h	(89)	71:29	37	25	—	—	486
d	74b, DCE, rt, 18 h	(56)	87:13	59	49	1R,2R	1R,2S	487
l	74b, DCE, rt, 18 h	(63)	81:19	73	84	1R,2R	1R,2S	487
d	74c, DCE, rt, 18 h	(63)	88:12	87	86	1R,2R	1R,2S	487
l	74c, DCE, rt, 18 h	(64)	77:23	90	99	1R,2R	1R,2S	487
d	75, CH$_2$Cl$_2$, rt	I (67)	88:12	87	81	1S,2S	1S,2R	488
l	75, CH$_2$Cl$_2$, rt	I (66)	85:15	89	84	1S,2S	1S,2R	488
d	79, DCE, rt, 23 h	(73)	76:24	5	23	1R,2R	1R,2S	491
l	79, DCE, rt, 23 h	(74)	78:22	30	34	1R,2R	1R,2S	491
d	82, DCE, rt, 23 h	(74)	82:18	8	10	1S,2S	1R,2S	491

TABLE V. ASYMMETRIC CYCLOPROPANATION OF ALKENES USING CHIRAL COPPER CATALYSTS (*Continued*)

Carbenoid Precursor	Substrate	Conditions	Product(s) and Yield(s) (%)	Refs.

Menthyl stereochemistry: *d*, *l*

Substrate: CH₂=C(t-Bu)CH₃ (isopropenyl t-Bu)

Conditions: **30b**, PhNHNH₂, DCE, rt, 24 h

Products:

I (t-Bu cyclopropane-CO₂R) + II (t-Bu cyclopropane-CO₂R), R = menthyl

		ee (%)		abs stereochem	
I+II	I:II	I	II	I	II
(55)	98:2	77	—	—	—
(60)	95:5	80	91	—	—

Ref. 124

Menthyl stereochemistry: *d*, *l*

Substrate: n-C₆H₁₃–CH=CH₂

Conditions: **30b**, PhNHNH₂, DCE, rt, 24 h

Products: n-C₆H₁₃–cyclopropane–CO₂R (I) + n-C₆H₁₃–cyclopropane–CO₂R (II), R = menthyl

		ee (%)		abs stereochem	
I+II	I:II	I	II	I	II
(72)	90:10	75	45	—	—
(76)	94:6	99	30	—	—

Ref. 124

Menthyl stereochemistry: *d*, *l*

Substrate: n-C₃H₇–CH=CH–C₃H₇-n

Conditions: **30b**, PhNHNH₂, DCE, rt, 24 h

Product: C₃H₇-n / n-C₃H₇ cyclopropane–CO₂R, R = menthyl

ee (%)	abs stereochem
(50) 85	—
(52) 88	—

Ref. 124

Menthyl stereochemistry: *l*-menthyl

Substrate: ethylidenecyclohexane

Conditions: **36a**, CH₂Cl₂, 0°, 14 h

Products: spiro[cyclohexane-cyclopropane]–CO₂R (I) + spiro[cyclohexane-cyclopropane]–CO₂R (II), R = menthyl

(54) I:II = 86:14

I ee = 82%

Ref. 125

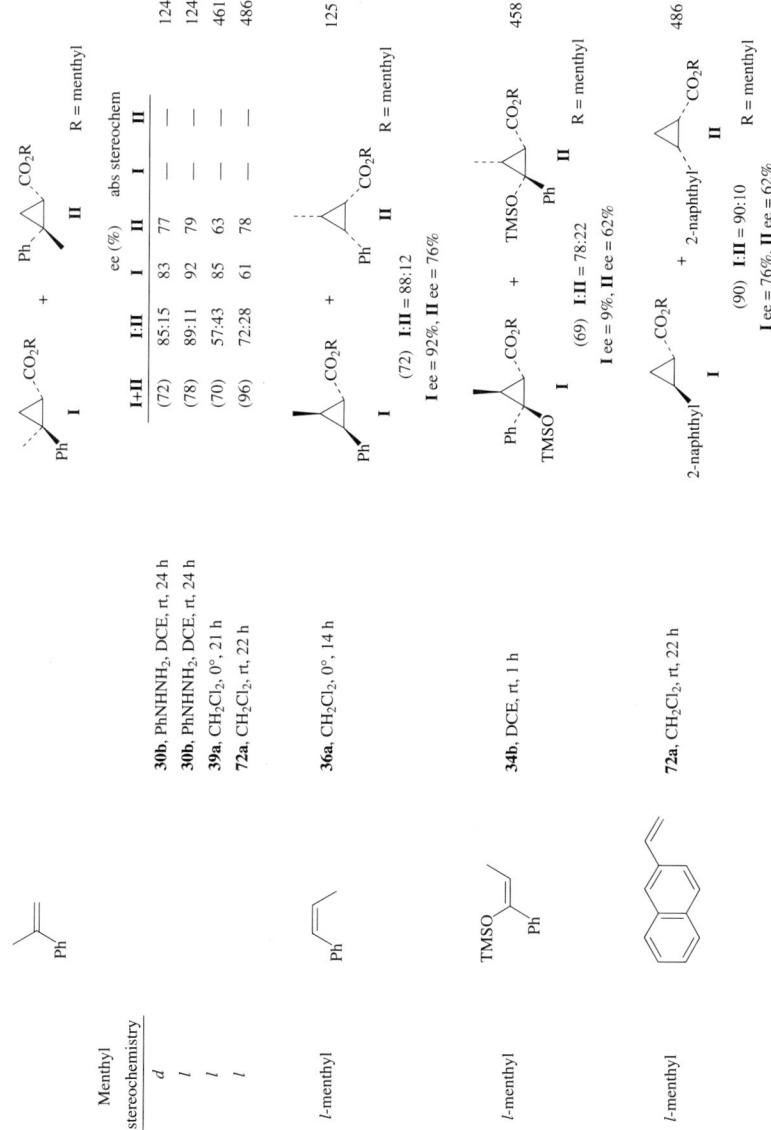

TABLE V. ASYMMETRIC CYCLOPROPANATION OF ALKENES USING CHIRAL COPPER CATALYSTS (*Continued*)

Carbenoid Precursor	Substrate	Conditions	Product(s) and Yield(s) (%)			Refs.

Menthyl stereochemistry				R = menthyl		
	Ph₂C=CH₂ (Ph, Ph)		Ph—CO₂R cyclopropane (Ph, Ph)			
				ee (%)	abs stereochem	
l		**31d**, PhNHNH₂, CH₂Cl₂, rt	(58)	82	1*R*	496
l		**72a**, CH₂Cl₂, rt, 22 h	(69)	74	1*S*	486
l		**72a**, CH₂Cl₂, 0°, 22 h	(75)	90	1*S*	486
l		**72a**, CH₂Cl₂, −20°, 22 h	(80)	95	1*S*	486
l		**72a**, CH₂Cl₂, −40°, 22 h	(83)	98	1*S*	486
l		**72a**·Cu(MeCN)₄PF₆, CH₂Cl₂, rt, 22 h	(80)	93	1*S*	486
l		**72a**·Cu(MeCN)₄PF₆, CH₂Cl₂, −20°, 22 h	(71)	96	1*S*	486

Menthyl stereochemistry				R = menthyl Ar = 4-R¹C₆H₄		
	(4-R¹-C₆H₄)₂C=CH₂	**72a**, CH₂Cl₂, rt, 22 h	Ar—CO₂R cyclopropane			486
	R¹			ee (%)	abs stereochem	
l	Cl		(82)	96	—	
l	OMe		(43)	94	—	

| *l*-menthyl | Ph,Ph-CH=CH-C₃H₇-*n* | **36a**, CH₂Cl₂, 0°, 14 h | I (Ph,Ph,C₃H₇-*n*,CO₂R) + II (Ph,Ph,C₃H₇-*n*,CO₂R) R = menthyl
(52) **I:II** = 98:2
I ee = 84% | | | 125 |

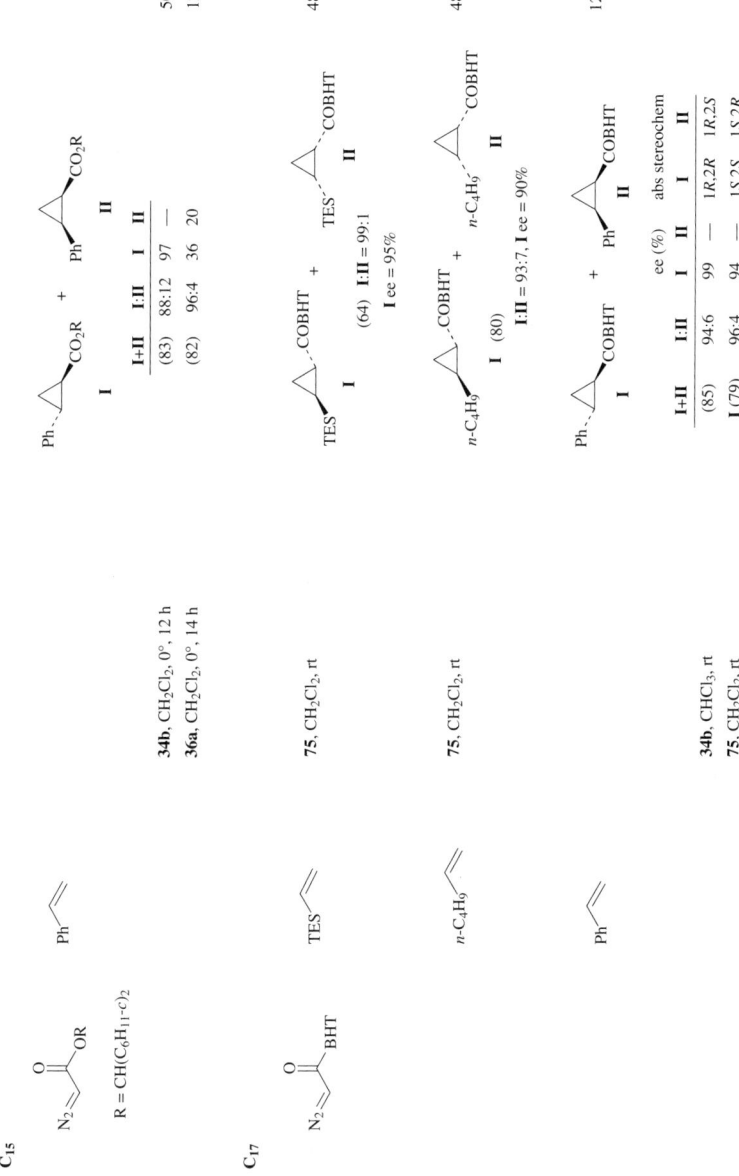

TABLE V. ASYMMETRIC CYCLOPROPANATION OF ALKENES USING CHIRAL COPPER CATALYSTS (*Continued*)

Carbenoid Precursor	Substrate	Conditions	Product(s) and Yield(s) (%)	Refs.
	4-R¹-C₆H₄-CH=CH₂	**75**, CH₂Cl₂, rt	**I** (aryl-COBHT, trans) + **II** (aryl-COBHT, cis) 		I+II	I:II	ee (%) I	ee (%) II	abs stereochem I	abs stereochem II	 	CF₃	(81)	94:6	96	—	—	—	 	OMe	(90)	94:6	87	—	—	—		488
	Ph-CH₂-CH=CH₂	**75**, CH₂Cl₂, rt	Ph-CH₂-△-COBHT (**I**, 78) + Ph-CH₂-△-COBHT (**II**) I:II = 94:6, I ee = 91%	488
C_{18} N₂=CH-C(O)-OR R = (−)-8-phenyl-menthyl	Ph-CH=CH₂	**39a**, CH₂Cl₂, 0°, 21 h	Ph-△-CO₂R (**I**) + Ph-△-CO₂R (**II**) (80) I:II = 80:20 I ee = 96% (1R,2R), II ee = 91% (1R,2S)	468, 461

B. *Non-C_2-Symmetric Complexes (See Chart 3 for Catalyst Structures)*

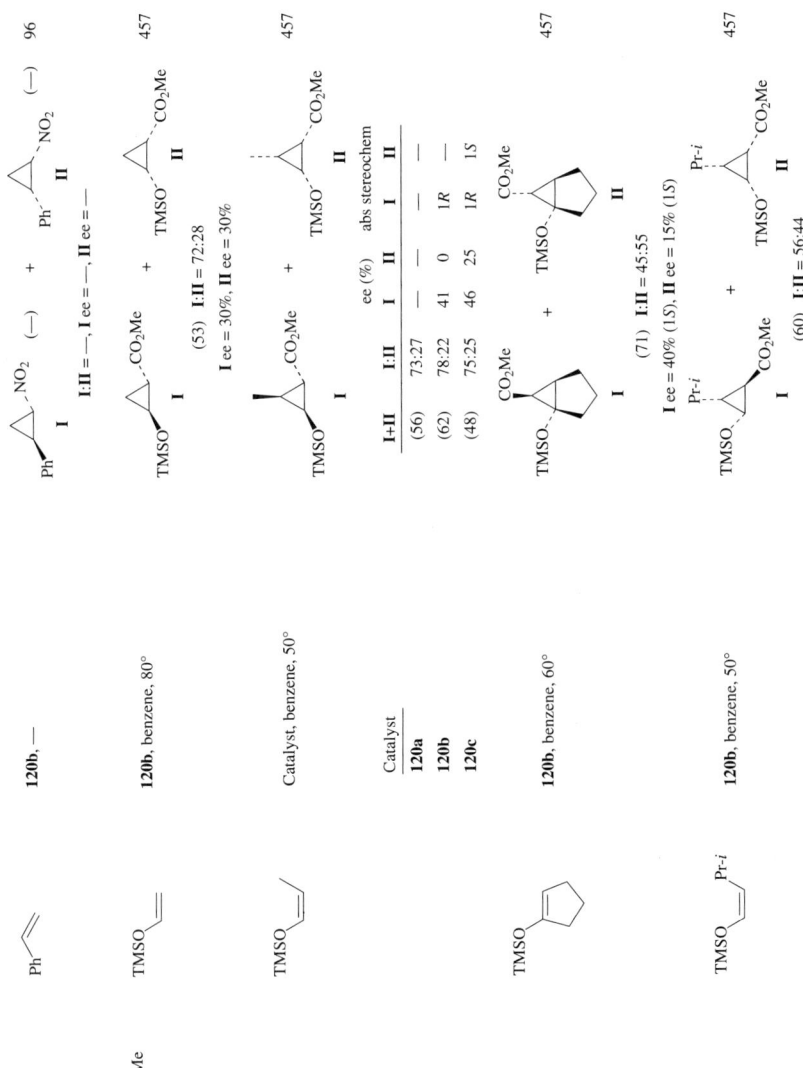

TABLE V. ASYMMETRIC CYCLOPROPANATION OF ALKENES USING CHIRAL COPPER CATALYSTS (*Continued*)

Carbenoid Precursor	Substrate	Conditions	Product(s) and Yield(s) (%)							Refs.
TMSO–C(=CH–t-Bu)–R		Catalyst, DCE, 50°, 2 h	t-Bu–△(CO₂Me)(TMSO) **I** + TMSO–△(CO₂Me)(t-Bu)–R **II**							502
					ee (%)		abs stereochem			
	R	Catalyst	(I+II)	I:II	I	II	I	II		
	H	120b	(64)	47:53	60	76	—	—		
	H	95a	(68)	41:59	61	68	—	—		
	Me	120b	(42)	38:62	30	8	—	—		
	Me	95a	(26)	38:62	32	6	—	—		
TMSO–C(=CH–Ar)–R		Catalyst, DCE, 50°, 2 h	Ar–△(CO₂Me)(TMSO)–R **I** + TMSO–△(CO₂Me)(Ar)–R **II**							
					ee (%)		abs stereochem			
	R Ar	Catalyst	(I+II)	I:II	I	II	I	II		
	H Ph	120b	(59)	47:53	33	38	—	—		502
	H Ph	95a	(46)	44:56	44	63	—	—		502, 460
	H Ph	95b	(60)	47:53	21	35	—	—		502
	Me Ph	120b	(35)	52:48	68	71	—	—		502
	Me Ph	95a	(39)	47:53	75	82	—	—		502, 460
	Me Ph	95b	(40)	62:38	53	33	—	—		502
	Me 4-FC₆H₄	120b	(58)	61:39	47	54	—	—		502
	Me 4-FC₆H₄	95a	(44)	59:41	61	69	—	—		502

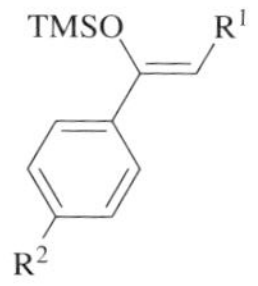

95a, DCE, 50°, 5 h

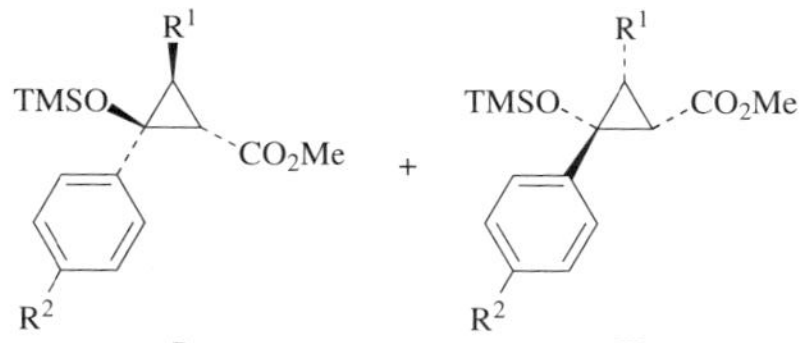

R¹	R²	I+II	I:II	ee (%) I	ee (%) II	abs stereochem I	abs stereochem II
H	OMe	(63)	36:64	65	73	—	—
H	Me	(67)	42:58	59	74	—	—
H	CF$_3$	(61)	48:52	33	60	—	—
H	NO$_2$	(41)	46:54	41	60	—	—
Me	OMe	(40)	59:41	53	60	—	—
Me	Me	(49)	56:44	69	78	—	—
Me	CF$_3$	(32)	57:43	74	78	—	—
Me	NO$_2$	(49)	61:39	64	74	—	—

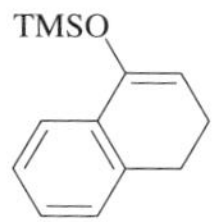

460

Catalyst, DCE, 50°, 2 h

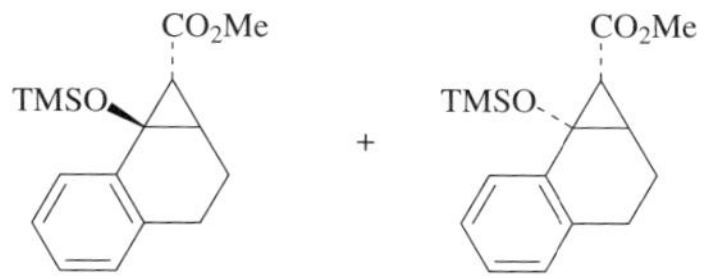

Catalyst	I+II	I:II	ee (%) I	ee (%) II	abs stereochem I	abs stereochem II
120b	(55)	35:65	54	67	—	—
95a	(51)	31:69	65	80	—	—
95b	(39)	32:68	44	53	—	—

502

TABLE V. ASYMMETRIC CYCLOPROPANATION OF ALKENES USING CHIRAL COPPER CATALYSTS (*Continued*)

Carbenoid Precursor	Substrate	Conditions	Product(s) and Yield(s) (%)					Refs.

Carbenoid Precursor: N_2=CHCO$_2$Et

Substrate 1: TMSO-C(Ph)=CMe$_2$ type (TMSO, Ph on alkene)

Conditions: Catalyst, DCE, 50°, 2 h

Products: I (Ph, TMSO, CO$_2$Me cyclopropane) + II (diastereomer)

Catalyst	I+II	I:II	ee (%) I	ee (%) II	abs stereochem I	abs stereochem II	Refs.
120b	(47)	38:62	47	34	—	—	502
95a	(37)	33:67	49	46	—	—	

Substrate 2: R-CH$_2$-CH=CMe$_2$

Conditions: 91e, neat, rt, 5 h

Products: I (R-CH$_2$-cyclopropane-CO$_2$Et) + II

R	I+II	I:II	ee (%) I	ee (%) II	abs stereochem I	abs stereochem II	Refs.
CH$_2$Br	(47)	21:79	—	—	—	—	116
CH$_2$Cl	(73)	16:84	51	90	—	1R	
CHCl$_2$	(71)	12:88	31	85	—	1R	
CCl$_3$	(59)	15:85	11	91	—	1R	

Substrate 3: Ph-CH=CH$_2$ (styrene)

Products: I (Ph-cyclopropane-CO$_2$Et cis) + II (trans)

Conditions	I+II	I:II	ee (%) I	ee (%) II	abs stereochem I	abs stereochem II	Refs.
90, 60°	(72)	70:30	6	—	1R,2R	—	114
ent-90, 60°	(—)	—	6	—	1S,2S	—	114
91e, CH$_3$CN, PhNHNH$_2$, 50°	(30)	74:26	83	88	—	—	494
91l, CH$_3$CN, PhNHNH$_2$, 50°	(33)	74:26	74	53	—	—	494
91m, CH$_3$CN, PhNHNH$_2$, 50°	(37)	71:29	86	82	—	—	494
91n, CH$_3$CN, PhNHNH$_2$, 50°	(42)	75:25	77	87	—	—	494

99, neat, 60°, 6 h	(72)	70:30	6	6	—	—	113
120d, CH$_3$CN, PhNHNH$_2$, 50°	(50)	85:15	67	63	—	—	494
120d, CH$_2$Cl$_2$, PhNHNH$_2$, 50°	(35)	84:16	41	37	—	—	494
120d, DCE, PhNHNH$_2$, 50°	(35)	79:21	29	32	—	—	494
120d, toluene, PhNHNH$_2$, 50°	(45)	79:21	8	5	—	—	494
126a, neat, 50°, 1 h	(45)	70:30	18	0	1S,2S	—	484
126b, neat, 50°, 1 h	(50)	72:28	4	0	1S,2S	—	484
127a, neat, 60°, 2 h	(46)	70:30	57	75	1R,2R	1R,2S	485
127b, neat, 60°, 2 h	(40)	65:35	57	75	1R,2R	1R,2S	485
128a, neat, 60°, 2 h	(52)	72:28	64	74	1R,2R	1R,2S	485
128b, neat, 60°, 2 h	(59)	71:29	66	70	1R,2R	1R,2S	485
129, neat, 60°, 2 h	(58)	69:31	59	66	1R,2R	1R,2S	485
131, —	(—)	—	<2	—	—	—	481
132, —	(—)	—	<2	—	—	—	481
133a, —	(—)	—	3-4	—	—	—	481
133b, —	(—)	—	3-4	—	—	—	481
134a, —	(—)	—	10-11	—	—	—	481
134b, —	(—)	—	0	—	—	—	481
134c, —	(—)	—	0	—	—	—	481
134d, —	(—)	—	0	—	—	—	481
135a, —	(—)	—	10-11	—	—	—	481
135b, —	(—)	—	0	—	—	—	481
135c, —	(—)	—	0	—	—	—	481
135d, —	(—)	—	0	—	—	—	481
136, CH$_2$Cl$_2$, rt, 2 h	(62)	64:36	2	2	1S,2S	1R,2S	474
137, CH$_2$Cl$_2$, rt, 2 h	(54)	58:42	20	8	1R,2R	1S,2R	474
138, CH$_2$Cl$_2$, rt, 24 h	(6)	59:41	1	21	1R,2R	1R,2S	474
139, CH$_2$Cl$_2$, rt, 10 h	(13)	69:31	8	15	1R,2R	1S,2R	474
140, CH$_2$Cl$_2$, rt, 2 h	(61)	70:30	9	9	1R,2R	1R,2S	474
141a, CH$_2$Cl$_2$, 50°, 2 h	(54)	76:24	5	10	1S,2S	1S,2R	503
141b, CH$_2$Cl$_2$, 50°, 2 h	(53)	72:28	2	12	1R,2R	1R,2S	503
141c, CH$_2$Cl$_2$, 50°, 2 h	(54)	79:21	6	18	1R,2R	1R,2S	503

TABLE V. ASYMMETRIC CYCLOPROPANATION OF ALKENES USING CHIRAL COPPER CATALYSTS (Continued)

Carbenoid Precursor	Substrate	Conditions	Product(s) and Yield(s) (%)						Refs.
		141d, CH$_2$Cl$_2$, 50°, 2 h	(45)	73:27	5	3	1S,2S	1R,2S	503
		141e, CH$_2$Cl$_2$, 50°, 2 h	(55)	72:28	6	1	1S,2S	1R,2S	503
		141f, CH$_2$Cl$_2$, 50°, 2 h	(46)	80:20	58	46	1S,2S	1S,2R	503
		141g, CH$_2$Cl$_2$, 50°, 2 h	(37)	78:22	53	35	1S,2S	1S,2R	503
		141h, CH$_2$Cl$_2$, 50°, 2 h	(65)	64:36	13	10	1R,2R	1R,2S	503
		142a, CH$_2$Cl$_2$, 50°, 2 h	(45)	74:26	40	10	1S,2S	1S,2R	503
		142b, CH$_2$Cl$_2$, 50°, 2 h	(48)	79:21	52	32	1S,2S	1S,2R	503
		143a, DCE, PhNHNH$_2$, rt	(92)	67:33	48	84	1S,2S	1S,2R	504
		144, —	(—)	—	1-8	1-8	—	—	505
		145, —	(—)	—	1-8	1-8	—	—	505
		146, —	(—)	—	1-8	1-8	—	—	505
		147a, DCE, reflux, 8 h	(23)	69:31	10	3	—	—	506
		147b, CHCl$_3$, reflux, 8 h	(71)	69:31	38	23	—	—	506
		147b, DCE, reflux, 8 h	(65)	69:31	28	22	—	—	506
		147b, benzene, reflux, 8 h	(69)	72:28	35	25	—	—	506
		147c, CHCl$_3$, reflux, 8 h	(63)	68:32	34	22	—	—	506
		147c, benzene, reflux, 8 h	(75)	69:31	32	28	—	—	506
		147d, CHCl$_3$, reflux, 8 h	(77)	65:35	45	26	—	—	506
		147e, CHCl$_3$, reflux, 8 h	(91)	72:28	27	22	—	—	506
		148a, rt, 2 h	(—)	62:38	4-7	4-7	—	—	507
		ent-148a, rt, 2 h	(—)	62:38	4-7	4-7	—	—	507
		148b, rt, 2 h	(—)	62:38	4-7	4-7	—	—	507
		ent-148b, rt, 2 h	(—)	62:38	4-7	4-7	—	—	507
		148c, rt, 2 h	(—)	62:38	4-7	4-7	—	—	507
		148d, rt, 2 h	(—)	62:38	4-7	4-7	—	—	507
		148e, rt, 2 h	(—)	62:38	4-7	4-7	—	—	507
		148f, rt, 2 h	(—)	62:38	4-7	4-7	—	—	507
		148g, rt, 2 h	(—)	62:38	4-7	4-7	—	—	507

Entry	(Yield)	Ratio					Ref
148h, rt, 2 h	(—)	62:38	4-7	4-7			507
148i, rt, 2 h	(—)	62:38	4-7	4-7			507
148j, rt, 2 h	(—)	62:38	4-7	4-7			507
148k, rt, 2 h	(—)	62:38	4-7	4-7			507
148l, rt, 2 h	(—)	62:38	4-7	4-7			507
149a, DCE, rt, 24 h	(70)	61:39	10	8	1S,2S	1S,2R	508
149b, DCE, rt, 24 h	(50)	59:41	13	21	1S,2S	1S,2R	508
149c, DCE, rt, 24 h	(59)	61:39	5	10	1S,2S	1S,2R	508
149d, DCE, rt, 24 h	(53)	60:40	5	11	1S,2S	1S,2R	508
149e, DCE, rt, 24 h	(53)	58:42	10	9	1S,2S	1S,2R	508
149f, DCE, rt, 24 h	(32)	59:41	10	9	1S,2S	1S,2R	508
149g, DCE, rt, 24 h	(64)	58:42	7	9	1S,2S	1S,2R	508
152a, DCE, rt, 24 h	(63)	65:35	18	19	1S,2S	1S,2R	465
152b, DCE, rt, 24 h	(37)	63:37	1	0	1R,2R	—	465
152c, DCE, rt, 24 h	(84)	62:38	38	5	1S,2S	1S,2R	465
152d, DCE, rt, 24 h	(76)	62:38	4	0	1R,2R	—	465
153a, DCE, rt, 24 h	(76)	62:38	4	1	1R,2R	1R,2S	465
153b, DCE, rt, 24 h	(74)	63:37	4	7	1R,2R	1R,2S	465
154a, 55°, 2 h	(53)	24:76	40	62	1R,2R	1R,2S	509
154b, CH$_2$Cl$_2$, 12 h	(46)	40:60	81	85	1R,2R	1R,2S	493
155a, DCE, rt, 24 h	(52)	44:56	31	51	1R,2R	1R,2S	510
155a, DCE, 0°, 24 h	(64)	37:63	37	57	1R,2R	1R,2S	510
155a, DCE, −78°, 24 h	(48)	37:63	40	60	1R,2R	1R,2S	510
155b, CH$_2$Cl$_2$, 12 h	(49)	69:31	25	18	1R,2R	1R,2S	493
155c, CH$_2$Cl$_2$, 12 h	(49)	62:38	9	4	1R,2R	1S,2R	493
156a, neat, rt, 3 h	(89)	65:35	23	16			511
156a/MeCN, neat, rt, 2.5 h	(81)	60:40	<5	<10			511
156a/MeCN, MeCN, 50°, 7 h	(73)	80:20	—	—			511

215

TABLE V. ASYMMETRIC CYCLOPROPANATION OF ALKENES USING CHIRAL COPPER CATALYSTS (*Continued*)

Carbenoid Precursor	Substrate	Conditions	Product(s) and Yield(s) (%)			Refs.
		Zeolite-**156b**, CH$_2$Cl$_2$, rt, 6 h	(100)	53:47	<5 —	512a
		Zeolite-**156b**/MeCN, neat, rt, 96 h	(85)	66:34	<5 —	512a
		Zeolite-**156b**/MeCN, MeCN, 50°, 47 h	(81)	78:22	12 7 —	512a
		157a, neat, rt, 4 h	(80)	62:38	<1 <1 —	511
		157b, neat, rt, 3 h	(82)	67:33	<5 —	511
		Zeolite-**157c**, neat, rt, 5 h	(32)	48:52	11 5 —	512a
		Zeolite-**157d**, neat, rt, 51 h	(48)	43:57	<5 —	512a
		158a, neat, rt, 1 h	(91)	65:35	0 0 —	511
		Zeolite-**158b**, neat, rt, 20 h	(66)	55:45	<5 —	512a

Ph₂C=CH₂ → Ph,Ph-cyclopropane-CO$_2$Et

Catalyst, neat, heat, 2 h

Catalyst	Temp	ee (%)	abs stereochem	Refs.
99	55°	11	1S	513
100	55°	1	1R	513
101	55°	9	1S	513
102	50°	4	1S	513
103	60°	1	1R	513
104	55°	1	1S	513
105a	80°	0	—	513
105b	55°	—	—	513
106	55°	1	1R	513
107	55°	3	1S	513
108	55°	2	1S	513
109	55°	10	1S	513
110a	55°	9	1R	513
110b	55°	1	1R	513
111	55°	1	1R	513
112	55°	7	1R	513
113a	55°	4	1S	513
113b	55°	6	1R	513

114	60°	0	—	513
115a	55°	2	1R	513
115b	55°	13	1R	513
116	55°	3	1S	513
117a	55°	4	1S	513
117b	55°	3	1S	513
118a	65°	13	1S	513
118b	55°	4	1S	513
119a	60°	52	1S	513
119b	55°	45	1S	513
119c	55°	66	1S	513
119d,	50°	38	1S	513
121	60°	1	1S	513
122	55°	6	1R	513
123	80°	5	1S	513
124	55°	16	1S	513
125	55°	2	1S	513
148a	rt	3-14	—	507[a]
ent-148a	rt	3-14	—	507[a]
148b	rt	3-14	—	507[a]
ent-148b	rt	3-14	—	507[a]
148c	rt	3-14	—	507[a]
148d	rt	3-14	—	507[a]
148e	rt	3-14	—	507[a]
148f	rt	3-14	—	507[a]
148g	rt	3-14	—	507[a]
148h	rt	3-14	—	507[a]
148i	rt	3-14	—	507[a]
148j	rt	3-14	—	507[a]
148k	rt	14	—	507[a]

TABLE V. ASYMMETRIC CYCLOPROPANATION OF ALKENES USING CHIRAL COPPER CATALYSTS (*Continued*)

Carbenoid Precursor	Substrate	Conditions	Product(s) and Yield(s) (%)	Refs.

C₆ carbenoid: N₂=C(NO₂)(CO₂Et)

Substrate: MeO₂C-CH=C(iPr)-CH₂-CH=C(Me)-Me (geranate-type)

Products I and II: cyclopropanes with MeO₂C group, CO₂Et

		I+II	I:II	ee (%) I	ee (%) II	abs stereochem I	abs stereochem II	
	ent-**91c**, cyclohexane, 40°	(88)	64:36	3	14	—	—	512b
	ent-**91c**, neat, 40°	(70)	38:62	17	78	—	—	
	91c, cyclohexane, 40°	(86)	60:40	9	25	—	—	
	91c, neat, 40°	(71)	37:63	29	80	—	—	
	98, cyclohexane, 40°	(68)	62:38	<2	<2	—	—	
	98, neat, 40°	(72)	64:36	12	11	—	—	

Substrate: Ph-CH=CH₂ (styrene)
Products: Ph-cyclopropyl-NO₂/CO₂Et (I) + II

	120b, —	%ee = —	96

C₆ carbenoid: N₂=CH-CO₂Bu-*t*

Substrate: Ph-CH=CH₂

Products I and II: Ph-cyclopropyl-CO₂Bu-*t*

	I+II	I:II	ee (%) I	ee (%) II	abs stereochem I	abs stereochem II	
143a, DCE, PhNHNH₂, rt	(56)	86:14	47	87	1*S*,2*S*	1*S*, 2*R*	504
147b, CHCl₃, reflux, 8 h	(55)	80:20	31	21	—	—	506
147d, CHCl₃, reflux, 8 h	(38)	74:26	52	34	—	—	506

C₈ carbenoid: 2-diazo-5,5-dimethyl-1,3-cyclohexanedione

Substrate: Ph-CH=CH₂

Product: spirocyclopropane-dione with Ph

Catalyst, benzene, reflux

Catalyst	Time		ee (%)
92a	24 h	(36)	92
92b	32 h	(21)	73
92c	24 h	(48)	100
93	10 h	(43)	98

141

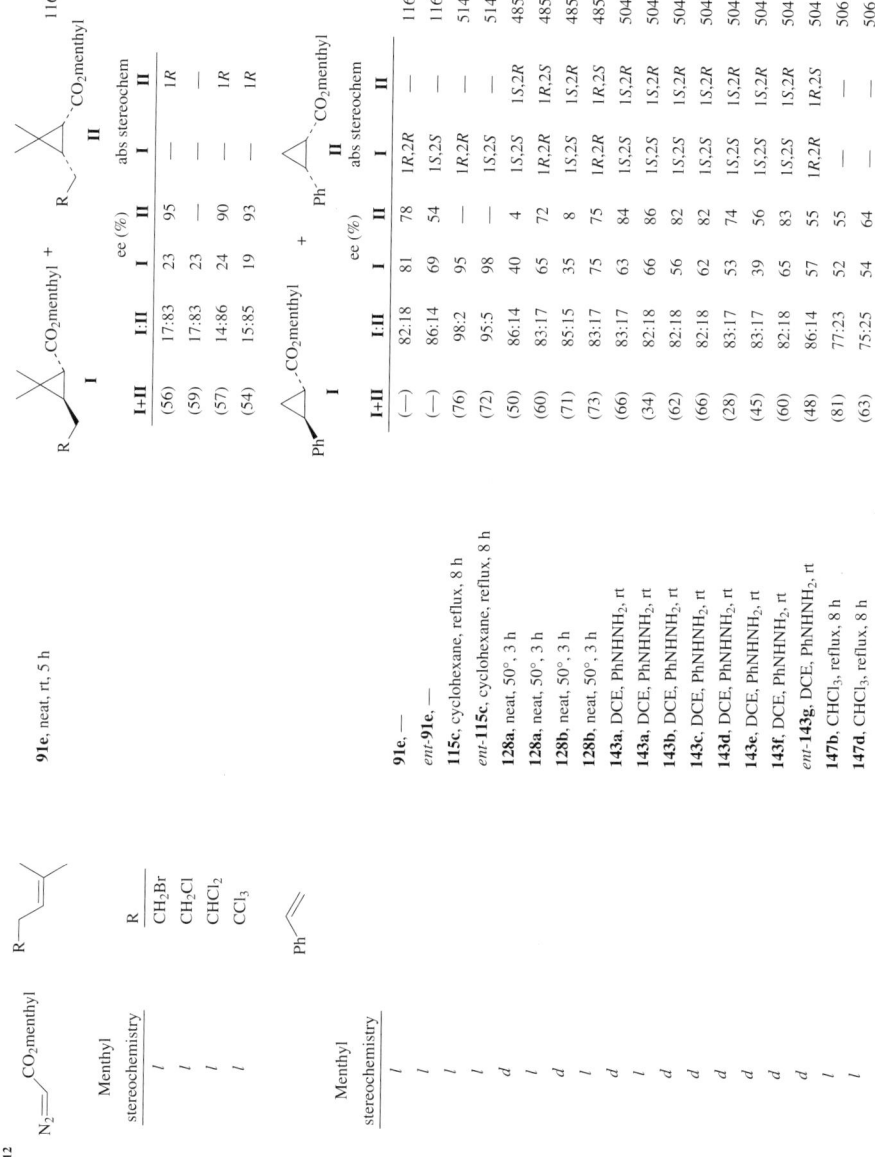

C$_{12}$									
N$_2$=CO$_2$menthyl									

	Menthyl stereochemistry	R			I+II	I:II	ee (%) I	ee (%) II	abs stereochem I	abs stereochem II	
		CH$_2$Br			(56)	17:83	23	95	—	1R	116
	l	CH$_2$Cl			(59)	17:83	23	54	—	—	116
	l	CHCl$_2$			(57)	14:86	24	90	—	1R	514
	l	CCl$_3$			(54)	15:85	19	93	—	1R	514

	Menthyl stereochemistry		I+II	I:II	ee (%) I	ee (%) II	abs stereochem I	abs stereochem II	
91e, —	l		(—)	82:18	81	78	1R,2R	—	116
ent-91e, —	l		(—)	86:14	69	54	1S,2S	—	116
115c, cyclohexane, reflux, 8 h	d		(76)	98:2	95	—	1R,2R	—	514
ent-115c, cyclohexane, reflux, 8 h	l		(72)	95:5	98	—	1S,2S	—	514
128a, neat, 50°, 3 h	d		(50)	86:14	40	4	1S,2S	1S,2R	485
128a, neat, 50°, 3 h	l		(60)	83:17	65	72	1R,2R	1R,2S	485
128b, neat, 50°, 3 h	d		(71)	85:15	35	8	1S,2S	1S,2R	485
128b, neat, 50°, 3 h	l		(73)	83:17	75	75	1R,2R	1R,2S	485
143a, DCE, PhNHNH$_2$, rt	d		(66)	83:17	63	84	1S,2S	1S,2R	504
143a, DCE, PhNHNH$_2$, rt	l		(34)	82:18	66	86	1S,2S	1S,2R	504
143b, DCE, PhNHNH$_2$, rt	d		(62)	82:18	56	82	1S,2S	1S,2R	504
143c, DCE, PhNHNH$_2$, rt	d		(66)	82:18	62	82	1S,2S	1S,2R	504
143d, DCE, PhNHNH$_2$, rt	d		(28)	83:17	53	74	1S,2S	1S,2R	504
143e, DCE, PhNHNH$_2$, rt	d		(45)	83:17	39	56	1S,2S	1S,2R	504
143f, DCE, PhNHNH$_2$, rt	d		(60)	82:18	65	83	1S,2S	1S,2R	504
ent-143g, DCE, PhNHNH$_2$, rt	d		(48)	86:14	57	55	1R,2R	1R,2S	504
147b, CHCl$_3$, reflux, 8 h	l		(81)	77:23	52	55	—	—	506
147d, CHCl$_3$, reflux, 8 h	l		(63)	75:25	54	64	—	—	506

TABLE V. ASYMMETRIC CYCLOPROPANATION OF ALKENES USING CHIRAL COPPER CATALYSTS (*Continued*)

Carbenoid Precursor	Substrate	Conditions	Product(s) and Yield(s) (%)	Refs.
d,l-menthyl	D,Ph,R¹,R² alkene	*ent*-91d, —	I + II; I:II 85:15 (50-60)/(50-60); ee(%) I 70-80, II 70-80; abs stereochem I 1S,2S,3S / 1S,2S,3R	515
		ent-91d, —	I + II; I:II = 85:15; I ee = 90% (1S,2S,3S), II ee = 90%	516
		ent-91d, —	(—) I:II = —; I ee = 90% (1S,2S,3R), II ee = —	517
Menthyl stereochemistry: *l* / *l*	n-C₆H₁₃ CH=CH₂	91e, — / *ent*-91e, —	I + II; I:II 78:22 / 83:17; ee(%) I 84/76, II 64/46; abs stereochem I 1R,2R / 1S,2S	116 / 116

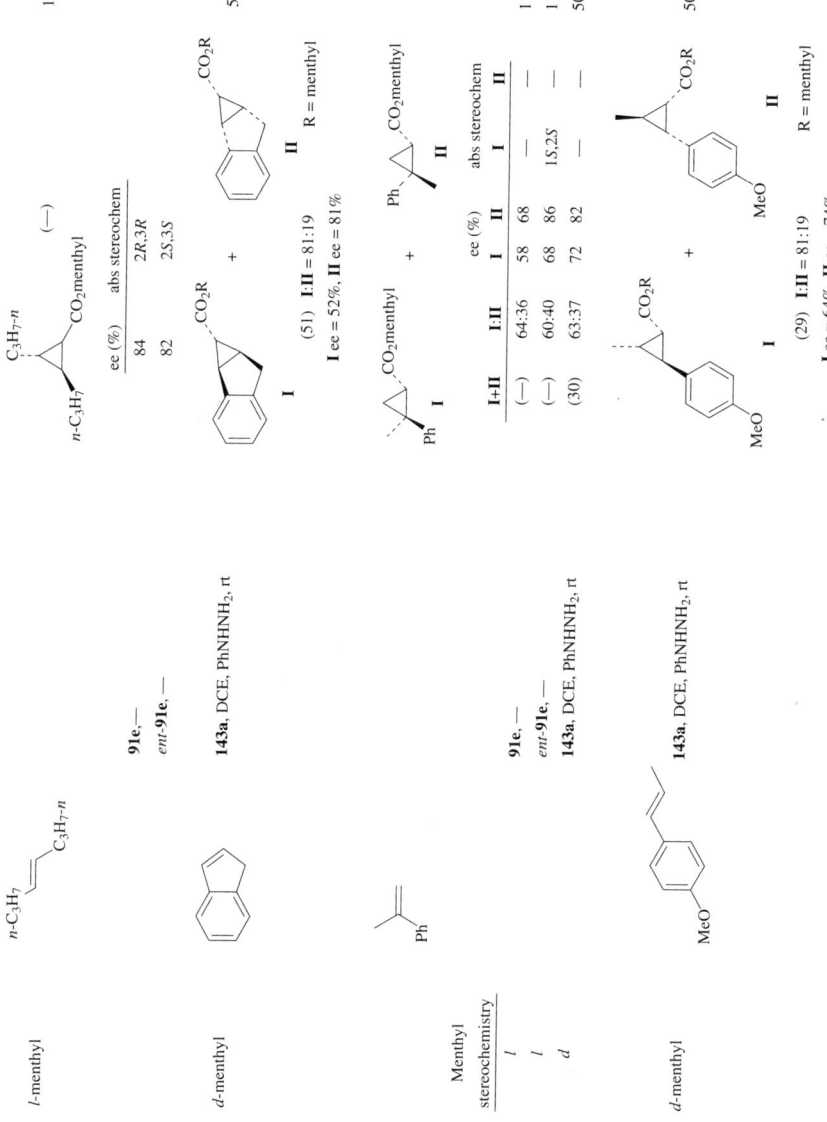

TABLE V. ASYMMETRIC CYCLOPROPANATION OF ALKENES USING CHIRAL COPPER CATALYSTS (*Continued*)

Carbenoid Precursor	Substrate	Conditions	Product(s) and Yield(s) (%)	Refs.
l-menthyl, MeO₂C–CH=C(iPr)–CH₂–CH=C(Me)₂ structure	MeO₂C–C(iPr)=CH–CH₂–... (prenyl)	**91e**, cyclohexane, 35°	MeO₂C–...–cyclopropane–CO₂R (**I**) + MeO₂C–...–cyclopropane–CO₂R (**II**), R = *l*-menthyl; (70) **I:II** = 44:66; **I** ee = >95%, **II** ee = >95%	512b
d-menthyl	Ph-C(=CH₂)-Ph	**143a**, DCE, PhNHNH₂, rt	Ph,Ph-cyclopropane–CO₂menthyl-*d* (27) 59% ee	504
C_{15} N_2=CH–CO₂DCM	Ph–CH=CH₂	Catalyst, CHCl₃, reflux, 8 h	Ph–cyclopropane–CO₂DCM (**I**) + Ph–cyclopropane–CO₂DCM (**II**)	506

Catalyst	**I + II**	**I:II**	ee (%) **I**	ee (%) **II**
147b	(64)	86:14	45	47
147d	(54)	86:14	64	63

Carbenoid Precursor	Substrate	Conditions	Product(s) and Yield(s) (%)	Refs.
C_{16} 2,6-di-*t*-Bu-C₆H₃–O–C(=O)–CH=N₂	Ph–CH=CH₂	**143a**, DCE, PhNHNH₂, rt	Ph–cyclopropane–CO₂Ar (**I**) + Ph–cyclopropane–CO₂Ar (**II**); Ar = 2,6-(*t*-Bu)₂C₆H₃; (73) **I:II** = 93:7; **I** ee = 9%, **II** ee = —	504
C_{17} N_2=CH–COBHT	Ph–CH=CH₂	**155a**, DCE, rt, 24 h	Ph–cyclopropane–COBHT (**I**) + Ph–cyclopropane–COBHT (**II**); (94) **I:II** = 96:4; **I** ee = 10%, **II** ee = —	510

[a] The authors do not indicate whether or not the reaction was run neat.

TABLE VI. ASYMMETRIC CYCLOPROPANATION OF ALKENES USING CHIRAL RHODIUM CATALYSTS

Carbenoid Precursor	Substrate	Conditions	Product(s) and Yield(s) (%)	Refs.

(See Chart 4 for Catalyst Structures)

C₃

$N_2\text{=}\diagup\text{CO}_2\text{Me}$

| | R—CH=CH₂ | Catalyst, CH₂Cl₂, reflux | R—△—CO₂Me (I) + R—△—CO₂Me (II) | 309 |

			ee (%)	abs stereochem	
R	Catalyst	I+II	I:II	I	II	I	II
n-Pr	Rh₂(5S-MEPY)₄ (**171a**)	(55)	58:42	44	40	—	—
n-Pr	Rh₂(4S-PHOX)₄ (**173**)	(21)	38:62	42	56	—	—
CH₂OAc	Rh₂(5S-MEPY)₄ (**171a**)	(46)	55:45	56	44	—	—
CH₂OAc	Rh₂(4S-PHOX)₄ (**173**)	(11)	47:53	4	10	—	—
OAc	Rh₂(5S-MEPY)₄ (**171a**)	(62)	53:47	10	55	—	—
OAc	Rh₂(4S-PHOX)₄ (**173**)	(14)	44:56	56	68	—	—
O₂CPh	Rh₂(5S-MEPY)₄ (**171a**)	(40)	50:50	22	54	—	—
O₂CPh	Rh₂(4S-PHOX)₄ (**173**)	(22)	38:62	32	54	—	—

| | (cyclopentenyl-OTMS) | Catalyst, CH₂Cl₂, reflux | TMSO—[bicyclic]—CO₂Me (I) + TMSO—[bicyclic]—CO₂Me (II) | 309 |

Catalyst		
Rh₂(5S-MEPY)₄ (**171a**)	(36) **I:II** = 43:53; **I** ee = 50%, **II** ee = 14%	
Rh₂(4S-PHOX)₄ (**173**)	(5) **I:II** = 57:43; **I** ee = —%, **II** ee = —%	

| | (dihydropyran) | Catalyst, CH₂Cl₂, reflux | O—[bicyclic]—CO₂Me (I) + O—[bicyclic]—CO₂Me (II) | 309 |

Catalyst		
Rh₂(5S-MEPY)₄ (**171a**)	(18) **I:II** = 87:13; **I** ee = 10%, **II** ee = 49%	
Rh₂(4S-PHOX)₄ (**173**)	(31) **I:II** = 71:29; **I** ee = 14%, **II** ee = 76%	

TABLE VI. ASYMMETRIC CYCLOPROPANATION OF ALKENES USING CHIRAL RHODIUM CATALYSTS (*Continued*)

Carbenoid Precursor	Substrate	Conditions	Product(s) and Yield(s) (%)	Refs.
	TMSO-C(=CH₂)-Ph	Catalyst, CH₂Cl₂, reflux	TMSO⋯△⋯CO₂Me / Ph (**I**) + Ph⋯△⋯CO₂Me / TMSO (**II**)	309
		Catalyst: Rh₂(5S-MEPY)₄ (**171a**) Rh₂(4S-PHOX)₄ (**173**)	(80) **I:II** = 45:55; **I** ee = 30%, **II** ee = 14% (44) **I:II** = 47:53; **I** ee = 14%, **II** ee = 10%	
	Ph-CH=CH-CH₃	Catalyst, CH₂Cl₂, reflux	Ph⋯△⋯CO₂Me (**I**) + Ph⋯△⋯CO₂Me (**II**)	309
		Catalyst: Rh₂(5S-MEPY)₄ (**171a**) Rh₂(4S-PHOX)₄ (**173**)	(12) **I:II** = 99:1; **I** ee = —%, **II** ee = —% (10) **I:II** = 89:11; **I** ee = —%, **II** ee = —%	
C₄ N₂=CH-CO₂Et	R-CH=CH₂		R⋯△⋯CO₂Et (**I**) + R⋯△⋯CO₂Et (**II**)	

R	Conditions	I+II	I:II	ee (%) I	ee (%) II	abs stereochem I	abs stereochem II	Refs.
OEt	**178b**, CH₂Cl₂, rt	(—)	55:45	10	15	—	—	518
OAc	Rh₂(5S-MEPY)₄ (**171a**), CH₂Cl₂, reflux	(75)	52:48	24	60	—	—	309
OAc	Rh₂(4S-PHOX)₄ (**173**), CH₂Cl₂, reflux	(26)	44:56	50	48	—	—	309

| | Ph-CH=CH₂ | | Ph⋯△⋯CO₂Et (**I**) + Ph⋯△⋯CO₂Et (**II**) | |

	I+II	I:II	ee (%) I	ee (%) II	abs stereochem I	abs stereochem II	ref
Rh$_2$(S-TBSP)$_4$ (161b), pentane	(80)	55:45	6	30	1R,2R	1R,2S	128, 58
Rh$_2$(S-BDME)$_4$ (166), CH$_2$Cl$_2$, rt, 15 h	(98)	31:69	57	53	1S,2S	1S,2R	129
168a, neat, 30°, 3 h	(76)	57:43	12	6	—	—	127
168a, neat, 50°, 3 h	(72)	56:44	10	5	—	—	127
168a, neat, 70°, 3 h	(65)	52:48	9	4	—	—	127
168b, neat, 30°, 3 h	(71)	52:48	7	4	—	—	127
168c, neat, 30°, 3 h	(71)	51:49	9	6	—	—	127
168d, neat, 30°, 3 h	(71)	51:49	8	4	—	—	127
168e, neat, 30°, 3 h	(69)	52:48	6	3	—	—	127
168f, neat, 30°, 3 h	(70)	54:46	3	4	—	—	127
168g, neat, 30°, 3 h	(69)	52:48	5	2	—	—	127
168h, neat, 30°, 3 h	(48)	—	7	3	—	—	127
168i, H$_2$O, 30°, 3 h	(28)	54:46	10	6	—	—	127
168j, H$_2$O, 30°, 3 h	(33)	55:45	8	7	—	—	127
168k, H$_2$O, 30°, 3 h	(36)	47:53	5	4	—	—	127
168l, H$_2$O, 30°, 3 h	(12)	58:42	4	3	—	—	127
168m, acetone, 30°, 3 h	(8)	63:37	5	3	—	—	127
168n, CH$_2$Cl$_2$, 0°, 1 h	(78)	55:45	2	2	—	—	519
Rh$_2$(5S-MEPY)$_4$ (171a), CH$_2$Cl$_2$, reflux	(59)	56:44	58	33	1S,2S	1S,2R	108, 309
Rh$_2$(5S-DMAP)$_4$ (171c), CH$_2$Cl$_2$, reflux	(47)	66:34	22	38	1S,2S	1S,2R	520
Rh$_2$(4S-PHOX)$_4$ (173), CH$_2$Cl$_2$, reflux, 13 h	(41)	34:66	24	57	1R,2R	1R,2S	521, 309
Rh$_2$(S-PTPI)$_4$ (175), CH$_2$Cl$_2$, reflux, 3 h	(50)	62:38	17	28	1S,2S	1S,2R	519
Rh$_2$(S-PTPI)$_4$ (175), Et$_2$O, 40°, 3 h	(63)	57:43	51	49	1S,2S	1S,2R	522
Rh$_2$(4S-BNAZ)$_4$ (176a), CH$_2$Cl$_2$, rt	(74)	41:59	23	29	1S,2S	1S,2R	523
Rh$_2$(4S-IBAZ)$_4$ (176b), CH$_2$Cl$_2$, rt	(62)	36:64	47	73	1S,2S	1S,2R	523
Rh$_2$(4S-IBAZ)$_4$ (176b), CH$_2$Cl$_2$, reflux	(62)	38:62	47	68	1S,2S	1S,2R	523
177, AgOTf, THF, rt, 6 h	(44)	48:52	85	82	1S,2S	1S,2R	473
178a, CH$_2$Cl$_2$, 0°, 24 h	(—)	30:70	10	—	—	1S,2R	524, 100
178b, CH$_2$Cl$_2$, rt	(—)	29:71	—	15	—	1R,2S	518
189b (Chart 5), CH$_2$Cl$_2$, rt, 24 h	(—)	60:40	14	5	1S,2S	1S,2R	455

TABLE VI. ASYMMETRIC CYCLOPROPANATION OF ALKENES USING CHIRAL RHODIUM CATALYSTS (*Continued*)

Carbenoid Precursor	Substrate	Conditions	Product(s) and Yield(s) (%)	Refs.
	(isopropenyl-Ph)	Rh₂(5S-MEPY)₄ (**171a**), CH₂Cl₂, reflux	Ph-cyclopropane-CO₂Et (**I**) + Ph-cyclopropane-CO₂Et (**II**) (44) **I:II** = 55:45; **I** ee = 15%, **II** ee = 5%	309

			I+II	I:II	ee (%) I	ee (%) II	abs stereochem I	abs stereochem II	
	Ph-CH=CH-CH₃	Rh₂(5S-MEPY)₄ (**171a**), CH₂Cl₂, reflux	(44)	44:56	12	16	—	—	309
		Rh₂(4S-PHOX)₄ (**173**), CH₂Cl₂, reflux	(19)	36:64	24	18	—	—	309
		178a, CH₂Cl₂, 0°, 24 h	(—)	11:89	50	20	—	—	524, 100
		178b, CH₂Cl₂, rt	(—)	16:84	20	25	—	—	518
		178b, CH₂Cl₂, 0°	(—)	7:93	20	0	—	—	518

			I+II	I:II	ee (%) I	ee (%) II	abs stereochem I	abs stereochem II	
	Bn-CH=CH₂	**178a**, CH₂Cl₂, 0°, 24 h	(—)	19:81	60	45	—	—	524, 100
		178b, CH₂Cl₂, rt	(—)	50:50	10	10	—	—	518
		178b, CH₂Cl₂, 0°	(—)	20:80	40	20	—	—	518

| | 2-vinylnaphthalene | Rh₂(4S-PHOX)₄ (**173**), CH₂Cl₂, reflux | 2-naphthyl-cyclopropane-CO₂Et (**I**) + 2-naphthyl-cyclopropane-CO₂Et (**II**) (32) **I:II** = 42:58 | 309 |

TABLE VI. ASYMMETRIC CYCLOPROPANATION OF ALKENES USING CHIRAL RHODIUM CATALYSTS (*Continued*)

Carbenoid Precursor	Substrate	Conditions	Product(s) and Yield(s) (%)	Refs.
PhCO$_2$–CH=N$_2$ (implied)	PhCO$_2$–CH=CH$_2$	Catalyst, CH$_2$Cl$_2$, reflux	PhCO$_2$⟋△⟍CO$_2$Bu-*t* **I** + PhCO$_2$⟋△⟍CO$_2$Bu-*t* **II** Catalyst Rh$_2$(5*S*-MEPY)$_4$ (**171a**) (24) **I:II** = 63:37; **I** ee = 28%, **II** ee = 78% Rh$_2$(4*S*-PHOX)$_4$ (**173**), 13 h (10) **I:II** = 53:47; **I** ee = 76%, **II** ee = 56%	309
	Ph–CH=CH–CH$_3$	Rh$_2$(4*S*-PHOX)$_4$ (**173**), CH$_2$Cl$_2$, reflux	Ph⟋△⟍CO$_2$Bu-*t* **I** + Ph⟋△⟍CO$_2$Bu-*t* **II** (17) **I:II** = 98:2; **I** ee = 8%, **II** ee = —%	309
	2-naphthyl–CH=CH$_2$	Rh$_2$(*S*-BDME)$_4$ (**166**), CH$_2$Cl$_2$, rt, 15 h Rh$_2$(4*S*-PHOX)$_4$ (**173**), CH$_2$Cl$_2$, reflux	2-naphthyl⟋△⟍CO$_2$Bu-*t* **I** + 2-naphthyl⟋△⟍CO$_2$Bu-*t* **II** (59) **I:II** = 48:52; **I** ee = 35%, **II** ee = 87% (25) **I:II** = 37:63	129 309
	biphenyl–CH=CH$_2$	Rh$_2$(4*S*-PHOX)$_4$ (**173**), CH$_2$Cl$_2$, reflux	biphenyl⟋△⟍CO$_2$Bu-*t* **I** + biphenyl⟋△⟍CO$_2$Bu-*t* **II** (28) **I:II** = 37:63	309
MeO$_2$C–C(N$_2$)–CH=CH–CH$_3$	EtO–CH=CH$_2$	Rh$_2$(*S*-DOSP)$_4$ (**161a**), pentane, −78°	EtO⟋△⟍(CH=CH–CH$_3$)(CO$_2$Me) (75) ee = 93%	213

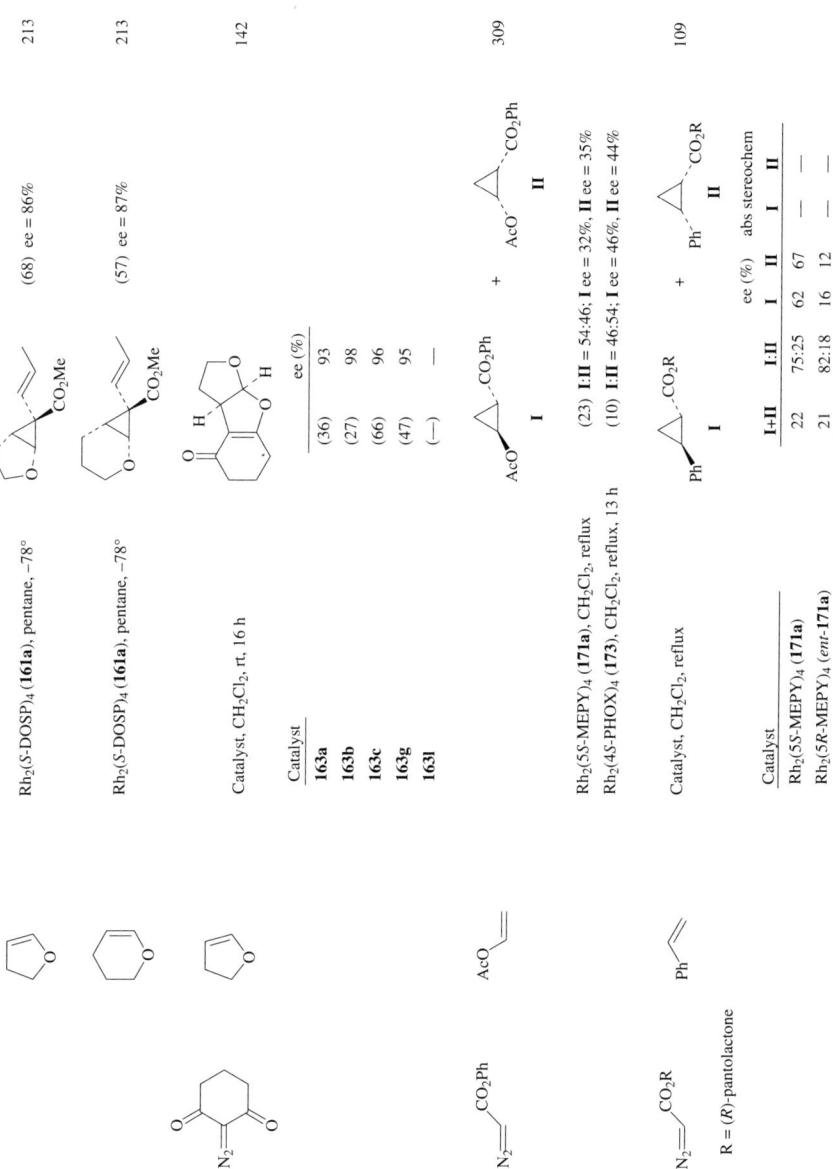

TABLE VI. ASYMMETRIC CYCLOPROPANATION OF ALKENES USING CHIRAL RHODIUM CATALYSTS (*Continued*)

Carbenoid Precursor	Substrate	Conditions	Product(s) and Yield(s) (%)	Refs.
N₂=C(CO₂Me)(cyclopentenyl)	2,3-dihydrofuran	Rh₂(S-DOSP)₄ (**161a**), pentane, −78°	(49) ee = 76%	213
	3,4-dihydro-2H-pyran	Rh₂(S-DOSP)₄ (**161a**), pentane, −78°	(53) ee = 88%	213
C₉ N₂=C(CO₂Me)(Ph)	R-CH=CH₂	Rh₂(S-DOSP)₄ (**161a**), pentane	**I** (R, Ph, CO₂Me cyclopropane) + **II** (R, Ph, CO₂Me cyclopropane)	73

R	I+II	I:II	ee (%) I	ee (%) II
Et	(86)	93:7	80	—
n-Bu	(86)	93:7	77	—
i-Pr	(63)	87:13	74	—
CH₂OAc	(85)	97:3	80	—

Substrate: RO-CH=CH₂

Product: **I** (RO, Ph, CO₂Me) + **II** (RO, Ph, CO₂Me)

R	I+II	I:II	ee (%) I	ee (%) II	abs stereochem I	abs stereochem II	Conditions	Refs
Et	(88)	97:3	66	—	1R,2S	—	Rh₂(S-DOSP)₄ (**161a**), pentane	73
Et	(86)	>95:5	79	—	1S,2R	—	Rh₂(S-biTISP)₂ (**169c**), CH₂Cl₂, −50°	525
n-Bu	(84)	97:3	64	—	1R,2S	—	Rh₂(S-DOSP)₄ (**161a**), pentane	73

Catalyst, conditions	I+II	I:II	ee (%) I	ee (%) II	abs stereochem I	abs stereochem II	Ref
Rh₂(S-TBSP)₄ (161b), pentane	(90)	98:2	87	—	1R,2S	—	73
Rh₂(S-TBSP)₄ (161b), pentane	(73)	96:4	85	—	1R,2S	—	74
Rh₂(S-TBSP)₄ (161b), CH₂Cl₂	(77)	97:3	61	—	1R,2S	—	74
Rh₂(S-TBSP)₄ (161b), SC-CHF₃ (100 bar), 30°, 1h	(—)	—	40	—	—	—	526
Rh₂(S-TBSP)₄ (161b), SC-CHF₃ (52 bar), 30°, 1h	(—)	—	77	—	—	—	526
Rh₂(S-TBSP)₄ (161b), SC-CO₂ (110 bar), 34°, 1h	(—)	—	80	—	—	—	526
Rh₂(S-TBSP)₄ (161b), SC-CO₂ (79 bar), 34°, 1h	(—)	—	83	—	—	—	526
161e, CH₂Cl₂	(45)	97:3	60	—	1R,2S	—	74
164a, CH₂Cl₂, reflux	(71)	98:2	35	—	—	—	145
164a, benzene, reflux	(98)	>99:1	40	—	—	—	145
164a, pentane, reflux	(96)	>99:1	49	—	—	—	145
164a, cyclohexane, reflux	(88)	98:2	51	—	—	—	145
164b, CH₂Cl₂, reflux	(70)	99:1	36	—	—	—	145
168n, pentane	(82)	96:4	16	—	1S,2R	—	74
170, CH₂Cl₂, reflux	(86)	>99:1	31	—	—	—	145
Rh₂(5S-MEPY)₄ (171a), CH₂Cl₂	(27)	97:3	49	—	1R,2S	—	74
Rh₂(4S-MEOX)₄ (172c), CH₂Cl₂	(57)	96:4	41	—	1R,2S	—	74
Rh₂(4S-MBOIM)₄ (174a), CH₂Cl₂	(73)	96:4	48	—	1R,2S	—	74
Rh₂(4S-TBOIM)₄ (174b), CH₂Cl₂	(63)	95:5	77	—	1R,2S	—	74
Rh₂(4S-TBOIM)₄ (174b), pentane	(69)	94:6	75	—	1R,2S	—	74

TABLE VI. ASYMMETRIC CYCLOPROPANATION OF ALKENES USING CHIRAL RHODIUM CATALYSTS (*Continued*)

Carbenoid Precursor	Substrate	Conditions	Product(s) and Yield(s) (%)							Refs.

| | 4-R-C6H4-CH=CH2 (R = Cl, OMe) | Rh2(S-TBSP)4 (**161b**), pentane | I (Ph, CO2Me, Ar) + II (Ph, CO2Me, Ar) | | | | | | | 73 |

				I+II	I:II	ee (%) I	ee (%) II	abs stereochem I	abs stereochem II	
	R = Cl			(84)	98:2	85	—	1R,2S	—	
	R = OMe			(82)	98:2	88	—	1R,2S	—	

| | CH2=C(Me)Ph | | I (Ph, Me, CO2Me, Ph) + II (Ph, Ph, CO2Me) | | | | | | | |

				I+II	I:II	ee (%) I	ee (%) II	abs stereochem I	abs stereochem II	
		Rh2(S-TBSP)4 (**161b**), pentane		(88)	60:40	85	81	—	—	74
		161e, CH2Cl2		(86)	61:39	63	73	—	—	74
		164a, CH2Cl2, reflux		(96)	73:27	43	50	—	—	145
		164a, benzene, reflux		(61)	72:28	49	55	—	—	145
		164a, pentane, reflux		(94)	72:28	64	62	—	—	145
		164a, cyclohexane, reflux		(78)	72:28	65	63	—	—	145
		164b, CH2Cl2, reflux		(98)	74:26	42	51	—	—	145
		170, CH2Cl2, reflux		(94)	74:26	26	45	—	—	145
		Rh2(4S-TBOIM)4 (**174b**), CH2Cl2		(91)	69:31	36	14	—	—	74

232

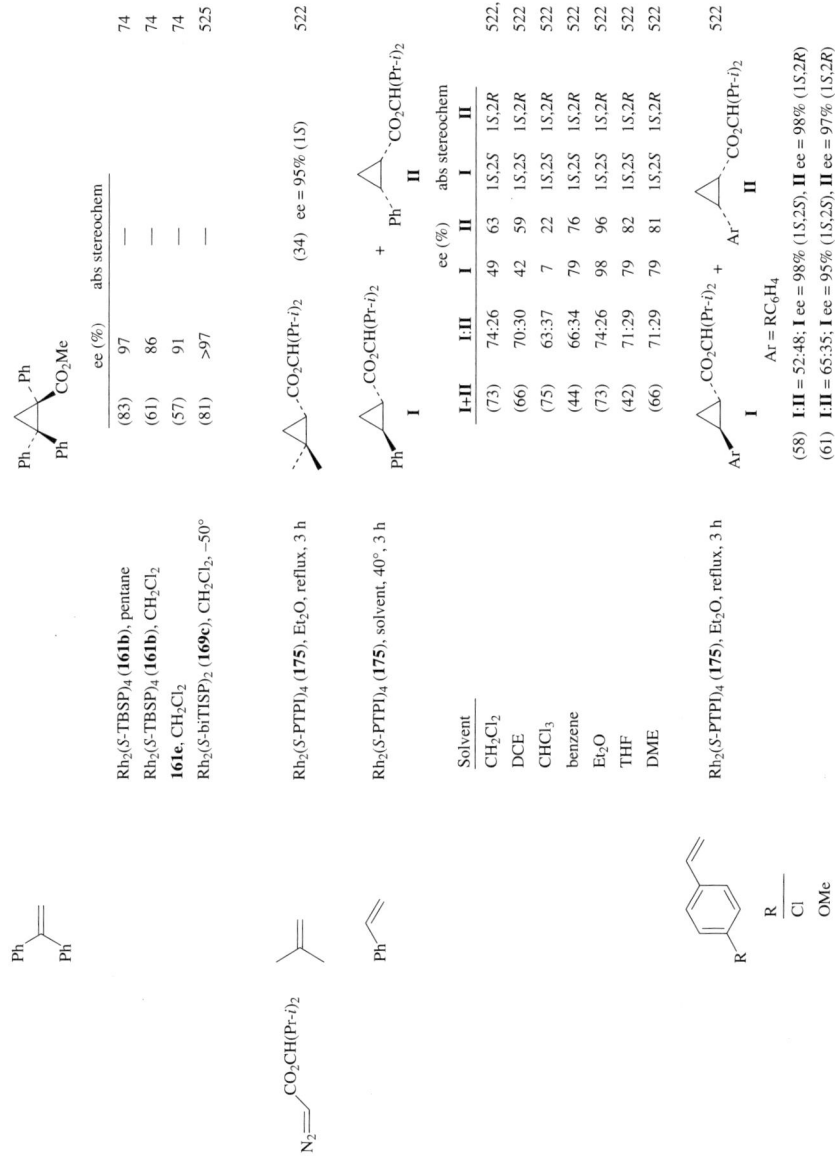

	ee (%)	abs stereochem	
Rh$_2$(S-TBSP)$_4$ (161b), pentane	(83) 97	—	74
Rh$_2$(S-TBSP)$_4$ (161b), CH$_2$Cl$_2$	(61) 86	—	74
161e, CH$_2$Cl$_2$	(57) 91	—	74
Rh$_2$(S-biTISP)$_2$ (169c), CH$_2$Cl$_2$, −50°	(81) >97	—	525

Rh$_2$(S-PTPI)$_4$ (175), Et$_2$O, reflux, 3 h	(34)	ee = 95% (1S)	522

Rh$_2$(S-PTPI)$_4$ (175), solvent, 40°, 3 h

		ee (%)		abs stereochem		
Solvent	I+II	I:II	I	II	I	II
CH$_2$Cl$_2$	(73)	74:26	49	63	1S,2S	1S,2R	522, 519
DCE	(66)	70:30	42	59	1S,2S	1S,2R	522
CHCl$_3$	(75)	63:37	7	22	1S,2S	1S,2R	522
benzene	(44)	66:34	79	76	1S,2S	1S,2R	522
Et$_2$O	(73)	74:26	98	96	1S,2S	1S,2R	522
THF	(42)	71:29	79	82	1S,2S	1S,2R	522
DME	(66)	71:29	79	81	1S,2S	1S,2R	522

Rh$_2$(S-PTPI)$_4$ (175), Et$_2$O, reflux, 3 h

Ar = RC$_6$H$_4$

R	
Cl	(58) I:II = 52:48; I ee = 98% (1S,2S), II ee = 98% (1S,2R)
OMe	(61) I:II = 65:35; I ee = 95% (1S,2S), II ee = 97% (1S,2R)

522

TABLE VI. ASYMMETRIC CYCLOPROPANATION OF ALKENES USING CHIRAL RHODIUM CATALYSTS (*Continued*)

Carbenoid Precursor	Substrate	Conditions	Product(s) and Yield(s) (%)	Refs.
$N_2\diagdown CO_2R$ $R = CMe(Pr-i)_2$	$n\text{-}C_6H_{13}$ CH=CH$_2$	Rh$_2$(S-PTPI)$_4$ (**175**), Et$_2$O, reflux, 3 h	(52) **I:II** = 71:29 **I** ee = 89% (1S,2S), **II** ee = 82% (1S,2R) $n\text{-}C_6H_{13}\text{-}\triangle\text{-}CO_2R$ (**I**) + $n\text{-}C_6H_{13}\text{-}\triangle\text{-}CO_2R$ (**II**) $R = CH(Pr-i)_2$	522
	Ph C(CH$_3$)=CH$_2$	Rh$_2$(S-PTPI)$_4$ (**175**), Et$_2$O, reflux, 3 h	(80) **I:II** = 54:46 **I** ee = 94% (1S,2S), **II** ee = 95% (1S,2R) Ph-△-CO$_2$R (**I**) + Ph-△-CO$_2$R (**II**) $R = CH(Pr-i)_2$	522
	Ph$_2$C=CH$_2$	Rh$_2$(S-PTPI)$_4$ (**175**), Et$_2$O, reflux, 3 h	(77) ee = 95% (1S) Ph$_2$-△-CO$_2$R $R = CH(Pr-i)_2$	522
	Ph CH=CH$_2$	Rh$_2$(S-PTPI)$_4$ (**175**), Et$_2$O, reflux, 3 h	(92) **I:II** = 62:38 **I** ee = 65% (1S,2S), **II** ee = 52% (1S,2R) Ph-△-CO$_2$R (**I**) + Ph-△-CO$_2$R (**II**)	519
	Ph CH=CH$_2$	Rh$_2$(5S-MEPY)$_4$ (**171a**), CH$_2$Cl$_2$, reflux	(51) **I:II** = 73:27 **I** ee = 67% (1S,2S), **II** ee = 83% (1S,2R) Ph-△-CO$_2$R (**I**) + Ph-△-CO$_2$R (**II**)	108

C$_{10}$

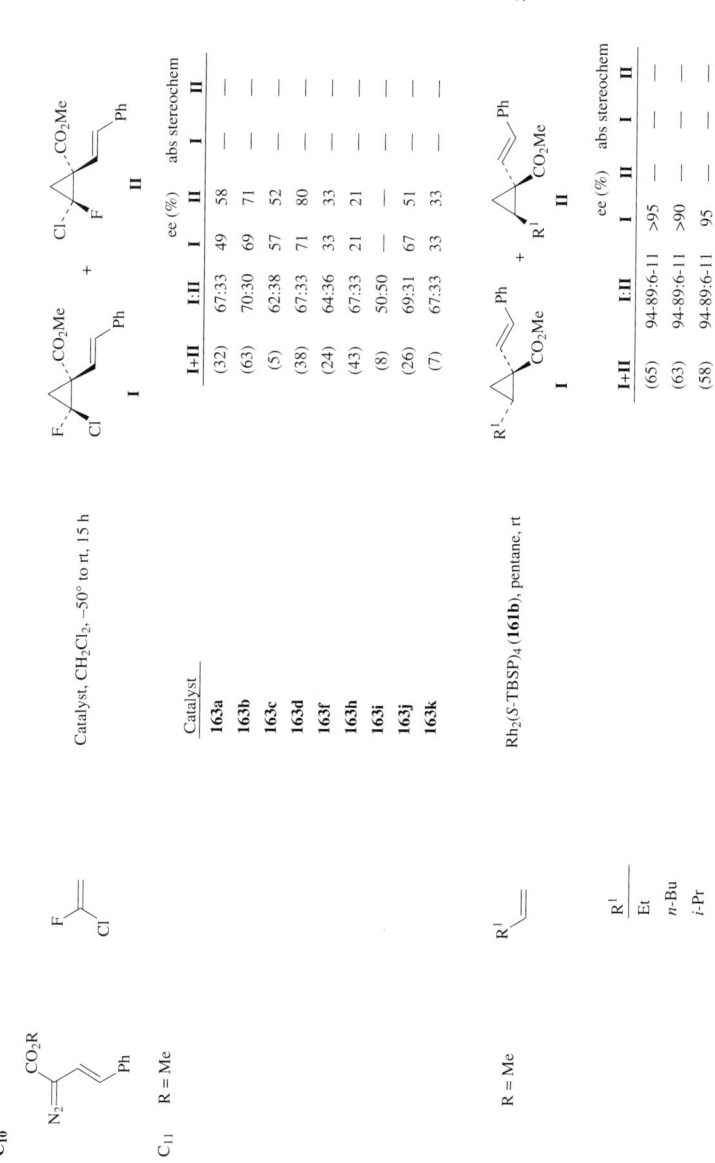

TABLE VI. ASYMMETRIC CYCLOPROPANATION OF ALKENES USING CHIRAL RHODIUM CATALYSTS (*Continued*)

Carbenoid Precursor	Substrate	Conditions	Product(s) and Yield(s) (%)							Refs.
			I+II	**I:II**	**I**	**II**	**I**	**II**		
					ee (%)		abs stereochem			
R = Me	R¹O—=		R¹O⋯△—CH=CH—Ph, CO₂Me (**I**) + R¹O⋯△—CH=CH—Ph, CO₂Me (**II**)							
	R¹									
	Ac	Rh₂(S-DOSP)₄ (**161a**), pentane, −78°	(26)	>95:5	95	—	1R,2S	—		58
	Et	Rh₂(S-DOSP)₄ (**161a**), pentane, −78°	(65)	>95:5	93	—	1R,2S	—		58
	Ac	Rh₂(S-TBSP)₄ (**161b**), pentane, rt	(40)	>95:5	76	—	1R,2S	—		58, 128
	Et	Rh₂(S-TBSP)₄ (**161b**), pentane, rt	(83)	>95:5	59	—	1R,2S	—		58, 128
	Et	**163a**, pentane, rt, 15 h	(78)	>95:5	47	—	—	—		527
	Et	**163b**, pentane, rt, 15 h	(78)	>95:5	52	—	—	—		527
	Et	**163c**, pentane, rt, 15 h	(76)	>95:5	23	—	—	—		527
	Et	**163d**, pentane, rt, 15 h	(70)	>95:5	60	—	—	—		527
	Et	**163f**, pentane, rt, 15 h	(69)	>95:5	19	—	—	—		527
	Et	**163h**, pentane, rt, 15 h	(70)	>95:5	19	—	—	—		527
	Et	**163i**, pentane, rt, 15 h	(70)	>95:5	14	—	—	—		527
	Et	**163j**, pentane, rt, 15 h	(71)	>95:5	47	—	—	—		527
	Et	Rh₂(S-biTISP)₂ (**169c**), CH₂Cl₂, −50°	(53)	>95:5	87	—	1S,2R	—		525
R = Me	⟨O⟩ (dihydrofuran)	Rh₂(S-DOSP)₄ (**161a**), pentane, −78°	(84) ee = 86%							58
R = Me	isobutylene	Rh₂(S-DOSP)₄ (**161a**), pentane	(52) ee = 95%							58
R = Me	propene	Rh₂(S-DOSP)₄ (**161a**), pentane	(80)							58

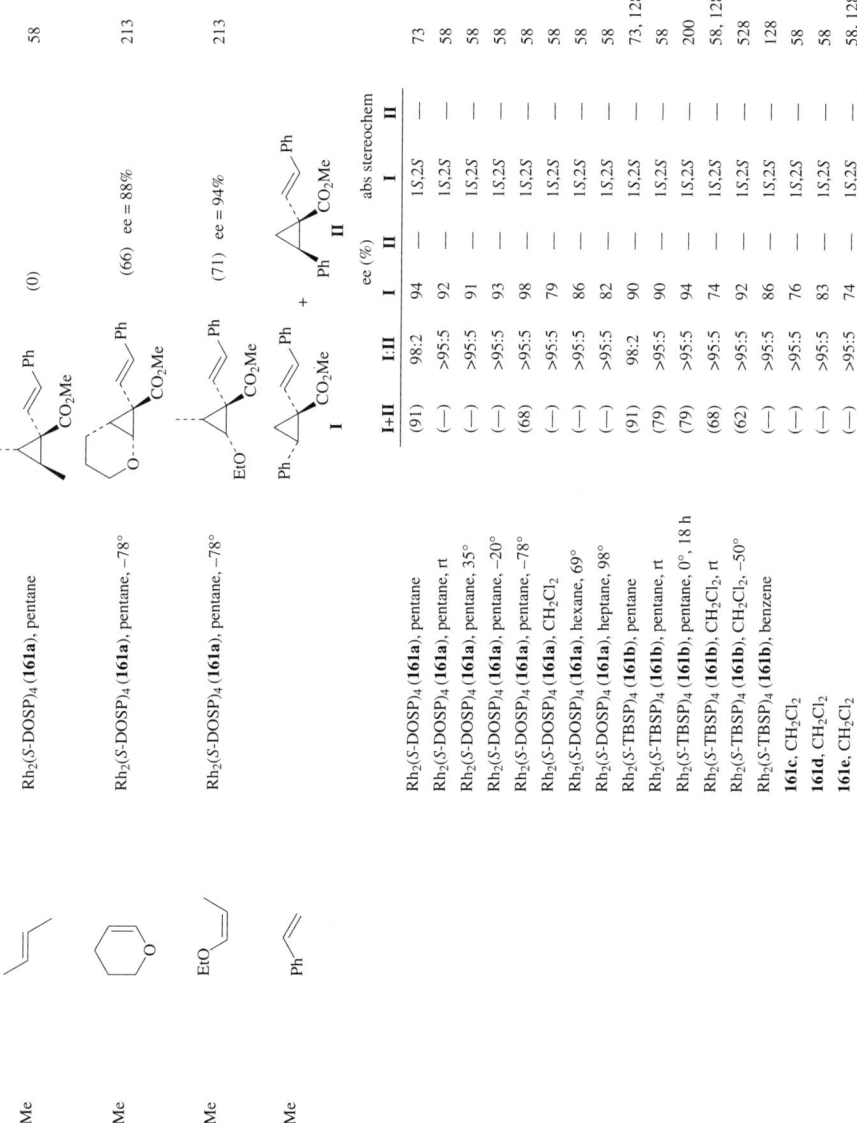

	I+II	I:II	ee (%) I	ee (%) II	abs stereochem I	abs stereochem II	
Rh$_2$(S-DOSP)$_4$ (161a), pentane	(91)	98:2	94	—	1S,2S	—	73
Rh$_2$(S-DOSP)$_4$ (161a), pentane, rt	(—)	>95:5	92	—	1S,2S	—	58
Rh$_2$(S-DOSP)$_4$ (161a), pentane, 35°	(—)	>95:5	91	—	1S,2S	—	58
Rh$_2$(S-DOSP)$_4$ (161a), pentane, −20°	(—)	>95:5	93	—	1S,2S	—	58
Rh$_2$(S-DOSP)$_4$ (161a), pentane, −78°	(68)	>95:5	98	—	1S,2S	—	58
Rh$_2$(S-DOSP)$_4$ (161a), CH$_2$Cl$_2$	(—)	>95:5	79	—	1S,2S	—	58
Rh$_2$(S-DOSP)$_4$ (161a), hexane, 69°	(—)	>95:5	86	—	1S,2S	—	58
Rh$_2$(S-DOSP)$_4$ (161a), heptane, 98°	(—)	>95:5	82	—	1S,2S	—	58
Rh$_2$(S-TBSP)$_4$ (161b), pentane	(91)	98:2	90	—	1S,2S	—	73, 128
Rh$_2$(S-TBSP)$_4$ (161b), pentane, rt	(79)	>95:5	90	—	1S,2S	—	58
Rh$_2$(S-TBSP)$_4$ (161b), pentane, 0°, 18 h	(79)	>95:5	94	—	1S,2S	—	200
Rh$_2$(S-TBSP)$_4$ (161b), CH$_2$Cl$_2$, rt	(68)	>95:5	74	—	1S,2S	—	58, 128
Rh$_2$(S-TBSP)$_4$ (161b), CH$_2$Cl$_2$, −50°	(62)	>95:5	92	—	1S,2S	—	528
Rh$_2$(S-TBSP)$_4$ (161b), benzene	(—)	>95:5	86	—	1S,2S	—	128
161c, CH$_2$Cl$_2$	(—)	>95:5	76	—	1S,2S	—	58
161d, CH$_2$Cl$_2$	(—)	>95:5	83	—	1S,2S	—	58
161e, CH$_2$Cl$_2$	(—)	>95:5	74	—	1S,2S	—	58, 128

TABLE VI. ASYMMETRIC CYCLOPROPANATION OF ALKENES USING CHIRAL RHODIUM CATALYSTS (Continued)

Carbenoid Precursor	Substrate	Conditions	Product(s) and Yield(s) (%)			Refs.
		161e, benzene	(—)	>95:5	87	—	128
		162a, CH$_2$Cl$_2$	(—)	>95:5	75	1S,2S	58
		162b, CH$_2$Cl$_2$	(—)	>95:5	61	1S,2S	58
		162c, CH$_2$Cl$_2$	(—)	>95:5	30	1S,2S	58
		163a, pentane, rt, 15 h	(75)	>95:5	64	1S,2S	527
		163b, pentane, rt, 15 h	(91)	>95:5	76	1S,2S	527
		163c, pentane, rt, 15 h	(68)	>95:5	59	1S,2S	527
		163d, pentane, rt, 15 h	(84)	>95:5	48	1S,2S	527
		163e, pentane, rt, 15 h	(43)	>95:5	36	1S,2S	527
		163f, pentane, rt, 15 h	(52)	>95:5	40	1S,2S	527
		163h, pentane, rt, 15 h	(74)	>95:5	7	1S,2S	527
		163i, pentane, rt, 15 h	(13)	>95:5	20	1R,2R	527
		163j, pentane, rt, 15 h	(55)	>95:5	64	1S,2S	527
		163k, pentane, rt, 15 h	(71)	>95:5	55	1R,2R	527
		164a, pentane	(—)	>95:5	81	1S,2S	58
		165, pentane	(—)	>95:5	81	1S,2S	58
		167, CH$_2$Cl$_2$, rt	(81)	>95:5	59	1R,2R	528
		167, CH$_2$Cl$_2$, −50°	(52)	>95:5	83	1R,2R	528
		167, pentane, rt	(78)	>95:5	56	1R,2R	528
		168o, CH$_2$Cl$_2$	(—)	>95:5	30	1S,2S	58
		168p, CH$_2$Cl$_2$	(—)	>95:5	6	1S,2S	58
		Rh$_2$(S-biTBSP)$_2$ (**169a**), CH$_2$Cl$_2$, rt	(52-92)	>95:5	63	1R,2R	525
		Rh$_2$(S-biTBSP)$_2$ (**169a**), hexane, rt	(52-92)	>95:5	55	1R,2R	525
		Rh$_2$(S-biDOSP)$_2$ (**169b**), CH$_2$Cl$_2$, rt	(52-92)	>95:5	68	1R,2R	525
		Rh$_2$(S-biDOSP)$_2$ (**169b**), hexane, rt	(52-92)	>95:5	53	1R,2R	525
		Rh$_2$(S-biTISP)$_2$ (**169c**), CH$_2$Cl$_2$, rt	(52-92)	>95:5	90	1R,2R	525
		Rh$_2$(S-biTISP)$_2$ (**169c**), CH$_2$Cl$_2$, −50°	(52-92)	>95:5	98	1R,2R	525
		Rh$_2$(S-biTISP)$_2$ (**169c**), hexane, rt	(52-92)	>95:5	74	1R,2R	525

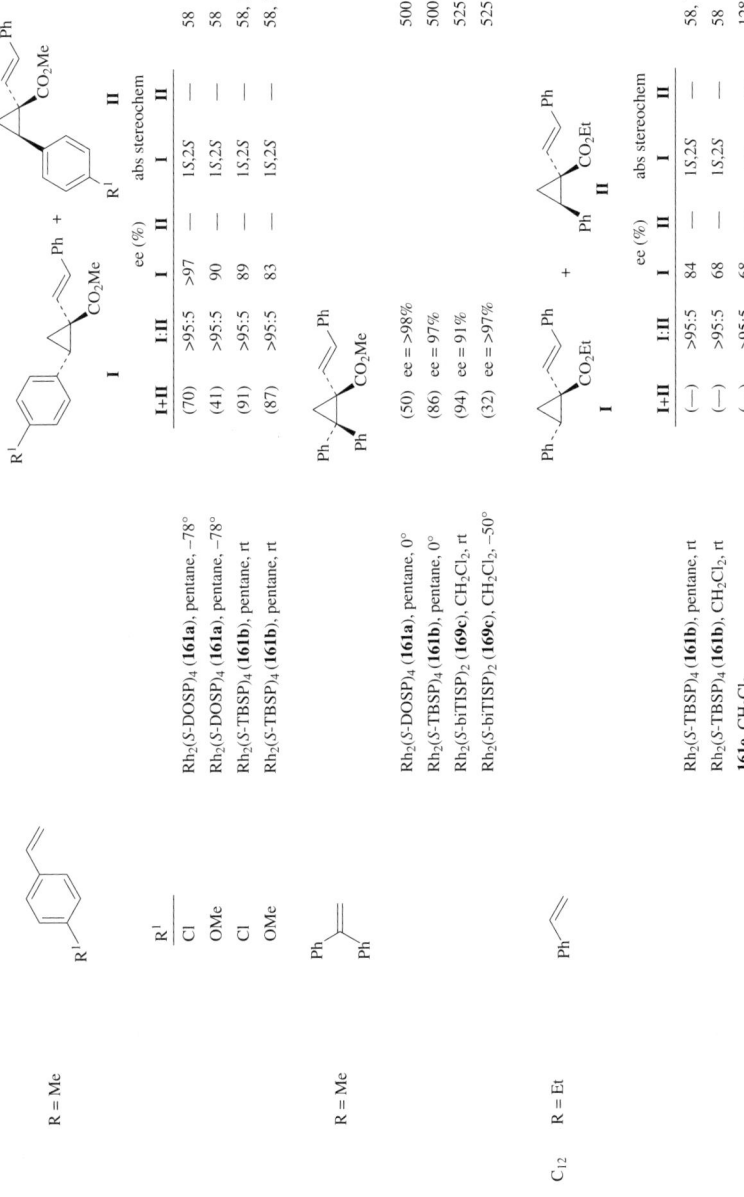

R = Me

R¹				I+II	I:II	ee (%) I	ee (%) II	abs stereochem I	abs stereochem II	
Cl	Rh₂(S-DOSP)₄ (**161a**), pentane, –78°			(70)	>95:5	>97	—	1S,2S	—	58
OMe	Rh₂(S-DOSP)₄ (**161a**), pentane, –78°			(41)	>95:5	90	—	1S,2S	—	58
Cl	Rh₂(S-TBSP)₄ (**161b**), pentane, rt			(91)	>95:5	89	—	1S,2S	—	58, 128
OMe	Rh₂(S-TBSP)₄ (**161b**), pentane, rt			(87)	>95:5	83	—	1S,2S	—	58, 128

R = Me

Rh₂(S-DOSP)₄ (**161a**), pentane, 0°				(50)	ee = >98%		500
Rh₂(S-TBSP)₄ (**161b**), pentane, 0°				(86)	ee = 97%		500
Rh₂(S-biTISP)₂ (**169c**), CH₂Cl₂, rt				(94)	ee = 91%		525
Rh₂(S-biTISP)₂ (**169c**), CH₂Cl₂, –50°				(32)	ee = >97%		525

C₁₂ R = Et

				I+II	I:II	ee (%) I	ee (%) II	abs stereochem I	abs stereochem II	
Rh₂(S-TBSP)₄ (**161b**), pentane, rt				(—)	>95:5	84	—	1S,2S	—	58, 128
Rh₂(S-TBSP)₄ (**161b**), CH₂Cl₂, rt				(—)	>95:5	68	—	1S,2S	—	58
161e, CH₂Cl₂				(—)	>95:5	68	—	—	—	128

239

TABLE VI. ASYMMETRIC CYCLOPROPANATION OF ALKENES USING CHIRAL RHODIUM CATALYSTS (Continued)

	Carbenoid Precursor	Substrate	Conditions	Product(s) and Yield(s) (%)					Refs.
						ee (%)		abs stereochem	
				I+II	I:II	I	II	I	II	
C_{13}	R = i-Pr		$Rh_2(S\text{-TBSP})_4$ (**161b**), pentane, rt	(—)	>95:5	76	—	1S,2S	—	58, 128
			$Rh_2(S\text{-TBSP})_4$ (**161b**), CH_2Cl_2, rt	(—)	>95:5	43	—	1S,2S	—	58
			161e, CH_2Cl_2	(—)	>95:5	43	—	—	—	128
						ee (%)		abs stereochem	
				I+II	I:II	I	II	I	II	
C_{14}	R = t-Bu		$Rh_2(S\text{-TBSP})_4$ (**161b**), pentane, rt	(—)	>95:5	50	—	1S,2S	—	58, 128
			$Rh_2(S\text{-TBSP})_4$ (**161b**), CH_2Cl_2, rt	(38)	>95:5	9	—	1S,2S	—	58
			161e, CH_2Cl_2	(—)	>95:5	9	—	—	—	128
			161e, benzene	(—)	>95:5	28	—	—	—	128
			167, CH_2Cl_2, rt	(59)	>95:5	49	—	1R,2R	—	528

C_{12}

$N_2\diagdown\!\!\!\diagup CO_2R$, R = menthyl

$R^1\diagdown\!\!\triangle\!\!\diagup CO_2R$ + $R^1\diagdown\!\!\triangle\!\!\diagup CO_2R$
       I   **II**

						ee (%)		abs stereochem		
Menthyl stereochemistry	R^1			I+II	I:II	I	II	I	II	
d	OEt		$Rh_2(5S\text{-MEPY})_4$ (**171a**), CH_2Cl_2, reflux	(48)	66:34	37	49	—	—	309
l	CH_2OAc			(26)	70:30	38	26	—	—	
d	Bu			(65)	91:9	25	75	—	—	
d	t-Bu			(62)	80:20	53	82	—	—	
l	CH_2OAc		$Rh_2(4S\text{-PHOX})_4$ (**173**), CH_2Cl_2, reflux	(5)	51:49	34	50	—	—	

240

R = menthyl

Menthyl stereochemistry		I+II	I:II	ee (%) I	ee (%) II	abs stereochem I	abs stereochem II	
d	Rh$_2$(S-BDME)$_4$ (166), CH$_2$Cl$_2$, rt, 15 h	(100)	37:63	45	99	1S,2S	1S,2R	129
d	Rh$_2$(5S-MEPY)$_4$ (171a), CH$_2$Cl$_2$, reflux	(69)	67:33	48	86	1S,2S	1S,2R	108, 109
l	Rh$_2$(5S-MEPY)$_4$ (171a), CH$_2$Cl$_2$, reflux	(47)	67:33	56	79	1S,2S	1S,2R	108, 109
d	Rh$_2$(5R-MEPY)$_4$ (ent-171a), CH$_2$Cl$_2$, reflux	(32)	51:49	36	79	1R,2R	1R,2S	109
l	Rh$_2$(5R-MEPY)$_4$ (ent-171a), CH$_2$Cl$_2$, reflux	(38)	56:44	31	88	1R,2R	1R,2S	109
d	Rh$_2$(5S-PYCA)$_4$ (171b), DCE, reflux	(79)	—	50	54	—	—	529
l	Rh$_2$(5S-DMAP)$_4$ (171c), CH$_2$Cl$_2$, reflux	(27)	63:37	2	13	1S,2S	1S,2R	520
d	Rh$_2$(5S-DMAP)$_4$ (171c), DCE, reflux	(75)	64:36	1	15	1S,2S	1S,2R	520
d	Rh$_2$(5S-DMAP)$_4$ (171c), DCE, reflux	(83)	58:42	7	33	1S,2S	1S,2R	520
d	Rh$_2$(4R-BNOX)$_4$ (172a), CH$_2$Cl$_2$, reflux	(46)	67:33	34	62	1S,2S	1S,2R	108
l	Rh$_2$(4R-BNOX)$_4$ (172a), CH$_2$Cl$_2$, reflux	(51)	62:38	4	25	1S,2S	1S,2R	108
d	Rh$_2$(4S-BNOX)$_4$ (ent-172a), CH$_2$Cl$_2$, reflux	(50)	63:37	4	24	1R,2R	1R,2S	108
l	Rh$_2$(4S-BNOX)$_4$ (ent-172a), CH$_2$Cl$_2$, reflux	(56)	61:39	34	63	1R,2R	1R,2S	108
d	Rh$_2$(4S-IPOX)$_4$ (173), CH$_2$Cl$_2$, reflux	(74)	75:25	2	4	1R,2R	1R,2S	108
l	Rh$_2$(4S-IPOX)$_4$ (173), CH$_2$Cl$_2$, reflux	(55)	70:30	34	56	1R,2R	1R,2S	108
d	Rh$_2$(4S-PHOX)$_4$ (173), CH$_2$Cl$_2$, reflux, 13 h	(24)	59:41	4	6	1S,2S	1R,2S	521, 309
l	Rh$_2$(4S-PHOX)$_4$ (173), CH$_2$Cl$_2$, reflux, 13 h	(19)	27:73	40	72	1R,2R	1R,2S	521, 309
d	175, CH$_2$Cl$_2$, reflux, 3 h	(79)	72:18	89	83	1S,2S	1S,2R	519
l	175, CH$_2$Cl$_2$, reflux, 3 h	(62)	52:48	73	79	1S,2S	1S,2R	519
l	Rh$_2$(4S-BNAZ)$_4$ (176a), CH$_2$Cl$_2$, reflux	(85)	43:57	24	44	1S,2S	1S,2R	524
d	Rh$_2$(4S-IBAZ)$_4$) (176b), CH$_2$Cl$_2$, rt	(61)	44:56	62	92	1S,2S	1S,2R	524
l	Rh$_2$(4S-IBAZ)$_4$) (176b), CH$_2$Cl$_2$, rt	(56)	38:62	62	88	1S,2S	1S,2R	524

TABLE VI. ASYMMETRIC CYCLOPROPANATION OF ALKENES USING CHIRAL RHODIUM CATALYSTS (*Continued*)

Carbenoid Precursor	Substrate	Conditions	Product(s) and Yield(s) (%)	Refs.

R = menthyl, substrate: CH$_2$=C(CH$_3$)Ph

Products: Ph-cyclopropane-CO$_2$R (**I**) + Ph-cyclopropane-CO$_2$R (**II**)

Menthyl stereochemistry	Conditions	(I+II)	I:II	ee (%) I	ee (%) II	abs stereochem I	abs stereochem II	Refs.
d	Rh$_2$(5S-MEPY)$_4$ (**171a**), CH$_2$Cl$_2$, reflux	(32)	59:41	58	78	—	—	309
l	Rh$_2$(5S-MEPY)$_4$ (**171a**), CH$_2$Cl$_2$, reflux	(59)	61:39	5	9	—	—	309
d	Rh$_2$(S-PTPI)$_4$) (**175**), CH$_2$Cl$_2$, reflux, 3 h	(63)	64:36	90	74	—	—	519

R = menthyl, substrate: CH$_2$=CPh$_2$

Product: Ph$_2$C-cyclopropane-CO$_2$R

Menthyl stereochemistry	Conditions		ee (%)	Refs.
l	Rh$_2$(5S-MEPY)$_4$ (**171a**), CH$_2$Cl$_2$, reflux	(46)	4	309
d	Rh$_2$(S-PTPI)$_4$) (**175**), CH$_2$Cl$_2$, reflux, 3 h	(81)	84	519

C$_{14}$ N$_2$=CHCOR, R = (+)-2-phenylcyclohexyloxy, substrate: CH$_2$=CHCH$_2$Ph

Products: Ph-CH$_2$-cyclopropane-COR (**I**) + Ph-CH$_2$-cyclopropane-COR (**II**)

Conditions	(I+II)	I:II	ee (%) I	ee (%) II	abs stereochem I	abs stereochem II	Refs.
Rh$_2$(5S-MEPY)$_4$ (**171a**), CH$_2$Cl$_2$, reflux	(55)	84:16	48	37	—	—	109
Rh$_2$(5R-MEPY)$_4$ (*ent*-**171a**), CH$_2$Cl$_2$, reflux	(48)	82:18	70	80	—	—	109

C_{15}

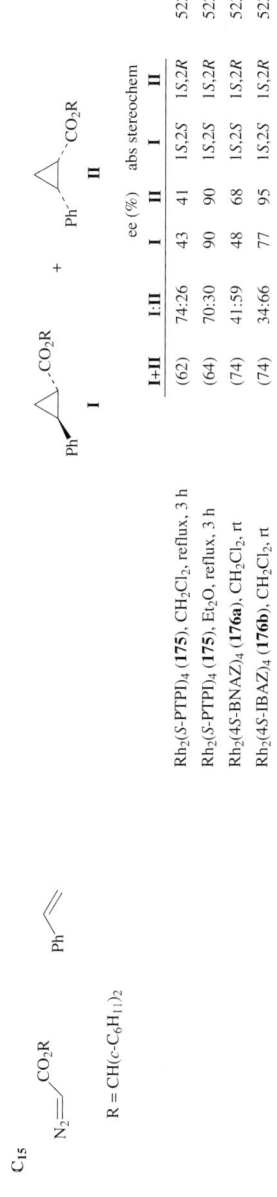

R = CH(c-C_6H_{11})$_2$

	I+II	I:II	ee (%) I	ee (%) II	abs stereochem I	abs stereochem II	
Rh$_2$(S-PTPI)$_4$ (**175**), CH$_2$Cl$_2$, reflux, 3 h	(62)	74:26	43	41	1S,2S	1S,2R	522
Rh$_2$(S-PTPI)$_4$ (**175**), Et$_2$O, reflux, 3 h	(64)	70:30	90	90	1S,2S	1S,2R	522
Rh$_2$(4S-BNAZ)$_4$ (**176a**), CH$_2$Cl$_2$, rt	(74)	41:59	48	68	1S,2S	1S,2R	523
Rh$_2$(4S-IBAZ)$_4$ (**176b**), CH$_2$Cl$_2$, rt	(74)	34:66	77	95	1S,2S	1S,2R	523

TABLE VII. ASYMMETRIC CYCLOPROPANATION OF ALKENES USING MISCELLANEOUS CHIRAL CATALYSTS

(See Chart 5 for Catalyst Structures)

C₃

Carbenoid Precursor	Substrate	Conditions	Product(s) and Yield(s) (%)	Refs.
N_2=C(CN)(CN)	Ph—	**179**, neat, 35°	Ph-△-(CN)(CN) (20) ee = 5% (2S)	530, 88
N_2=CHCO$_2$Me	MeO$_2$C—	**179**, neat, rt	MeO$_2$C-△-CO$_2$Me (11) ee = 33% (1S,2S)	530
	Ph—		Ph-△-CO$_2$Me (I) + Ph-△-CO$_2$Me (II)	

				ee (%)		abs stereochem		
		I+II	I:II	I	II	I	II	
	179, neat, 0°	(94)	41:59	61	—	—	—	87
	182g, CH$_2$Cl$_2$, rt, 18 h	(74)	90:10	88	70	1R,2R	1R,2S	104
	182g, CH$_2$Cl$_2$, 30°, 10 h	(82)	89:11	92	97	1R,2R	1R,2S	531
	193a, CH$_2$Cl$_2$, 30°, 10 h	(79)	89:11	71	48	1R,2R	1R,2S	531
	193b, CH$_2$Cl$_2$, 30°, 10 h	(88)	83:17	86	63	1R,2R	1R,2S	531

Carbenoid Precursor	Substrate	Conditions	Product(s) and Yield(s) (%)	Refs.
N_2=CHCO$_2$Et	NC—	**179**, neat, 49°	NC-△-CO$_2$Et (I) + NC-△-CO$_2$Et (II) (13) I:II = —; I ee = —, II ee = —	530

C₄

Carbenoid Precursor	Substrate	Conditions	Product(s) and Yield(s) (%)	Refs.
	n-C$_4$H$_9$—	**196**, DCE, reflux, 12 h	n-C$_4$H$_9$-△-CO$_2$Et (I) + n-C$_4$H$_9$-△-CO$_2$Et (II) (73) I:II = 66:34 I ee = 72% (1S,2S), II ee = 74% (1S,2R)	532

	I+II	I:II	ee (%)		abs stereochem			
					I		II	
			I	II	I	II	I	II	
179, neat, 10°	(91)	50:50	70	68	1S,2S	1S,2R		88
179, neat, 99°	(95)	57:43	33	16	1S,2S	1S,2R		500, 87
179, neat, rt	(93)	49:51	75	61	1S,2S	1S,2R		500
179, neat, 0°	(92)	46:64	75	67	1S,2S	1S,2R		500
179, neat, −15°	(80)	49:51	76	72	1S,2S	1S,2R		500
179, acetone, 0°	(81)	57:43	—	—	—	—		500
179, ethyl acetate, 0°	(92)	50:50	—	—	—	—		500
179, di-n-butyl ether, 0°	(88)	40:60	—	—	—	—		500
179, n-hexane, 0°	(67)	46:54	—	—	—	—		500
179, acetophenone, 0°	(95)	49:51	—	—	—	—		500
180, neat, 35°, 16 h	(50–60)	64:36	20	75	1S,2S	1S,2R		85
182a, CH$_2$Cl$_2$, rt, 18 h	(50)	91:9	82	54	1S,2S	1S,2R		106
182b, CH$_2$Cl$_2$, rt, 18 h	(82)	94:6	91	79	1R,2R	1R,2S		106
182c, CH$_2$Cl$_2$, 30°, 18 h	(66)	90:10	68	24	1R,2R	1R,2S		106
182d, CH$_2$Cl$_2$, 35°, 18 h	(38)	89:11	69	41	1S,2S	1S,2R		106
ent-182d, CH$_2$Cl$_2$, rt, 10 h	(30)	86:14	76	—	1R,2R	—		534
182e, CH$_2$Cl$_2$, rt, 18 h	(2)	88:12	80	53	1R,2R	1R,2S		533
182e, CH$_2$Cl$_2$, 40°, 18 h	(45)	90:10	78	55	1R,2R	1R,2S		533
182f, CH$_2$Cl$_2$, rt, 18 h	(29)	92:8	86	70	1R,2R	1R,2S		533
182f, CH$_2$Cl$_2$, 40°, 18 h	(66)	90:10	84	68	1R,2R	1R,2S		533
182g, CH$_2$Cl$_2$, rt, 18 h	(73)	91:9	89	79	1R,2R	1R,2S		533, 104, 535
182g, CH$_2$Cl$_2$, 40°, 18 h	(90)	93:7	89	82	1R,2R	1R,2S		533
182h, CH$_2$Cl$_2$, rt, 18 h	(60)	92:8	90	80	1R,2R	1R,2S		533
182h, CH$_2$Cl$_2$, 40°, 18 h	(84)	90:10	87	72	1R,2R	1R,2S		533
182i, CH$_2$Cl$_2$, rt, 18 h	(74)	91:9	93	87	1R,2R	1R,2S		533
182i, CH$_2$Cl$_2$, 40°, 18 h	(81)	90:10	85	68	1R,2R	1R,2S		533

TABLE VII. ASYMMETRIC CYCLOPROPANATION OF ALKENES USING MISCELLANEOUS CHIRAL CATALYSTS (Continued)

Carbenoid Precursor	Substrate	Conditions	Product(s) and Yield(s) (%)				Refs.
		184, CH$_2$Cl$_2$, 5 h	(98)	86:14	—	—	536
		185, CH$_2$Cl$_2$, rt, 10 h	(51)	91:9	60	1S,2S	534
		186, CH$_2$Cl$_2$, rt, 10 h	(47)	89:11	41	1R,2R	534
		187, neat, rt, 5 h	(—)	90:10	14	34	84
		188, CH$_2$Cl$_2$, rt, 18 h	(83)	95:5	87	1S,2S	370
		188, CH$_2$Cl$_2$, 0°, 18 h	(63)	96:4	91	1S,2S	370
		188, DCE, rt	(100)	96:4	87	1R,2R	83
		ent-**188**, DCE, rt	(100)	96:4	16	1S,2S	83
		188, DCE, 0°	(100)	95:5	91	1R,2R	83
		188, benzene, rt	(100)	91:9	84	1R,2R	83
		189a, neat, rt	(85)	80:20	46	1R,2R	537
		189a, CH$_2$Cl$_2$, rt, 24 h	(—)	86:14	58	1R,2R	455
		189c, CH$_2$Cl$_2$, rt, 24 h	(—)	87:13	15	1R,2R	455
		189d, CH$_2$Cl$_2$, rt, 24 h	(—)	70:30	0	11	455
		190a, CH$_2$Cl$_2$, rt, 24 h	(47)	65:35	6	1R,2R	538
		190b, CH$_2$Cl$_2$, rt, 24 h	(32)	65:35	4	1R,2R	538
		190c, CH$_2$Cl$_2$, rt, 24 h	(50)	65:35	8	1R,2R	538
		190d, CH$_2$Cl$_2$, rt, 24 h	(60)	65:35	6	1R,2R	538
		190e, CH$_2$Cl$_2$, rt, 24 h	(49)	65:35	4	1R,2R	538
		190f, CH$_2$Cl$_2$, rt, 24 h	(56)	65:35	3	1R,2R	538
		190g, CH$_2$Cl$_2$, rt, 24 h	(49)	65:35	8	1R,2R	538
		191, CH$_2$Cl$_2$, rt, 24 h	(63)	65:35	4	1R,2R	538
		193b, CH$_2$Cl$_2$, 30°, 10 h	(93)	89:11	90	1R,2R	531
		194, CH$_2$Cl$_2$, rt, 24 h	(0)	—	—	—	539
		195, CH$_2$Cl$_2$, rt, 24 h	(0)	—	—	—	539
		196, DCE, reflux, 12 h	(71)	68:32	87	1S,2S	532
		197, CH$_2$Cl$_2$, rt, 24 h	(21)	60:40	14	35	539
		197, CH$_2$Cl$_2$, −76°, 24 h	(4)	91:9	18	—	539
		197/AgOTf, CH$_2$Cl$_2$, rt, 24 h	(84)	59:41	17	40	539
		197/[NEt$_4$]Cl, CH$_2$Cl$_2$, rt, 24 h	(15)	60:40	13	32	539

R		I+II	I:II	ee (%) I	ee (%) II	abs stereochem I	abs stereochem II	
Cl	**179**, neat, 0°	(96)	45:55	—	—	—	—	87
Me	**179**, neat, 0°	(91)	50:50	—	—	—	—	87
OMe	**179**, neat, 0°	(96)	49:51	—	—	—	—	87
Cl	**188**, CH$_2$Cl$_2$, rt, 18 h	(66)	96:4	90	4	—	—	370
Me	**188**, CH$_2$Cl$_2$, rt, 18 h	(78)	95:5	81	9	—	—	370
OMe	**188**, CH$_2$Cl$_2$, rt, 18 h	(61)	94:6	85	8	—	—	370
F	**189a**, neat, rt	(92)	92:8	50	—	—	—	537
Cl	**189a**, neat, rt	(93)	90:10	52	—	—	—	537
Br	**189a**, neat, rt	(93)	86:14	45	—	—	—	537
Me	**189a**, neat, rt	(92)	91:9	46	—	—	—	537
OMe	**189a**, neat, rt	(>95)	86:14	47	—	—	—	537

180, neat, 35°, 16 h   (50–60) **I:II** = 64:36   **I** ee = 97%, **II** ee = 18%   85

		I+II	I:II	ee (%) I	ee (%) II	abs stereochem I	abs stereochem II	
	179, neat, 0°	(97)	—	—	—	—	—	530
	188, CH$_2$Cl$_2$, rt, 18 h	(69)	75:25	87	35	—	—	530
								370

TABLE VII. ASYMMETRIC CYCLOPROPANATION OF ALKENES USING MISCELLANEOUS CHIRAL CATALYSTS (*Continued*)

Carbenoid Precursor	Substrate	Conditions	Product(s) and Yield(s) (%)	Refs.
	indene	**196**, DCE, reflux, 12 h	(75) **I:II** = 72:28 **I** ee = 89%, **II** ee = 75%	532
	Ph-CH=CH-CH₃	**196**, DCE, reflux, 12 h	(67) **I:II** = 70:30 **I** ee = 64% (1S,2S), **II** ee = 60% (1S,2R)	532
	MeO₂C-C(=CH₂)-Ph	**179**, neat, 0°	(92) **I:II** = — **I** ee = 37% (1R,2S), **II** ee = 71% (1S,2S)	530
	Ph₂C=CH₂		ee (%) abs stereochem	
		179, —	(70) 37 1S	88
		179, neat, 5°	(95) 70 1S	87, 530
		188, CH₂Cl₂, rt, 18 h	(76) 81 —	370
		196, DCE, reflux, 12 h	(65) 81 1S	532
C₅ N₂=CH-CO₂Pr-*i*	Ph-CH=CH₂	**179**, neat, 0° **193b**, CH₂Cl₂, 30°, 10 h	(91) **I:II** = 53:47; **I** ee = 84% (80) **I:II** = 92:8; **I** ee = 90% (1R,2R), **II** ee = 68% (1R,2S)	87 531

C_6

N$_2$=CHCO$_2$Bu-i + Ph-CH=CH$_2$ → Ph△CO$_2$Bu-i (I) + Ph△''CO$_2$Bu-i (II)

179, neat, 0° — (94) **I:II** = 48:52; **I** ee = 80% — 87

N$_2$=CHCO$_2$Bu-t + Ph-CH=CH$_2$ → Ph△CO$_2$Bu-t (I) + Ph△''CO$_2$Bu-t (II)

	I+II	I:II	ee (%) I	ee (%) II	abs stereochem I	abs stereochem II	
181a, CH$_2$Cl$_2$, rt, 24 h	(0)	—	—	—	—	—	540, 541, 542
181b, CH$_2$Cl$_2$, rt, 24 h	(0)	—	—	—	—	—	540
181c, CH$_2$Cl$_2$, rt, 24 h	(0)	—	—	—	—	—	541
181d, CH$_2$Cl$_2$, rt, 24 h	(0)	—	—	—	—	—	540
181e, CH$_2$Cl$_2$, rt, 24 h	(0)	—	—	—	—	—	541, 542
181f, CH$_2$Cl$_2$, rt, 24 h	(76)	98:2	73	—	1S,2S	—	541, 540, 542
181g, CH$_2$Cl$_2$, rt, 24 h	(85)	96:4	89	93	1S,2S	1S,2R	540, 542
181h, CH$_2$Cl$_2$, rt, 24 h	(76)	95:5	75	—	1S,2S	—	541, 540, 542
181i, CH$_2$Cl$_2$, rt, 24 h	(55)	94:6	83	42	1S,2S	1S,2R	540, 542
181j, CH$_2$Cl$_2$, rt, 24 h	(79)	95:5	64	51	1S,2S	1S,2R	540, 541, 542
ent-**181j**, CH$_2$Cl$_2$, rt, 24 h	(69)	96:4	74	—	1R,2R	—	541
181k, CH$_2$Cl$_2$, rt, 24 h	(83)	95:5	66	82	1S,2S	1S,2R	540, 542
181l, CH$_2$Cl$_2$, rt, 24 h	(80)	96:4	93	91	1S,2S	1S,2R	540, 542
182g, CH$_2$Cl$_2$, rt, 18 h	(65)	97:3	94	87	1R,2R	1R,2S	104
183, DCE, 65°, 24 h	(53)	67:33	<10	<10	—	—	103
193b, CH$_2$Cl$_2$, 30°, 10 h	(82)	93:7	68	42	1R,2R	1S,2R	531

TABLE VII. ASYMMETRIC CYCLOPROPANATION OF ALKENES USING MISCELLANEOUS CHIRAL CATALYSTS (*Continued*)

Carbenoid Precursor	Substrate	Conditions	Product(s) and Yield(s) (%)	Refs.
	4-Cl-C₆H₄-CH=CH₂	Catalyst, CH₂Cl₂, rt, 24 h	**I** (4-Cl-C₆H₄, CO₂Bu-*t* cyclopropane) + **II** (4-Cl-C₆H₄, CO₂Bu-*t* cyclopropane) Catalyst | I+II | I:II | ee(%) I | ee(%) II | abs stereochem I | abs stereochem II *ent*-**181j** | (67) | 94:6 | 71 | — | — | — **181l** | (86) | 97:3 | 96 | — | — | —	541 540, 542
	2-naphthyl-CH=CH₂	Catalyst, CH₂Cl₂, rt, 24 h	**I** (2-naphthyl, CO₂Bu-*t*) + **II** (2-naphthyl, CO₂Bu-*t*) Catalyst | I+II | I:II | ee I | ee II | abs I | abs II *ent*-**181j** | (66) | 88:12 | 70 | — | — | — **181l** | (87) | 95:5 | 92 | — | — | —	541 540, 542
C₇				
N₂=CH-CO₂R R = *neo*pentyl	Ph-CH=CH₂	**179**, neat, 0°	**I** (Ph, CO₂R) + **II** (Ph, CO₂R) (87) **I:II** = 70:30 **I** ee = 88% (1*S*,2*S*), **II** ee = 81% (1*S*,2*R*)	530, 87
C₈				
N₂=CH-COPh	Ph-CH=CH₂	**179**, neat, 50°	**I** (Ph, COPh) + **II** (Ph, COPh) (44) **I:II** = — **I** ee = 20% (1*S*,2*S*), **II** ee = —	530
N₂=CH-CO₂R R = *c*-C₆H₁₁	Ph-CH=CH₂	**179**, neat, 0°	**I** (Ph, CO₂R) + **II** (Ph, CO₂R) (72) **I:II** = 59:41; **I** ee = 78%	87

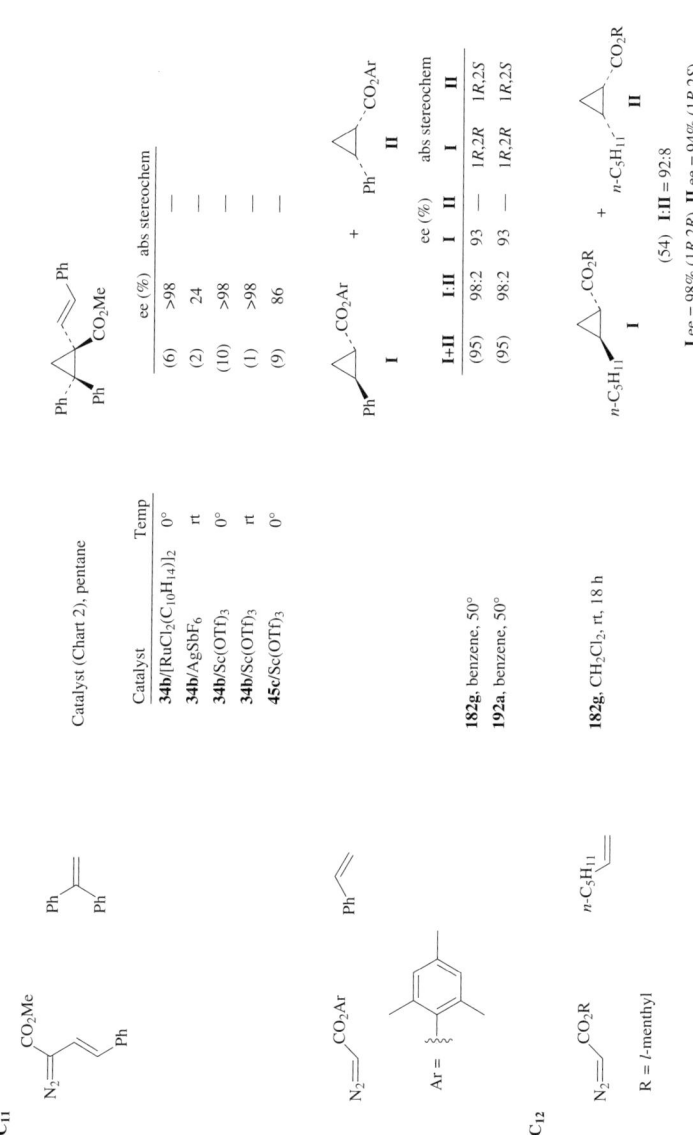

TABLE VII. ASYMMETRIC CYCLOPROPANATION OF ALKENES USING MISCELLANEOUS CHIRAL CATALYSTS (Continued)

Carbenoid Precursor: $N_2=\!\!\!=\!\!\!\diagup\!\!\!^{CO_2R}$

Substrate: $Ph\!\!-\!\!\!\diagup\!\!=$

Product(s): $Ph\cdots\!\triangle\!\!-\!CO_2R$ (I) + $Ph\cdots\!\triangle\!\!-\!CO_2R$ (II)

R = menthyl

Menthyl stereochemistry	Conditions	Yield (%) I+II	I:II	ee (%) I	ee (%) II	abs stereochem I	abs stereochem II	Refs.
l	182e, CH$_2$Cl$_2$, 40°, 18 h	(79)	94:6	84	38	1R,2R	1R,2S	533
l	182f, CH$_2$Cl$_2$, 40°, 18 h	(89)	96:4	90	67	1R,2R	1R,2S	533
l	182g, CH$_2$Cl$_2$, 40°, 18 h	(93)	97:3	93	79	1R,2R	1R,2S	533
l	182h, CH$_2$Cl$_2$, 40°, 18 h	(93)	97:3	94	83	1R,2R	1R,2S	533
l	182i, CH$_2$Cl$_2$, 40°, 18 h	(95)	96:4	97	85	1R,2R	1R,2S	533
d	182g, CH$_2$Cl$_2$, rt, 18 h	(82)	97:3	87	97	1R,2R	1R,2S	104, 535
l	182g, CH$_2$Cl$_2$, rt, 18 h	(83)	97:3	96	80	1R,2R	1R,2S	104, 535
d	182a, CH$_2$Cl$_2$, rt, 18 h	(77)	98:2	95	74	1S,2S	1S,2R	106
l	182a, CH$_2$Cl$_2$, rt, 18 h	(77)	96:4	72	86	1S,2S	1S,2R	106
d	182b, CH$_2$Cl$_2$, rt, 18 h	(86)	97:3	80	93	1R,2R	1R,2S	106
l	182b, CH$_2$Cl$_2$, rt, 18 h	(89)	99:1	96	66	1R,2R	1R,2S	106
d	182c, CH$_2$Cl$_2$, 30°, 18 h	(78)	93:7	71	86	1R,2R	1R,2S	106
l	182c, CH$_2$Cl$_2$, 30°, 18 h	(82)	98:2	93	64	1R,2R	1R,2S	106
d	182d, CH$_2$Cl$_2$, 35°, 18 h	(51)	95:5	84	14	1S,2S	1S,2R	106
l	182d, CH$_2$Cl$_2$, 35°, 18 h	(56)	97:3	14	66	1S,2S	1S,2R	106
d	193b, CH$_2$Cl$_2$, 30°, 10 h	(81)	96:4	55	69	1R,2R	1R,2S	531
l	193b, CH$_2$Cl$_2$, 30°, 10h	(84)	99:1	94	64	1R,2R	1R,2S	531

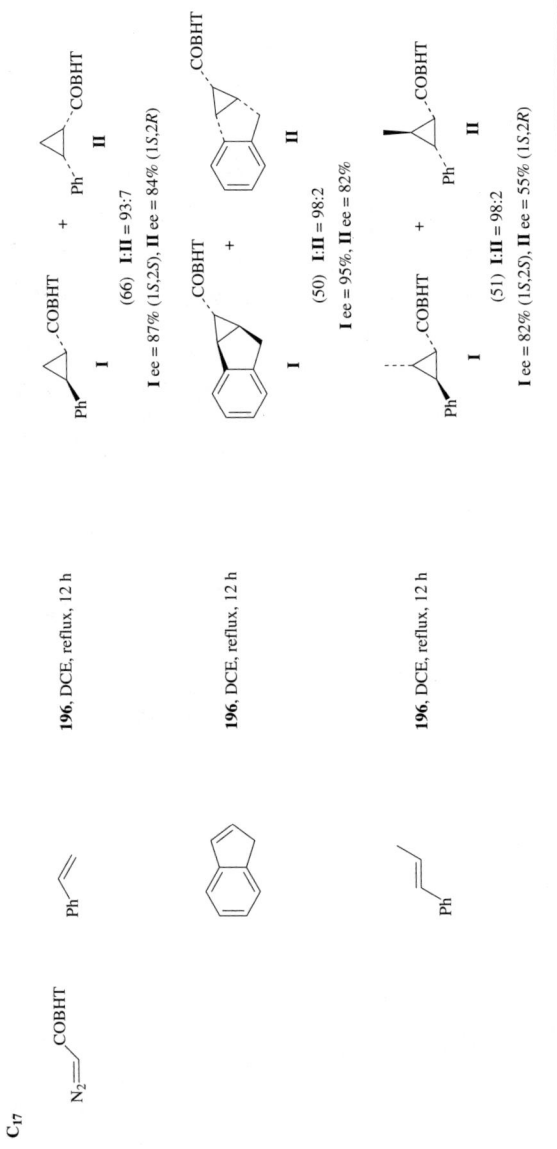

TABLE VIII. REACTION OF CARBENOIDS WITH DIENES

Carbenoid Precursor	Substrate	Conditions	Product(s) and Yield(s) (%)	Refs.
C₂ $N_2\!\!=\!\!\!\diagup\!\!\!^{CO_2R}$				
C₄ R = Et	Cl–CH=CH–CH=CH₂	Catalyst, neat, rt, 8 h	**I** (cyclopropane with CO₂Et and CH=CH–Cl) + **II** (cyclopropane with CO₂Et and CH=CH–Cl isomer)	
			Catalyst — I:II	
			Rh₂(OAc)₄ (73) 62:38	71, 151
			Rh₆(CO)₁₆ (21) 67:33	71, 151
			CuCl•P(OPr-i)₃ (17) 72:28	71, 151
			PdCl₂•2PhCN (<8) —	71
R = Et	(Z)-Cl–CH=CH–CH=CH₂	Rh₂(OAc)₄, neat, rt, 8 h	**I** + **II** (49) I:II = 62:38	71, 543
R = Et	CH₂=C(Cl)–CH=CH₂	Catalyst, neat, rt, 8 h	**I**: R¹ = CO₂Et, R² = H; **II**: R¹ = H, R² = CO₂Et + **III**: R¹ = CO₂Et, R² = H; **IV**: R¹ = H, R² = CO₂Et	
			Catalyst — I:II:III:IV	
			Rh₂(OAc)₄ (76) 37:33:17:13	71, 151
			Rh₆(CO)₁₆ (80) 37:33:19:11	71, 151
			CuCl•P(OPr-i)₃ (33) 48:37:10:5	71, 151
			PdCl₂•2PhCN (4) —	71

TABLE VIII. REACTION OF CARBENOIDS WITH DIENES (Continued)

Carbenoid Precursor	Substrate	Conditions	Product(s) and Yield(s) (%)					Refs.
		Catalyst (Chart 4), CH$_2$Cl$_2$, rt	I (CO$_2$R cyclopropane, vinyl) + II (CO$_2$R cyclopropane, vinyl)					
						ee (%)		
		Catalyst	I+II	I:II	I	II		
C$_3$ R = Me		Rh$_2$(OAc)$_4$	(60)	43:57	—	—		287
Me		Rh$_2$(5S-MEPY)$_4$ (171a)	(27)	48:52	56	60		
Me		Rh$_2$(4S-PHOX)$_4$ (173)	(45)	43:57	58	47		
C$_6$ t-Bu		Rh$_2$(OAc)$_4$	(37)	44:56	—	—		
t-Bu		Rh$_2$(5S-MEPY)$_4$ (171a)	(17)	43:57	74	—		
t-Bu		Rh$_2$(4S-PHOX)$_4$ (173)	(20)	52:48	46	—		
C$_8$ Ph		Rh$_2$(OAc)$_4$	(77)	42:58	—	—		
Ph		Rh$_2$(5S-MEPY)$_4$ (171a)	(38)	51:49	26	—		
Ph		Rh$_2$(4S-PHOX)$_4$ (173)	(60)	51:49	48	—		
C$_{12}$ l-menthyl		Rh$_2$(OAc)$_4$	(90)	37:63	—	—		
l-menthyl		Rh$_2$(5S-MEPY)$_4$ (171a)	(41)	38:62	72	60		
l-menthyl		Rh$_2$(4S-PHOX)$_4$ (173)	(31)	49:51	54	52		
C$_{12}$ d-menthyl		Rh$_2$(OAc)$_4$	(64)	38:62	—	—		
d-menthyl		Rh$_2$(5S-MEPY)$_4$ (171a)	(22)	45:55	76	50		
d-menthyl		Rh$_2$(4S-PHOX)$_4$ (173)	(29)	43:57	14	8		
d-menthyl		179 (Chart 5)	(87)	—	—	—		530
d-menthyl		28a (Chart 2)	(~60)	63:37	97 (S)	97 (S)		69, 118

Catalyst (Chart 4), neat, rt, 4 or 8 h

I: $R^1 = CO_2R, R^2 = H$ **III**: $R^1 = CO_2R, R^2 = H$
II: $R^1 = H, R^2 = CO_2R$ **IV**: $R^1 = H, R^2 = CO_2R$

	R	Catalyst	Time		I:II:III:IV	
C_3	Me	$Rh_2(OAc)_4$	4 h	(99)	30:32:25:13	544
	Me	**168q**	4 h	(99)	37:26:14:23	544, 71
C_4	Et	$Rh_2(OAc)_4$	8 h	(93)	28:29:24:12	151, 543, 544, 71
	Et	**168q**	4 h	(98)	37:27:14:22	544
	Et	$Rh_6(CO)_{16}$	8 h	(70)	32:29:25:14	151
	Et	$Cu(OTf)_2$	4 h	(57)	33:33:22:12	543, 80, 71
	Et	$CuCl·P(OPr-i)_3$	8 h	(81)	37:28:27:8	71, 151
	Et	$CuCl·P(OPh)_3$	8 h	(73)	36:30:27:7	71
	Et	Cu-bronze	8 h	(20)	35:30:28:7	71
	Et	$Cu(acac)_2$	8 h	(55)	38:29:25:8	71
	Et	$Pd(OAc)_2$	4 h	(37)	10:12:46:32	543
	Et	$PdCl_2·2PhCN$	8 h	(24)	16:18:33:33	71, 151
	n-Bu	$Rh_2(OAc)_4$	4 h	(88)	61(**I+II**):27:12	544
	n-Bu	**168q**	4 h	(99)	62(**I+II**):17:21	544
	t-Bu	$Rh_2(OAc)_4$	8 h	(85)	27:36:28:9	544
	t-Bu	**168q**	4 h	(95)	9:52:24:15	544
C_4	Et	$Rh_2(OPiv)_4$, rt		(93)	61(**I+II**):39(**III+IV**)	80
	Et	**179** (Chart 5)		(87)	—	530
	Et	$(CO)_5W=C(OMe)Ph$		(25)	—	321
	Et	$RuCl_2(PPh_3)_3$		(26)	73.5:14.5:8:4	52
	Et	$RuCl_2(PPh_3)_3/C_6H_6$		(25)	74:14:8:4	52
	Et	$RuH_3[Si(OEt)_3](PPh_3)_2$		(19)	78:11.5:7:3.5	52
	Et	$Ru[Si(OEt)_3]_2(PPh_3)_2$		(23)	81:10:6:3	52
C_6	n-Bu	$Pd(OAc)_2$, rt		(48)	23(**I+II**):77(**III+IV**)	80

TABLE VIII. REACTION OF CARBENOIDS WITH DIENES (Continued)

Carbenoid Precursor	Substrate	Conditions	Product(s) and Yield(s) (%)	Refs.
C_4 R = Et	(butadiene)	Catalyst, neat, rt, 4 h	**I** (CO$_2$Et, cyclopropane-vinyl) + **II** (CO$_2$Et, cyclopropane-vinyl)	
		Catalyst	I+II I:II	
		Rh$_2$(OAc)$_4$	(96) >80:20	543
		Cu(OTf)$_2$	(82) >80:20	543
		Pd(OAc)$_2$	(48) >75:25	74
R = Et	(pentadiene)		**I**: $R^1 = CO_2Et, R^2 = H$ **III**: $R^1 = CO_2Et, R^2 = H$ **II**: $R^1 = H, R^2 = CO_2Et$ **IV**: $R^1 = H, R^2 = CO_2Et$	
			I+II:III+IV	
		Rh$_2$(OAc)$_4$, neat, rt, 8 h	(61) 54:34:12 (**I:II:III+IV**)	71, 151
		Rh$_6$(CO)$_{16}$, neat, rt, 8 h	(72) 89:11	151
		CuCl•P(OPr-i)$_3$ neat, rt, 8 h	(23) 91:9	151
		PdCl$_2$•2PhCN neat, rt, 8 h	(31) 77:23	151
		Rh$_2$(OPiv)$_4$, rt	(96) 76:24	80
		Cu(OTf)$_2$, rt	(82) 80:20	80
		Pd(OPiv)$_2$, rt	(48) 73:27	80

R = Et

I: R² = CO₂Et, R³ = H III: R² = CO₂Et, R³ = H
II: R² = H, R³ = CO₂Et IV: R² = H, R³ = CO₂Et

R¹		I:II:III+IV	
		I:II:III:IV (I:II:III:IV)	
Me	Rh₂(OAc)₄, neat, rt, 8 h	(90) 52:37:4:7	71, 151
Me	Rh₆(CO)₁₆, neat, rt, 8 h	(63) 52:40:8	71, 151
Me	CuCl·P(OPr-i)₃ neat, rt, 8 h	(57) 56:38:16	71, 151
Me	PdCl₂·2PhCN neat, rt, 8 h	(22) 40:33:27	71, 151
Me	Cu	(—) —	545
TMS	Cu	(—) —	545

TABLE VIII. REACTION OF CARBENOIDS WITH DIENES (Continued)

Carbenoid Precursor	Substrate					Conditions	Product(s) and Yield(s) (%)					Refs.
							I + II					
							(I+II)	I:II	ee (%) I	ee (%) II		
	R	R^1	R^2	R^3	R^4							
C_3	Me	H	H	H	TMS	120b (Chart 3)	(64)	56:44	58	74		502
	Me	H	H	H	TMS	95a (Chart 3)	(59)	58:42	61	72		502
	Me	H	H	H	TMS	34b (Chart 2)	(72)	55:45	74	>95		458
	Me	H	H	H	TBDMS	Cu(acac)$_2$, EtOAc	(64)	67:33	—	—		546
	Me	H	H	H	TBDMS	34b (Chart 2)	(50)	62:38	69	>95		458
	Me	H	Me	H	TBDMS	Cu(acac)$_2$, EtOAc	(52)	—	—	—		546
	Me	H	Me	Me	TMS	120b (Chart 3)	(53)	58:42	64	67		502
	Me	H	Me	H	TMS	95a (Chart 3)	(27)	56:44	53	62		502
	Me	H	H	Me	TMS	34b (Chart 2)	(28)	>90:10	64	—		458
	Me	H	H	Me	TMS	95a (Chart 3)	(35)	64:36	72	72		458
	Me	Me	H	H	TMS	Cu(acac)$_2$, EtOAc	(61)	—	—	—		547
	Me	CH$_2$OTBDMS	H	H	TMS	Rh$_2$(OAc)$_4$, CH$_2$Cl$_2$	(73)	—	—	—		547
	Me	Ph	H	H	TBDMS	Cu(acac)$_2$, EtOAc	(80)	72:28	—	—		506
	Me	Ph	H	H	TMS	Cu(acac)$_2$, EtOAc	(73)	48:52	23	42		502
	Me	Ph	H	H	TMS	95a (Chart 3)	(67)	52:48	56	70		502
	Me	p-(CF$_3$)C$_6$H$_4$	H	H	TBDMS	Cu(acac)$_2$, EtOAc	(84)	72:28	—	—		546
	Me	Me	H	H	TBDMS	Cu(acac)$_2$, EtOAc	(76)	65:35	—	—		546
	Me	MeO	H	H	TBDMS	Cu(acac)$_2$, EtOAc	(72)	75:25	—	—		546
C_4	Et	H	H	H	Me	Rh$_2$(OAc)$_4$, neat, rt, 8 h	(88)	—	—	—		71, 151
	Et	H	H	H	Me	Rh$_6$(CO)$_{16}$, neat, rt, 8 h	(63)	—	—	—		151
	Et	H	H	H	Me	CuCl•P(OPr-i)$_3$ neat, rt, 8 h	(38)	—	—	—		151

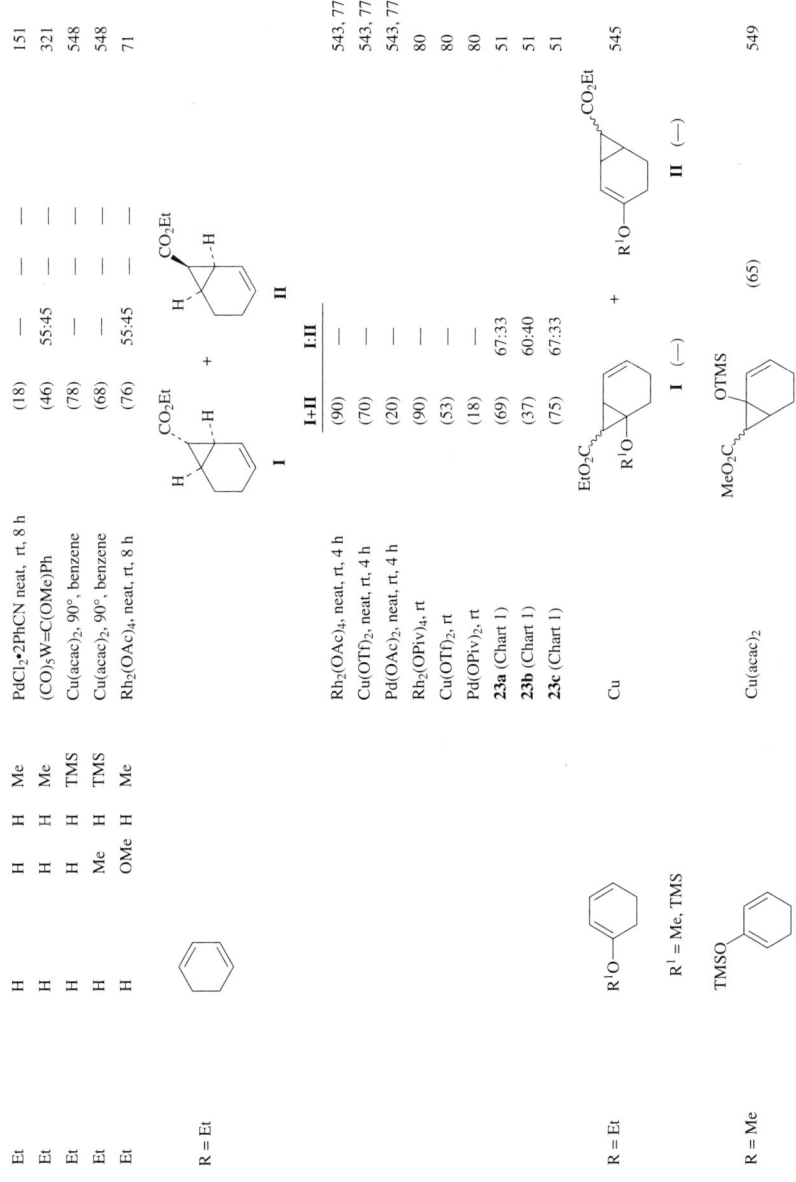

TABLE VIII. REACTION OF CARBENOIDS WITH DIENES (Continued)

Carbenoid Precursor	Substrate	R	Conditions	Product(s) and Yield(s) (%)		ee (%)		abs stereochem		Refs.
				I+II	I:II	I	II	I	II	
C₃		Me	Rh₂(O₂CC₆H₄CF₃-3)₄, rt	(69)	44:56	—	—	—	—	544
		Me	Rh₂(OAc)₄, neat, rt, 4 h	(56)	52:48	—	—	—	—	544
C₄		Et	Rh₆(CO)₁₆, neat, rt, 8 h	(77)	—	—	—	—	—	543
		Et	CuCl•P(OPr-i)₃ neat, rt, 8 h	(35)	—	—	—	—	—	543
		Et	Rh₂(TFA)₄, neat, rt, 4 h	(29)	52:48	—	—	—	—	544
		Et	Rh₂(O₂CC₆F₅)₄, neat, rt, 8 h	(64)	52:48	—	—	—	—	543
		Et	Rh₂(OPiv)₄, neat, rt, 8 h	(56)	47:53	—	—	—	—	543
		Et	Rh₂[O₂C-2,4-Cl₂C₆H(NO₂)₂-3,5)]₄	(15)	47:53	—	—	—	—	544
		Et	Rh₂[O₂CC₆H₃(OMe)₂-2,6]₄, rt	(68)	50:50	—	—	—	—	544
		Et	168q (Chart 4), neat, rt, 4 h	(58)	45:55	—	—	—	—	544
		Et	Rh₂(O₂CC₆H₄CF₃-3)₄, rt	(86)	42:58	—	—	—	—	544
		Et	Cu	(45)	57:43	—	—	—	—	463
		Et	95a (Chart 3)	(33)	60:40	22	15	1S,3S	1S,3R	463
		Et	94 (Chart 3)	(3)	59:41	5	17	1R,3R	1S,3R	463
		Et	97 (Chart 3)	(15)	58:42	10	29	1R,3R	1S,3R	463
		Et	96 (Chart 3)	(16)	57:43	16	2	1R,3R	1S,3R	463
		Et	151a (Chart 3), 53°	(22)	60:40	17	5	1S,3S	1R,3S	551
		Et	150a (Chart 3), 50°	(44)	58:42	7	5	1S,3S	1S,3R	551
		Et	151b (Chart 3), 50°	(49)	64:36	6	5	1R,3R	1R,3S	551
		Et	150b (Chart 3), 50°	(44)	60:40	3	10	1S,3S	1R,3S	551
		Et	151c (Chart 3), 50°	(17)	56:44	4	9	1R,3R	1R,3S	551
		Et	150c (Chart 3), 50°	(43)	57:43	12	16	1R,3R	1R,3S	551

	Et	Cu-(L-Ala)$_2$	(46)	55:45	<3	—	336
	Et	NaCuX-zeolite-(L-Ala)$_2$	(59)	55:45	<3	—	336
	Et	NaCuX-zeolite-(L-Val)$_2$	(57)	53:47	<3	—	336
	Et	NaCuX-zeolite-(L-Pro)$_2$	(58)	55:45	<3	—	336
	Et	NaCuX-zeolite-[(R)-1,2-propanediamine]$_2$	(54)	54:46	<3	—	336
C$_6$	n-Bu	Rh$_2$(OAc)$_4$, neat, rt, 4 h	(58)	46:54	—	—	544
	n-Bu	Rh$_2$(O$_2$CC$_6$H$_4$CF$_3$-m)$_4$, rt	(96)	40:60	—	—	544
	t-Bu	Rh$_2$(OAc)$_4$, neat, rt, 8 h	(56)	44:56	—	—	543
	t-Bu	Rh$_2$(TFA)$_4$, neat, rt, 8 h	(45)	56:44	—	—	543
	t-Bu	Rh$_2$(O$_2$CC$_6$F$_5$)$_4$, neat, rt, 8 h	(56)	53:47	—	—	543
	t-Bu	Rh$_2$(OPiv)$_4$, neat, rt, 8 h	(32)	44:56	—	—	543
	t-Bu	Cu powder/CuSO$_4$, 110°	(16)	—	—	—	552
	t-Bu	Cu powder/CuSO$_4$, 110°	(23)	55:45	—	—	553
C$_8$	CH$_2$CH$_2$NEt$_2$	36a (Chart 2)	(60)	97:3	92	—	125
	CH$_2$CH$_2$SBu	36a (Chart 2)	(60)	88:12	92	—	125
C$_9$	2,4-dimethyl-3-pentyl	36a (Chart 2)	(62)	99:1	92	85	125
C$_{12}$	l-menthyl						
C$_{15}$	dicyclohexylmethyl						

C$_4$ R = Et

Catalyst, neat, rt, 8 h

I (CO$_2$Et, methylpropenyl cyclopropane) + II (CO$_2$Et, isopropenyl cyclopropane)

Catalyst	I+II	I:II
Rh$_2$(OAc)$_4$	(85)	—
Rh$_6$(CO)$_{16}$	(76)	—
CuCl·P(OPr-i)$_3$	(35)	—

543

C$_{12}$

R				
d-menthyl	(77)	I:II = 63:37; I 97% ee (1S), II 97% ee (1S)		69
l-menthyl	(86)	I:II = 79:21; I 98% ee (1R,3R), II 79% ee (1R,3S)		106

I (CO$_2$R cyclopropane) + II (CO$_2$R cyclopropane)

TABLE VIII. REACTION OF CARBENOIDS WITH DIENES (Continued)

Carbenoid Precursor	Substrate	Conditions	Product(s) and Yield(s) (%)	Refs.
C$_4$				
R = Et	∽∽		I (CO$_2$Et, propenyl cyclopropane) + II (CO$_2$Et, propenyl cyclopropane)	
			I+II I:II	
		Cu	(57) 57:43	150
		Rh$_2$(OAc)$_4$, rt, 4 h	(76) —	543
		Cu(OTf)$_2$, rt, 8 h	(85) —	543
		CuCl•P(OPr-i)$_3$ neat, rt, 8 h	(30) —	543
			R^1,R^2-cyclopropane + propenyl + R^1,R^2-cyclopropane	
R = Et	∽∽ (Z,Z)		I: R^1 = CO$_2$Et, R^2 = H III: R^1 = CO$_2$Et, R^2 = H II: R^1 = H, R^2 = CO$_2$Et IV: R^1 = H, R^2 = CO$_2$Et	
			I+II+III+IV (I:II:III:IV)	
		Cu	(51) 41:31:17:10	150
		Rh$_2$(OAc)$_4$, neat, rt, 8 h	(87) 76:24	543
		Rh$_6$(CO)$_{16}$, neat, rt, 8 h	(81) 76:24	543
		CuCl•P(OPr-i)$_3$ neat, rt, 8 h	(60) 50:50	543
R = Et	∽∽ (Z)	Cu	I (CO$_2$Et) + II (CO$_2$Et) (59) I:II = 68:32	150

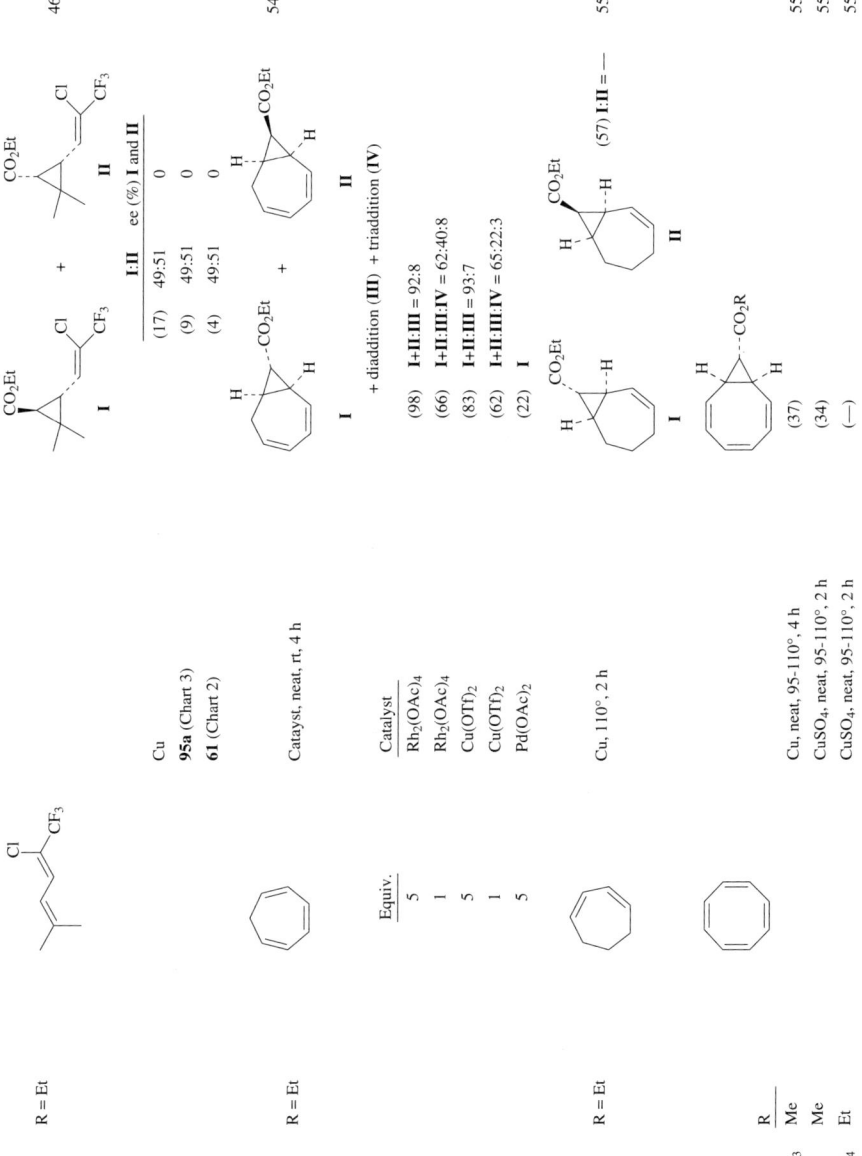

TABLE VIII. REACTION OF CARBENOIDS WITH DIENES (Continued)

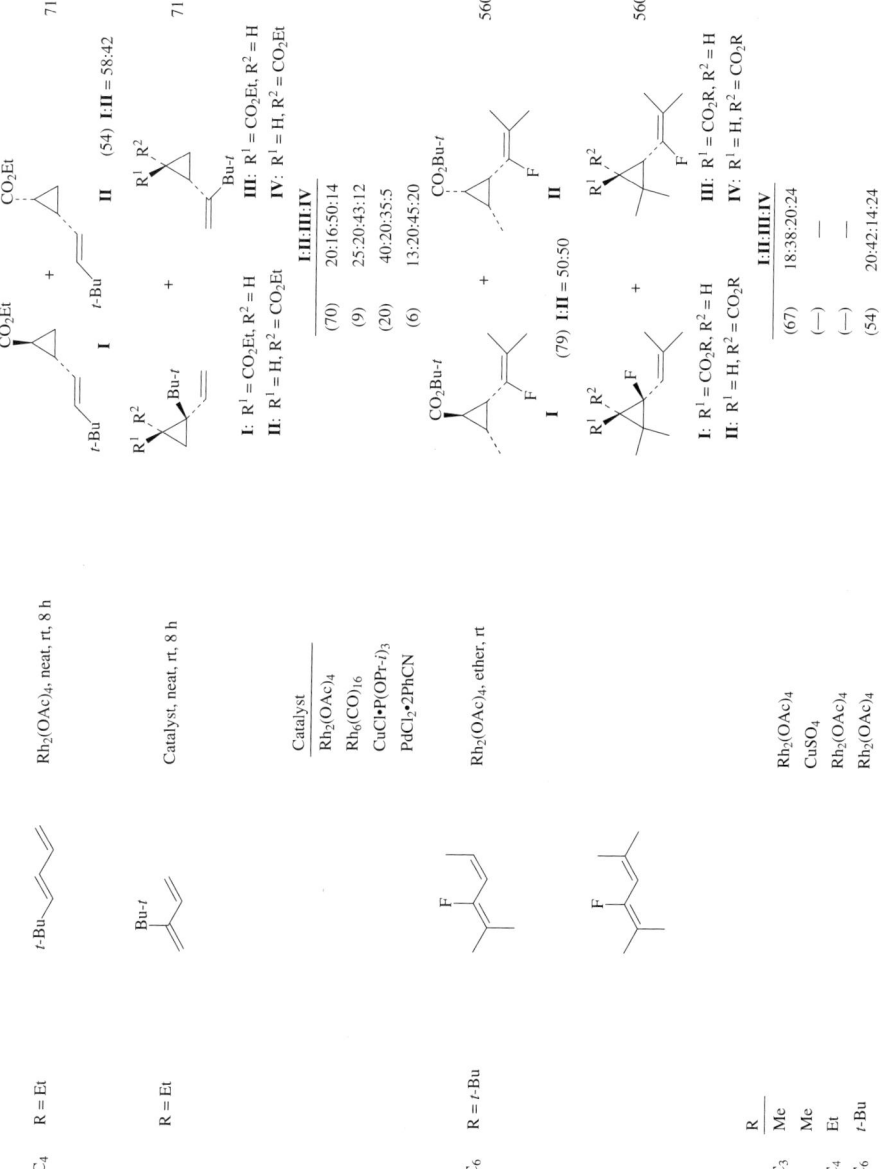

TABLE VIII. REACTION OF CARBENOIDS WITH DIENES (Continued)

	Carbenoid Precursor	Substrate	Conditions	Product(s) and Yield(s) (%) I+II	I:II	ee (%) I	ee (%) II	abs stereochem I	abs stereochem II	Refs.
	R									
C_3	Me		$Rh_2(OAc)_4$, neat, rt, 4 h	(100)	61:39	—	—	—	—	544
	Me		168q (Chart 4), neat, rt, 4 h	(100)	50:50	—	—	—	—	544
C_4	CH_2CN		$Cu/CuSO_4$, 80°	(75)	—	—	—	—	—	349
	CH_2CH_2Cl		Cu, 115-120°	(53)	55:45	—	—	—	—	341
	CH_2CH_2Br		$Cu/CuSO_4$, 110°	(50)	—	—	—	—	—	342
	Et		Cu-bronze	(34)	73:27	—	—	—	—	71
	Et		$Cu(OTf)_2$	(93)	70:30	—	—	—	—	71, 77
	Et		$Cu(acac)_2$	(76)	64:36	—	—	—	—	71
	Et		$Rh_2(OAc)_4$, 60°	(50)	—	—	—	—	—	544, 543, 560
	Et		$Rh_2(OAc)_4$, neat, rt, 4 h	(100)	59:41	—	—	—	—	544, 71, 77, 78, 94
	Et		$Rh_2(acetamide)_4$,	(—)	60:40	—	—	—	—	78
	Et		$Rh_2(TFA)_4$, neat, rt, 4 h	(85)	60:40	—	—	—	—	544
	Et		$Rh_2[O_2C-2,4-Cl_2-3,5-(O_2N)_2C_6H]_4$	(53)	56:44	—	—	—	—	544
	Et		$Rh_2[O_2C-2,6-(MeO)_2C_6H_3]_4$, rt	(92)	57:43	—	—	—	—	544
	Et		168q (Chart 4), neat, rt, 4 h	(99)	49:51	—	—	—	—	544
	Et		$Rh_2(O_2CC_6H_4CF_3\text{-}m)_4$, rt	(100)	57:43	—	—	—	—	71, 318, 543
	Et		$Rh_6(CO)_{16}$, neat, rt, 8 h	(87)	69:31	—	—	—	—	543
	Et		$CuCl \cdot P(OPr\text{-}i)_3$, rt, 8 h	(55)	73:27	—	—	—	—	71, 77, 318, 543
	Et		$CuCl \cdot P(OPh)_3$, rt, 8 h	(66)	73:27	—	—	—	—	71
	Et		$Rh_2(O_2C\text{-polyethylene})_4$	(56)	70:30	—	—	—	—	367
	Et		$PdCl_2 \cdot 2PhCN$	(20)	70:30	—	—	—	—	71, 318
	Et		$Pd(OAc)_2$	(35)	—	—	—	—	—	77
	Et		$CuSO_4$	(—)	61:39	—	—	—	—	569, 561

	R	Catalyst/Conditions	(Yield)	Ratio	ee	Config	Ref
	Et	23a (Chart 1)	(96)	58:42	—	—	51
	Et	23b (Chart 1)	(61)	60:40	—	—	51
	Et	23c (Chart 1)	(97)	56:44	—	—	51
	Et	130a (Chart 3), 5°	(64)	55:45	~8	—	562
	Et	130b (Chart 3), 5°	(74)	57:43	~10	—	562
	Et	130c (Chart 3), 5°	(71)	55:45	~1	—	562
	Et	91e (Chart 3), neat, 20°	(54)	51:49	68	62 1S,3S 1S,3R	112
	Et	91a (Chart 3), neat, 40°, 5 h	(64)	62:38	—	—	41
	Et	91b (Chart 3), neat, 40°, 5 h	(60)	59:41	—	—	41
	Et	91c (Chart 3), neat, 40°, 5 h	(54)	58:42	—	—	41
	Et	ent-91f (Chart 3), neat, 40°, 5 h	(58)	62:38	—	—	41
	Et	ent-91g (Chart 3), neat, 40°, 5 h	(53)	58:42	—	—	41
	Et	ent-91h (Chart 3), neat, 40°, 5 h	(52)	57:43	—	—	41
	Et	91i (Chart 3), neat, 40°, 5 h	(59)	60:40	—	—	41
	Et	91j (Chart 3), neat, 40°, 5 h	(46)	58:42	—	—	41
	Et	91k (Chart 3), neat, 40°, 5 h	(51)	67:33	—	—	41
C_6	n-Bu	Rh$_2$(OAc)$_4$, neat, rt, 4 h	(100)	57:43	—	—	544
	n-Bu	168q (Chart 4), neat, rt, 4 h	(97)	45:55	—	—	544
	t-Bu	Rh$_2$(OAc)$_4$, neat, rt, 4 h	(100)	64:36	—	—	544
	t-Bu	168q (Chart 4), neat, rt, 4 h	(99)	44:56	—	—	544
C_8	cyclohexyl	91e (Chart 3), neat, 20°	(74)	75:25	75	46 1S,3S 1S,3R	112
	CH$_2$CH$_2$SBu-n	91e (Chart 3), neat, 20°	(71)	58:42	70	58 1S,3S 1S,3R	112
	CH$_2$CH$_2$NEt$_2$	Cu/CuSO$_4$, 110°	(40)	55:45	—	—	553
	2,3-Me$_2$-2-butyl	Cu/CuSO$_4$, 110°	(23)	55:45	—	—	552
C_9	2,3-Me$_2$-2-pentyl	91e (Chart 3), neat, 20°	(71)	78:22	85	43 1S,3S 1S,3R	112
	2,3-Me$_2$-2-pentyl	36a (see Chart 2), 0°- rt	(78)	93:7	94	— 1R,3R	125
	2,3-Me$_2$-2-pentyl	Rh$_2$(OAc)$_4$, rt, 7 h	(88)	65:35	—	—	78
C_{10}	2,3,4-Me$_3$-3-pentyl	Rh$_2$(acetamide)$_4$, rt, 7 h	(95)	59:41	—	—	78
	2,3,4-Me$_3$-3-pentyl	91e (Chart 3), neat, 20°	(64)	92:8	88	— 1S,3S	112
	2,3,4-Me$_3$-3-pentyl	36a (Chart 2), 0°- rt	(76)	85:15	88	— 1R,3R	125
C_{11}	α,α-Me$_2$-benzoyl	91e (Chart 3), neat, 20°	(60)	56:44	71	— 1S,3S	112
	2-Me-3-(i-Pr)-3-heptyl	Rh$_2$(OAc)$_4$, rt, 7 h	(73)	60:40	—	—	78
	2-Me-3-(i-Pr)-3-heptyl	Rh$_2$(acetamide)$_4$, rt, 7 h	(62)	60:40	—	—	78
C_{12}	l-adamantyl	91e (Chart 3), neat, 20°	(82)	84:16	85	46 1S,3S 1S,3R	112
	d-neomenthyl	91e (Chart 3), neat, 20°	(77)	89:11	87	— 1S,3S	112

TABLE VIII. REACTION OF CARBENOIDS WITH DIENES (Continued)

	Carbenoid Precursor	Substrate	Conditions	Product(s) and Yield(s) (%)						Refs.
	d-menthyl		91e (Chart 3), neat, 20°	(64)	72:28	90	59	1S,3S	1S,3R	112
	dl-menthyl		91e (Chart 3), neat, 20°	(67)	81:19	90	75	1S,3S	1S,3R	112
	l-menthyl		91d (Chart 3), neat, 20°	(67)	89:11	87	25	1S,3S	1S,3R	112
	l-menthyl		91h (Chart 3), neat, 20°	(42)	91:9	86	22	1S,3S	1S,3R	112
	l-menthyl		Cu powder	(69)	76:24	0.7	0	—	—	112
	l-menthyl		31b (Chart 2), CH_2Cl_2, 0°- rt	(61)	84:16	24	16	1R,3S	1R,3R	125
	l-menthyl		34b (Chart 2), CH_2Cl_2, 0°- rt	(60)	84:16	24	20	1R,3S	1R,3R	125
	l-menthyl		37 (Chart 2), CH_2Cl_2, 0°- rt	(60)	88:12	40	25	1R,3S	1R,3R	125
	l-menthyl		36b (Chart 2), CH_2Cl_2, 0°- rt	(68)	90:10	72	60	1R,3S	1R,3R	125
	l-menthyl		36c (Chart 2), CH_2Cl_2, 0°- rt	(65)	90:10	82	65	1R,3S	1R,3R	125
	l-menthyl		37 (Chart 2), CH_2Cl_2, 0°- rt	(58)	80:20	90	80	1R,3S	1R,3R	125
	l-menthyl		36a (Chart 2), CH_2Cl_2, 0°- rt	(70)	86:14	90	78	1R,3S	1R,3R	125
	d-menthyl		36a (Chart 2), CH_2Cl_2, 0°- rt	(72)	92:8	92	84	1R,3S	1R,3R	125
	l-menthyl		66 (Chart 2), $C_2H_4Cl_2$, 0°	(47)	88:12	74	—	1R,3R	—	483
	l-menthyl		36a (Chart 2), CH_2Cl_2, 0°- rt	(78)	95:5	94	—	1R,3R	—	125
C_{15}	dicyclohexylmethyl		$Rh_2(OAc)_4$, rt, 7 h	(80)	94:6	—	—	—	—	78
	2,6-(t-Bu)$_2$-4-MeC$_6$H$_2$		Rh_2(acetamide)$_4$, rt, 7 h	(75)	98:2	—	—	—	—	78
C_{17}	2,6-(t-Bu)$_2$-4-MeC$_6$H$_2$		37 (Chart 2), CH_2Cl_2, 0°- rt	(60)	92:8	92	—	1R,3R	—	125
	2,6-(t-Bu)$_2$-4-MeC$_6$H$_2$		36a (Chart 2), CH_2Cl_2, 0°- rt	(75)	94:6	94	—	1R,3R	—	125
C_4	R = Et	[TMS-substituted diene]	Cu(acac)$_2$, benzene	[product I: CO$_2$Et, TMS cyclopropane] + [product II: CO$_2$Et, TMS vinyl cyclopropane] (53) I:II = —						563
	R = Et	[benzodioxole diene]	$Rh_2(OAc)_4$, Et_2O, rt, 4 h	I [CO$_2$Et fused dioxole] + II [CO$_2$Et fused dioxole isomer] (70) I:II = 1.2						564, 565

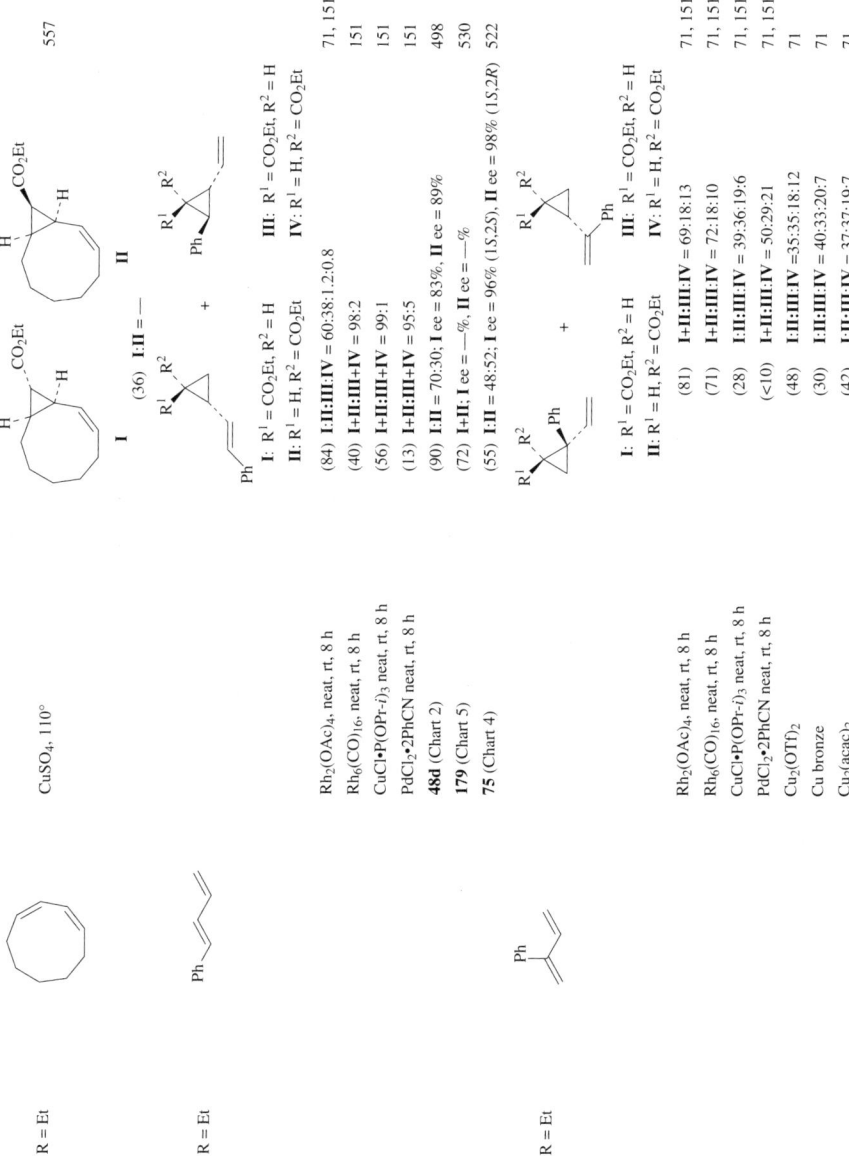

TABLE VIII. REACTION OF CARBENOIDS WITH DIENES (Continued)

Carbenoid Precursor	Substrate	Conditions	Product(s) and Yield(s) (%)	Refs.
R = Et		CuBr	(70)	558
R = Et		Catalyst, neat, rt, 4 h	**I**: $R^1 = H, R^2 = CO_2Et$ **II**: $R^1 = CO_2Et, R^2 = H$ **III**: $R^1 = H, R^2 = CO_2Et$ **IV**: $R^1 = CO_2Et, R^2 = H$	543
			I:II:III:IV	
		$Rh_2(OAc)_4$	(72) 30:28:21:21	
		$Cu(OTf)_2$	(38) 30:30:25:15	
		$Pd(OAc)_2$	(17) 31:31:23:15	
R = Et		Catalyst, neat, rt, 4 h	**I**: $R^1 = CO_2Et, R^2 = H$ **II**: $R^1 = H, R^2 = CO_2Et$ **III**: $R^1 = CO_2Et, R^2 = H$ **IV**: $R^1 = H, R^2 = CO_2Et$	
			I:II:III:IV	
		$CuSO_4$	(50) 27:13:30:30	566
		$Rh_2(OAc)_4$, neat, rt, 4 h	(87) 26:19:29:26	543
		$Cu(OTf)_2$, neat, rt, 4 h	(72) 25:18:32:35	543
		$Pd(OAc)_2$, neat, rt, 4 h	(33) 42:33:16:9	543

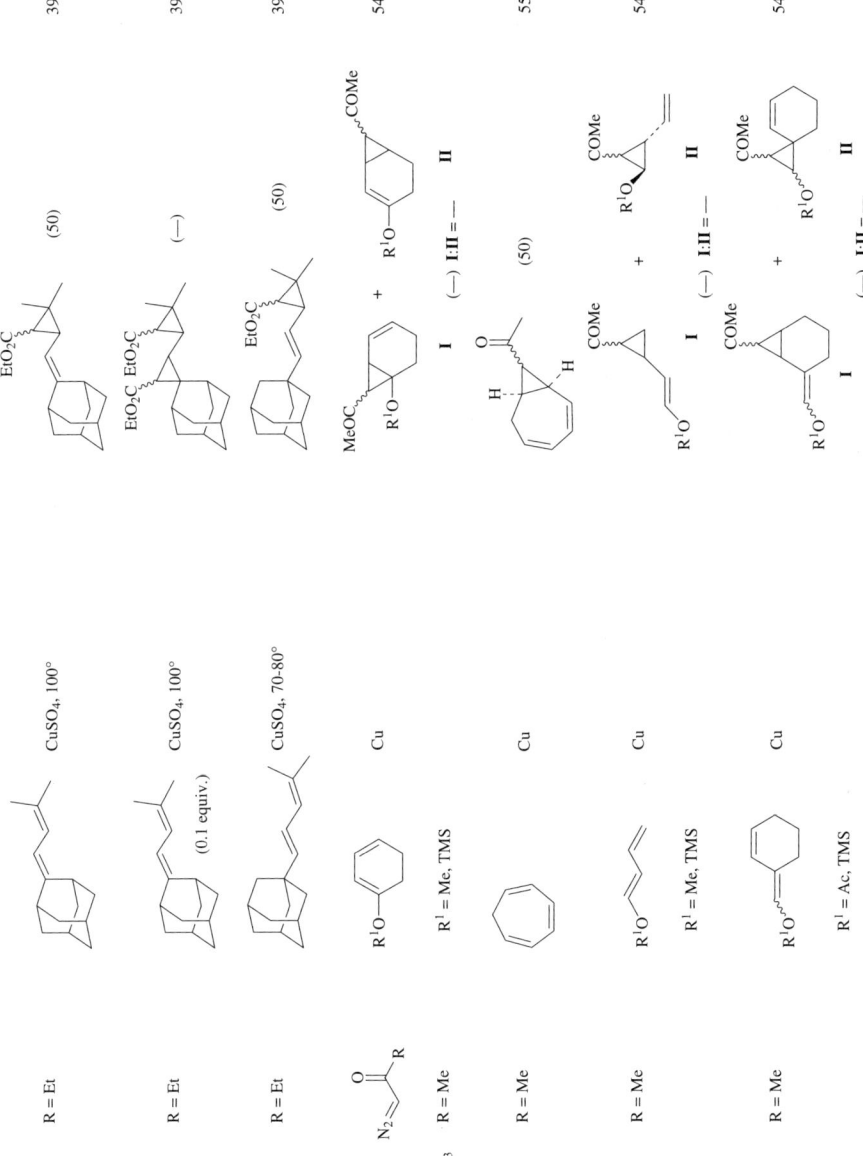

TABLE VIII. REACTION OF CARBENOIDS WITH DIENES (*Continued*)

Carbenoid Precursor	Substrate	Conditions	Product(s) and Yield(s) (%)	Refs.
C_6 R = *t*-Bu	OMe-diene	CuSO$_4$, hexane, 45°	**I** + **II** (43)	567
R = *t*-Bu	R^{1}O-diene	CuSO$_4$, hexane, 45°	**I** **II+III** Me: (42) (0) Et: (41) (15) **II**: R^2 = COBu-*t*, R^3 = H **III**: R^2 = H, R^3 = COBu-*t*	567
C_8 R = Ph	cycloheptatriene	Cu	(55)	558
C_8 R = C$_6$H$_{13}$	butadiene	Rh$_2$(OAc)$_4$, CH$_2$Cl$_2$	**I** **I+II** (83); **I:II** ~67:33 **II**	568
C_5 R = Et (N$_2$CHCOCO$_2$R)	butadiene	Rh$_2$(OAc)$_4$, CH$_2$Cl$_2$	(35) + (26) + (4)	568

TABLE VIII. REACTION OF CARBENOIDS WITH DIENES (*Continued*)

Carbenoid Precursor	Substrate	Conditions	Product(s) and Yield(s) (%)	Refs.
R = Me	TBSO⟶OMe	Rh$_2$(OAc)$_4$	TBSO-cyclopentene-CO$_2$Me, CO$_2$Me, OMe (31)	53
R = Me	TBSO-(1-cyclohexenyl)	Rh$_2$(OAc)$_4$	TBSO-bicyclic-CO$_2$Me, CO$_2$Me (35)	53
R = Me	TBSO⟶Ph	Rh$_2$(OAc)$_4$	TBSO-cyclopentene-CO$_2$Me, CO$_2$Me, Ph (43)	53
$\underset{R^2}{\diagdown}\!\!=\!\!\diagdown\!\!\underset{N_2}{\diagup}\!\!\text{COR}^1$ R^1 = OMe, R^2 = Ph	Cl-diene	Rh$_2$(S-DOSP)$_4$ (**161a**, Chart 4), pentane, −78°	Cl-cycloheptadiene-CO$_2$Me, Ph (69) 82% ee	157
C$_{11}$	⟶OAc	1. Rh$_2$(OAc)$_4$, CH$_2$Cl$_2$ 2. Kugelrohr	cycloheptatriene-CO$_2$Et, R^2	156

	R^1	R^2	
C$_9$	OEt	CO$_2$Et	(50)
C$_{12}$	OEt	SO$_2$Ph	(67)

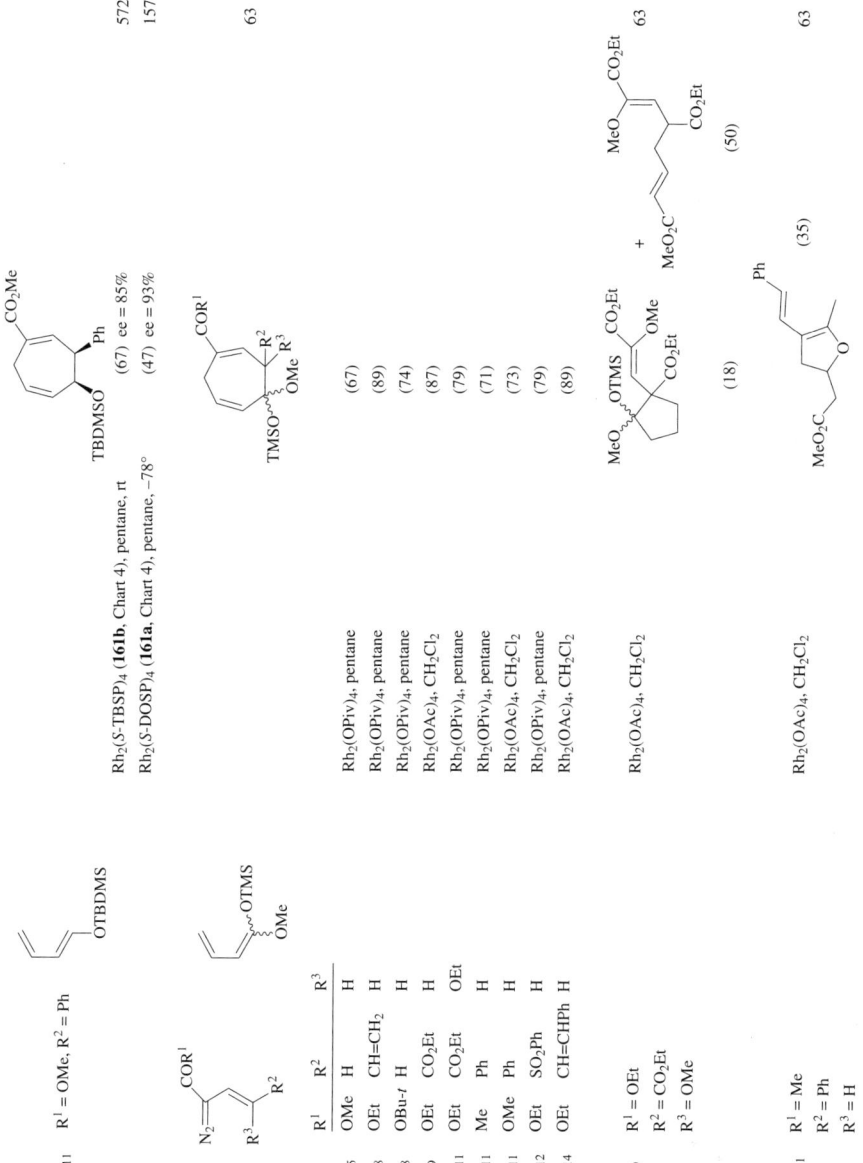

TABLE VIII. REACTION OF CARBENOIDS WITH DIENES (Continued)

Carbenoid Precursor	Substrate	Conditions	Product(s) and Yield(s) (%)	Refs.
C_{12} R^1 = OEt, R^2 = SO_2Ph, R^3 = H	(OTMS/OMe diene)	$Rh_2(OAc)_4$, CH_2Cl_2	cycloheptadiene product (23) + cyclopentane product (47)	63
				64
	R^1 / R^2 / R^3			
C_6	OMe / Me / H	$Rh_2(OOct)_4$, hexane	(>67)	
C_8	OBu-t / H / H	$Rh_2(OPiv)_4$, hexane	(60)	
C_8	OEt / CO_2Et / H	$Rh_2(OAc)_4$, CH_2Cl_2	(91)	
C_{12}	OEt / SPh / H	$Rh_2(OPiv)_4$, hexane	(>46)	
C_{14}	OBu-t / Ph / H	$Rh_2(OAc)_4$, CH_2Cl_2	(91)	
C_5 R^1 = OMe, R^2 = H, R^3 = OMe		1. $Rh_2(OPiv)_4$, hexane 2. DDQ·TsOH	(32)	64
		1. $Rh_2(OPiv)_4$, hexane 2. DDQ·TsOH		64
	R^1 / R^2 / R^3			
C_5	OMe / Me / H		(58)	
C_8	OMe / SEt / H		(39)	
C_{11}	OMe / Ph / H		(64)	

| C_{12} | R^1 = OEt
R^2 = SPh
R^3 = H | 1. $Rh_2(OPiv)_4$, hexane
2. DDQ·TsOH | (11) + (77) | 64 |

C_5

	R		I:II	
C_5	OMe	$Rh_2(OAc)_4$, pentane	(85) 95:5	159
	OMe	$Rh_2(OAc)_4$, benzene	(74) 84:16	159
	OMe	$Rh_2(OAc)_4$, CH_2Cl_2	(81) 67:33	159
	OMe	$Rh_2(TFA)_4$, pentane	(69) 46:54	159
	OMe	$Rh_2(TFA)_4$, CH_2Cl_2	(60) 32:68	159
	OMe	$Rh_2(OPiv)_4$, pentane	(86) 98:2	159
	OMe	$Rh_2(OPiv)_4$, CH_2Cl_2	(86) 90:10	159
C_8	OBu-t	$Rh_2(OAc)_4$, CH_2Cl_2	(—) 52:48	451
C_{19}	BHT	$Rh_2(OAc)_4$, CH_2Cl_2	(—) 0:100	451

TABLE VIII. REACTION OF CARBENOIDS WITH DIENES (Continued)

	Carbenoid Precursor			Substrate	Conditions			Product(s) and Yield(s) (%)		Refs.
	R^1	R^2	R^3		Catalyst (Chart 4)	Solvent	Temp	(yield)	ee (%)	
C_5	Me	H	H		$Rh_2(OAc)_4$	pentane	—	(75)	—	159
	OMe	H	H		$Rh_2(S\text{-}TBSP)_4$ (**161b**)	pentane	rt	(76)	63	572
	OMe	H	H		$Rh_2(S\text{-}DOSP)_4$ (**161a**)	pentane	−78°	(70)	63	157
C_6	OMe	Me	H		$Rh_2(S\text{-}TBSP)_4$ (**161b**)	pentane	rt	(66)	64	572
	OMe	Me	H		$Rh_2(S\text{-}DOSP)_4$ (**161a**)	pentane	−78°	(74)	62	157
	OMe	H	Me		$Rh_2(S\text{-}TBSP)_4$ (**161b**)	pentane	rt	(75)	83	572
	OMe	H	Me		$Rh_2(S\text{-}DOSP)_4$ (**161a**)	pentane	−78°	(47)	91	157
C_7	OMe	H	$CH=CH_2$		$Rh_2(S\text{-}TBSP)_4$ (**161b**)	pentane	rt	(64)	91	572
	OMe	H	$CH=CH_2$		$Rh_2(S\text{-}DOSP)_4$ (**161a**)	pentane	−78°	(92)	58	157
C_8	OEt	H	$CH=CH_2$		$Rh_2(OAc)_4$	pentane	—	(83)	—	159
	OBu-t	H	H		$Rh_2(OAc)_4$	pentane	—	(72)	—	159
	OBu-t	H	H		$Rh_2(S\text{-}TBSP)_4$ (**161b**)	pentane	rt	(50)	5	572
C_{10}	OEt	H	CO_2Et		$Rh_2(OAc)_4$	neat	—	(98)	—	156, 573
	OEt	H	CO_2Et		$Rh_2(S\text{-}TBSP)_4$ (**161b**)	pentane	rt	(98)	10	572
C_{11}	Me	H	Ph		$Rh_2(OAc)_4$	neat	—	(88)	—	574
	OMe	OTBDMS	H		$Rh_2(OAc)_4$	neat	—	(66)	—	156
	OMe	OTBDMS	H		$Rh_2(S\text{-}TBSP)_4$ (**161b**)	pentane	rt	(66)	42	572
	OMe	OTBDMS	H		$Rh_2(S\text{-}DOSP)_4$ (**161a**)	pentane	−78°	(97)	74	157
	OMe	H	Ph		$Rh_2(OAc)_4$	neat	—	(73)	—	156, 573
	OMe	H	Ph		$Rh_2(S\text{-}TBSP)_4$ (**161b**)	pentane	rt	(92)	75	572
C_{12}	OEt	H	Ph		$Rh_2(S\text{-}DOSP)_4$ (**161a**)	pentane	−78°	(77)	93	157
	OEt	H	SO_2Ph		$Rh_2(OAc)_4$	neat	—	(80)	—	156
C_{13}	OMe	H	$CH=CHPh$		$Rh_2(S\text{-}DOSP)_4$ (**161a**)	pentane	−78°	(80)	90	157
C_{14}	OEt	H	$CH=CHPh$		$Rh_2(OAc)_4$	neat	—	(72)	—	156
C_{19}	$OC_6H_2Me_3\text{-}2,4,4$	OTBDMS	H		$Rh_2(OAc)_4$	neat	—	(80)	—	574

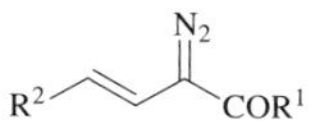

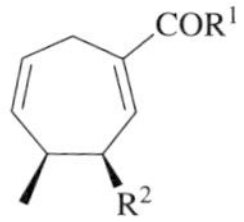

	R¹	R²
C₁₁	OMe	Ph
	OMe	Ph
C₁₃	OMe	CH=CHPh
C₁₄	OEt	CH=CHPh

Catalyst (Chart 4)	Solvent	Temp	(%)	ee (%)	
Rh₂(S-TBSP)₄ (**161b**)	pentane	rt	(41)	90	572
Rh₂(S-DOSP)₄ (**161a**)	pentane	−78°	(51)	98	157
Rh₂(S-DOSP)₄ (**161a**)	pentane	−78°	(62)	98	157
Rh₂(OAc)₄	CH₂Cl₂	—	(75)	—	156

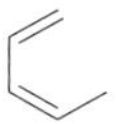

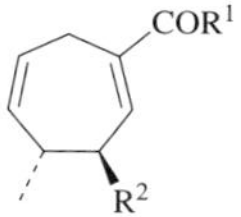

	R¹	R²
C₁₁	OMe	Ph
	OMe	Ph
C₁₃	OMe	CH=CHPh
C₁₄	OEt	CH=CHPh

Catalyst (Chart 4)	Solvent	Temp	(%)	ee (%)	
Rh₂(S-TBSP)₄ (**161b**)	pentane	rt	(79)	90	572
Rh₂(S-DOSP)₄ (**161a**)	pentane	−78°	(47)	96	157
Rh₂(S-DOSP)₄ (**161a**)	pentane	−78°	(82)	95	157
Rh₂(OAc)₄	CH₂Cl₂	—	(68)	—	156

R¹ = OEt
R² = CH=CHPh

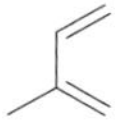

Rh₂(OAc)₄, CH₂Cl₂

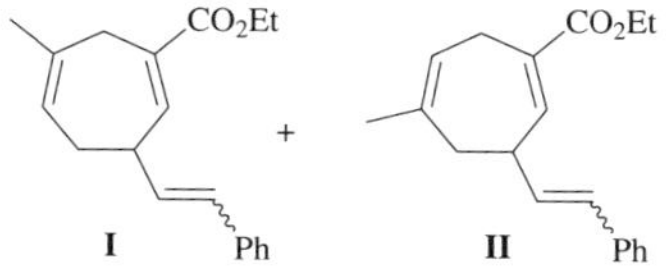

I + II = (72); **I : II** = 86:14   156

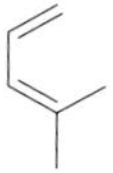

	R¹	R²
C₉	OEt	CO₂Et
C₁₄	OEt	CH=CHPh

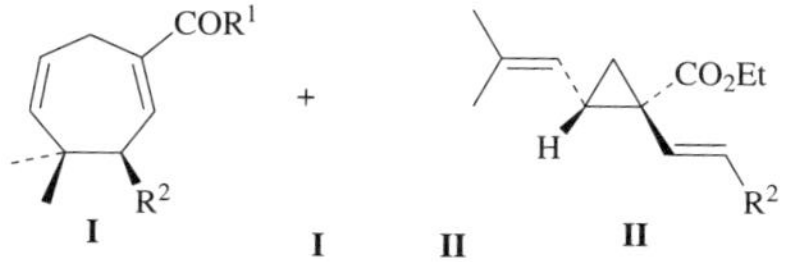

Rh₂(OAc)₄, CH₂Cl₂

I	II
(58)	(14)
(80)	(0)

156

TABLE VIII. REACTION OF CARBENOIDS WITH DIENES (*Continued*)

Carbenoid Precursor	Substrate	Conditions	Product(s) and Yield(s) (%)	Refs.
C$_{14}$ R^1 = OEt, R^2 = CH=CHPh		Rh$_2$(OAc)$_4$, CH$_2$Cl$_2$	(53)	156
C$_{13}$ R^1 = OMe, R^2 = CH=CHPh		Rh$_2$(S-DOSP)$_4$ (**161a**, Chart 4), hexanes, reflux	(41) ee = 48%	158
			I + II	
			I **II** ee (%)	
C$_9$ OEt CO$_2$Et		Rh$_2$(OAc)$_4$, CH$_2$Cl$_2$	(49) (13) —	156
C$_{11}$ OMe Ph		Rh$_2$(S-DOSP)$_4$ (**161a**, Chart 4), pentane, −78°	(45) (0) 91	157
C$_{14}$ OEt CH=CHPh		Rh$_2$(OAc)$_4$, CH$_2$Cl$_2$	(42) (0) —	156
			I + II	
			I **II** ee (%)	
C$_9$ OEt CO$_2$Et		Rh$_2$(OAc)$_4$, CH$_2$Cl$_2$	(63) (14) —	156
C$_{13}$ OMe CH=CHPh		Rh$_2$(S-DOSP)$_4$ (**161a**, Chart 4), pentane, −78°	(87) (0) 97	157
C$_{14}$ OMe CH=CHPh		Rh$_2$(OAc)$_4$, CH$_2$Cl$_2$	(59) (0) —	156

	R¹	R²
C₅	OMe	H
C₆	OMe	Me
C₇	OMe	CH=CHPh
C₁₁	OMe	Ph
C₁₃	OMe	CH=CHPh

	ee (%)
(41)	73
(87)	98
(56)	96
(83)	98
(84)	93

Rh₂(S-DOSP)₄ (**161a**, Chart 4), pentane, −78°

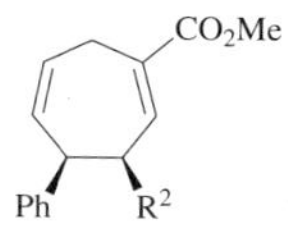

157

	R¹	R²	R³
C₅	OMe	H	H
	OMe	H	H
	OMe	H	H
C₆	Et	H	H
	Et	H	H
C₁₁	OMe	H	Me
	OMe	H	Ph
	OMe	H	Ph
	OMe	OTBDMS	H
	OMe	OTBDMS	H

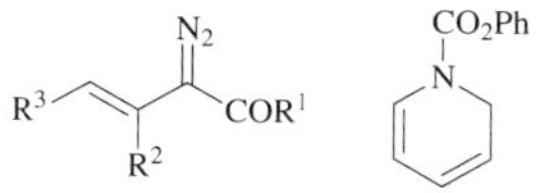

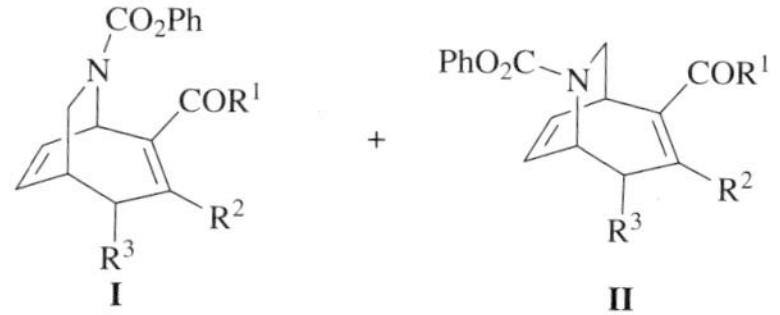

Catalyst, hexane

Catalyst	I	II
Rh₂(OOct)₄	(30)	(29)
Rh₂(OPiv)₄	(56)	(12)
Rh₂(TPA)₄	(50)	(3)
Rh₂(OOct)₄	(11)	(25)
Rh₂(OPiv)₄	(27)	(12)
Rh₂(OPiv)₄	(58)	(3)
Rh₂(OOct)₄	(30)	(9)
Rh₂(OPiv)₄	(69)	(trace)
Rh₂(OOct)₄	(17)	(24)
Rh₂(OPiv)₄	(34)	(3)

575

TABLE VIII. REACTION OF CARBENOIDS WITH DIENES (Continued)

Carbenoid Precursor	Substrate	Conditions	Product(s) and Yield(s) (%)	Refs.
C₅, C₆, C₁₁: N₂=C(COR¹)(CH=CHR²) R¹=OMe, R²=H R¹=OMe, R²=Me R¹=OMe, R²=Ph	1-(CO₂Ph)-3-methylpyridine	Rh₂(OPiv)₄, hexane	bicyclic product with PhO₂C-N, CO₂Me, R² groups (65) (50) (71)	575
C₁₀, C₁₁, C₁₂, C₁₃: N₂=C(COR¹)(=C(OR²)(COR¹)) R¹=OEt, R²=Me R¹=OEt, R²=Et R¹=OEt, R²=i-Pr R¹=OEt, R²=t-Bu	cyclopentadiene	Rh₂(OAc)₄, neat	spiro bicyclic product with CO₂Et, R²O, CO₂Et (89) (77) (80) (86)	156, 573
C₇: N₂=C(CO₂Me)(2-methyl-3-oxocyclopent-1-enyl)	diene with R¹, R² substituents R¹ | R² H | OAc H | OTMS OTMS | H OTBDMS | H OTMS | OMe	Rh₂(OAc)₄, CH₂Cl₂, 40°	azulenone with CO₂Me, R¹, R² (67) (86) (53) (94) (59)	576

284

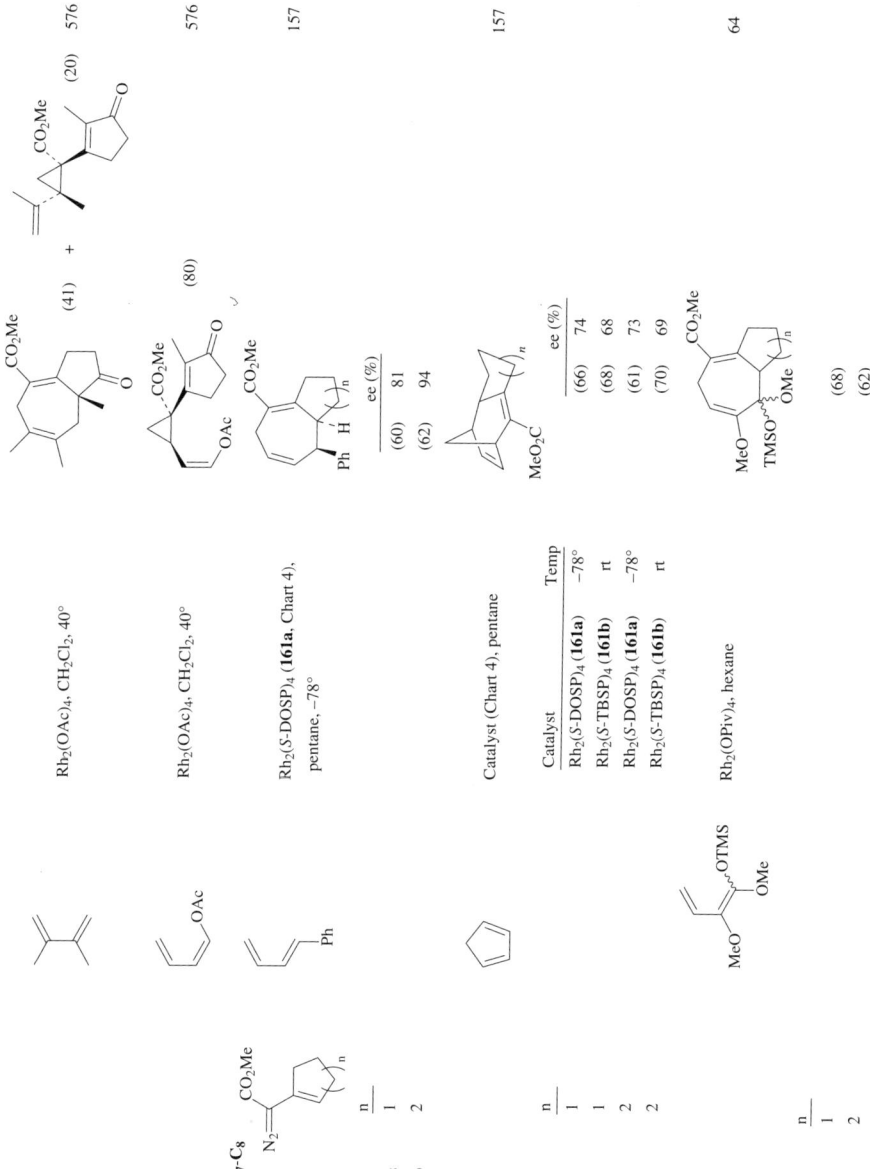

TABLE VIII. REACTION OF CARBENOIDS WITH DIENES (*Continued*)

Carbenoid Precursor	Substrate	Conditions	Product(s) and Yield(s) (%)	Refs.
C$_{10}$		1. Rh$_2$(OOct)$_4$, hexane, reflux 2. Kugelrohr	**I** (49)	158
		1. Rh$_2$(S-DOSP)$_4$ (**161a**, Chart 4), hexane, reflux 2. Kugelrohr	**I** (—) ee = 4%	158
C$_{13}$		Catalyst, CH$_2$Cl$_2$	**I** + **II**	148

Catalyst	Temp	**I+II**	de (%)
Rh$_2$(OAc)$_4$	40°	(60)	91
Rh$_2$(OPiv)$_4$	0°	(63)	69
Rh$_2$(OHex)$_4$	0°	(58)	>90
Rh$_2$(OHex)$_4$	40°	(80)	>90
Rh$_2$[(−)-mandelate]$_4$	40°	(60)	38
Rh$_2$[(+)-mandelate]$_4$	40°	(56)	4

C₁₆

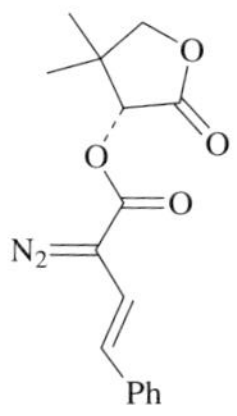

Rh₂(OAc)₄, CH₂Cl₂, 40°

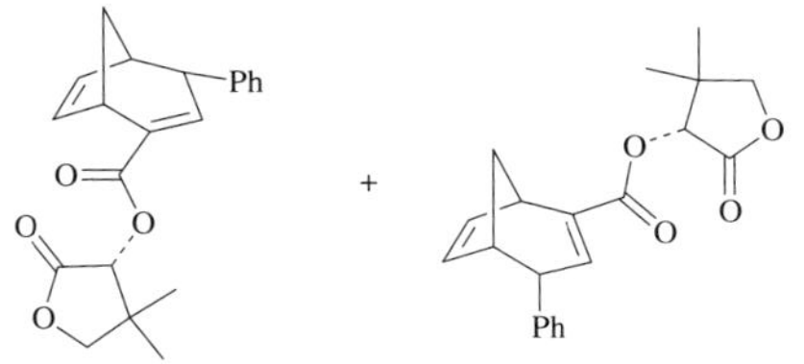

(87) **I+II** de = 76%

148

TABLE IX. REACTION OF CARBENOIDS WITH FURANS

Carbenoid Precursor	Substrate	Conditions	Product(s) and Yield(s) (%)	Refs.
C₄				
$N_2\text{=}\!\!\!\!\text{—}\!\!\text{CO}_2\text{Et}$	(furan)	$Rh_2(OAc)_4$	(65) **I:II:III:IV** = 18:10:4:1	172
		$Rh_2(OAc)_4$, neat, rt, 18 h	(66) **I:II:III:IV** = 17:10:5:1	173
		$Rh_2(OAc)_4$, neat, 0°, 1 h	(—) **I:II:III** = 6:1:1	161
		$CuSO_4$, neat, rt, 2 h	(22) **I**, (30) **II+III**	162
		$CuSO_4$, neat, 80°, 2 h	(5) **I**, (53) **II+III**	162
		$Cu(acac)_2$, neat, 0°, 24 h	(—) **I:II:III** = 6:1:1	161
$N_2\text{=}\!\!\!\!\text{—}\!\!\text{CO}_2\text{Et}$ (2.5 eq)	(2,5-disubstituted furan, R¹, R²: H,H; Me,H; Me,Me)	$Rh_2(OAc)_4$, CH_2Cl_2, rt, 3 h	(35), (28), (80) / (14), (0), (0)	577

Conditions	Products	Refs.
Rh₂(OAc)₄	(62) **I:II:III:IV** = 19:6:5:1	172
Rh₂(OAc)₄, neat, rt, 18 h	(54) **I:II:III:IV** = 31:12:10:1	173
Rh₂(OAc)₄, CH₂Cl₂, rt, 10 h	(92) **I:II** = 1.5:1	171
Rh₂(OAc)₄, CH₂Cl₂, rt, 10 h	(12–17)	578
1. Rh₂(OAc)₄, CH₂Cl₂, rt, 10 h 2. I₂	(—)	171
Rh₂(OAc)₄	**I+II+III** (75) **I:II:III** = 22:15:1	172
Rh₂(OAc)₄, neat, rt, 18 h	**I+II+III** (78) **I:II:III** = 9:6:1	173
CuSO₄, neat, 80°, 2 h	**I** (72)	162
1. Cu-Bronze, neat, 91°, 4 h 2. Silica gel	**I** (38); **III** (6)	160
1. Cu-Bronze, neat, 91°, 3 h 2. Alumina	**II** (36)	160

TABLE IX. REACTION OF CARBENOIDS WITH FURANS (Continued)

Carbenoid Precursor	Substrate	Conditions	Product(s) and Yield(s) (%)	Refs.

Substrate: RO₂C-furan (2-furoate ester)

Products: I (cyclopropanated bicycle with CO₂Et and RO₂C) + II (ring-opened diene with CO₂Et and CO₂R, ketone)

R	Conditions	I	II	Ref.
Me	Rh₂(OAc)₄	(—)	(0)	172
Me	Rh₂(OAc)₄, CH₂Cl₂, 15 h	(55)	(0)	173
Me	CuSO₄	(0)	(36)	162
Et	Rh₂(OAc)₄, neat, rt	(22)	(0)	578

Substrate: MeO₂C-CH=CH-furan

Conditions	I	Ref.
Rh₂(OAc)₄	(—)	172
Rh₂(OAc)₄, CH₂Cl₂, rt, 15 h	(51)	173

Substrate: 2-R⁴-furan with R²R¹C=C(R³)- substituent

Carbenoid: EtO₂C-CR¹=... giving cyclopropane product I and ring-opened dienone II

Conditions: CuSO₄, 50–70°   Refs. 174, 330

R¹	R²	R³	R⁴	I	II
Me	Me	H	H	(29)	(47)
Me	H	H	H	(26)	(46)
H	H	H	H	(52)	(5)
H	H	H	Me	(68)	(5)
H	H	H	Cl	(36)	(0)
H	H	H	Br	(22)	(0)
Me	H	Me	H	(—)	(58)

Substrate	Conditions	Product(s) (Yield %)	Refs.
2-furyl-furan	1. Rh₂(OAc)₄, CH₂Cl₂, rt, 12 h 2. I₂, CH₂Cl₂, rt, 18 h	furyl dienone-CO₂Et (40)	54
benzofuran	Rh₂(OAc)₄, CH₂Cl₂, rt, 1 h	benzofuran-CO₂Et adduct (40)	161
furan-CO₂Et	Rh₂(OAc)₄, CH₂Cl₂, rt, 3 h	oxabicycle-CO₂Et (76)	577
furyl-ethyl	1. Rh₂(OAc)₄, CH₂Cl₂, rt, 26 h 2. I₂, rt, 1 h	furyl dienone (52)	173
furyl-CO₂Et	CuSO₄	EtO₂C-dienone (37)	162
furyl-C₈H₁₇-n	Rh₂(OAc)₄	mixture I + II	172
	1. Rh₂(OAc)₄ 2. Silica gel	(—) I:II = 1.5:1 I (60)	172, 173

291

TABLE IX. REACTION OF CARBENOIDS WITH FURANS (Continued)

Carbenoid Precursor	Substrate	Conditions	Product(s) and Yield(s) (%)	Refs.
C5 R = Me (acrylate diazo)	furan	Rh2(OOct)4, hexane, reflux, 1 h	(±)-**I** (51) + **II** (10-15)	177
		Rh2(S-TBSP)4 (**161b**, Chart 4), hexane, reflux, 1 h	(1S,5S)-**I** (64) ee = 80% + **II** (15-20)	177
C9 R = (CH(Me)CO2Et)		Rh2(OOct)4, hexane, reflux, 1 h	**I** (63) de = 57%	177
		Rh2(S-TBSP)4 (**161b**, Chart 4), hexane, reflux, 1 h	**I** (51) de = 68%	177
		Rh2(R-TBSP)4 (ent-**161b**, Chart 4), hexane, reflux, 1 h	**I** (44) de = 0%	177
C5 R = Me (OTBDMS diazo)	furan	Rh2(OOct)4, hexane, reflux, 1 h	(±)-**I** (90)	177
		Rh2(S-TBSP)4 (**161b**, Chart 4), hexane, reflux, 1 h	(1S,5S)-**I** (94) ee = 46%	177
C9 R = (CH(Me)CO2Et)		Rh2(OOct)4, hexane, reflux, 1 h	**I** (72) de = 79%	177
		Rh2(S-TBSP)4 (**161b**, Chart 4), hexane, reflux, 1 h	**I** (97) de = 80%	177
		Rh2(R-TBSP)4 (ent-**161b**, Chart 4), hexane, reflux, 1 h	**I** (99) de = 53%	177

C10

Substrate: 4,4-dimethyl-γ-butyrolactone with R substituent (R = CO2R group with OTBDMS)

Product I: bicyclic ether with CO2R and OTBDMS groups

Conditions	Product (yield)	de	ref
Rh2(OOct)4, hexane, reflux, 1 h	I (82)	de = 94%	177
Rh2(S-TBSP)4 (161b, Chart 4), hexane, reflux, 1 h	I (83)	de = 20%	177
Rh2(R-TBSP)4 (ent-161b, Chart 4), hexane, reflux, 1 h	I (67)	de = 70%	177

C10–C13

Substrate: R^1O-diazo ester with OTBDMS and R^2 group + furan (R^3, R^4)

Product I: oxabicyclic with CO_2R^1, OTBDMS, R^2, R^3, R^4

Conditions: Rh2(OOct)4, hexane, reflux, 1 h — ref 177

R^1	R^2	R^3	R^4	(yield)	de (%)	abs stereochem
ethyl (S)-lactate	Me	H	H	(62)	90	1S
(R)-pantolactone	Me	H	H	(75)	95	1R
ethyl (S)-lactate	H	Me	H	(81)	75	1S
(R)-pantolactone	H	Me	H	(91)	83	1R
ethyl (S)-lactate	Me	Me	H	(91)	84	1S
(R)-pantolactone	Me	Me	H	(69)	94	1R
ethyl (S)-lactate	H	COMe	H	(74)	79	1S
(R)-pantolactone	H	COMe	H	(65)	94	1R
ethyl (S)-lactate	Me	COMe	H	(71)	80	1S
(R)-pantolactone	H	Me	CO2Me	(65)	82	1R

C5

Substrate: methyl 2-diazo-2-(ethoxycarbonyl)propanoate analog (N_2=C(Me)CO2Et) + furan

Products: I (cyclopropanated dihydrofuran with CO2Et) + II (dienal with CO2Et, CHO)

Conditions	(yield)	I:II	ref
Rh2(OAc)4, neat	(—)	I:II = 8:1	172
Rh2(OAc)4, neat, rt, 18 h	(60)	I:II = 8:1	173

TABLE IX. REACTION OF CARBENOIDS WITH FURANS (Continued)

Carbenoid Precursor	Substrate	Conditions	Product(s) and Yield(s) (%)	Refs.
EtO2C-C(=O)-CH=N2	2-methylfuran	Rh2(OAc)4, neat	(—) I:II = 3:1	172
		Rh2(OAc)4, neat, rt, 18 h	(69) I:II = 3:1	173
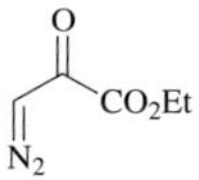 (MeO2C)2C=N2	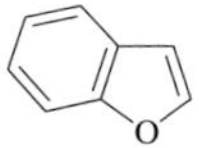 benzofuran	Cu-bronze, neat, 110°, 5 h	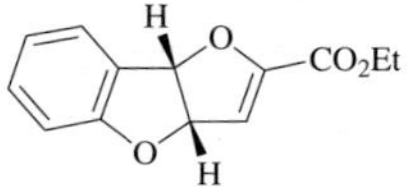 (27)	130
	2-R-furan (R = H, Me, CH=CMe2)	Rh2(OAc)4, neat, rt	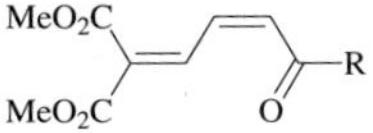 (100)	579

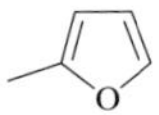

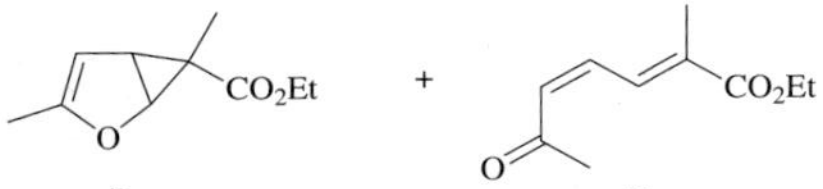

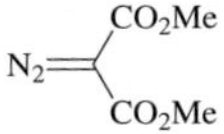

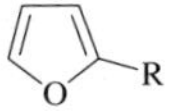

C6

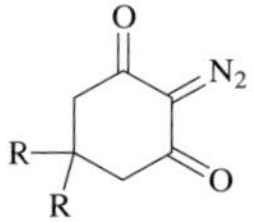

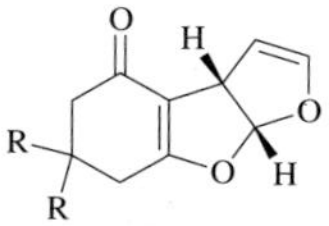

			ee (%)	
R = H		Rh2(OAc)4, C6H5F, rt, 15 h	(92) —	67
H		Rh2(OAc)4, neat, rt, 24 h	(48) —	580
H		Rh2(BNP)4, neat, rt	(44) 50	176
C8 Me		Rh2(OAc)4, C6H5F, rt, 15 h	(56) —	67
Me		Rh2(OPiv)4, C6H5F, rt, 15 h	(44) —	67
Me		Rh2(OAc)4, neat, rt, 24 h	(56) —	580
Me		Rh2(BNP)4, neat, rt	(50) 49	176

294

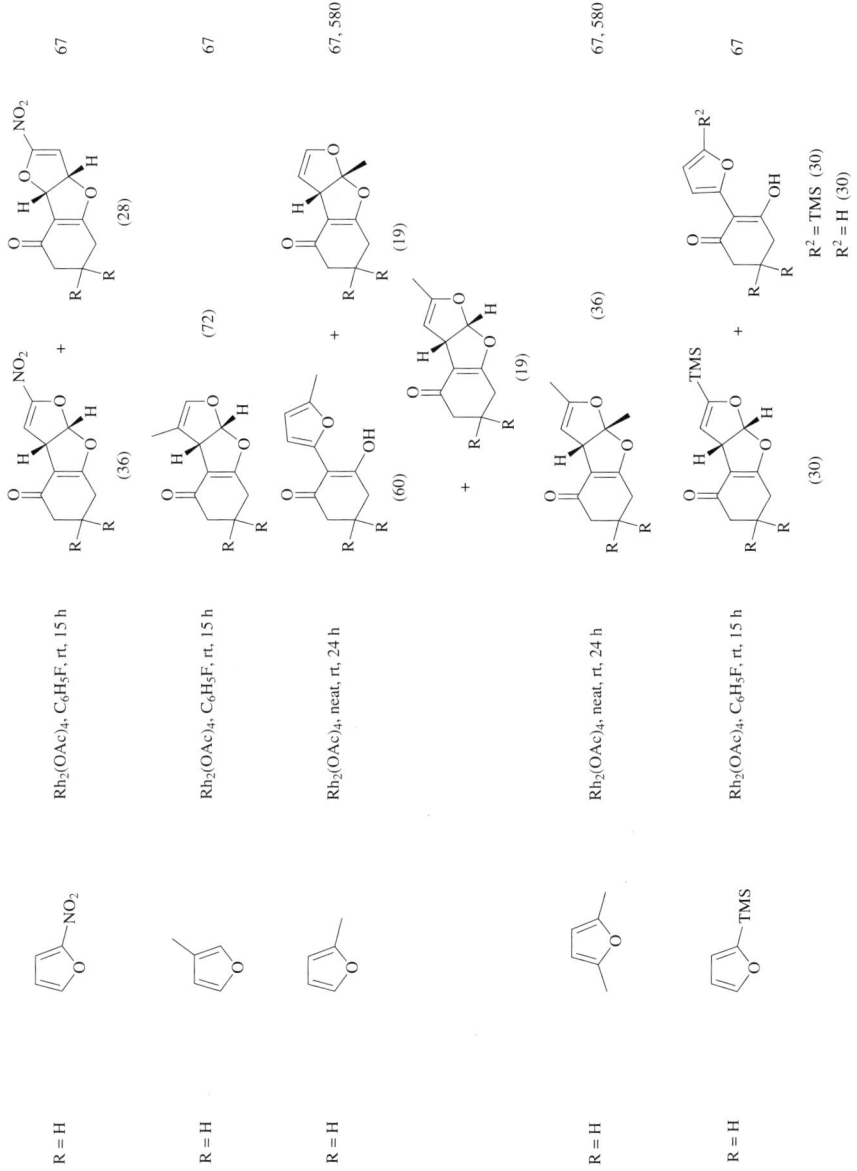

TABLE IX. REACTION OF CARBENOIDS WITH FURANS (*Continued*)

Carbenoid Precursor	Substrate	Conditions	Product(s) and Yield(s) (%)	Refs.
R = H (EtO2C, CO2Et furan)		Rh2(OAc)4, C6H5F, rt, 15 h	(fused bicyclic product) (26)	67
R = H (benzofuran)		Rh2(OAc)4, C6H5F, rt, 15 h Rh2(OAc)4, neat, rt, 24 h	I + II I (52) II (14) I (36) II (19)	67 580
(C7 diazo ketoester with CO2Me)	furan	Rh2(OAc)4 or Rh2(OAc)4, neat, rt, 18 h	I + II + III (—) I:II:III = 7:1:1	172, 173
		1. Rh2(OAc)4, neat, rt, 18 h 2. I2, CH2Cl2, rt, 6 h	IV (68)	173
		Rh2(OAc)4, neat, rt, 2 h	I (75–80) + II (20–25)	163, 168

296

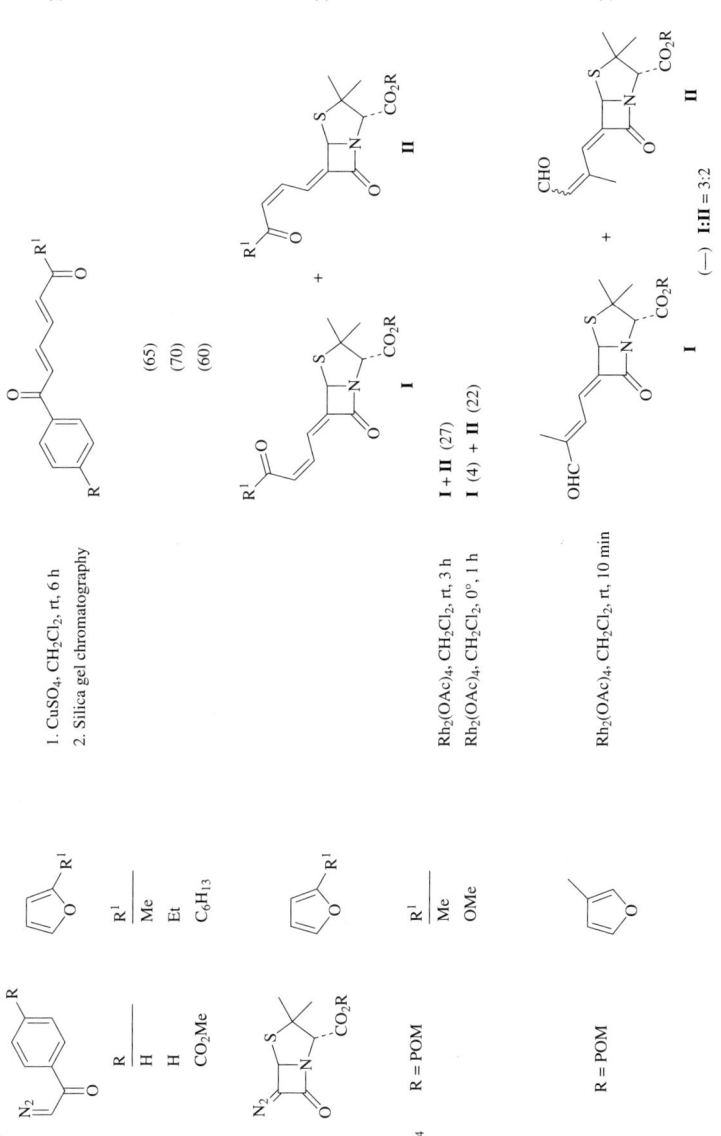

TABLE IX. REACTION OF CARBENOIDS WITH FURANS (Continued)

Carbenoid Precursor	Substrate	Conditions	Product(s) and Yield(s) (%)	Refs.
C_{15} R = POM	benzofuran	$Rh_2(OAc)_4$, CH_2Cl_2, rt, 1 h	(15)	161
C_{15} R = PNB	2,5-dimethylfuran	$Rh_2(OAc)_4$, CH_2Cl_2, 0°, 2 h	(2.5)	161
C_{21} R = CHPh$_2$	furan	$Rh_2(OAc)_4$, neat, 0°, 1 h	I + II (100) **I:II** = 2:1	161

298

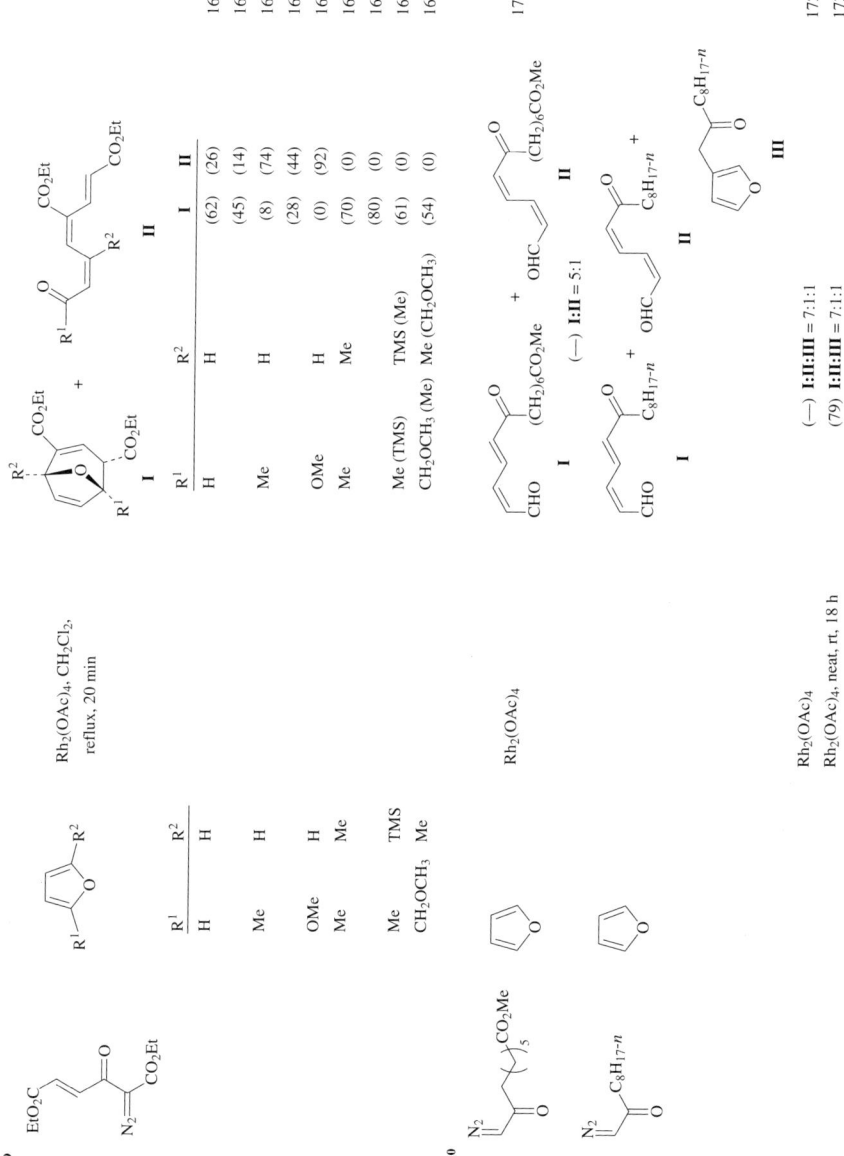

TABLE IX. REACTION OF CARBENOIDS WITH FURANS (Continued)

Carbenoid Precursor	Substrate	Conditions	Product(s) and Yield(s) (%)	Refs.
C₅H₁₁-n with N₂/C=O alkyne	furan	Rh₂(OAc)₄, neat, rt, 1 h	CHO-diene-yne-ketone (—)	169
(CH₂)₃CO₂Me with N₂/C=O alkyne	furan	Rh₂(OAc)₄, neat, rt, 2 h	CHO-diene-yne-ketone-CO₂Me (75-80) + CHO-diene-yne-ketone-CO₂Me (20-25)	164, 168
(CH₂)₅CO₂Et with N₂/C=O (C₁₃)	2-n-C₆H₁₃-furan	1. Rh₂(OAc)₄, CH₂Cl₂, rt, 12 h; 2. I₂, CH₂Cl₂, rt, 2 h	enone-(CH₂)₂-CO₂Et (46)	170
alkyne-ketone-N₂	furan	Rh₂(OAc)₄	CHO-dienyne-ketone (—)	164, 168
(C₁₇) aryl diazo with alkyne-Ph	2-R-furan (R = H, Me)	Rh₂(OAc)₄, neat, rt, 30 min	indanone-Ph-enone-R (R=H, 91; R=Me, 94)	581

TABLE X. REACTION OF CARBENOIDS WITH PYRROLES

Carbenoid Precursor	Substrate	Conditions	Product(s) and Yield(s) (%)			Refs.
			I+II	I:II		
C_2 $N_2=\!\!\!\!=\!\!\!\!<\!\!\!\!\!\!\begin{array}{c}\\ CO_2R\end{array}$						
C_4 R = Et	N-Me pyrrole					
		Cu powder, 110°	(31)	81:19		180, 181
		Cu-bronze, 100°	(30–40)	84–79:21–16		180, 181
		Cu_2O, 100°	(45)	80:20		180, 181
		$CuSO_4$, 80°	(58)	65:35		180, 181
		$Cu(OAc)_2$, 80°	(45)	81:19		180
		$Cu(NO_3)_2$, 55°	(54)	71:29		180
		CuCN, 70°	(43)	76:24		180
		$Cu(BF_4)_2$, 50°	(66)	54:46		180, 181
		$Cu(OSO_2CF_3)_2$, 40°	(63)	58:42		180, 181
		CuCl, 75°	(48)	80:20		180, 181
		$CuCl\cdot P(OEt)_3$, 60°	(51)	81:19		180
		CuBr, 60°	(27)	80:20		180
		$CuBr\cdot P(OEt)_3$, 65°	(34)	82:18		180
		CuI, 75°	(34)	81:19		180
		$CuI\cdot(POEt)_3$, 65°	(27)	84:16		180
		CuF_2, 80°	(43)	85:15		180
		$CuCl_2$, 75°	(42)	82:18		180
		$CuBr_2$, 60°	(33)	81:19		180
		$Cu(OMe)_2$, 60°	(45)	83:17		180
		$Pd(OAc)_2$, 30°	(35)	66:34		180, 181
		$(\pi\text{-}C_3H_5PdCl)_2$, 35°	(42)	67:33		180, 181
		$DiphosNiCl_2$, 90°	(13)	67:33		180
		ZnI_2, 110°	(–3)	—		180
		$LiClO_4$, 110°	(15)	—		180
		$Rh_2(OAc)_4$, 80°	(trace)	—		180

Product I: 2-(CH$_2$CO$_2$Et)-N-Me-pyrrole; Product II: 3-(CH$_2$CO$_2$Et)-N-Me-pyrrole

TABLE X. REACTION OF CARBENOIDS WITH PYRROLES (Continued)

Carbenoid Precursor	Substrate	Conditions	Product(s) and Yield(s) (%)	Refs.
R = Et	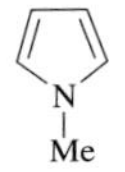 N-Me pyrrole	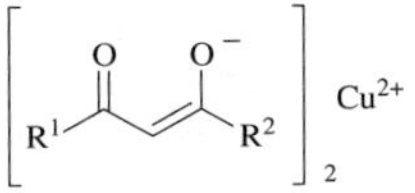 [R¹C(O)=C(R²)O⁻]₂ Cu²⁺	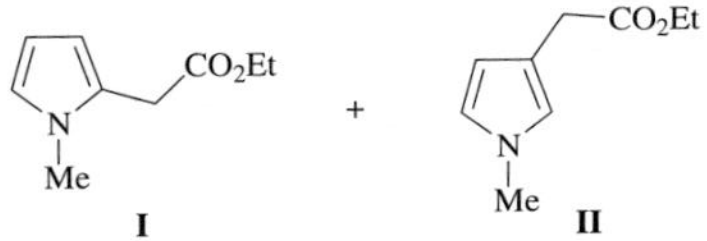 I + II (2-CH₂CO₂Et and 3-CH₂CO₂Et N-methylpyrroles)	180

R¹	R²	I+II	I:II
Me	Me	(34-42)	91-93:9-7
Me	Ph	(44-45)	90-91:10-9
Me	CF₃	(54)	80:20
CF₃	CF₃	(55)	78-80:22-20

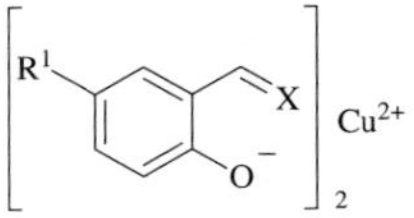

				180

salicylaldimine Cu²⁺ complex

R¹	X	I+II	I:II
H	O	(54)	83:17
Cl	O	(60)	80:20
OMe	O	(51)	83:17
NO₂	O	(63)	74:26
H	NH	(50)	92:8
H	NOH	(51)	82:18
H	NPr-i	(50)	94:6
H	NMePh	(47)	94:6
H	NC₆H₄Cl-4	(52)	91:9
NO₂	NH	(44)	85:15

		I+II	I:II	180
		(42)	89:11	

Cu salen complex

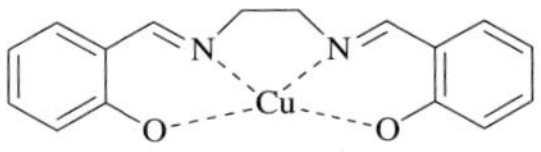

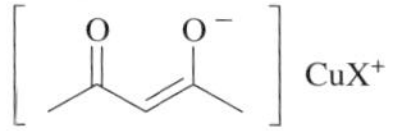

$$\left[\begin{array}{c}\text{O}^-\\ \\ \text{O}\end{array}\right]\text{CuX}^+$$

X
OSO_2CF_3
OMe

R = Et

Pyrrole (N–R) + diazo → I (2-CH₂CO₂Et pyrrole) + II (3-CH₂CO₂Et pyrrole)

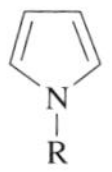

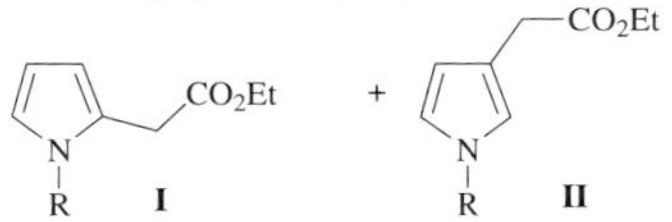

R	Catalyst	I+II	I:II	Ref.
H	Cu(acac)₂	(58)	81:19	180
H	Cu-bronze	(37)	90:10	180
H	Cu(acac)₂	(53)	94:6	
H	Cu-bronze	(42)	92:8	
H	Cu(OSO₂CF₃)₂	(52)	73:27	
Pr-i	Cu(acac)₂	(30)	60:40	
Pr-i	Cu(OSO₂CF₃)₂	(20)	46:54	
Bu-t	Cu(acac)₂	(34)	<2:98	
Ph	CuCl·P(OEt)₃	(<2)	—	
Ph	Cu(OSO₂CF₃)₂	(5)	33:67	
Ph	Cu(acac)₂	(trace)	—	
Ph	Cu-bronze	(trace)	—	

R = Et

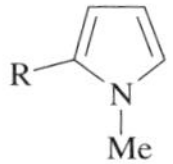

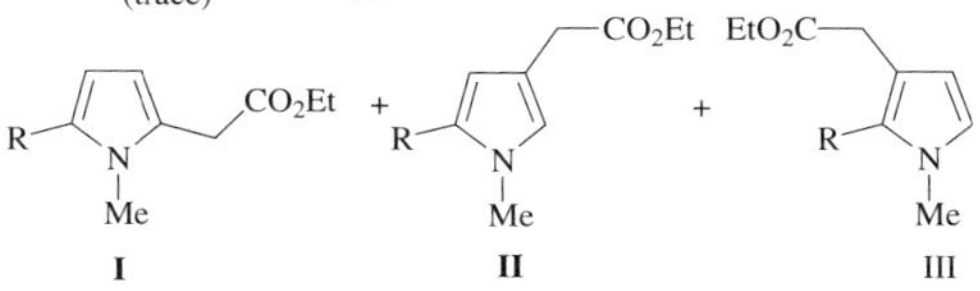

N-Methyl-2-R-pyrrole + diazo → I + II + III (180)

R	Catalyst	(yield)	I:II:III
Me	Cu(acac)₂	(34)	78:3:19
Me	Cu(OSO₂CF₃)₂	(64)	50:25:25
4-ClC₆H₄CO	CuCl·P(OEt)₃	(trace)	—
4-ClC₆H₄CO	Cu(OSO₂CF₃)₂	(~5)	—
CH₂CO₂Et	Cu(acac)₂	(60)	80:5:15
CH₂CO₂Et	Cu(OSO₂CF₃)₂	(56)	56:22:22

TABLE X. REACTION OF CARBENOIDS WITH PYRROLES (Continued)

Carbenoid Precursor	Substrate	Conditions	Product(s) and Yield(s) (%)	Refs.
R = Et	N-Me pyrrole (4-methyl)	Cu(acac)$_2$ Cu(OSO$_2$CF$_3$)$_2$	I + II + III I:II:III (65) 57:14:29 (74) 40:30:30	180
R = Et	pyrrole with BnO$_2$C, Me, Me, N-Me substituents	Cu-bronze	(43)	66
R = Bn	pyrrole with MeO$_2$C, propyl, t-BuO$_2$C, N-Me	Cu powder, 90°	(25)	582
R = Et	pyrrole with o-NO$_2$-C$_6$H$_4$, Me, n-Bu, N-Me	Rh$_2$(OAc)$_4$	(27)	583
C$_3$ R = Me	N-BOC pyrrole	Rh$_2$(OAc)$_4$	bicyclic MeO$_2$C / N-BOC product (45)	183

R = Et		Cu-bronze	(11) + (36)	584
C₄ R = Et		Cu-bronze	(55)	584
R = Et	0.33 eq	Cu powder	(41)	585
R = Et	0.75 eq	Cu powder	(16)	585
R = Et		Cu powder		217

I: $R^2 = Ph$, $R^3 = CH_2CO_2Et$
II: $R^2 = CH_2CO_2Et$, $R^3 = Ph$

R^1	R^2	R^3	**I**	**II**
H	Ph	H	(16)	(0)
H	H	Ph	(0)	(25)
Me	Ph	H	(19)	(0)
Me	H	Ph	(0)	(31)

TABLE X. REACTION OF CARBENOIDS WITH PYRROLES (Continued)

Carbenoid Precursor	Substrate	Conditions	Product(s) and Yield(s) (%)	Refs.
R = Et		Cu powder		

R^1	R^2	R^3		
H	H	H	(51)	217
Me	H	H	(39)	217
Me	H	4-OC$_6$H$_5$	(12-61)	216
Me	H	2-Cl	(12-61)	216
Me	H	3-Cl	(12-61)	216
Me	H	4-Cl	(12-61)	216
Me	H	2,4-Cl$_2$	(12-61)	216
Me	H	3,4-Cl$_2$	(12-61)	216
Me	H	2,5-Cl$_2$	(12-61)	216
Me	H	2,3,4-Cl$_3$	(12-61)	216
Me	H	4-Bu-t	(12-61)	216
Me	H	4-NO$_2$	(12-61)	216
Me	H	4-NH$_2$	(12-61)	216
Me	H	4-F	(12-61)	216
Me	H	4-OMe	(12-61)	216
Me	4-Cl	H	(12-61)	216
H	H	4-Cl	(12-61)	216
Me	H	4-OH	(12-61)	216

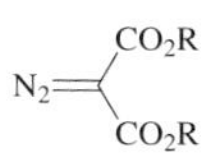

C_5, R = Me

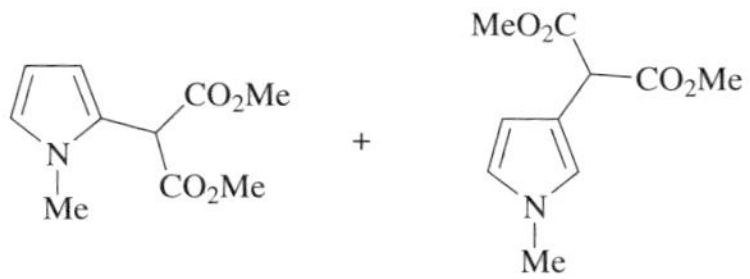

Substrate: N-methylpyrrole with diazomalonate (CO2R, CO2R, N2)

I: 2-substituted N-methylpyrrole with CH(CO2Me)2
II: 3-substituted N-methylpyrrole with CH(CO2Me)2

Catalyst	(I+II)	I:II
Cu-bronze	(~30)	87:13
Cu(acac)₂	(65)	90:10
Cu(F₆-acac)₂	(83)	88:12
Cu(4-chlorosalicylate)₂	(69)	84:16
Cu[(N-i-Pr)salicylaldiminate]₂	(76)	92:8
Cu(OSO₂CF₃)₂	(70)	83:17
Rh₂(OAc)₄	(92)	83:17

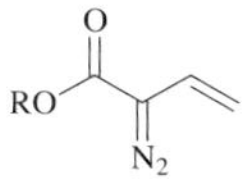

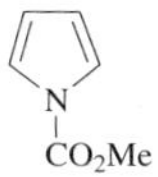

66

C_5, R = Me

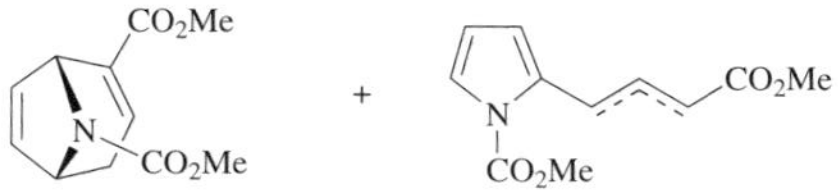

Substrate: N-(methoxycarbonyl)pyrrole with methyl 2-diazo-3-butenoate

I: bicyclic cycloadduct
II: 2-(3-methoxycarbonyl-1-propenyl)-N-(methoxycarbonyl)pyrrole

Catalyst	I	I:II
Rh₂(OAc)₄, CH₂Cl₂	(35)	55:45
Rh₂(mandelate)₄, CH₂Cl₂	(20)	40:60
Rh₂(TFA)₄, CH₂Cl₂	(16)	15:85
Rh₂(OHex)₄, CH₂Cl₂	(34)	86:14
Rh₂(OHex)₄, benzene	(68)	>95:5
Rh₂(OHex)₄, hexane	(75)	>95:5
Rh₂(OPiv)₄, hexane	(67)	>95:5
Rh₂(TFA)₄, hexane	(29)	26:74

185

TABLE X. REACTION OF CARBENOIDS WITH PYRROLES (Continued)

Carbenoid Precursor	Substrate	Conditions	Product(s) and Yield(s) (%)	Refs.
R = Me	N-R[1] pyrrole	Rh$_2$(S-TBSP)$_4$ (161b), hexane	product (bicyclic, CO$_2$Me, R[1]): R[1] / ee (%) CO$_2$Bu-t (42) / 51 CO$_2$Me (44) / 42 Ac (46) / 17 SO$_2$Me (34) / 29	175
C$_8$ R = Me R = CH$_2$CO$_2$Et	N-BOC pyrrole	Rh$_2$(S-TBSP)$_4$ (161b), hexane	(42) (62) ee = 51%	175, 586 586
C$_5$ R = Me C$_8$ R = CH$_2$CO$_2$Et	2-Me N-BOC pyrrole	Rh$_2$(S-TBSP)$_4$ (161b), hexane Rh$_2$(OOct)$_4$, hexane Rh$_2$(OOct)$_4$, hexane	I + II + III + IV I / II / III / IV (24) (6) (19) (21) (38) (16) (10) (27) (56) (8) (8) (12)	175 586 586

C5 R = Me		Rh2(OOct)4, hexane	(56)	587
C6 R = Et			(33)	

C5 R = Me		Rh2(OOct)4, hexane	(73) + (−)	185

R¹	R²			
C12 Et	SO2Ph	Rh2(OAc)4	(61)	184
C11 Me	Ph		(53)	
C14 Et	CH=CHPh		(18)	

TABLE X. REACTION OF CARBENOIDS WITH PYRROLES (Continued)

Carbenoid Precursor	Substrate	Conditions	Product(s) and Yield(s) (%)			Refs.

C_9-C_{10} carbenoid precursor: $R^1O-C(=O)-C(=N_2)-CH=CH_2$

Substrate: BOC-pyrrole with R^2, R^3, R^4 substituents

Product: bicyclic azabicyclo with CO_2R^1, BOC, R^2, R^3, R^4

Conditions: $Rh_2(OOct)_4$, hexane, reflux

R^1	R^2	R^3	R^4	de (%)	abs stereochem
ethyl (S)-lactate	H	H	H	(82) 66	1R
(R)-pantolactone	H	H	H	(64) 69	1S
ethyl (S)-lactate	Me	H	H	(54) 59	1R
ethyl (S)-lactate	CH$_2$OTBDMS	H	H	(62) 70	1R
ethyl (S)-lactate	Ph	H	H	(64) 53	1R
ethyl (S)-lactate	Ac	H	H	(30) 67	1R
ethyl (S)-lactate	Me	H	Me	(33) 25	1R
ethyl (S)-lactate	H	Me	H	(19) 52	1R
ethyl (S)-lactate	—(CH$_2$)$_4$—		H	(48) 55	1R

References: 175, 186; 175, 186; 175; 175; 175; 175; 175; 175; 175

C_9-C_{10} carbenoid precursor: $R^1O-C(=O)-C(=N_2)-C(OTBDMS)=CH_2$

Product: bicyclic azabicyclo with CO_2R^1, OTBDMS, BOC, R^2, R^3, R^4

Conditions: $Rh_2(OOct)_4$, hexane, reflux

R^1	R^2	R^3	R^4	de (%)	abs stereochem
ethyl (S)-lactate	H	H	H	(64) 66	1R
(R)-pantolactone	H	H	H	(66) 68	1S
ethyl (S)-lactate	Me	H	H	(55) 58	1R
ethyl (S)-lactate	Ph	H	H	(74) 52	1R
(R)-pantolactone	Ph	H	H	(56) 52	1S
ethyl (S)-lactate	Ac	H	H	(58) 79	1R
(R)-pantolactone	Ac	H	H	(69) 78	1S
ethyl (S)-lactate	Me	H	Me	(30) 52	1R
(R)-pantolactone	—(CH$_2$)$_4$—		H	(31) 37	1S

References: 175

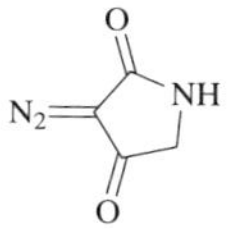	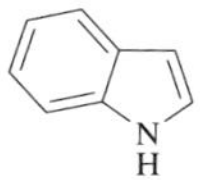	$Rh_2(OAc)_4$	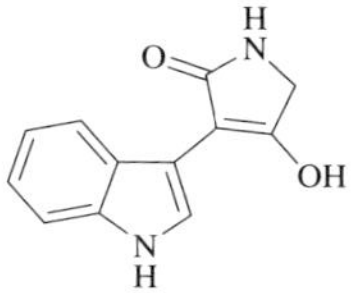 (65)	588
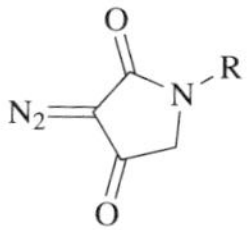	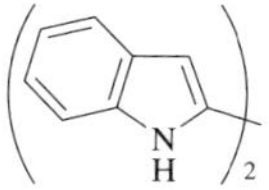	$Rh_2(OAc)_4$	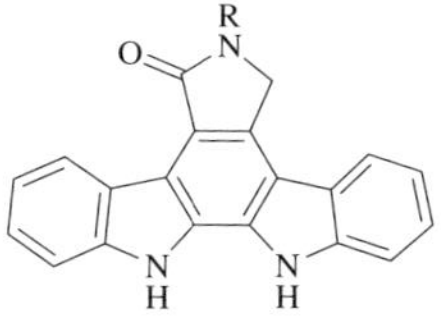	588

	R		
	H		(25)
C_8	*t*-Bu		(40)
C_{11}	4-$MeOC_6H_4$		(55)
	Bn		(58)
C_{12}	3,4-$(MeO)_2C_6H_3$		(62)

C_6

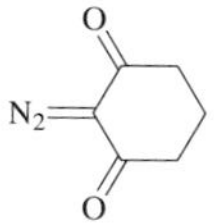	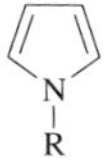	$Rh_2(OAc)_4$, C_6H_5F	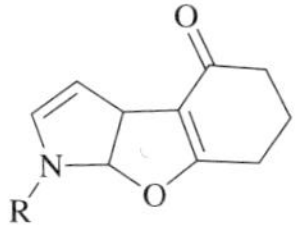	67
	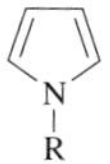	$Rh_2(OAc)_4$, solvent	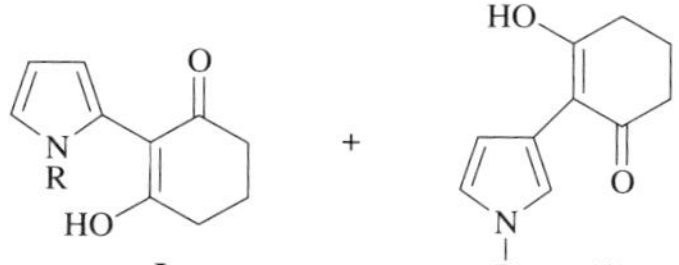	67

R	
CO_2Et	(34)
Ac	(32)

I + **II**

R	Solvent	I	II
Me	C_6F_6	(60)	(29)
Ts	C_6F_6	(—)	(59)
Ph	C_6H_5F	(—)	(72)
4-FC_6H_4	C_6F_6	(30)	(59)
4-$MeOC_6H_4$	C_6H_5F	(—)	(66)

TABLE X. REACTION OF CARBENOIDS WITH PYRROLES (Continued)

Carbenoid Precursor	Substrate	Conditions	Product(s) and Yield(s) (%)	Refs.
C_6-C_8				
2,2-disubstituted 2-diazo-1,3-cyclohexanedione (R = H, Me)	N-Ac indole	$Rh_2(OAc)_4$, C_6H_5F	fused tetracyclic product; R = H (54), Me (54)	580, 67
C_8				
$PhCOCHN_2$	pyrrole (N-H)	Cu powder	1-phenyl-2-(pyrrol-2-yl)ethanone (32)	589
C_9				
diethyl 2-diazo-3-butenedioate (EtO_2C-$CH=CH$-$C(N_2)CO_2Et$)	N-R pyrrole; R = CO_2Me, CO_2Et, $CO_2CH_2CH_2TMS$	$Rh_2(OAc)_4$	azabicyclic diester (62), (54), (71)	184
C_{10}				
diethyl 2-diazo-3-methoxy-3-pentenedioate	N-CO_2Me pyrrole	$Rh_2(OAc)_4$	polycyclic adduct (33)	184

REFERENCES

[1] Dave, V.; Warnhoff, E. W. *Org. React.* **1970**, *18*, 217.
[2] Adams, J.; Spero, D. M. *Tetrahedron* **1991**, *47*, 1765.
[3] Calter, M. A. *Curr. Org. Chem.* **1997**, *1*, 37.
[4] Davies, H. M. L. *Tetrahedron* **1993**, *49*, 5203.
[5] Davies, H. M. L. *Aldrichimica Acta* **1997**, *30*, 107.
[6] Davies, H. M. L. *Curr. Org. Chem.* **1998**, *2*, 463.
[7] Davies, H. M. L. In *Advances in Cycloaddition*; Haramata, M. E., Ed.; JAI Press: Greenwich, CT, 1999; Vol. 5, pp. 119–164.
[8] Doyle, M. P. *Chem. Rev.* **1986**, *86*, 919.
[9] Doyle, M. P. *Acc. Chem. Res.* **1986**, *19*, 348.
[10] Doyle, M. P. *Recl. Trav. Chim. Pays-Bas* **1991**, *110*, 305.
[11] Doyle, M. P. In *Catalytic Asymmetric Synthesis*; Ojima, I., Ed.; VCH: New York, 1993; pp. 63–99.
[12] Doyle, M. P.; McKervey, M. A.; Ye, T. *Modern Catalytic Methods for Organic Synthesis with Diazo Compounds: From Cyclopropanes to Ylides*; Wiley: New York, 1998.
[13] Doyle, M. P.; Forbes, D. C. *Chem. Rev.* **1998**, *98*, 911.
[14] Doyle, M. P.; Protopopova, M. N. *Tetrahedron* **1998**, *54*, 7919.
[15] Maas, G. *Top. Curr. Chem.* **1987**, *137*, 75.
[16] Nefedov, O. M.; Shapiro, E. A.; Dyatkin, A. B. In *Supplement B: The Chemistry of Acid Derivatives*; Patai, S., Ed.; Wiley: New York, 1992; Ch. 25.
[17] Padwa, A.; Hornbuckle, S. F. *Chem. Rev.* **1991**, *91*, 263.
[18] Padwa, A.; Krumpe, K. E. *Tetrahedron* **1992**, *48*, 5385.
[19] Padwa, A.; Weingarten, M. D. *Chem. Rev.* **1996**, *96*, 223.
[20] Ye, T.; McKervey, M. A. *Chem. Rev.* **1994**, *94*, 1091.
[21] Demonceau, A.; Noels, A. F.; Hubert, A. J. *Aspects Homogeneous Catal.* **1988**, *6*, 199; *Chem. Abstr.* **1988**, *109*, 189473q.
[22] Regitz, M.; Maas, G. In *Aliphatic Diazo Compounds- Properties and Synthesis*; Academic: New York, 1986.
[23] *Carbocyclic Three- and Four- Membered Ring Compounds, Methods of Organic Chemistry (Houben-Weyl)*; de Meijere, A., Ed.; Georg Thieme Verlag: New York, 1997; Vol. E 17a.
[24] Reissig, H.-U. In *Stereoselective Synthesis, Methods of Organic Chemistry (Houben-Weyl)*, Helmchen, G., Hoffmann, R. W., Mulzer, J., Schaumann, E., Eds.; Georg Thieme Verlag: New York, 1995; Vol. E 21c, pp. 3179–3270.
[25] Noels, A. F.; Demonceau, A. In *Applied Homogeneous Catalysis with Organometallic Compounds*; Cornils, B., Herrmann, W. A., Eds.; VCH: New York, 1996, pp. 733–747.
[26] Marchand, A. P.; Brockway, N. M. *Chem. Rev.* **1974**, *74*, 431.
[27] Burke, S. D.; Grieco, P. A. *Org. React.* **1979**, *26*, 361.
[28] *Carbene (Carbenoide), Methoden der Organischen Chemie (Houben-Weyl)*, Regitz, M., Ed.; Georg Thieme Verlag: New York, 1989; Vol. E 19b.
[29] Doyle, M. P.; McKervey, M. A. *J. Chem. Soc., Chem. Commun.* **1997**, 983.
[30] Singh, V. K.; Arpita, D.; Sekar, G. *Synthesis* **1997**, 137.
[31] Tomilov, Yu. V.; Dokitchev, V. A.; Dzhemilev, U. M.; Nefedov, O. M. *Russ. Chem. Rev.* **1993**, *62*, 799; *Chem. Abstr.* **1994**, *121*, 178781.
[32] Doyle, M. P.; Griffin, J. H.; Bagheri, V.; Dorow, R. L. *Organometallics* **1984**, *3*, 53.
[33] De Meijere, A.; Schulz, T. J.; Kostikov, R. R.; Graupner, F.; Murr, T.; Bielfeldt, T. *Synthesis* **1991**, 547.
[34] Hoveyda, A. H.; Evans, D. A.; Fu, G. C. *Chem. Rev.* **1993**, *93*, 1307.
[35] Molander, G. A.; Harring, L. S. *J. Org. Chem.* **1989**, *54*, 3525.
[36] Simmons, H. E.; Cairns, T. L.; Vladuchick, S. A.; Hoiness, C. M. *Org. React.* **1973**, *20*, 1.
[37] Brookhart, M.; Studabaker, W. B. *Chem. Rev.* **1987**, *87*, 411.
[38] Dötz, K. H. *Angew. Chem., Int. Ed. Engl.* **1984**, *23*, 587.
[39] Pirrung, M. C.; Morehead, A. T., Jr. *J. Am. Chem. Soc.* **1996**, *118*, 8162.

[40] Pirrung, M. C.; Morehead, A. T., Jr. *J. Am. Chem. Soc.* **1994**, *116*, 8991.
[41] Alonso, M. E.; Hernandez, M. I.; Gomez, M.; Jano, P.; Pekerar, S. *Tetrahedron* **1985**, *41*, 2347.
[42] Alonso, M. E.; Garcia, M. C. *Tetrahedron* **1989**, *45*, 69.
[43] Nishiyama, H.; Aoki, K.; Itoh, H.; Iwamura, T.; Sakata, N.; Kurihara, O.; Motoyama, Y. *Chem. Lett.* **1996**, 1071.
[44] Park, S.-B.; Sakata, N.; Nishiyama, H. *Chem. Eur. J.* **1996**, *2*, 303.
[45] Park, S.-B.; Nishiyama, H.; Itoh, Y.; Itoh, K. *J. Chem. Soc., Chem. Commun.* **1994**, 1315.
[46] Sheehan, S. M.; Padwa, A.; Snyder, J. P. *Tetrahedron Lett.* **1998**, *39*, 949.
[47] Taber, D. F.; You, K. K.; Rheingold, A. L. *J. Am. Chem. Soc.* **1996**, *118*, 547.
[48] Taber, D. F.; Malcolm, S. C. *J. Org. Chem.* **1998**, *63*, 3717.
[49] Ref. 12, pp. 163–220.
[50] Hanks, T. W.; Jennings, P. W. *J. Am. Chem. Soc.* **1987**, *109*, 5023.
[51] Demonceau, A.; Saive, E.; de Froidmont, Y.; Noels, A. F.; Hubert, A. J.; Chizhevsky, I. T.; Lobanova, I. A.; Bregadze, V. I. *Tetrahedron Lett.* **1992**, *33*, 2009.
[52] Demonceau, A.; Abreu Dias, E.; Lemoine, C. A.; Stumpf, A. W.; Noels, A. F.; Pietraszuk, C.; Gulinski, J.; Marciniec, B. *Tetrahedron Lett.* **1995**, *36*, 3519.
[53] Schnaubelt, J.; Marks, E.; Reissig, H.-U. *Chem. Ber.* **1996**, *129*, 73.
[54] Wenkert, E.; Guo, M.; Pizzo, F.; Ramachandran, K. *Helv. Chim. Acta* **1987**, *70*, 1429.
[55] O'Bannon, P. E.; Dailey, W. P. *J. Org. Chem.* **1989**, *54*, 3096.
[56] Brown, K. C.; Kodadek, T. *J. Am. Chem. Soc.* **1992**, *114*, 8336.
[57] Wolf, J. R.; Hamaker, C. G.; Djukic, J.-P.; Kodadek, T.; Woo, L. K. *J. Am. Chem. Soc.* **1995**, *117*, 9194.
[58] Davies, H. M. L.; Bruzinski, P. R.; Lake, D. H.; Kong, N.; Fall, M. J. *J. Am. Chem. Soc.* **1996**, *118*, 6897.
[59] Davies, H. M. L.; Clark, T. J.; Church, L. A. *Tetrahedron Lett.* **1989**, *30*, 5057.
[60] Wulfman, D. S.; McDaniel, R. S., Jr.; Peace, B. W. *Tetrahedron* **1976**, *32*, 1241.
[61] Peace, B. W.; Wulfman, D. S. *Synthesis* **1973**, 137.
[62] Wenkert, E. *Acc. Chem. Res.* **1980**, *13*, 27.
[63] Davies, H. M. L.; Clark, T. J.; Kimmer, G. F. *J. Org. Chem.* **1991**, *56*, 6440.
[64] Davies, H. M. L.; Clark, T. J. *Tetrahedron* **1994**, *50*, 9883.
[65] Alonso, M. E.; Morales, A.; Chitty, A. W. *J. Org. Chem.* **1982**, *47*, 3747.
[66] Maryanoff, B. E. *J. Org. Chem.* **1982**, *47*, 3000.
[67] Pirrung, M. C.; Zhang, J.; Lackey, K.; Sternbach, D. D.; Brown, F. *J. Org. Chem.* **1995**, *60*, 2112.
[68] Wenkert, E.; Alonso, M. E.; Buckwalter, B. L.; Chou, K. J. *J. Am. Chem. Soc.* **1977**, *99*, 4778.
[69] Fritschi, H.; Leutenegger, U.; Pfaltz, A. *Helv. Chim. Acta* **1988**, *71*, 1553.
[70] Pfaltz, A. In *Modern Synthetic Methods*; Scheffold, R., Ed.; Springer-Verlag: Berlin, 1989, Vol. 5, pp. 199–248.
[71] Doyle, M. P.; Dorow, R. L.; Buhro, W. E.; Griffin, J. H.; Tamblyn, W. H.; Trudell, M. L. *Organometallics* **1984**, *3*, 44.
[72] Alonso, M. E.; Jano, P.; Hernandez, M. I.; Greenberg, R. S.; Wenkert, E. *J. Org. Chem.* **1983**, *48*, 3047.
[73] Davies, H. M. L.; Bruzinski, P. R.; Fall, M. J. *Tetrahedron Lett.* **1996**, *37*, 4133.
[74] Doyle, M. P.; Zhou, Q.-L.; Charnsangavej, C.; Longoria, M. A.; McKervey, M. A.; Garcia, C. F. *Tetrahedron Lett.* **1996**, *37*, 4129.
[75] Ref. 12, pp. 61–111.
[76] Paulissen, R.; Reimlinger, H.; Hayez, E.; Hubert, A. J.; Teyssie, P. *Tetrahedron Lett.* **1973**, 2233.
[77] Hubert, A. J.; Noels, A. F.; Anciaux, A. J.; Teyssie, P. *Synthesis* **1976**, 600.
[78] Doyle, M. P.; Bagheri, V.; Wandless, T. J.; Harn, N. K.; Brinker, D. A.; Eagle, C. T.; Loh, K. L. *J. Am. Chem. Soc.* **1990**, *112*, 1906.
[79] Majchrzak, M. W.; Kotelko, A.; Lambert, J. B. *Synthesis* **1983**, 469.
[80] Anciaux, A. J.; Hubert, A. J.; Noels, A. F.; Petiniot, N.; Teyssie, P. *J. Org. Chem.* **1980**, *45*, 695.

[81] Demonceau, A.; Simal, F.; Noels, A. F.; Vinas, C.; Nunez, R.; Teixidor, F. *Tetrahedron Lett.* **1997**, *38*, 4079.
[82] Bianchini, C.; Glendenning, L. *Chemtracts: Inorg. Chem.* **1994**, *6*, 52.
[83] Frauenkron, M.; Berkessel, A. *Tetrahedron Lett.* **1997**, *38*, 7175.
[84] Galardon, E.; Le Maux, P.; Simonneaux, G. *J. Chem. Soc., Chem. Commun.* **1997**, 927.
[85] Jommi, G.; Pagliarin, R.; Rizzi, G.; Sisti, M. *Synlett* **1993**, 833.
[86] Katsuki, T. Jpn. Kokai Tokkyo Koho, 1997; *Chem. Abstr.* **1997**, *126*, 293110.
[87] Nakamura, A.; Konishi, A.; Tsujitani, R.; Kudo, M.; Otsuka, S. *J. Am. Chem. Soc.* **1978**, *100*, 3449.
[88] Tatsuno, Y.; Konishi, A.; Nakamura, A.; Otsuka, S. *J. Chem. Soc., Chem. Commun.* **1974**, 588.
[89] Pazynina, G. V.; Kaliya, O. L.; Luk'yanets, E. A.; Bolesov, I. G. *Zh. Org. Khim.* **1987**, *23*, 813; *Chem. Abstr.* **1988**, *108*, 111521.
[90] Smith, D. A.; Reynolds, D. N.; Woo, L. K. *J. Am. Chem. Soc.* **1993**, *115*, 2511.
[91] Demonceau, A.; Lemoine, C. A.; Noels, A. F. *Tetrahedron Lett.* **1996**, *37*, 1025.
[92] Seitz, W. J.; Saha, A. K.; Casper, D.; Hossain, M. M. *Tetrahedron Lett.* **1992**, *33*, 7755.
[93] Wulfman, D. S.; Peace, B. W.; McDaniel Jr., R. S. *Tetrahedron* **1976**, *32*, 1251.
[94] Doyle, M. P.; Van Leusen, D.; Tamblyn, W. H. *Synthesis* **1981**, 787.
[95] Wenkert, E.; Greenberg, R. S.; Raju, M. S. *J. Org. Chem.* **1985**, *50*, 4681.
[96] O'Bannon, P. E.; Dailey, W. P. *Tetrahedron* **1990**, *46*, 7341.
[97] Abramovitch, R. A.; Roy, J. *J. Chem. Soc., Chem. Commun.* **1965**, 542.
[98] Lewis, R. T.; Motherwell, W. B. *Tetrahedron Lett.* **1988**, *29*, 5033.
[99] Callot, H. J.; Metz, F. *Tetrahedron* **1985**, *41*, 4495.
[100] Maxwell, J. L.; O'Malley, S.; Brown, K. C.; Kodadek, T. *Organometallics* **1992**, *11*, 645.
[101] Callot, H. J.; Metz, F.; Piechocki, C. *Tetrahedron* **1982**, *38*, 2365.
[102] Callot, H. J.; Piechocki, C. *Tetrahedron Lett.* **1980**, *21*, 3489.
[103] Nishiyama, H.; Park, S.-B.; Haga, M.; Aoki, K.; Itoh, K. *Chem. Lett.* **1994**, 1111.
[104] Nishiyama, H.; Itoh, Y.; Matsumoto, H.; Park, S.-B.; Itoh, K. *J. Am. Chem. Soc.* **1994**, *116*, 2223.
[105] Nishiyama, H. *Asahi Garasu Zaidan Josei Kenkyu Seika Hokoku* **1994**, 169; *Chem. Abstr.* **1995**, *122*, 249245x.
[106] Nishiyama, H.; Itoh, Y.; Sugawara, Y.; Matsumoto, H.; Aoki, K.; Itoh, K. *Bull. Chem. Soc. Jpn.* **1995**, *68*, 1247.
[107] Nishiyama, H. *Zh. Org. Khim.* **1996**, *32*, 180; *Chem. Abstr.* **1996**, *125*, 328115.
[108] Doyle, M. P.; Brandes, B. D.; Kazala, A. P.; Pieters, R. J.; Jarstfer, M. B.; Watkins, L. M.; Eagle, C. T. *Tetrahedron Lett.* **1990**, *31*, 6613.
[109] Doyle, M. P.; Protopopova, M. N.; Brandes, B. D.; Davies, H. M. L.; Huby, N. J. S.; Whitesell, J. K. *Synlett* **1993**, 151.
[110] Doyle, M. P.; Dorow, R. L.; Terpstra, J. W.; Rodenhouse, R. A. *J. Org. Chem.* **1985**, *50*, 1663.
[111] Krieger, P. E.; Landgrebe, J. A. *J. Org. Chem.* **1978**, *43*, 4447.
[112] Aratani, T.; Yoneyoshi, Y.; Nagase, T. *Tetrahedron Lett.* **1977**, 2599.
[113] Nozaki, H.; Takaya, H.; Moriuti, S.; Noyori, R. *Tetrahedron* **1968**, *24*, 3655.
[114] Nozaki, H.; Moriuti, S.; Takaya, H.; Noyori, R. *Tetrahedron Lett.* **1966**, 5239.
[115] Aratani, T.; Yoneyoshi, Y.; Nagase, T. *Tetrahedron Lett.* **1975**, *1707*.
[116] Aratani, T.; Yoneyoshi, Y.; Nagase, T. *Tetrahedron Lett.* **1982**, *23*, 685.
[117] Aratani, T. *Pure Appl. Chem.* **1985**, *57*, 1839.
[118] Fritschi, H.; Leutenegger, U.; Pfaltz, A. *Angew. Chem., Int. Ed. Engl.* **1986**, *25*, 1005.
[119] Fritschi, H.; Leutenegger, U.; Siegmann, K.; Pfaltz, A.; Keller, W.; Kratky, C. *Helv. Chim. Acta* **1988**, *71*, 1541.
[120] Leutenegger, U.; Umbricht, G.; Fahrni, C.; Von Matt, P.; Pfaltz, A. *Tetrahedron* **1992**, *48*, 2143.
[121] Muller, D.; Umbricht, G.; Weber, B.; Pfaltz, A. *Helv. Chim. Acta* **1991**, *74*, 232.
[122] Pfaltz, A. *Bull. Soc. Chim. Belg.* **1990**, *99*, 729.
[123] Pfaltz, A. *Acc. Chem. Res.* **1993**, *26*, 339.
[124] Lowenthal, R. E.; Abiko, A.; Masamune, S. *Tetrahedron Lett.* **1990**, *31*, 6005.
[125] Lowenthal, R. E.; Masamune, S. *Tetrahedron Lett.* **1991**, *32*, 7373.

[126] Evans, D. A.; Woerpel, K. A.; Hinman, M. M.; Faul, M. M. *J. Am. Chem. Soc.* **1991**, *113*, 726.
[127] Brunner, H.; Kluschanzoff, H.; Wutz, K. *Bull. Soc. Chim. Belg.* **1989**, *98*, 63.
[128] Davies, H. M. L.; Hutcheson, D. K. *Tetrahedron Lett.* **1993**, *34*, 7243.
[129] Ishitani, H.; Achiwa, K. *Synlett* **1997**, 781.
[130] Wenkert, E.; Alonso, M. E.; Buckwalter, B. L.; Sanchez, E. L. *J. Am. Chem. Soc.* **1983**, *105*, 2021.
[131] Wenkert, E.; Alonso, M. E.; Gottlieb, H. E.; Sanchez, E. L.; Pellicciari, R.; Cogolli, P. *J. Org. Chem.* **1977**, *42*, 3945.
[132] Wenkert, E.; Ananthanarayan, T. P.; Ferreira, V. F.; Hoffmann, M. G.; Kim, H. S. *J. Org. Chem.* **1990**, *55*, 4975.
[133] Lee, Y. R. *Tetrahedron* **1995**, *51*, 3087.
[134] Lee, Y. R.; Morehead Jr., A. T. *Tetrahedron* **1995**, *51*, 4909.
[135] Bien, S.; Segal, Y. *J. Org. Chem.* **1977**, *42*, 1685.
[136] Pirrung, M. C.; Lee, Y. R. *J. Am. Chem. Soc.* **1995**, *117*, 4814.
[137] Alonso, M. E.; Jano S. P.; Hernandez, M. I. *J. Org. Chem.* **1980**, *45*, 5299.
[138] Pirrung, M. C.; Lee, Y. R. *Tetrahedron Lett.* **1994**, *35*, 6231.
[139] Pirrung, M. C.; Lee, Y. R. *J. Chem. Soc., Chem. Commun.* **1995**, 673.
[140] Peace, B. W.; Carman, F.; Wulfman, D. S. *Synthesis* **1971**, 658.
[141] Matlin, S. A.; Lough, W. J.; Chan, L.; Abram, D. M. H.; Zhou, Z. *J. Chem. Soc., Chem. Commun.* **1984**, 1038.
[142] Ishitani, H.; Achiwa, K. *Heterocycles* **1997**, *46*, 153.
[143] Davies, H. M. L.; Hu, B. *Tetrahedron Lett.* **1992**, *33*, 453.
[144] Davies, H. M. L.; Hu, B. *Heterocycles* **1993**, *35*, 385.
[145] Starmans, W. A. J.; Thijs, L.; Zwanenburg, B. *Tetrahedron* **1998**, *54*, 629.
[146] Davies, H. M. L.; Rusiniak, L. *Tetrahedron Lett.* **1998**, *39*, 8811.
[147] Davies, H. M. L.; Cantrell, W. R., Jr. *Tetrahedron Lett.* **1991**, *32*, 6509.
[148] Davies, H. M. L.; Huby, N. J. S.; Cantrell, W. R., Jr.; Olive, J. L. *J. Am. Chem. Soc.* **1993**, *115*, 9468.
[149] Reissig, H.-U. *Top. Curr. Chem.* **1988**, *144*, 73.
[150] Mazzocchi, P. H.; Tamburin, H. J. *J. Org. Chem.* **1973**, *38*, 2221.
[151] Doyle, M. P.; Dorow, R. L.; Tamblyn, W. H.; Buhro, W. E. *Tetrahedron Lett.* **1982**, *23*, 2261.
[152] Burgess, K. *J. Org. Chem.* **1987**, *52*, 2046.
[153] Hudlicky, T.; Fan, R.; Reed, J. W.; Gadamasetti, K. G. *Org. React.* **1992**, *41*, 1.
[154] Piers, E. In *Comprehensive Organic Synthesis*; Trost, B. M., Ed.; Pergamon: Oxford, 1991; Vol. 5, pp. 971–998.
[155] Davies, H. M. L.; Smith, H. D.; Korkor, O. *Tetrahedron Lett.* **1987**, *28*, 1853.
[156] Davies, H. M. L.; Clark, T. J.; Smith, H. D. *J. Org. Chem.* **1991**, *56*, 3817.
[157] Davies, H. M. L.; Stafford, D. G.; Doan, B. D.; Houser, J. H. *J. Am. Chem. Soc.* **1998**, *120*, 3326.
[158] Davies, H. M. L.; Doan, B. D. *J. Org. Chem.* **1998**, *63*, 657.
[159] Davies, H. M. L.; Saikali, E.; Clark, T. J.; Chee, E. H. *Tetrahedron Lett.* **1990**, *31*, 6299.
[160] Wenkert, E.; Bakuzis, M. L. F.; Buckwalter, B. L.; Woodgate, P. D. *Synth. Commun.* **1981**, *11*, 533.
[161] Matlin, S. A.; Chan, L.; Catherwood, B. *J. Chem. Soc., Perkin Trans. 1* **1990**, 89.
[162] Nefedov, O. M.; Shostakovskii, V. M.; Samoilova, M. Y.; Kravchenko, M. I. *Bull. Acad. Sci. USSR Div. Chem. Sci.* **1972**, 2342; *Chem. Abstr.* **1973**, *78*, 43165.
[163] Rokach, J.; Adams, J.; Perry, R. *Tetrahedron Lett.* **1983**, *24*, 5185.
[164] Adams, J.; Rokach, J. *Tetrahedron Lett.* **1984**, *25*, 35.
[165] Ong, C. W.; Chen, C. M.; Juang, S. S. *J. Org. Chem.* **1994**, *59*, 7915.
[166] Davies, H. M. L.; Clark, D. M.; Alligood, D. B.; Eiband, G. R. *Tetrahedron* **1987**, *43*, 4265.
[167] Davies, H. M. L.; Clark, D. M.; Smith, T. K. *Tetrahedron Lett.* **1985**, *26*, 5659.
[168] Adams, J.; Leblanc, Y.; Rokach, J. *Tetrahedron Lett.* **1984**, *25*, 1227.
[169] Leblanc, Y.; Fitzsimmons, B. J.; Adams, J.; Perez, F.; Rokach, J. *J. Org. Chem.* **1986**, *51*, 789.
[170] Sheu, J. H.; Yen, C. F.; Huang, H. C.; Hong, Y. L. V. *J. Org. Chem.* **1989**, *54*, 5126.

[171] Wenkert, E.; Khatuya, H.; Klein, P. S. *Tetrahedron Lett.* **1999**, *40*, 5171.
[172] Wenkert, E. In *New Trends in Natural Products Chemistry, Studies in Organic Chemistry*; Rahman, A., Quesne, P. W., Eds.; Elsevier: Amsterdam, 1986; Vol. 26, pp. 557–563.
[173] Wenkert, E.; Guo, M.; Lavilla, R.; Porter, B.; Ramachandran, K.; Sheu, J. H. *J. Org. Chem.* **1990**, *55*, 6203.
[174] Nefedov, O. M.; Shostakovsky, V. M.; Vasilvizky, A. E. *Angew. Chem., Int. Ed. Engl.* **1977**, *16*, 646.
[175] Davies, H. M. L.; Matasi, J. J.; Hodges, L. M.; Huby, N. J. S.; Thornley, C.; Kong, N.; Houser, J. H. *J. Org. Chem.* **1997**, *62*, 1095.
[176] Pirrung, M. C.; Zhang, J. *Tetrahedron Lett.* **1992**, *33*, 5987.
[177] Davies, H. M. L.; Ahmed, G.; Churchill, M. R. *J. Am. Chem. Soc.* **1996**, *118*, 10774.
[178] Mann, J. *Tetrahedron* **1986**, *42*, 4611.
[179] Davies, H. M. L. In *Advances in Nitrogen Heterocycles*; Moody, C. J., Ed.; JAI Press: London, 1995; Vol. 1, pp. 1–18.
[180] Maryanoff, B. E. *J. Org. Chem.* **1979**, *44*, 4410.
[181] Maryanoff, B. E. *J. Heterocycl. Chem.* **1977**, *14*, 177.
[182] Tanny, S. R.; Grossman, J.; Fowler, F. W. *J. Am. Chem. Soc.* **1972**, *94*, 6495.
[183] Bubert, C.; Reiser, O. *Tetrahedron Lett.* **1997**, *38*, 4985.
[184] Davies, H. M. L.; Young, W. B.; Smith, H. D. *Tetrahedron Lett.* **1989**, *30*, 4653.
[185] Davies, H. M. L.; Saikali, E.; Young, W. B. *J. Org. Chem.* **1991**, *56*, 5696.
[186] Davies, H. M. L.; Huby, N. J. S. *Tetrahedron Lett.* **1992**, *33*, 6935.
[187] Burgess, K.; Ho, K. K.; Moye-Sherman, D. *Synlett* **1994**, 575.
[188] Stammer, C. H. *Tetrahedron* **1990**, *46*, 2231.
[189] Hudlicky, T.; Kutchan, T. M.; Naqvi, S. M. *Org. React.* **1985**, *33*, 247.
[190] Reissig, H.-U. In *The Chemistry of the Cyclopropyl Group*; Rappoport, Z., Ed.; Wiley: New York, 1987; Vol. 1, pp. 375–444.
[191] Tidwell, T. T. In *The Chemistry of the Cyclopropyl Group*; Rappoport, Z., Ed.; Wiley: New York, 1987; Vol. 1, Ch. 10.
[192] Vehre, R.; De Kimpe, N. In *The Chemistry of the Cyclopropyl Group*; Rappoport, Z., Ed.; Wiley: New York, 1987; Vol. 1, pp. 445–564.
[193] Wong, H. N. C.; Hon, M.-Y.; Tse, C.-W.; Yip, Y.-C.; Tanko, J.; Hudlicky, T. *Chem. Rev.* **1989**, *89*, 165.
[194] Elliott, M.; Janes, N. F. *Chem. Soc. Rev.* **1978**, *7*, 473.
[195] Pellicciari, R.; Natalini, B.; Marinozzi, M.; Monahan, J. B.; Snyder, J. P. *Tetrahedron Lett.* **1990**, *31*, 139.
[196] Paulini, K.; Reissig, H.-U. *J. Prakt. Chem./Chem.-Ztg.* **1995**, *337*, 55; *Chem. Abstr.* **1995**, *122*, 265954h.
[197] Danishefsky, S. J.; Singh, R. K. *J. Org. Chem.* **1975**, *40*, 3807.
[198] Danishefsky, S. J.; Rovnyak, G. *J. Org. Chem.* **1975**, *40*, 114.
[199] Danishefsky, S. J. *Acc. Chem. Res.* **1979**, *12*, 66.
[200] Corey, E. J.; Gant, T. G. *Tetrahedron Lett.* **1994**, *35*, 5373.
[201] Reissig, H.-U.; Reichelt, I. *Tetrahedron Lett.* **1984**, *25*, 5879.
[202] Reissig, H.-U. *Tetrahedron Lett.* **1985**, *26*, 3943.
[203] Schnaubelt, J.; Zschiesche, R.; Reissig, H.-U.; Linder, H. J.; Richter, J. *Justus Liebigs Ann. Chem.* **1993**, 61.
[204] Schnaubelt, J.; Reissig, H.-U. *Synlett* **1995**, 452.
[205] Schnaubelt, J.; Ullmann, A.; Reissig, H.-U. *Synlett* **1995**, 1223.
[206] Ullmann, A.; Reissig, H.-U.; Rademacher, O. *Eur. J. Org. Chem.* **1998**, 2541.
[207] Ullmann, A.; Schnaubelt, J.; Reissig, H.-U. *Synthesis* **1998**, 1052.
[208] Frey, B.; Hünig, S.; Koch, M.; Reissig, H.-U. *Synlett* **1991**, 854.
[209] Zschiesche, R.; Grimm, E. L.; Reissig, H.-U. *Angew. Chem., Int. Ed. Engl.* **1986**, *25*, 1086.
[210] Marino, J. P.; Silveira, C.; Comasseto, J.; Petragnani, N. *J. Org. Chem.* **1987**, *52*, 4139.
[211] Marino, J. P.; Laborde, E. *J. Am. Chem. Soc.* **1985**, *107*, 734.
[212] Davies, H. M. L.; Hu, B. *J. Org. Chem.* **1992**, *57*, 3186.

[213] Davies, H. M. L.; Kong, N.; Churchill, M. R. *J. Org. Chem.* **1998**, *63*, 6586.
[214] Piers, E.; Moss, N. *Tetrahedron Lett.* **1985**, *26*, 2735.
[215] Pirrung, M. C.; Lee, Y. R. *Tetrahedron Lett.* **1996**, *37*, 2391.
[216] Laufer, S. A.; Augustin, J.; Dannhardt, G.; Kiefer, W. *J. Med. Chem.* **1994**, *37*, 1894.
[217] Dannhardt, G.; Lehr, M. *Arch. Pharm.* (Weinheim, Ger.) **1988**, *321*, 159.
[218] Fairfax, D. J.; Austin, D. J.; Xu, S. L.; Padwa, A. *J. Chem. Soc., Perkin Trans. 1* **1992**, 2837.
[219] Moriarty, R. M.; Bailey, B. R., III; Prakash, O.; Prakash, I. *J. Am. Chem. Soc.* **1985**, *107*, 1375.
[220] Moriarty, R. M.; May, E. J.; Guo, L.; Prakash, O. *Tetrahedron Lett.* **1998**, *39*, 765.
[221] Müller, P.; Fernandez, D. *Helv. Chim. Acta* **1995**, *78*, 947.
[222] Müller, P.; Fernandez, D.; Nury, P.; Rossier, J.-C. *J. Phys. Org. Chem.* **1998**, *11*, 321.
[223] Shioiri, T.; Aoyama, T.; Mori, S. *Org. Synth.* **1990**, *68*, 1.
[224] Vallgarda, J.; Hacksell, U. *Tetrahedron Lett.* **1991**, *32*, 5625.
[225] Vallgarda, J.; Appelberg, U.; Csoregh, I.; Hacksell, U. *J. Chem. Soc., Perkin Trans. 1* **1994**, 461.
[226] Denmark, S. E.; Stavenger, R. A.; Faucher, A.-M.; Edwards, J. P. *J. Org. Chem.* **1997**, *62*, 3375.
[227] Seitz, W. J.; Hossain, M. M. *Tetrahedron Lett.* **1994**, *35*, 7561.
[228] Salomon, R. G.; Salomon, M. F.; Heyne, T. R. *J. Org. Chem.* **1975**, *40*, 756.
[229] Ref. 12, pp. 238–288.
[230] Corey, E. J.; Myers, A. G. *J. Am. Chem. Soc.* **1985**, *107*, 5574.
[231] Danishefsky, S. J.; McKee, R.; Singh, R. K. *J. Am. Chem. Soc.* **1977**, *99*, 7711.
[232] Fox, M. E.; Li, C.; Marino, J. P.; Overman, L. E. *J. Am. Chem. Soc.* **1999**, *121*, 5467.
[233] Doyle, M. P.; Kalinin, A. V. *J. Org. Chem.* **1996**, *61*, 2179.
[234] Doyle, M. P.; Austin, R. E.; Bailey, A. S.; Dwyer, M. P.; Dyatkin, A. B.; Kalinin, A. V.; Kwan, M. M. Y.; Liras, S.; Oalmann, C. J.; Pieters, R. J.; Protopopova, M. N.; Raob, C. E.; Roos, G. H. P.; Zhou, Q. L.; Martin, S. F. *J. Am. Chem. Soc.* **1995**, *117*, 5763.
[235] Doyle, M. P.; Protopopova, M. N.; Poulter, C. D.; Rogers, D. H. *J. Am. Chem. Soc.* **1995**, *117*, 7281.
[236] Doyle, M. P.; Zhou, Q.-L.; Dyatkin, A. B.; Ruppar, D. A. *Tetrahedron Lett.* **1995**, *36*, 7579.
[237] Doyle, M. P.; Pieters, R. J.; Martin, S. F.; Austin, R. E.; Oalmann, C. J.; Mueller, P. *J. Am. Chem. Soc.* **1991**, *113*, 1423.
[238] Doyle, M. P.; Peterson, C. S.; Zhou, Q.-L.; Nishiyama, H. *J. Chem. Soc., Chem. Commun.* **1997**, 211.
[239] Charette, A. B.; Cote, B.; Marcoux, J.-F. *J. Am. Chem. Soc.* **1991**, *113*, 8166.
[240] Charette, A. B.; Cote, B. *J. Org. Chem.* **1993**, *58*, 933.
[241] Hoemann, M. Z.; Agrios, K. A.; Aube, J. *Tetrahedron* **1997**, *53*, 11087.
[242] Kasdorf, K.; Liotta, D. C. *Chemtracts* **1997**, *10*, 533.
[243] Imai, N.; Sakamoto, K.; Takahashi, H.; Kobayashi, S. *Tetrahedron Lett.* **1994**, *35*, 7045.
[244] Takahashi, H.; Yohioka, M.; Ohno, M.; Kobayashi, S. *Tetrahedron Lett.* **1992**, *33*, 2575.
[245] Semmelhack, M. F.; Tamura, R. *J. Am. Chem. Soc.* **1983**, *105*, 6750.
[246] Buchert, M.; Reissig, H.-U. *Tetrahedron Lett.* **1988**, *29*, 2319.
[247] Buchert, M.; Reissig, H.-U. *Chem. Ber.* **1992**, *125*, 2723.
[248] Buchert, M.; Hoffmann, M.; Reissig, H.-U. *Chem. Ber.* **1995**, *128*, 605.
[249] Hoffmann, M.; Buchert, M.; Reissig, H.-U. *Angew. Chem., Int. Ed. Engl.* **1997**, *36*, 283.
[250] Fernandez, M. D.; Alcaraz, C.; de Frutos, M. P.; Marco, J. L.; Bernabe, M.; Foces-Foces, C.; Cano, F. H. *Tetrahedron* **1994**, *50*, 12443.
[251] Schöllkopf, U.; Hupfeld, B.; Küper, S.; Egert, E.; Dyrbusch, M. *Angew. Chem., Int. Ed. Engl.* **1988**, *27*, 433.
[252] Romo, D.; Romine, D. L.; Midura, W.; Meyers, A. I. *Tetrahedron* **1990**, *46*, 4851.
[253] Yamazaki, S.; Takada, T.; Imanishi, T.; Moriguchi, Y.; Yamabe, S. *J. Org. Chem.* **1998**, *63*, 5919.
[254] Yamazaki, S.; Kumagai, H.; Yamabe, S.; Yamamoto, K. *J. Org. Chem.* **1998**, *63*, 3371.
[255] Yamazaki, S.; Tanaka, M.; Yamaguchi, A.; Yamabe, S. *J. Am. Chem. Soc.* **1994**, *116*, 2356.

[256] Dorizon, P.; Su, G.; Ludvig, G.; Nikitina, L.; Paugam, R.; Ollivier, J.; Salaün, J. *J. Org. Chem.* **1999**, *64*, 4712.
[257] Arndt, F. *Org. Synth. Coll. Vol. 2* **1943**, 163.
[258] de Boer, T. J.; Backer, H. J. *Org. Synth. Coll. Vol. 4* **1963**, 250.
[259] Black, T. W. *Aldrichimica Acta* **1983**, *16*, 3.
[260] Moore, J. A.; Reed, D. E. *Org. Synth. Coll. Vol. 5* **1973**, 351.
[261] Moss, S. *Chem. & Industry* **1994**, 122.
[262] Ref. 12, pp. 1–60.
[263] Spencer, H. *Chem. Brit.* **1981**, *17*, 106.
[264] Rewicki, D.; Tuchscherer, C. *Angew. Chem., Int. Ed. Engl.* **1972**, *11*, 44.
[265] Bollinger, F. W.; Tuma, L. D. *Synlett* **1996**, 407.
[266] Hazen, G. G.; Weinstock, L. M.; Connell, R.; Bollinger, F. W. *Synth. Commun.* **1981**, *11*, 947.
[267] Hazen, G. G.; Bollinger, F. W.; Roberts, F. E.; Russ, W. K.; Seman, J. J.; Staskiewicz, S. *Org. Synth.* **1996**, *73*, 144.
[268] Baum, J. S.; Shook, D. A.; Davies, H. M. L.; Smith, H. D. *Synth. Commun.* **1987**, *17*, 1709.
[269] Davies, H. M. L.; Cantrell, W. R., Jr.; Romines, K. R.; Baum, J. S. *Org. Synth.* **1992**, *70*, 93.
[270] Regitz, M. *Synthesis* **1972**, 351.
[271] Taber, D. F.; Gleave, D. M.; Heer, R. J.; Moody, K.; Hennessy, M. J. *J. Org. Chem.* **1995**, *60*, 2283.
[272] Taber, D. F.; You, K.; Song, Y. *J. Org. Chem.* **1995**, *60*, 1093.
[273] Danheiser, R. L.; Miller, R. F.; Brisbois, R. G. *Org. Synth.* **1996**, *73*, 134.
[274] House, H. O.; Blankley, C. J. *J. Org. Chem.* **1968**, *33*, 47.
[275] Corey, E. J.; Myers, A. G. *Tetrahedron Lett.* **1984**, *25*, 3559.
[276] Ouihia, A.; René, L.; Guilheim, J.; Pascard, C.; Badet, B. *J. Org. Chem.* **1993**, *58*, 1641.
[277] Marino, J. P., Jr.; Osterhout, M. H.; Padwa, A. *J. Org. Chem.* **1995**, *60*, 2704.
[278] Karady, S.; Amato, J. S.; Reamer, R. A.; Weinstock, L. M. *J. Am. Chem. Soc.* **1981**, *103*, 6765.
[279] Davies, H. M. L.; Hougland, P. W.; Cantrell, W. R., Jr. *Synth. Commun.* **1992**, *22*, 971.
[280] Fink, J.; Regitz, M. *Synthesis* **1985**, 569.
[281] Searle, N. E. *Org. Synth. Coll. Vol. 4* **1963**, 424.
[282] Womack, E. B.; Nelson, A. B. *Org. Synth. Coll. Vol. 3* **1955**, 392.
[283] Creary, X. *Org. Synth.* **1986**, *64*, 207.
[284] Wheeler, T. N.; Meinwald, J. *Org. Synth.* **1972**, *52*, 53.
[285] Blankley, J.; Sauter, F. J.; House, H. O. *Org. Synth. Coll. Vol. 5* **1973**, 259.
[286] Gerkin, R. M.; Rickborn, B. *J. Am. Chem. Soc.* **1967**, *89*, 5850.
[287] Muller, P.; Baud, C.; Ene, D.; Motallebi, S.; Doyle, M. P.; Brandes, B. D.; Dyatkin, A. B.; See, M. M. *Helv. Chim. Acta* **1995**, *78*, 459.
[288] O'Bannon, P. E.; Dailey, W. P. *Tetrahedron Lett.* **1988**, *29*, 987.
[289] Dowd, P.; Kaufman, C.; Paik, Y. H. *Tetrahedron Lett.* **1985**, *26*, 2283.
[290] Milner, D. J. *J. Organomet. Chem.* **1984**, *262*, 85.
[291] Lin, Y.-L.; Turos, E. *J. Am. Chem. Soc.* **1999**, *121*, 856.
[292] Kunz, T.; Jonowitz, A.; Reissig, H.-U. *Synthesis* **1990**, 43.
[293] Saigo, K.; Okagawa, S.; Nohira, H. *Bull. Chem. Soc. Jpn.* **1981**, *54*, 3603.
[294] Kunkel, E.; Reichelt, I.; Reissig, H.-U. *Justus Liebigs Ann. Chem.* **1984**, 512.
[295] Graziano, M. L.; Iesce, M. R. *Synthesis* **1985**, 762.
[296] Saigo, K.; Kurihara, H.; Miura, H.; Hongu, A.; Kubota, N.; Nohira, H.; Hasegawa, M. *Synth. Commun.* **1984**, *14*, 787.
[297] Augusti, R.; Eberlin, M. N.; Kascheres, C. *J. Heterocycl. Chem.* **1995**, *32*, 1355.
[298] Doyle, M. P.; Davidson, J. G. *J. Org. Chem.* **1980**, *45*, 1538.
[299] Doyle, M. P.; Dorow, R. L.; Tamblyn, W. H. *J. Org. Chem.* **1982**, *47*, 4059.
[300] Paulini, K.; Reissig, H.-U. *Liebigs Ann. Chem.* **1991**, 455.
[301] Dolgii, I. E.; Shapiro, E. A.; Nefedov, O. M. *Bull. Acad. Sci. USSR Div. Chem. Sci. (Engl. Transl.)* **1983**, 2168; *Chem. Abstr.* **1984**, *100*, 85285.
[302] Doyle, M. P.; Tamblyn, W. H.; Bagheri, V. *J. Org. Chem.* **1981**, *46*, 5094.
[303] Lai, M. T.; Liu, H. W. *J. Am. Chem. Soc.* **1990**, *112*, 4034.

[304] Baldwin, J. E.; Widdison, W. C. *J. Am. Chem. Soc.* **1992**, *114*, 2245.
[305] Ivshin, V. P.; Komelin, M. S.; Kozhevnikova, T. V.; Morozova, N. S. *J. Org. Chem. USSR (Engl. Transl.)* **1982**, *18*, 696; *Chem. Abstr.* **1982**, *97*, 455326.
[306] Wiberg, K. B.; Kass, S. R.; Bishop, K. C. *J. Am. Chem. Soc.* **1985**, *107*, 996.
[307] Reichelt, I.; Reissig, H.-U. *Chem. Ber.* **1983**, *116*, 3895.
[308] Maas, G.; Werle, T.; Alt, M.; Mayer, D. *Tetrahedron* **1993**, *49*, 881.
[309] Mueller, P.; Baud, C.; Ene, D.; Motallebi, S.; Doyle, M. P.; Brandes, B. D.; Dyatkin, A. B.; See, M. M. *Helv. Chim. Acta* **1995**, *78*, 459.
[310] Nishiyama, S.; Ueki, S.; Watanabe, T.; Yamamura, S.; Kato, K.; Takita, T. *Tetrahedron Lett.* **1991**, *32*, 2141.
[311] LeGoaller, R.; Pierre, J.-L. *Can. J. Chem.* **1977**, *55*, 757.
[312] Reissig, H.-U.; Hirsch, E. *Angew. Chem., Int. Ed. Engl.* **1980**, *19*, 813.
[313] Reissig, H.-U.; Reichelt, I.; Kunz, T. *Org. Synth.* **1992**, *71*, 189.
[314] Shimamoto, K.; Ishida, M.; Shinozaki, H.; Ohfune, Y. *J. Org. Chem.* **1991**, *56*, 4167.
[315] Andrist, A. H.; Agnello, R. M.; Wolfe, D. C. *J. Org. Chem.* **1978**, *43*, 3422.
[316] Salomon, R. G.; Kochi, J. K. *J. Am. Chem. Soc.* **1973**, *95*, 3300.
[317] Green, J.; Sinn, E.; Woodward, S.; Butcher, R. *Polyhedron* **1993**, *12*, 991.
[318] Doyle, M. P.; Tamblyn, W. H.; Buhro, W. E.; Dorow, R. L. *Tetrahedron Lett.* **1981**, *22*, 1783.
[319] Zhu, Z.; Espenson, J. H. *J. Am. Chem. Soc.* **1996**, *118*, 9901.
[320] Boeckman, R. K.; Bruza, K. J. *Tetrahedron* **1981**, *37*, 3997.
[321] Doyle, M. P.; Griffin, J. H.; Da Conceicao, J. *J. Chem. Soc., Chem. Commun.* **1985**, 328.
[322] Seitz, W. J.; Saha, A. K.; Hossain, M. M. *Organometallics* **1993**, *12*, 2604.
[323] Doyle, M. P.; Loh, K. L.; DeVries, K. M.; Chinn, M. S. *Tetrahedron Lett.* **1987**, *28*, 833.
[324] Timmers, C. M.; Leeuwenburgh, M. A.; Verheijen, J. C.; van der Marel, G. A.; van Boom, J. H. *Tetrahedron: Asymmetry* **1996**, *7*, 49.
[325] Demonceau, A.; Noels, A. F.; Hubert, A. J. *Tetrahedron* **1990**, *46*, 3889.
[326] Hagen, M.; Luning, U. *Chem. Ber.* **1997**, *130*, 231.
[327] Jendralla, H. *Chem. Ber.* **1982**, *115*, 201.
[328] Goldschmidt, Z.; Finkel, D. *J. Chem. Soc., Perkin Trans. 1* **1983**, 45.
[329] Wenkert, E.; Mueller, R. A.; Reardon, E. J. J.; Sathe, S. S.; Scharf, D. J.; Tosi, G. *J. Am. Chem. Soc.* **1970**, *92*, 7428.
[330] Nefedov, O. M.; Shostakovskii, V. M.; Vasil'vitskii, A. E.; Kravchenko, M. I. *Bull. Acad. Sci. USSR Div. Chem. Sci. (Engl. Transl.)* **1980**, 425; *Chem. Abstr.* **1980**, *93*, 71417.
[331] Hoberg, J. O.; Claffey, D. J. *Tetrahedron Lett.* **1996**, *37*, 2533.
[332] Henry, K. J., Jr.; Fraser-Reid, B. *Tetrahedron Lett.* **1995**, *36*, 8901.
[333] de Meijere, A.; Kozhushkov, S. I.; Spaeth, T.; Zefirov, N. S. *J. Org. Chem.* **1993**, *58*, 502.
[334] Ollivier, J.; Salaun, J. *J. Chem. Soc., Chem. Commun.* **1985**, 1269.
[335] Callot, H. J.; Schaeffer, E. *Nouv. J. Chim.* **1980**, *4*, 311.
[336] Oudejans, J. C.; Kaminska, J.; Kock-Van Dalen, A. C.; Van Bekkum, H. *Recl. Trav. Chim. Pays-Bas* **1986**, *105*, 421.
[337] Marvell, E. N.; Rusay, R. *J. Org. Chem.* **1977**, *42*, 3336.
[338] Demonceau, A.; Lemoine, C. A.; Noels, A. F.; Chizhevsky, I. T.; Sorokin, P. V. *Tetrahedron Lett.* **1995**, *36*, 8419.
[339] Fraile, J. M.; Garcia, J. I.; Mayoral, J. A. *J. Chem. Soc., Chem. Commun.* **1996**, 1319.
[340] Simal, F.; Jan, D.; Demonceau, A.; Noels, A. F. *Tetrahedron Lett.* **1999**, *40*, 1653.
[341] Shapiro, E. A.; Lun'kova, G. V.; Dolgii, I. E.; Nefedov, O. M. *Bull. Acad. Sci. USSR Div. Chem. Sci. (Engl. Transl.)* **1984**, 2317; *Chem. Abstr.* **1985**, *102*, 131555.
[342] Shapiro, E. A.; Romanova, T. N.; Dolgii, I. E.; Nefedov, O. M. *Bull. Acad. Sci. USSR Div. Chem. Sci. (Eng. Transl.)* **1984**, 2323; *Chem. Abstr.* **1985**, *102*, 184741.
[343] Gajewski, J. J.; Burka, L. T. *J. Am. Chem. Soc.* **1972**, *94*, 8860.
[344] Alonso, M. E.; Gomez, M.; de Sierraalta, S. P.; Jano, P. S. *J. Heterocyclic Chem.* **1982**, *19*, 369.
[345] Dappen, M. S.; Pellicciari, R.; Natalini, B.; Monahan, J. B.; Chiorri, C.; Cordi, A. A. *J. Med. Chem.* **1991**, *34*, 161.

[346] Pellicciari, R.; Natalini, B.; Marinozzi, M.; Sadeghpour, B. M.; Cordi, A. A.; Lanthorn, T. H.; Hood, W. F.; Monahan, J. B. *Farmaco* **1991**, *46*, 1243; *Chem. Abstr.* **1992**, *117*, 82891.
[347] Epshtein, A. E.; Dolgil, I. E.; Limanov, V. E.; Skvortsova, E. K.; Nefedov, O. M. *Bull. Acad. Sci. USSR Div. Chem. Sci. (Eng. Transl.)* **1978**, 438; *Chem. Abstr.* **1978**, *88*, 190135.
[348] Perez, P. J.; Brookhart, M.; Templeton, J. L. *Organometallics* **1993**, *12*, 261.
[349] Shapiro, E. A.; Dyatkin, A. B.; Nefedov, O. M. *Bull. Acad. Sci. USSR Div. Chem. Sci. (Eng. Transl.)* **1992**, 272; *Chem. Abstr.* **1993**, *118*, 147203.
[350] Shapiro, E. A.; Romanova, T. N.; Dolgii, I. E.; Nefedov, O. M. *Bull. Acad. Sci. USSR Div. Chem. Sci. (Engl. Transl.)* **1984**, *11*, 2436; *Chem. Abstr.* **1985**, *102*, 184741.
[351] Demonceau, A.; Simal, F.; Noels, A. F.; Vinas, C.; Nunez, R.; Teixidor, F. *Tetrahedron Lett.* **1997**, *38*, 7879.
[352] Kunz, H.; Lindig, M. *Chem. Ber.* **1983**, *116*, 220.
[353] Lam, J.; Johnson, B. L. *Aust. J. Chem.* **1972**, *25*, 2269.
[354] Olteanu, E.; Caproiu, M. T.; Draghici, C. *Rev. Roum. Chim.* **1996**, *41*, 953; *Chem. Abstr.* **1997**, *127*, 148849.
[355] Salomon, R. G.; Salomon, M. F.; Kachinski, J. L. C. *J. Am. Chem. Soc.* **1977**, *99*, 1043.
[356] Stewart, F. F.; Jennings, P. W. *J. Am. Chem. Soc.* **1991**, *113*, 7037.
[357] Simal, F.; Demonceau, A.; Noels, A. F. *Tetrahedron Lett.* **1998**, *39*, 3493.
[358] Vincens, M.; Dumont, C.; Vindal, M. *Bull. Soc. Chim. Fr.* **1974**, *12*, 2811.
[359] Wenkert, E.; Hudlicky, T. *J. Org. Chem.* **1988**, *53*, 1953.
[360] Banthorpe, D.; Christou, P. N. *J. Chem. Soc., Perkin Trans. 1* **1981**, 105.
[361] Alonso, M. E.; Garcia, M. C. *J. Org. Chem.* **1985**, *50*, 988.
[362] Zaitseva, G. S.; Novikova, O. P.; Livantsova, L. I.; Kisin, A. V.; Baukov, Y. I. *J. Org. Chem. USSR (Engl. Transl.)* **1988**, *58*, 1495; *Chem. Abstr.* **1990**, *113*, 172142.
[363] Tadic-Biadatti, M. H.; Newcomb, M. *J. Chem. Soc., Perkin Trans. 2* **1996**, 1467.
[364] Simal, F.; Demonceau, A.; Noels, A. F.; Knowles, D. R. T.; O'Leary, S.; Maitlis, P. M.; Gusev, O. *J. Organomet. Chem.* **1998**, *558*, 163.
[365] Diaz-Requejo, M. M.; Perez, P. J.; Brookhart, M.; Templeton, J. L. *Organometallics* **1997**, *16*, 4399.
[366] Galardon, E.; LeMaux, P.; Toupet, L.; Simonneaux, G. *Organometallics* **1998**, *17*, 565.
[367] Bergbreiter, D. E.; Morvant, M.; Chen, B. *Tetrahedron Lett.* **1991**, *32*, 2731.
[368] Boverie, S.; Simal, F.; Demonceau, A.; Noels, A. F.; Eremenko, I. L.; Sidorov, A. A.; Nefedov, S. E. *Tetrahedron Lett.* **1997**, *38*, 7543.
[369] Bertani, R.; Michelin, R. A.; Mozzon, M.; Traldi, P.; Seraglia, R.; da Silva, M. F. C. G.; Pombeiro, A. J. L. *Organometallics* **1995**, *14*, 551.
[370] Lo, W.-C.; Che, C.-M.; Cheng, K.-F.; Mak, T. C. W. *J. Chem. Soc., Chem. Commun.* **1997**, 1205.
[371] Falk, H.; Suste, A. *Monatsh. Chem.* **1994**, *125*, 325.
[372] Cetinkaya, B.; Ozdemir, I.; Dixneuf, P. H. *J. Organomet. Chem.* **1997**, *534*, 153.
[373] Gross, Z.; Simkhovich, L.; Galili, N. *J. Chem. Soc., Chem. Commun.* **1999**, 599.
[374] Maxwell, J. L.; Brown, K. C.; Bartley, D. W.; Kodadek, T. *Science* **1992**, *256*, 1544.
[375] Shapiro, E. A.; Eismont, M. Y.; Pereverzeva, Y. O.; Nefedov, A. O.; Strashnenko, A. V.; Kostyrko, I. N.; Roslavtseva, S. A. *Bull. Acad. Sci. USSR Div. Chem. Sci. (Engl. Transl.)* **1990**, 573; *Chem. Abstr.* **1990**, *113*, 97312.
[376] Kusuyama, Y.; Tokami, K. *Magn. Reson. Chem.* **1992**, *30*, 361.
[377] Wimalasena, K.; May, S. W. *J. Am. Chem. Soc.* **1987**, *109*, 4036.
[378] Molchanov, A. P.; Serkina, T. G.; Badovskaya, L. A.; Kostikov, R. R. *J. Org. Chem. USSR (Engl. Transl.)* **1993**, *28*, 1874; *Chem. Abstr.* **1993**, *119*, 271053.
[379] Ceccherelli, P.; Coccia, R.; Curini, M.; Pellicciari, R. *Gazz. Chim. Ital.* **1983**, *113*, 453; *Chem. Abstr.* **1984**, *100*, 175154.
[380] Piers, E.; Maxwell, A. R.; Moss, N. *Can. J. Chem.* **1985**, *63*, 555.
[381] Piers, E.; Jung, G. L.; Moss, N. *Tetrahedron Lett.* **1984**, *25*, 3959.
[382] Komendantov, M. I.; Pronyaev, V. N.; Bekmukhametov, R. R. *J. Org. Chem. USSR (Engl. Transl.)* **1979**, *15*, 284; *Chem. Abstr.* **1979**, *91*, 4968.

[383] Kremer, K. A. M.; Helquist, P. *J. Organomet. Chem.* **1985**, *285*, 231.
[384] Demonceau, A.; Noels, A. F.; Saive, E.; Hubert, A. J. *J. Mol. Cat.* **1992**, *76*, 123.
[385] Wilson, S. R.; Zucker, P. A. *J. Org. Chem.* **1988**, *53*, 4682.
[386] Brown, S. P.; Bal, B. S.; Pinnick, H. W. *Tetrahedron Lett.* **1981**, *22*, 4891.
[387] Newcomb, M.; Chestney, D. L. *J. Am. Chem. Soc.* **1994**, *116*, 9753.
[388] Luhowy, R.; Keehn, P. M. *J. Am. Chem. Soc.* **1977**, *99*, 3797.
[389] Kirmse, W.; Hellwig, G.; Chiem, P. V. *Chem. Ber.* **1986**, *119*, 1511.
[390] Radunz, H.-E.; Reissig, H.-U.; Schneider, G.; Riethmüller, A. *Justus Liebigs Ann. Chem.* **1990**, 705.
[391] Arenare, L.; Caprariis, P. D.; Marinozzi, M.; Natalini, B.; Pellicciari, R. *Tetrahedron Lett.* **1994**, *35*, 1425.
[392] Gream, G. E.; Pincombe, C. F. *Aust. J. Chem.* **1974**, *27*, 543.
[393] Wenkert, E.; Arrhenius, T. S.; Bookser, B.; Guo, M.; Mancini, P. *J. Org. Chem.* **1990**, *55*, 1185.
[394] Speicher, A.; Eicher, T. *Synthesis* **1995**, 998.
[395] Ochiai, M.; Sumi, K.; Fujita, E.; Shiro, M. *Tetrahedron Lett.* **1982**, *23*, 5419.
[396] Kostikov, R. R.; Boganov, S. E.; Molchanov, A. P.; Slobodin, Y. M. *Russ. J. Org. Chem. (Engl. Transl.)* **1988**, *24*, 1084; *Chem. Abstr.* **1989**, *110*, 114340.
[397] Rilling, H. C.; Poulter, C. D.; Epstein, W. W.; Larsen, B. *J. Am. Chem. Soc.* **1971**, *93*, 1783.
[398] Molchanov, A. P.; Filippov, G. Y.; Kostikov, R. R. *Russ. J. Org. Chem. (Engl. Transl.)* **1994**, *30*, 1412; *Chem. Abstr.* **1995**, *123*, 339199.
[399] Kutney, J. P.; Choudhury, M. K.; Decesare, J. M.; Jacobs, H.; Singh, A. K.; Worth, B. R. *Can. J. Chem.* **1981**, *59*, 3162.
[400] Evans, D. A.; Sims, C. L.; Andrews, G. C. *J. Am. Chem. Soc.* **1977**, *99*, 5453.
[401] Shul'ts, E. E.; Vafina, G. F.; Spirikhin, L. V.; Tolstikov, G. A. *J. Org. Chem. USSR (Engl. Transl.)* **1991**, *27*, 643.
[402] Vig, O. P.; Trehan, I. R.; Kad, G. L.; Bedi, A. L. *Indian J. Chem. Sect. B* **1978**, *16B*, 455.
[403] Marino, J. P.; Pradilla, R. F.; Laborde, E. *J. Org. Chem.* **1984**, *49*, 5279.
[404] Vig, O. P.; Kad, G. L.; Bedi, A. L.; Kumar, S. D. *Indian J. Chem. Sect. B* **1978**, *16B*, 452.
[405] Weber, E.; Hecker, M.; Csoeregh, I.; Czugler, M. *J. Am. Chem. Soc.* **1989**, *111*, 7866.
[406] Warner, P.; Sutherland, R. *J. Org. Chem.* **1992**, *57*, 6294.
[407] Fakhretdinov, R. N.; Marvanov, R. M.; Dzhemilev, U. M.; Tolstikov, G. A. *Bull. Acad. Sci. USSR Div. Chem. Sci. (Engl. Transl.)* **1986**, 2555; *Chem. Abstr.* **1987**, *106*, 32431.
[408] Zindel, J.; Maitra, S.; Lightner, D. A. *Synthesis* **1996**, 1217.
[409] Feldman, K. S.; Vong, A. K. K. *Tetrahedron Lett.* **1990**, *31*, 823.
[410] Biggs, T. N.; Swenton, J. S. *J. Org. Chem.* **1992**, *57*, 5568.
[411] Melnick, M. J.; Bisaha, S. N.; Gammill, R. B. *Tetrahedron Lett.* **1990**, *31*, 961.
[412] Marino, J. P.; Long, J. K. *J. Am. Chem. Soc.* **1988**, *110*, 7916.
[413] Pellicciari, R.; Cecchetti, S.; Natalini, B.; Roda, A.; Grigolo, B.; Fini, A. *J. Med. Chem.* **1984**, *27*, 746.
[414] Proudfoot, J. R.; Djerassi, C. *J. Chem. Soc., Perkin Trans. 1* **1987**, 1283.
[415] Coates, R. M.; Sandefur, L. O.; Smillie, R. D. *J. Am. Chem. Soc.* **1975**, *97*, 1619.
[416] Wenkert, E.; Buckwalter, B. L.; Craviro, A. A.; Sanchez, E. L.; Sathe, S. S. *J. Am. Chem. Soc.* **1978**, *100*, 1267.
[417] Tsuge, O.; Kanemasa, S.; Otsuka, T.; Suzuki, T. *Bull. Chem. Soc. Jpn.* **1988**, *61*, 2897.
[418] McMurry, J. E.; Glass, T. E. *Tetrahedron Lett.* **1971**, 2575.
[419] De Kimpe, N.; Verhre, R.; De Buyck, L.; Schamp, N. *Synth. Commun.* **1978**, *8*, 75.
[420] Shi, G.; Xu, Y. *J. Org. Chem.* **1990**, *55*, 3383.
[421] Lewis, R. T.; Motherwell, W. B.; Shipman, M.; Salwin, A. M. Z.; Williams, D. J. *Tetrahedron* **1995**, *51*, 3289.
[422] Tsuge, O.; Kanemasa, S.; Suzuki, T.; Matsuda, K. *Bull. Chem. Soc. Jpn.* **1986**, *59*, 2851.
[423] Franck-Neumann, M.; Geoffroy, P.; Winling, A. *Tetrahedron Lett.* **1995**, *36*, 8213.
[424] House, H. O.; Fischer, W. F.; Gall, M.; McLaughlin, T. E.; Peet, N. P. *J. Org. Chem.* **1971**, *36*, 3429.

[425] Tishchenko, I. G.; Kulinkovich, O. G.; Glazkov, Y. V. *J. Org. Chem. USSR* **1975**, 579; *Chem. Abstr.* **1975**, *83*, 28027.
[426] Jeganathan, A.; Richardson, S. K.; Mani, R. S.; Haley, B. E.; Watt, D. S. *J. Org. Chem.* **1986**, *51*, 5362.
[427] Smeets, F. L. M.; Thijs, L.; Zwanenburg, B. *Tetrahedron* **1980**, *36*, 3269.
[428] Lund, E. A.; Kennedy, I. A.; Fallis, A. G. *Tetrahedron Lett.* **1993**, *34*, 6841.
[429] Lund, E. A.; Kennedy, I. A.; Fallis, A. G. *Can. J. Chem.* **1996**, *74*, 2401.
[430] Eberlin, M. N.; Kascheres, C. *J. Org. Chem.* **1988**, *53*, 2084.
[431] Gefflaut, T.; Perie, J. *Synth. Commun.* **1994**, *24*, 29.
[432] Nickels, H.; Dürr, H.; Toda, F. *Chem. Ber.* **1986**, *119*, 2249.
[433] O'Bannon, P. E.; Dailey, W. P. *Tetrahedron Lett.* **1989**, *30*, 4197.
[434] Sezer, O.; Daut, A.; Anac, O. *Helv. Chim. Acta* **1995**, *78*, 2036.
[435] Alonso, M. E.; Pekerar, S. V.; Borgo, M. L. *Mag. Res. Chem.* **1990**, *28*, 956.
[436] Chelucci, G.; Saba, A. *Tetrahedron Lett.* **1995**, *36*, 4673.
[437] Blanchard, L. A.; Schneider, J. A. *J. Org. Chem.* **1986**, *51*, 1372.
[438] Li, K.; Du, W.; Que, N. L. S.; Liu, H. *J. Am. Chem. Soc.* **1996**, *118*, 8763.
[439] Shostakovskii, V. M.; Zlatkina, V. L.; Vasil'vitskii, A. E.; Nefedov, G. M. *Proc. Acad. Sci. USSR, Gen. Chem.* **1983**, 1877.
[440] Danishefsky, S. J.; Etheredge, S. J.; Dynak, J.; McCurry, P. *J. Org. Chem.* **1974**, *39*, 2658.
[441] Huval, C. C.; Singleton, D. A. *J. Org. Chem.* **1994**, *59*, 2020.
[442] Alonso, M. E.; Fernandez, R. *Tetrahedron* **1989**, *45*, 3313.
[443] Wulfman, D. S.; McGibboney, B. G.; Steffen, E. K.; Thinh, N. V.; McDaniel, R. S., Jr.; Peace, B. W. *Tetrahedron* **1976**, *32*, 1257.
[444] Mazzocchi, P. H.; Lustig, R. S. *J. Am. Chem. Soc.* **1975**, *97*, 3707.
[445] Dehmlow, E. V.; Birkhahn, M. *Tetrahedron* **1988**, *44*, 4363.
[446] Marinozzi, M.; Natalini, B.; Constantino, G.; Pellicciari, R.; Bruno, V.; Nicoletti, F. *Farmaco* **1996**, *51*, 121; *Chem. Abstr.* **1996**, *124*, 290233.
[447] Piers, E.; Hall, T.-W. *Can. J. Chem.* **1980**, *58*, 2613.
[448] Gallucci, R. R.; Jones, M. Jr. *J. Am. Chem. Soc.* **1976**, *98*, 7704.
[449] Padwa, A.; Coats, S. J.; Hadjiarapoglou, L. *Heterocycles* **1995**, *41*, 1631.
[450] Rosenfeld, M. J.; Shankar, B. K. R.; Shechter, H. *J. Org. Chem.* **1988**, *53*, 2699.
[451] Davies, H. M. L.; Hu, B.; Saikali, E.; Bruzinski, P. R. *J. Org. Chem.* **1994**, *59*, 4535.
[452] Davies, H. M. L.; Hu, B. *J. Org. Chem.* **1992**, *57*, 4309.
[453] Seyferth, D.; Marmor, R. S.; Hilbert, P. *J. Org. Chem.* **1971**, *36*, 1379.
[454] Scherer, H.; Hartmann, A.; Regitz, M.; Tunggal, B. D.; Gunther, H. *Chem. Ber.* **1972**, *105*, 3357.
[455] Gross, Z.; Galili, N.; Simkhovich, L. *Tetrahedron Lett.* **1999**, *40*, 1571.
[456] Davies, H. M. L.; Ahmed, G.; Calvo, R. L.; Churchill, M. R.; Churchill, D. G. *J. Org. Chem.* **1998**, *63*, 2641.
[457] Kunz, T.; Reissig, H.-U. *Tetrahedron Lett.* **1989**, *30*, 2079.
[458] Schumacher, R.; Dammast, F.; Reissig, H.-U. *Chem. Eur. J.* **1997**, *3*, 614.
[459] Ebinger, A.; Heinz, T.; Umbricht, G.; Pfaltz, A. *Tetrahedron* **1998**, *54*, 10469.
[460] Schumacher, R.; Reissig, H.-U. *Liebigs Ann./Recl.* **1997**, 521.
[461] Bedekar, A. V.; Koroleva, E. B.; Andersson, P. G. *J. Org. Chem.* **1997**, *62*, 2518.
[462] Kwong, H.-L.; Lee, W.-S.; Ng, H.-F.; Chiu, W.-H.; Wong, W.-T. *J. Chem. Soc., Dalton Trans.* **1998**, 1043.
[463] Laidler, D. A.; Milner, D. J. *J. Organomet. Chem.* **1984**, *270*, 121.
[464] Roje, M.; Vinkovic, V.; Sunjic, V.; Solladie-Cavallo, A.; Diep-Vohuule, A.; Isarno, T. *Tetrahedron* **1998**, *54*, 9123.
[465] Dakovic, S.; Liscic-Tumir, L.; Kirin, S. I.; Vinkovic, V.; Raza, Z.; Suste, A.; Sunjic, V. *J. Mol. Catal. A: Chem.* **1997**, *118*, 27.
[466] Fraile, J. M.; Garcia, J. I.; Mayoral, J. A.; Tarnai, T. *Tetrahedron: Asymmetry* **1998**, *9*, 3997.
[467] Fraile, J. M.; Garcia, J. I.; Mayoral, J. A.; Tarnai, T. *Tetrahedron: Asymmetry* **1997**, *8*, 2089.
[468] Bedekar, A. V.; Andersson, P. G. *Tetrahedron Lett.* **1996**, *37*, 4073.

[469] Harm, A. M.; Knight, J. G.; Stemp, G. *Synlett* **1996**, 677.
[470] Harm, A. M.; Knight, J. G.; Stemp, G. *Tetrahedron Lett.* **1996**, *37*, 6189.
[471] Imai, Y.; Zhang, W.; Kida, T.; Nakatsuji, Y.; Ikeda, I. *Tetrahedron Lett.* **1997**, *38*, 2681.
[472] Gupta, A. D.; Bhuniya, D.; Singh, V. K. *Tetrahedron* **1994**, *50*, 13725.
[473] Christenson, D. L.; Tokar, C. J.; Tolman, W. B. *Organometallics* **1995**, *14*, 2148.
[474] Chelucci, G.; Antonietta Cabras, M.; Saba, A. *J. Mol. Catal. A: Chem.* **1995**, *95*, L7.
[475] Uozumi, Y.; Kyota, H.; Kishi, E.; Kitayama, K.; Hayashi, T. *Tetrahedron: Asymmetry* **1996**, *7*, 1603.
[476] Rippert, A. J. *Helv. Chim. Acta* **1998**, *81*, 676.
[477] Meyers, A. I.; Price, A. *J. Org. Chem.* **1998**, *63*, 412.
[478] Tanner, D.; Harden, A.; Johansson, F.; Wyatt, P.; Andersson, P. G. *Acta Chem. Scand.* **1996**, *50*, 361.
[479] Tanner, D.; Andersson, P. G.; Harden, A.; Somfai, P. *Tetrahedron Lett.* **1994**, *35*, 4631.
[480] Tanner, D.; Johansson, F.; Harden, A.; Andersson, P. G. *Tetrahedron* **1998**, *54*, 15731.
[481] Brunner, H.; Altmann, S. *Chem. Ber.* **1994**, *127*, 2285.
[482] Basu, B.; Frejd, T. *Acta Chem. Scand.* **1996**, *50*, 316.
[483] Kanemasa, S.; Hamura, S.; Harada, E.; Yamamoto, H. *Tetrahedron Lett.* **1994**, *35*, 7985.
[484] Brunner, H.; Goldbrunner, J. *Chem. Ber.* **1989**, *122*, 2005.
[485] Brunner, H.; Wutz, K. *New J. Chem.* **1992**, *16*, 57.
[486] Suga, H.; Fudo, T.; Ibata, T. *Synlett* **1998**, 933.
[487] Kim, S.-G.; Cho, C.-W.; Ahn, K. H. *Tetrahedron: Asymmetry* **1997**, *8*, 1023.
[488] Lo, M. M.-C.; Fu, G. C. *J. Am. Chem. Soc.* **1998**, *120*, 10270.
[489] Brunner, H.; Schiessling, H. *Bull. Soc. Chim. Belg.* **1994**, *103*, 119.
[490] Hoarau, O.; Ait-Haddou, H.; Castro, M.; Balavoine, G. G. A. *Tetrahedron: Asymmetry* **1997**, *8*, 3755.
[491] Raza, Z.; Dakovic, S.; Vinkovic, V.; Sunjic, V. *Croat. Chem. Acta* **1996**, *69*, 1545; *Chem. Abstr.* **1997**, *126*, 212258.
[492] Reetz, M. T.; Bohres, E.; Goddard, R. *J. Chem. Soc., Chem. Commun.* **1998**, 935.
[493] Keyes, M. C.; Chamberlain, B. M.; Caltagirone, S. A.; Halfen, J. A.; Tolman, W. B. *Organometallics* **1998**, *17*, 1984.
[494] Cai, L.; Mahmoud, H.; Han, Y. *Tetrahedron: Asymmetry* **1999**, *10*, 411.
[495] Ito, K.; Katsuki, T. *Synlett* **1993**, 638.
[496] Yang, Z.-C.; Zhong, M.; Chen, L. *Acta Chim. Sinica* **1994**, *52*, 1218; *Chem. Abstr.* **1995**, *122*, 290747.
[497] Ito, K.; Tabuchi, S.; Katsuki, T. *Synlett* **1992**, 575.
[498] Ito, K.; Katsuki, T. *Tetrahedron Lett.* **1993**, *34*, 2661.
[499] Ito, K.; Yoshitake, M.; Katsuki, T. *Heterocycles* **1996**, *42*, 305.
[500] Moye-Sherman, D.; Welch, M. B.; Reibenspies, J.; Burgess, K. *J. Chem. Soc., Chem. Commun.* **1998**, 2377.
[501] Shu, F. C.; Zhou, Q. L. *Synth. Commun.* **1999**, *29*, 567.
[502] Dammast, F.; Reissig, H.-U. *Chem. Ber.* **1993**, *126*, 2727.
[503] Brunner, H.; Berghofer, J. *J. Organomet. Chem.* **1995**, *501*, 161.
[504] Ichiyanagi, T.; Shimizu, M.; Fujisawa, T. *Tetrahedron* **1997**, *53*, 9599.
[505] Brunner, H.; Hassler, B. *Z. Naturforsch., B: Chem. Sci.* **1998**, *53*, 126.
[506] Wu, X. Y.; Li, X. H.; Zhou, Q. L. *Tetrahedron: Asymmetry* **1998**, *9*, 4143.
[507] Brunner, H.; Hassler, B. *Z. Naturforsch., B: Chem. Sci.* **1998**, *53*, 476.
[508] Sunjic, V.; Sepac, D.; Kojic-Prodic, B.; Kiralj, R.; Mlinaric-Majerski, K.; Vinkovic, V. *Tetrahedron: Asymmetry* **1993**, *4*, 575.
[509] Brunner, H.; Singh, U. P.; Boeck, T.; Altmann, S.; Scheck, T.; Wrackmeyer, B. *J. Organomet. Chem.* **1993**, *443*, C16.
[510] Tokar, C. J.; Kettler, P. B.; Tolman, W. B. *Organometallics* **1992**, *11*, 2737.
[511] Carmona, A.; Corma, A.; Iglesias, M.; Sanchez, F. *Inorg. Chim. Acta* **1996**, *244*, 239.
[512a] Carmona, A.; Corma, A.; Iglesias, M.; Sanchez, F. *Inorg. Chim. Acta* **1996**, *244*, 79.
[512b] Becalski, A.; Cullen, W. R.; Fryzuk, M. D.; Herb, G.; James, B. R.; Kutney, J. P.; Piotrowska, K.; Tapiolas, D. *Can. J. Chem.* **1988**, *66*, 3108.

513. Brunner, H.; Miehling, W. *Monatsh. Chem.* **1984**, *115*, 1237.
514. Cho, N. S.; Shin, D. H.; Lee, C. C.; Ra, D. Y. *Bull. Korean Chem. Soc.* **1988**, *9*, 195; *Chem. Abstr.* **1989**, *110*, 213094.
515. Baldwin, J. E.; Chang, G. E. C. *Tetrahedron* **1982**, *38*, 825.
516. Baldwin, J. E.; Patapoff, T. W.; Barden, T. C. *J. Am. Chem. Soc.* **1984**, *106*, 1421.
517. Baldwin, J. E.; Barden, T. C. *J. Am. Chem. Soc.* **1984**, *106*, 5312.
518. O'Malley, S.; Kodadek, T. *Organometallics* **1992**, *11*, 2299.
519. Watanabe, N.; Matsuda, H.; Kuribayashi, H.; Hashimoto, S. *Heterocycles* **1996**, *42*, 537.
520. Doyle, M. P.; Winchester, W. R.; Simonsen, S. H.; Ghosh, R. *Inorg. Chim. Acta* **1994**, *220*, 193.
521. Doyle, M. P.; Winchester, W. R.; Protopopova, M. N.; Muller, P.; Bernardinelli, G.; Ene, D.; Motallebi, S. *Helv. Chim. Acta* **1993**, *76*, 2227.
522. Kitagaki, S.; Matsuda, H.; Watanabe, N.; Hashimoto, S. *Synlett* **1997**, 1171.
523. Doyle, M. P.; Zhou, Q. L.; Simonsen, S. H.; Lynch, V. *Synlett* **1996**, 697.
524. O'Malley, S.; Kodadek, T. *Tetrahedron Lett.* **1991**, *32*, 2445.
525. Davies, H. M. L.; Panaro, S. A. *Tetrahedron Lett.* **1999**, *40*, 5287.
526. Wynne, D. C.; Jessop, P. G. *Angew. Chem., Int. Ed. Engl.* **1999**, *38*, 1143.
527. Yoshikawa, K.; Achiwa, K. *Chem. Pharm. Bull.* **1995**, *43*, 2048.
528. Davies, H. M. L.; Kong, N. *Tetrahedron Lett.* **1997**, *38*, 4203.
529. Doyle, M. P.; Eismont, M. Y.; Bergbreiter, D. E.; Gray, H. N. *J. Org. Chem.* **1992**, *57*, 6103.
530. Nakamura, A.; Konishi, A.; Tatsuno, Y.; Otsuka, S. *J. Am. Chem. Soc.* **1978**, *100*, 3443.
531. Nishiyama, H.; Soeda, N.; Naito, T.; Motoyama, Y. *Tetrahedron-Asymmetry* **1998**, *9*, 2865.
532. Song, J. H.; Cho, D. J.; Jeon, S. J.; Kim, Y. H.; Kim, T. J.; Jeong, J. H. *Inorg. Chem.* **1999**, *38*, 893.
533. Park, S.-B.; Murata, K.; Matsumoto, H.; Nishiyama, H. *Tetrahedron: Asymmetry* **1995**, *6*, 2487.
534. Davies, I. W.; Gerena, L.; Cai, D.; Larsen, R. D.; Verhoeven, T. R.; Reider, P. J. *Tetrahedron Lett.* **1997**, *38*, 1145.
535. Park, S.-B.; Sakata, N.; Nishiyama, H. *Chem. Eur. J.* **1996**, *2*, 303.
536. Ko, P.-H.; Chen, T.-Y.; Zhu, J.; Cheng, K.-F.; Peng, S.-M.; Che, C.-M. *J. Chem. Soc., Dalton Trans.* **1995**, 2215.
537. Galardon, E.; Roue, S.; Le Maux, P.; Simonneaux, G. *Tetrahedron Lett.* **1998**, *39*, 2333.
538. Navarro, R.; Urriolabeitia, E. P.; Cativiela, C.; Diaz-de-Villegas, M. D.; Lopez, M. P.; Alonso, E. *J. Mol. Catal. A: Chem.* **1996**, *105*, 111.
539. Lee, H. M.; Bianchini, C.; Jia, G.; Barbaro, P. *Organometallics* **1999**, *18*, 1961.
540. Fukuda, T.; Katsuki, T. *Tetrahedron* **1997**, *53*, 7201.
541. Fukuda, T.; Katsuki, T. *Synlett* **1995**, 825.
542. Ito, Y. N.; Katsuki, T. *Bull. Chem. Soc. Jpn.* **1999**, *72*, 603.
543. Anciaux, A. J.; Demonceau, A.; Noels, A. F.; Warin, R.; Hubert, A. J.; Teyssie, P. *Tetrahedron* **1983**, *39*, 2169.
544. Demonceau, A.; Noels, A. F.; Anciaux, A. J.; Hubert, A. J.; Teyssie, P. *Bull. Soc. Chim. Belg.* **1984**, *93*, 949; *Chem. Abstr.* **1985**, *102*, 204113.
545. Wenkert, E.; Goodwin, T. E.; Ranu, B. C. *J. Org. Chem* **1977**, *42*, 2137.
546. Hoffman, B.; Reissig, H.-U. *Chem. Ber.* **1994**, *127*, 2315.
547. Frey, B.; Schnaubelt, J.; Reissig, H.-U. *Eur. J. Org. Chem.* **1999**, 1377.
548. Kuehne, M. E.; Pitner, J. B. *J. Org. Chem.* **1989**, *54*, 4553.
549. Grimm, E. L.; Zschiesche, R.; Reissig, H.-U. *J. Org. Chem.* **1985**, *50*, 5543.
550. Holland, D.; Milner, D. J. *J. Chem. Res. (S)* **1979**, 317.
551. Holland, D.; Laidler, D. A.; Milner, D. J. *J. Mol. Catal.* **1981**, *11*, 119.
552. Shapiro, E. A.; Romanova, T. N.; Bunkina, T. A.; Dolgii, I. E.; Nefedov, I. M. *Bull. Acad. Sci. USSR Div. Chem. Sci. (Engl. Transl.)* **1987**, 2318; *Chem. Abstr.* **1988**, *109*, 73036.
553. Shapiro, E. A.; Romanova, T. N.; Dolgii, I. E.; Nefedov, O. M. *Bull. Acad. Sci. USSR Div. Chem. Sci. (Eng. Transl.)* **1988**, 2322; *Chem. Abstr.* **1988**, *109*, 37498.
554. Moshe, A. *Eur. J. Med. Chem. Chim. Ther.* **1981**, *16*, 199.
555. Boche, G.; Martens, D. *Chem. Ber.* **1979**, *112*, 175.

[556] Jones, M., Jr.; Reich, S. D.; Scott, L. T. *J. Am. Chem. Soc.* **1970**, *92*, 3118.
[557] Dauben, W. G.; Michno, D. M. *J. Am. Chem. Soc.* **1981**, *103*, 2284.
[558] Decock-le Reverend, B.; Durand, M.; Merenyi, R. *Bull. Soc. Chim. Fr.* **1978**, 369.
[559] Tomilov, Y. V.; Tsvetkova, N. M.; Nefedov, O. M. *Russ. Chem. Bull.* **1997**, *46*, 507; *Chem. Abstr.* **1997**, *127*, 277999.
[560] Cottens, S.; Schlosser, M. *Tetrahedron* **1988**, *44*, 7127.
[561] Kelly, L. F. *J. Chem. Educ.* **1987**, *64*, 1061.
[562] Wright, M. E.; Svejda, S. A.; Jin, M.-J.; Peterson, M. A. *Organometallics* **1990**, *9*, 136.
[563] Sanda, F.; Murata, J.; Endo, T. *Macromolecules* **1997**, *30*, 160.
[564] Mahon, M. F.; Molloy, K.; Pittol, C. A.; Pryce, R. J.; Roberts, S. M.; Ryback, G.; Sik, V.,; Williams, J. O.; Winders, J. A. *J. Chem. Soc., Perkin Trans. 1* **1991**, 1255.
[565] Downing, W.; Latouche, R.; Pittol, C. A.; Pryce, R. J.; Roberts, S. M.; Ryback, G.; Williams, J. O. *J. Chem. Soc., Perkin Trans. 1* **1990**, 2613.
[566] De Smet, A.; Anteunis, M.; Tavernier, D. *Bull. Soc. Chim. Belg.* **1975**, *84*, 67.
[567] Kheruze, Y. I.; Petrov, A. A. *J. Org. Chem. USSR (Engl. Transl.)* **1974**, 1423.
[568] Wenkert, E.; Greenberg, R. S.; Kim, H. S. *Helv. Chim. Acta* **1987**, *70*, 2159.
[569] Mueller, L. G.; Lawton, R. G. *J. Org. Chem.* **1979**, *44*, 4741.
[570] Mandel'shtam, T. V.; Kristol, L. D.; Bogdanova, L. A.; Ratnikova, T. N. *Russ. J. Org. Chem. (Engl. Transl.)* **1968**, *4*, 963; *Chem. Abstr.* **1968**, *69*, 43498.
[571] Birkhahn, M.; Dehmlow, E. V.; Bogge, H. *Angew. Chem., Int. Ed. Engl.* **1987**, *26*, 72.
[572] Davies, H. M. L.; Peng, Z.-Q.; Houser, J. H. *Tetrahedron Lett.* **1994**, *35*, 8939.
[573] Davies, H. M. L.; Oldenburg, C. E. M.; McAfee, M. J.; Nordahl, J. G.; Henretta, J. P.; Romines, K. R. *Tetrahedron Lett.* **1988**, *29*, 975.
[574] Davies, H. M. L.; Houser, J. H.; Thornley, C. *J. Org. Chem.* **1995**, *60*, 7529.
[575] Davies, H. M. L.; Hodges, L. M.; Thornley, C. T. *Tetrahedron Lett.* **1998**, *39*, 2707.
[576] Cantrell, W. R., Jr.; Davies, H. M. L. *J. Org. Chem.* **1991**, *56*, 723.
[577] Wenkert, E.; Khatuya, H. *Helv. Chim. Acta* **1998**, *81*, 2370.
[578] Saltykova, L. E.; Vasil'vitskii, A. E.; Shostakovskii, V. M.; Nefedov, O. M. *Bull. Acad. Sci. Div. Chem. Sci. (Engl. Transl.)* **1988**, 2557; *Chem. Abstr.* **1989**, *111*, 23308.
[579] Nefedov, O. M.; Vasil'vitskii, A. E.; Zlatkina, V. L.; Yufit, D. S.; Struchkov, Y. T.; Shostakovskii, V. M. *Bull. Acad. Sci. Div. Chem. Sci. (Engl. Transl.)* **1988**, 2107; *Chem. Abstr.* **1989**, *110*, 212512.
[580] Pirrung, M. C.; Zhang, J.; McPhail, A. T. *J. Org. Chem.* **1991**, *56*, 6269.
[581] Padwa, A.; Krumpe, K. E.; Kassir, J. M. *J. Org. Chem.* **1992**, *57*, 4940.
[582] Battersby, A. R.; Block, M. H.; Fookes, C. J. R.; Harrison, P. J.; Henderson, G. B.; Leeper, F. J. *J. Chem. Soc., Perkin Trans. 1* **1992**, 2175.
[583] Alazard, J.-P.; Millet-Paillusson, C.; Guenard, D.; Thal, C. *Bull. Soc. Chim. Fr.* **1996**, *133*, 251.
[584] Keller, H.; Langer, E.; Lehner, H. *Monatsh. Chem.* **1977**, *108*, 123.
[585] Bergman, J.; Koch, E.; Pelcman, B. *Tetrahedron* **1995**, *51*, 5631.
[586] Davies, H. M. L.; Matasi, J. J.; Thornley, C. *Tetrahedron Lett.* **1995**, *36*, 7205.
[587] Davies, H. M. L.; Saikali, E.; Huby, N. J. S.; Gilliatt, V. J.; Matasi, J. J.; Sexton, T.; Childers, S. R. *J. Med. Chem.* **1994**, *37*, 1262.
[588] Wood, J. L.; Stolz, B. M.; Dietrich, H.-J. *J. Am. Chem. Soc.* **1995**, *117*, 10413.
[589] Cragg, J. E.; Herbert, R. B.; Jackson, F. B.; Moody, C. J.; Nicolson, I. T. *J. Chem. Soc., Perkin Trans. 1* **1982**, 2477.

CHAPTER 2

OXIDATION OF PHENOLIC COMPOUNDS WITH ORGANOHYPERVALENT IODINE REAGENTS

ROBERT M. MORIARTY AND OM PRAKASH[#]

Department of Chemistry, The University of Illinois at Chicago, Chicago, Illinois 60607–7061

CONTENTS

	PAGE
ACKNOWLEDGMENTS	329
INTRODUCTION	329
MECHANISM	331
SCOPE AND LIMITATIONS	332
Oxidation of Phenols to Quinones and Related Compounds	332
Formation of Quinones	332
Formation of *p*-Quinols, *p*-Quinol Ethers, and Quinone Monoacetals	334
Intramolecular Participation in the Oxidation of Phenols	337
Formation of Heterocyclic Spiroquinones	337
Formation of Oxygen Heterocyclic Compounds	342
Coumaran-3-ones	342
3,4-Dihydrocoumarins	343
Intramolecular Carbon-Carbon Bond Formation	343
Phenolic Oxidative Coupling and Its Synthetic Applications	343
Formation of Iodonium Ylides and Salts and Their Synthetic Applications	349
Miscellaneous	351
p-Thiocyanation of Phenols	351
Oxidation of Flavonols	352
Cleavage of Amino-Terminal Tyrosyl-Peptide Bonds	352
Oxidation of 3′-Desmethylpapaverine	353
Oxidative Phenol-Propenylbenzene Cycloadditions Leading to *trans*-2,3-Dihydrobenzofurans	353
Synthesis of Dynemicin A	354
COMPARISON WITH OTHER METHODS	356
Oxidation of Phenols to Quinones and Related Compounds	356

[#]On leave of absence from Kurukshetra University, Kurukshetra, Haryana 136119, INDIA.

Organic Reactions, Vol. 57, Edited by Larry E. Overman et al.
ISBN 0-471-43511-2 © 2001 Organic Reactions, Inc. Published by John Wiley & Sons, Inc.

Intramolecular Participation in the Oxidation of Phenols. 357
Intramolecular Carbon-Carbon Bond Formation: Phenolic Oxidative Coupling . . 357
EXPERIMENTAL PROCEDURES 359
  Practical Aspects and Availability of Organohypervalent Iodine Reagents . . . 359
  5-Hydroxy-1,4-naphthoquinone (Juglone) [Oxidation of a Phenol to a *p*-Quinone . 359
  4-Hydroxy-4-methylcyclohexa-2,5-dienone [Oxidation of a *para*-Substituted
    Phenol to a *p*-Quinol] 360
  4,4-Dimethoxy-3,5-dimethylcyclohexa-2,5-dienone [Oxidation of a Monophenol
    to a *p*-Quinone Monoacetal] 360
  1-Oxaspiro[4,5]deca-6,9-dien-8-one [Formation of a Spiroquinone by Intramolecular
    Participation of a Primary Alcoholic Group]. 360
  (1*S*,8*R*,9*R*,10*R*)-1'-Methyl-8'-phenyl-1'-decalyl 7,9-Dibromo-8-methoxy-6-oxo-1-oxa-
    2-azaspiro[4,5]deca-2,7,9-triene-3-carboxylate [Asymmetric Induction in the
    Oxidative Cyclization of an *o*-Phenolic Oxime Ester] 361
  1-Benzoyloxy-7-methyl-*cis*-azabicyclo[4.3.0]non-2-en-4-one [Formation of a
    Hexahydroindol-6-one by Oxidation of an *N*-Alkyl-*N*-benzoyltyramine] . . . 362
  5,6,7,8,9,10-Hexahydropyrido[2,3-*g*]quinoline-5,10-dione-9-spiro-4'-cyclohexa-
    2,5'-dien-1'-one [Oxidative Cyclization of an *O*-Silylated Phenol to an
    Azacarbocyclic Spirodienone] 362
  5,7-Dinitro-1,3-benzoxathiole-2-thione [Synthesis of an Oxathiole by
    Photochemical Decomposition of a Phenolic Iodonium Ylide in the Presence
    of Carbon Disulfide]. 363
    Step I. Formation of 2,4-Dinitro-6-phenyliodonium Phenolate 363
    Step II. Formation of 5,7-Dinitro-1,3-benzoxathiole-2-thione 363
  4-Thiocyanato-1-naphthol [*p*-Thiocyanation of a Phenol] 364
  p-(Methoxymethyl)phenol [Cleavage of an Amino-Terminal Tyrosyl-Peptide Bond] . 364
  (±)-*trans*-2-(3,4-Methylenedioxyphenyl)-2,3-dihydro-6-allyloxy-5-methoxy-3-
    methylbenzofuran [Formation of a Dihydrobenzofuran by Oxidative Cycloaddition
    of a Phenol to an Electron-rich Styrene]. 364
TABULAR SURVEY 365
  Table IA. Formation of Quinones by Oxidation of Monophenols 366
  Table IB. Formation of Quinones by Oxidation of Bisphenols 371
  Table II. Formation of *p*-Quinols by Oxidation of Phenols and Phenol Tripropylsilyl
    Ethers 374
  Table III. Formation of Quinone Acetals by Oxidation of Phenols 376
  Table IV. Formation of *p*-Quinol Ethers by Oxidation of Phenols 379
  Table V. *ipso*-Fluorination of 4-Alkylphenols 382
  Table VI. Formation of Spiro-Heterocyclic Compounds 383
  Table VII. Oxidation of *N*-Acyltyramines 390
  Table VIII. Formation of Hydroindolenones by Oxidative Cyclization of Phenols . 391
  Table IX. Formation of Oxygen Heterocyclic Compounds by Oxidative Cyclization of
    o-Acylphenols 392
  Table X. Intramolecular Carbon-Carbon Bond Formation: Phenolic Oxidative
    Coupling. 394
  Table XI. Oxidative Cyclization of *O*-Silylated Phenols to Azacarbocyclic
    Spirodienones. 399
  Table XII. Formation of Iodonium Ylides and Salts 401
  Table XIII. *p*-Thiocyanation of Phenols 405
  Table XIV. Formation of 2,3-Dimethoxy-3-hydroxyflavanones by Oxidation of
    Flavonols and Analogs 407
  Table XV. Cleavage of Amino-Terminal Tyrosylpeptides 409
  Table XVI. Synthesis of 2,3-Dihydrobenzofurans by Oxidative Cycloaddition of
    Phenols to Propenylbenzenes 410
REFERENCES 412

ACKNOWLEDGMENTS

Thanks are due to Dr. Neena Rani, Mr. Jerry Kosmeder, Dr. Vijay Sharma, and Dr. Hitesh Batra, UIC, Chicago, and Ritu Prakash for their valuable help in the preparation of this manuscript. Financial support from the U. S. National Science Foundation through grant CHE 9520157 is gratefully acknowledged.

INTRODUCTION

The oxidation of phenols is a key biochemical process in oxidative phosphorylation, and it is also important in numerous biosynthetic pathways.[1-12] Controlled oxidative transformations of substituted phenols are found in many synthetic organic sequences. Accordingly, a large number of oxidizing agents for phenols have been developed. Useful reagents include Fremy's salt $(KSO_3)_2NO$,[13] and a wide variety of other redox metal-based oxidants such as Pb(IV),[14] Mn(III),[15] Tl(III),[16,17] Cu(II),[18] and Fe(III).[19-25] Recently, organohypervalent iodine reagents have emerged as particularly useful agents for oxidizing phenols.[26-47a] In the area of natural product synthesis, organohypervalent iodine reagents have been used extensively to effect oxidative intramolecular bicyclization and spirocyclization. The focus of this chapter is on synthetic applications. The chapter also covers formation and applications of phenolic iodonium ylides.[28,40]

The processes that occur during the oxidation of phenols with organohypervalent iodine reagents can be divided into two categories. The first category involves ligand exchange of the phenolic proton with the hypervalent iodine reagent to generate the O-I(III) intermediate **1**, which is subsequently transformed to various products, depending upon the reaction conditions (Eq. 1).

$$\text{4-Z-C}_6\text{H}_4\text{OH} \xrightarrow{\text{PhIX}_2} [\text{4-Z-C}_6\text{H}_4\text{-O-I(Ph)X}]_{\mathbf{1}} \longrightarrow \text{Products} + \text{PhI} \quad (\text{Eq. 1})$$

These transformations include oxidation of phenols to quinones and related compounds (Eq. 2), formation of spirocyclic and oxygen heterocyclic compounds via oxidative intramolecular participation reactions (Eqs. 3 and 4), and intramolecular carbon-carbon bond formation via phenolic oxidative coupling (Eq. 5). All of these reactions are driven by the reduction of iodine(III) [or iodine(V)] to iodine(I) (iodobenzene).

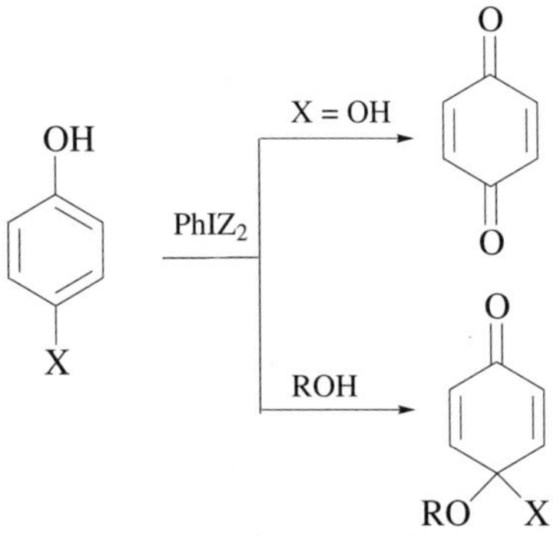

(Eq. 2)

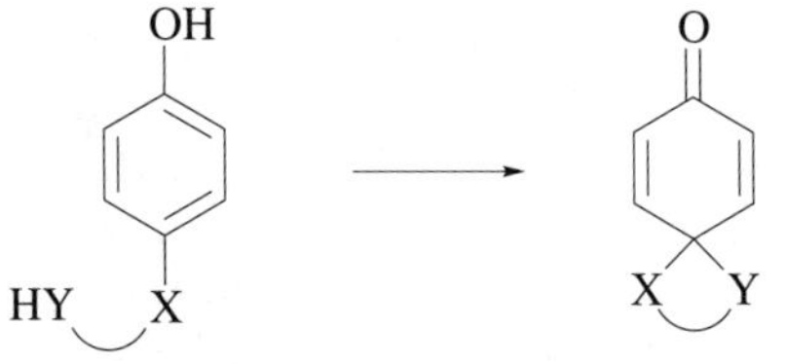

(Eq. 3)

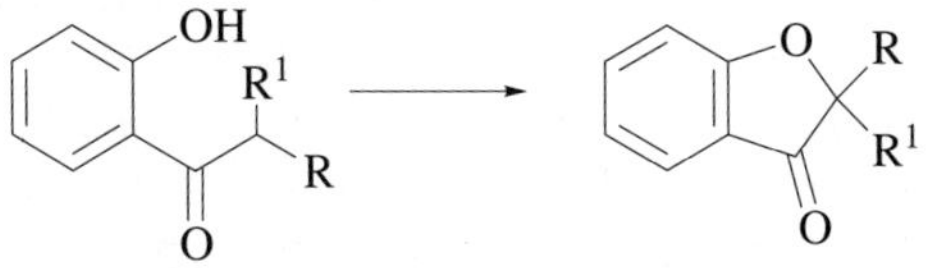

(Eq. 4)

(Eq. 5)

A second class of reactions that is discussed in this chapter includes formation of stable iodonium salts and ylides, which are important intermediates in organic synthesis. These reactions occur via carbon-I(III) bond formation without loss of iodobenzene (Eq. 6).

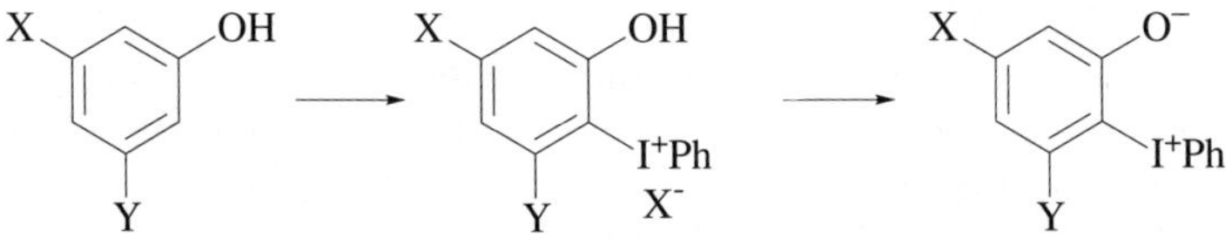

(Eq. 6)

MECHANISM

Oxidation of phenols provides products that come from a formal two-electron oxidation. These may be formed by a sequential one-electron or two-electron process.

(Eq. 7)

Two pathways (Routes A and B, Eq. 7) have been proposed for the two-electron oxidation of phenols to quinones and related compounds.[48] Common to both pathways is an aryloxyiodonium(III) intermediate **1**, which is formed by ligand exchange between a phenolic hydroxy group and the organohypervalent iodine reagent $PhIX_2$. In route A, **1** reacts with nucleophiles such as an alcohol or water to form product **2**. Route B involves dissociation of **1** to the phenoxenium ion **1a**, which then reacts with nucleophiles to give the products.

Product **2** can undergo a proton shift (where Z = H) to yield a *para*-substituted phenol (**3**); where Z is hydroxy or alkoxy, a quinone **4** or quinone monoacetal **5** results. Spirobicyclization can be accommodated within this mechanism as resulting from intramolecular participation in the reductive elimination of iodobenzene. In Eq. 8 the group Z attached by a three-atom chain at the para position acts as the internal nucleophile.

[Eq. 8 scheme showing conversion of 6 → 7 → 8 and 9]

Thus, either intramolecular spirocyclization (**7** → **8**) or bimolecular attack by a nucleophile or nucleophilic solvent (**7** → **9**) can occur. With an appropriately *meta*-substituted phenol, bicyclization by attack either at the ortho or para position is possible.

SCOPE AND LIMITATIONS

Oxidation of Phenols to Quinones and Related Compounds

Early studies on the oxidation of phenols with hypervalent iodine reagents reported the formation of resinous products,[48a] and it is only because of recent developments that hypervalent iodine mediated oxidative transformations have become synthetically attractive. Recent work in this area has shown that product distribution among *p*-quinones, *o*-quinones, quinone acetals, and quinone ethers is dependent on: (1) the nature of the phenolic substrates and hypervalent iodine reagents, (2) the nature of the solvent (nucleophilicity and polarity), and (3) the ratio of oxidizing agent to substrate.

Formation of Quinones. Oxidation of various phenols with organohypervalent iodine reagents under suitable conditions leads to the formation of quinones. Although these transformations have been effected by using various iodine(III) reagents, iodobenzene diacetate [IBD; PhI(OAc)$_2$] and iodobenzene bis(trifluoroacetate) [IBTA; PhI(O$_2$CCF$_3$)$_2$] are the most efficient. IBD is preferred over IBTA because of its lower toxicity and lower price. The structure of the phenols, the stoichiometry of the reactants, and the reaction conditions affect the course of these oxidations. Phenols containing groups such as alkoxy and *N*-acylamino at the para position are oxidized to *p*-quinones with one equivalent of the I(III) reagent in the presence of water (Eq. 9),[49] whereas phenols without substituents at the para position consume 2 equivalents of oxidant to afford the corresponding

[Eq. 9 scheme: 4-methoxyphenol → IBTA (1 eq), MeCN, H$_2$O → intermediate → p-benzoquinone (86%)]

p-quinones (Eq. 10).[50] Water acts as an external nucleophile in these reactions, and hydroquinones are intermediates.[51-53]

Eq. 10: 1,5-dihydroxynaphthalene → (via 1,4,5-trihydroxynaphthalene intermediate with IBTA (1 eq)) → IBTA (1 eq), MeCN, H$_2$O → juglone (58%)

A recent example is the oxidation of pentachlorophenol to tetrachloro-1,4-benzoquinone (**11**) by using IBTA in the presence of acetate buffer (Eq. 11). In this reaction, in situ-generated **10** undergoes loss of one molecule of hydrochloric acid to give **11**. This process offers an analytical method for trace quantities of environmental polychlorophenols.[54]

*Eq. 11: pentachlorophenol → IBTA, 0.1 M Acetate buffer → intermediate **10** → –HCl → tetrachloro-1,4-benzoquinone **11***

In general, oxidation of phenols yields mixtures of isomeric p-benzoquinones and o-benzoquinones. However, steric hindrance by the I(III) group such as I(Ph)(O$_2$CCX$_3$) can favor formation of p-quinones as the major or sole products.

Organohypervalent iodine reagents can also be used for the oxidation of hydrophenols; the reaction requires one equivalent of oxidant. Examples of such oxidations, including additional examples of monophenols, are given in Eqs. 12–15.

*Eq. 12: 5,8-dihydroxy-2,3-dichloronaphthoquinone → IBTA, MeCOMe → bisquinone **12** (81%)*

Some noteworthy features exemplifying the scope and limitations of this approach are:

(1) Quinones such as **12** (Eq. 12), which are useful precursors in the site-selective cycloaddition reactions of bisquinones, are readily accessible through this approach.[55,56]

[Eq. 13: Dihydroxyphthalic anhydride → p-benzoquinone-2,3-dicarboxylic acid anhydride, IBTA / C₆H₆, (68%)]

[Eq. 14: 5,8-dihydroxyquinoline → quinoline-5,8-dione, IBTA (2.2 eq.), (88%)]

[Eq. 15: 2,2-dimethyl-4,5-dihydroxychroman → quinone, IBD / MeCN, H₂O, (69%)]

(2) The preparation of *p*-benzoquinone-2,3-dicarboxylic acid anhydride using IBTA (Eq. 13) is a notable example, as even fuming nitric acid fails to accomplish this conversion.[57]

(3) Phenols containing nitrogen heterocycles are smoothly converted to the corresponding *p*-quinones, leaving the nitrogen unaffected (Eq. 14).[58]

(4) The presence of an oxidizable non-phenolic hydroxy group does not interfere with the oxidation process (Eq. 15).[59]

(5) An efficient synthesis of 5-hydroxy-1,4-naphthoquinone, juglone (Eq. 10),[50] which is a useful intermediate for the synthesis of tetracyclines[60] and antitumor antibiotics,[61] is available.

(6) Phenols containing electron-withdrawing substituents (NO_2, $COCH_3$, CO_2R) do not give quinones by this method; rather the corresponding iodonium ylides or salts are formed (see Eq. 51).

Formation of *p*-Quinols, *p*-Quinol Ethers, and Quinone Monoacetals. Oxidation of certain *p*-substituted phenols with IBTA in the presence of water affords *p*-quinols. The yields of this reaction can be improved by using the silyl ether derivatives of the phenols. For example, oxidation of the tripropylsilyl ether of *p*-cresol with IBTA in aqueous acetonitrile gives 4-hydroxy-4-methylcyclohexadienone in 73% yield, whereas a 48% yield is obtained from the oxidation of *p*-cresol itself (Eq. 16).[62]

Since phenolic oxidations proceed via an electrophilic intermediate of type **1**, the oxidation of phenols has been extensively investigated in the presence of other nucleophiles such as alcohols. These studies have identified an elegant and gen-

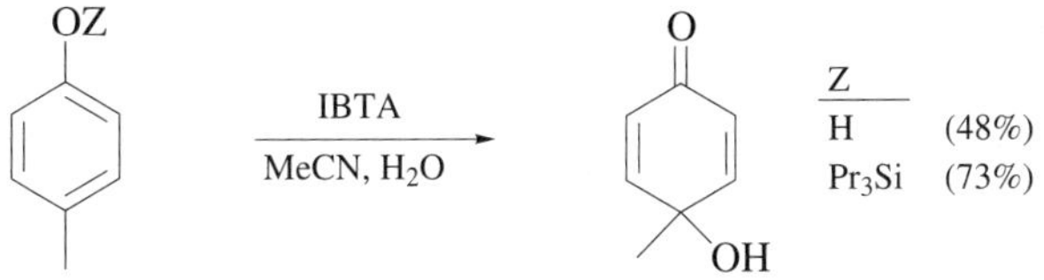

$$\text{4-Z-O-C}_6\text{H}_4\text{-CH}_3 \xrightarrow[\text{MeCN, H}_2\text{O}]{\text{IBTA}} \text{4-methyl-4-hydroxycyclohexa-2,5-dienone}$$

Z	
H	(48%)
Pr$_3$Si	(73%)

(Eq. 16)

eral way to synthesize *p*-quinone ethers and 4-bis(alkoxy)cyclohexadienones. Both the stoichiometry and structure of the phenols play a role in determining the course of the reaction. For example, phenols with no substituent at the para position afford quinone monoacetals in good yields upon oxidation with 2 equivalents of IBD in methanol (Eq. 17).[53] On the other hand, oxidation of *p*-alkoxyphenols with one equivalent of the iodine(III) reagent in the presence of methanol or other alcohols results in the formation of corresponding *p*-quinone monoacetals.[53,63] Thus, oxidation of *p*-methoxyphenol with one equivalent of IBD in methanol or ethanol leads to the formation of quinone monoacetals in high yields (Eq. 18).[53,63]

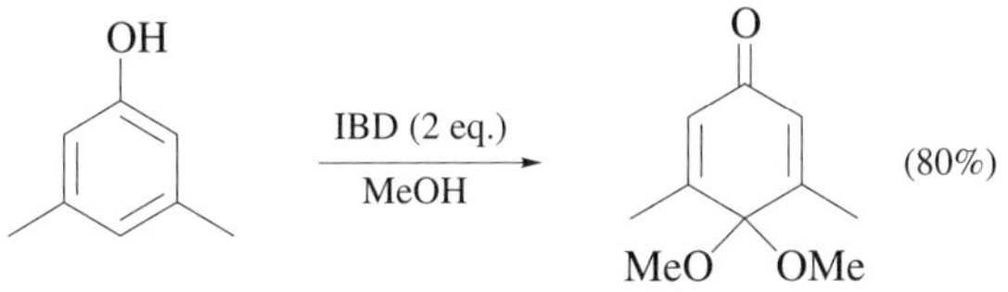

<!-- Eq. 17: 3,5-dimethylphenol + IBD (2 eq.)/MeOH → 4,4-dimethoxy-3,5-dimethyl-2,5-cyclohexadienone (80%) -->

(Eq. 17)

<!-- Eq. 18: 4-methoxyphenol + IBD/ROH → 4-methoxy-4-alkoxy-2,5-cyclohexadienone -->

R	
Me	(99%)
Et	(78%)

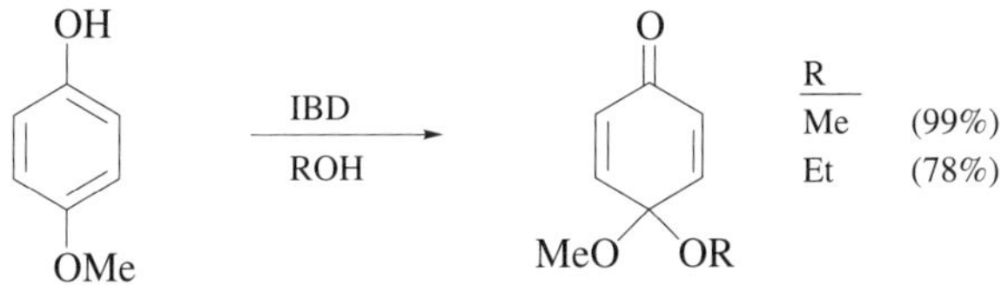

(Eq. 18)

When the alcohol is dienic, nucleophilic addition is followed by an intramolecular Diels-Alder reaction upon the thus-formed *p*-quinone monoacetal. For example, when methanol is replaced by sorbyl alcohol in the above reaction (Eq. 18), in situ intramolecular Diels-Alder reactions proceeding by way of the quinone monoacetal take place.[63] The products from this reaction are found to vary according to the workup employed. Thus, if sodium bicarbonate is added prior to the removal of excess sorbyl alcohol by distillation, **13** is obtained. But if distillation is carried out prior to base addition, the reaction affords only spiroacetal **14**. While the yields of products are modest, these transformations represent a facile, one-pot assembly of a highly functionalized ring system in a single step (Eq. 18a).

Oxidation of phenols bearing an alkoxy substituent at the ortho position results in formation of the corresponding *o*-quinone acetals, which are unstable with respect to dimerization.[63a] *o*-Benzoquinones are generated in situ as dienes in inter- and intramolecular Diels-Alder reactions. Thus, masked *o*-benzoquinone

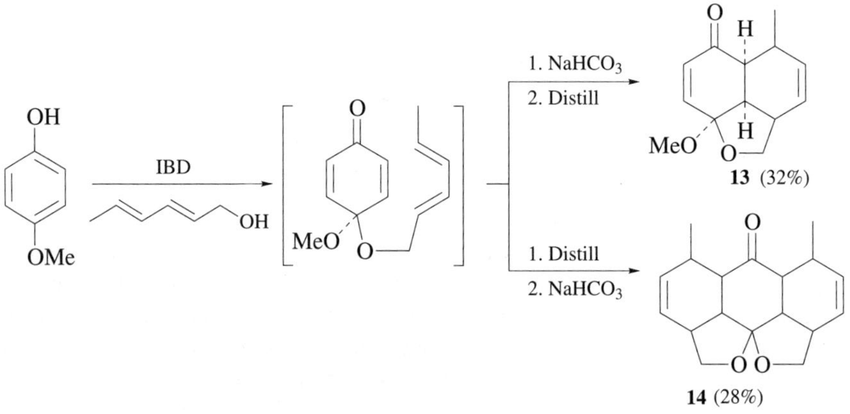

(Eq. 18a)

15, generated in situ by the oxidation of methyl vanillate with IBD in dichloromethane containing allyl alcohol, undergoes an intramolecular Diels-Alder reaction to yield **16** (Eq. 19).[64,64a]

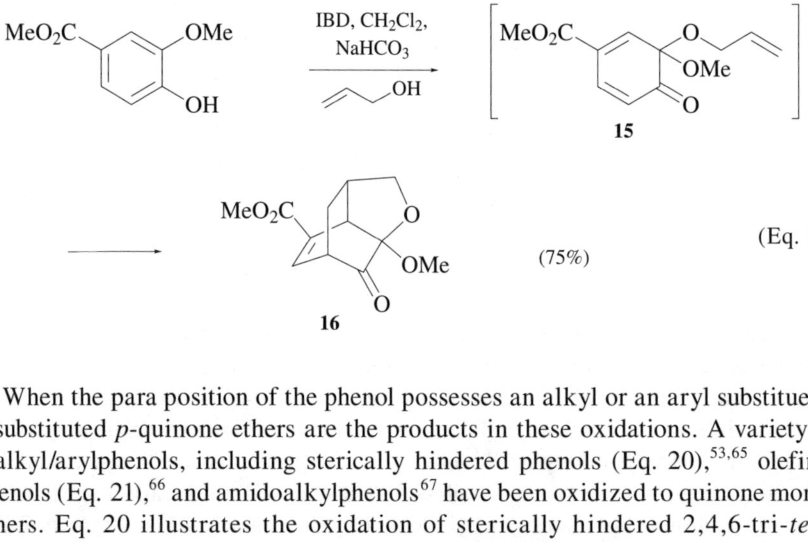

(Eq. 19)

When the para position of the phenol possesses an alkyl or an aryl substituent, *p*-substituted *p*-quinone ethers are the products in these oxidations. A variety of *p*-alkyl/arylphenols, including sterically hindered phenols (Eq. 20),[53,65] olefinic phenols (Eq. 21),[66] and amidoalkylphenols[67] have been oxidized to quinone monoethers. Eq. 20 illustrates the oxidation of sterically hindered 2,4,6-tri-*tert*-butylphenol in the presence of a phenolic nucleophile **17**, and forms the basis of a useful method for the preparation of thyronine derivatives.[68,69]

Oxidation of α-naphthols with organoiodine(III) reagents affords *p*-naphthoquinone monoacetals, whereas oxidation of β-naphthols leads to stable *o*-naphthoquinone monoacetals (Eq. 22).[70,71] Both *o*- and *p*-quinone acetals, accessible through I(III) mediated oxidation of phenols, find application in a wide variety of syntheses of hydroxyanthraquinones, benz[*a*]anthraquinones, and related natural products such as anthracyclines.[71–72a]

[Eq. 20], [Eq. 21], [Eq. 22] schemes shown.

The *ipso* fluorination of *p*-alkylphenols occurs by using IBTA in the presence of pyridinium polyhydrogen fluoride as a source of fluoride anion. The reaction is useful for the synthesis of polycyclic 4-fluorocyclohexa-2,5-dienones (Eq. 23). The presence of alcohols and ketones, which are potentially oxidizable with IBTA, does not interfere in this reaction.[73]

[Eq. 23] scheme shown.

Intramolecular Participation in the Oxidation of Phenols

Formation of Heterocyclic Spiroquinones. Intramolecular participation leads to spiro compounds when appropriate nucleophilic substituents are present at the ortho or para position of the phenolic compound (Eq. 8). The oxidation of benzylidene-1,1′-bis(2-naphthols) **18** with IBD in benzene or pyridine results in formation of the less hindered phenylnaphtho[2,1-*b*]furan-2(1*H*)-spiro-1′(2′*H*)-naphthalen-2′-one **20**. This is an example of stereoselective spirocyclization of a phenol where the phenolic hydroxy group acts as the intramolecular nucleophile

(Eq. 24).[74,75] Although the stereoselectivity of these reactions is attributed to the formation of cyclic intermediate **19**, it is not certain that such an intermediate actually exists, since other bis(naphthols) and related compounds with phenolic hydroxy groups in unfavorable positions undergo similar oxidative spirocyclization.

Intramolecular spirocyclization involving the carboxy group as an internal nucleophile occurs when 3-(4-hydroxyphenyl)propanoic acids are oxidized with IBTA or other I(III) reagents (Eq. 25).[76] In most cases non-nucleophilic solvents are required to prevent solvent participation.

Organoiodine(III)-induced cyclization is applicable to *N*-acyltyrosine derivatives (Eq. 26).[76-78] This approach has been used in the total synthesis of aranorosin,[79,80] a natural product that is isolated from the fungal strain *Pseudoarachniotus roseus* and is associated with antibiotic, antifungal, and antitumor properties.[81-83]

Oxygen atom participation in the hypervalent iodine oxidation of phenols can also occur with alcoholic groups (Eq. 27[84,84a]) and oximino groups (Eqs. 28 and 29). While oxidative cyclization of 4-hydroxyphenyl ketoximes **21** smoothly gives the corresponding spirocyclic isoxazolines **22** (Eq. 28), 2-hydroxyphenyl ketoxime **23** produces a Diels-Alder dimer **25** of the expected spirocyclic isoxazoline **24** (Eq. 29).[85] When a chiral *p*-oximino ester is used, asymmetric induction at the para position to the carboalkoxy carbon occurs.[86,87]

Oxidation of *N*-acyltyramines **26** with IBTA proceeds with intramolecular participation of the amido group oxygen (**26** → **27**) in solvents of low nucleophilicity. In nucleophilic solvents such as methanol and acetic acid, addition of the solvent at the para position occurs (**26** → **28**) (Eq. 30).[67]

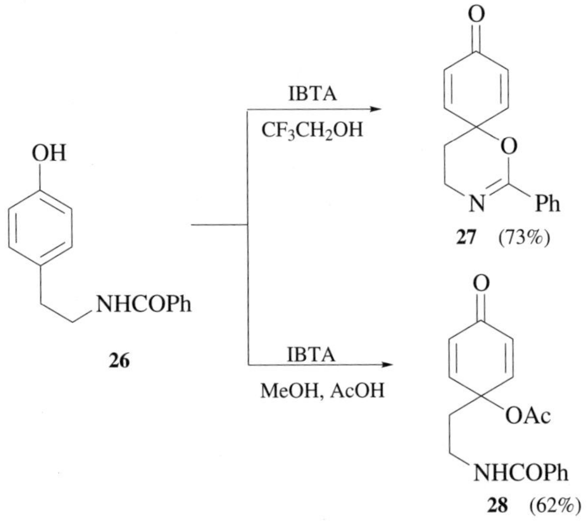

(Eq. 30)

A more complicated example of intramolecular participation in the oxidation of phenols is presented in Eq. 31. In this reaction, oxidation of *N*-methyl-*N*-benzoyltyramine (**29**) with IBTA in the presence of water results in the formation of azabicyclic compound **30**.

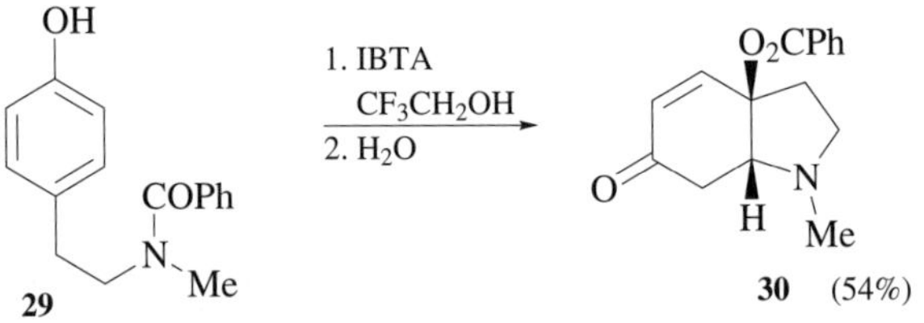

(Eq. 31)

The transformation **29** → **30** is accounted for by intramolecular cyclization to yield the intermediate imidate salt **31**, which is neutralized by water (**31** → **32**) during workup. Ring opening (**32** → **33**) followed by intramolecular Michael addition (**33** → **30**) completes the sequence (Eq. 32).

This strategy of oxidative cyclization followed by intramolecular Michael addition has been used to prepare the core hydrindole ring system **35** of the *Stemona* alkaloids. Oxidation of *N*-protected tyrosine **34** with IBD/MeOH in the presence of sodium bicarbonate can proceed under three different conditions as shown in

Eq. 33. These results illustrate an exceptional diastereotopic group-selective intramolecular conjugate addition.[77]

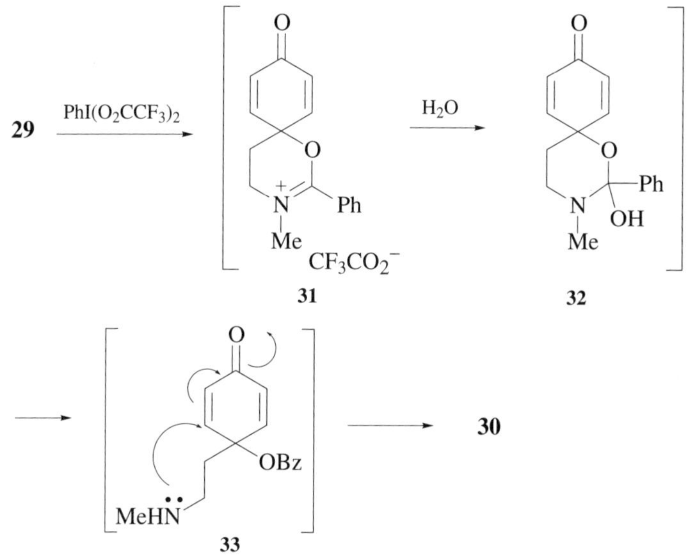

(Eq. 32)

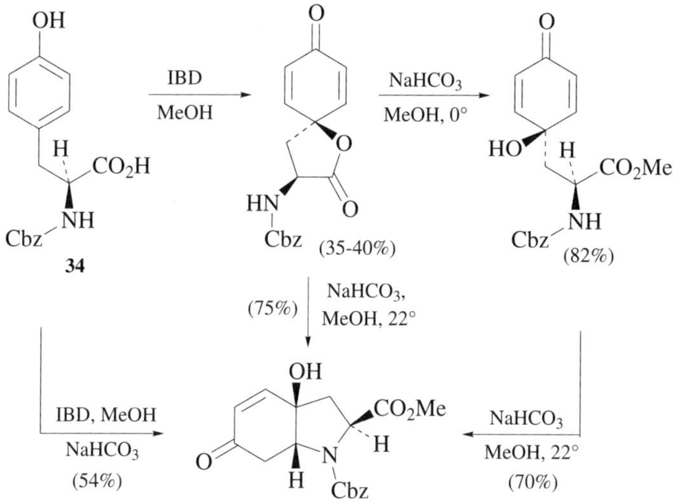

(Eq. 33)

In a related study, it has been observed that oxidation of N-benzyl-N-benzyloxycarbonyltyrosine (**36**) with 3 equivalents of IBD in acetonitrile followed by quenching with aqueous sodium halide (chloride or bromide) solution gives the corresponding dihalodienone lactone **37** (Eq. 34).[88] With 3-(4-hydroxyphenyl)propanoic acid, oxidation proceeds in the similar manner to afford dihalodienone lactones,

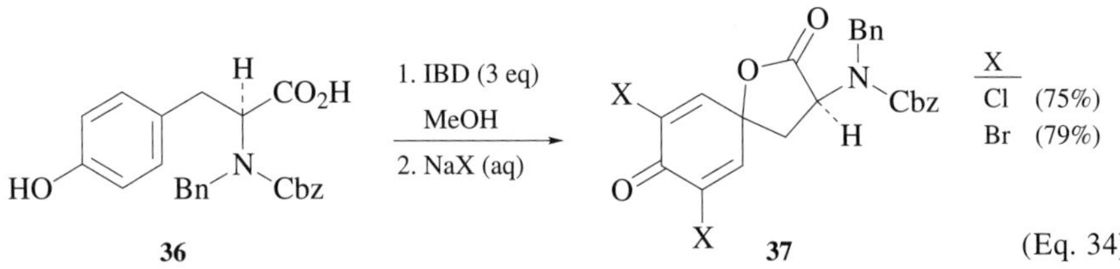

[Structure of **36**: 4-hydroxyphenyl CH₂CH(CO₂H)N(Bn)(Cbz)]
1. IBD (3 eq), MeOH
2. NaX (aq)
→ [Structure **37**: dihalodienone spirolactone with N(Bn)(Cbz)]

X	
Cl	(75%)
Br	(79%)

(Eq. 34)

In contrast, oxidation of oxazoline analogs of phenolic compounds with IBD in trifluoroethanol leads to spirocyclic amides (Eq. 34a).[88a] The preferential formation of spirolactones (Eqs. 33, 34) is likely due to an electronic effect.

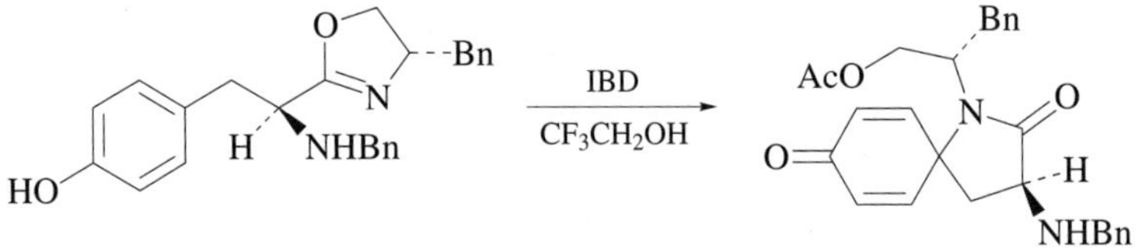

[Oxazoline substrate] —IBD, CF₃CH₂OH→ [spirocyclic amide product] (47%)

(Eq. 34a)

Formation of Oxygen Heterocyclic Compounds

Coumaran-3-ones. Hypervalent iodine oxidation of *o*-hydroxyacetophenones and related compounds under suitable conditions offers convenient syntheses of coumaran-3-ones (Eqs. 35–37).[89–92] Although these oxidative cyclizations

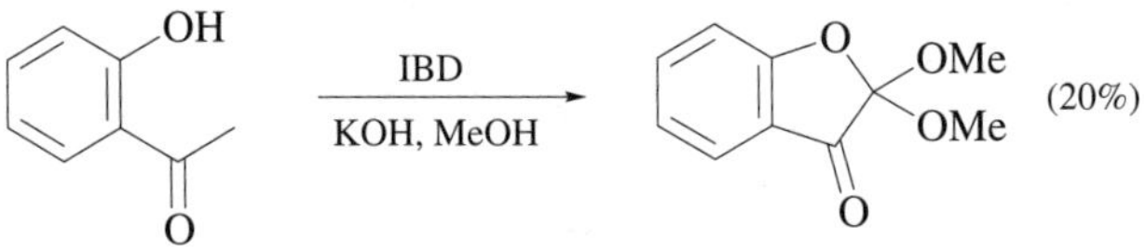

[*o*-hydroxyacetophenone] —IBD, KOH, MeOH→ [2,2-dimethoxy coumaran-3-one] (20%)

(Eq. 35)

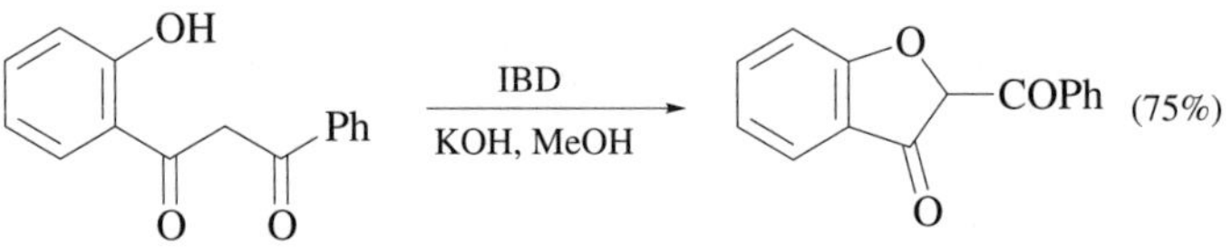

[*o*-hydroxy dibenzoylmethane] —IBD, KOH, MeOH→ [2-benzoyl coumaran-3-one] (75%)

(Eq. 36)

[bis-OTMS enol ether substrate] —(PhIO)ₙ, BF₃·Et₂O, H₂O→ [coumaran-3-one] (31%)

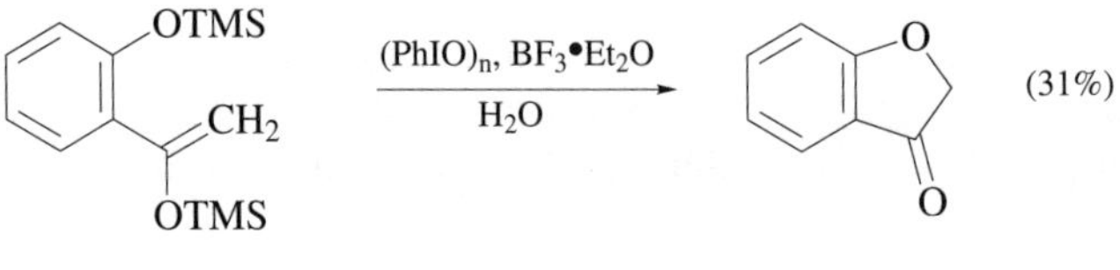

(Eq. 37)

have previously been proposed to occur via intermediates of type **38**, which contain a C-iodine(III) bond (Eq. 38, route **a**), an alternative route involving oxidation of the phenolic group to form intermediates **39** and **40** also explains these results (Eq. 38, route **b**).

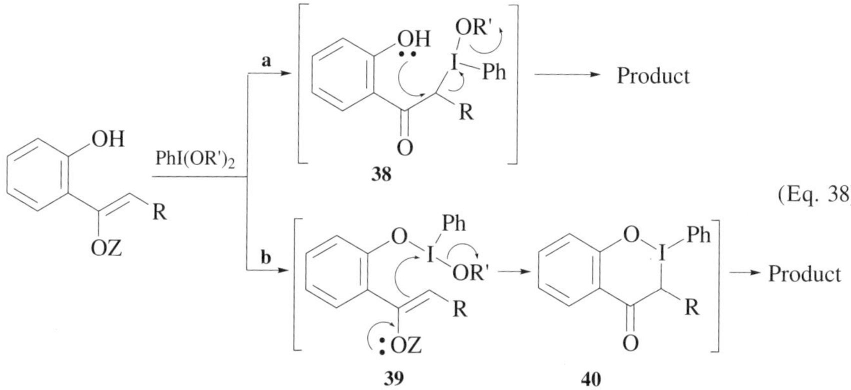

(Eq. 38)

3,4-Dihydrocoumarins.
Oxidative cyclization of N-acyl-3-(3,4-dihydroxyphenyl)-L-alanines **41** provides a one-step synthesis of optically active 3,4-dihydrocoumarins **42** (Eq. 39).[93] The mechanistic pathway for this cyclization is similar to the one given in Eq. 8. In this case, intramolecular bicyclization occurs involving nucleophilic attack by the carboxy group (**43** → **44**) (Eq. 40).

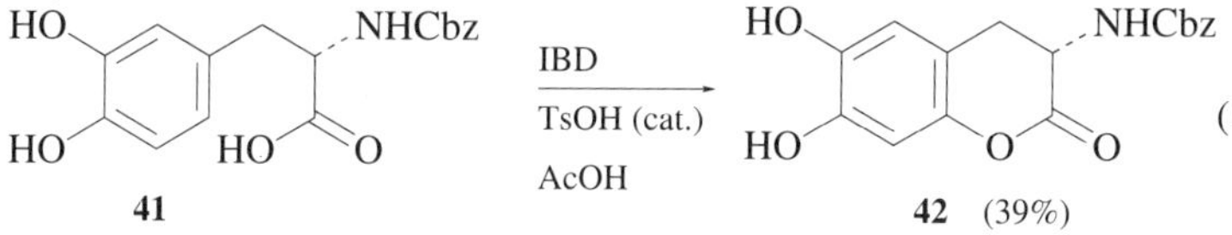

(Eq. 39)

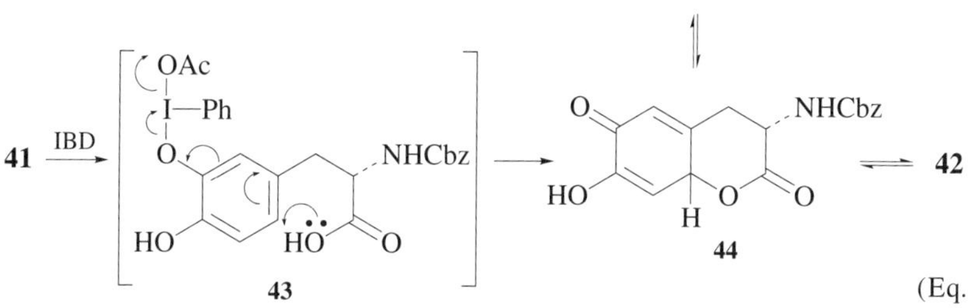

(Eq. 40)

Intramolecular Carbon-Carbon Bond Formation

Phenolic Oxidative Coupling and Its Synthetic Applications. A fascinating feature of hypervalent iodine reagents is their ability to effect oxidative carbon-carbon coupling of phenols. Such coupling is a key biosynthetic step to many natural products including several antibiotics and alkaloids.[5] A general mechanistic scheme for the coupling reaction is outlined in Eq. 41. This coupling process depends upon prior formation of an O-I(III) intermediate **45**, which can undergo intramolecular ortho-para or para-para carbon-carbon coupling (**45** → **46**) upon reaction with a second equivalent of the phenol.

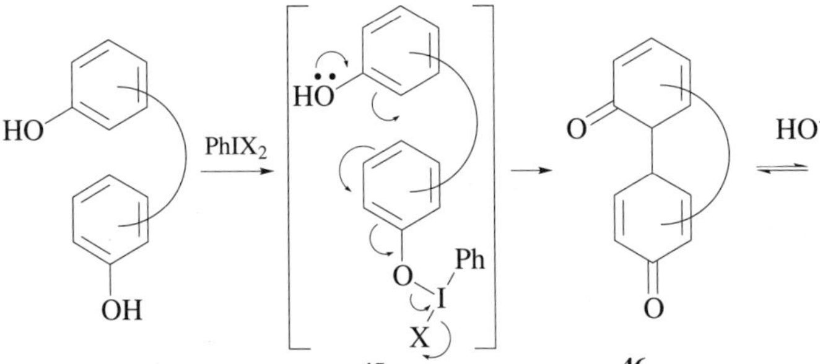

45 **46** (Eq. 41)

Phenolic oxidative coupling finds extensive use in biomimetic syntheses of a number of morphine alkaloids and analogs.[94–95] A key step in the biosynthetic pathway for morphine alkaloids is formation of salutaridine by intramolecular oxidative cyclization of reticuline.[95] Various organohypervalent iodine reagents that have been used in the conversion of reticuline derivatives **47** to salutaridine derivatives **48** include IBD, IBTA, bis(trichloroacetoxy)iodobenzene, and different tetraethylammonium (acyloxy iodate) derivatives (Eq. 42).[96–98] This conversion involves ortho-para coupling with the creation of one new asymmetric center. A further noteworthy example of this approach is the asymmetric synthesis of (9*R*)-salutaridine derivative (+)-**50** from the oxidation of (−)-**49** (Eq. 43).[99]

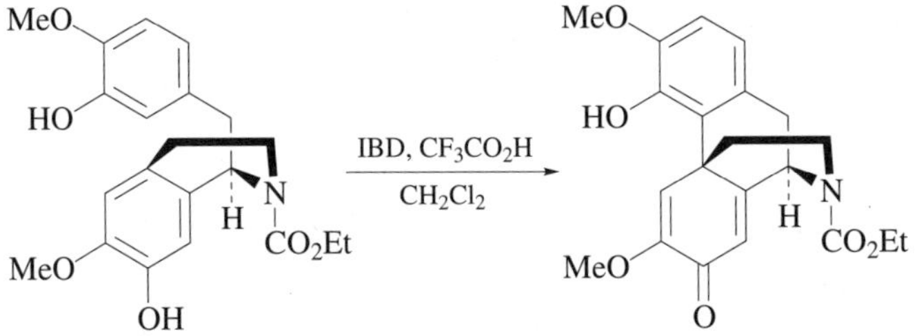

47 **48** (22%) (Eq. 42)

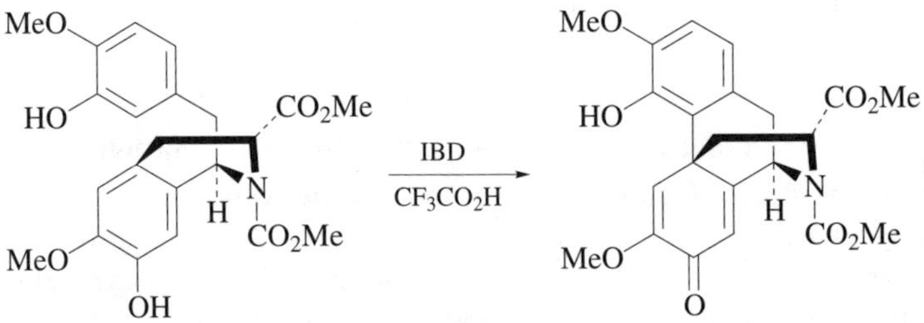

(−)-**49** 54% ee (+)-**50** (25%) 58% ee (Eq. 43)

Oxidation of conformationally rigid molecules such as bisphenol **51** with IBD gives a mixture of C-C coupling products related to aporphine (isoboldine analog **52**, major product) and morphinane alkaloids (**53**, minor product) (Eq. 44).[100]

(Eq. 44)

These results show that both coupling sites on the isoquinoline moiety are accessible to a pseudoaxial benzyl group. The larger percentage of isoboldine-type product **52**, which results from para-para coupling at the apparent expense of the ortho-para coupled product **53**, is likely the result of a steric factor. The benzyl group in an axial position may hinder the formation of ortho-para coupled product because of steric interaction with the C(5) hydrogens of the isoquinoline moiety.

The phenolic oxidative coupling using IBTA as oxidant has been extended to the synthesis of 6a-epipretazettine (**58**) (Eq. 45).[101] The acetal **54** is converted to oxidative coupling product **55** with IBTA (2 equiv) in the presence of propylene oxide (10 equiv) in 13% yield. This step is followed by intramolecular Michael addition (**56** → **57**).

To control the selectivity of such oxidative coupling reactions, the starting phenols can be converted to the corresponding OTMS derivatives, which are subsequently oxidized to the desired products. Oxidation of O-silylated phenols **59a** and **59b** with IBTA in 2,2,2-trifluoroethanol (TFE) leads to the formation of azacarbocyclic spirodienones **60** and **61** in good yields.[102,103] Oxidation of the corresponding methyl ether derivative **59c** under similar conditions affords a mixture of **60** and rearranged products **62** and **63** (Eq. 46).[102] A remarkable use of this approach is the successful synthesis of the anticancer marine alkaloid discorhabdin C.[103,104]

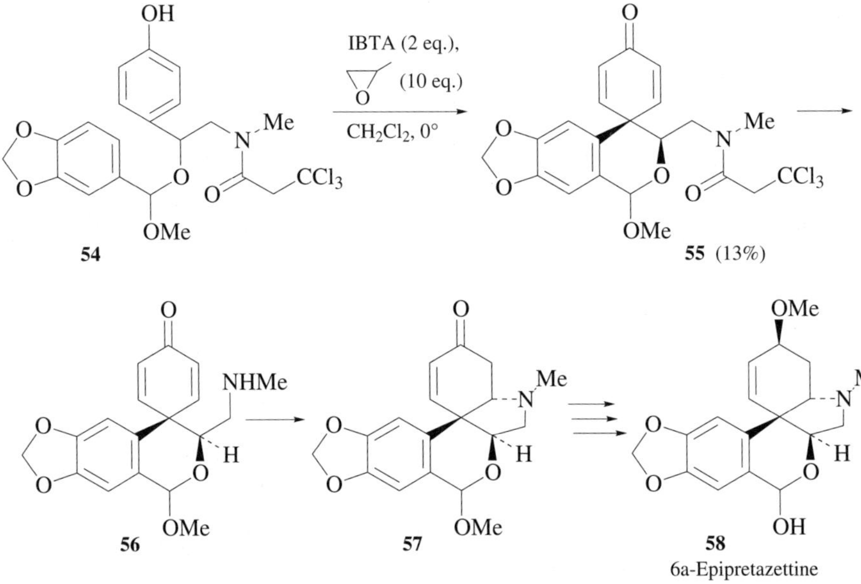

(Eq. 45)

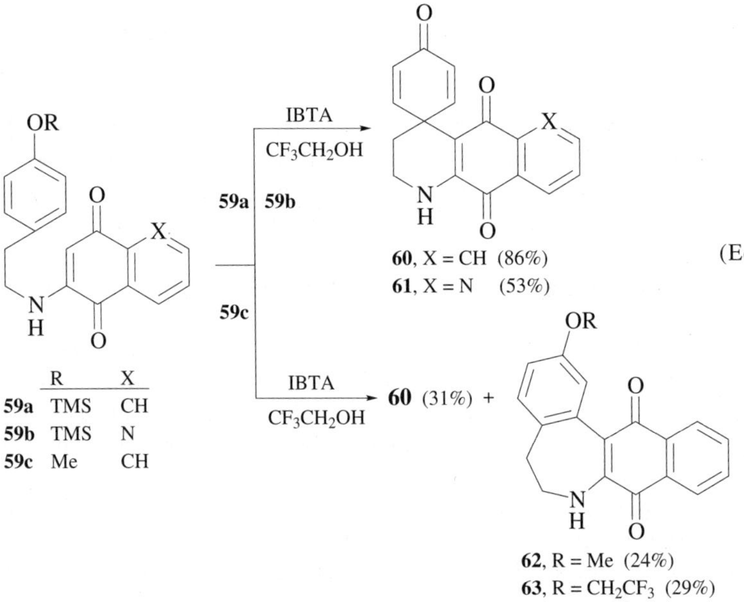

(Eq. 46)

The conversion **59a → 60** presumably proceeds by initial formation of an intermediate (**64 → 65**). The cleavage of the OTMS group by trifluoroacetate gives the product (**65 → 60**) (Eq. 47).

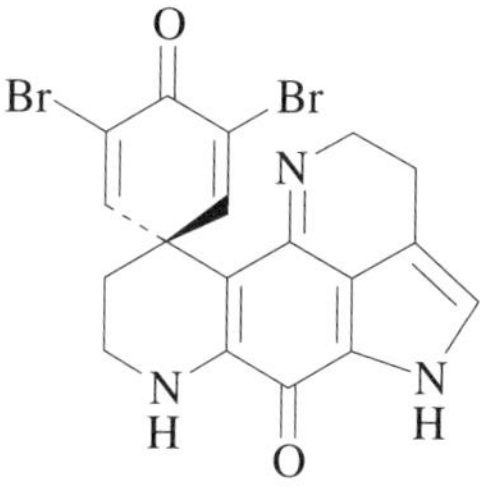

Discorhabdin C

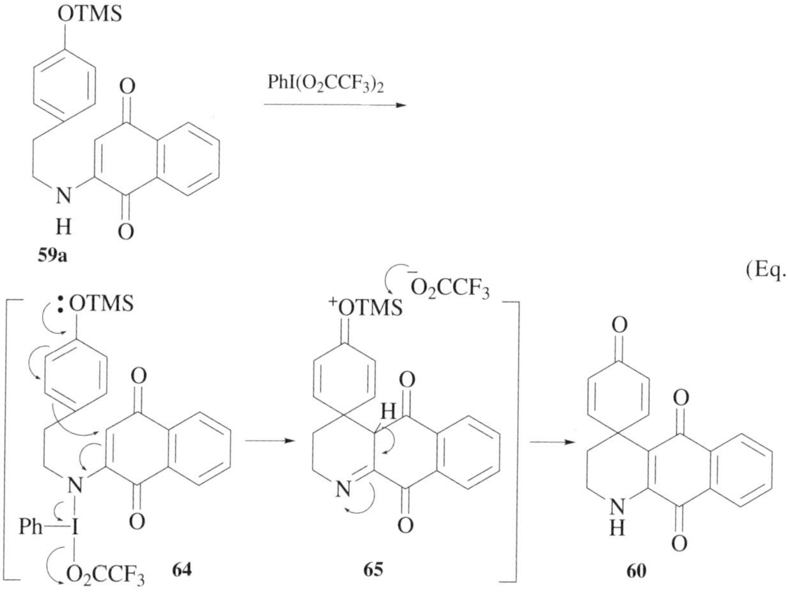

(Eq. 47)

On oxidation with IBTA in TFE, phenolic dibenzylbutyrolactone **66** gives as the major product either the dibenzocyclooctadiene **67** (R = H) (Eq. 48, route **a**) or the spirodienone **68** (route **b**), depending upon the reaction time. When the reaction is left for 24 hours, the major product (48%) is **67** (R = H). When the reaction time is only 1 hour, the major product (47%) is **68** together with a minor amount (3%) of the p-quinol monoether **69c**. The use of methanol instead of TFE in this reaction affords a mixture of products containing p-quinol **69a** and monoether **69b** in addition to **67** (R = Me) and **68** (route **c**).[105,106] Route **a** of the oxidative cyclization has been employed to effect the asymmetric synthesis of isostegnane derivatives.[107]

Intramolecular carbon-carbon bond formation is further exemplified by Eq. 49 in which the aromatic ring attacks the para position of the phenol ring.[108] The cyclization of styryl type systems (**70** → **71**, Eq. 50) is a logical extension of the process of Eq. 49.[66] However, there is a second mechanistic possibility involving initial attack of IBD at the styryl double bond.

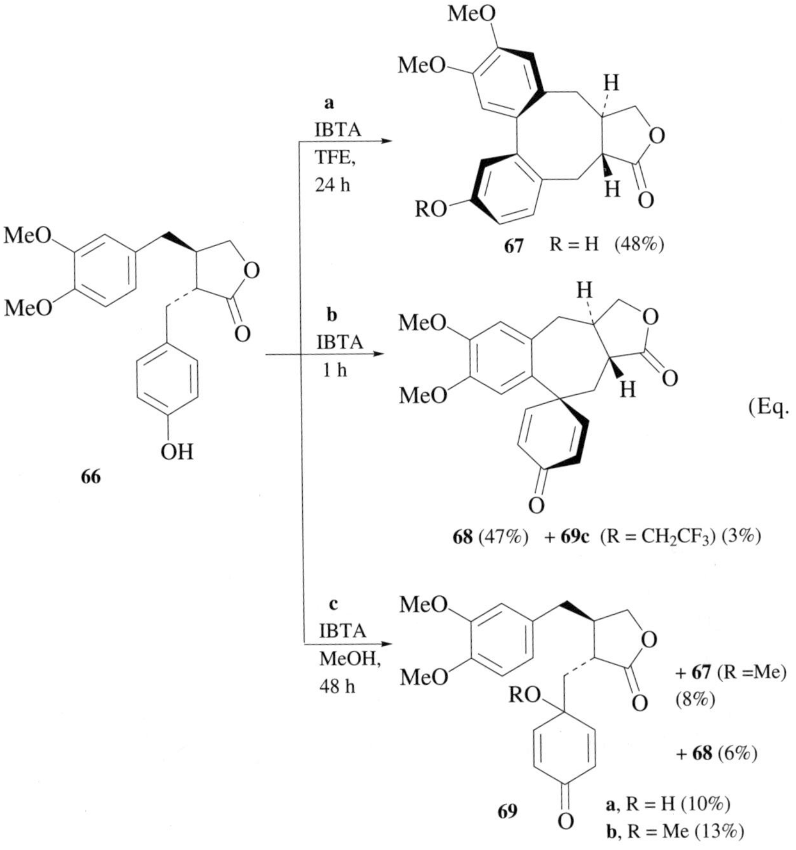

(Eq. 48)

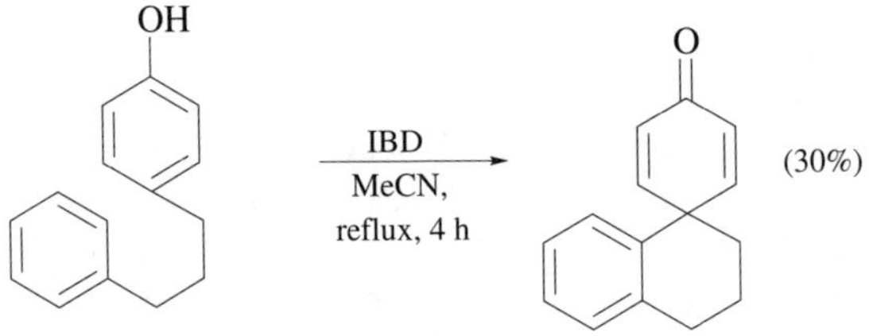

(Eq. 49)

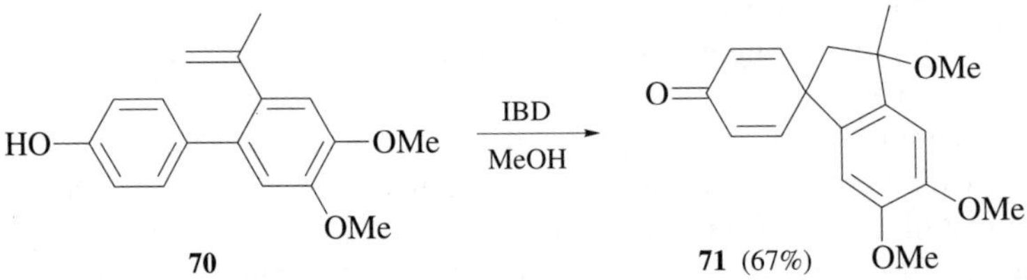

(Eq. 50)

Formation of Iodonium Ylides and Salts and their Synthetic Applications. Phenols containing at least one electron-accepting group in the para position and one free ortho position react with organoiodine(III) reagents to yield stable iodonium ylides **73**.[109–112] The first step in ylide formation is the generation of iodine(III) intermediate **72** (iodonium salt). This subsequently loses trifluoroacetic acid under the influence of heat or base to give the stable ylide **73** (Eq. 51).[109] These iodonium ylides are important intermediates in the synthesis of heterocyclic compounds by cycloaddition reactions, in the formation of o-iodoethers by rearrangements, and in the formation of o-substituted phenols.

(Eq. 51)

A variety of five-membered heterocyclic compounds can be synthesized by cycloaddition reactions of phenolic iodonium ylides with compounds containing double[113,114] or triple bonds.[114,115] The iodonium ylide **73** reacts under photochemical conditions with carbon disulfide, alkene **74**, and alkynes **75** to afford 5,7-dinitro-1,3-benzoxathiole-thione (**76**),[113] 2,3-dihydrobenzo[b]furan **77**,[114] and benzo[b]furan **78**,[114] respectively (Eq. 52).

(Eq. 52)

Another characteristic feature of iodonium ylides is their reaction with electrophiles to afford products bearing substituents in place of the phenyliodonium functionality. The electrophile first gives iodonium salts, which normally undergo subsequent nucleophilic substitution with the counteranion. For example, reaction of the ylide of 2-hydroxy-1,4-naphthoquinone (**79**) with benzoyl chloride produces 2-benzoyloxy-3-chloro-1,4-naphthoquinone (**81**) (Eq. 53).[115] The intermediate iodonium salt **80** could not be isolated. However, when the counteranion is not nucleophilic, it is possible to isolate the intermediate iodonium salts (e.g. **82**). In the presence of nucleophiles such as phenol and aniline, these salts can be converted to various functionalized phenolic compounds (Eq. 54).[116]

The iodonium ylides of phenols react with nucleophiles in two different ways. The first path outlined in Eq. 55 illustrates attack of the nucleophile on iodine to give an adduct.[117] This iodine adduct formation normally occurs under photochemical conditions.

The second path involves atttack of a nucleophile at carbon, thereby giving the S_NAr substitution product with the expulsion of iodobenzene (Eq. 56).[118] Non-photochemical reactions in the presence of strong nucleophiles favor a SNAr substitution route.

$$73 \xrightarrow[\text{MeOH}]{\text{MeONa}} \text{[2,4-dinitro-6-methoxyphenol]} \quad (98\%) \quad \text{(Eq. 56)}$$

Phenolic iodonium ylides undergo thermal 1,4-formal rearrangement to o-iodoaryl ethers.[110–122] Examples of such rearrangements are shown in Eq. 57[112,119] and Eq. 58.[120,121] Equation 57 illustrates the rearrangement of phenolic ylide **83** containing an enolizable ketone, which remains unaffected in IBD-KOH/MeOH. The conversions **84 → 86** and **86 → 87** represent the formation of the stabilized monocarbonyl iodonium ylide **86** and its rearrangement to iodoether **87** (Eq. 58). Intermediate iodonium salt **85** is also stable.

(Eq. 57)

(Eq. 58)

Miscellaneous

***p*-Thiocyanation of Phenols.** A variety of phenols and α-naphthols undergo *p*-selective thiocyanation in the presence of (dichloroiodo)benzene and lead dithiocyanate in dry dichloromethane (Eq 59). The ligand exchange reaction of $PhICl_2$ and $Pb(SCN)_2$ generates in situ [(bisthiocyanato)iodo]benzene, which effects the oxidation of the phenolic group analogously to other iodine(III) reagents

$$\text{(Eq. 59)}$$

such as IBTA or IBD. The reaction is completed by the nucleophilic attack of thiocyanate anion at the para position.[123,123a]

Oxidation of Flavonols. Flavonols **88** behave like phenols upon oxidation with hypervalent iodine reagents. Thus, oxidation of **88** with HTIB or IBD in methanol proceeds with the introduction of two methoxy groups into the carbon-carbon double bond, thereby giving 2,3-dimethoxy-3-hydroxyflavanones **91** (Eq. 60).[124,125] The use of periodic acid as an oxidant gives similar results.[126] Iodine(III) intermediate **89** of this reaction gives α-diketone **90** by Michael-type addition of methanol. The latter affords product **91** by nucleophilic addition of methanol at the C(3) position.

$$\text{(Eq. 60)}$$

Cleavage of Amino-Terminal Tyrosyl-Peptide Bonds. Tyrosine (**92**) (Eq. 61) and tyrosyl dipeptides **97** (Eq. 62) undergo cleavage of their carbon-NH$_2$ bonds when oxidized with IBD in methanol/potassium hydroxide.[127] These reactions are explained on the basis of phenolic group oxidations discussed earlier. The first step of the reaction gives O-I(III) intermediate **93**, which then yields the spirolactone **94** by intramolecular participation of carboxylate anion. Decarboxylation followed by oxidation of the resulting imine **95** by the second molecule of IBD finally affords the cleavage product **96** (Eq. 61).

Cleavage of the amino-terminal tyrosyl-peptide bond in **97** also occurs via initial ligand exchange involving the phenolic hydroxy group and PhI(OMe)$_2$ to give intermediate **98**. This intermediate undergoes reductive cleavage of iodobenzene via fragmentation initiated by the amino group, providing **99**. Finally, nucleophilic attack of MeOH gives product **100** (Eq. 62). Cleavage of tyrosyl proteins using *o*-iodosobenzoic acid has also been reported.[128,129]

Oxidation of 3′-Desmethylpapaverine. 3′-Desmethylpapaverine (**101**) on oxidation with two equivalents of IBD in methanol at room temperature gives a novel product **104** in 16% yield.[130] A possible pathway for this reaction involves the formation of *o*-quinone acetal **102** by oxidation of the phenolic group. This intermediate is converted to **103** which finally gives the product **104** by nucleophilic attack of methanol (Eq. 63).

Oxidative Phenol-Propenylbenzene Cycloadditions Leading to *trans*-2,3-Dihydrobenzofurans. Oxidation of various 4-methoxyphenols and naphthols with IBTA in the presence of electron-rich styrene derivatives such as 1,2-dimethoxy-4-propenylbenzene leads to formal cycloaddition of the alkene across the oxygen and C(2) of the phenol, thereby affording a *trans*-dihydrobenzofuran structure related to the neolignan family of natural products (Eq. 64).[131] This reaction permits a facile preparation of **105**, which can be converted to (±)-kadsurenone and (±)-denudatin B. Similar cycloaddition reactions have also

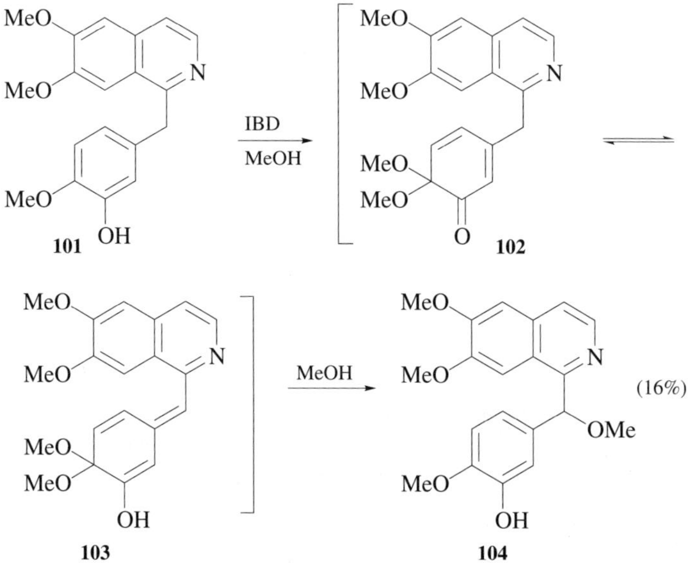

(Eq. 63)

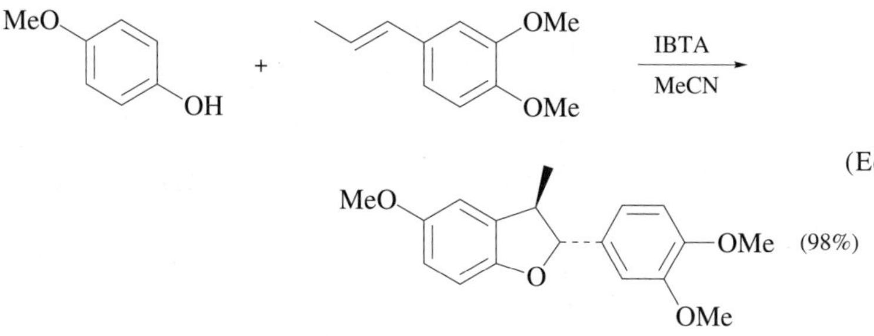

(Eq. 64)

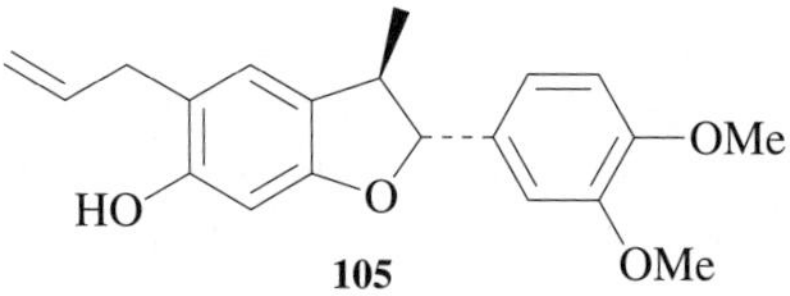

been effected by using electrochemical methods. However, for the reaction of 2-naphthol and propenylbenzene, anodic oxidation gives a much lower yield of the product compared to IBTA oxidation.

Synthesis of Dynemicin A. Iodine(III)-induced oxidative conversion of phenols to quinone imines has been employed in the total synthesis of dynemicin A, a metabolite isolated from *Micromonospora cherisina*, which has DNA cleaving capability and demonstrated in vitro antitumor properties. Phenolic intermediate **107**, generated in situ by removal of some of the protecting groups from **106**, undergoes oxidation to quinone imine **108** with IBD (Eq. 65).[132] Compound **108** is the key intermediate for subsequent elaboration to dynemicin A.

(Eq. 65)

COMPARISON WITH OTHER METHODS

There are several alternative methods for effecting most of the major processes reviewed in this chapter with the exception of ylide formation, which is a unique property of organohypervalent iodine reagents. In most cases, organohypervalent iodine based methodology is more effective owing to its ease, simplicity, selectivity, and efficiency. However, hypervalent iodine-mediated oxidations do have some limitations, such as low yields of products in a few cases and relatively high cost of the organohypervalent iodine reagents (although the iodobenzene generated as byproduct of the reaction can be recycled). Some examples of direct comparison of iodine(III)-based reactions with other common methods are presented according to the type of reaction accomplished.

Oxidation of Phenols to Quinones and Related Compounds. Quinones, quinone monoacetals, and related compounds are generally prepared by (1) chemical oxidation of phenols or 4-alkoxyphenols with various metal oxidants,[133,134] Fremy's salt,[13] silver oxide,[135–138] or 2,3-dichloro-5,6-dicyano-1,4-benzoquinone (DDQ);[139] (2) electrochemical oxidation of 4-methoxyphenols[140–142] or their trimethylsilyl ethers;[143] or (3) monohydrolysis of bis(acetals).[144] Although the first group of reagents offers the most facile and shortest route to quinones and their derivatives, there are some limitations, especially with phenols that have acid-sensitive groups. For example, oxidation of 4-methoxy-2-(tetrahydropyranyloxy)methylphenol (**109**) with various oxidizing agents is problematic. While oxidation of **109** with DDQ in methanol gives a complex mixture, use of ammonium ceric(IV) nitrate[145] in the presence of water or potassium carbonate, respectively, results in the formation of the *p*-quinone alcohol **110** or the desired *p*-quinone **111**, but in unsatisfactory yield (69%). With Fremy's salt in potassium dihydrogen phosphate solution, oxidation of **109** does not give *p*-quinone **111** but rather *o*-quinone **112**. Since oxidation of **109** to **111** has been most effectively accomplished by using IBTA in acetonitrile/water in the presence of potassium carbonate, the I(III) method is obviously advantageous. Of course, the use of other oxidants may serve as a complementary approach for a specific purpose (e.g. **109** to **112**) (Eq. 66).

(Eq. 66)

Another example of phenolic oxidation where an I(III) reagent is of special interest is the formation of unsymmetrical monoacetals **113** in the reaction of *p*-alkoxyphenols with IBTA in methanol. This result is quite different from that observed in thallium trinitrate oxidation,[17,146] which gives the symmetrical dimethylacetal **114** (Eq. 67).

(Eq. 67)

Intramolecular Participation in the Oxidation of Phenols. Of the various examples of the oxidative cyclization involving intramolecular participation of oxygen or nitrogen, one that deserves special comment is the conversion of bisnaphthol **115** to spiran **116**. In this reaction, no oxidizing agent except IBD is successful (Eq. 68).[74]

(Eq. 68)

Intramolecular Carbon-Carbon Bond Formation: Phenolic Oxidative Coupling. Intramolecular C-C bond formation in phenolic oxidation using I(III) reagents provides a nonmetallic and safer alternative to existing methods, although yields in most reactions are low. The study of the oxidation of conformationally rigid 1-benzyltetrahydroisoquinolines **51** (Eq. 44) with various oxidizing agents shows that IBD and vanadium oxychloride may be the reagents of choice for preparing aporphine-type products **52**. Oxidation of **51** with thallium(III) trifluoroacetate gives **52** in very low yield (15%). The morphinane-type product **53** is available only from IBD oxidation, although the yield is very poor (4%) (Eq. 69).[100]

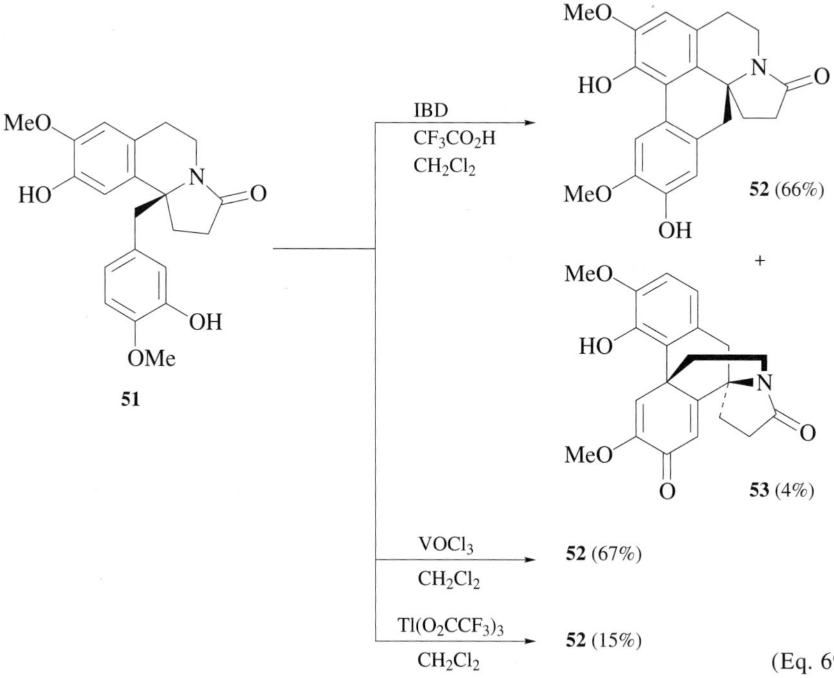

(Eq. 69)

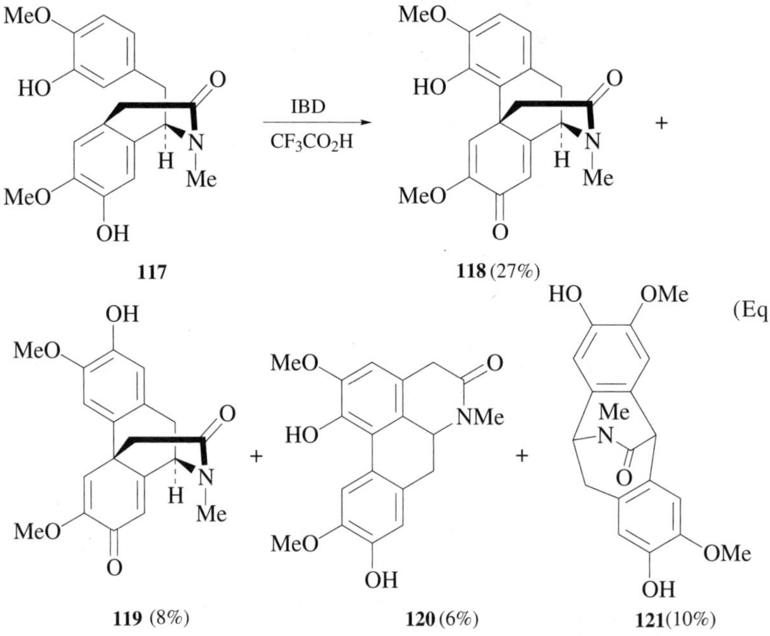

(Eq. 70)

Oxidation of 3-oxoreticuline (**117**) with IBD in the presence of TFA gives 16-oxosalutaridine (**118**) in 27% yield, together with 6-oxopallidine (**119**) (8%), 5-

oxoisoboldine (**120**) (6%), and 13-oxothalidine (**121**) (10%) (Eq. 70). When oxidative coupling of **117** is carried out with vanadium oxychloride as oxidant, products **119**, **120**, and **121** are obtained in yields of 26%, 11%, and 9%, respectively, but none of the salutaridine derivative **118** is formed. When the same reaction is carried out with a slight excess of DDQ in dichloromethane, the only product obtained is **121** (85%) (Eq. 71).[98] Thus, it is possible to choose oxidizing agents that give a more selective outcome.

$$\mathbf{117} \begin{cases} \xrightarrow{\text{VOCl}_3,\ \text{CH}_2\text{Cl}_2} \mathbf{119}\ (26\%) + \mathbf{120}\ (11\%) + \mathbf{121}\ (9\%) \\ \xrightarrow{\text{DDQ},\ \text{CH}_2\text{Cl}_2} \mathbf{121}\ (85\%) \end{cases} \quad (\text{Eq. 71})$$

EXPERIMENTAL PROCEDURES

Practical Aspects and Availability of Organohypervalent Iodine Reagents. All organohypervalent iodine reagents are solids that are fairly stable at room temperature and generally insensitive to atmospheric oxygen and moisture. Most reagents have relatively low toxicity and can be handled easily. IBD,[147–150] IBTA,[151] and HTIB[152,153] are stable and commercially available, or can be prepared by standard procedures. Iodosobenzene[149] can be prepared by hydrolysis of either (dichloroiodo)benzene[149] or IBD and should be stored in a refrigerator in dark containers. The preparation of iodine(V) reagent iodylbenzene can best be effected by direct oxidation of iodobenzene with aqueous hypochlorite and phase-transfer reagents.[154]

$$\text{1,5-dihydroxynaphthalene} \xrightarrow[\text{MeCN, H}_2\text{O}]{\text{IBTA (2.2 eq)}} \text{Juglone}\ (58\%)$$

5-Hydroxy-1,4-naphthoquinone (Juglone) [Oxidation of a Phenol to a *p*-Quinone].[50] A solution of 1,5-dihydroxynaphthalene (160 mg, 1 mmol) in acetonitrile-water (2/1, v/v) (12 mL) was added dropwise to [bis(trifluoroacetoxy)iodo]benzene (IBTA) (946 mg, 2.2 mmol). The resulting solution was stirred at 0° under nitrogen for one hour. The solvent was removed in vacuo, and the resulting residue was extracted with dichloromethane (3 × 15 mL). The combined organic extracts were dried (MgSO$_4$) and evaporated in vacuo to yield the crude product, which was purified by column chromatography to give 101 mg (58%) of pure juglone, mp 154–155°; IR 1645, 1670, 3300–3600 cm^{-1}; ^{1}H NMR (CDCl$_3$, 60 MHz) δ 6.8 (s, 2 H), 7.2 (m, 1 H), 7.5 (s, 2 H).

4-Hydroxy-4-methylcyclohexa-2,5-dienone [Oxidation of a para-Substituted Phenol to a *p*-Quinol].[62] To a stirred solution of *p*-cresol (108 mg, 1 mmol) in acetonitrile-water (3/1, v/v) (4 mL) was added IBTA (473 mg, 1.1 mmol) at 0° and stirring was continued for about 15 minutes. The reaction was quenched by addition of water (4 mL) and the resulting mixture was extracted with dichloromethane (4 × 5 mL). The combined organic extracts were dried (Na_2SO_4) and evaporated under reduced pressure to give a brown residue, which was purified by column chromatography on silica gel (ether-light petroleum) to give 60 mg (48%) of the title product as colorless needles, mp 75–77° (from $CHCl_3$-hexane); ^{1}H NMR ($CDCl_3$, 60 MHz) δ 1.44 (s, 3 H), 3.48 (bs, 1 H), 6.05 (d, J = 10 Hz, 2 H), 6.89 (d, J = 10 Hz, 2 H).

The same procedure was used for effecting the oxidation of tripropylsilyl ethers of para-substituted phenols to *p*-quinols. Oxidations required 0.5–2.0 hours.

4,4-Dimethoxy-3,5-dimethylcyclohexa-2,5-dienone [Oxidation of a Monophenol to a *p*-Quinone Monoacetal].[53] To a stirred solution of 3,5-dimethylphenol (620 mg, 5 mmol) in dry methanol (10 mL) was added dropwise at room temperature a solution of iodobenzene diacetate (IBD) (3.22 g, 10 mmol) in methanol (25 mL) via a double-ended needle. The solution became reddish and was stirred under nitrogen for about 40 minutes. Methanol was removed in vacuo to give a yellow oil, which was purified by column chromatography on silica gel (230–40 mesh) using light petroleum (bp 40–60°)-dichloromethane as eluants to afford 728 mg (80%) of the title product as colorless prisms, mp 59–61°; IR (KBr) 1650, 1685 cm^{-1}; ^{1}H NMR ($CDCl_3$) δ 1.9 (s, 3 H), 3.02 (s, 6 H), 6.28 (s, 2 H); ^{13}C NMR ($CDCl_3$) δ 16.2, 50.8, 98.1, 131.8, 155.0, 184.7; MS m/z 182 (0.5), 167 (100), 135 (33), 127 (41), 123 (53).

1-Oxaspiro[4,5]deca-6,9-dien-8-one [Formation of a Spiroquinone by Intramolecular Participation of a Primary Alcoholic Group].[84] To a stirred solution of p-(3-hydroxypropyl)phenol (152 mg, 1 mmol) and pyridine (0.3 mL) in acetonitrile (10 mL) at 0° was added a solution of IBTA (430 mg, 1 mmol) in acetonitrile (2 mL). The mixture was stirred at room temperature for 10 minutes, diluted with water, and extracted with diethyl ether (3 × 10 mL). The combined organic extracts were washed with saturated aqueous sodium chloride solution, dried (MgSO$_4$), and concentrated in vacuo. The residue was purified by column chromatography on silica gel using hexanes-ethyl acetate to give 89 mg (59%) of the title product as a syrup; IR (CHCl$_3$) 1630, 1670, 1690 cm^{-1}; ^{1}H NMR (CDCl$_3$) δ 2.0–2.4 (m, 4 H), 4.06 (t, J = 6 Hz, 2 H), 6.08 (d, J = 10 Hz, 2 H), 6.76 (d, J = 10 Hz, 2 H).

(1S,8R,9R,10R)-1'-Methyl-8'-phenyl-1'-decalyl 7,9-Dibromo-8-methoxy-6-oxo-1-oxa-2-azaspiro[4.5]deca-2,7,9-triene-3-carboxylate [Asymmetric Induction in the Oxidative Cyclization of an o-Phenolic Oxime Ester].[86,87] (−)-Camphorsulfonic acid (1.28 g, 5.5 mmol) was added to a suspension of iodosobenzene (1.22 g, 5.5 mmol) in dichloromethane (60 mL) at room temperature and the mixture was stirred for 2 hours. The resulting clear solution was cooled to −78° and a solution of o-phenolic oxime-ester **122** (3.06 g, 5 mmol) in dichloromethane (60 mL) was added. The mixture was allowed to warm to 0°. After addition of water, the mixture was stirred for 30 minutes at room temperature and extracted with diethyl ether (3 × 100 mL). The combined extracts were dried (MgSO$_4$) and concentrated in vacuo. The residue was purified by column chromatography on silica gel using benzene as eluant to afford 2.53 g (83%) of **123** [70–80% de; the diastereomeric excess was estimated on the basis of the

500-MHz ^{1}H NMR spectrum (CDCl$_3$), in which a peak due to a vinyl proton (10-H) of each diastereomer appears as a singlet at δ 6.71 and 6.67 (91:1)]; mp 175–177°; IR (CHCl$_3$) 1600, 1710, 2590, 3370 cm^{-1}; ^{1}H NMR (CDCl$_3$) δ 1.46 (s, 3 H), 1.21–1.80 (m, 12 H), 2.10–2.94 (m, 3 H), 2.17 (major product) (d, J = 17.6 Hz, 1 H), 3.00 (major product) (d, J = 17.6 Hz, 1 H), 2.72 (minor product) (d, J = 17.7 Hz, 1 H), 2.86 (minor product) (d, J = 17.7 Hz, 1 H), 4.15 (s, 3 H), 6.65 (major product) (s, 1 H), 6.66 (minor product) (s, 1 H), 7.1–7.3 (m, 5 H); ^{13}C NMR (CDCl$_3$) (major product) δ 21.2, 22.9, 26.3, 33.8, 34.5, 37.1, 37.7, 40.7, 44.4, 46.5, 49.6, 62.0, 86.7, 90.4, 107.2, 119.1, 125.6, 127.8, 128.4, 137.1, 147.3, 150.9, 157.2, 163.1, 188.9; MS (FAB) m/z 609 (M$^-$ + 4), 607 (M$^-$ + 2), 605 (M$^-$).

1-Benzoyloxy-7-methyl-*cis*-azabicyclo[4.3.0]non-2-en-4-one [Formation of a Hexahydroindol-6-one by Oxidation of an *N*-Alkyl-*N*-benzoyltyramine].[67] To a solution of *N*-methyl-*N*-benzoyltyramine (21.1 mg, 0.08 mmol) in trifluoroethanol (1 mL) was added IBTA (43 mg, 0.1 mmol). The mixture was stirred at room temperature for 30 minutes and then neutralized with solid sodium bicarbonate. The mixture was concentrated in vacuo and the resulting residue was dissolved in ethyl acetate and filtered. The filtrate was evaporated in vacuo and the residue was purified by column chromatography on silica gel to give 12.1 mg (54%) of pure product as a hygroscopic colorless oil; IR 1600, 1685, 1715 cm^{-1}; ^{1}H NMR (CDCl$_3$) δ 2.3–2.4 (m, 1 H), 2.34 (s, 3 H), 2.4–2.55 (m, 2 H), 2.71 (dd, J = 2, 17 Hz, 1 H), 2.95–2.99 (m, 1 H), 3.05 (m, dd, J = 5, 17 Hz, 1 H) 3.1–3.2 (m, 1 H), 6.03 (d, J = 10 Hz, 1 H), 7.06 (dd, J = 2, 10 Hz, 1 H), 7.46 (t, J = 7 Hz, 2 H), 7.59 (t, J = 7 Hz, 1 H), 8.01 (d, J = 8 Hz, 2 H); HRMS 271.1204.

5,6,7,8,9,10-Hexahydropyrido[2,3-*g*]quinoline-5,10-dione-9-spiro-4'-cyclohexa-2',5'-dien-1'-one (61) [Oxidative Cyclization of an *O*-Silylated Phenol to an Azacarbocyclic Spirodienone].[104] To a stirred suspension of 6-[2-(4-hydroxyphenyl)ethylamino]quinoline-5,8-dione (26.5 mg; 0.900 mmol) in dichloromethane (4 mL) was added dropwise *O*-trimethylsilyl ketene acetal (75 mg, 0.469 mmol) at room temperature over 3 hours under nitrogen. The mixture was concentrated in vacuo to give the *O*-silylated phenol of **59b**, which was dissolved in 2,2,2-trifluoroethanol (5 mL). To the resulting solution was added IBTA (46.5 mg, 0.108 mmol). After 15 minutes, the reaction mixture was concentrated in vacuo, and the residue was purified by column chromatography to give 14 mg (53%) of azacarbocyclic spiro dienone **61** as a red crystalline solid, mp 235–237° (from chloroform-hexanes); UV (MeOH) 231 (ε 24,400), 249 (21,100), 464 (2640) nm; IR 1600, 1660, 3400 cm^{-1}; ^{1}H NMR (CDCl$_3$) δ 1.98 (t, J = 6 Hz, 2 H)), 3.60–3.68 (m, 2 H), 6.33 (br s, 1 H), 6.39 (d, J = 10 Hz, 2 H), 6.92 (d, J = 10 Hz, 2 H), 7.55 (m, 1 H), 8.34 (d, J = 8 Hz, 1 H), 8.97 (d, J = 4 Hz, 1 H).

5,7-Dinitro-1,3-benzoxathiole-2-thione [Synthesis of an Oxathiole by Photochemical Decomposition of a Phenolic Iodonium Ylide in the Presence of Carbon Disulfide].

Step I. Formation of 2,4-Dinitro-6-phenyliodonium Phenolate (**73**).[118] A solution of 2,4-dinitrophenol (1.84 g, 10 mmol) in acetonitrile (10 mL) was added to a solution of IBTA (4.3 g, 10 mmol) in acetonitrile (10 mL). After 72 hours at room temperature, **73** (1.35 g) separated from the mixture as a yellow crystalline solid, mp 195–196°; λ_{max} (Me$_2$SO) 372 nm (log ε 4.15); IR 3086, 1585, 1560, 1555, 1260, 1090, 715 cm^{-1}; ^{1}H NMR [(CD$_3$)$_2$SO] δ 7.78 (m, 3 H), 8.37–8.48 (m, 2 H), 8.98 (s, 2 H); MS m/z 386 (M$^+$, 93), 204 (30), 139, (95), 93 (100).

The filtrate was concentrated under reduced pressure without heating, and fresh acetonitrile (10 mL) was added. After 24 hours, a second crop of **73** (580 mg) was collected and upon repetition of the procedure the total yield of **73** amounted to 65–70%.

Step II. Formation of 5,7-Dinitro-1,3-benzoxathiole-2-thione.[113] A suspension of iodonium ylide **73** (200 mg) in acetonitrile (20 mL), carbon disulfide (20 mL), and benzene (6 mL) was irradiated with a low-pressure Hg lamp (400 W, Pyrex vessel) for 9 hours with occasional shaking. Partial removal of solvents led to the precipitation of some unreacted **73** (50 mg). The filtrate was chromatographed on silica gel using hexanes-chloroform as eluant to give 73 mg (73%, based on **73** consumed) of the title product **76** as an orange crystalline solid, mp 158–159°(from chloroform-hexane); IR (Nujol) 3095, 1600, 1540 cm^{-1}; UV (EtOH) λ_{max} 237 nm (ε 14,790), 265 (12,880), 298 (14,120), 355 (10,710); ^{1}H NMR (CDCl$_3$) δ 8.44 (d, J = 2 Hz, 1 H), 8.85 (d, J = 2 Hz, 1 H); MS m/z 258 (M$^+$, 45), 168 (45), 124 (48), 93 (90), 82 (85), 62 (100).

4-Thiocyanato-1-naphthol [p-Thiocyanation of a Phenol].[123] To an ice-cooled suspension of lead dithiocyanate (485 mg, 1.5 mmol) in dry dichloromethane (10 mL) was added (dichloroiodo)benzene (330 mg, 1.2 mmol). The resulting mixture was stirred at the same temperature for 20 minutes, and then 1-naphthol (144 mg, 1.0 mmol) was added. The mixture was stirred for 1 hour, filtered, and silica gel (2 g) was added to the filtrate. Filtration followed by concentration in vacuo gave crude product. Purification by column chromatography on silica gel using ethyl acetate-hexanes as eluant afforded 166 mg (88%) of the title product as a solid, mp 113°.

p-(Methoxymethyl)phenol [Cleavage of an Amino-Terminal Tyrosyl-Peptide Bond].[127] Potassium hydroxide (283 mg, 5 mmol) was dissolved in methanol (50 mL) and the solution was cooled to 0°. L-Tyrosyl-L-alanine (252 mg, 1 mmol) was added to the stirred solution. IBD (322 mg, 1.0 mmol) was added over a period of 1.5 hours and stirring was continued for 1.5 hours. The reaction mixture was then acidified with acetic acid and extracted with chloroform (3 × 20 mL). The combined organic extracts were dried (MgSO$_4$), and concentrated in vacuo. The crude residue was purified by crystallization from ether-hexane (1:1) to give 98 mg (71%) of the product, mp 76–79°; ^{1}H NMR (CDCl$_3$) δ 3.20 (s, 3 H), 4.30 (s, 2 H), 6.80–7.40 (m, 4 H).

($\pm$)-$trans$-2-(3,4-Methylenedioxyphenyl)-2,3-dihydro-6-allyloxy-5-methoxy-3-methylbenzofuran [Formation of a Dihydrobenzofuran by Oxidative Cycloaddition of a Phenol to an Electron-rich Styrene] (Eq. 73).[131] To a solution of 3-allyloxy-4-methoxyphenol (500 mg, 2.78 mmol) and (E)-isosafrole (1.98 g, 11.1 mmol) in acetonitrile (2 mL) at 0° was added IBTA (1.43 g, 3.34 mmol). The temperature was kept at 0° for 30 minutes and the mixture was concentrated in vacuo. Purification of the crude product by column chromatography on silica gel using ethyl acetate-petroleum ether as eluant afforded 652 mg (69%) of the title product as a colorless crystalline solid (from ether − petroleum ether), mp 59–61°; IR (KBr) 1495, 1240, 1215, 1175, 1015 cm^{-1}; ^{1}H NMR (CDCl$_3$) δ 1.36 (d, J = 6.7 Hz, 3 H), 3.28–340 (m, 1 H), 3.85 (s, 3 H), 4.58 (dt, J = 5.4, 1.5 Hz, 2 H), 5.03 (d, J = 8.7 Hz, 1 H), 5.25–5.46 (m, 2 H), 5.96 (s, 2 H), 5.96–6.17 (m, 1 H), 6.51 (s, 1 H), 6.70 (d, J = 1 Hz, 1 H), 6.77–6.92 (m, 3 H).

TABULAR SURVEY

The tables are arranged in parallel with the text and in the order of complexity of the substrates. Numbers in parentheses are yields of isolated pure products, whereas a dash indicates that no yield is reported. Where isolated yields and yields by GLC or NMR are reported, we give only the former. Where isolated yields are not reported and yields based on NMR/GLC are available, we give the latter along with a footnote. Numbers without parentheses are ratios of products.

The following abbreviations are used in the tables:

Ac	acetyl
Bn	benzyl
Boc	*tert*-butoxycarbonyl
Cbz	benzyloxycarbonyl
CSA	camphorsulfonic acid
DMF	*N,N*-dimethylformamide
DMSO	dimethyl sulfoxide
HTIB	[hydroxy(tosyloxy)iodo]benzene
IBD	iodobenzene diacetate
IBTA	iodobenzene bis(trifluoroacetate)
MOM	methoxymethyl
Py	pyridine
TBAF	tetrabutylammonium fluoride
TEOC	trimethylsilylethoxycarbonyl
TFA	trifluoroacetic acid
TFE	2,2,2,-trifluoroethanol
THF	tetrahydrofuran
THP	tetrahydropyranyl
Ts	*p*-toluenesulfonyl

TABLE IA. FORMATION OF QUINONES BY OXIDATION OF MONOPHENOLS

Substrate	Conditions	Product(s) and Yield(s) (%)	Refs.
C_6			
phenol	4-HO$_2$C-C$_6$H$_4$-I(OAc)$_2$ (4 eq)	ortho-benzoquinone (30–40)	48
2,6-dichloro-4-chlorophenol	PhI(O$_2$CCF$_3$)$_2$	2,6-dichloro-1,4-benzoquinone (—)	54
2,6-dibromo-4-bromophenol	4-HO$_2$C-C$_6$H$_4$-I(OAc)$_2$ (4 eq)	2,6-dibromo-1,4-benzoquinone (—)	48
2,3,4,6-tetrachlorophenol	PhI(O$_2$CCF$_3$)$_2$	tetrachloro-1,4-benzoquinone (—)	54
pentachlorophenol	PhI(O$_2$CCF$_3$)$_2$	tetrachloro-1,4-benzoquinone (—)	54
C_7			
o-cresol	4-HO$_2$C-C$_6$H$_4$-I(OAc)$_2$ (4 eq)	2-methyl-1,4-benzoquinone (30–40)	48

C$_{7-16}$

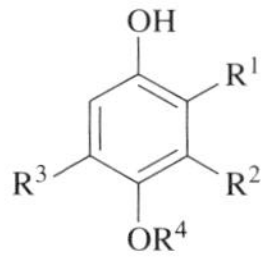

Starting material: phenol with OH, R^1, R^2, OR^4, R^3 substituents

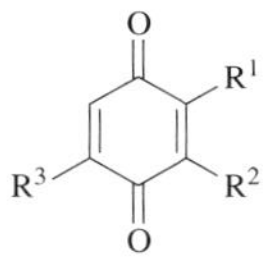

Conditions: IBTA, H$_2$O, MeCN, rt

Product: ortho-quinone with R^1, R^2, R^3 substituents (86-100)

R^1	R^2	R^3	R^4	Refs.
H	H	H	Me	49
CH$_2$OMOM	H	H	Me	49
CH$_2$OTHP	H	H	Me	49
H	OMe	OMe	Me	49
H	OMe	H	Me	49
CH$_2$OH	H	H	Me	49
–CO(CH$_2$)$_3$–		H	Et	49
H	H	H	Me	49
(CH$_2$)$_2$NHCOCF$_3$	H	H	Me	49
(CH$_2$)$_2$NHCOCF$_3$	H	OMe	Me	155
(CH$_2$)$_2$NHTEOC	H	OMe	Me	155

C$_8$

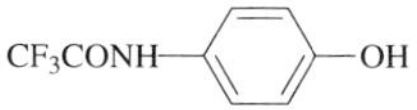

Starting material: CF$_3$CONH–C$_6$H$_4$–OH (para)

Conditions: IBTA, H$_2$O, MeCN, rt

Product: 1,4-benzoquinone (80) — Ref. 49

C$_{9-10}$

Starting material: phenol with OH, R^1, R^2, R^3, R^4 substituents

Product: para-quinone with R^1, R^2, R^3, R^4 substituents

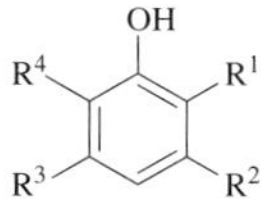

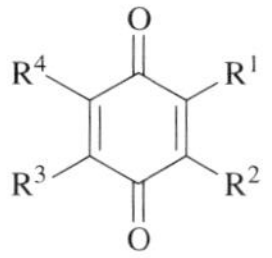

R^1	R^2	R^3	R^4	Conditions	Yield	Refs.
Me	Me	H	Me	(PhIO)$_n$, BF$_3$•Et$_2$O, rt, 20 min.	(70)	156
Me	Me	H	Me	(PhIO)$_n$ (2 eq), AcOH, 2–5 min.	(75)	156
Me	Me	H	Me	m-HO$_2$CC$_6$H$_4$IO, [RuCl$_2$(PPh$_3$)$_3$] (1 mol%)	(74)	156

TABLE IA. FORMATION OF QUINONES BY OXIDATION OF MONOPHENOLS (*Continued*)

Substrate								Conditions	Product(s) and Yield(s) (%)	Refs.
R^1	R^2	R^3	R^4							
Me	Me	Me	H					(PhIO)$_n$, BF$_3$•Et$_2$O, rt, 20 min.	(42)	156
Me	Me	Me	H					(PhIO)$_n$ (2 eq), AcOH, 2–5 min.	(73)	156
Me	Me	Me	Me					(PhIO)$_n$, BF$_3$•Et$_2$O, rt, 20 min.	(58)	156
Me	Me	Me	Me					(PhIO)$_n$ (2 eq), AcOH, 2–5 min.	(80)	156
Me	Me	Me	Me					*m*-HO$_2$CC$_6$H$_4$IO, [RuCl$_2$(PPh$_3$)$_3$] (1 mol%)	(43)	156

C$_{9-11}$

A	B	C	D	R	Conditions	Product(s) and Yield(s) (%)	Refs.
CH	CH	CH	CH	H	IBTA (2.2 eq), H$_2$O, MeCN	(73)	58
CH	CH	CH	CH	Me	(PhIO)$_n$, BF$_3$•Et$_2$O, rt, 20 min.	(61)	156
CH	CH	CH	CH	Me	(PhIO)$_n$ (2 eq), AcOH, 2–5 min.	(50)	156
CH	CH	CH	CH	Me	*m*-HO$_2$CC$_6$H$_4$IO, [RuCl$_2$(PPh$_3$)$_3$] (1 mol%)	(30)	156
CH	CH	CH	CH	Me	(PhIO)$_n$ (2 eq), [RuCl$_2$(PPh$_3$)$_3$] (1 mol%), CH$_2$Cl$_2$, 2 min.	(18)	156
CH	CH	CH	COH	H	IBTA (2.2 eq), H$_2$O, MeCN, 0°	(58)	50,58
CH	CH	CH	COH	H	IBD (2.2 eq), H$_2$O, MeCN, 0°	(26)	50
CH	CH	CH	COH	H	(PhIO)$_n$, MeOH, rt	(47)	50
CH	CH	CH	COH	H	PhIO$_2$, VO$_2$ acac (cat.), PhMe	(30)	50
CH	CH	CH	COH	H	C$_6$F$_5$I(O$_2$CCF$_3$)$_2$, H$_2$O, MeCN	(76)	50
CH	CH	CH	COH	H	C$_6$F$_{13}$I(O$_2$CCF$_3$)$_2$, H$_2$O, MeCN	(91)	50
CH	CH	CH	COMe	H	IBTA (2.2 eq), H$_2$O, MeCN, 0°	(80)	58
CH	CH	CH	N	H	IBTA (2.2 eq), H$_2$O, MeCN, 0°	(88)	58
CH	CH	N	CH	H	IBTA (2.2 eq), H$_2$O, MeCN, 0°	(80)	58
CH	CH	CMe	N	H	IBTA (2.2 eq), H$_2$O, MeCN, 0°	(82)	58
N	CMe	CMe	N	H	IBTA (2.2 eq), H$_2$O, MeCN, 0°	(70)	58

	Reagent	Product (yield %)
C_{10} (2-naphthol)	PhIO$_2$/AcOH	1,2-naphthoquinone (65), 51
C_{10-11} (substituted naphthol)	PhIO$_2$, H$_2$O, MeCN	ortho-quinone I + para-quinone II, 157

R^1	R^2	R^3	R^4	I	II
H	H	H	H	(11)	(63)
Me	H	H	H	(27)	(66)
H	OMe	H	Me	(15)	(59)
H	H	OMe	H	(18)	(62)
H	OMe	H	H	(17)	(68)
H	H	H	OMe	(16)	(46)
Me$_2$COH	H	H	H	(34)	(59)

C_{11}: IBD, H$_2$O, MeCN → chromanone product (69), 59

C_{12}: IBTA, H$_2$O, MeCN → naphthoquinone product (96-100), 49

R	
H	(96-100), 49
Br	(96-100), 49

TABLE IA. FORMATION OF QUINONES BY OXIDATION OF MONOPHENOLS (*Continued*)

Substrate	Conditions	Product(s) and Yield(s) (%)	Refs.
C₁₃₋₁₄			
(furoquinoline-OH, R = H or Me)	IBTA, H$_2$O, MeCN, rt	(furoquinoline-dione)	
R = H	−10°	(78)	158
R = Me	0°	(98)	158
C$_{14}$ (2-t-Bu-4-t-Bu-phenol)	PhIO$_2$, AcOH	(ortho-quinone with t-Bu groups) (82)	51
C$_{15}$ (tetrahydronaphthalenol with Me and i-Pr)	PhIO$_2$, H$_2$O, MeCN	(para-quinone + ortho-quinone) (—)	157

TABLE IB. FORMATION OF QUINONES BY OXIDATION OF BISPHENOLS

Substrate	Conditions	Product(s) and Yield(s) (%)	Refs.
C₆ HO–C₆H₄–OH	IBD, MeOH, rt	p-benzoquinone (94)	52, 53
	IBTA, H₂O, MeCN, rt	(91)	49
	HTIB, MeCN, CH₂Cl₂, rt	(80)	52, 53
	(PhIO)ₙ, Et₂O	(70)	156
	(PhIO)ₙ, [RuCl₂(PPh₃)₃] (0.2 mol%), Et₂O	(93)	156
C₈ dichloro-dicyano-bisphenol	IBTA, MeCN	dichlorodicyanobenzoquinone (90)	151
C₈₋₁₄ dihydroxy-phthalimide (X = O, NPh)	IBTA, C₆H₆	quinone (68)	57
	IBTA, CCl₄	(97)	57
C₉ trimethylhydroquinone	IBD, MeOH, rt	trimethylbenzoquinone (100)	52, 53
	HTIB, MeCN, CH₂Cl₂, rt	(84)	52, 53

TABLE IB. FORMATION OF QUINONES BY OXIDATION OF BISPHENOLS (*Continued*)

Substrate	Conditions	Product(s) and Yield(s) (%)	Refs.
C$_{10}$ 4-*t*-Bu-catechol	IBD, MeOH, rt	4-*t*-Bu-*o*-benzoquinone (99)	52, 53
1,4-dihydroxynaphthalene	(PhIO)$_n$, MeCOMe (PhIO)$_n$, [RuCl$_2$(PPh$_3$)$_3$] (1.7 mol%), MeCOMe	1,4-naphthoquinone (92) (85)	156 156
2,3-X$_2$-5,8-dihydroxy-1,4-naphthoquinone (X = H, Cl)	IBTA, MeCOMe, rt, 1 h IBTA, MeCOMe, rt, 1 h	2,3-X$_2$-1,4,5,8-naphthodiquinone (81) (81)	55 55
C$_{11}$ 4-*t*-Bu-3-methylcatechol	(PhIO)$_n$, CH$_2$Cl$_2$, rt, 1.5 h (PhIO)$_n$, [RuCl$_2$(PPh$_3$)$_3$] (0.3 mol%), CH$_2$Cl$_2$	4-*t*-Bu-3-methyl-*o*-benzoquinone (70) (98)	156 156

Substrate	Conditions	Product(s) and Yield(s) (%)	Refs.
C_{12} 3-tert-butyl-6-methylcatechol	(PhIO)$_n$, CH$_2$Cl$_2$, rt, 1.5 h (PhIO)$_n$, [RuCl$_2$(PPh$_3$)$_3$] (0.2 mol%), CH$_2$Cl$_2$	3-tert-butyl-6-methyl-o-benzoquinone (48) (42)	156 156
C_{12} 4,4'-biphenol	IBD, MeOH, rt	4,4'-diphenoquinone (70) (94)^a	52, 53 52, 53
C_{14} 3,5-di-tert-butylcatechol	IBD, MeOH, rt (PhIO)$_n$, CH$_2$Cl$_2$, rt, 1.5 h (PhIO)$_n$, [RuCl$_2$(PPh$_3$)$_3$] (0.5 mol%), CH$_2$Cl$_2$	3,5-di-tert-butyl-o-benzoquinone (91) (98) (86)	52, 53 156 156
C_{15} 1,4-dihydroxyanthraquinone	IBTA, MeCOMe, rt, 1 h	anthraquinone derivative (66)	55
C_{15} bridged dihydroxyanthraquinone	IBTA, MeCOMe, rt, 1 h	bridged quinone (58)	56

^{a} This yield is based upon HPLC.

TABLE II. FORMATION OF *p*-QUINOLS BY OXIDATION OF PHENOLS AND PHENOL TRIPROPYLSILYL ETHERS

Substrate							Conditions	Product(s) and Yield(s) (%)	Refs.

C_{7-19}

R	R^1	R^2	R^3	R^4	R^5				
H	H	H	Me	H	H		IBTA, MeCN, H_2O, 0°	(48)	62
Pr_3Si	H	H	Me	H	H			(73)	
H	H	H	Et	H	H			(48)[a]	
H	H	H	Bn	H	H			(38)[b]	
H	Me	H	Me	H	H			(60)	
Pr_3Si	Me	H	Me	H	H			(73)	
H	H	Me	Me	H	H			(67)	
Pr_3Si	H	Me	Me	H	H			(75)	
H	Me	H	Me	H	Me			(67)	
Pr_3Si	Me	H	Me	H	Me			(78)	
H	H	Me	Me	Me	H			(63)	
Pr_3Si	H	Me	Me	Me	H			(70)	
H	H	H	CH_2CO_2Me	H	H			(27)	
Pr_3Si	H	H	CH_2CO_2Me	H	H			(59)	
H	H	H	Me	H	CO_2Et			(30)	
H	Br	H	Me	H	Br			(78)	
H	Br	H	CH_2CO_2Et	H	Br			(67)	
H	*t*-Bu	H	*t*-Bu	H	*t*-Bu			(76)	
H	*t*-Bu	H	Me	H	*t*-Bu			(78)	

C_{11-14}

[naphthalene with OR and methyl substituents] — IBTA, MeCN, H₂O, 0° → [4-methyl-4-hydroxy-naphthalenone] 62

R	
H	(59)
Pr₃Si	(44)

[a] 4-Acetylphenol was also obtained in 15% yield.
[b] Oxidation of 4-benzylphenol gave 2-benzyl-1,4-benzoquinone as the only identifiable product (38%).

TABLE III. FORMATION OF QUINONE ACETALS BY OXIDATION OF PHENOLS

C_{6-16}

Substrate					Conditions	Product(s) and Yield(s) (%)							Refs.
R^1	R^2	R^3	R^4	R^5		R	R^1	R^2	R^4	R^5	R^6		
H	H	H	H	H	IBD, MeOH, rt	Me	H	H	H	H	Me	(68)	53
H	H	H	H	Bn	IBD, MeOH, rt	Me	H	H	H	Bn	Me	(83)	53
H	H	OMe	H	H	IBD, MeOH, rt	Me	H	H	H	H	Me	(99)	53
H	H	OMe	H	H	IBD, EtOH, rt	Et	H	H	H	H	Me	(78)	63
H	H	OMe	H	H	IBD, n-PrOH, rt	n-Pr	H	H	H	H	Me	(77)	63
H	H	OMe	H	H	IBD, i-PrOH, rt	i-Pr	H	H	H	H	Me	(59)	63
H	H	OMe	H	H	IBD, i-BuOH, rt	i-Bu	H	H	H	H	Me	(59)	63
H	H	OMe	H	H	IBD, t-BuOH, rt	t-Bu	H	H	H	H	Me	(0)	63
H	H	OMe	OMe	H	IBTA, K_2CO_3, MeOH, MeCN, 0°–rt	Me	H	H	OMe	H	Me	(90)	84
H	H	OEt	H	H	IBTA, K_2CO_3, MeOH, MeCN, 0°–rt	Me	H	H	H	H	Et	(90)	84
H	Me	H	Me	H	IBD, MeOH, rt	Me	H	Me	Me	H	Me	(80)	53
Me	H	OMe	OMe	H	IBTA, K_2CO_3, MeOH, MeCN, 0°–rt	Me	Me	H	OMe	H	Me	(85)	84
OMe	H	OMe	OMe	H	IBTA, K_2CO_3, MeOH, MeCN, 0°–rt	Me	OMe	H	OMe	H	Me	(86)	84
H	OMe	OMe	OMe	$CH_2CH=CH_2$	IBTA, K_2CO_3, MeOH, MeCN, 0°–rt	Me	H	OMe	OMe	$CH_2CH=CH_2$	Me	(85)	84
OMe	OMe	OMe	OMe	H	IBTA, K_2CO_3, MeOH, MeCN, 0°–rt	Me	OMe	OMe	OMe	H	Me	(quant)	84
OMe	OMe	OMe	OMe	H	IBTA, K_2CO_3, EtOH, MeCN, 0°–rt	Et	OMe	OMe	OMe	H	Et	(quant)	84
OMe	OMe	OMe	OMe	H	IBTA, K_2CO_3, i-PrOH, MeCN, 0°–rt	i-Pr	OMe	OMe	OMe	H	i-Pr	(97)	84
H	OMe	H	H	$(CH_2)_2$NHTEOC	IBTA, MeOH, MeCN, rt	Me	H	OMe	H	$(CH_2)_2$NHTEOC	Me	(—)	155
OMe	OMe	H	H	$(CH_2)_2$NHTEOC	IBTA, MeOH, MeCN, rt	Me	OMe	OMe	H	$(CH_2)_2$NHTEOC	Me	(quant)	155
OMe	OMe	H	H	$(CH_2)_2$NHCOCF$_3$	IBTA, MeOH, MeCN, rt	Me	OMe	OMe	H	$(CH_2)_2$NHCOCF$_3$	Me	(quant)	155

C₇	IBTA, K₂CO₃, ROH, MeCN	R: Me (83), Et (99), i-Pr (80) — 84, 84, 84
C₈	IBD (1.5 eq), MeOH	4:1 mixture — 72
C₈₋₁₁	IBD (1.0 eq), MeOH; IBD (1.0 eq), MeOH; IBD (1.5 eq), MeOH	R: Me — I (high); CH₂CH=CH₂ — I (high); t-Bu — I (43) + II (55) — 72, 72, 72
C₁₀₋₁₄	IBD, MeOH, 0° to rt	

TABLE III. FORMATION OF QUINONE ACETALS BY OXIDATION OF PHENOLS (Continued)

Substrate	Conditions	Product(s) and Yield(s) (%)	Refs.
X — H		(76)	71
Br		(66)	71
7-OMe		(88)	71
5-CO$_2$Me		(43)	71
6-CH(Me)CO$_2$Me		(52)	71
C$_{11}$ (1,4-dihydroxy-methoxynaphthalene)	IBD, ROH	R = Me (74), Et (63)	63, 63
C$_{15}$ (2-phenyl-6-hydroxychromone)	IBD, ROH	III: R=Me (20), Et (18); IV: R=Me (31), Et (34)	159, 159
C$_{16}$ (chalcone derivative)	IBTA, MeOH, 0°	(67)	160

Ar = p-ClC$_6$H$_4$

TABLE IV. FORMATION OF p-QUINOL ETHERS BY OXIDATION OF PHENOLS

Substrate	Conditions	Product(s) and Yield(s) (%)	Refs.

C_{7-14}

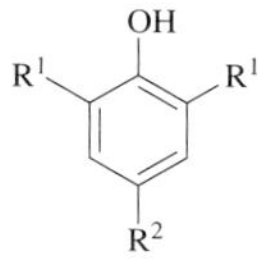

Substrate: 2,6-R¹,4-R²-phenol (OH, R¹, R², R¹)

Conditions: IBD, R³OH, Solvent

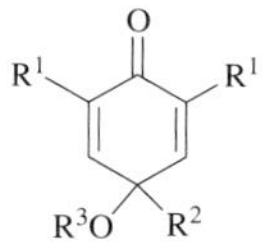

Product: 4-R³O-4-R²-2,6-R¹-cyclohexa-2,5-dienone

R¹	R²	R³	Solvent, etc.	Yield	Ref.
H	NHAc	Me	MeOH	(75)	63
H	Bn	Me	MeOH	(65)	53
Br	Me	Me	CH$_2$Cl$_2$, MeOH	(63)	65
Br	Me	Et	CH$_2$Cl$_2$, EtOH	(42)	65
Me	Me	Me	MeOH	(72)	53
t-Bu	t-Bu	Me	MeOH	(94)	53
t-Bu	t-Bu	Me	CH$_2$Cl$_2$, MeOH	(37)	65
t-Bu	t-Bu	Ph	CH$_2$Cl$_2$ overnight	(16)	65
t-Bu	t-Bu	4-MeC$_6$H$_4$	CH$_2$Cl$_2$ overnight	(23)	65
t-Bu	t-Bu	4-MeOC$_6$H$_4$	CH$_2$Cl$_2$ overnight	(11)	65
t-Bu	t-Bu	4-O$_2$NC$_6$H$_4$	CH$_2$Cl$_2$ overnight	Complex mixture	65
t-Bu	t-Bu	2,6-Br$_2$-4-MeC$_6$H$_2$	CH$_2$Cl$_2$ overnight	(59)	65
t-Bu	t-Bu	2,4,6-Me$_3$C$_6$H$_2$	CH$_2$Cl$_2$ overnight	Complex mixture	65

C_{9-10}

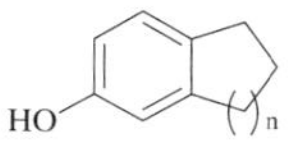

Substrate: bicyclic phenol with fused (CH$_2$)$_n$ ring

Conditions: IBD, MeOH

Product: bicyclic MeO-dienone

n	Yield	Ref.
1	(—)	71
2	(67)	71

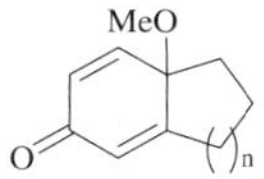

379

TABLE IV. FORMATION OF *p*-QUINOL ETHERS BY OXIDATION OF PHENOLS (*Continued*)

Substrate	Conditions	Product(s) and Yield(s) (%)	Refs.

C$_{12-15}$

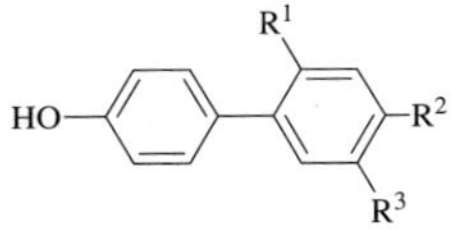

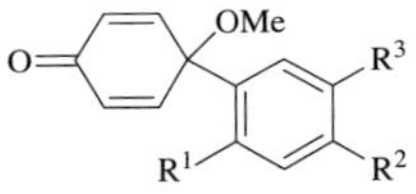

R^1	R^2	R^3			
H	H	H	IBD, MeOH	(71)	66
CH=CH$_2$	H	H		(46)	66
CH=CH$_2$	OMe	H		(63)	66
CH=CH$_2$	H	OMe		(34)	66
CH=CH$_2$	–[OCH$_2$O]–			(9)	66

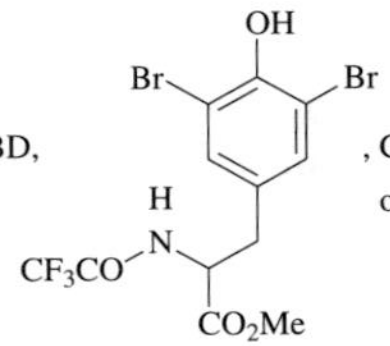

C$_{18}$

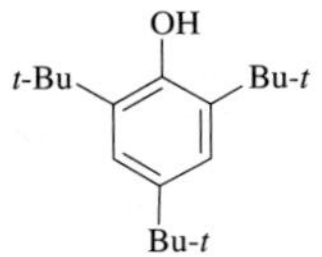

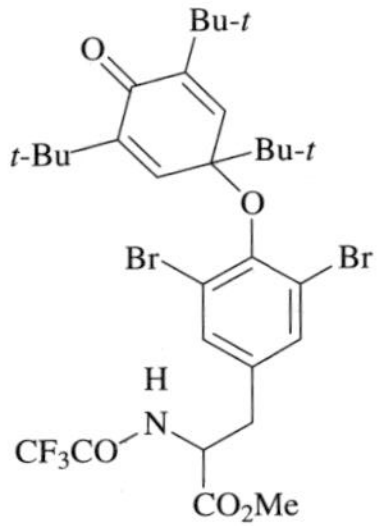

(65) 65

2,4,6-tri-*t*-Bu-phenol, IBD, CH$_2$Cl$_2$, 1 h	bis-peroxide dimer (40)	65
C$_{20}$ diol-phenol, IBTA, MeOH, 48 h / IBTA, TFE, 1 h	spirolactone R = Me (13)^a / R = CH$_2$CF$_3$ (3)^a	106
C$_{28}$ menthyl diol-phenol, IBTA, TFE, 24 h	menthyl spirolactone (2)^a	107

^{a} Several other products were also formed.

TABLE V. *ipso*-FLUORINATION OF 4-ALKYLPHENOLS

Substrate	Conditions	Product(s) and Yield(s) (%)	Refs.
C₇ (4-methylphenol)	Py-HF (30 : 70 w/w), IBTA	(68) + (8)	73
C₈ (4-ethylphenol)	Py-HF (30 : 70 w/w), IBTA	(61)	73
C₉ (indanol)	Py-HF (30 : 70 w/w), IBTA	(42)	73
C₁₀ (tetrahydronaphthol)	Py-HF (30 : 70 w/w), IBTA	(66)	73
C₁₈ (steroid)	Py-HF (30 : 70 w/w), IBTA	R = H (77); R = OH (58)	73, 73

TABLE VI. FORMATION OF SPIRO-HETEROCYCLIC COMPOUNDS

Substrate			Conditions	Product(s) and Yield(s) (%)	Refs.
C_{8-14}					
R	R^1	R^2			
H	H	H	IBTA, MeCN, reflux	(20)	85
Me	H	H	IBTA, MeCN, 0°	(63)	85
Me	Br	Br	IBTA, MeCN, 0°	(58)	85
OMe	OMe	H	IBTA, MeCN, 0°	(39)	86
Et	H	H	IBTA, MeCN, 0°	(68)	85
Et	Br	Br	IBTA, MeCN, 0°	(52)	85
OEt	H	H	IBTA, MeCN, 0°	(68)	86
CO_2Me	H	H	IBTA, MeCN, 0°	(65)	85
CO_2Me	Br	Br	IBTA, MeCN, 0°	(58)	85
CMe_3	H	H	IBTA, MeCN, 0°	(93)	85
CMe_3	Br	Br	IBTA, MeCN, 0°	(89)	85
C_6H_5	H	H	IBTA, MeCN, 0°	(86)	85
C_6H_5	Br	Br	IBTA, MeCN, 0°	(78)	85
4-BrC_6H_4	H	H	IBTA, MeCN, 0°	(82)	85
4-BrC_6H_4	Br	Br	IBTA, MeCN, 0°	(77)	85
C_{8-16}					

TABLE VI. FORMATION OF SPIRO-HETEROCYCLIC COMPOUNDS (*Continued*)

Substrate		Conditions	Product(s) and Yield(s) (%)	Refs.
X	Y			
O	OH	IBTA, Py, MeCN, 0°-rt	(80)	84
CH_2	OH	IBTA, Py, MeCN, 0°-rt	(86)	84
CH_2	OH	IBTA, MeCN, 0°	(83)	76
CH_2	OH	IBTA, Py, MeCN, 0°	(45-65)	76
CH_2	OH	IBD, MeCN, 0°	(27)	76
CH_2	OH	4-MeC$_6$H$_4$I(O$_2$CCF$_3$)$_2$, MeCN, 0°	(69)	76
CH_2	OH	4-MeC$_6$H$_4$I(O$_2$CCF$_3$)$_2$, MeCN, Py, 0°	(35)	76
CH_2	OH	4-ClC$_6$H$_4$I(O$_2$CCF$_3$)$_2$, MeCN, 0°	(52)	76
CH_2	OH	4-ClC$_6$H$_4$I(O$_2$CCF$_3$)$_2$, MeCN, Py, 0°	(17)	76
CH_2	OH	4-O$_2$NC$_6$H$_4$I(O$_2$CCF$_3$)$_2$, MeCN, 0°	(65)	76
CH_2	OH	4-O$_2$NC$_6$H$_4$I(O$_2$CCF$_3$)$_2$, MeCN, Py, 0°	(56)	76
CH_2	OH	1,4-[I(O$_2$CCF$_3$)$_2$]$_2$C$_6$H$_4$, MeCN	(67)	76
CH_2	OH	HTIB, MeCN	(53)	76
CH_2	NHBn	IBTA, Py, MeCN, 0°-rt	(69)	84

C$_9$

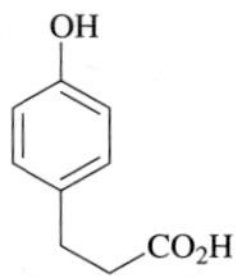

[structure: 4-hydroxyphenylpropanoic acid with OH on para position and CH$_2$CH$_2$CO$_2$H chain]

IBD (3 eq), MeCN, 0°

[spiro product structure with two X substituents]

X	
Cl	(49)[a]
Br	(72)[b]

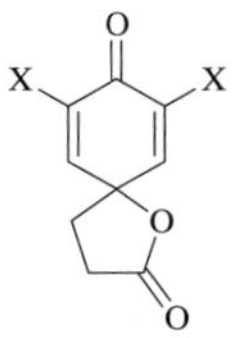

Refs. 88, 88

Substrate	Conditions	Product	Yield (%)
C$_{10-14}$ HO-C$_6$H$_4$-(CH$_2$)$_3$-OH	IBTA, Py, MeCN, 0°-rt	spiro dienone-tetrahydrofuran (59)	84
oxime substrate (various R, R^1, R^2)	IBD, MeCN, 0°	isoxazoline-spirodienone	

R	R^1	R^2	Yield (%)
OMe	Br	H	86 (46)
OMe	H	OMe	86 (40)
OMe	Br	OMe	86 (70)
OBu-t	Br	OMe	86 (72)
NH(CH$_2$)$_3$OMe	H	OMe	86 (45)
NH(CH$_2$)$_3$OMe	Br	OMe	86 (64)

C$_{11-17}$ HO-C$_6$H$_4$-CH$_2$-CH(R)-CO$_2$H

R	Conditions	Yield (%)
NHAc	IBTA, MeCN, 0°	76 (28)
NHAc	IBTA, MeOH	76 (37)
NHAc	IBTA, MeCN, Py, 0–10°	78 (60–70)

TABLE VI. FORMATION OF SPIRO-HETEROCYCLIC COMPOUNDS (*Continued*)

Substrate	Conditions	Product(s) and Yield(s) (%)	Refs.
NHCOCF$_3$	IBTA, MeCN, Py, 0–10°, 18 h	(60)	78
NHCO$_2$Bn	IBTA, MeCN, 0°	(38)	76
NHCO$_2$Bn	IBTA, MeOH	(41)	76
NHCO$_2$Bn	IBTA, MeCN, Py, 0–10°, 18 h	(60–70)	78
NHCO$_2$Bn	IBD, MeOH, 0°	(35–40)	80
NHBoc	IBTA, MeCN, 0°	(38)	76
NHBoc	IBTA, MeOH	(34)	76
NHBoc	IBTA, MeCN, Py, 0–10°, 18 h	(67)	78
NHBoc	IBD, MeOH	(68)	77
	IBTA, MeCN, Py, 0–10°, 18 h	(60–70)	78
C$_{21-27}$	IBD, C$_6$H$_6$	(34)^c	
		R = H (18)^c	75
		R = Ph (62)^c	75
C$_{23}$	IBD, C$_6$H$_6$	(34)^c	75

Substrate	Conditions	Product	Yield
C24 (bisphenol with Ph)	IBD, AcOH	spirocyclic dienone with Ph	(43)[c,d] 75
Tyrosine derivative (Bn, NCbz, CO2H)	IBD (3 eq), MeCN, 0°	spirolactone with Bn/NCbz	(32)[e] 88
	IBD (3 eq), MeCN, 0°		(41)[f] 88
	IBD (3 eq), MeCN, −5°		(43)[f,a] 88
	IBTA (1.1 eq), H2O, MeCN, rt		(63) 88
Tyrosine derivative	IBD (3 eq), MeCN, −5°	dihalo spirolactone	X = Cl (74)[a] 88
			X = Br (79)[b] 88

TABLE VI. FORMATION OF SPIRO-HETEROCYCLIC COMPOUNDS (Continued)

Substrate	Conditions	Product(s) and Yield(s) (%)	Refs.
C_{25}	IBD, AcOH	(28)[c]	75
C_{26} R = (−)-8-phenylmenthyl	IBD IBTA (PhIO)$_n$, (+)-CSA (PhIO)$_n$, (−)-CSA	(31) 33% de (88) 67% de (70) 82% de (87) 82% de	86,87 86,87 86,87 86,87
C_{27-28} R H	IBD, C_6H_6	(84)[c]	74

R			
H	IBD, Py	(79)[c,g]	74
H	IBD, AcOH	(67)[c,g]	74
2-F	IBD, C$_6$H$_6$	(96)[c,g]	74
2-F	IBD, Py	(82)[c,g]	74
2-F	IBD, AcOH	(56)[c,g]	74
3-F	IBD, C$_6$H$_6$	(86)[c]	74
3-F	IBD, Py	(85)[c,g]	74
3-F	IBD, AcOH	(71)[c,g]	74
4-F	IBD, C$_6$H$_6$	(87)[c]	74
4-F	IBD, Py	(80)[c,g]	74
4-F	IBD, AcOH	(79)[c]	74
2-OMe	IBD, C$_6$H$_6$	(89)[c,g]	74
2-OMe	IBD, Py	(70)[c,g]	74
2-OMe	IBD, AcOH	(34)[c,g]	74
3-OMe	IBD, C$_6$H$_6$	(66)[c,g]	74
3-OMe	IBD, Py	(81)[c,g]	74
3-OMe	IBD, AcOH	(76)[c,g]	74
4-OMe	IBD, C$_6$H$_6$	(61)[c,g]	74
4-OMe	IBD, Py	(80)[c,g]	74
4-OMe	IBD, AcOH	(54)[c,g]	74

[a] The reaction was worked up with sat. aq. NaCl.
[b] The reaction was worked up with sat. aq. NaBr.
[c] The product was obtained as a racemic mixture.
[d] The product was isolated as the dimer.
[e] The reaction was worked up with sat. aq. Na$_2$S$_2$O$_3$.
[f] The reaction was worked up with aq. 10% citric acid.
[g] The other diastereomer was also formed as a minor product.

TABLE VII. OXIDATION OF N-ACYLTYRAMINES

Substrate	Conditions	Product(s) and Yield(s) (%)	Refs.
C_{10-15} HO-C$_6$H$_4$-CH$_2$CH$_2$-NHCOR		**I** (spiro cyclohexadienone with R^{1}O); **II** (spiro cyclohexadienone oxazine with R)	
R = Me	IBTA, MeOH	**I**, R = Me; R^1 = Me (76)	67
Me	IBTA, EtOH	**I**, R = Me; R^1 = Et (47)	67
Me	IBTA, i-PrOH	**I**, R = Me; R^1 = i-Pr (22)	67
Me	IBTA, AcOH	**I**, R = Me; R^1 = Ac (20)	67
Me	IBTA, CH$_2$Cl$_2$, K$_2$CO$_3$	**II**, R = Me (29)	67
t-Bu	IBTA, MeOH	**I**, R = t-Bu; R^1 = Me (64) + **II**, R = t-Bu (18)	67
t-Bu	IBTA, AcOH	**I**, R = t-Bu; R^1 = Ac (44) + **II**, R = t-Bu (8)	67
t-Bu	IBTA, CF$_3$CH$_2$OH	**II**, R = t-Bu (75)	67
t-Bu	IBTA, CH$_2$Cl$_2$, K$_2$CO$_3$	**II**, R = t-Bu (24)	67
Ph	IBTA, MeOH	**I**, R = Ph; R^1 = Me (61) + **II**, R = Ph (27)	67
Ph	IBTA, AcOH	**I**, R = Ph; R^1 = Ac (62)	67
Ph	IBTA, CF$_3$CH$_2$OH	**II**, R = Ph (73)	67
Ph	IBTA, CH$_2$Cl$_2$, K$_2$CO$_3$	**II**, R = Ph (38)	67
2,6-(MeO)$_2$C$_6$H$_3$	IBTA, MeOH	**I**, R = 2,6-(MeO)$_2$C$_6$H$_3$; R^1 = Me (68)	67
2,6-(MeO)$_2$C$_6$H$_3$	IBTA, AcOH	**I**, R = 2,6-(MeO)$_2$C$_6$H$_3$; R^1 = Ac (57)	67
2,6-(MeO)$_2$C$_6$H$_3$	IBTA, CF$_3$CH$_2$OH	**II**, R = 2,6-(MeO)$_2$C$_6$H$_3$ (74)	67
2,6-(MeO)$_2$C$_6$H$_3$	IBTA, CH$_2$Cl$_2$, K$_2$CO$_3$	**II**, R = 2,6-(MeO)$_2$C$_6$H$_3$ (17)	67

TABLE VIII. FORMATION OF HYDROINDOLENONES BY OXIDATIVE CYCLIZATION OF PHENOLS

Substrate	Conditions	Product(s) and Yield(s) (%)	Refs.
C_{14-17}			
(4-HO-C₆H₄-CH₂-CH(NHR)-CO₂H)		(bicyclic product with HO, CO₂Me, N-R)	
R = Boc	IBD, MeOH, NaHCO₃	(54)	77
R = Cbz	IBD, MeOH, NaHCO₃	(54)	77
C_{16-17}			
(4-HO-C₆H₄-CH₂-CH₂-NRCOPh)		(bicyclic product with O₂CPh, N-R)	
R = Me	IBTA, CF₃CH₂OH, NaHCO₃	(54)	67
R = Et	IBTA, CF₃CH₂OH, NaHCO₃	(48)	67

TABLE IX. FORMATION OF OXYGEN HETEROCYCLIC COMPOUNDS BY OXIDATIVE CYCLIZATION OF o-ACYLPHENOLS

Substrate	Conditions	Product(s) and Yield(s) (%)	Refs.
Coumaran-3-ones			
C_8 (2-hydroxyacetophenone)	IBD, KOH, MeOH, 0–5° 1 h, rt 2 h	2,2-dimethoxy-benzofuran-3(2H)-one (20)	89
C_9 (2-hydroxypropiophenone)	IBD, KOH, MeOH, 0–5° 1 h, rt 2 h	2-methyl-2-methoxy-benzofuran-3(2H)-one (35)	89
C_{10}	IBD, KOH, MeOH, 0–5° 1 h, rt 2 h	2,2-dimethoxy-4,6-dimethyl-benzofuran-3(2H)-one (21)	89
C_{14} (bis-OTMS enol ether)	(PhIO)$_n$, BF$_3$·Et$_2$O, ether, H$_2$O, −40° 1 h, −40° to rt 1 h, rt 0.5 h	benzofuran-3(2H)-one (31) + 2-hydroxyphenacyl alcohol (25) + [2-hydroxyphenyl ketone]$_2$ (5)	92
C_{15}	IBD, KOH, MeOH, 0–5° 1 h, rt 2 h	2-benzyl-2-methoxy-benzofuran-3(2H)-one (40)	89

C$_{15-19}$

[starting material: 2-hydroxyphenyl 1,3-diketone with ArCO-CH2-CO- group, R1, R2 substituents]

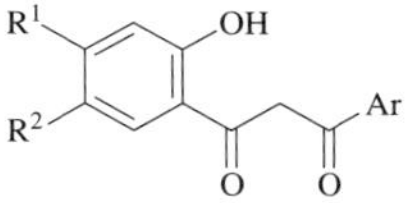

1. IBD, KOH, MeOH, 0–5° 1 h, rt 2 h
2. HCl (6 N)

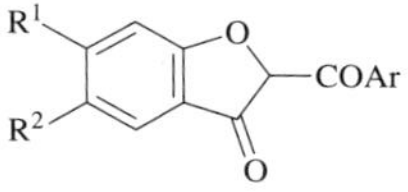

[product: benzofuran-3(2H)-one with COAr at 2-position, R1, R2 on benzene ring]

R^1	R^2	Ar		
H	H	Ph	(75)	90
H	Cl	Ph	(—)	90
Cl	Me	Ph	(—)	90
H	OMe	Ph	(—)	90
H	COMe	Ph	(82)	91
H	COMe	p-MeC$_6$H$_4$	(80)	91
H	COMe	p-MeOC$_6$H$_4$	(86)	91
H	COEt	p-MeC$_6$H$_4$	(79)	91
H	COEt	p-MeOC$_6$H$_4$	(82)	91

3,4-Dihydrocoumarins

C$_{11-17}$

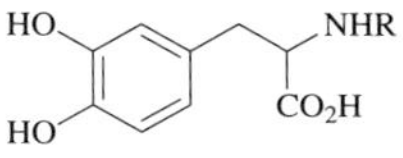

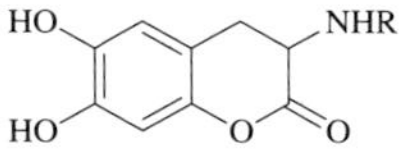

[starting material: 3,4-dihydroxyphenylalanine derivative with NHR, CO$_2$H]

[product: 6,7-dihydroxy-3,4-dihydrocoumarin with NHR at 3-position]

R		
Ac	IBD, AcOH	(9)	93
Boc	IBD, AcOH	(31)	93
Cbz	IBD, p-TsOH (cat.), AcOH	(39)	93

TABLE X. INTRAMOLECULAR CARBON-CARBON BOND FORMATION: PHENOLIC OXIDATIVE COUPLING

Substrate	Conditions	Product(s) and Yield(s) (%)	Refs.
C_{15-17}			
(structure with OH, R^1, R^2, R^3)	IBD, MeCN, reflux, 4 h	(structure) (30)	108

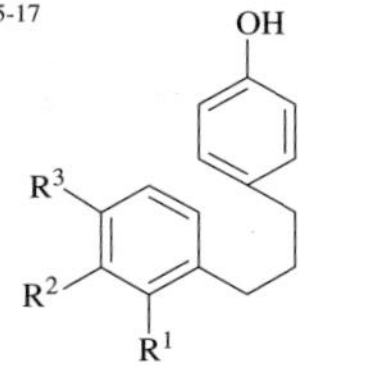

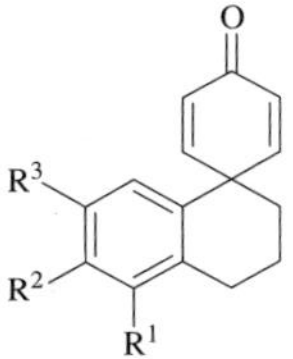

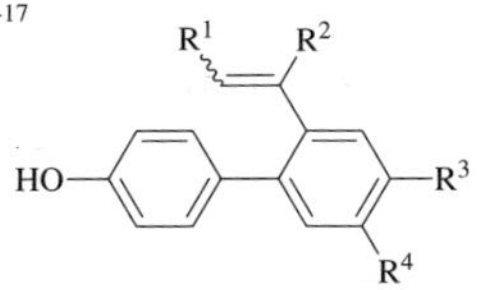

R^1	R^2	R^3
H	H	H
H	H	OMe
H	OMe	H
OMe	H	OMe
H	OMe	OMe

C_{15-17}

(substrate structure with HO, R^1, R^2, R^3, R^4)   IBD, MeOH   (spiro product with R^1, R^2, OMe, R^3, R^4)

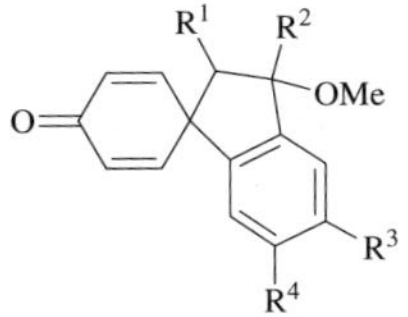

R^1	R^2	R^3	R^4	Yield	Ref.
H	H	H	OMe	(56)	66
H	Me	H	H	(67)	66
H	Me	OMe	H	(48)	66
H	Me	OMe	OMe	(76)	66
H	H	-[OCH$_2$O]-		(34)	66
H	Me	-[OCH$_2$O]-		(79)	66
Me	H	-[OCH$_2$O]-		(75)[a]	66
-(CH$_2$)$_3$-		H	H	(50)	66

394

C19 structure (with NMe, MeO, HO, OH, OMe)		IBD, CF$_3$CO$_2$H, CH$_2$Cl$_2$		(27)	98

R^1	R^2	X			
H	CO$_2$Me	H	IBD, CF$_3$CO$_2$H, CH$_2$Cl$_2$	(26)	99
H	CO$_2$Et	H	IBD, AcOH, CH$_2$Cl$_2$	(14-32)	96
H	CO$_2$Et	H	IBTA, CF$_3$CO$_2$H, CH$_2$Cl$_2$	(14-32)	96
H	CO$_2$Et	H	PhI(O$_2$CCCl$_3$)$_2$, CCl$_3$CO$_2$H, CH$_2$Cl$_2$	(14-32)	96
H	CO$_2$Et	H	IBD, CF$_3$CO$_2$H, CH$_2$Cl$_2$	(22)	98
H	CO$_2$Et	Cl	IBD, AcOH. CH$_2$Cl$_2$	(14-32)	96
H	CO$_2$Et	Cl	IBTA, CF$_3$CO$_2$H, CH$_2$Cl$_2$	(14-32)	96
H	CO$_2$Et	Cl	PhI(O$_2$CCCl$_3$)$_2$, CCl$_3$CO$_2$H, CH$_2$Cl$_2$	(14-32)	96
H	CO$_2$Et	Cl	Et$_4$N$^+$ (IOAc$_2$)$^-$, CH$_2$Cl$_2$, C$_5$H$_5$N$^+$H·O$_2$CCF$_3$	(—)	96
H	CO$_2$Et	Cl	Et$_4$N$^+$ [I(CO$_2$CF$_3$)$_2$]$^-$, CH$_2$Cl$_2$	(25-58)	96
H	CO$_2$Et	Cl	Et$_4$N$^+$ [I(CO$_2$CCl$_3$)$_2$]$^-$, CH$_2$Cl$_2$	(25-58)	96
H	CO$_2$Et	Br	IBTA, O$_2$NC$_6$H$_5$, CH$_2$Cl$_2$, −78°	(28)	97
H	CO$_2$Et	Br	IBD, AcOH. CH$_2$Cl$_2$	(14-32)	96
H	CO$_2$Et	Br	IBTA, CF$_3$CO$_2$H, CH$_2$Cl$_2$	(14-32)	96
H	CO$_2$Et	Br	PhI(O$_2$CCCl$_3$)$_2$, CCl$_3$CO$_2$H, CH$_2$Cl$_2$	(14-32)	96

TABLE X. INTRAMOLECULAR CARBON-CARBON BOND FORMATION: PHENOLIC OXIDATIVE COUPLING (Continued)

Substrate			Conditions	Product(s) and Yield(s) (%)	Refs.
R^1	R^2	X			
H	CO_2Et	Br	Et_4N^+ $(IOAc_2)^-$, CH_2Cl_2, $C_6H_5N^+H$ $^-O_2CCF_3$	(—)	96
H	CO_2Et	Br	Et_4N^+ $[I(CO_2CF_3)_2]^-$, CH_2Cl_2	(25-58)	96
H	CO_2Et	Br	Et_4N^+ $[I(CO_2CCl_3)_2]^-$, CH_2Cl_2	(25-58)	96
H	CHO	H	IBD, AcOH, CH_2Cl_2	(14-32)	96
H	CHO	H	IBTA, CF_3CO_2H, CH_2Cl_2	(14-32)	96
H	CHO	H	$PhI(O_2CCCl_3)_2$, CCl_3CO_2H, CH_2Cl_2	(14-32)	96
H	CHO	Cl	IBD, AcOH, CH_2Cl_2	(14-32)	96
H	CHO	Cl	IBTA, CF_3CO_2H, CH_2Cl_2	(14-32)	96
H	CHO	Cl	$PhI(O_2CCCl_3)_2$, CCl_3CO_2H, CH_2Cl_2	(14-32)	96
H	CHO	Cl	Et_4N^+ $(IOAc_2)^-$, CH_2Cl_2, $C_6H_5N^+H$ $^-O_2CCF_3$	(25-58)	96
H	CHO	Cl	Et_4N^+ $[I(CO_2CF_3)_2]^-$, CH_2Cl_2	(25-58)	96
H	CHO	Cl	Et_4N^+ $[I(CO_2CCl_3)_2]^-$, CH_2Cl_2	(25-58)	96
H	CHO	Br	IBD, AcOH, CH_2Cl_2	(14-32)	96
H	CHO	Br	IBTA, CF_3CO_2H, CH_2Cl_2	(14-32)	96
H	CHO	Br	$PhI(O_2CCCl_3)_2$, CCl_3CO_2H, CH_2Cl_2	(14-32)	96
H	CHO	Br	Et_4N^+ $(IOAc_2)^-$, CH_2Cl_2, $C_6H_5N^+H$ $^-O_2CCF_3$	(25-58)	96
H	CHO	Br	Et_4N^+ $[I(CO_2CF_3)_2]^-$, CH_2Cl_2	(25-58)	96
H	CHO	Br	Et_4N^+ $[I(CO_2CCl_3)_2]^-$, CH_2Cl_2	(25-58)	96
H	$COCF_3$	Br	IBTA, CH_2Cl_2, $-40°$	(21)	97
H	$COCF_3$	Br	o-$MeC_6H_4I(O_2CCF_3)_2$, CH_2Cl_2, $-40°$	(10)	97
H	$COCF_3$	Br	m-$MeC_6H_4I(O_2CCF_3)_2$, CH_2Cl_2, $-40°$	(12)	97
H	$COCF_3$	Br	p-$MeC_6H_4I(O_2CCF_3)_2$, CH_2Cl_2, $-40°$	(12)	97
H	$COCF_3$	Br	m-$MeOC_6H_4I(O_2CCF_3)_2$, CH_2Cl_2, $-40°$	(14)	97
H	$COCF_3$	Br	m-$O_2NC_6H_4I(O_2CCF_3)_2$, CH_2Cl_2, $-40°$	(8)	97
H	$COCF_3$	Br	p-$ClC_6H_4I(O_2CCF_3)_2$, CH_2Cl_2, $-40°$	(9)	97
H	$COCF_3$	Br	$(PhIO)_n$, CF_3CO_2H, $-40°$	(11)	97
CO_2Me	CO_2Me	H	IBD, CF_3CO_2H, CH_2Cl_2	(25)	99

C$_{20}$ structure	IBD, CF$_3$CO$_2$H, CH$_2$Cl$_2$	(66) + (4) 100
C$_{20-21}$ structure	IBTA (2 eq), CH$_2$Cl$_2$, propylene oxide (10 eq), −10°, 0.5 h	R: Me (—) 101; CH$_2$CCl$_3$ (13) 101
structure	IBTA, TFE, 1 h IBTA, TFE, 24 h	X: H (47) 106; OMe (13) 106

TABLE X. INTRAMOLECULAR CARBON-CARBON BOND FORMATION: PHENOLIC OXIDATIVE COUPLING (*Continued*)

Substrate	Conditions	Product(s) and Yield(s) (%)	Refs.
C$_{21}$			
	IBTA, TFE, 24 h	(40)	106
C$_{30}$			
	IBTA, TFE, 0.5 h	(—)^a	107
	IBTA, TFE, 24 h	(+) + (+) (44)	107

^{a} This product was not isolated, but HPLC analysis indicated that it was the major product.

TABLE XI. OXIDATIVE CYCLIZATION OF O-SILYLATED PHENOLS TO AZACARBOCYCLIC SPIRODIENONES

Substrate	Conditions	Product(s) and Yield(s) (%)	Refs.
C$_{17}$ (phenol substrate structure)	1. TMSO-C(OMe)=CHMe 2. IBTA, TFE, rt	spirodienone product (53)	103,104
C$_{17-18}$ (phenol substrate structure)	1. TMSO-C(OMe)=CHMe 2. IBTA, TFE, rt	spirodienone product	

R^1	R^2	X	Y		
H	H	CH	CH	(86)	102
Br	H	CH	CH	(42)	102
H	H	N	CH	(53)	103,104
H	H	CH	N	(64)	103,104
H	Me	N	N	(57)	103,104

TABLE XI. OXIDATIVE CYCLIZATION OF O-SILYLATED PHENOLS TO AZACARBOCYCLIC SPIRODIENONES (Continued)

Substrate	Conditions	Product(s) and Yield(s) (%)	Refs.
C₁₈	1. TMSO-C(OMe)=CHMe 2. IBTA, TFE, rt	(42)	103,104
C₂₀₋₂₄ R R¹ R² H H (CH₂)₂NHCOCF₃ H H (CH₂)₂NHTEOC Et Me CH₂OAc	1. TMSO-C(OMe)=CHMe 2. IBTA, TFE, rt	(58) (62) (86)	104 104 103,104
C₂₂	1. TMSO-C(OMe)=CHMe 2. IBTA, TFE, rt	(71)	103,104

TABLE XII. FORMATION OF IODONIUM YLIDES AND SALTS

Substrate	Conditions	Product(s) and Yield(s) (%)	Refs.
C_6			
2-Cl-4-NO_2-phenol	1. 2-Cl-C$_6$H$_4$-I(OAc)$_2$, AcOH 2. Product dried over KOH	(2-Cl-C$_6$H$_4$)I$^+$-(3-Cl-5-NO_2-2-O$^-$-C$_6$H$_2$) (—)	111
2,4-dinitrophenol	$C_6F_5I(O_2CCF_3)_2$, MeCN	$C_6F_5I^+$-(2,4-$(NO_2)_2$-6-O$^-$-C$_6$H$_2$) (10)	118
C_{6-9} 2-Y-4-X-phenol		Ph-I$^+$-(2-Y-4-X-6-O$^-$-C$_6$H$_2$)	

X	Y			
NO_2	H	IBD, AcOH	(84)	109
NO_2	H	IBD, AcOH	(—)	110
NO_2	NO_2	IBTA, MeCN	(65–70)	118
CHO	H	IBD, AcOH	(—)	110
COMe	H	IBD, AcOH	(—)	110
CO_2Et	H	IBD, AcOH	(—)	110
CO_2Me	NO_2	IBTA, MeCN	(30)	118

TABLE XII. FORMATION OF IODONIUM YLIDES AND SALTS (Continued)

Substrate	Conditions	Product(s) and Yield(s) (%)	Refs.

C_{6-12}

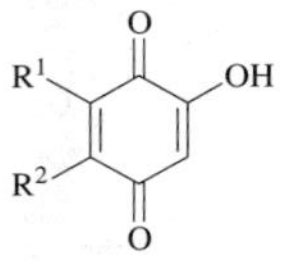

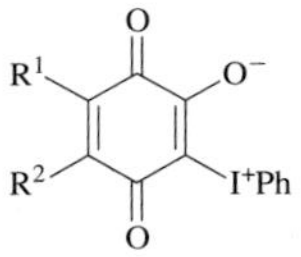

Substrate: 2-hydroxy-1,4-benzoquinone with R^1, R^2 substituents

Product: 2-oxido-6-(phenyliodonio)-1,4-benzoquinone with R^1, R^2 substituents

R^1	R^2			
H	H	IBD, CH$_2$Cl$_2$	(—)	121
H	Br	IBD, CH$_2$Cl$_2$	(—)	121
H	Me	IBD, CH$_2$Cl$_2$	(91)	121
H	Ph	IBD, CH$_2$Cl$_2$	(93)	121
Me	Me	IBD, CH$_2$Cl$_2$	(60)	121

C_{8-12}

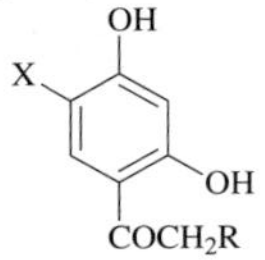

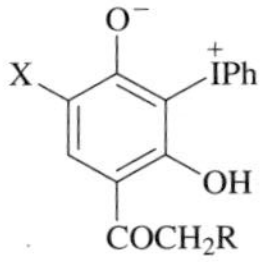

Substrate: 4-X-substituted resorcinol with COCH$_2$R group

Product: iodonium ylide (phenyliodonio-oxido) with COCH$_2$R and X

R	X			
H	H	IBD, KOH, MeOH	(40)	112,119
H	H	TsN=IPh, CH$_2$Cl$_2$	(30)	119
H	Br	IBD, KOH, MeOH	(35)^a	112
H	NO$_2$	IBD, KOH, MeOH	(45)	112,119
H	NO$_2$	TsN=IPh, CH$_2$Cl$_2$	(30)	119
H	COMe	IBD, KOH, MeOH	(40)	112,119
H	COMe	TsN=IPh, CH$_2$Cl$_2$	(28)	119
Me	NO$_2$	IBD, KOH, MeOH	(48)	161
Me	COEt	IBD, KOH, MeOH	(60)^a	161

C$_9$	RC$_6$H$_4$I(OAc)$_2$, MeOH, rt; R = H, Me, OMe	(60-65) 122
C$_{9-15}$	RC$_6$H$_4$I(OAc)$_2$	
X = O, R = H		(93) 162,163
X = NH, R = H		(90) 162
X = NH, R = Me		(84) 162
X = NH, R = OMe		(92) 162
X = NMe, R = H		(93) 162
X = NPh, R = H		(93) 162
C$_{10}$	IBD, CHCl$_3$, 0°	(92) 115
	HTIB, MeOH, rt	(81) 164

TABLE XII. FORMATION OF IODONIUM YLIDES AND SALTS (Continued)

Substrate	Conditions	Product(s) and Yield(s) (%)	Refs.
C_{10-11}			
![substrate with OH, Me, N, R]		![product with OH, +IPh, -OTs, Me, N, R]	
R			
H	HTIB, MeCN, CH_2Cl_2	(85)	120
Cl	HTIB, MeCN, CH_2Cl_2	(90)	120
Me	HTIB, MeCN, CH_2Cl_2	(82)	120
OMe	HTIB, MeCN, CH_2Cl_2	(80)	120
![substrate with OH, Me, N, R]		![product with O−, +IPh, Me, N, R]	
R			
H	1. HTIB, MeCN, CH_2Cl_2 2. K_2CO_3	(75)	120
Cl	1. HTIB, MeCN, CH_2Cl_2 2. K_2CO_3	(83)	120
Me	1. HTIB, MeCN, CH_2Cl_2 2. K_2CO_3	(73)	120
OMe	1. HTIB, MeCN, CH_2Cl_2 2. K_2CO_3	(68)	120

[a] The iodonium ylide was unstable, and the product was isolated as a rearranged iodoether.

TABLE XIII. *p*-THIOCYANATION OF PHENOLS

Substrate	Conditions	Product(s) and Yield(s) (%)	Refs.
C_{6-14}			
(phenol with R^1, R^2, R^3)	PhICl$_2$, Pb(SCN)$_2$, dry CH$_2$Cl$_2$, 0°	(product with SCN para to OH)	123

R^1	R^2	R^3	Yield
H	H	H	(93)
CN	H	H	(61)
Me	H	Me	(78)
t-Bu	H	*t*-Bu	(97)
Cl	H	Cl	(64)
H	Me	H	(95)
Me	Me	Me	(94)
CH$_2$=CH-CH$_2$	Me	Me	(77)
COMe	Me	Me	(78)

C_{10-16}

(naphthol with R^1, R^2) — PhICl$_2$, Pb(SCN)$_2$, dry CH$_2$Cl$_2$, 0° — 123

R^1	R^2	Yield
H	H	(88)
Me	Me	(67)
COMe	H	(97)
CO$_2$Et	Me	(97)
CO$_2$Et	CO$_2$Et	(58)
CO$_2$Et	CH$_2$OAc	(88)

TABLE XIII. *p*-THIOCYANATION OF PHENOLS (*Continued*)

Substrate	Conditions	Product(s) and Yield(s) (%)	Refs.
C_{10}			
5-hydroxy-1-methyl-3,4-dihydroquinolin-2(1H)-one	PhICl$_2$, Pb(SCN)$_2$, dry CH$_2$Cl$_2$, 0°	8-SCN derivative (65)	123

TABLE XIV. FORMATION OF 2,3-DIMETHOXY-3-HYDROXYFLAVANONES BY OXIDATION OF FLAVONOLS AND ANALOGS

C_{13-19}

Substrate			Conditions	Product(s) and Yield(s) (%)	Refs.
Ar	R^1	R^2			
Ph	H	H	HTIB, MeOH	(65)	124
Ph	OMe	H	HTIB, MeOH	(72)	124
Ph	H	COEt	HTIB, MeOH	(90-92)	125
Ph	H	COEt	IBD, MeOH	(90-92)	125
Ph	H	COEt	IBTA, MeOH	(92)	125
4-ClC$_6$H$_4$	H	H	HTIB, MeOH	(73)	124
4-ClC$_6$H$_4$	H	COEt	IBTA, MeOH	(92)	125
4-ClC$_6$H$_4$	H	COEt	HTIB, MeOH	(92)	125
4-ClC$_6$H$_4$	H	COEt	IBD, MeOH	(92)	125
4-MeC$_6$H$_4$	H	COEt	IBTA, MeOH	(90)	125
4-MeC$_6$H$_4$	H	COEt	IBD, MeOH	(90-92)	125
4-MeC$_6$H$_4$	H	COEt	HTIB, MeOH	(90-92)	125
4-MeOC$_6$H$_4$	H	H	HTIB, MeOH	(75)	124
4-MeOC$_6$H$_4$	H	COEt	IBTA, MeOH	(90)	125
4-MeOC$_6$H$_4$	H	COEt	IBD, MeOH	(90)	125
4-MeOC$_6$H$_4$	OMe	H	HTIB, MeOH	(45)	124
3,4-(MeO)$_2$C$_6$H$_3$	H	H	HTIB, MeOH	(77)	124
2-furyl	H	COEt	IBTA, MeOH	(80)	125
2-thienyl	H	COEt	IBTA, MeOH	(85)	125

TABLE XIV. FORMATION OF 2,3-DIMETHOXY-3-HYDROXYFLAVANONES BY OXIDATION OF FLAVONOLS AND ANALOGS (*Continued*)

Substrate	Conditions	Product(s) and Yield(s) (%)	Refs.
C₁₉ (2-phenyl-3-hydroxy-4H-naphtho[2,1-b]pyran-4-one)	HTIB, MeOH	2,3-dimethoxy analog (71)	124

TABLE XV. CLEAVAGE OF NH$_2$-TERMINAL TYROSYL PEPTIDES

Substrate	Conditions	Product(s) and Yield(s) (%)	Refs.
C$_8$ (tyramine)	IBD, KOH, MeOH, 0°	4-HOC$_6$H$_4$CH$_2$OMe (65)	127
C$_9$ (tyrosine X = OH or NH$_2$)	IBD, KOH, MeOH, 0°	4-HOC$_6$H$_4$CH$_2$CN (52), (61)	127
C$_{11-18}$ (Tyr-NHCHR-CO$_2$H)		4-HOC$_6$H$_4$CH$_2$OMe	
R = H	IBD, KOH, MeOH, 0°	(62)	127
Me	IBD, KOH, MeOH, 0°	(71)	127
Pr-i	IBD, KOH, MeOH, 0°	(65)	127
CH$_2$C$_6$H$_4$OH-p	IBD, KOH, MeOH, 0°	(78)	127
C$_{20}$ (Trp-Tyr)	IBD, KOH, MeOH, 0°	3-(MeOCH$_2$)indole (62)	126

TABLE XVI. SYNTHESIS OF 2,3-DIHYDROBENZOFURANS BY OXIDATIVE CYCLOADDITIONS OF PHENOLS TO PROPENYLBENZENES

Substrate	Conditions	Product(s) and Yield(s) (%)	Refs.

C_{7-10}

Substrate **I**: phenol with OMe, R^1, R^2, R^3; Substrate **II**: R^4-CH=CH-Ar

Conditions: IBTA, MeCN

Product: 2,3-dihydrobenzofuran with OMe, R^1, R^2, R^3, R^4, Ar substituents

Ref. 131

R^1	R^2	R^3	R^4	Ar	Molar ratio of **I:II**	Yield (%)
H	H	H	Me[t]	1,2-(MeO)$_2$C$_6$H$_3$	1.1	(67)
H	H	H	Me[t]	1,2-(MeO)$_2$C$_6$H$_3$	4.0	(71)
H	H	H	Me[c]	1,2-(MeO)$_2$C$_6$H$_3$	1.5	(50)
H	H	H	H	4-MeOC$_6$H$_4$	1.0	(57)
H	OMe	H	Me[t]	1,2-(MeO)$_2$C$_6$H$_3$	1.0	(26)
H	OMe	H	Me[c]	1,2-(MeO)$_2$C$_6$H$_3$	1.5	(23)
H	OMe	H	Me[t]	1,2-(MeO)$_2$C$_6$H$_3$	3.6	(64)
H	OMe	H	H	4-MeOC$_6$H$_4$	4.0	(34)
H	Me	H	Me[t]	1,2-(MeO)$_2$C$_6$H$_3$	1.1	(81)
Allyl	H	H	Me[t]	1,2-(MeO)$_2$C$_6$H$_3$	3.0	(68)
Allyl	H	Cl	Me[t]	1,2-(MeO)$_2$C$_6$H$_3$	4.0	(53)

C_{10}

Substrate: 2-allyloxy-4-hydroxy-1-methoxybenzene + 5-propenyl-1,3-benzodioxole

Conditions: IBTA, MeCN

Product: dihydrobenzofuran with OMe, OAllyl, methylenedioxyphenyl, methyl substituents (69)

Ref. 131

C_{7-10}

Reactants	Conditions	Product	Ref.
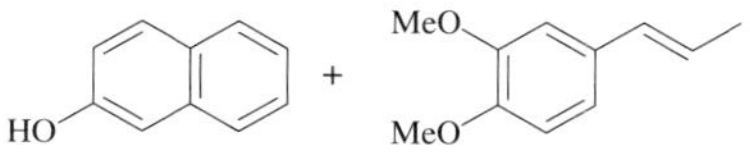 2-naphthol + (E)-1-(3,4-dimethoxyphenyl)propene	IBTA, MeCN	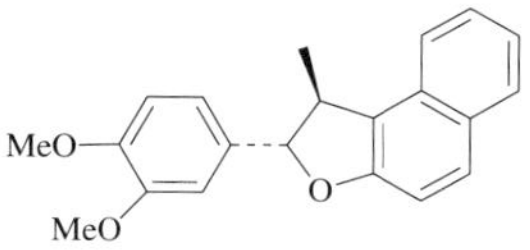 (39) dihydrofuran product	131
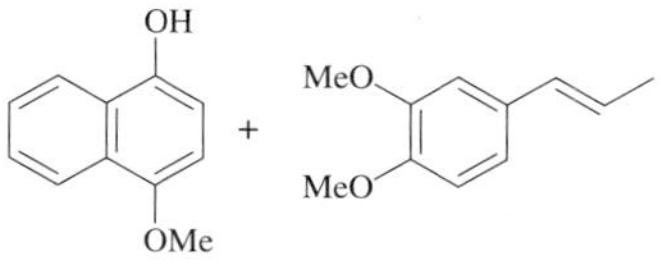 1,4-dihydroxy-4-methoxynaphthalene + (E)-1-(3,4-dimethoxyphenyl)propene	IBTA, MeCN	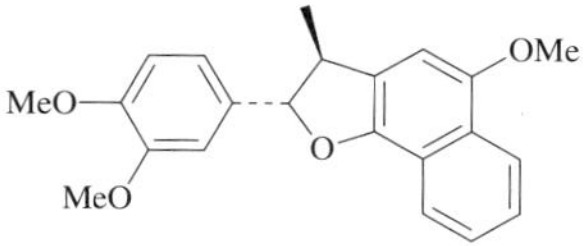 (44) OMe-substituted dihydrofuran product	131

---

[c] Denotes a *cis* isomer.
[t] Denotes a *trans* isomer.

REFERENCES

[1] Humphries, S. G. In *Biogenesis of Natural Products*; Bernfeld, P., Ed.; McMillan: New York, 1963, p. 617.
[2] Musso, H. *Angew. Chem., Int. Ed. Engl.* **1963**, *2*, 723.
[3] Musso, H. *Angew. Chem.* **1963**, *75*, 965.
[4?] Scott, A. I. *Quart. Rev.* **1965**, *19*, 1.
[5] *Oxidative Coupling of Phenols;* Taylor, W. I., Battersby, A. R., Eds.; Arnold: London, 1967.
[6] Mihailovic, M. L. J.; Cekovic, Z. In *The Chemistry of the Hydroxyl Group, Part 1*; Patai, S., Ed.; Wiley: London, 1971.
[7] Kametani, T.; Fukumoto, K. *Synthesis* **1972**, 657.
[8] *Oxidation in Organic Chemistry, Part B*; Trahanovsky, W. S., Ed.; Academic: New York, 1973, p. 97.
[9] Haslam, E. *The Shikimate Pathway*; Butterworths: London, 1974.
[10] Higuchi, T. In *Biosynthesis and Biodegradation of Wood Components*; Higuchi, T., Ed.; Academic: London, 1985, Ch. 7.
[11] Ayers, D. C.; Loike, J. D. In *Lignans: Chemical, Biological, and Clinical Properties*; Cambridge University Press: Cambridge, 1990, Ch. 7.
[12] Whiting, D. A. In *Comprehensive Organic Chemistry;* Stoddart, J. F., Ed.; Pergamon: Oxford, Vol. 1, 1979, p. 707.
[13] Deya, P. M.; Dopico, M.; Raso, A. G.; Morey, J.; Saa, J. M. *Tetrahedron* **1987**, *43*, 3523.
[14] Taub, D.; Kuo, C. H.; Slates, H. L.; Wendler, N. L. *Tetrahedron* **1963**, *19*, 1.
[15] Dewar, M. J. S.; Nakaya, T. *J. Am. Chem. Soc.* **1968**, *90*, 7134.
[16] Schwartz, M. A.; Mami, I. *J. Am. Chem. Soc.* **1975**, *97*, 1239.
[17] McKillop, A.; Perry, D. H.; Edwards, M.; Antus, S.; Farkas, L.; Nogradi, M.; Taylor, E. C. *J. Org. Chem.* **1976**, *41*, 282.
[18] Brussee, J.; Jansen, A. C. A. *Tetrahedron Lett.* **1983**, *24*, 3261.
[19] Barton, D. H. R.; Deflorin, A. M.; Edwards, O. E. *J. Chem. Soc.* **1956**, 530.
[20] Thyagarajan, B. S. *Chem. Rev.* **1958**, *58*, 439.
[21] Battersby, A. R.; Brown, T. H.; Clements, J. H. *J. Chem. Soc.* **1965**, 4550.
[22] Hewgill, F. R.; Middleton, B. S. *J. Chem. Soc. C* **1967**, 2316.
[23] Pelter, A.; Bradshaw, J.; Warren, R. F. *Phytochem.* **1971**, *10*, 835.
[24] Balogh, V.; Fetizon, M.; Golfier, M. *J. Org. Chem.* **1971**, *36*, 1339.
[25] Tobinaga, S.; Kotani, E. *J. Am. Chem. Soc.* **1972**, *94*, 309.
[26] Varvoglis, A. *Chem. Soc. Rev.* **1981**, *10*, 377.
[27] Umemoto, T. *Yuki, Gosei Kagaku Kyokaishi* **1983**, *41*, 251; *Chem. Abstr.* **1983**, *98*, 214835y.
[28] Koser, G. F. In *The Chemistry of Functional Groups, Suppl. D*; Patai, S., Rappoport, Z., Eds.; Wiley: New York, 1983, Ch. 18 and 25.
[29] Nguyen, T. T.; Martin, J. C. In *Comprehensive Heterocyclic Chemistry*; Katritzky, A. R., Rees, C. W., Eds.; Pergamon: Oxford, 1984, Vol. 1, p. 563.
[30] Varvoglis, A. *Synthesis* **1984**, 709.
[31] Ochiai, M.; Nagao, Y. *Yuki, Gosei Kagaku Kyokaishi* **1986**, *44*, 660; *Chem. Abstr.* **1987**, *106*, 84682s.
[32] Moriarty, R. M.; Prakash, O. *Acc. Chem. Res.* **1986**, *19*, 244.
[33] Merkushev, E. B. *Russ. Chem. Rev. (Engl. Transl.)* **1987**, *56*, 826.
[34] Moriarty, R. M.; Vaid, R. K.; Koser, G. F. *Synlett* **1990**, 365.
[35] Moriarty, R. M.; Vaid, R. K. *Synthesis* **1990**, 431.
[36] Varvoglis, A. *The Organic Chemistry of Polycoordinated Iodine*; VCH: New York, 1992.
[36a] Kita, Y.; Tohma, H.; Yakura, T. *Trends in Organic Chemistry* **1992**, *3*, 113.
[37] Stang, P. J. *Angew. Chem., Int. Ed. Engl.* **1992**, *31*, 274.
[38] Prakash, O.; Saini, N.; Sharma, P. K. *Synlett* **1994**, 221.
[39] Prakash, O.; Singh, S. P. *Aldrichimica Acta* **1994**, *27*, 15.
[40] Koser, G. F. In *The Chemistry of Halides, Pseudo-Halides and Azides, Suppl. D*; Patai, S., Rappoport, Z., Eds.; Wiley: New York, 1995, Ch. 21.

[41] Prakash, O.; Saini, N.; Tanwar, M. P.; Moriarty, R. M. *Cont. Org. Synth.* **1995**, *2*, 121.
[42] Prakash, O. *Aldrichimica Acta* **1995**, *28*, 63.
[43] Stang, P. J.; Zhdankin, V. V. *Chem. Rev.* **1996**, *96*, 1123.
[44] Varvoglis, A. *Hypervalent Iodine in Organic Synthesis*; Academic: New York, 1996.
[45] Varvoglis, A. *Tetrahedron* **1997**, *53*, 1179.
[46] Moriarty, R. M.; Prakash, O. In *Advances in Heterocyclic Chemistry*; Katritzky, A. R., Ed.; Academic: New York, 1998, Vol. 69, Ch. 1.
[46a] Moriarty, R. M.; Prakash, O. *Org. React.* **1999**, *54*, 273.
[47] Kitamura, T.; Fujiwara, Y. *Org. Prep. Proced. Int.* **1997**, *29*, 411.
[47a] Wirth, T.; Hirt, U. H. *Synthesis* **1999**, 1271.
[48] Kurti, L.; Herczegh, P; Visy, J.; Simonyi, M.; Antus, S.; Pelter, A. *J. Chem. Soc., Perkin Trans. 1* **1999**, 379.
[48a] Siegel, A.; Antony, F. *Monatsh. Chem.* **1955**, *86*, 292.
[49] Tamura, Y.; Yakura, T.; Tohma, H.; Kikuchi, K.; Kita, Y. *Synthesis* **1989**, 126.
[50] Barret, R.; Daudon, M. *Synth. Commun.* **1990**, *20*, 2907.
[51] Barton, D. H. R.; Godfrey, C. R. A.; Morzycki, J. W.; Mortherwell, W. B.; Stobie, A. *Tetrahedron Lett.* **1982**, *23*, 957.
[52] Pelter, A.; Elgendy, S. *Tetrahedron Lett.* **1988**, *29*, 677.
[53] Pelter, A.; Elgendy, S. M. A. *J. Chem. Soc., Perkin Trans. 1* **1993**, 1891.
[54] Saby, C.; Luong, J. H. T. *J. Chem. Soc., Chem. Commun.* **1997**, 1197.
[55] Yoshino, S.; Hayakawa, K.; Kanematsu, K. *J. Org. Chem.* **1981**, *46*, 3841.
[56] Hayakawa, K.; Aso, M.; Kanematsu, K. *J. Org. Chem.* **1985**, *50*, 2036.
[57] Kanematsu, K.; Morita, S.; Fukushima, S.; Osawa, E. *J. Am. Chem. Soc.* **1981**, *103*, 5211.
[58] Barret, R.; Daudon, M. *Tetrahedron Lett.* **1990**, *31*, 4871.
[59] Saitz, B. C.; Valderrama, J. A.; Tapia, R.; Farina, F.; Paredes, M. C. *Synth. Commun.* **1992**, *22*, 955.
[60] Barton, D. H. R.; Magnus, P. D.; Quinney, J. C. *J. Chem. Soc., Perkin Trans. 1* **1975**, 1610.
[61] Simoneau, B.; Brassard, P. *Tetrahedron* **1986**, *42*, 3767.
[62] McKillop, A.; McLaren, L.; Taylor, R. J. K. *J. Chem. Soc., Perkin Trans. 1* **1994**, 2047.
[63] Fleck, A. E.; Hobart, J. A.; Morrow, G. W. *Synth. Commun.* **1992**, *22*, 179.
[63a] Kurti, L.; Sazilagyi, L.; Antus, S.; Nogradi, M. *Eur. J. Org. Chem.* **1999**, 2579.
[64] Chu, C.-S.; Lee, T.-H.; Liao, C.- C. *Synlett* **1994**, 635.
[64a] Lee, T.-H; Liao, C.-C; Liu, W.-C. *Tetrahedron Lett.* **1996**, *37*, 5897.
[65] Lewis, N.; Wallbank, P. *Synthesis* **1987**, 1103.
[66] Callinan, A.; Chen, Y.; Morrow, G. W.; Swenton, J. S. *Tetrahedron Lett.* **1990**, *31*, 4551.
[67] Kita, Y.; Tohma, H.; Kikuchi, K.; Inagaki, M.; Yakura, T. *J. Org. Chem.* **1991**, *56*, 435.
[68] Matsuura, T.; Cahnmann, H. J. *J. Am. Chem. Soc.* **1960**, *82*, 2055.
[69] Matsuura, T.; Nishinaga, A. *J. Org. Chem.* **1962**, *27*, 3072.
[70] Mallik, U. K.; Mallik, A. K. *Indian J. Chem.* **1992**, *30B*, 611.
[71] Mal, D.; Roy, H. N.; Hazra, N. K.; Adhikari, S. *Tetrahedron* **1997**, *53*, 2177.
[72] Mitchell, A. S.; Russell, R. A. *Tetrahedron Lett.* **1993**, *34*, 545.
[72a] Mitchell, A. S.; Russell, R. A. *Tetrahedron* **1997**, *53*, 4387.
[73] Karam, O.; Jacquesy, J.-C.; Jouannetaud, M. P. *Tetrahedron Lett.* **1994**, 35, 2541.
[74] Bennett, D. J.; Dean, F. M.; Herbin, G. A.; Matkin, D. A.; Price, A. W.; Robinson, M. L. *J. Chem. Soc., Perkin Trans. 1* **1980**, 1978.
[75] Dean, F. M.; Herbin, A.; Matkin, D. A.; Price, A. W.; Robinson, M. L. *J. Chem. Soc., Perkin Trans. 1* **1980**, 1986.
[76] McKillop, A.; McLaren, L.; Taylor, R. J. K.; Watson, R. J.; Lewis, N. *Synlett* **1992**, 201.
[76a] McKillop, A.; McLaren, L.; Taylor, R. J. K; Watson, R. J.; Lewis, N. *J. Chem. Soc., Perkin Trans. 1* **1996**, 1386.
[77] Wipf, P.; Kim, Y. *Tetrahedron Lett.* **1992**, *33*, 5477.
[78] Rao, A. V. R.; Gurjar, M. K.; Sharma, P. A. *Tetrahedron Lett.* **1991**, *32*, 6613.
[79] Wipf, P.; Kim, Y. *J. Org. Chem.* **1993**, *58*, 1649.

[80] Wipf, P.; Kim, Y.; Fritch, P. C. *J. Org. Chem.* **1993**, *58*, 7195.
[81] Fehlhaber, H. W.; Kogler, H.; Mukhopadhyay, T.; Vijayakumar, E. K. S.; Ganguli, B. N. *J. Am. Chem. Soc.* **1988**, *110*, 8242.
[82] Roy, K.; Mukhopadhyay, T.; Reddy, G. C. S.; Desikan, K. R.; Rupp, R. H.; Ganguli, B. N. *J. Antibiot.* **1988**, *41*, 1780.
[83] Fehllhaber, H. W.; Kogler, H.; Mukhopadhyay, T.; Vijaykumar, E. K. S.; Roy, K.; Ganguli, B. N. *J. Antibiot.* **1988**, *41*, 1785.
[84] Tamura, Y.; Yakura, T.; Haruta, J.; Kita, Y. *J. Org. Chem.* **1987**, *52*, 3927.
[84a] Pelter, A.; Hussain, A.; Smith, G.; Ward, R. S. *Tetrahedron* **1997**, *53*, 3879.
[85] Kacan, M.; Koyuncu, D.; McKillop, A. *J. Chem. Soc., Perkin Trans. 1* **1993**, 1771.
[86] Murakata, M.; Yamada, K.; Hoshino, O. *J. Chem. Soc., Chem. Commun.* **1994**, 443.
[87] Murakata, M.; Tamura, M.; Hoshino, O. *J. Org. Chem.* **1997**, *62*, 4428.
[88] Hara, H.; Inoue, T.; Nakamura, H.; Endoh, M.; Hoshino, O. *Tetrahedron Lett.* **1992**, *33*, 6491.
[88a] Braun, N. A.; Ciufolini, M. A.; Peters, K.; Peters, E.-M. *Tetrahedron Lett.* **1998**, *39*, 4667.
[89] Moriarty, R. M.; Prakash, O.; Prakash, I.; Musallam, H. A. *J. Chem. Soc., Chem. Commun.* **1984**, 1342.
[90] Prakash, O.; Goyal, S.; Pahuja, S.; Singh, S. P. *Synth. Commun.* **1990**, *20*, 1409.
[91] Khanna, M. S.; Sangeeta; Garg, C. P.; Kapoor, R. P. *Synth. Commun.* **1992**, *22*, 2555.
[92] Moriarty, R. M.; Prakash, O.; Duncan, M. P. *Synth. Commun.* **1986**, *16*, 1239.
[93] Kaiser, A.; Koch, W.; Scheer, M.; Wolcke, U. *Helv. Chim. Acta* **1970**, *53*, 1708.
[94] Dhingra, O. P. In *Oxidation in Organic Chemistry, Part D*; Trahanovsky, W. S., Ed., Academic: New York, 1982, p. 97.
[95] Barton, D. H. R.; Bhakuni, D. S.; James, R.; Kirby, G. W. *J. Chem. Soc. C* **1967**, 128.
[96] Szantay, C.; Blasko, G.; Barczai-Beke, M.; Pechy, P.; Dornyei, G. *Tetrahedron Lett.* **1980**, *21*, 3509.
[97] White, J. D.; Caravatti, G.; Kline, T. B.; Edstrom, E.; Rice, K. C.; Brossi, A. *Tetrahedron* **1983**, *39*, 2393.
[98] Vanderlaan, D. G.; Schwartz, M. A. *J. Org. Chem.* **1985**, *50*, 743.
[99] Schwartz, M. A.; Pham, P. T. K. *J. Org. Chem.* **1988**, *53*, 2318.
[100] Burnett, D. A.; Hart, D. J. *J. Org. Chem.* **1987**, *52*, 5662.
[101] White, J. D.; Chong, W. K. M.; Thirring, K. *J. Org. Chem.* **1983**, *48*, 2300.
[102] Kita, Y.; Yakura, T.; Tohma, H.; Kikuchi, K.; Tamura, Y. *Tetrahedron Lett.* **1989**, *30*, 1119.
[103] Kita, Y.; Tohma, H.; Inagaki, M.; Hatanaka, K.; Kikuchi, K.; Yakura, T. *Tetrahedron Lett.* **1991**, *32*, 2035.
[104] Kita, Y.; Tohma, H.; Inagaki, M.; Hatanaka, K.; Yakura, T. *J. Am. Chem. Soc.* **1992**, *114*, 2175.
[105] Pelter, A.; Ward, R. S.; Abd-El-Ghani, A. *J. Chem. Soc., Perkin Trans. 1* **1992**, 2249.
[106] Ward, R. S.; Pelter, A.; Abd-El-Ghani, A. *Tetrahedron* **1996**, *52*, 1303.
[107] Pelter, A.; Ward, R. S.; Abd-El-Ghani, A. *Tetrahedron Asymmetry* **1994**, *5*, 329.
[108] Rama Krishna, K. V.; Sujatha, K.; Kapil, R. S. *Tetrahedron Lett.* **1990**, *31*, 1351.
[109] Fox, A. R.; Pausacker, K. H. *J. Chem. Soc.* **1957**, 295.
[110] Kokil, P. B.; Nair, P. M. *Tetrahedron Lett.* **1977**, 4113.
[111] Page, S. W.; Mazzola, E. P.; Mighell, A. D.; Himes, V. L.; Hubbard, C. R. *J. Am. Chem. Soc.* **1979**, *101*, 5858.
[112] Prakash, O.; Tanwar, M. P.; Goyal, S.; Pahuja, S. *Tetrahedron Lett.* **1992**, *33*, 6519.
[113] Papadopoulou, M.; Spyroudis, S.; Varvoglis, A. *J. Org. Chem.* **1985**, *50*, 1509.
[114] Spyroudis, S. P. *J. Org. Chem.* **1986**, *51*, 3453.
[115] Hatzigrigoriou, E.; Spyroudis, S.; Varvoglis, A. *Justus Liebigs Ann. Chem.* **1989**, 167.
[116] Pongratz, E.; Kappe, T. *Monatsh. Chem.* **1984**, *115*, 231.
[117] Spyroudis, S. P. *Justus Liebigs Ann. Chem.* **1986**, 947.
[118] Spyroudis, S.; Varvoglis, A. *J. Chem. Soc., Perkin Trans. 1* **1984**, 135.
[119] Spyroudis, S.; Tarantili, P. *Tetrahedron* **1994**, *50*, 11541.
[120] Prakash, O.; Kumar, D.; Saini, R. K.; Singh, S. P. *Tetrahedron Lett.* **1994**, *35*, 4211.
[121] Papoutsis, I.; Spyroudis, S.; Varvoglis, A. *Tetrahedron Lett.* **1994**, *35*, 8449.

[122] Georgantji, A.; Spyroudis, S. *Tetrahedron Lett.* **1995**, *36*, 443.
[123] Kita, Y.; Okuno, T.; Egi, M.; Iio, K.; Takeda, Y.; Akai, S. *Synlett* **1994**, 1039.
[123a] Kita, Y.; Takeda, Y.; Okuno, T.; Egi, M.; Iio, K.; Kawagauchi, K.; Akai, S. *Chem. Pharm. Bull.* **1997**, *45*, 1887.
[124] Moriarty, R. M.; Prakash, O.; Musallam, H. A.; Mahesh, V. K. *Heterocycles* **1986**, *24*, 1641.
[125] Khanna, M. S.; Sangeeta; Garg, C. P.; Kapoor, R. P. *Synth. Commun.* **1992**, *22*, 893.
[126] Smith, M. A.; Webb, R. A.; Cline, L. J. *J. Org. Chem.* **1965**, *30*, 995.
[127] Moriarty, R. M.; Sultana, M.; Ku, Y-Y. *J. Chem. Soc., Chem. Commun.* **1985**, 974.
[128] Mahoney, W. C.; Smith, P. K.; Hermodson, M. A. *Biochemistry* **1981**, *20*, 443.
[129] Fontana, A.; Dalzoppo, D.; Grandi, C.; Zambonin M. *Biochemistry* **1981**, *20*, 6997.
[130] Reddy, G. S. *Tetrahedron Lett.* **1995**, *36*, 1001.
[131] Gates, B. D.; Dalidowicz, P.; Tebben, A.; Wang, S.; Swenton, J. S. *J. Org. Chem.* **1992**, *57*, 2135.
[132] Shair, M. D.; Yoon, T.; Danishefsky, S. J. *Angew. Chem., Int. Ed. Engl.* **1995**, *34*, 1721.
[133] Cason, J. *Org. React.* **1948**, *4*, 305.
[134] McKillop, A.; Swann, B. P.; Taylor, E. C. *Tetrahedron* **1970**, *26*, 4031.
[135] Boldt, P. *Chem. Ber.* **1967**, *100*, 1270.
[136] Snyder, C. D.; Rapoport, H. *J. Am. Chem. Soc.* **1974**, *96*, 8046.
[137] Syper, L.; Kloc, K.; Mlochowski, J.; Szulc, Z. *Synthesis* **1979**, 521.
[138] Luly, J. R.; Rapoport, H. *J. Org. Chem.* **1981**, *46*, 2745.
[139] Becker, H.-D. *J. Org. Chem.* **1965**, *30*, 982.
[140] Nilsson, A.; Ronlan, A.; Parker, V. D. *Tetrahedron Lett.* **1975**, 1107.
[141] Foster, C. H.; Payne, D. A. *J. Am. Chem. Soc.* **1978**, *100*, 2834.
[142] Chen, C.-P.; Swenton, J. S. *J. Chem. Soc., Chem. Commun.* **1985**, 1291.
[143] Stewart, R. F.; Miller, L. L. *J. Am. Chem. Soc.* **1980**, *102*, 4999.
[144] Swenton, J. S. *Acc. Chem. Res.* **1983**, *16*, 74.
[145] Jacob, P., III; Callery, P. S.; Shulgin, A. T.; Castagnoli, N., Jr. *J. Org. Chem.* **1976**, *41*, 3627.
[146] Crouse, D. J.; Wheeler, M. M.; Goemann, M.; Tobin, P. S.; Basu, S. K.; Wheeler, D. M. S. *J. Org. Chem.* **1981**, *46*, 1814.
[147] Sharefkin, J. G.; Saltzman, H. *Org. Synth.* **1963**, *43*, 62; *Coll. Vol. V* **1973**, 660.
[148] Pausacker, K. H. *J. Chem. Soc.* **1953**, 107.
[149] Lucas, H. J.; Kennedy, E. R. *Org. Synth. Coll. Vol. III* **1955**, 482.
[150] McKillop, A.; Kemp, D. *Tetrahedron* **1989**, *45*, 3299.
[151] Spyroudis, S.; Varvoglis, A. *Synthesis* **1975**, 445.
[152] Neiland, O.-Y.; Karele, B.-Y. *J. Org. Chem. USSR (Engl. Tranl.)* **1970**, *6*, 889.
[153] Koser, G. F.; Wettach, R. H. *J. Org. Chem.* **1977**, *42*, 1476.
[154] Bayraktaroglou, T. O.; Gooding, M. A.; Khatib, S. F.; Lee, H.; Kourouma, M.; Landolt, R. G. *J. Org. Chem.* **1993**, *58*, 1264.
[155] Kita, Y.; Tohma, H.; Inagaki, M.; Hatanaka, K. *Heterocycles* **1992**, *33*, 503.
[156] Müller, P.; Gilabert, D. M. *Chimia* **1986**, *40*, 127.
[157] Murali, D.; Rao, G. S. K. *Indian J. Chem.* **1987**, *26B*, 668.
[158] Nebois, P.; Cherkaoui, O.; Benameur, L.; Fillion, H. *Tetrahedron* **1994**, *50*, 8457.
[159] Sharma, V. Ph.D. Dissertation, 1998, Kurukshetra University, Kurukshetra, Haryana, India.
[160] Thakkar, K.; Cushman, M. *J. Org. Chem.* **1995**, *60*, 6499.
[161] Prakash, O.; Sharma, V.; Tanwar, M. P. *Can. J. Chem.* **1999**, *77*, 1.
[162] Kappe, T.; Korbuly, G.; Stadlbauer, W. *Chem. Ber.* **1978**, *111*, 3857.
[163] Hanefeld, W.; Spangenberg, B. *Arch. Pharm.* (Weinheim, Ger.) **1987**, *320*, 666.
[164] Prakash, O. Kurukshetra University, Kurukshetra, Haryana, India, unpublished results.

CHAPTER 3

SYNTHETIC USES OF TOSYLMETHYL ISOCYANIDE (TosMIC)

DAAN VAN LEUSEN AND ALBERT M. VAN LEUSEN

Department of Organic and Molecular Inorganic Chemistry, Groningen University, Nijenborgh 4, 9747 AG Groningen, The Netherlands

CONTENTS

	PAGE
ACKNOWLEDGMENTS	419
INTRODUCTION	420
Overview of TosMIC Applications	421
General Aspects	423
Preparation of TosMIC	424
Analogs of TosMIC	425
Homologs of TosMIC	425
REDUCTIVE CYANATION; ONE-CARBON HOMOLOGATION OF KETONES AND ALDEHYDES	428
Ketones	428
Aldehydes	431
Dicarbonyl Compounds; Less Reactive Ketones	432
α,β-Unsaturated Aldehydes	433
SYNTHESIS OF OXAZOLINES AND OXAZOLES FROM CARBONYL COMPOUNDS	433
Oxazolines and Products Derived Therefrom	434
Stereochemistry	436
Oxazoles	437
SYNTHESIS OF 1-FORMAMIDO-1-TOSYLALKENES AND 1-ISOCYANO-1-TOSYLALKENES	441
Scope and Limitations	442
TOSMIC SUBSTITUTED AT THE METHYLENE GROUP	443
Monosubstituted TosMIC Derivatives	443
Disubstituted TosMIC Derivatives	445
TosMIC as a Connective Reagent	448
Acylated TosMIC Derivatives	451
SYNTHESIS OF IMIDAZOLES AND 1,2,4-TRIAZOLES	451
Imidazoles	451
1,2,4-Triazoles	455

Organic Reactions, Vol. 57, Edited by Larry E. Overman et al.
ISBN 0-471-43511-2 © 2001 Organic Reactions, Inc. Published by John Wiley & Sons, Inc.

SYNTHESIS OF THIAZOLES FROM THIOCARBONYL COMPOUNDS 456
SYNTHESIS OF PYRROLES AND INDOLES 457
   Pyrroles from Michael Acceptors 458
   Pyrroles from 1-Isocyano-1-tosyl-1-alkenes 463
   2,3-Dialkenylpyrroles and Indoles 464
SYNTHESIS OF KETONES BY HYDROLYSIS OF TosMIC DERIVATIVES 465
   Ketones 466
   Diketones 467
   Tri- and Tetraketones 468
   Enones 469
   α-Hydroxy Ketones and α-Hydroxy Aldehydes 471
REDUCTION OF TosMIC DERIVATIVES 473
   Alkanes 473
   Alkyl and Aralkyl Isocyanides 473
   N-Methylamines. 473
   β-Hydroxy N-Methylamines 474
MISCELLANEOUS 474
   α-Additions to the Isocyano Carbon of TosMIC 474
   Cycloadditions to the Isocyano Carbon of TosMIC 475
   TosMIC as a Metal Ligand 477
   Analogs of TosMIC 477
EXPERIMENTAL PROCEDURES 478
   N-(p-Tolylsulfonylmethyl)formamide and p-Tolylsulfonylmethyl Isocyanide
      (Two-step Synthesis of TosMIC) 478
   2-Cyanoadamantane (Reductive Cyanation of a Ketone) 479
   2-$tert$-Butyl-5-cyanomethylthiophene (Reductive Cyanation of an Aldehyde) . 479
   4-Phenyl-5-tosyl-2-oxazoline (Reaction of an Aldehyde with TosMIC) . . . 480
   (2'R)- and (2'S)-3-Methoxy-17-(2-methoxy-3-oxazolin-4-yl)-androsta-3,5,16-triene
      (Reaction of a Monosubstituted TosMIC Derivative with an Aldehyde and
      Methanol) 480
   N-(Cyclohexylidenetosylmethyl)formamide (Reaction of TosMIC with a Ketone in
      a Nonprotic Solvent) 481
   (Isocyanotosylmethylidene)cyclohexane (Dehydration of an α,β-Unsaturated
      Formamide) 481
   $trans$-5-$tert$-Butyl-5-methyl-4-tosyl-2-oxazoline and (E)-N-(2,3,3-Trimethyl-1-tosyl-1-
      butenyl)formamide (Reaction of TosMIC with a Sterically Screened Ketone by the
      Two-step/Two base Approach) 482
   1-Tosylcyclobutyl Isocyanide (Intramolecular Dialkylation of TosMIC) . . . 483
   9-Isocyano-9-tosylheptadecane and n-Heptadecane (Dialkylation of TosMIC and
      Reduction of the Product; TosMIC as CH_2 Connector) 483
   5-(2-Hydroxyphenyl)-1-neopentylimidazole (Reaction of TosMIC with an Aldimine) . 484
   3-Cyano-4-phenylpyrrole (Reaction of a Michael Acceptor with TosMIC) . . 484
   3-Methyl-4-phenylpyrrole (Reaction of a Michael Donor with a 1-Isocyano-1-
      tosylalkene) 485
   (E)-4-Benzoyl-2-(cyclopent-1-enyl)-1-methyl-3-(2-phenylethenyl)pyrrole and
      3-Benzoyl-7,8-dihydro-1-methyl-5-phenyl-6H-cyclopent[g]indole (Formation of a
      2,3-Dialkenylpyrrole from a TosMIC Derivative and a Michael Acceptor, Followed by
      Ring Closure and Dehydrogenation to an Indole) 485
   (Z)-19-Isocyano-19-tosyltricos-9-ene and (Z)-Tricos-9-en-19-one (Alkylation of a
      Monosubstituted TosMIC Derivative and Hydrolysis of a Disubstituted TosMIC
      Derivative. TosMIC as CO Connector) 486
   1,3-Diphenyl-1-hydroxypropan-2-one (Synthesis of a Tosyloxazoline, Followed by
      Hydrolysis to a Hydroxyketone) 487

N-Methyl-2-(3,4-dimethoxyphenyl)-2-hydroxyethylamine (Reaction of TosMIC with an Aldehyde, Followed by Reduction with Lithium Aluminum Hydride) . . . 487
TABULAR SURVEY 488
   Table I-A. Cyanides from Ketones by Reductive Cyanation 490
   Table I-B. Cyanides from Aldehydes by Reductive Cyanation 501
   Table I-C. Carboxylic Acids and Esters from Ketones or Aldehydes via Reductive Cyanation 503
   Table II. Oxazolines from Ketones or Aldehydes 504
   Table III-A. Oxazoles from Aldehydes 516
   Table III-B. Tosyloxazoles from Carboxylic Acid Derivatives 524
   Table IV. 1-Formamido-1-tosylalkenes from Aldehydes or Ketones 525
   Table V-A. 1-Isocyano-1-tosylalkenes by Dehydration of 1-Formamido-1-tosylalkenes . 530
   Table V-B. 1-Isocyano-1-tosylalkenes by in situ Dehydration of 1-Formamido-1-tosylalkenes 533
   Table V-C. 1-Isocyano-1-tosylalkenes from Aldehydes or Ketones by Peterson Olefination 537
   Table VI. Monoalkylation of TosMIC 538
   Table VII-A. Disubstituted Derivatives of TosMIC 544
   Table VII-B. Disubstituted Derivatives of TosMIC from 1-Isocyano-1-tosylalkenes . 564
   Table VII-C. α-Isocyano-α-tosyl Ketones from TosMIC Homologs and Acid Chlorides . 569
   Table VIII-A. Imidazoles from Aldimines 570
   Table VIII-B. 4-Tosylimidazoles from Imidoyl Chlorides or Isothiocyanates . . . 580
   Table VIII-C. Imidazoles from 1-Isocyano-1-tosylalkenes 583
   Table IX. 1,2,4-Triazoles from Diazonium Salts 586
   Table X. Thiazoles from Thiocarbonyl Compounds 587
   Table XI-A1. Pyrroles from TosMIC or TosMIC Homologs TosCHRNC and Michael Acceptors 590
   Table XI-A2. Pyrroles from TosMIC Homologs TosC($=$CR$_2$)NC and Michael Acceptors 607
   Table XI-B. Pyrroles from 1-Isocyano-1-tosylalkenes and Michael Donors . . . 612
   Table XII. Ketones by Acid Hydrolysis of Disubstituted TosMIC Derivatives . . 613
   Table XIII. 1,2-Diketones from Acid Chlorides 637
   Table XIV. α-Hydroxy Ketones and α-Hydroxy Aldehydes by Hydrolysis of Oxazolines 640
   Table XV. Alkanes, N-Methylamines, and Isocyanides by Reduction of TosMIC Derivatives 645
   Table XVI. β-Hydroxy-N-Methylamines by Reduction of 4-Tosyloxazolines . . 649
   Table XVII. α-Additions to the Isocyano Carbon of TosMIC 653
   Table XVIII. Cycloadditions to the Isocyano Carbon of TosMIC 656
REFERENCES 659

ACKNOWLEDGMENTS

The authors wish to dedicate this Chapter to the workers in the area of TosMIC chemistry. It has been a pleasing challenge—although a time-consuming one—to collect their many contributions and to put these in perspective. Where we have succeeded, this has become possible only through their efforts. As for shortcomings and mistakes of the present survey, the reader has to put the blame on us.

INTRODUCTION

TosMIC is the acronym for 4-tolylsulfonylmethyl isocyanide or tosylmethyl isocyanide.* It is the best-known member of a series of about 25 sulfonyl-substituted methyl isocyanides RSO_2CH_2NC collected in Chart 1 (p. 426). TosMIC is a multipurpose synthetic reagent. It is by far the most versatile synthon derived from methyl isocyanide.[1-5] This chapter provides a complete account of the synthetic uses of TosMIC based on a literature search closed in January 1996, and supplemented with further data available to the authors. Also included are applications of some of the more important TosMIC homologs (TosCHRNC).

TosMIC

TosMIC is the only commercially available sulfonylmethyl isocyanide. It is a stable, colorless, practically odorless solid, which can be stored at room temperature without decomposition.

Four brief review papers on the chemistry of TosMIC have appeared, in 1980,[6] 1987,[7] 1993,[8] and 1995.[9] Several reviews on the chemistry of isocyanides in general are available.[10-25]

An effort has been made to make this chapter as complete as possible, in coverage both by tables and by references, with respect to the immediate objective: a survey of the "Synthetic Uses of TosMIC." Emphasis is on the primary products derived from reactions of TosMIC and related isocyanides, with limited reference to the further utilization of these products. Thus, the next section gives a complete account of reductive cyanation, the conversion of aldehydes or ketones into cyanides with the use of TosMIC, but for obvious reasons synthetic applications of the product cyanides are not treated in this chapter. One of the subsequent sections describes the application of TosMIC to the synthesis of pyrroles, including 2,3-divinylpyrroles, which are important precursors in a new synthesis of indoles. This latter transformation, made possible by the easy availability of the precursors through TosMIC chemistry, is afforded brief coverage.

As a rule, reference to preliminary papers has been omitted when the same information is available from the corresponding full papers. The patent literature is covered highly selectively and is cited only when providing new and relevant information. Negative results of reactions are reported in the text or tables only when at least some relevant information is available with respect to the conditions of such unsuccessful attempts.

---

*Throughout this chapter, the tosyl group (4-tolylsulfonyl, *p*-tolylsulfonyl) is abbreviated as Tos, both in names and in formulas, rather than as Ts as is recommended by IUPAC. Not only is the name TosMIC generally accepted, but the Tos abbreviation is also used in nearly all the papers on which this chapter is based.

Overview of TosMIC Applications

The synthetic applications of TosMIC are diverse. Even with the same substrate molecule, TosMIC can be used to prepare quite different products simply by varying the reaction conditions. The main purpose of this overview section is to provide a quick insight into the major synthetic applications of TosMIC. Each type of reaction is illustrated by an example of wider applicability. Boldface symbols **C** and **N** are used—only in this overview section—to show which atoms of the TosMIC molecule (and TosMIC analog or homolog) end up in what positions of the final products. Where appropriate, hydrogen atoms are depicted explicitly to emphasize that the use of TosMIC leads to products in which certain positions are intrinsically unsubstituted.

1. Reductive Cyanation of Ketones and Aldehydes (Eq. 1)

cyclohexanone → (TosCH₂N=C) → 1-cyano-1-H-cyclohexane (80%)   (Eq. 1)

2. Synthesis of Oxazoles from Aldehydes (Eq. 2) and Oxazolines from Ketones (Eq. 3)

PhCHO → (TosCH₂N=C) → 5-phenyl-oxazole (H**C**=**N**, **C**H, O, Ph) (91%)   (Eq. 2)

cyclohexanone → (TosCH₂N=C) → spiro-oxazoline (Tos, H–**C**–**N**, **C**H, O) (60%)   (Eq. 3)

3. Synthesis of Imidazoles from TosMIC and Imines (Eq. 4) or by Using Knoevenagel-type Condensation Products of TosMIC (Eq. 5)

CH₂=CH–N(Et) → (TosCH₂N=C) → imidazole (H**C**=**N**, **C**H, Me, N-Et) (70%)   (Eq. 4)

MeNH₂ + Ph–**C**(H)=**C**(Tos)(N=C) → imidazole (H**C**=**N**, **C**H, Ph, N-Me) (87%)   (Eq. 5)

4. Synthesis of Thiazoles from Thionoesters (Eq. 6)

$$\text{furan-C(=S)-S-CH}_2\text{CO}_2\text{H} \xrightarrow{\text{TosCH}_2\text{N=C}} \text{furan-thiazole(Tos)} \quad (67\%) \quad (\text{Eq. 6})$$

5. Knoevenagel-type Condensation Products from Aldehydes or Ketones (Eq. 7)

$$\text{cyclohexanone} \xrightarrow{\text{TosCH}_2\text{N=C}} \text{cyclohexylidene=C(N=C)Tos} \quad (58\%) \quad (\text{Eq. 7})$$

6. Synthesis of Pyrroles from TosMIC and Michael Acceptors (Eq. 8)

$$\text{PhCH}_2\text{O-CH}_2\text{-CH=CH-C(O)CH}_3 \xrightarrow{\text{TosCH}_2\text{N=C}} \text{pyrrole} \quad (83\%) \quad (\text{Eq. 8})$$

$$\text{PhC(O)Et} + \text{Ph-C(H)=C(N=C)Tos} \longrightarrow \text{pyrrole} \quad (73\%) \quad (\text{Eq. 9})$$

7. Synthesis of Indoles; Application of Knoevenagel-type Condensation Products of TosMIC (Eq. 10)

$$\text{Ph-CH=CH-CH=CH-C(O)Ph} + \text{CH}_3\text{-C(CH}_3\text{)=C(N=C)Tos} \xrightarrow{\text{four steps}} \text{indole} \quad (90\%) \quad (\text{Eq. 10})$$

8. Phase Transfer Monoalkylation of TosMIC; Preparation of TosMIC Homologs (Eq. 11)

$$\text{BnBr} \xrightarrow[\text{Base}]{\text{TosCH}_2\text{N=C}} \text{TosCH(Bn)N=C} \quad (80\%) \quad \text{(Eq. 11)}$$

9. TosMIC as a Connective Reagent; Dialkylation of TosMIC (Eq. 12)

muscalure - pheromone of housefly (Eq. 12)

10. Synthesis of Ketones; Umpolung of Formaldehyde Reactivity (Eq. 13)

triamcinolone acetonide (Eq. 13)

The above sequence of examples is not necessarily paralleled by the order of the following sections in which the examples are discussed in detail. The sequence of the foregoing examples emphasizes product-correspondence, such as the various azoles (Eqs. 2–6, 8–10), whereas the order of sections to follow is based primarily on mechanistic correlations.

General Aspects

The majority of the synthetic applications of TosMIC derive from the following fundamental properties:

1. TosMIC contains an activated methylene group and readily forms a stabilized, nucleophilic carbanion, which will react with a variety of electrophiles.
2. The divalent isocyano carbon usually acts as an electrophilic center, which can participate in ring closing reactions.

3. The activating tosyl group frequently operates as a moderately good leaving group in a 1,2-elimination step to produce sulfur-free products.
4. The geminal isocyano and tosyl groups in TosMIC (and in TosMIC analogs or homologs) can be looked upon as a special type of N,S-acetal, and in fact can be made to react accordingly.

The near quantitative formation of 5-phenyloxazole[26] from TosMIC and benzaldehyde with potassium carbonate in methanol at reflux (Eq. 14) serves to illustrate the above points 1, 2, and 3. When the reaction is carried out at room temperature, a 4-tosyloxazoline is formed; at reflux temperature subsequent base-induced elimination of 4-toluenesulfinic acid leads to 5-phenyloxazole.

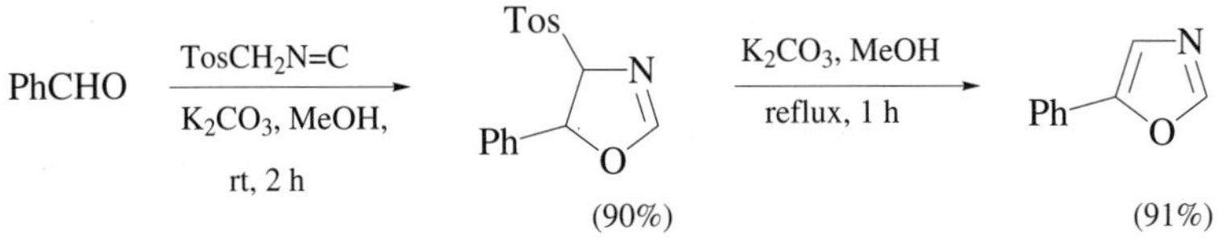

$$\text{PhCHO} \xrightarrow[\substack{K_2CO_3,\text{ MeOH},\\ \text{rt, 2 h}}]{\text{TosCH}_2\text{N=C}} \underset{(90\%)}{\text{[4-Tos-oxazoline, Ph]}} \xrightarrow[\text{reflux, 1 h}]{K_2CO_3,\text{ MeOH}} \underset{(91\%)}{\text{[5-Ph-oxazole]}}$$

(Eq. 14)

Basic aspects of the chemistry of TosMIC, as exemplified by Eq. 14, are summarized in Scheme 1 (top).[7] The various reaction modes shown in this scheme may, but need not, take place in a combined fashion. For example, it is possible to alkylate the methylene group of TosMIC, once or twice, without affecting the isocyano and tosyl groups. The conversion of formaldehyde into TosMIC, and its reversal (bottom), constitutes an umpolung of the formaldehyde molecule.

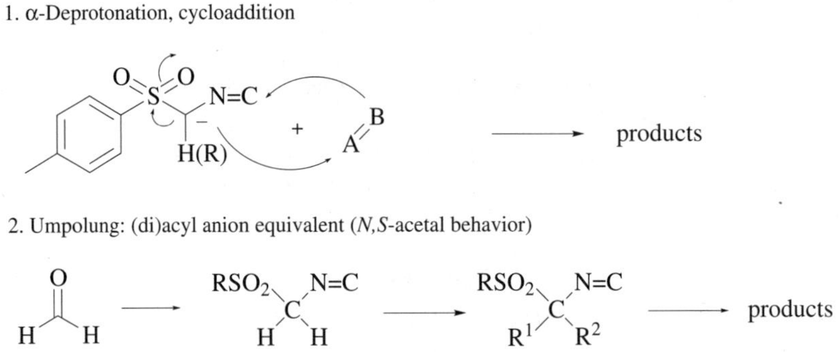

1. α-Deprotonation, cycloaddition

2. Umpolung: (di)acyl anion equivalent (N,S-acetal behavior)

Scheme 1

Preparation of TosMIC

TosMIC was first obtained in 1967 as the unexpected result of the photolysis of tosyldiazomethane in liquid hydrogen cyanide.[27] TosMIC (CAS Registry Number 36635–61–7) is commercially available; alternatively it can be readily prepared in two steps from the commercial products sodium p-toluenesulfinate, formaldehyde, and formamide (Eq. 16). A version of this synthesis is described in Organic Syntheses.[28] The overall yield has been further improved to 75% by modifying the conditions of the Mannich reaction.[29,30] The procedure in the experimental section combines the best of the previous procedures into a simple preparation of TosMIC in 55% overall yield.[31]

The same approach has been used for the synthesis of labeled TosMIC compounds: Tos^{14}CH$_2$NC (CAS Registry Number 62796–16–1),[32] TosCH$_2$N^{13}C (CAS Registry Number 60684–36–8),[33,34] and TosCH$_2$^{15}NC (CAS Registry Number 160999–37–1).[34]

A few alternative preparations of TosMIC are known. Two of these are worth mentioning here because they have been used in the synthesis of TosMIC analogs RSO$_2$CH$_2$NC: the reaction of 4-toluenesulfonyl fluoride with lithiomethyl isocyanide;[34–37] and the oxidation of 4-CH$_3$C$_6$H$_4$SCH$_2$NHCHO, followed by dehydration.[35,38–40] Alternatives to the latter dehydration using phosphorus oxychloride and triethylamine have been reported.[41,42]

Analogs of TosMIC

Chart 1 depicts the TosMIC analogs RSO$_2$CH$_2$NC with R other than 4-tolyl. By and large these compounds show the same chemistry as TosMIC, but they have been used much less frequently for synthetic purposes than TosMIC proper. In general, the crystalline compounds are more stable to storage. The chemistry of the TosMIC analogs of Chart 1 is not covered exhaustively in this chapter. One analog, however, is worth mention: (−)-S-phenyl-N-tosylsulfonimidoylmethyl isocyanide (**8**), is a chiral TosMIC analog that bears its stereogenic center close to the prime reaction site, the methylene group.[40,43]

Polymeric TosMIC analogs have been prepared in high yields by free-radical polymerization of (4-ethenylphenyl)sulfonylmethyl isocyanide (Eq. 15).[44] The polymeric TosMIC analogs have been used for the reductive cyanation of ketones

(Eq. 15)

and for the synthesis of imidazoles. The eliminated (polymeric) sulfinic acids can be reused.[44] Further aspects of the chemistry of TosMIC analogs are discussed at the end of this chapter.

Homologs of TosMIC

Most synthetic applications of TosMIC are based on an initial reaction of TosMIC anion (TosC$^-$HNC). Obviously, related anions may be derived from monosubstituted TosMIC derivatives TosCHR^{1}NC **13**. Such homologs **13** can be used synthetically in most of the synthetic transformations of TosMIC except

Chart 1. Analogs of TosMIC

RSO$_2$CH$_2$N=C

R	mp
Me (1)[30,181]	49-51°
n-Bu[35]	46-47°
t-Bu[35]	104-105°
n-C$_{10}$H$_{21}$[295]	65-68°
Bn[35]	103-106°

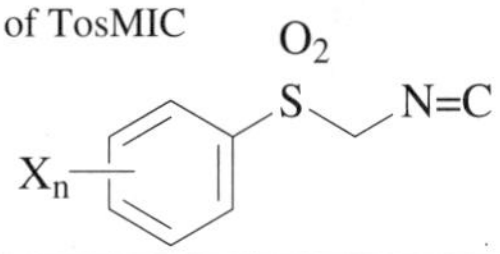

X$_n$—C$_6$H$_4$—SO$_2$CH$_2$N=C

X	mp	X	mp	X	mp
4-H[35,294]	88°	4-Et[92]	—	4-MeO[181]	95-96°
4-Cl[181]	124-126°	2,3,4,5,6-Me$_5$[181]	146-148°	4-EtO[92]	—
4-Me[28–35]	116-117°	2-MeO[40]	93-94°	4-CH$_2$=CH[44]	63-64°

Chiral Alkyl

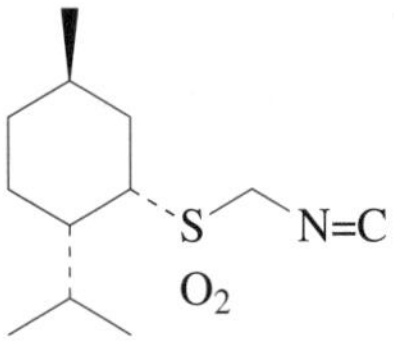

(+)-neomenthyl, mp 68° **2**[39,40]

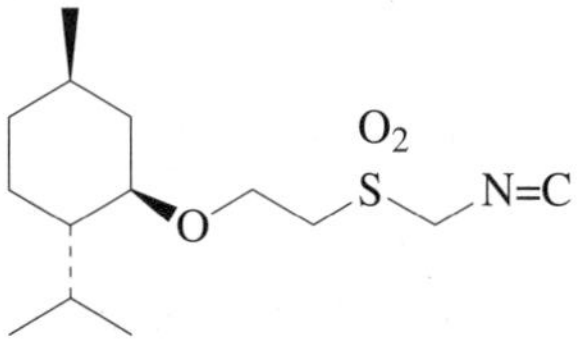

(−)-2-menthoxyethyl (oil) **3**[40]

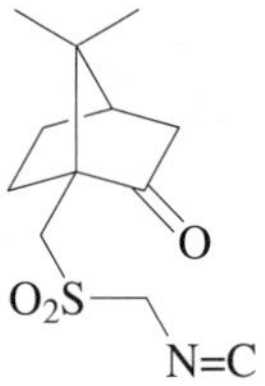

(+)-camphor (oil) **4**[40]

Chiral Aryl

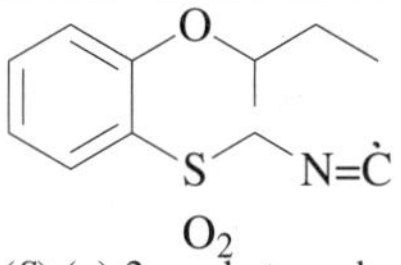

(S)-(−)-2-sec-butoxyphenyl (oil) **5**[40]

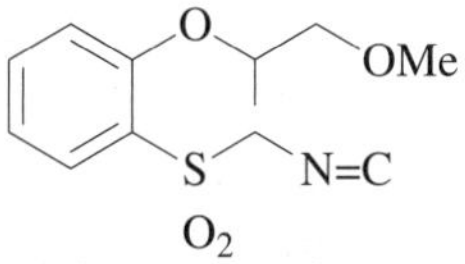

(±)-2-(1-methoxy-2-propoxy)phenyl (oil) **6**[40]

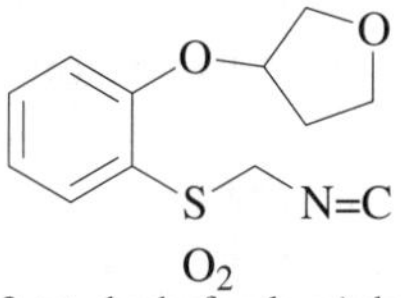

(R)-(−)-2-(3-tetrahydrofuryloxy)phenyl, mp 107-108° **7**[40]

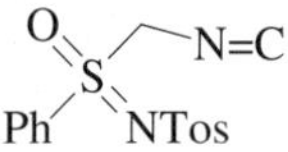

(−)-S-phenyl-N-tosylsulfonimidoyl mp 96° **8**[40,43]

Other

Et$_2$NSO$_2$CH$_2$N=C

mp 69-70° **9**[180]

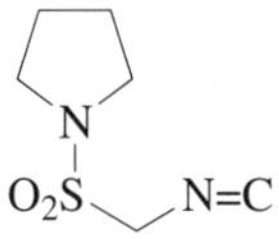

mp 119-120° **10**[180]

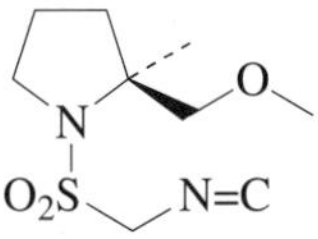

oil **11**[180]

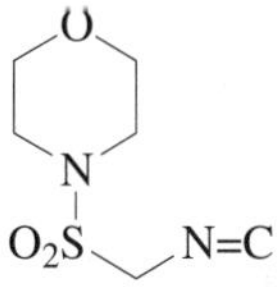

mp 89-90° **12**[180]

those requiring both acidic hydrogens, notably reductive cyanation (Eq. 1) and the Knoevenagel-type condensation reaction (Eq. 7).

Since disubstituted TosMIC derivatives TosCR^{1}R^{2}NC **14** cannot form carbanions, these compounds are treated as TosMIC-derived products. They are crucial intermediates in the TosMIC-based synthesis of ketones (Eq. 13) and in the connective reagent application (Eq. 12). Several compounds TosCHR^{1}NC **13** have been prepared by a Mannich reaction followed by dehydration (Eq. 16).

TosH + R¹CHO + H₂NCHO →[Mannich reaction]

$$\text{Tos}-\underset{\underset{H}{|}}{N}(R^1)(CHO) \xrightarrow[\text{base}]{POCl_3} \text{Tos}-C(R^1)H-N=C \quad \mathbf{13}$$

(Eq. 16)

R¹ = H   75% [28-30]
R¹ = Me   68% [90]
R¹ = Et   64% [90]
R¹ = Ph   72% [90,37]
R¹ = C₆H₄CHO-3   37% [82]

Most of the alkyl-substituted compounds **13** (R^1 = alkyl) are prepared by phase transfer catalyzed (PTC) monoalkylation of TosMIC (Eq. 17, Table VI). This is the method of choice, since TosMIC is commercially available. Aryl derivatives (R^1 = aryl) are accessible via the Mannich reaction (Eq. 16), or by reaction of arenesulfonyl fluorides with lithiobenzyl isocyanides.[37]

$$\text{Tos}-CH_2-N=C \xrightarrow[\text{PTC}]{R^1X,\ \text{base}} \text{Tos}-CH(R^1)-N=C$$

(Eq. 17)

The (formal) Knoevenagel condensation products of ketones and aldehydes with TosMIC (Eq. 7) deserve special attention. When a γ carbon in these condensation products bears a hydrogen, as for example in cyclohexanone derivative **15**, the tautomeric hydrogen shift to **16** provides a TosMIC derivative with an alkenyl substituent (Eq.18). In most reactions, the β,γ-unsaturated compounds are

[Structures: cyclohexylidene(Tos)(N=C) **15** → base → 1-cyclohexenyl-CH(Tos)(N=C) **16**]

(Eq. 18)

[Structures: anion **17** shown in two resonance forms]

the thermodynamically more stable tautomers. The α,β-unsaturated condensation product of TosMIC and acetone is one of the few established exceptions to this rule. Reactions of anion **17** obtained by deprotonating either **15** or **16** and related compounds further extends the chemistry of TosMIC derivatives.

Intermediates of type **17** play a crucial role in reactions such as Eqs. 10 and 13. Since Knoevenagel condensation products such as **15** and **16** find other applications as well, they are considered primarily to be the products of TosMIC reactions. The sodium borohydride reduction of the α,β double bond of compounds **15** provides another entry into monosubstituted TosMIC derivatives **13**.[45]

A related series of synthons exists that can act as TosMIC equivalents: tosylmethyl substituted carbodiimides ($TosCH_2N=C=NR$),[46] imidates and thioimidates [$TosCH_2N=C(XCH_3)R$; $X = O, S$],[47–51] and imino(dithio)carbonates [$TosCH_2N=C(XR)_2$; $X = O, S$].[47,51] These synthons have been used occasionally for the synthesis of azoles with a substituent at position 2.[52]

The following sections are arranged according to the various types of products that have been realized with the use of TosMIC, since structurally different products may be derived from the same substrate molecule, simply by varying the reaction conditions.

REDUCTIVE CYANATION; ONE-CARBON HOMOLOGATION OF KETONES AND ALDEHYDES

Ketones

Ketones are converted in good yields to cyanides upon reaction with TosMIC and base in nonprotic solvents (Eq. 19).[32,53,54]

$$\underset{R^2}{\overset{R^1}{>}}=O \quad \xrightarrow[\text{base}]{TosCH_2N=C} \quad \underset{R^2}{\overset{R^1}{>}}\underset{CN}{\overset{H}{<}} \qquad \text{(Eq. 19)}$$

The reductive cyanation of ketones is carried out in one operation, although the reaction involves several stages. The reaction of adamantanone with ^{14}C-labeled TosMIC shows that the cyano carbon of the product is derived from the TosMIC methylene group (Eq. 20).[32]

$$\text{adamantanone} \quad \xrightarrow[\text{t-BuOK}]{Tos^{14}CH_2N=C} \quad \text{adamantyl-}^{14}CN \quad (55\%) \qquad \text{(Eq. 20)}$$

The vertical track of Scheme 2 shows the probable mechanism of the reductive cyanation reaction.[32,55] Nucleophilic attack of TosMIC anion at the ketone carbonyl leads to oxazoline anion **18**. In nonprotic solvents a hydrogen shift to **19** is followed by electrocyclic ring opening to **20**, which at or above room temperature loses 4-toluenesulfinate anion to afford *N*-formylketenimine **21** (which has not been isolated). The formyl group of **21** is lost upon subsequent attack of a nucleophile (for example alkoxide) to give anion **22**. Workup with water provides cyanide **23**, obtained as an equilibrium mixture of diastereomers when ketones with stereogenic centers are used.

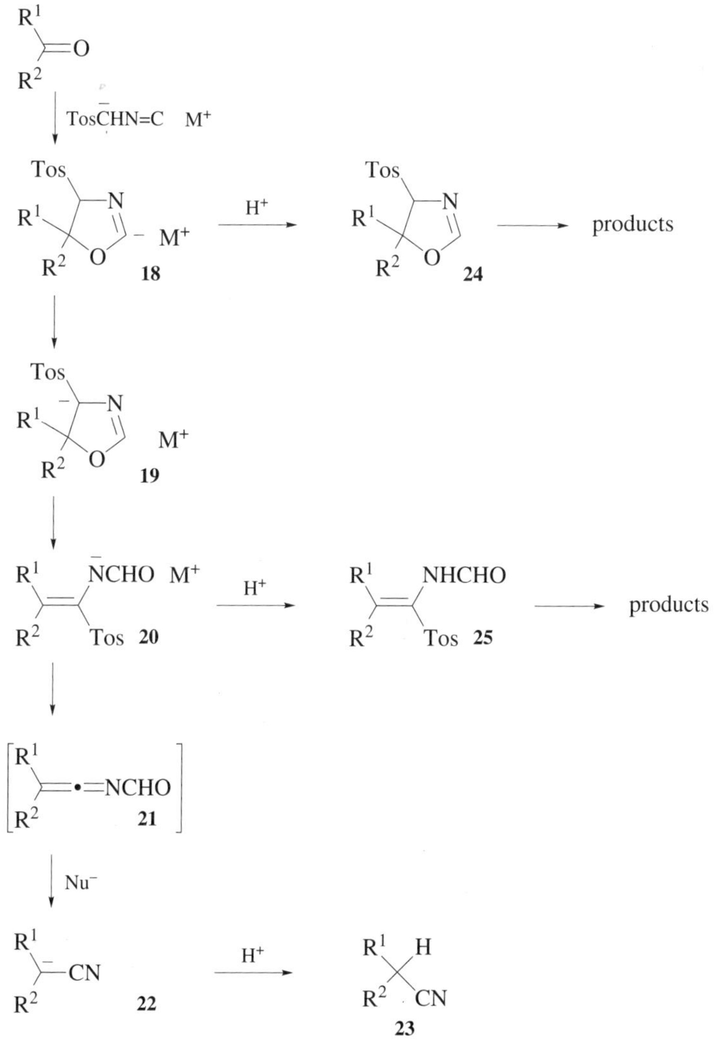

Scheme 2

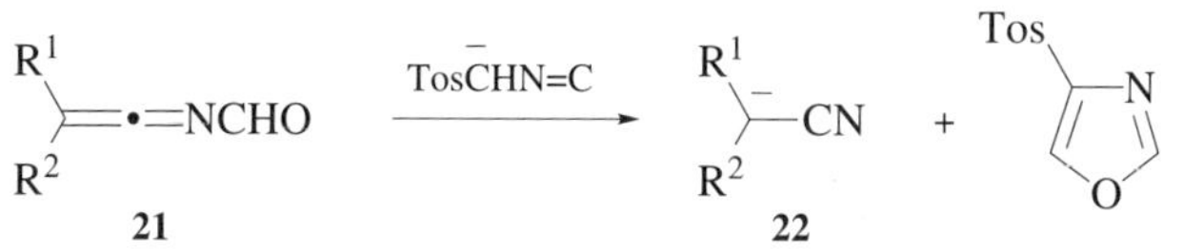

Occasionally, TosMIC anion has been found to act as the nucleophile in the deformylation of **21** to **22**, resulting in the formation of 4-tosyloxazole (Eq. 21).[32,56] Since this reaction causes an undesirable consumption of a second equivalent of TosMIC, the reductive cyanation is best carried out in tetrahydrofuran (THF) or 1,2-dimethoxyethane (DME) containing 1 to 2 equivalents of methanol or ethanol, producing a formate ester, rather than 4-tosyloxazole.[32,56] The presence of small amounts of alcohol in the otherwise nonprotic solvents does not significantly promote the undesirable reactions **18** → **24** and **20** → **25**, and reductive cyanation proceeds efficiently.

Two aspects of Scheme 2 are important in relation to later sections of this chapter:

(1) Reductive cyanation of ketones and aldehydes is the end point of a series of sequential reactions ultimately leading to cyanides (Eqs. 20 and 21).

(2) By changing the reaction conditions, quite different products may be formed even from the same starting materials.

At this stage of the discussion it is necessary only to realize that under protic conditions intermediates **18** or **20** may be sidetracked into compounds **24** and **25**, respectively. Both 4-tosyloxazolines **24** and unsaturated tosylformamides **25** are important products for further synthetic applications, as will be shown later.

Reductive cyanation has been applied to many different ketones. The reaction fails when the carbonyl carbon is severely sterically screened from nucleophilic attack by TosMIC anion, as exemplified by di-*tert*-butyl ketone, which is recovered unchanged even under forcing conditions (170 hours at 45°),[32] and when the ketone is easily deprotonated, exemplified by benzyl phenyl ketone.

The reductive cyanation of ketones is usually carried out in DME with 1.1 equivalents of TosMIC and 2 equivalents or more of potassium *tert*-butoxide. Reaction temperatures and reaction times strongly depend on the reactivity of the ketone, and range from 0 to 50° and 1 to 72 hours, respectively.[32] Occasionally, THF, dimethyl sulfoxide (DMSO), and hexamethylphosphoramide (HMPA) have been used as solvents or cosolvents. Most probably, all reactions carried out in DME could be carried out equally well in the cheaper THF.[55] An excess of base is used to suppress cyclodimerization reactions of TosMIC, which are initiated by the addition of TosMIC anion to the isocyano carbon of TosMIC.[32,37,56]

In a few cases potassium *tert* butoxide may be replaced by sodium ethoxide. *n*-Butyllithium cannot be used in reductive cyanation reactions, although this base is quite useful in other applications of TosMIC; the proton shift in 2-lithiooxazoline **18** to isomer **19** (Scheme 2, M = Li) apparently is retarded such that isomer **18** will react with a second molecule of ketone (Eq. 22).[55]

Selective reductive cyanation of ketones is possible in the presence of other functional groups that do not normally react with TosMIC: ethers, thioethers, acetals, aromatic halides, aliphatic hydroxy groups, esters, lactones, urethanes, aromatic nitro groups, and isolated C-C multiple bonds survive. Reductive cyanation has been successfully applied to a stable radical species (Eq. 23).[57]

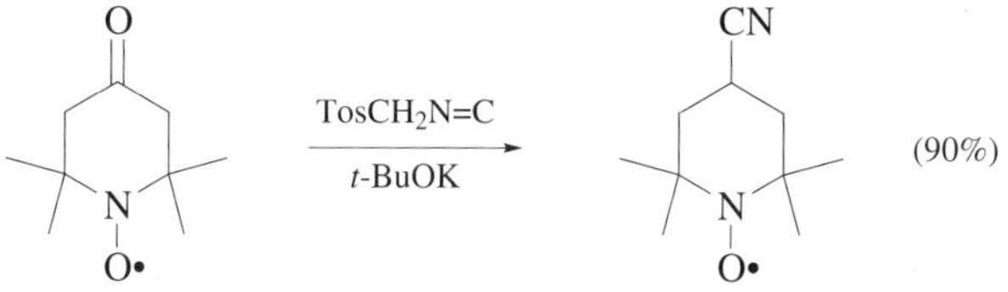

(Eq. 23)

The one-step reductive cyanation of ketones with the use of TosMIC is much more convenient (especially as far as handling of reagents is concerned) than a three-step approach that was developed almost simultaneously.[58] The latter process, detailed in *Organic Syntheses* for the conversion of cyclohexanone to cyclohexyl cyanide,[59] makes use of methyl hydrazine carboxylate, hydrogen cyanide, and bromine. A more recent, interesting alternative involves the samarium (II) iodide reduction of α-cyanophosphates, formed in situ from ketones (or aldehydes) by reaction with diethyl cyanophosphate and lithium cyanide.[60,61]

Aldehydes

Reductive cyanation is also applicable to aldehydes, although it is less frequently employed (Scheme 2, $R^1 = H$). A slight modification of reaction conditions is needed to obtain good results (see below). When the reductive cyanation of aldehydes is carried out under conditions typically employed for ketones, yields are low. Benzaldehyde, for example, gives benzyl cyanide in only 15% yield.[62] Byproducts to be expected are oxazoles and/or oxazolines. As noted earlier, a crucial step in reductive cyanation is electrocyclic ring opening of **19** to **20** (Scheme 2), which requires a proton shift from **18** to **19**. With aldehydes, a competing proton shift becomes possible from **18a** (i.e. **18** where $R^1 = H$) to **26**, which initiates the elimination of Tos⁻ to give oxazoles **27** (Scheme 3).

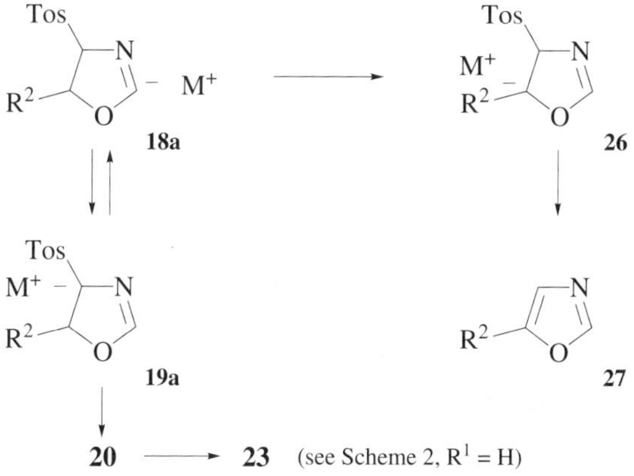

Scheme 3

In protic solvents the formation of oxazoles **27** is well established in the reaction of TosMIC with aromatic aldehydes (Scheme 3). The acidity of the hydrogen at C5 in anion **18a**, when R^2 is aromatic, is such that the hydrogen shift **18a** → **26** can compete with the proton shift **18a** → **19a**. When the C5 hydrogen is less acidic, as is the case for aliphatic aldehydes, oxazoles are not formed.[63] In methanol at reflux, the tosyl group in oxazoline **24** (Scheme 2) is replaced by MeO (Eqs. 30 and 32).

In actual practice the problem of side reactions in the reductive cyanation of aldehydes is largely overcome by carrying out the reaction in two operations (see Experimental Procedures). First, the aldehyde is allowed to react with TosMIC anion at low temperatures (-50 to $-20°$) under nonprotic conditions to carry the reaction all the way to intermediate **20** (Scheme 2, $R^1 = H$). The reaction is then completed by heating, after addition of excess methanol.[62,64] Whether this two-step procedure is beneficial to the lower yielding reductive cyanations of ketones remains to be seen.

Dicarbonyl Compounds; Less Reactive Ketones

A double reductive cyanation has been carried out with symmetrical dialdehydes (Eq. 24).[65]

It is not known whether a single reductive cyanation of just one of the aldehyde functions of the same substrate is possible in an efficient manner. However, several reductive monocyanations of diones have been described albeit only for unsymmetrical starting materials (Eq 25).[66]

Reductive cyanations of less reactive ketones such as benzophenone and hindered ketones such as camphor require longer reaction times (Eq. 26).[32]

Monocyanation of 1,4-cyclohexanedione has been described only for the monoketal derivative.[67] The reactivity of the keto groups at C3 and C17 in steroids with two double bonds in the A ring is sufficiently different to allow a selective reductive cyanation at C17 (Eq. 27).[32] However, without the 1,2 double bond of the A ring, protection of the C3 keto group becomes necessary. The less reactive steroidal C11 carbonyl does not need protection, as demonstrated by the selective cyanation of 3,3-(ethylenedioxy)androst-5-ene-11,17-dione.[56]

$$\text{[bicyclic ketone]} \xrightarrow{\text{TosCH}_2\text{N=C, }t\text{-BuOK}} \text{[bicyclic-CN]} \quad \text{(Eq. 26)}$$

DME, rt, 1.5 h (<5%)
HMPA, 45°, 17 h (73%)
DMSO, 45°, 70 h (80%)

$$\text{[steroid-17-one]} \xrightarrow{\text{TosCH}_2\text{N=C}} \text{[steroid-17-CN]} \quad (47\%) \quad \text{(Eq. 27)}$$

α,β-Unsaturated Aldehydes

Although α,β-unsaturated ketones undergo conjugate addition of TosMIC anion to give pyrroles (Eq. 8), the higher electrophilicity of the aldehyde carbon permits reductive cyanation of α,β-unsaturated aldehydes, as exemplified by the reductive cyanation of citral (Eq. 28).[62]

$$\text{citral-CHO} \xrightarrow{\text{TosCH}_2\text{N=C}} \text{citral-CN} \quad (58\%) \quad \text{(Eq. 28)}$$

In closing this section, it may be noted that an efficient reversal of Eq. 19, an oxidative decyanation, has been reported for methyl cyanides with at least one aryl substituent in the form of a phase transfer catalyzed reaction with oxygen.[68]

SYNTHESIS OF OXAZOLINES AND OXAZOLES FROM CARBONYL COMPOUNDS

Base-promoted reactions of TosMIC with ketones or aldehydes in alcoholic solvents provide five-membered heterocycles containing nitrogen and oxygen. Three different types of heterocycles may be formed in alcoholic solution from TosMIC and ketones: 4-tosyl-2-oxazolines **28**, 2-alkoxy-3-oxazolines **29**, and 4-alkoxy-2-oxazolines **30** (R^3 = H). The same type of heterocycles **28**, **29**, and **30**

28, **29**, **30**, **31**

R^1, R^2, R^3 = H, alkyl, aryl; R^4 = alkyl

($R^1 = R^3 = H$) result from reacting TosMIC with aldehydes, but in this case the formation of oxazoles **31** is also possible (Eq. 29).

$$\underset{\underset{R^2}{\overset{R^3}{\diagup}}\diagdown_O\diagup^{\overset{Tos}{\diagdown}N}}{} \xrightarrow[\text{reflux}]{K_2CO_3, \text{ MeOH,}} \underset{\underset{R^2}{\overset{R^3}{\diagup}}\diagdown_O\diagup^N}{\underset{\textbf{31}}{}} \quad \text{(Eq. 29)}$$

4-Tosyloxazolines **28** are the initial products in all of these reactions, and are formed simply by stirring TosMIC and the carbonyl compound with a catalytic amount of base in methanol or ethanol at room temperature. At elevated temperatures and/or prolonged reaction times in alcohol, tosyloxazolines **28** are converted to one of the compounds **29** to **31**. Which of these is formed depends on the nature of the substituents R^1, R^2, and R^3.

Oxazolines and Products Derived Therefrom

Elimination of 4-toluenesulfinic acid is obviously not possible with tosyloxazolines **28** derived from ketones. Instead, the tosyl group may be replaced by an alkoxy group to give 2-alkoxyoxazolines **29** or 4-alkoxyoxazolines **30** by reaction with the alcoholic solvent. These reactions are assumed to take place in a combined addition-elimination process to give **29**, or by two such processes to give isomer **30** (Eq. 30).

(Eq. 30)

On the basis of available information, 4-alkoxy-2-oxazolines **30** appear to be the main (or sole) products when $R^3 = H$, whereas 2-alkoxy-3-oxazolines **29** are formed exclusively when $R^3 = $ alkyl. Whereas the reductive cyanation of ketones or aldehydes requires that the methylene group of TosMIC be unsubstituted, this is no longer a necessary condition for the synthesis of oxazolines **28**, **29**, and **30** and oxazoles **31**. Monosubstituted TosMIC homologs give the same series of products **28** to **31** with $R^3 \neq H$. For example, the reaction of 2-phenyl-1-tosylethyl isocyanide gives a 4-tosyloxazoline of type **28** (Eq. 31).[69]

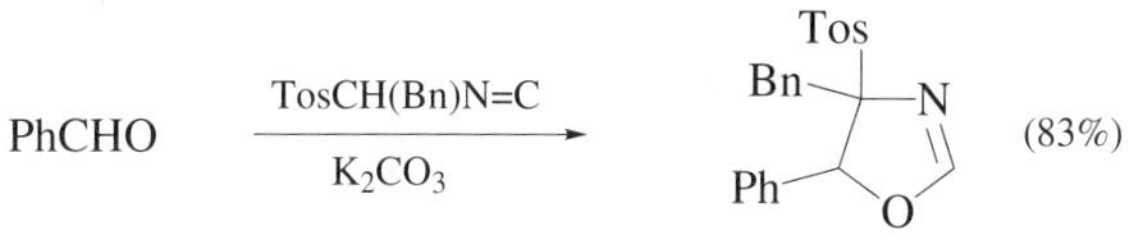

PhCHO $\xrightarrow[K_2CO_3]{TosCH(Bn)N=C}$ [Bn-N, Tos, Ph-O oxazoline] (83%) (Eq. 31)

The next example, Eq. 32, amounts to conversion of a 17-oxosteroid to a 2-methoxyoxazoline of type **29** using TosMIC, formaldehyde, and methanol.[70] First, 3-methoxyandrosta-3,5-dien-17-one is converted to **32**, the formal Knoevenagel condensation product of TosMIC and the 17-oxosteroid. Condensation products of type **32** can be used effectively as monosubstituted TosMIC homologs through γ deprotonation, as discussed earlier. Thus, a 4-tosyloxazoline **28** is formed in the reaction of steroid **32** with formaldehyde (Eq. 32).

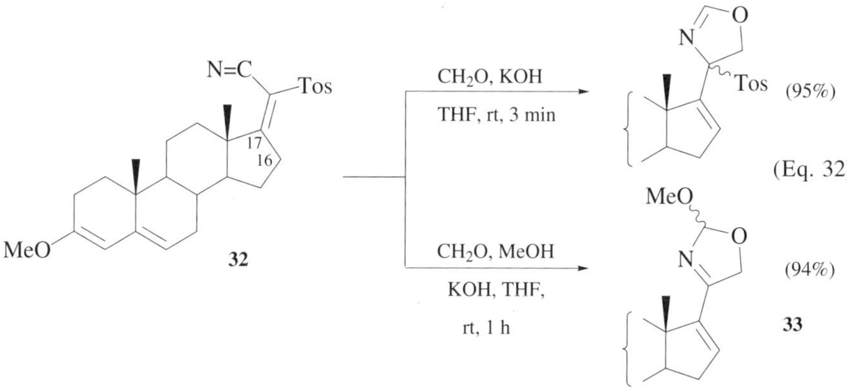

(Eq. 32)

When the reaction of **32** with formaldehyde is carried out in the presence of 5 equivalents of methanol, the initially formed 4-tosyloxazoline is converted in situ into 2-methoxyoxazoline **33** (Eq. 32) by an addition-elimination process as described in Eq. 30.

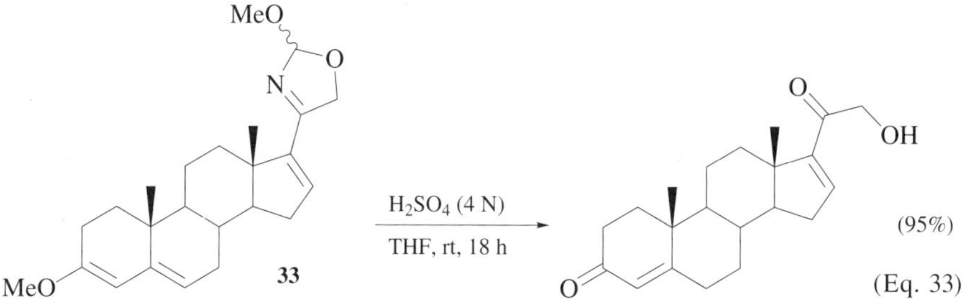

(Eq. 33)

The synthetic utility of oxazolines **28 − 30**, in addition to being intermediates in the synthesis of oxazoles when $R^1 =$ H, relies above all on their propensity to hydrolysis. Thus, acid hydrolysis of 2-methoxyoxazoline **33** leads to a 17-(hydroxyacetyl)steroid in high yield (Eq. 33).[70]

The reactions shown in Eqs. 32 and 33 form an efficient method for the introduction of 17-(hydroxyacetyl) side chains into 17-oxosteroids.[70,71]

In a similar fashion, acid hydrolysis of 4-alkoxy-2-oxazolines **30** leads to α-hydroxy aldehydes (Eq. 34).[72]

$$\text{Ph-CO-CH}_3 \xrightarrow[\text{EtOH}]{\text{TosCH}_2\text{N=C}} \text{Ph-(EtO)-oxazoline} \xrightarrow{\text{H}_3\text{O}^+} \text{Ph-C(OHC)(OH)-} \quad (70\%)$$

(60%), E:Z = 4:1 (Eq. 34)

Although 4-tosyloxazolines **28** can also be hydrolyzed by acid to afford α-hydroxy carbonyl compounds,[73] the 2- and 4-alkoxyoxazolines **29** and **30** are preferred precursors since they are more stable.[70]

Reduction of 4-tosyloxazolines **28** with LiAlH$_4$ leads to β-hydroxy N-methylamines. This reaction, which so far has been little investigated, applies to products derived from both ketones and aldehydes. The synthesis of β-hydroxy N-methylamines is exemplified by Eqs. 35[74] and 37.[75]

(Eq. 35)

Stereochemistry

Several aspects of stereochemistry are involved in the reactions leading to oxazolines. The anion of TosMIC will attack sterically screened carbonyl groups from the least hindered side. For example, the $(-)aR$ chromium complex of 2-methylbenzaldehyde is attacked by TosMIC anion exclusively from the side opposite to the (tricarbonyl)chromium group (Eq. 35),[74] and norbornanone is attacked exclusively from the exo side.[72]

4-Tosyloxazolines **28** may form cis/trans mixtures when $R^1 \neq R^2$. In the case of 4-tosyloxazolines derived from TosMIC and aldehydes, the dominant or exclusive products have the trans configuration (as in Eq. 35),[36,75] facile epimerization at C4 favoring the more thermodynamically stable product.[76] A comparable situation exists for tosyloxazolines derived from TosMIC and ketones (**28**, R^3 = H). For example, only *trans*-5-*tert*-butyl-5-methyl-4-tosyl-2-oxazoline is formed in the reaction of TosMIC and *tert*-butyl methyl ketone (Eq. 36).[55,77] Stereoselection in the formation of 2-alkoxy- and 4-alkoxyoxazolines **29** and **30** is less pronounced.

$$\text{t-Bu-CO-Me} \xrightarrow{\text{TosCH}_2\text{N=C}, \text{ t-BuOLi}} \text{trans-5-t-Bu-5-Me-4-Tos-2-oxazoline} \quad (99\%) \quad \text{(Eq. 36)}$$

In a catalytic, asymmetric process related to Eq. 35, several aldehydes have been converted to *trans*-4-tosyloxazolines, with enantiomeric excesses ranging from 73 to 86%, by reaction with TosMIC under the influence of a chiral silver (I) catalyst derived from silver triflate and a *N,N,N′,N′*-tetraalkylethylenediamino-substituted bis(diphenylphosphino)ferrocene ligand (Eq. 37).[75]

$$\text{PhCHO} \xrightarrow{\text{TosCH}_2\text{N=C, AgOTf, ferrocene ligand}} \text{4-Tos-5-Ph-oxazoline} \quad 4R,5R, (96\%) \; 83\% \text{ ee}, \text{ trans:cis} = 100:1 \quad \text{(Eq. 37)}$$

$$\xrightarrow{\text{LiAlH}_4} \text{Ph-CH(OH)-CH}_2\text{-NHMe} \quad R \; (90\%)$$

Oxazoles

Two types of oxazoles are accessible from TosMIC or monosubstituted TosMIC homologs. Reaction with aldehydes and base gives mono- or disubstituted oxazoles with a 5-aryl, 5-alkyl, or 5-hydrogen substituent, depending on the type of aldehyde used. A substituent R at the TosMIC methylene group leads to a second substituent on the oxazole ring at C4 (Eq. 38).[26,69,78]

$$\text{PhCHO} \xrightarrow[\text{reflux, 1-2 h}]{\text{TosCHRN=C, K}_2\text{CO}_3\text{, MeOH}} \text{4-R-5-Ph-oxazole}$$

R	
H	(91%)
Me	(75%)
Et	(73%)

(Eq. 38)

Reaction of TosMIC with acid chlorides or esters leads in a similar fashion to oxazoles which, however, always carry a tosyl group at C4 (Eq. 39).[26]

$$\text{PhCOCl} \xrightarrow[n\text{-BuLi, THF}]{\text{TosCH}_2\text{N}=\text{C}} \underset{\text{Ph}}{\underset{\text{O}}{\text{Tos}}}\!\text{N} \quad (65\%) \qquad \text{(Eq. 39)}$$

4-Tosyloxazoles are obtained similarly from carboxylic anhydrides and from selenol esters. Esters of aliphatic alcohols react with TosMIC only when two equivalents of n-butyllithium in THF are used (Eq. 40), the more nucleophilic dianion being necessary for attack at the ester carbonyl.[79]

$$\text{Ph}\!\!\!\!\sim\!\!\!\!\text{CO}_2\text{Me} \xrightarrow[\substack{n\text{-BuLi (2 eq),}\\-70 \text{ to } 0°, 2\text{ h}}]{\text{TosCH}_2\text{N}=\text{C},} \text{Ph}\!\!\sim\!\!\!\overset{\text{Tos}}{\underset{\text{O}}{\text{N}}} \quad (53\%) \qquad \text{(Eq. 40)}$$

The example of Eq. 40 also shows that under the conditions given no reaction takes place at the conjugated double bond of the ester. With the softer monoanion of TosMIC, however, the reaction would take place exclusively at the C-C double bond. The latter reactions provide a highly useful synthesis of pyrroles, as is discussed in a later section.

In all cases, the oxazole ring carbon C2, originating from the isocyano group, remains unsubstituted (Eqs. 38–40). However, oxazoles with amino or alkoxy substituents at C2 are available by a related process using TosCH$_2$N=C=NR[46] and TosCH$_2$N=C(OR)[52] synthons, respectively, instead of TosMIC.

4-Tosyloxazolines of type **28** (R^1 = H) are the first-formed intermediates of Eq. 38 en route to oxazoles. Although oxazoles may be formed directly by base-induced elimination of TosH as suggested in Eq. 30, it is not at all unlikely that alkoxyoxazolines **30** and/or **29** (Eq. 30) are intermediates when the synthesis of oxazoles is carried out in alcoholic solvents. The oxazoles are formed eventually by elimination of R^4OH from **29** or **30** (R^1 = H). Several species of each of these potential intermediates have been isolated and characterized.

Whether alkoxyoxazolines **29** and **30** really occur as intermediates in the synthesis of oxazoles seems to depend in part on the relative acidities of the C4 and C5 hydrogens. For example, 5-hexyloxazole is formed only in 13% yield from TosMIC and n-heptanal,[31] whereas a 72% yield of 5-hexyl-4-methyloxazole is obtained when TosMIC is replaced by 1-tosylethyl isocyanide.[31] The low yield of 5-hexyloxazole is assumed to be a reflection of a larger difference in acidity between the hydrogens at C4 and C5 in the 4-tosyloxazoline precursor (see Eq. 30), as compared to the corresponding species derived from benzaldehyde (Eq. 38). In the latter reaction the increased acidity of the C5 hydrogen in the 4-tosyloxazoline precursor facilitates the base-induced elimination of TosH. The low yield of

5-hexyloxazole is improved by utilizing the two-step approach of Eq. 41. The acidity of the C4 hydrogen is first decreased by replacing the tosyl group of the oxazoline precursor by a methoxy group, followed by elimination of methanol by t-BuOK in diethyl ether.[31]

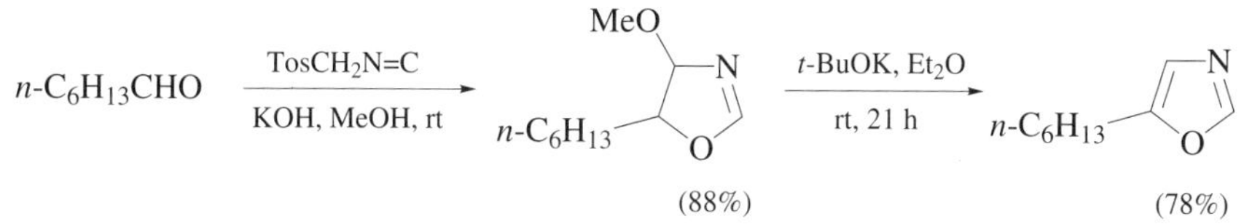

$$n\text{-}C_6H_{13}CHO \xrightarrow[\text{KOH, MeOH, rt}]{TosCH_2N=C} \underset{(88\%)}{n\text{-}C_6H_{13}\text{-oxazoline(MeO)}} \xrightarrow[\text{rt, 21 h}]{t\text{-BuOK, Et}_2O} \underset{(78\%)}{n\text{-}C_6H_{13}\text{-oxazole}}$$

(Eq. 41)

α,β-Unsaturated aldehydes react with TosMIC or monosubstituted TosMIC homologs at the aldehyde group to give oxazoles (Eq. 42).[80] In a similar fashion, oxazoles are formed in reactions with α-keto aldehydes and with a hydrazone of glyoxal (Eq. 43).[63]

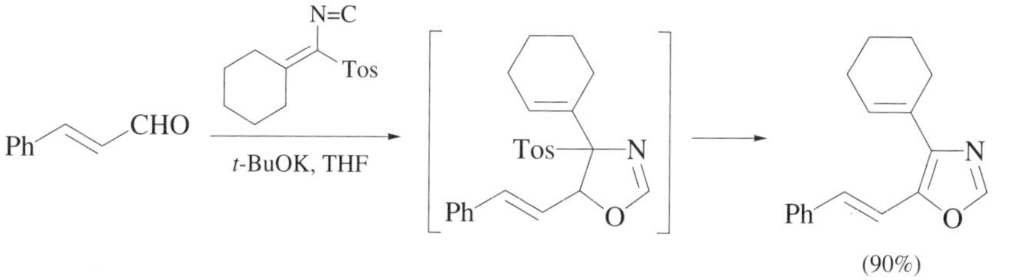

(Eq. 42)

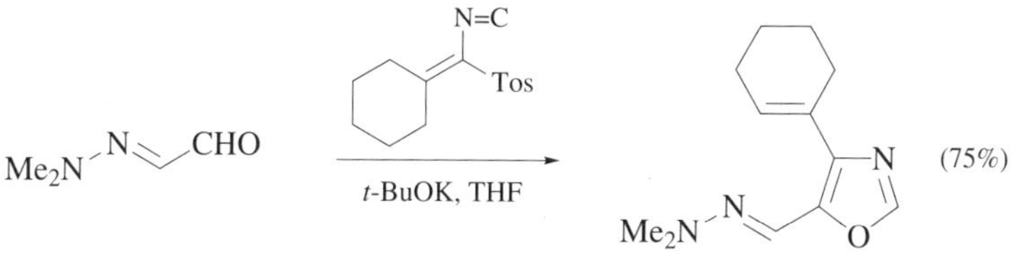

(Eq. 43)

The reaction of Eq. 44 is one of the few examples of the conversion of formaldehyde into an oxazole.[70]

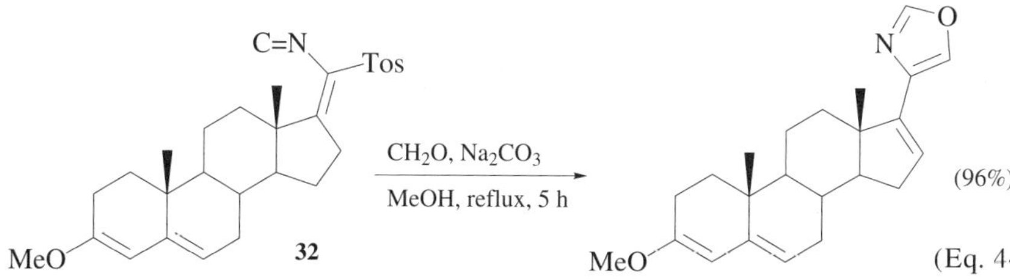

(Eq. 44)

Only two aldehydes are reported to behave differently. Whereas 2-indolecarboxaldehyde forms an oxazole with TosMIC,[81] the reaction with 2-pyrrolecarboxaldehyde gives 3-tosylpyrrolo[1,2-c]pyrimidine (in low yield), through the participation of the pyrrole NH group in the ring-closing reaction (Eq. 45).[78] A

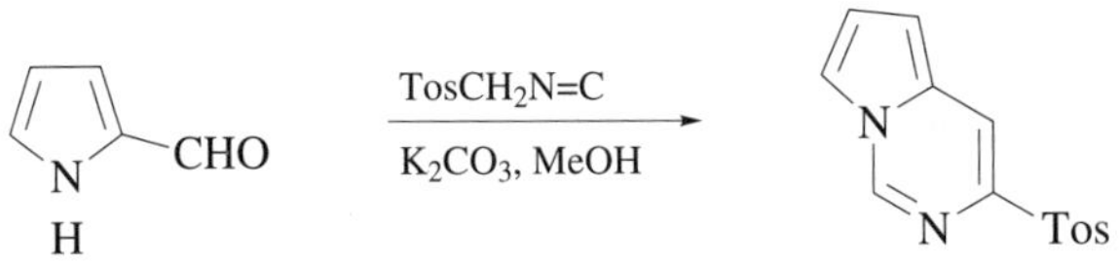

$$\text{pyrrole-2-CHO} \xrightarrow[K_2CO_3, \text{MeOH}]{TosCH_2N=C} \text{pyrrolo[1,2-c]pyrimidine-Tos} \quad (20\%) \quad \text{(Eq. 45)}$$

similar pyrimidine, instead of an oxazole, is formed from 2-imidazolecarboxaldehyde and TosMIC (Eq. 46).[81] However, when the reaction conditions of Eq. 46 are changed to K_2CO_3 in methanol, products of ring opening of the ex-

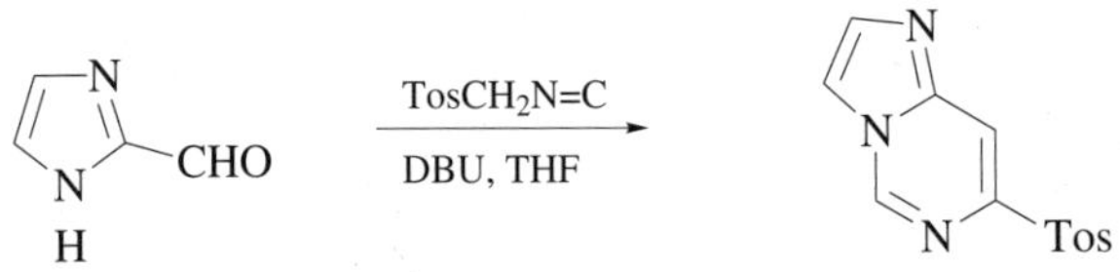

$$\text{imidazole-2-CHO} \xrightarrow[DBU, THF]{TosCH_2N=C} \text{imidazo-pyrimidine-Tos} \quad (14\%) \quad \text{(Eq. 46)}$$

pected 4-tosyloxazolines are formed (cf. Scheme 2): an unsaturated formamide **34** when the reaction is carried out at room temperature, and an unusual tosyl substituted primary enamine **35** at reflux temperature.[81]

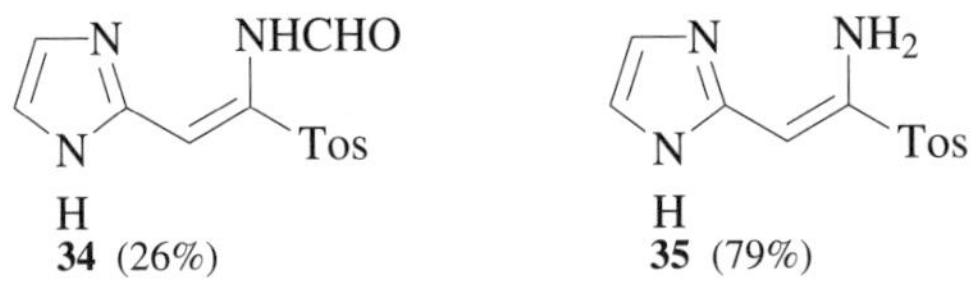

34 (26%)   **35** (79%)

Unusual examples of TosMIC oxazole syntheses are shown in Eqs. 47[82] and 48.[83,84]

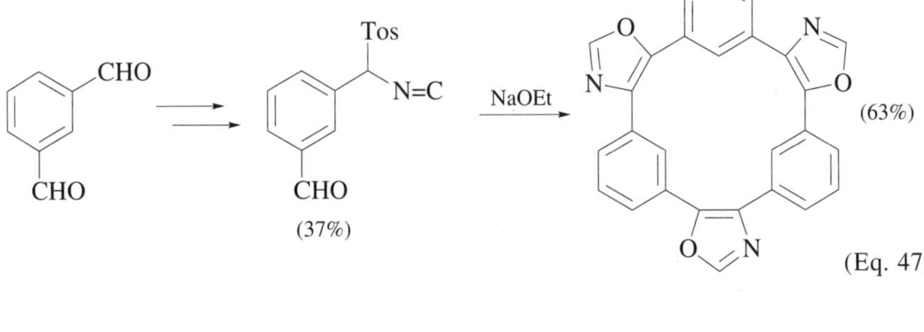

$$\text{isophthalaldehyde} \longrightarrow \text{Tos-CH(N=C)-C_6H_4-CHO} \xrightarrow{NaOEt} \text{tetraoxazole macrocycle} \quad (63\%)$$

(37%)

(Eq. 47)

SYNTHESIS OF 1-FORMAMIDO-1-TOSYLALKENES AND 1-ISOCYANO-1-TOSYLALKENES

Reaction of ketones or aldehydes with TosMIC and base, at low temperatures in non-protic solvents, leads to 1-tosyl-substituted α,β-unsaturated formamides **25**.[36,63,85,86] Subsequent dehydration of these unsaturated formamides gives 1-tosyl-substituted α,β-unsaturated isocyanides **36**, which are used extensively both as homologs of TosMIC and as Michael acceptors.

The overall result of Eq. 49 amounts to a formal Knoevenagel condensation between cyclohexanone and TosMIC. Similar results are obtained by an application of the Peterson olefination conditions to TosMIC (Eq. 50). The latter approach is restricted to aldehydes and reactive ketones such as cyclobutanone.[87]

The mechanistic rationale for the formation of these unsaturated formamides was discussed earlier (Scheme 2). To form unsaturated TosMIC derivatives **36**, a dehydration step is needed; the two steps may be combined into a one-pot procedure.[54,86] Conversions to unsaturated tosyl formamides **25** are best carried out with 1 equivalent of TosMIC and 1.1 equivalents of t-BuOK in tetrahydrofuran at -70 to $-40°$ for 30 minutes, followed by a quench with water and acetic acid.[63]

Scope and Limitations

The formal Knoevenagel condensation of TosMIC according to Eq. 49 is feasible with many different ketones and aldehydes. The limitations to the first reaction step are the same as in the reductive cyanation: no reaction occurs with severely sterically hindered ketones or with readily enolizable carbonyl compounds. With the less reactive, sterically screened carbonyl compounds exemplified by *tert*-butyl methyl ketone, the strenuous conditions necessary to realize the first reaction step also facilitate subsequent steps, leading to reductive cyanation. This problem was solved by carrying out the conversion to the unsaturated formamide in two separate steps (Eq. 51).[55] First, 5-*tert*-butyl-5-methyl-4-tosyl-

2-oxazoline (**37**) is formed by reaction in tetrahydrofuran, using lithium *tert*-butoxide. The ring opening to (E)-N-(2,3,3-trimethyl-1-tosyl-1-butenyl)formamide (**38**) is then carried out under mild conditions using potassium *tert*-butoxide in 1,2-dimethoxyethane at $-50°$. This approach is also recommended for other less reactive carbonyl compounds.

When cis/trans isomers are possible, as with aldehydes and unsymmetrical ketones, single stereoisomers are formed almost without exception. The bulkier groups are assumed to be trans to the tosyl group, as in the established structure[88] of the isocyanide derived from pivalaldehyde.[86]

The Peterson olefination approach is applicable only to aldehydes and activated ketones, such as cyclobutanone. With aldehydes, the best results are obtained with non-enolizable compounds such as cinnamaldehyde (Eq. 52).[87]

Two examples of a direct condensation of amido acetals and TosMIC have been reported briefly (Eq. 53).[89]

$$\begin{array}{ccc} R^1 & & R^2 \\ H & Me & (87\%) \\ -(CH_2)_3- & & (98\%) \\ R^3 \text{ not specified} & & \end{array}$$

TosCH$_2$N=C with R^1, R^{2}N(Me), (OR^3)$_2$ gives the vinyl product R^1(R^{2}N(Me))C=C(Tos)(N=C).

(Eq. 53)

TosMIC SUBSTITUTED AT THE METHYLENE GROUP

Both mono- and disubstituted TosMIC derivatives are well known, substituents most commonly being alkyl or aralkyl.

Monosubstituted TosMIC Derivatives

Normally, the reaction of TosMIC with one equivalent of an alkylating agent leads to mixtures of starting material with mono- and dialkylated products. When monosubstitution of TosMIC is desired, phase transfer catalysis (PTC) is the method of choice, as in Eq. 54. Benzyltriethylammonium chloride (BTEAC),

$$\text{TosCH}_2\text{N=C} \xrightarrow[\text{BTEAC, CH}_2\text{Cl}_2]{\text{MeI, NaOH (30\%)}} \text{TosCH(Me)N=C} \quad (95\%)$$

(Eq. 54)

tetrabutylammonium bromide (TBAB), and tetrabutylammonium iodide (TBAI) have been commonly employed for these reactions, in a mixture of dichloromethane and 20–50% aqueous sodium hydroxide. High yields are normally obtained from primary bromides or iodides, and also from benzyl bromide or allyl chloride (Eq. 55).[73,90,91]

$$\text{TosCH}_2\text{N=C} \xrightarrow[\text{PTC}]{\text{CH}_2=\text{CHCH}_2\text{Cl}} \text{TosCH(N=C)(CH}_2\text{CH=CH}_2\text{)} \quad (91\%)$$

(Eq. 55)

1-Tosylethyl isocyanide has been prepared by monomethylation of TosMIC (Eq. 54).[90,91] The same compound is also available from the Mannich route (Eq. 16).[90]

Benzylic-type monoalkylation has been used in a double fashion to form compound **40** as part of a synthesis of cyclophanes (Eq. 56). A TosMIC analog (4-ethylphenylsulfonylmethyl isocyanide) was used in this sequence.[92]

[Structures for Eq. 56: compound 39 (2,6-bis(bromomethyl)pyridine) + 4-EtC$_6$H$_4$SO$_2$CH$_2$N=C under PTC gives compound 40, then with H$_3$O$^+$ gives macrocyclic bis-pyridine diketone (17%)]

(Eq. 56)

In addition to 1-tosyl-3-butenyl isocyanide (Eq. 55), only two other examples are known of monosubstituted TosMIC derivatives specifically functionalized at the β position: (1) with a trimethylsilyl group[91,93] and (2) with an ester group.[91]

The yields of monoalkylations with secondary alkyl halides tend to be lower than with primary halides (Eq. 57).

$$\text{TosCH}_2\text{N=C} \xrightarrow{i\text{-PrX}} \text{TosCH(Pr-}i\text{)N=C}$$

	X	
PTC	Cl	(40%)[90]
NaH, DMSO, Et$_2$O	Br	(71%)[73]
PTC	I	(42%)[73]

(Eq. 57)

Better results are sometimes obtained with secondary alkyl bromides or iodides when sodium hydride in a mixture of dimethyl sulfoxide and diethyl ether is employed (Eq. 58).[73]

$$\text{TosCH}_2\text{N=C} \xrightarrow[\text{NaH, DMSO, Et}_2\text{O}]{\text{C}_6\text{H}_{11}\text{X}} \underset{\text{C}_6\text{H}_{11}}{\text{TosCHN=C}}$$

X	
Br	(0%)
I	(30%)

(Eq. 58)

As expected, the reaction of *tert*-butyl bromide or iodide with TosMIC does not lead to 2,2-dimethyl-1-tosylpropyl isocyanide. This compound has been prepared in two different ways by the Mannich approach (Eqs. 59 and 60).[94]

$$4\text{-MeC}_6\text{H}_4\text{SH} + t\text{-BuCHO} + \text{H}_2\text{NCHO} \xrightarrow[90°, 2\text{ h}]{\text{HCO}_2\text{H}}$$

4-MeC$_6$H$_4$S–CH(Bu-t)–NHCHO $\xrightarrow[-30 \text{ to } 20°, 17\text{ h}]{3\text{-ClC}_6\text{H}_4\text{CO}_3\text{H}}$ Tos–CH(Bu-t)–NHCHO (20%) (Eq. 59)

$$\text{TosH} + t\text{-BuCHO} + \text{H}_2\text{NCHO} \xrightarrow[90°, 2\text{ h}]{\text{HCO}_2\text{H}} \text{Tos–CH(Bu-}t\text{)–NHCHO} \quad (20\%)$$

$\xrightarrow[\text{THF}, 0°, 1\text{ h}]{\text{POCl}_3, i\text{-Pr}_2\text{NH}}$ Tos–CH(Bu-t)–N=C (80%) (Eq. 60)

Sodium borohydride reduction of the formal Knoevenagel condensation products of TosMIC and aldehydes or ketones also leads to monosubstituted TosMIC derivatives (Eq. 61).[45,95,96]

(MeO)$_2$C$_6$H$_3$–CH=C(Tos)–N=C $\xrightarrow{\text{NaBH}_4}$ (MeO)$_2$C$_6$H$_3$–CH$_2$–CH(Tos)–N=C (75%) (Eq. 61)

Disubstituted TosMIC Derivatives

Dialkylation of TosMIC is readily achieved by using two equivalents of alkylating agent. Such reactions are usually carried out by using sodium hydride in a dimethyl sulfoxide − diethyl ether mixture (Eqs. 62 and 63).[35,97]

TosCH$_2$N=C $\xrightarrow[\text{DMSO, Et}_2\text{O}]{\text{MeI (2 eq), NaH}}$ Tos–C(Me)$_2$–N=C (90%) (Eq. 62)

TosCH$_2$N=C $\xrightarrow[\substack{\text{2. MeI (1.5 eq), NaH,} \\ \text{DMSO, Et}_2\text{O}}]{\substack{\text{1. (Z)-EtCH=CH(CH}_2)_3\text{I,} \\ \text{PTC}}}$ Tos–C(Me)((CH$_2$)$_3$CH=CHEt-(Z))–N=C (72%) (Eq. 63)

Dialkylated compounds cannot be synthesized by the Mannich approach because of the instability of the formamido-precursors TosCR^{1}R^{2}NHCHO when both R^1 and R^2 are not hydrogen.[98] The only exceptions are 1-tosyl-1-formamidocyclopropanes.[99]

The best results in the dialkylation of TosMIC are usually obtained with primary alkyl bromides or iodides, and with allylic or benzylic chlorides or bromides. Although PTC conditions are used in particular for monoalkylations, the same conditions have been used occasionally for the introduction of a second alkyl group. Under appropriate conditions even a less reactive halide, such as the chlorine in Eq. 64, may remain untouched.[100] Nevertheless, alkyl chlorides have been used successfully at slightly elevated temperatures and/or prolonged reaction times.[101]

$$\text{Tos}\underset{\text{N=C}}{\diagdown}\xrightarrow[\text{NaH, DMSO, Et}_2\text{O}]{\text{Br(CH}_2)_3\text{Cl (1.5 eq)}} \text{Tos}\underset{\text{N=C}}{\overset{(\text{CH}_2)_3\text{Cl}}{\diagdown}} \quad (87\%) \quad \text{(Eq. 64)}$$

Dibromides or diiodides react with TosMIC to form ring systems by intramolecular dialkylation (Eq. 65). Simple 1,n-dibromides (or iodides) have been used to form 3- to 7-membered rings.[39,73,102]

$$\text{TosCH}_2\text{N=C} \xrightarrow[\text{NaH, DMSO, Et}_2\text{O}]{\text{Br(CH}_2)_2\text{Br}} \triangle\underset{\text{Tos}}{\overset{\text{N=C}}{\diagdown}} \quad (81\%) \quad \text{(Eq. 65)}$$

Monosubstituted TosMIC derivatives react with 1,n-dihalides to afford dialkylation products (Eq. 66).[100] Bridges of 2 to 9 atoms have thus been realized.

$$\text{Tos}\underset{\text{N=C}}{\diagdown}\xrightarrow[\text{NaH, DMSO, Et}_2\text{O}]{\text{Br(CH}_2)_3\text{Br}} \text{Tos}\underset{\text{N=C}}{\overset{}{\diagdown}}\diagdown\underset{\text{Tos}}{\overset{\text{N=C}}{\diagdown}} \quad (91\%) \quad \text{(Eq. 66)}$$

Lower yields obtained in the reactions with 1,2-dibromoethane or 1,2-diiodoethane may reflect steric hindrance, which is compounded with 2-phenyl-1-tosylethyl isocyanide. The latter fails to react with 1,2-dibromoethane and requires an iodide for introduction of a *sec*-butyl group (Eq. 67).[73,100]

$$\text{Tos}\underset{\text{N=C}}{\overset{\text{Bn}}{\diagdown}}\xrightarrow[\text{NaH, DMSO, Et}_2\text{O}]{\text{MeCHXEt}} \text{Tos}\underset{\text{N=C}}{\overset{\text{Bn}\quad\text{Et}}{\diagdown}} \quad \begin{array}{l}\text{X}\\\text{Br}\quad(0\%)\\\text{I}\quad(56\%)\end{array} \quad \text{(Eq. 67)}$$

Two unusual disubstituted TosMIC derivatives have been obtained from the reaction of dilithio-TosMIC with pyridine-*N*-oxide and with pyridazine-*N*-oxide, followed by benzylation (Eqs. 68 and 69).[79]

$$\text{TosCH}_2\text{N=C} \xrightarrow[\substack{\text{2. pyridine }N\text{-oxide}\\\text{3. BnBr (1 eq)}}]{\text{1. }n\text{-BuLi (2 eq)}} \text{Tos}\underset{\text{N=C}}{\overset{\text{Bn}}{\diagdown}}\diagup\!=\!\diagdown=\!\text{NOH} \quad \begin{array}{l}\text{(Eq. 68)}\\(64\%)\end{array}$$

$$\text{TosCH}_2\text{N=C} \xrightarrow[\substack{\text{2. pyridazine }N\text{-oxide}\\\text{3. BnBr (1 eq)}}]{\text{1. }n\text{-BuLi (2 eq)}} \text{Tos}\underset{\text{N=C}}{\overset{\text{Bn}}{\diagdown}}\diagup\!=\!\diagdown\!\!\equiv \quad (52\%) \quad \text{(Eq. 69)}$$

SYNTHETIC USES OF TOSYLMETHYL ISOCYANIDE (TosMIC)

A more general method for the synthesis of TosMIC derivatives **42** with a 1-alkenyl substituent involves alkylation of allylic anions **41**, formed by γ deprotonation of the (formal) Knoevenagel condensation products of TosMIC and ketones. These alkylations take place exclusively at C1.[101,103]

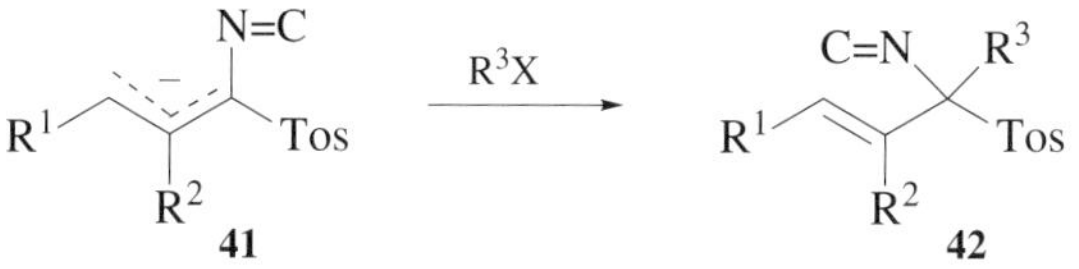

The allylic alkylation reaction **41** → **42** forms part of a widely applicable conversion of ketones to homologous enones, as in Eq. 70.[103]

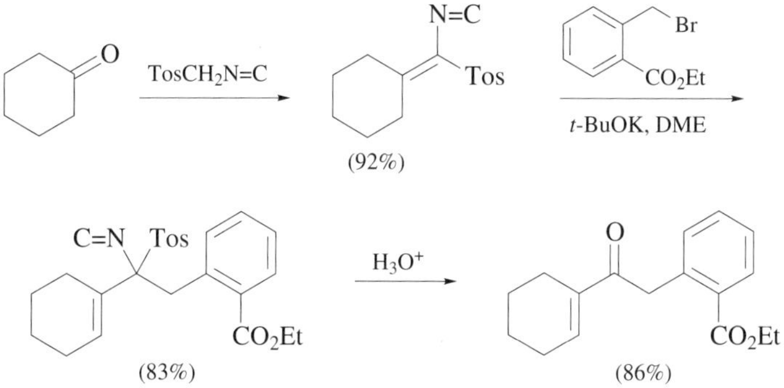

(Eq. 70)

The same approach, using dichloromethane in the alkylation step, is used for the introduction of functionalized acetyl side chains at C17 in 17-oxosteroids (Eq. 71)[101]

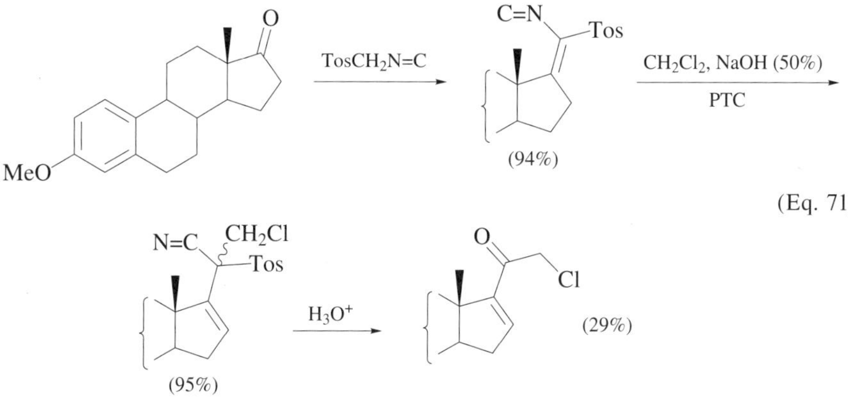

(Eq. 71)

A methylene transfer reaction, using dimethyloxosulfonium methylide, applied to the (formal) Knoevenagel condensation product of TosMIC and acetone leads to a cyclopropane derivative (Eq. 72).[102]

$$\text{[Structure: } \underset{\text{Tos}}{\overset{\text{N=C}}{\diagup\!\!\!\diagdown}} \xrightarrow[\text{DMSO, THF}]{\overset{+}{\text{Me}_2\text{S}}\overset{\text{O}^-}{\diagup}\text{CH}_2} \underset{\text{Tos}}{\overset{\text{N=C}}{\triangle}} \quad (85\%) \quad\quad \text{(Eq. 72)}$$

TosMIC as a Connective Reagent

In the reactions discussed so far, TosMIC is used to connect two (ar)alkyl halides, either inter- or intramolecularly, by a functionalized methylene bridge. Hence, TosMIC here fulfills the role of a connective or conjunctive reagent.[104] Practical applications of this type of TosMIC reaction derive from the various ways in which the geminal isocyano and tosyl groups in the products can be subsequently transformed. Most prominent among these transformations is acid hydrolysis to a keto group, as in Eqs. 13, 70, and 71. Details of this hydrolysis are discussed in the section on the TosMIC-based synthesis of ketones.

The remaining part of this section emphasizes the use of TosMIC to form methylene bridges. The geminal isocyano and tosyl groups are directly converted to a methylene group by reduction with lithium in liquid ammonia containing 3 equivalents of ethanol.[105,106] Thus, the combination of TosMIC dialkylation followed by Li/NH$_3$ reduction makes TosMIC a CH$_2$ connector of two alkyl halides, as in Eq. 73.[105]

$$\text{TosCH}_2\text{N=C} \xrightarrow[\text{PTC}]{n\text{-C}_8\text{H}_{17}\text{I}} \underset{\text{Tos}}{\overset{\text{C}_8\text{H}_{17}\text{-}n}{\text{CH}}}\text{N=C} \xrightarrow[\text{NaH, DMSO, Et}_2\text{O}]{\text{MeO-C}_6\text{H}_3(\text{OMe})\text{CH}_2\text{Br}}$$

[MeO, MeO-aryl]–C(Tos)(N=C)(C$_8$H$_{17}$-n) (90%) $\xrightarrow[\text{EtOH}]{\text{Li, NH}_3 \text{ (liq)}}$ [MeO, MeO-aryl]–CH$_2$–C$_8$H$_{17}$-n (90%) (Eq. 73)

This approach has been used for the synthesis of muscalure (a pheromone of the common housefly) by using TosMIC as a CH$_2$ connector: (1) between n-nonyl bromide and (Z)-tridec-4-enyl iodide (Eq. 74);[105,106] (2) between n-butyl iodide and the iodide derived from oleic acid.[107]

$$\underset{\text{Tos}}{\overset{\text{C}_9\text{H}_{19}\text{-}n}{\text{CH}}}\text{N=C} \xrightarrow{(Z)\text{-}n\text{-C}_8\text{H}_{17}\text{CH=CH(CH}_2)_3\text{I}} \underset{\text{Tos}}{\overset{n\text{-C}_9\text{H}_{19}}{\text{C}}}\overset{(\text{CH}_2)_3\text{CH=CHC}_8\text{H}_{17}\text{-}n\text{-}(Z)}{\text{N=C}} \quad (85\%)$$

$$\xrightarrow[\text{EtOH}]{\text{Li, NH}_3 \text{ (liq)}} (Z)\text{-}n\text{-C}_{13}\text{H}_{27}\text{CH=CHC}_8\text{H}_{17}\text{-}n \quad (90\%)$$

muscalure (Eq. 74)

Reduction of a monosubstituted TosMIC derivative forms a methyl group (Eq. 75).[105] Alternative reduction products which may be obtained from TosMIC derivatives are discussed later.

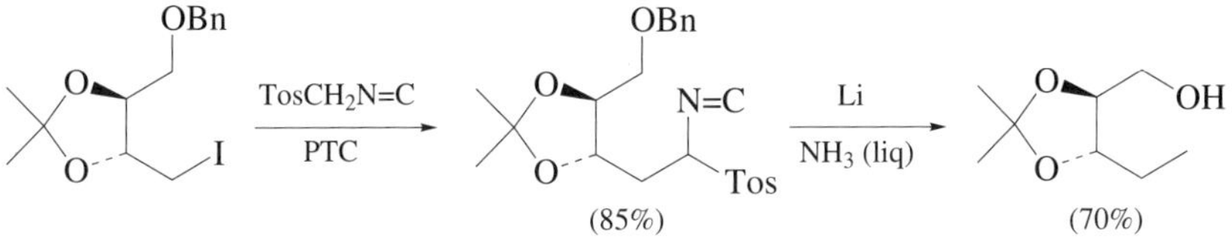

(Eq. 75)

The dibenzylation of TosMIC has been used extensively to form large ring systems, in particular various cyclophanes (Eqs. 76–78).[108–111]

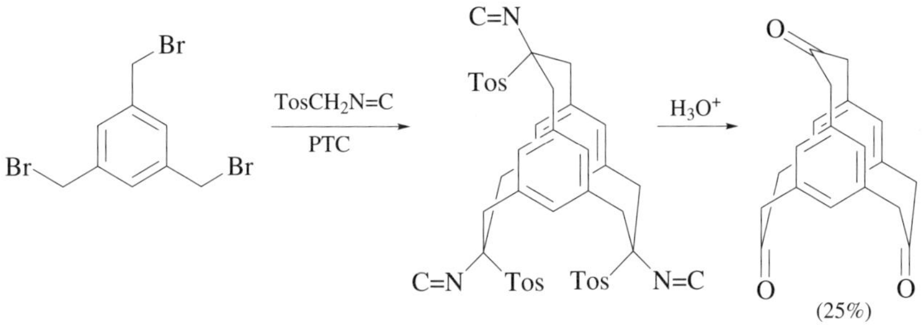

(Eq. 76)

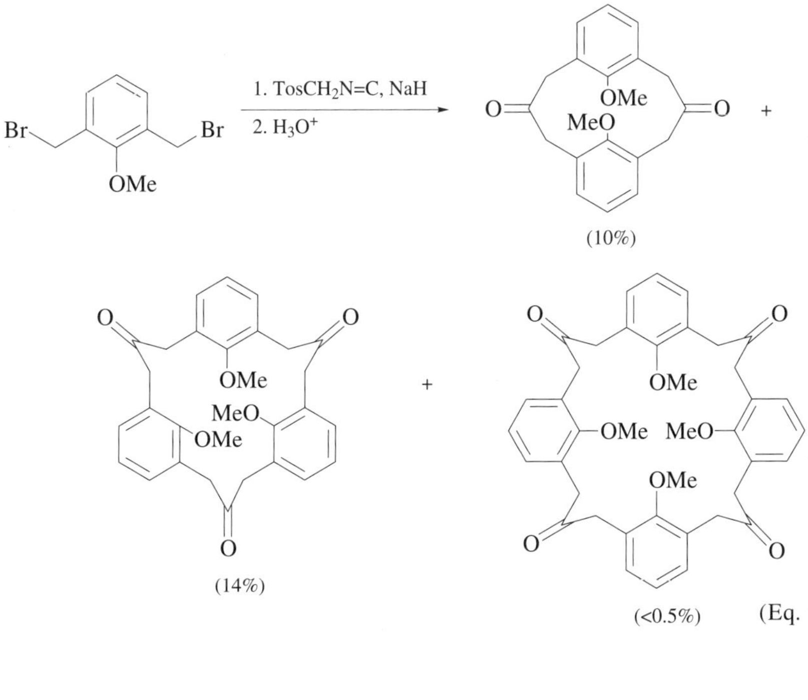

(Eq. 77)

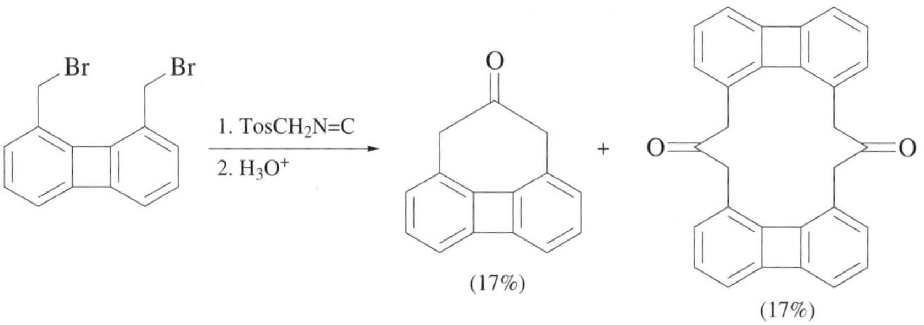

(Eq. 78)

Acylated TosMIC Derivatives

Reaction of monosubstituted TosMIC derivatives with acyl halides leads to disubstituted TosMIC compounds **43** carrying one acyl group, as in Eq. 79.[112] The products **43** are precursors of 1,2-diketones, which are formed upon acid hydrolysis. The acylation reaction is restricted to monosubstituted TosMIC derivatives, since the corresponding reaction with TosMIC leads to oxazoles.

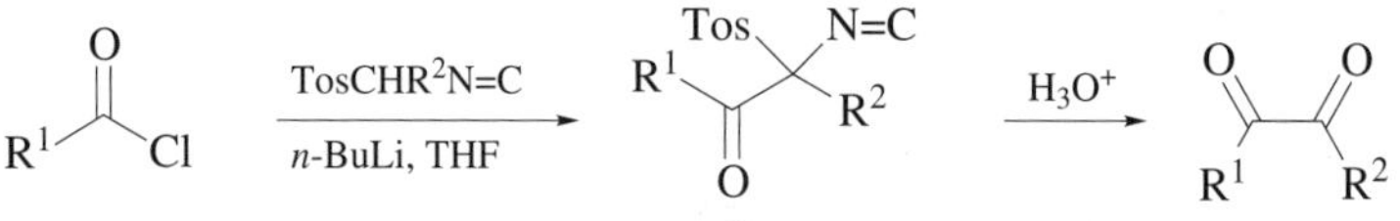

(Eq. 79)

SYNTHESIS OF IMIDAZOLES AND 1,2,4-TRIAZOLES

Imidazoles

The synthesis of imidazoles from imines and TosMIC or TosMIC homologs is closely related to the previously discussed synthesis of oxazoles from carbonyl compounds, but has not been as extensively investigated. No reports are available on reactions of ketimines, but the reaction of aldimines with TosMIC and base provides 1,5-disubstituted imidazoles in variable yields (Eq. 80).[37]

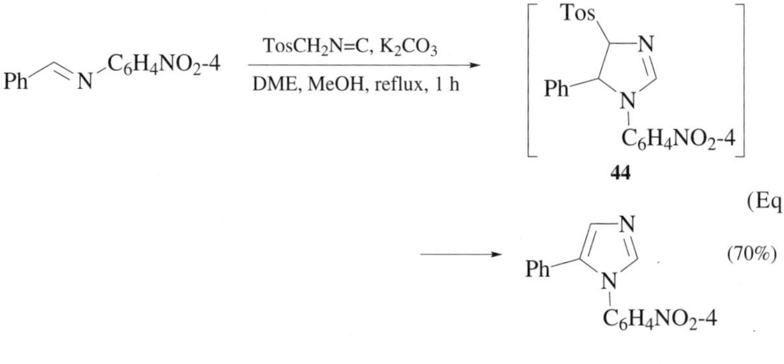

(Eq. 80)

1,4,5-Trisubstituted imidazoles are formed similarly when TosMIC is replaced by a homolog, as in Eq. 81.[37,69]

$$Ph\text{—}CH=N\text{—}C_6H_4NO_2\text{-}4 \xrightarrow[\text{NaH, DME, 0°, 1 h}]{\text{TosCH(Me)N=C}} \text{4-methyl-5-phenyl-1-(4-nitrophenyl)imidazole} \quad (75\%) \quad \text{(Eq. 81)}$$

Some 1,2,5-trisubstituted imidazoles have been prepared with the use of (formal) TosMIC derivatives TosCH$_2$N=C(SCH$_3$)R (R = C$_6$H$_5$, CH$_3$S), which provide a phenyl or a methylthio group at the 2-position.[47]

$$R\text{—}CH=N\text{—}R \xrightarrow[\text{t-BuOK, t-BuOH, DME}]{\text{TosCH}_2\text{N=C(SMe)}_2} \text{imidazole-SMe product} \quad (64\%) \quad R = 4\text{-ClC}_6H_4$$

Intermediate 4-tosylimidazolines have been isolated in a few cases only. 4-Tosylimidazoline **44**, for example, is obtained in 73% yield when the reaction of Eq. 80 is carried out with sodium hydride in 1,2-dimethoxyethane in the absence of methanol at $-20°$.[37] Under these conditions, however, no tosylimidazoline **45** is obtained from the reactants of Eq. 81. It is not known whether 4-toluenesulfinic acid is eliminated directly from 4-tosylimidazolines like **44**, or that methoxyimidazolines of type **46** and/or **47** are involved. In contrast to the oxazoline series, such methoxyimidazolines have not been identified.

45 (Tos, Ph, C$_6$H$_4$NO$_2$-4 substituted imidazoline) **46** (R, OMe imidazoline) **47** (MeO, R imidazoline)

N-(Trimethylsilyl)aldimines react with TosMIC derivatives to produce *N*-unsubstituted imidazoles (Eq. 82).[113]

$$\underset{\text{OHC}}{\overset{\text{TMS}}{\text{N}}}\text{—TMS} \xrightarrow[\text{THF}]{n\text{-BuLi}} \left[n\text{-Bu}\text{—}CH=N\text{—TMS}\right] \xrightarrow[\text{LiN(TMS)}_2]{\text{TosCH(Bn)N=C}} \text{4-}n\text{-Bu-5-Bn-imidazole} \quad (66\%) \quad \text{(Eq. 82)}$$

The C-N double bonds of quinoline, isoquinoline, quinazoline,[79] and 4*H*-pyrrolo[1,2-*a*][1,4]benzodiazepine[8] are also subject to the formation of imidazoles, as in Eq. 83, but only when the more nucleophilic reagent dilithio-TosMIC is used. This type of reaction fails with pyridine or pyridazine.[79]

$$\text{isoquinoline} \xrightarrow[n\text{-BuLi, (2 eq)}]{\text{TosCH}_2\text{N=C, LiBr}} \text{imidazo-fused product} \quad (44\%) \quad (\text{Eq. 83})$$

α,β-Unsaturated aldimines usually form imidazoles[114,115] (Eq. 84), although reaction with the C-C double bond occasionally leads to pyrrole formation.

$$\text{aldimine-Bu-}i \xrightarrow[\text{K}_2\text{CO}_3]{\text{TosCH}_2\text{N=C}} \text{imidazole-Bu-}i \quad (60\%) \quad (\text{Eq. 84})$$

Imidazoles are formed similarly by reaction of the 4-nitrophenylimine of cinnamaldehyde and of a hydrazone derivative of glyoxal with a TosMIC homolog derived from cyclohexanone (Eq. 85).[63,80] On the other hand, a pyrrole is formed from the phenylimine of cinnamaldehyde and an analogous 7-ring TosMIC homolog (Eq. 86).[80] The formation of pyrroles is avoided by replacing the C-C double bond by a triple bond (Eq. 87).[80]

$$\text{Me}_2\text{N-N=CH-CH=N-C}_6\text{H}_4\text{NO}_2\text{-4} \xrightarrow[\text{Triton B}]{\text{cyclohexanone N=C Tos}} \text{imidazole product} \quad (85\%) \quad (\text{Eq. 85})$$

$$\text{Ph-CH=CH-CH=N-Ph} \xrightarrow[t\text{-BuOK}]{\text{cycloheptanone C=N Tos}} \text{pyrrole product} \quad (70\%) \quad (\text{Eq. 86})$$

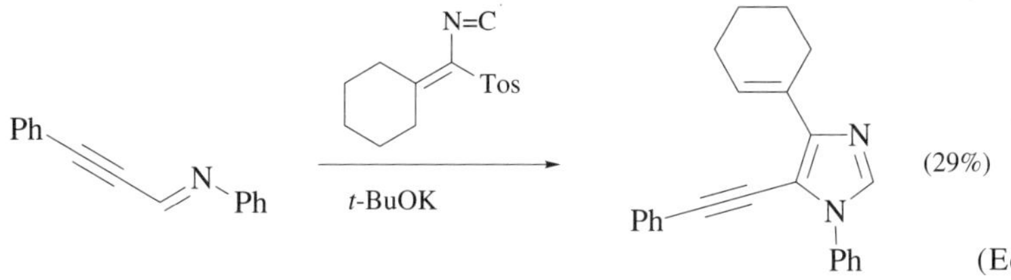

(Eq. 87)

A 1:1 mixture of imidazole and pyrrole is obtained in the reaction of the *N*-tosylimine of cinnamaldehyde (Eq. 88).[80]

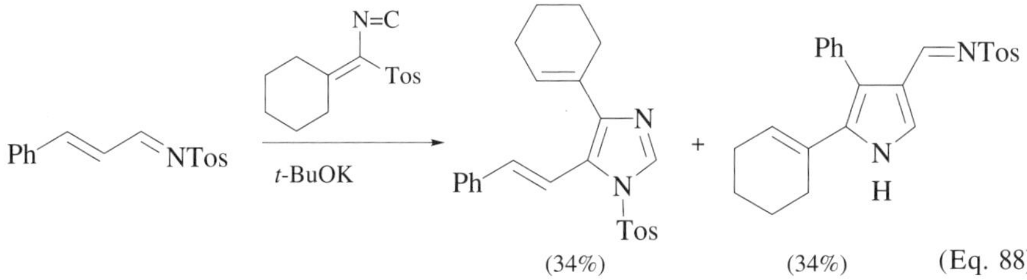

(Eq. 88)

An interesting application of the TosMIC imidazole synthesis is depicted in Eq. 89, where four imidazole rings are formed in one operation. A base-induced

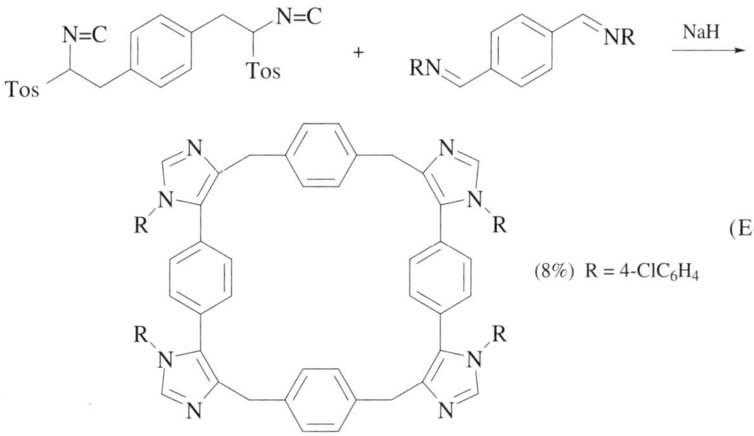

(Eq. 89) (8%) R = 4-ClC$_6$H$_4$

conversion of a bis(imine) derived from phthalic dicarboxaldehyde and the reaction product of 1,4-(dibromomethyl)benzene and two molecules of TosMIC gives a paracyclophane as a 2:2 reaction product.[116]

The majority of imidazoles have been prepared by the reaction of aldimines derived from aromatic aldehydes and aliphatic amines. This protocol also applies to aldimines derived from aliphatic aldehydes and aliphatic amines, as well as aromatic aldehydes and aromatic amines. However, the method seems to fail with aldimines derived from aliphatic aldehydes and aromatic amines.[77]

The reaction of imidoyl chlorides and TosMIC leads to 4-tosylimidazoles (Eq. 90).[37,77]

$$\underset{Ph}{\overset{Cl}{\diagup}}\!\!=\!\!N\text{-}C_6H_4NO_2\text{-}4 \quad \xrightarrow[NaH]{TosCH_2N=C} \quad \underset{Ph}{\text{Tos-imidazole}}\text{-}C_6H_4NO_2\text{-}4 \quad (81\%) \quad \text{(Eq. 90)}$$

Leaving groups other than chloride have been used occasionally: the diethyl phosphonate group in the reaction of a benzodiazepine derivative[117] and an amino fragment in the reaction of TosMIC with *sym*-triazine (Eq. 91).[118]

$$\text{sym-triazine} \quad \xrightarrow[EtONa]{TosCH_2N=C} \quad \text{4-Tos-imidazole-NH} \quad (28\%) \quad \text{(Eq. 91)}$$

Formimidates have been used similarly for the synthesis of 4-tosylimidazoles such as **48**. The tosylimidazolyl group in compound **48** may then be used as a leaving group to effect the net nucleophilic displacement of amino groups from the heterocycle (Eq. 92).[119]

$$\text{4-aminopyridine} \xrightarrow{HC(OEt)_3} \text{formimidate} \xrightarrow[NaH]{TosCH_2N=C} \textbf{48}\ (90\%) \xrightarrow{Nu^-} \text{4-Nu-pyridine}$$

Nu = EtO, *n*-BuS, PhS, H

(Eq. 92)

One example describes the formation of an imidazole from a nitrile and dilithio-TosMIC (Eq. 93).[79]

$$PhCN \quad \xrightarrow{TosCLi_2N=C} \quad \text{4-Tos-5-Ph-imidazole-NH} \quad (33\%) \quad \text{(Eq. 93)}$$

The reaction of TosMIC with isothiocyanates may lead to 4-tosylimidazoles, although 4-tosylthiazoles are also formed, as shown in Eq. 94. When the same reaction is carried out with sodium hydride in 1,2-dimethoxyethane at 0°, the tosylthiazole is obtained as the sole product in 76% yield.[120]

$$C_6H_{11}NCS \xrightarrow[n\text{-BuLi (2 eq)}]{TosCH_2N=C} \text{[tosylimidazole]} (40\%) + \text{[tosylthiazole]} (35\%)$$

(Eq. 94)

An alternative way to synthesize imidazoles involves the addition of amines to 1-isocyano-1-tosyl-1-alkenes (Eq. 95). The reaction is limited to the use of aliphatic primary amines and ammonia; aniline is not sufficiently nucleophilic to react.[86]

(Eq. 95)

It should be emphasized that the essential difference between the two synthetic approaches to imidazoles merely is a matter of sequence in which the three basic reaction partners—TosMIC, aldehyde, and amine—are brought together. In the approach of Eq. 80 an aldimine is prepared first (from an aldehyde and amine) which then reacts with TosMIC, whereas in the procedure of Eq. 95 TosMIC is reacted first with an aldehyde (e.g., benzaldehyde) to give **49** which is followed by reaction of the product with an amine. The two approaches, however, do not cover the same range of products. The first method allows, through the use of TosMIC homologs, the synthesis of 1,4,5-trisubstituted imidazoles (Eq. 81), which are not otherwise accessible. 5-Monosubstituted imidazoles are available by using ammonia in Eq. 95; the same imidazoles require the use of N-trimethylsilyl aldimines in the first approach of Eq. 82. As was mentioned before, the second approach (Eq. 95) does not seem to apply to aromatic amines, whereas the first one does. Where both methods have been applied to the synthesis of the same imidazole, the second approach seems to give the higher yields.[86] The aldimines used in the first approach (Eq. 80) may be formed in situ.[121]

1,2,4-Triazoles

One report describes the synthesis of 1,2,4-triazoles by a base-induced cycloaddition of TosMIC to diazonium salts (Eq. 96).[122] The reaction fails with

$$4\text{-MeOC}_6\text{H}_4\overset{+}{\text{N}}\equiv\text{N}\ \text{Cl}^- \xrightarrow[\text{DMSO, MeOH, H}_2\text{O}]{\text{TosCH}_2\text{N=C, K}_2\text{CO}_3}$$

[1,2,4-triazole with Tos at 3, N-C$_6$H$_4$OMe-4] (80%) + [1,2,4-triazole with Tos at 5, N-C$_6$H$_4$OMe-4] (12%)

(Eq. 96)

4-nitrophenyldiazonium tetrafluoroborate, which reacts with TosMIC anion via displacement of the diazonium group.[122]

SYNTHESIS OF THIAZOLES FROM THIOCARBONYL COMPOUNDS

Thiazoles are formed by base-induced cycloadditions of TosMIC to the C-S double bond of various thiocarbonyl compounds. The thiocarbonyl substrates used so far have led to 4-tosyl substituted thiazoles only, as in the reaction of a thionolactone (Eq. 97).[123]

[thionolactone] $\xrightarrow[\text{THF, }-78°,\ 1\text{ h}]{\text{TosCHLiN=C}}$ [HO—CH$_2$CH(Me)CH$_2$C(Tos)=C(N=C)SH intermediate] → [4-tosylthiazole product with HO-alkyl chain] (67%)

(Eq. 97)

In the mechanism shown in Eq. 97, the alkoxy group departs upon attack of TosMIC anion, prior to α-addition of the tautomeric thioenol function to the isocyano carbon; a concerted addition to the thiocarbonyl group is also possible.[123,124] The corresponding reaction of acyclic thiono esters (RC(=S)OR') and dithiono esters (RC(=S)SR') appears to fail, unless the leaving group in the dithio esters is activated by the use of R' = CH$_2$CO$_2$H or CH$_2$CO$_2$R''. Eq. 98 gives an example.[124]

$$4\text{-MeOC}_6\text{H}_4\text{-C(=S)-S-CH}_2\text{CO}_2\text{H} \xrightarrow[\substack{\text{KOH (3 eq), }t\text{-BuOH,} \\ \text{rt, 6-12 h}}]{\text{TosCH}_2\text{N=C (2 eq)}} \text{[2-(4-MeOC}_6\text{H}_4\text{)-4-Tos-thiazole]}\ (79\%)$$

(Eq. 98)

Addition of TosMIC anion to the thiono group of cumulated systems also leads to the formation of 4-tosylthiazoles. The reaction with carbon disulfide gives 5-thio-4-tosylthiazoles as in Eq. 99. Most 4-thio-5-tosylthiazoles **50** have only been isolated and characterized after in situ alkylation or acylation of the thiolate group.[125]

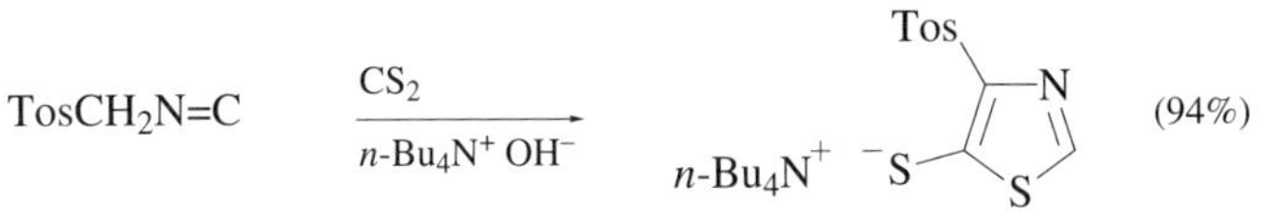

$$\text{TosCH}_2\text{N=C} \xrightarrow[n\text{-Bu}_4\text{N}^+ \text{OH}^-]{\text{CS}_2} n\text{-Bu}_4\text{N}^+ \begin{array}{c}\text{[Tos-thiazole-S}^-\text{]}\\ \mathbf{50}\end{array} \quad (94\%) \quad \text{(Eq. 99)}$$

A 4-unsubstituted thiazole is obtained in a process not based on the use of a thiocarbonyl substrate (Eq. 100).[126] Formamide **51** is dehydrated in the usual way to α,β-unsaturated tosyl isocyanide **52**, which by addition of hydrogen sulfide gives thiazole **53**.

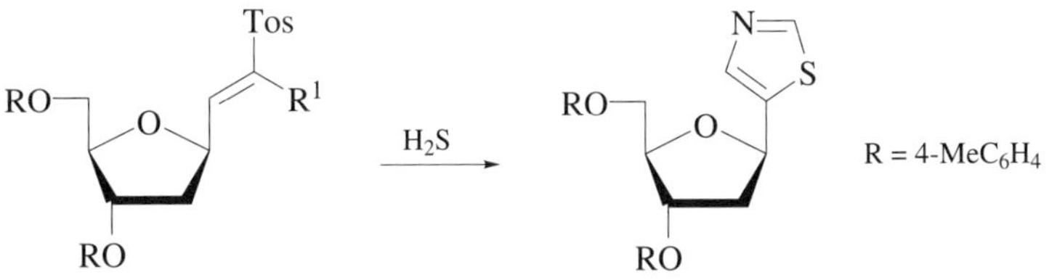

51 R^1 = NHCHO
52 R^1 = NC
53 (83%)

R = 4-MeC$_6$H$_4$

(Eq. 100)

No reactions of TosMIC with thioaldehydes or thioketones have been reported.

SYNTHESIS OF PYRROLES AND INDOLES

The preparation of pyrroles constitutes one of the major synthetic uses of TosMIC, and a variety of substituent types and patterns can be accessed. Three reactants are needed: TosMIC, an aldehyde, and a compound with an activated methyl or methylene group. In practice, the TosMIC-based synthesis of pyrroles is not carried out as a three-component reaction. Two reaction modes are possible (Scheme 4): (1) a base-induced reaction of TosMIC with a Michael acceptor molecule **54**, and (2) a base-induced reaction of an activated methyl compound with a 1-isocyano-1-tosyl-1-alkene **55**.

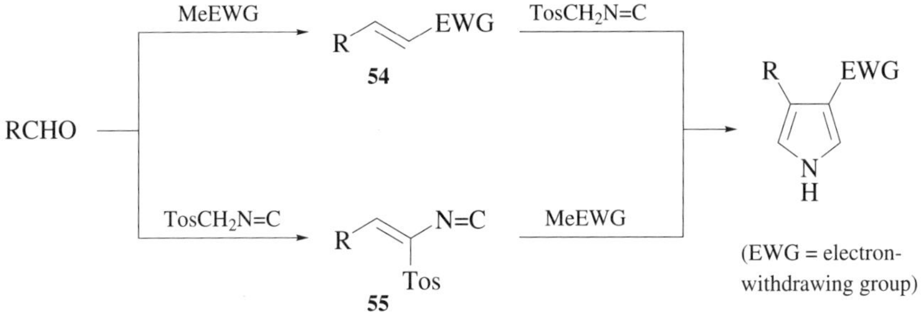

(EWG = electron-withdrawing group)

Scheme 4

2,3-Dialkenyl-substituted pyrroles are formed when the above reactions are extended to the use of dienic Michael acceptors and alkenyl substituted TosMIC homologs as reactants (Scheme 5). The six-π electron system of these 2,3-dialkenylpyrroles forms the basis of a novel synthesis of indoles through electrocyclic ring closure followed by dehydrogenation. The substituent on the TosMIC homolog thus becomes an integral part of the benzene moiety of the indole skeleton.

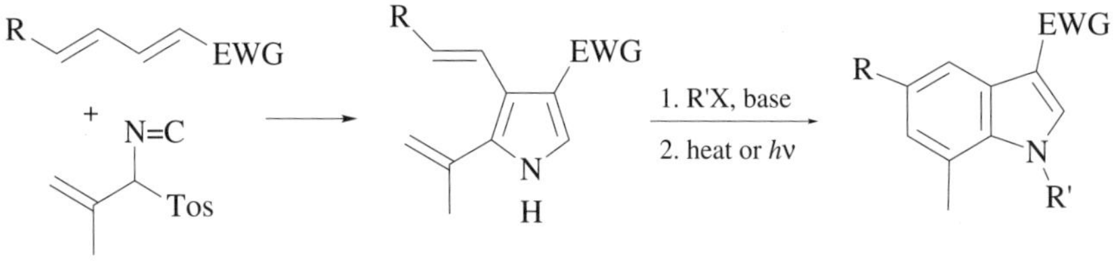

Scheme 5

Pyrroles from Michael Acceptors

A typical example of the TosMIC-based pyrrole synthesis (Eq. 101) shows that ring positions 1, 2, and 5 remain unsubstituted.[127]

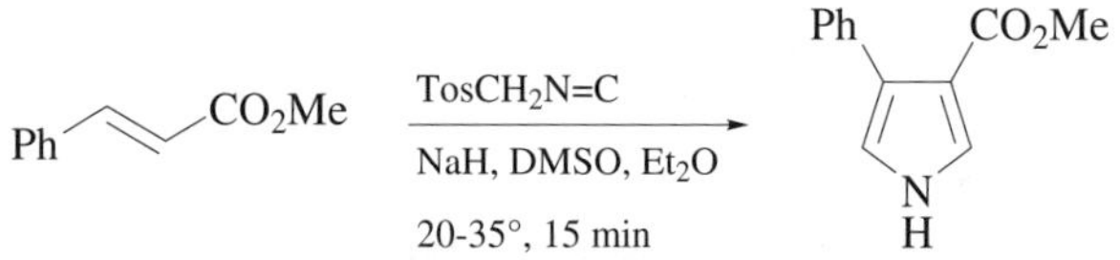

(70%)　(Eq. 101)

When TosMIC is replaced by a monosubstituted TosMIC homolog, an additional substituent is introduced in position 2 (Eq. 102).[69] In all cases, the 1 and 5 positions will carry no substituents.

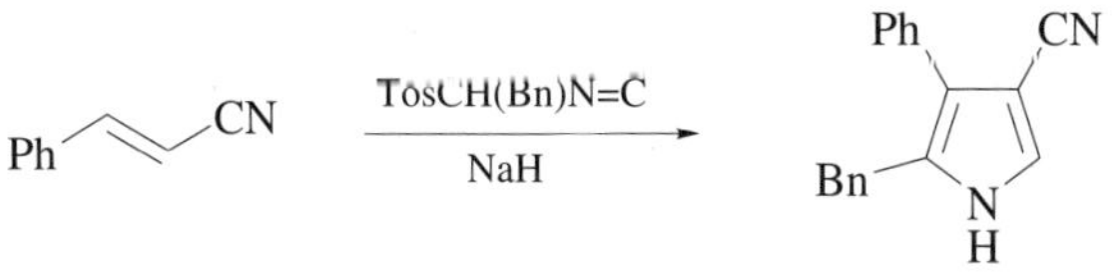

(80%)　(Eq. 102)

The (formal) condensation product **15** of TosMIC and cyclohexanone gives a 2,4-disubstituted pyrrole (Eq. 103).[128]

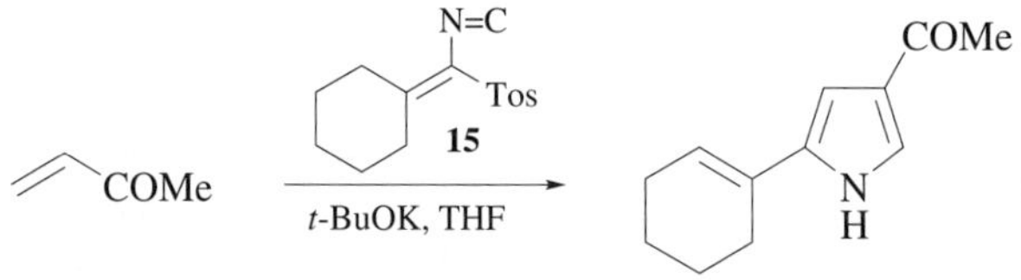

(96%)　(Eq. 103)

Scheme 6 offers two reasonable mechanisms, [1] and [2], to explain the results of Eqs. 101–103. In mechanism [1], after the key addition and cyclization steps,

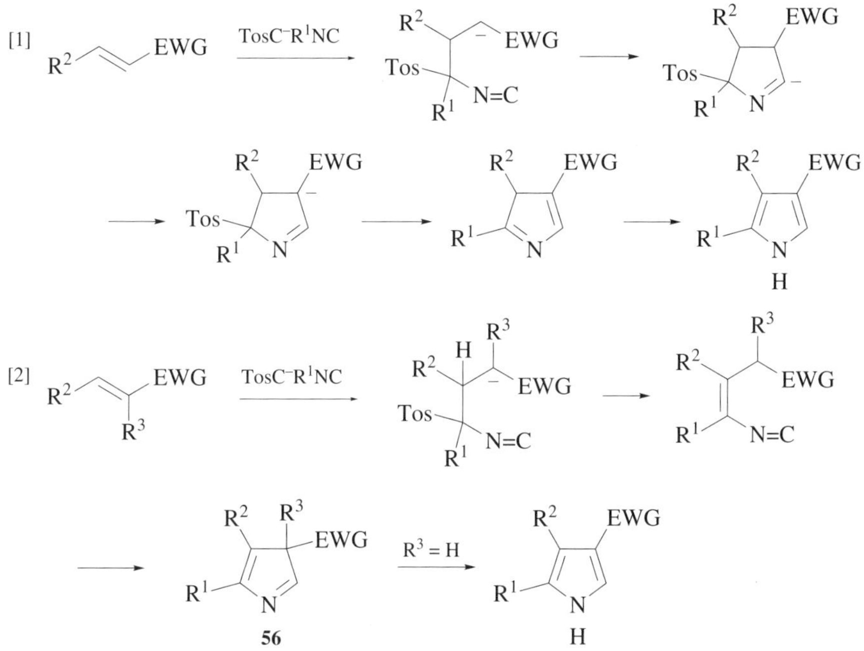

Scheme 6

prototropic shifts followed by a final aromatization through the elimination of toluenesulfinate anion lead to the observed products. In mechanism [2], elimination of sulfinate anion precedes the ring-closing reaction. Compound **56** with $R^1 = R^3 = H$, $R^2 = CH_3$, and EWG = CO_2Et is the only intermediate in the pyrrole synthesis of which identification has been claimed, although no details were given.[129]

The diverse Michael acceptors **54** used in reactions analogous to Eqs. 101–103 were usually the trans isomers, but cis isomers or cis-trans mixtures have also been employed.[63] In these Michael acceptors, the following electron-withdrawing groups have been used successfully: nitro, aroyl, alkanoyl, alkoxycarbonyl, arylsulfonyl and arylsulfinyl, cyano, carboxamido, and imidoyl. Only the carboxaldehyde function fails, since these substrates form oxazoles. α,β-Unsaturated aldimines can form both pyrroles and imidazoles, as previously discussed.

With extended Michael acceptors, addition can be directed to the α,β (Eq. 104) or γ,δ (Eq. 105) double bond, according to the reaction conditions.[79,91]

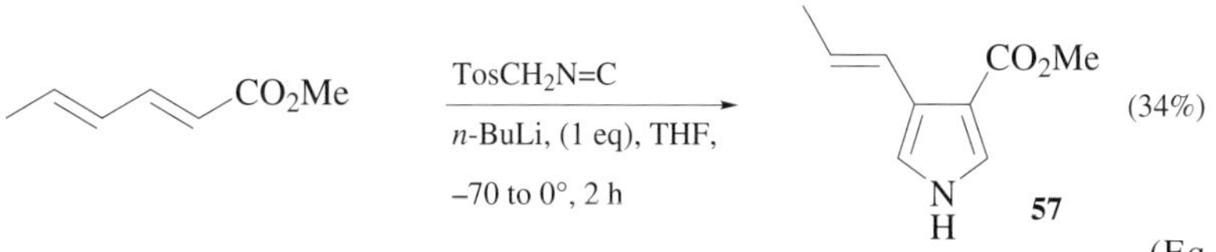

(Eq. 104)

[Eq. 105: methyl sorbate-type diene + TosCH$_2$N=C, NaH (1 eq), DMSO, Et$_2$O, rt, 1 h → 3-substituted pyrrole with CH=CH-CO$_2$Me side chain (56%) + **57** (2%)]

As part of the synthesis of a segment of antitumor agent CC-1065 the reaction of Eq. 105 was extended to the formation of a 3,3'-bipyrrole (Eq. 106).[91,129] Tri-, penta-, and heptapyrroles have been prepared in a comparable fashion.[130]

[Eq. 106: N-SO$_2$Ph pyrrole with CH=CH-CO$_2$Et + TosCH$_2$N=C, NaN(TMS)$_2$, THF → 3,3'-bipyrrole (85%)]

With 1-nitro-4-phenyl-1,3-butadiene, only one pyrrole is formed (Eq. 107). Variation of the base/solvent system did not lead to a pyrrole derived from the γ,δ double bond in this case.[131]

[Eq. 107: Ph-CH=CH-CH=CH-NO$_2$ + TosCH$_2$N=C, t-BuOK (2 eq), THF → pyrrole with styryl and NO$_2$ substituents (73%)]

1-Nitro-6-phenyl-1,3,5-hexatriene behaves correspondingly in the reaction with the TosMIC homolog **15**, affording a single pyrrole by addition to the α,β double bond (Eq. 108).[131]

[Eq. 108: Ph-CH=CH-CH=CH-CH=CH-NO$_2$ + cyclohexylidene(Tos)(N=C) **15**, t-BuOK (2 eq) → pyrrole with styryl, NO$_2$, and cyclohexenyl substituents (85%)]

1-(Phenylsulfonyl)-1,3-butadiene reacts with 1-tosylethyl isocyanide exclusively at the γ,δ double bond.[132,133]

[Eq. 109 scheme: PhSO₂-substituted diene + TosCH(Me)N=C, NaH, DMSO, THF → pyrrole with CH₂SO₂Ph and Me substituents, N–H (65%)]

(Eq. 109)

Further variation of the Michael acceptor molecule **54**, beyond R = alkyl, alkenyl, or aryl, is possible: pyrroles are also formed when R represents a second electron-withdrawing group. With two different electron-withdrawing groups, the reaction with TosMIC affords a single pyrrole (Eq. 110)[134] and with a TosMIC homolog, similar reactions can also be regiospecific. The direction of this process is governed by the more strongly electron-withdrawing substituent (Eq. 111).[63,93]

[Eq. 110 scheme: NC-CH=CH-CO₂Et + TosCH₂N=C, NaH, DMSO, Et₂O → 3-CN, 4-CO₂Et pyrrole, N–H (54%)]

(Eq. 110)

[Eq. 111 scheme: MeO₂C-CH=CH-COPh + TosCH(N=C)CH₂TMS; 1. Triton B (1.5 eq); 2. MeI, t-BuOK, THF → pyrrole bearing MeO₂C, COPh, TMS-CH₂, and N-Me (82%)]

(Eq. 111)

All examples discussed above involve reactions of TosMIC (or TosMIC homologs) with Michael acceptors carrying at least one hydrogen at each of the two sp^2 carbon atoms that will become part of the pyrrole ring. This, however, is not a necessary condition. Sometimes, it is advantageous to increase the Michael acceptor reactivity by using two electron-withdrawing groups at the same carbon atom, as in Eqs. 112[135] and 113.[136] Obviously, to form a pyrrole one of these groups has to be removed in the course of the reaction. This was shown to be possible for X = CO₂H, CO₂R, and CONH₂ (in **58**).[135–137]

[Eq. 112 scheme: 2,3-Cl₂C₆H₃-CH=C(CN)(CO₂H) + TosCH₂N=C, KOH, MeOH, CH₂Cl₂ → pyrrole with 2,3-Cl₂C₆H₃ and CN substituents, N–H (99%)]

(Eq. 112)

$$\text{Ph}\diagup\!\!\!\diagdown\substack{CO_2Et \\ CO_2Et} \xrightarrow[\text{NaOH, EtOH}]{TosCH_2N=C} \text{[3-Ph-4-CO}_2\text{Et pyrrole]} \quad (60\%) \quad (\text{Eq. 113})$$

$$R\diagup\!\!\!\diagdown\substack{X \\ EWG}$$

58

EWG = CN, CO_2Et

In a related process that uses 1,8-diazabicyclo[5.4.0]undec-7-ene (DBU), the nitro group is both the activating group and the leaving group.[138]

$$\text{4-MeOC}_6\text{H}_4\diagup\!\!\!\diagdown\substack{EWG \\ NO_2} \xrightarrow[\text{THF, }i\text{-PrOH}]{TosCH_2N=C,\ DBU} \text{[2-Tos-3-(4-MeOC}_6\text{H}_4\text{)-4-EWG pyrrole]} \quad (52\%) \quad (\text{Eq. 114})$$

Eq. 115[139] shows that two different electron-withdrawing groups at the same carbon may compete as leaving groups, in contrast to Eq. 112.

$$\diagup\!\!\!\diagdown\substack{(O)_n\text{SPh} \\ F} \xrightarrow[\text{NaH (2 eq), THF}]{TosCH_2N=C} \text{[2-Tos-4-F-3-Me pyrrole]} + \text{[2-Tos-3-Me-4-SPh(O)}_n\text{ pyrrole]} \quad (\text{Eq. 115})$$

n		
1	(24%)	(24%)
2	(8%)	(23%)

Few examples have been reported of pyrroles prepared from TosMIC and Michael acceptors with a C-C triple bond. Such reactions give 2-tosylpyrroles (Eq. 116).[140]

$$R-\!\!\equiv\!\!-CO_2Me \xrightarrow[\text{DBU, DMF}]{TosCH_2N=C} \text{[2-Tos-3-R-4-CO}_2\text{Me pyrrole]} + \text{[2-Tos-3-R-4-CO}_2\text{Me-5-vinyl pyrrole]}$$

R		
H	(12%)	(3%)
CO_2Me	(7%)	(10%)

(Eq. 116)

In a final example, two nitro groups are eliminated: one to form a pyrrole and one to form a triple bond (Eq. 117).[141]

$$\text{O}_2\text{N-C(Ph)=C(Ph)-CH=NO}_2 \xrightarrow[\text{rt, 30 min}]{\text{TosCH}_2\text{N=C, DBU, THF}} \text{2-Tos-3-Ph-4-(C≡CPh)-pyrrole} \quad (72\%) \quad \text{(Eq. 117)}$$

It is also possible to prepare 2,3,4-trisubstituted pyrroles using (formal) TosMIC derivatives that carry a substituent together with a leaving group at the "isocyano" carbon. The use of such synthons $\text{TosCH}_2\text{N=C(XCH}_3)\text{R}$ (X = O, S) for the synthesis of pyrroles[47–51] is not covered in detail in this chapter.

Pyrroles from 1-Isocyano-1-tosyl-1-alkenes

Being Michael acceptors, 1-isocyano-1-tosyl-1-alkenes **55** are susceptible to reaction with nucleophiles at the β carbon. This, in combination with the electrophilic nature of the isocyano carbon, provides an efficient synthesis of 3-nitropyrroles (Eq. 118).[136,142]

$$t\text{-Bu-CH=C(Tos)(N=C)} \xrightarrow[t\text{-BuOK, DME}]{\text{MeNO}_2} \text{2-}t\text{-Bu-3-NO}_2\text{-pyrrole} \quad (91\%) \quad \text{(Eq. 118)}$$

Extension of the method of Eq. 118 to synthesize pyrroles with substituents at C3 other than the nitro group is possible. Replacement of the 3-nitro function by the following groups has been realized: cyano, carboxylate, acetyl, benzoyl, and even methyl and phenyl.[136] It is, however, not possible to prepare a cyanopyrrole simply by replacing nitromethane (in Eq. 118) by acetonitrile. The CH acidity of acetonitrile needs to be enhanced by an acidifying auxiliary, for example, by using ethyl cyanoacetate (Eq. 119).[136] As before, the reaction is carried out in one operation; the auxiliary is removed by nucleophilic attack of ethoxide (Scheme 7).

$$2,3\text{-Cl}_2\text{C}_6\text{H}_3\text{-CH=C(Tos)(N=C)} \xrightarrow[\text{NaOH, EtOH}]{\text{EtO}_2\text{CCH}_2\text{CN}} \text{2-(2,3-Cl}_2\text{C}_6\text{H}_3)\text{-3-CN-pyrrole} \quad (93\%) \quad \text{(Eq. 119)}$$

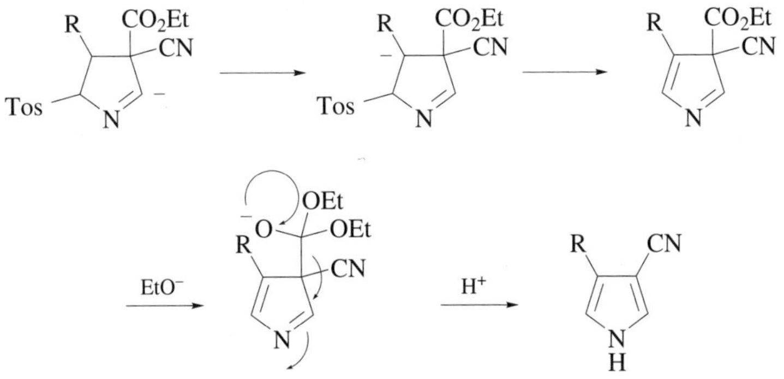

Scheme 7

3-Acetyl-4-phenylpyrrole is similarly obtained from acetylacetone.[136] The method also allows the introduction of substituents at C3 that are not of the electron-withdrawing type (Eq. 120).[136]

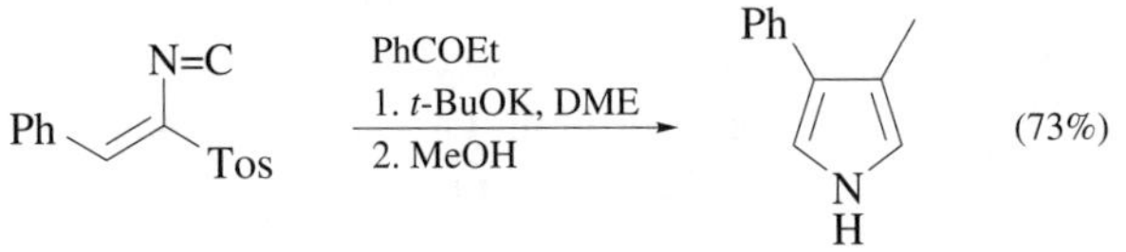

(73%)   (Eq. 120)

2,3-Dialkenylpyrroles and Indoles

Reactions of the extended anions derived from alkylidene TosMIC derivatives with Michael acceptors provide an efficient synthesis of 2,3-(dialk-1-enyl) pyrroles, as in Eq. 121.[128]

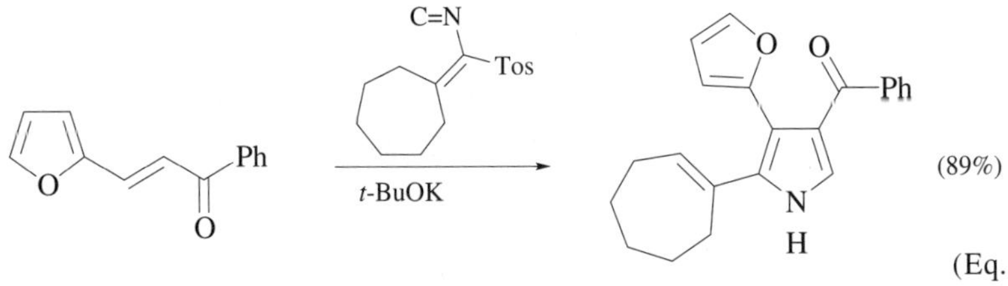

(89%)

(Eq. 121)

For obvious reasons the pyrrole ring in Eq. 121 is formed exclusively at the α,β double bond of the substrate molecule. The same is true for the reaction of the conjugated enyne acceptor of Eq. 122, which reflects reduced reactivity of the triple bond relative to the double bond.[63,143] Such products are nicely set up for six π electron cyclizations to form indoles. Eq. 123 gives an example of a thermal electrocyclic ring closure, followed by dehydrogenation with the use of 2,3-dichloro-5,6-dicyano-1,4-benzoquinone (DDQ).[128]

This method of synthesizing indoles is rather unusual in that the benzene ring is built onto a preformed pyrrole,[128] instead of the other way around.[144,145] The cyclization step may be carried out thermally[63,80,128,131,146] or photochemically.[128,143]

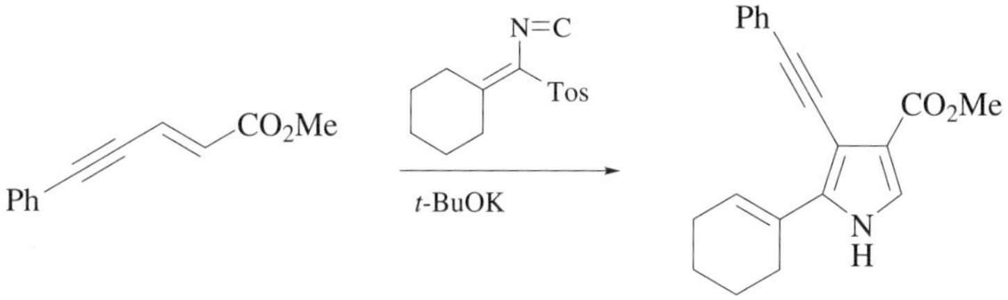

(Eq. 122)

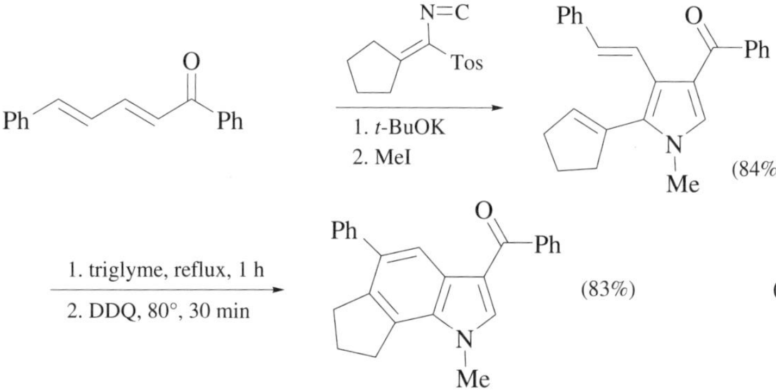

(Eq. 123)

Benzoxazoles and benzimidazoles are accessible similarly by electrocyclization and dehydrogenation of 2,3-(dialk-1-enyl)-oxazoles and -imidazoles, respectively. There is, however, only one brief report of this process, containing a limited number of examples.[80]

SYNTHESIS OF KETONES BY HYDROLYSIS OF TosMIC DERIVATIVES

Hydrolysis of TosMIC and similar compounds by acid ultimately produces aldehydes or ketones via initial hydration of the isocyanide. Eq. 124 provides an example. Compound **59**, obtained by a twofold alkylation of TosMIC, is hydrolyzed to ketone **60**.[107] This protocol shows TosMIC as a CO connective reagent. Wolff-Kishner reduction of ketone **60** gives muscalure. The application of TosMIC as a CO and CH_2 connective reagent is comparable to the use of other compounds with an activated methylene group, such as 1,3-dithiane[146a] and methylthiomethyl 4-tolyl sulfone.[147]

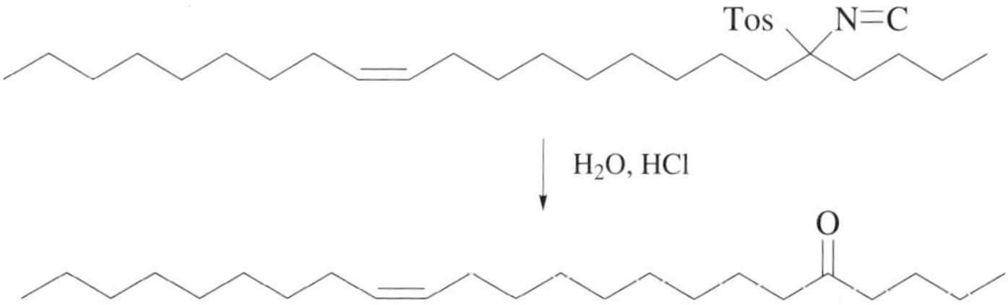

(Eq. 124)

This methodology is not restricted to the synthesis of monofunctional ketones. The same principles apply to the synthesis of enones, α-hydroxy ketones and aldehydes, 1,2-, 1,4-, 1,5-, and higher diketones, triketones, and tetraketones. Some specific examples of the TosMIC ketone synthesis are found in previous sections (Eqs. 13, 33, 34 56, 70, 71, and 76–79). Strikingly, only one aldehyde has been prepared by hydrolysis of a monosubstituted TosMIC derivative (apart from the synthesis of α-hydroxy aldehydes according to Eq. 143). Phenylacetaldehyde is obtained by acid hydration of 2-phenyl-1-tosylethyl isocyanide (**61**) and subsequent treatment of formamide **62** with aqueous base (Eq. 125).[73]

$$\underset{\underset{\mathbf{61}}{Bn\quad Tos}}{N=C} \xrightarrow{H_2O, HCl} \underset{\underset{\mathbf{62}}{Bn\quad Tos}}{NHCHO} \xrightarrow{H_2O, NaOH} BnCHO \quad (41\%)$$

(Eq. 125)

Ketones

Cycloalkylation of TosMIC is possible with dihaloalkanes such as 1,3-dibromopropane. This transformation forms the basis for an efficient synthesis of cyclobutanone.[148]

$$TosCH_2N=C \xrightarrow{Br(CH_2)_3Br} \underset{}{\square}-Tos \xrightarrow{H_2SO_4\ (50\%)}_{sulfolane} \underset{}{\square}=O \quad (84\%)$$

(Eq. 126)

The same approach leads to the synthesis of optically active 2-methylcyclobutanone, both from (S)-(+)-1,3-dibromobutane and TosMIC and from (±)-1,3-dibromobutane and (+)-neomenthylsulfonylmethyl isocyanide.[39] Other cyclic ketones have been synthesized by the same method, including examples of a 10-membered[149] and an 18-membered ring.[150] As previously noted, acid hydrolysis of the 3-membered ring compound prepared from TosMIC and 1,2-dibromoethane[73,99] gives formamide **63** and not cyclopropanone.[99]

$$\underset{Tos}{\triangle\!\!-N=C} \xrightarrow{H_2O,\ HCl} \underset{\underset{\mathbf{63}}{Tos}}{\triangle\!\!-NHCHO} \quad (79\%)$$

(Eq. 127)

The considerable difference in stability between the various types of geminal tosyl formamido compounds is understandable on the basis of the structure of the formyliminium species **64** formed upon spontaneous or acid-promoted ionization. Dialkyl substitution stabilises the carbocation to the point where the intermediates are not isolable, with the exception of the cyclopropyl case, where carbocation formation is disfavored.

The synthesis of progesterone from a 17-ketosteroid illustrates several aspects of TosMIC chemistry.[45] A condensation-dehydration sequence on the 17-keto 3,5-dienol ether is followed by reduction and alkylation to afford **65**; subsequent acid hydrolysis (Eq. 128) in the usual way gives progesterone in excellent overall yield.

This ketone synthesis has been used for the synthesis of chiral spiroketals, combining the hydrolysis and ketalization steps (olive fruit fly pheromones, Eq. 129).[151]

Diketones

Acid-promoted hydrolysis of products such as **66**, obtained from 1,n-dihaloalkanes and 2 equivalents of a monosubstituted TosMIC derivative, provides symmetrical 1,(n + 2)-diketones (Eq. 130). A related stepwise alkylation protocol leads to unsymmetrical diketones (Eq. 131).[100]

$$\underset{\text{Tos}}{\overset{\text{N=C}}{\searrow}} \xrightarrow{\text{Br(CH}_2)_3\text{Cl}} \underset{(87\%)}{\text{Tos}\searrow\overset{\text{N=C}}{\diagup}\text{Cl}} \xrightarrow{\text{NaI}} \underset{(89\%)}{\text{Tos}\searrow\overset{\text{N=C}}{\diagup}\text{I}}$$

$$\xrightarrow{\text{TosCH(Bn)N=C}} \underset{(80\%)}{\text{C=N}\diagup\text{Tos Tos}\diagdown\overset{\text{N=C}}{\diagup}\text{Bn}} \xrightarrow[\text{HCl}]{\text{H}_2\text{O}} \underset{(93\%)}{\overset{\text{O}\quad\quad\text{O}}{\diagdown\diagup\diagdown\diagup\diagdown}\text{Bn}}$$

(Eq. 131)

This methodology for preparing 1,5- and 1,6-diketones cannot always be applied to the synthesis of 1,3- and 1,4-diketones. Thus, reaction of 1,2-dibromoethane with two equivalents of 2-phenyl-1-tosylethyl isocyanide gives only **67**, with no disubstitution product. With the less sterically demanding 1-tosylethyl isocyanide, the reaction with 1,2-dibromoethane gives compound **68**, albeit in low yield (18%). Acid hydrolysis leads to 2,5-hexanedione (66% yield) in the only reported TosMIC-based synthesis of a 1,4-diketone.[100] The same approach fails altogether for the synthesis of 1,3-diketones.[100]

$$\underset{\text{Bn}}{\text{Tos}}\searrow\overset{\text{N=C}}{\diagup}\text{Br} \qquad \underset{\text{Tos}}{\text{Tos}}\searrow\overset{\text{N=C}}{\diagup}\text{N=C}$$
67 **68**

1,2-Diketones, on the other hand, are accessible by acylation of a monosubstituted TosMIC derivative, followed by acid hydrolysis (Eq. 132).[100,112]

$$\underset{\text{Bn}}{\overset{\text{N=C}}{\diagup}}\text{Tos} \xrightarrow[n\text{-BuLi}]{4\text{-O}_2\text{NC}_6\text{H}_4\text{COCl}} \underset{(77\%)}{\text{Tos}\overset{\text{N=C}}{\diagdown\diagup}\text{C}_6\text{H}_4\text{NO}_2\text{-4}} \xrightarrow[\text{HCl}]{\text{H}_2\text{O}} \underset{(65\%)\text{ overall}}{\text{Bn}\overset{\text{O}}{\diagdown\diagup}\text{C}_6\text{H}_4\text{NO}_2\text{-4}}$$

(Eq. 132)

Synthesis of 1,3-, 1,4-, and higher diketones from bromoketones requires protection of the keto group, as exemplified by the synthesis of jasmone via a 1,4-diketone (Eq. 133).[152]

Many symmetric as well as unsymmetric cyclic diketones, particularly in the cyclophane class, are available by a simple extension of this chemistry. Two molecules of TosMIC are first interconnected with a 1,n-dihalo compound in a double monoalkylation step. Reaction with a second dihalo compound, followed by acid hydrolysis, leads to the cyclic diketone (Eq. 134).[92,153]

Tri- and Tetraketones

Eq. 76 and 77 already showed that extrapolation of the principles of Eq. 132 can produce cyclic tri- and tetraketones. Another interesting example is the

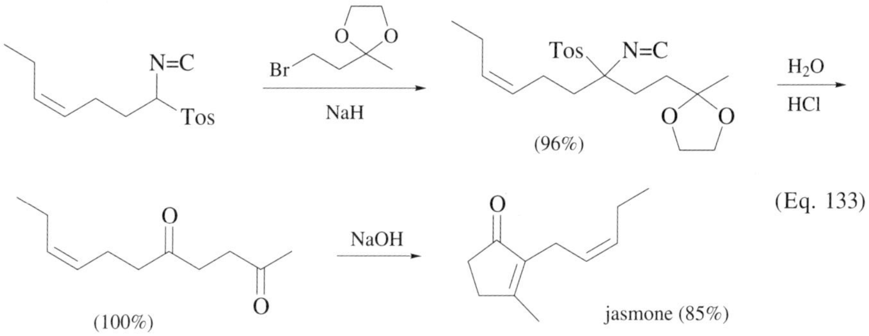

(Eq. 133)

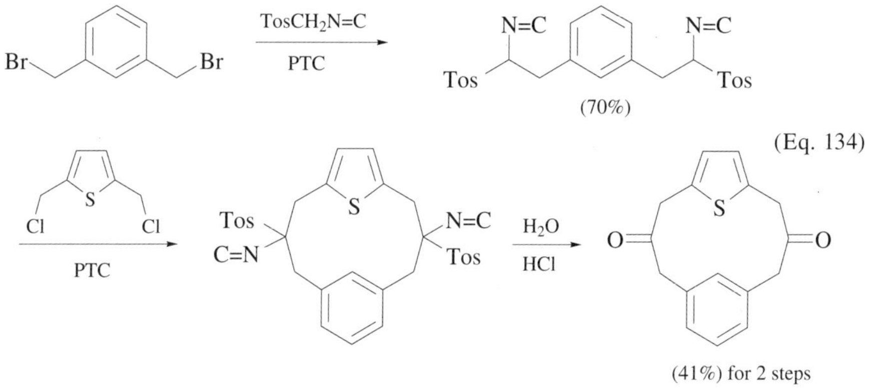

(Eq. 134)

threefold use of TosMIC as a CO connective reagent between two molecules of 1,3,5-tri(4-bromomethylphenyl)benzene.[154]

Cyclic tetraketones have been obtained as minor products in reactions that also form di- and triketones.[109,153,155,156]

Enones

Alkylation of 1-isocyano-1-tosyl-1-alkenes followed by acid hydrolysis affords enones. This method has been used extensively for the introduction of C17 side chains in steroids, as in Eq. 135.[101]

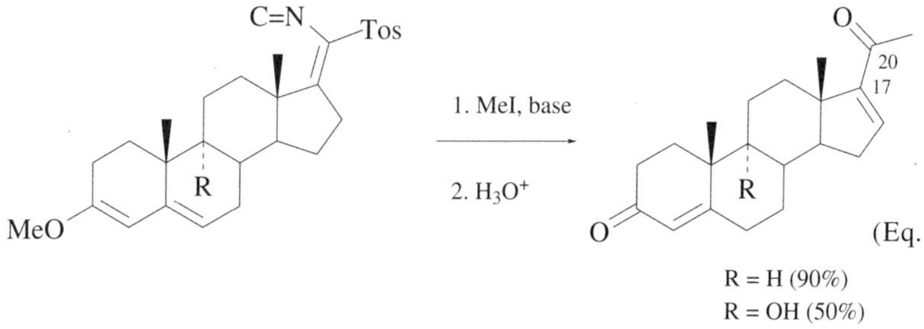

(Eq. 135)

R = H (90%)
R = OH (50%)

This route can also be used to synthesize nonconjugated olefins and bis(enones) (Eqs. 136 and 137).[101,103]

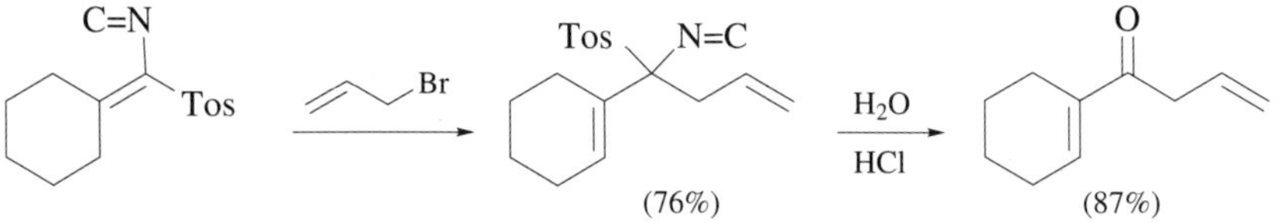

(Eq. 136)

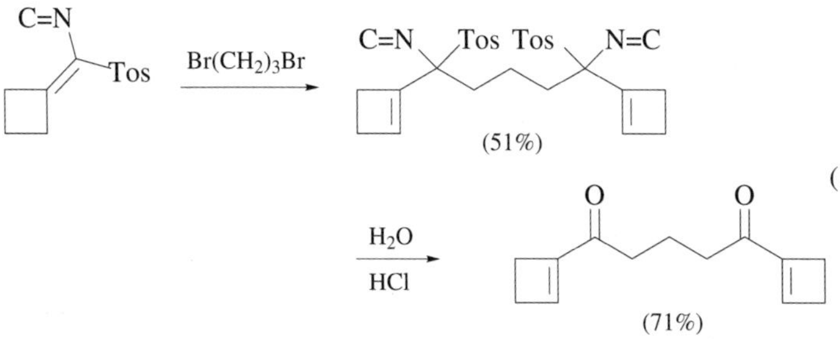

(Eq. 137)

When allyl bromide is replaced by α-bromoacetophenone or ethyl α-bromoacetate, the alkylation step succeeds, but the derived 1,3-diketone and β-keto ester decompose under the acid hydrolysis conditions.[103] α-Halo enones may be obtained using dibromomethane,[101] but the attempted synthesis of a β-bromo enone led to a divinyl ketone (Eq. 138).[103]

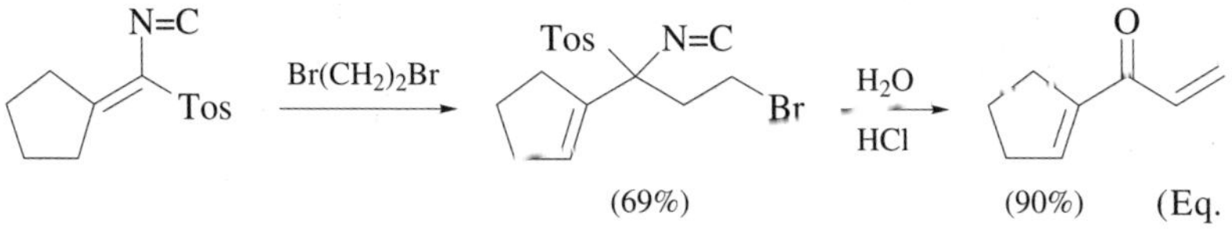

(Eq. 138)

The formation of a stable, conjugated enamide in the hydrolysis step has been reported (Eq. 139).[103]

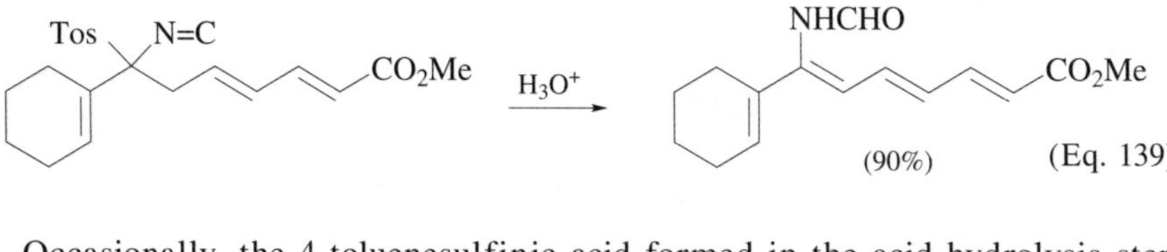

(Eq. 139)

Occasionally, the 4-toluenesulfinic acid formed in the acid hydrolysis step adds to the α,β double bond of the target enone, as exemplified by the formation of the sulfone **69** in high yield during an attempted synthesis of 16-dehydroprogesterone (Eq. 140).[101,157] When the reaction time was reduced to 30 minutes,

16-dehydroprogesterone is obtained in 75% yield. Further improvement to 90% was obtained by carrying out the hydrolysis in a mixture of dichloromethane and diethyl ether for 3 minutes with 40% aqueous perchloric acid.[101]

Alternatively, the isocyanide can first be oxidized with lead tetraacetate to an isocyanate, which is hydrolyzed on alumina (Eq. 141). To prevent oxidation in the A/B ring moiety, the dienol protecting group must be removed prior to the oxidative step.[101] A similar oxidative hydrolysis to form a saturated ketone, using peracetic acid or mercury(II) nitrate, has also been reported.[96]

α-Hydroxy Ketones and α-Hydroxy Aldehydes

Benzaldehyde and 2-phenyl-1-tosylethyl isocyanide have been made to react in a one-pot process to give 1,3-diphenyl-1-hydroxy-2-propanone (**70**) (Eq. 142). First, the two reactants are stirred in dry methanol with a catalytic amount of potassium carbonate for 5 minutes at 30°, then for another 5 minutes at 0° after addition of some concentrated hydrochloric acid.[73]

The course of the reaction of Eq. 142 is explained by a base-induced addition of 2-phenyl-1-tosylethyl isocyanide to benzaldehyde to give the (transient)[14] β-hydroxy isocyanide **71**. A fast intramolecular α-addition of the hydroxy group to the isocyano carbon gives 4-tosyloxazoline **72**, which by a base-induced addition-elimination process may give the 2-methoxyoxazoline **73**. Acid hydrolysis of any of these potential intermediates would lead to the hydroxy ketone **70**.

The reaction of TosMIC with ketones in alcoholic solvents, followed by hydrolysis, leads to α-hydroxy aldehydes (Eq. 143). Using thallium ethoxide as the base, the oxazoline **74** has been isolated and characterized. Hydrolysis with dilute hydrochloric acid in tetrahydrofuran at room temperature gives mostly the monomeric α-hydroxy aldehyde.[72]

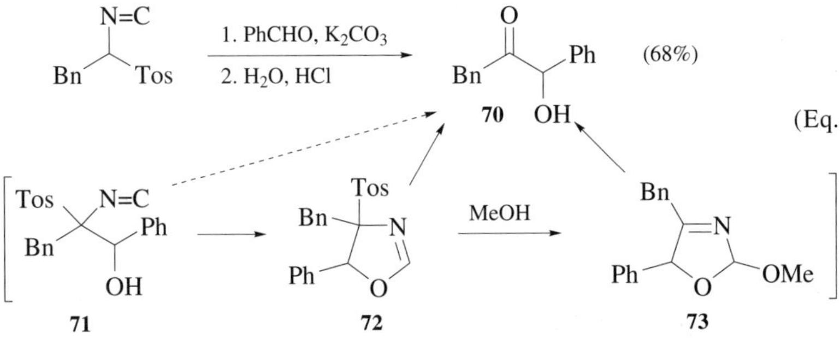

(Eq. 142)

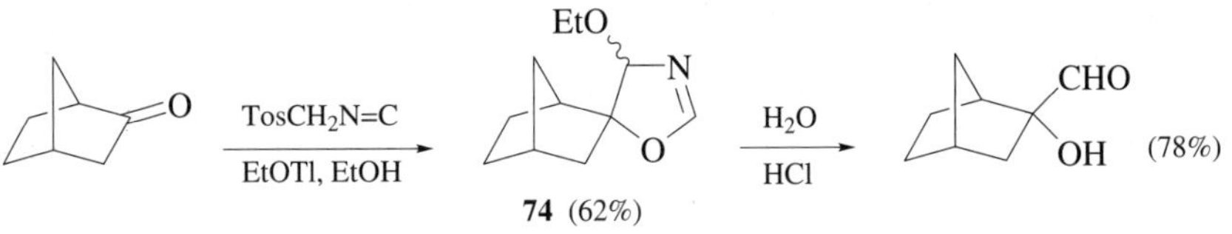

(Eq. 143)

The most prominent application of the α-hydroxy ketone synthesis, for the introduction of the corticoidal 17-hydroxyacetyl side chain in 17-ketosteroids, makes use of paraformaldehyde (Eq. 144).[70,158] The 2-methoxyoxazoline **75** may be isolated prior to acid hydrolysis to 21-hydroxy-3-methoxy-19-norpregna-1,3,5(10),16-tetraen-20-one (**76**).[70]

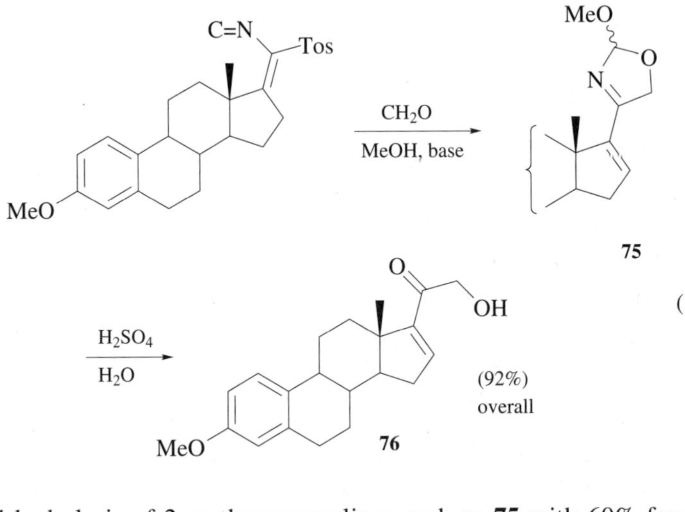

(Eq. 144)

Partial hydrolysis of 2-methoxyoxazolines such as **75** with 60% formic acid leads directly to 17-(formyloxy) acetyl side chains. The formyl protection allows the bis-hydroxylation of the C16-C17 double bond with potassium permanganate. This is a crucial step in a practical synthesis of the anti-inflammatory drug, triamcinolone acetonide (Eq. 13).[159]

REDUCTION OF TosMIC DERIVATIVES

Under appropriate conditions, the reduction of TosMIC derivatives may be controlled so as to produce alkanes, isocyanides, and methylamine derivatives.

Alkanes

The complete reductive removal of both the tosyl and the isocyano groups to give a methylene group, which was exemplified earlier in Eqs. 73–75, is rationalized in Scheme 8.[105] One-electron transfer, followed by loss of lithium cyanide

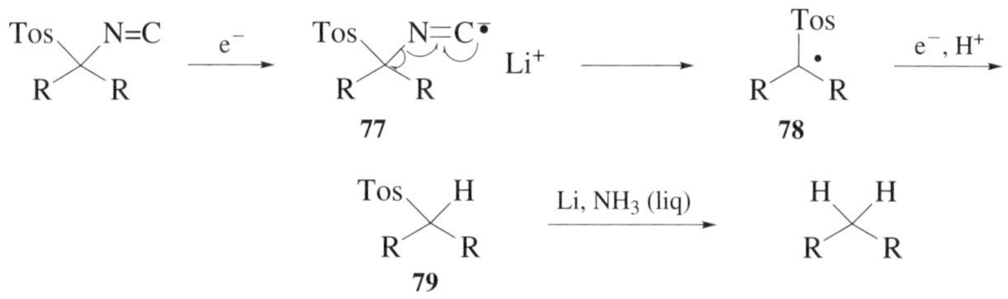

Scheme 8

from the radical anion **77** gives the radical **78**, which is further reduced and protonated to form the sulfone **79**. This sulfone is reduced in situ in a known process[159a] to the final product. Sulfone **79** has been identified as a possible intermediate upon use of an insufficient amount of lithium.[105] Isolated C-C double and triple bonds, ether, and acetal functions are unaffected,[105] but partial conversion of ester groups into amides requires an in situ reesterification step.[160]

Alkyl and Aralkyl Isocyanides

When the reduction of mono- and disubstituted derivatives of TosMIC is carried out by electrochemical means, only the sulfonyl group is removed (Eq. 145). This electroreductive detosylation, which is reported to proceed via a two-electron transfer process, has been used to prepare a number of symmetrical and unsymmetrical alkyl and aralkyl isocyanides.[161]

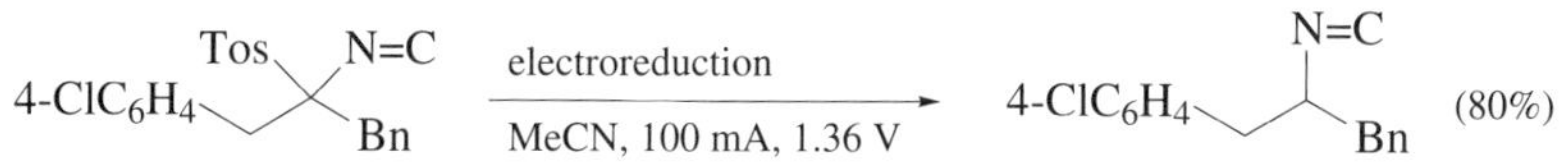

(Eq. 145)

N-Methylamines

Mono- and disubstituted TosMIC derivatives are reduced to N-methylamines by lithium aluminum hydride (Eq. 146); this reduction is assumed to take place

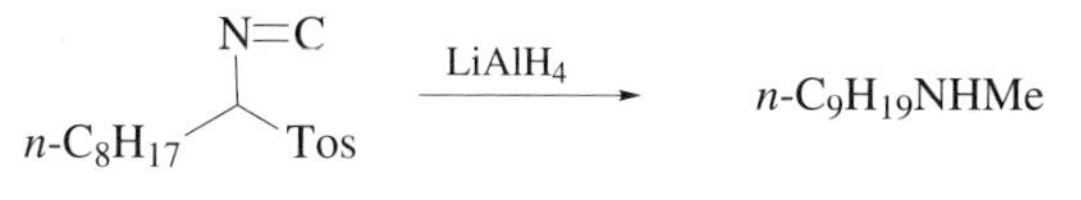

(Eq. 146)

via *N*-methylimines.[73] Sodium borohydride cannot be used, although this reagent will detosylate the acid hydration product of 2-phenyl-1-tosylethyl isocyanide. This is the only reported example of this type of reaction.[73]

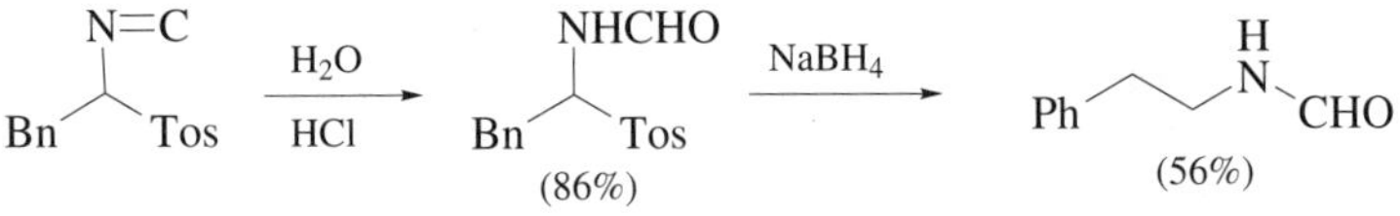

$$\underset{\underset{Bn}{}}{\overset{N\equiv C}{\diagdown}}\text{Tos} \xrightarrow{\underset{HCl}{H_2O}} \underset{\underset{Bn}{}}{\overset{NHCHO}{\diagdown}}\text{Tos} \xrightarrow{NaBH_4} Ph\diagdown\overset{H}{N}\diagdown CHO$$
(86%) (56%)

(Eq. 147)

β-Hydroxy *N*-methylamines

β-Hydroxy *N*-methylamines are obtained from 4-tosyloxazolines such as **80** by reductive ring opening with lithium aluminum hydride (Eq. 148). The 4-tosyloxazolines are usually reduced without purification.[73]

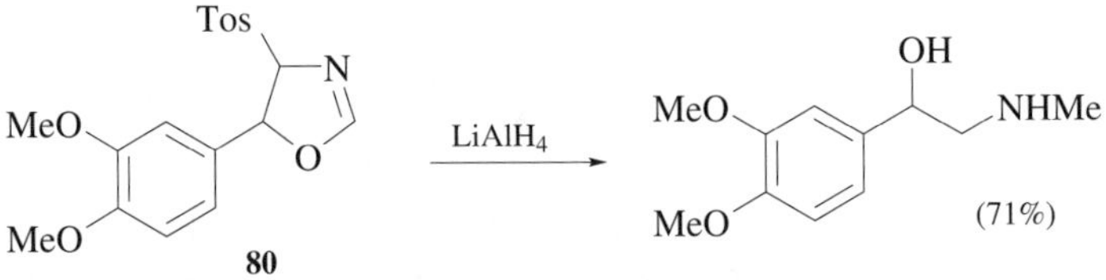

(Eq. 148)

So far, this conversion has been carried out only with TosMIC, not with TosMIC homologs. Optically active β-hydroxy *N*-methylamines are obtained from optically active 4-tosyloxazolines, as was shown previously in Eqs. 35 and 37.[74,75]

MISCELLANEOUS

Since the following reactions are not main stream synthetic uses of TosMIC, they are treated in an exemplary way only. For details the reader is referred to the references given in the text and Tables.

α-Additions to the Isocyano Carbon of TosMIC

Several reactions of TosMIC are known in which the tosylmethyl group remains unchanged, notably α-additions to the isocyano carbon. For example, TosMIC reacts with dry gaseous chlorine to give an isolable solid α-adduct **81**.[47,51] Replacement of one of the chlorines by methoxy provides the reagent **82** (Eq. 149), which has been used for the synthesis of azoles bearing a substituent at C2. Eq. 150 gives one example[52] of this type of reaction.[46–51,162]

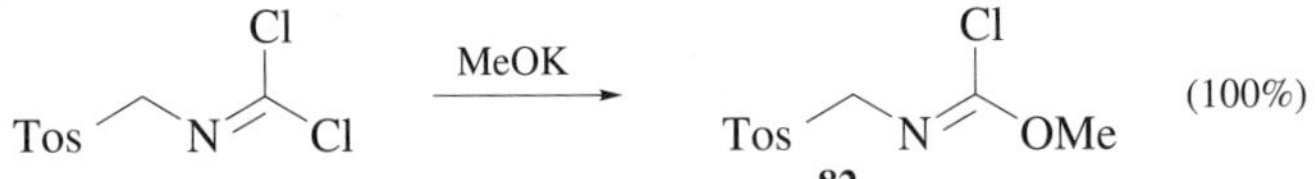

(Eq. 149)

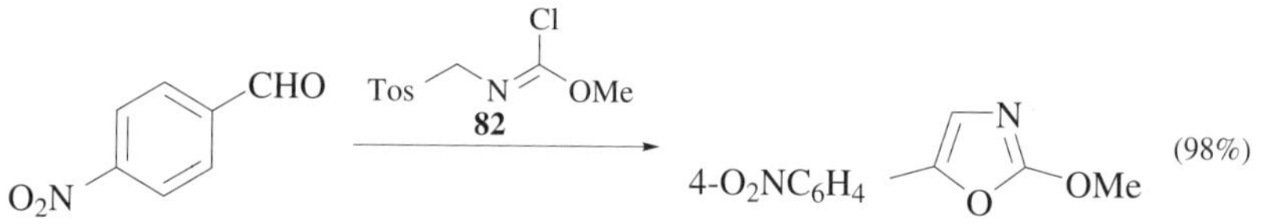

SYNTHETIC USES OF TOSYLMETHYL ISOCYANIDE (TosMIC)

(Eq. 150)

α-Additions of acid chlorides, anhydrides, and 1-hydroxy-2-thiopyridinone esters to TosMIC are also known.[163–166]. The (formal) condensation product **83** of TosMIC and veratraldehyde has been reacted with veratraldehyde acetal to give amide **84**, which was further transformed to papaverine.[95]

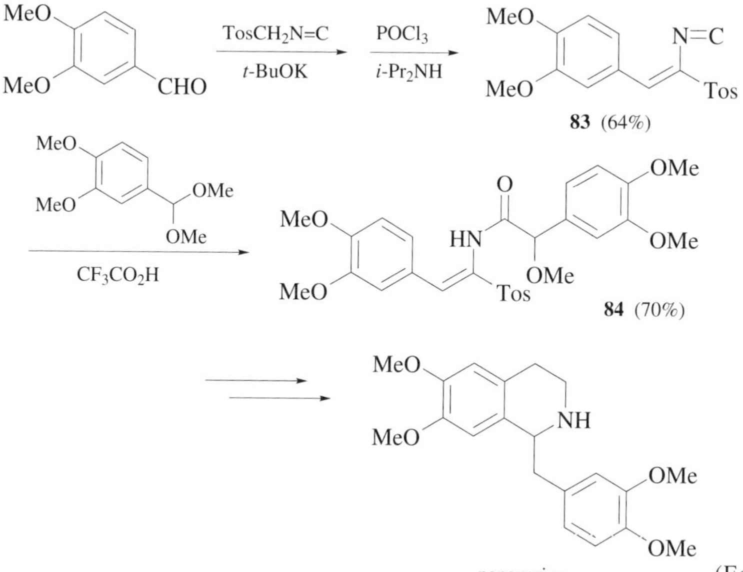

(Eq. 151)

TosMIC has been used as the isocyano component in a large series of Ugi four-component reactions[12,167] to form α-amino acid derivatives, using α-amino esters as the amine component, benzyloxycarbonyl-protected α-amino acids as the acidic component, and isobutyraldehyde.[168] The example of Eq. 152 shows that this route to dipeptide derivatives is not hampered by the presence of three bulky isopropyl groups.

Cycloadditions to the Isocyano Carbon of TosMIC

These reactions which form heterocyclic compounds by incorporation of the TosMIC isocyano carbon into a ring are essentially α-additions. They are, however, listed separately from the foregoing reactions, in view of the different product structures. Cycloadducts have been obtained from TosMIC by reactions with ketenes,[169] with dibenzoyldiazomethane,[170] and with a ketene N,N-acetal.[171]

[Eq. 152 scheme: (S)-BnO$_2$C-N(H)(Pr-i)-CO$_2$H + (S)-H$_2$N-CH(Pr-i)-CO$_2$Me + i-PrCHO + TosCH$_2$N=C]

(Eq. 152)

$\xrightarrow{\text{Et}_3\text{N}, \text{MeOH}}$ [product: i-Pr-(S)-C(=O)-N(CH(Pr-i)(S)-CO$_2$Me)-CH(NH-C(=O)-BnO$_2$C)-i-Pr$_{(R),(S)}$-C(=O)-N-CH$_2$-Tos] (76%)

Under neutral conditions dibenzoylacetylene reacts with two molecules of TosMIC according to Eq. 153.[172]

PhC(O)-C≡C-C(O)Ph $\xrightarrow{\text{TosCH}_2\text{N}=\text{C}}$ [bicyclic furo-furan product with Tos-N=CH, Ph, O, N-CH$_2$-Tos substituents] (34%)

(Eq. 153)

Various 3-membered ring compounds react in different ways with TosMIC.[173–176] Eq. 154 gives an example;[176] it is the only report of the formation of a pyridine ring with the use of TosMIC. The reaction is assumed to proceed via a vinylcarbene.

[cyclopropenyl cation with three SPr-i groups, ClO$_4^-$] $\xrightarrow{\text{TosCH}_2\text{N}=\text{C}, n\text{-BuLi, THF}}$ [vinylcarbene intermediate with Tos, i-PrS, i-PrS, i-PrS, N=C:] $\longrightarrow$ [pyridine with SPr-i, SPr-i, i-PrS, Tos substituents] (80%)

(Eq. 154)

A carbene-chromium complex has been used to form a ketenimine intermediate, which by a zinc chloride catalyzed cyclization gives a naphthalene derivative,[177] and a zirconacyclopentane intermediate has been used to convert a geminal diallyl compound into a cyclopentanone derivative via insertion of TosMIC into the zirconium complex.[178]

An 8-membered ring is formed by a base-induced intramolecular cyclization of the reaction product of 1,2-(dibromomethyl)benzene and two molecules of TosMIC (Eq. 155).[82]

[Eq. 155 scheme]

(Eq. 155)

TosMIC as a Metal Ligand

TosMIC forms stable, isolable complexes with various metal compounds. In these complexes, TosMIC coordinates with the metal through the isocyano carbon, as do many other isocyanides.[10] In certain copper complexes, the sulfonyl oxygens also appear to be involved in the coordination.[179]

Analogs of TosMIC

Synthetic applications of TosMIC analogs are limited in comparison with the parent compound. Two items are worthy of note: chiral analogs, and methylsulfonyl isocyanide. Of the chiral TosMIC analogs (**2–8, 11**), two have been prepared in an enantiomerically pure state (**3** and **4**).[40] Four isocyanomethylsulfonamides **9** to **12** are known; of these, the chiral proline derivative **11** was obtained as an impure oil only.[180] The scope of application of these chiral TosMIC analogs **2** to **8** has remained somewhat limited, since asymmetric inductions obtained with these synthons are modest. Reactions of isocyanides **2**, **3**, and **5-7** with acetophenone, as in Eq. 156, gave oxazolines **85** in diastereomeric excess ranging from 7 to 40%, and eventually α-hydroxy aldehydes **86** in enantiomeric excesses of 16 to 31%.[40]

[Eq. 156 scheme]

(Eq. 156)

Better results were obtained with sulfonimidoylmethyl isocyanide **8**, which, however, requires a more reactive substrate such as trifluoroacetophenone (Eq. 157).[40]

[Eq. 157 scheme: PhCOCF$_3$ → (94%) 80% de, reagents: Ti(OEt)$_4$, N-ethylpiperidine]

(Eq. 157)

Sulfonylmethyl isocyanide **4**, derived from (+)-10-camphorsulfonic acid, has the disadvantage of internal cyclization (Eq. 158) when the substrate molecule in an intermolecular process is not sufficiently reactive.[40,180]

$$\text{4} \xrightarrow{\text{base}} \text{product} \quad (80\%) \quad \text{(Eq. 158)}$$

The application of neomenthylsulfonylmethyl isocyanide (**2**) to the synthesis of 2-methylcyclobutanone in high optical purity (cf. Eq. 130) is based on the separation of diastereomers, not on asymmetric induction.[39]

Methylsulfonylmethyl isocyanide[30,54,181] (MesMIC) deserves to be further investigated for two reasons. Firstly, the "atom efficiency" of MesMIC in reactions is superior to TosMIC when the loss of methanesulfinic acid (molecular weight 80) is compared to 4-toluenesulfinic acid (molecular weight 156). Secondly, the significant solubility of MesMIC in water (10 g in 100 mL) offers the possibility of performing TosMIC-type reactions in aqueous solution. Only a few such reactions have been investigated, for example those of Eq. 159.[94]

$$\text{RCHO} \xrightarrow[\text{H}_2\text{O, rt, 5-30 min}]{\text{MsCH}_2\text{N=C, K}_2\text{CO}_3} \text{product} \quad \begin{array}{ll} R & \\ H & (81\%) \\ Ph & (90\%) \\ Bu\text{-}t & (90\%) \end{array} \quad \text{(Eq. 159)}$$

EXPERIMENTAL PROCEDURES

$$\text{TosNa} + \text{CH}_2\text{O} \xrightarrow[\text{H}_2\text{NCHO}]{\text{HCO}_2\text{H}} \text{Tos}\text{-CH}_2\text{-N(H)-CHO} \xrightarrow[\text{base}]{\text{POCl}_3} \text{Tos}\text{-CH}_2\text{-N=C}$$

***N*-(4-Tolylsulfonylmethyl)formamide[30] and 4-Tolylsulfonylmethyl Isocyanide (Two-step Synthesis of TosMIC, Eq. 16).[31,42]** A mixture of anhydrous sodium 4-toluenesulfinate (18.35 g, 100 mmol), formamide (30 mL, 750 mmol), paraformaldehyde (12.0 g, 400 mmol), and formic acid (19 mL, 700 mmol) was heated at 90° for 2 hours. The clear solution was cooled to room temperature and water (75 mL) was added. The mixture was stored in a refrigerator overnight. The solid was collected, washed with five 10-mL portions of ice-water, then dried in vacuum over calcium chloride at 60°, to give 17.2 g (81%) of *N*-(4-tolylsulfonylmethyl)formamide, mp 108–111°. A solution of this product (16.00 g, 75 mmol) in 100 mL of dry THF was cooled to −20° under nitrogen. Diisopropylamine (34 mL, 242 mmol) was added all at once, followed by POCl$_3$

(8.6 mL, 90 mmol) over 20 minutes, keeping the temperature between −10 and −20°. After stirring for 1 hour at 0°, the mixture was poured into a cold solution of potassium carbonate (20.7 g, 150 mmol) in 500 mL of water. After 10 minutes the solid was collected, washed with water (6 × 50 mL) and dried in vacuum over $CaCl_2$ to give 13.44 g of a brown solid, mp 96–110°. The crude product was dissolved in CH_2Cl_2 (50 mL) and filtered through a short column of silica (15 g), using CH_2Cl_2 (100 mL). The combined filtrates were concentrated to dryness. Crystallization from methanol (60 mL) gave 10.0 g (68%) of TosMIC, mp 116–117° (dec.). Some characteristic data of TosMIC are: IR (Nujol) 2150 (N=C), 1320 and 1155 cm^{-1} (SO_2); ^1H NMR ($CDCl_3$) δ 2.43 (s, CH_3), 4.53 (s, CH_2); ^{13}C NMR ($CDCl_3$) 61.1 (CH_2), 165.7 (N=C); estimated pK_a 13–15; CAS Registry Number 36635-61-7.

2-Cyanoadamantane (Reductive Cyanation of a Ketone).[28,32]

Solid potassium *tert*-butoxide (28.0 g, 0.24 mol, 95%) was added portionwise to a stirred and cooled solution of adamantanone (15.0 g, 0.10 mol) and TosMIC (25.0 g, 0.13 mol) in a mixture of 350 mL of 1,2-dimethoxyethane (DME) and 10 mL (0.17 mol) of absolute ethanol while keeping the temperature between 5 and 10°. Stirring was continued, first for 30 minutes without cooling, then for 30 minutes at 35–40°. The suspension thus obtained was cooled to room temperature with stirring. The precipitate (potassium *p*-toluenesulfinate) was removed and washed with DME. The combined DME solutions were concentrated to 25–35 mL and purified by flushing the concentrate over a 5-cm thick layer of alumina (ca. 200 g, on a Büchner funnel) with 250 mL of petroleum ether (bp 40–60°). Removal of the solvent provided 14–15 g (87–93%) of near-white 2-cyanoadamantane, mp 160–180° (sealed tube). Despite the wide melting range, this material is over 99.8% pure according to GLC (2-m ES-30 column, 190°). Charcoal treatment gave completely white material, melting in the same range.

2-*tert*-Butyl-5-cyanomethylthiophene (Reductive Cyanation of an Aldehyde).[64]

A solution of TosMIC (1.95 g, 10.0 mmol) in dry THF (10 mL) was added dropwise to a mixture of potassium *tert*-butoxide (1.12 g, 0.01 mol) in dry THF (10 mL) at −20° under argon. After the mixture was stirred for 30 minutes, 2-*tert*-butyl-5-thiophenecarboxaldehyde (1.68 g, 10.0 mmol) in dry THF (20 mL) was added dropwise. After the mixture was stirred for 30 minutes at −20°, 20 mL of methanol was added. Next, the mixture was refluxed for 15 minutes at

80° and then concentrated to dryness. The residue was treated with water (18 mL) and acetic acid (0.76 mL) and extracted with dichloromethane (3 × 10 mL). The product was obtained as an oil upon chromatography on silica with cyclohexane/ethyl acetate, 8:2. Yield 89%; IR (KBr) 2250 cm^{-1} (CN); ^{1}H NMR (80 MHz, CDCl$_3$) δ 6.50 (d, J = 3.7 Hz, 1 H), 6.15 (d, J = 3.7 Hz, 1 H), 3.75 (s, 2 H), 1.30 (s, 9 H). Anal. Calcd. for C$_{10}$H$_{13}$NS: C, 66.99; H, 7.30; N, 7.80. Found: C, 66.58; H, 7.30; N, 7.77.

5-Phenyl-4-tosyl-2-oxazoline (Reaction of an Aldehyde with TosMIC).[31] Potassium carbonate (14 mg, 2 mol %) was added to a stirred mixture of TosMIC (0.98 g, 5.0 mmol), benzaldehyde (0.53 g, 5.0 mmol), and methanol (12.5 mL) at room temperature. After 2 hours, water (5 mL) was added and the mixture was stirred for another hour at 0°. The solid was collected, washed with water, and dried in vacuum to give 1.36 g (90%) of product, mp 110–111°; IR (Nujol) 1611 (C=N), 1310, 1153 and 1114 cm^{-1} (SO$_2$); ^{1}H NMR (200 MHz, CDCl$_3$) δ 2.47 (s, 3 H), 5.05 (d, J = 6 Hz, 1 H), 6.07 (d, J = 6 Hz, 1 H), 7.23 (s, 1 H), 7.35–7.43 (m, 7 H), 7.84 and 7.88 (lower half of AB-q, 2 H); HRMS, m/z 301.077 (M+) (calculated 301.077).

(2′R)- and (2′S)-3-Methoxy-17-(2-methoxy-3-oxazolin-4-yl)-androsta-3,5,16-triene (33) (Reaction of a Monosubstituted TosMIC Derivative with an Aldehyde and Methanol).[70] Powdered potassium hydroxide (1.3 g, 22 mmol) was added to a stirred, ice cooled mixture of (E)-17-(isocyanotosylmethylene)-3-methoxyandrosta-3,5-diene^{54} (**32**, 2.38 g, 5.0 mmol), 37% aqueous formaldehyde (0.7 mL, 9.0 mmol), methanol (1 mL, 25 mmol), and THF (35 mL). The ice bath was removed, stirring was continued for 1 hour at room temperature, and Na$_2$SO$_4$ (5 g) was added to the reaction mixture. After 5 minutes the mixture was filtered through a layer of alumina (neutral, act. II/III; 2 × 8 cm i.d.) with CH$_2$Cl$_2$ to give 1.80 g (94%) of crude title product, mp 118–122°. Analytically pure product was obtained by crystallization from CH$_2$Cl$_2$/Et$_2$O, mp 123–128°; IR (KBr) 1655, 1630 cm^{-1} (C=N); ^{1}H NMR δ 0.8–2.6 (m), 1.04 (s), 1.02 (s), 3.35 and 3.36 (2 × s, ratio 2:1, CH$_3$OC$_{2'}$), 3.57 (s), 4.6–4.8 (m), 5.14 (s), 5.24 (m), 6.28 (m), 6.55 (m); ^{13}C NMR δ 168.9, 168.7, 155.4, 147.2, 141.4, 141.2, 122.4, 122.2, 117.6, 98.5, 73.4, 56.9, 54.3, 51.9, 48.7, 46.9, 35.4, 34.8, 33.6, 32.5, 31.5, 30.2, 25.2, 20.8, 18.8, 15.9. Anal. Calcd. for C$_{24}$H$_{33}$NO$_3$ (383.536): C, 75.16; H, 8.67; N, 3.65. Found: C, 75.1; H, 8.7; N, 3.9.

N-(Cyclohexylidenetosylmethyl)formamide (Reaction of TosMIC with a Ketone in a Nonprotic Solvent).[36,63] A solution of TosMIC (9.80 g, 50 mmol) in THF (50 mL) was added dropwise to a stirred solution of potassium tert-butoxide (6.26 g, 55 mmol) in THF (50 mL) at −35°. After 5 minutes, the reaction mixture was cooled to −70° and a solution of cyclohexanone (4.90 g, 50 mmol) in THF (50 mL) was added dropwise, then the temperature was allowed to rise to −40°. After 30 minutes, the reaction mixture was poured into ice-water (150 mL) and the mixture was neutralized with acetic acid, then extracted with Et$_2$O (100 mL). The organic layer was washed with sodium carbonate solution (20%, 50 mL), brine (30 mL), dried (Na$_2$SO$_4$), and concentrated. Crystallization from methanol gave 13.48 g (92%) of N-(cyclohexylidenetosylmethyl)formamide. Analytically pure material was obtained by crystallization from EtOH, mp 132°; IR (KBr) 3310 (NH), 1690 (C=O), 1630 (C=C), 1280 and 1135 cm^{-1} (SO$_2$); ^{1}H NMR (CDCl$_3$) δ 8.11 (s, CHO). Anal. Calcd. for C$_{15}$H$_{19}$NO$_3$S (293.4): C, 61.41; H, 6.53. Found: C, 61.36; H, 6.50.

(Isocyanotosylmethylidene)cyclohexane (Dehydration of an Unsaturated Formamide).[63,86] A solution of POCl$_3$ (2.05 mL, 22 mmol) in THF (20 mL) was added dropwise to a stirred solution of N-(cyclohexylidenetosylmethyl)formamide (5.90 g, 20 mmol) and diisopropylamine (8.0 mL, 57 mmol) in THF (60 mL). The temperature was maintained between −35 and −30°. After the addition, the temperature was allowed to rise to −20°. After 30 minutes, a solution of sodium carbonate (20%, 20 mL) was added at such a rate that the temperature was kept below 5°. After 1 hour, diethyl ether (70 mL) and water (70 mL) were added. The organic layer was washed with water (2 × 25 mL), with aqueous sodium carbonate solution (20%, 15 mL), and brine (20 mL), and dried

(Na_2SO_4). Removal of solvent (below 25°) gave 4.95 g (90%) of product which was analytically pure after one crystallization from methanol, mp 115° (dec.); IR (Nujol) 2180 (N=C), 1610 (C=C), 1360 and 1170 cm^{-1} (SO_2). Anal. Calcd. for $C_{15}H_{17}NO_2S$: C, 65.43; H, 6.22; N, 5.09, S, 11.64. Found: C, 65.0; H, 6.2; N, 5.0; S, 11.6.

In general, it may be advantageous to filter the Et_2O solution over a short (4-cm) column of alumina, to replace the process of drying over Na_2SO_4. One author[63] assumed that traces of $POCl_3$ in Et_2O solution may cause decomposition of the isocyanide, especially when the temperature increased during removal of the solvent.

trans-5-*tert*-Butyl-5-methyl-4-tosyl-2-oxazoline and (*E*)-*N*-(2,3,3-Trimethyl-1-tosyl-1-butenyl)formamide (Reaction of TosMIC with a Sterically Screened Ketone by the Two-step/Two-base Approach).[55]

n-Butyllithium (7 mL, 1.6 M solution in hexane, ca. 10 mmol) was added to a stirred solution of TosMIC (1.95 g, 10.0 mmol) in THF (25 mL) at −80°. *tert*-Butyl alcohol (1.0 mL, 11 mmol) was added to the mixture, followed after 5 minutes by 3,3-dimethyl-2-butanone (1.0 g, 10 mmol). The temperature was raised to −50° and the mixture was stirred for 1 hour at that temperature. The mixture was then poured into water (200 mL) and the solution was extracted with dichloromethane (3 × 50 mL). The combined extracts were dried ($MgSO_4$) and concentrated to give 2.91 g (99%) of the oxazoline, mp 123–127°. Analytically pure material was obtained after one crystallization from Et_2O, mp 127°; IR (Nujol) 1623 (C=N), 1322, 1155 cm^{-1} (SO_2); ^{1}H NMR ($CDCl_3$) δ 0.98 (s, 9 H), 1.85 (s, 3 H), 2.42 (s, 3 H), 4.77 (d,^{1,4} *J* = 2 Hz, 1 H), 6.92 (d,^{1,4} *J* = 2 Hz, 1 H), 7.20, 7.33, 7.72 and 7.85 (AB q, 4 H). Anal. Calcd. for $C_{15}H_{21}NO_3S$ (295.399): C, 60.99; H, 7.17; N, 4.74; S, 10.85. Found: C, 61.10; H, 7.15; N, 7.74; S, 10.91.

Potassium *tert*-butoxide (0.15 g, 1.3 mmol) was added to a stirred solution of the oxazoline (0.295 g, 1.00 mmol) in DME at −50°. After stirring for 10 minutes at −50°, the mixture was poured into water (50 mL) and was then extracted with EtOAc (3 × 15 mL). The combined extracts were washed with water, dried over Na_2SO_4, and concentrated to give 0.21 g (71%) of (*E*)-*N*-(2,3,3-trimethyl-1-tosyl-1-butenyl)formamide as an oil. Analytically pure material (0.18 g) was obtained by crystallization from Et_2O /petroleum ether, mp 112–113°; IR (neat) 3400 (NH), 1700 (C=O), 1310, 1150 cm^{-1} (SO_2); ^{1}H NMR ($CDCl_3$) δ 1.20 (s, 9 H), 2.0 (br s, 1.5 H), 2.25 (br s, 1.5 H), 2.40 (s, 3 H), 6.7–8.2 (m, 6 H). Anal. Calcd. for $C_{15}H_{21}NO_3S$ (295.399): C, 60.99; H, 7.17; N, 4.74; S, 10.85. Found: C, 60.93; H, 7.19; N, 7.73; S, 10.79.

TosCH₂N=C $\xrightarrow{\text{Br(CH}_2)_3\text{Br}}_{\text{NaH, DMSO, Et}_2\text{O}}$ [cyclobutyl with N=C and Tos substituents]

1-Tosylcyclobutyl Isocyanide (Intramolecular Dialkylation of TosMIC).[148] Sodium hydride (55–60% in mineral oil; 24 g, ~0.58 mol) was freed from mineral oil using dry petroleum ether (bp 40–60°) under nitrogen in a 2-L, three-necked flask, then dry DMSO (500 mL) and Et₂O (175 mL) were added. A solution of TosMIC (39.2 g, 0.20 mol) and 1,3-dibromopropane (48.5 g, 0.24 mol) in a mixture of dry DMSO (125 mL) and Et₂O (65 mL) was added over 45–60 minutes to the mechanically stirred suspension. During the addition the reaction mixture started to reflux and sodium bromide began to separate near the end. Stirring was continued for 1 hour, then water (150 mL) was added slowly. The mixture was extracted with Et₂O (2 × 250 mL, 1 × 150 mL and 1 × 100 mL), the combined extracts were washed with brine (3 × 100 mL), and dried over anhydrous Na₂SO₄. The Et₂O solution was concentrated to give a solid residue which was stirred with 100 mL of Et₂O for 10 minutes. After addition of petroleum ether (bp 40–60°, 100 mL) and overnight cooling at −20°, the product was collected; yield 33.3 g (71%), mp 95–97° (dec.). Analytically pure material was obtained from ethanol, mp 97°; IR (Nujol) 2190 (N=C), 1340, 1170 and 1150 cm⁻¹ (SO₂); ¹H NMR (CDCl₃) δ 2.48 (s, 3 H), 1.8–3.6 (m, 6 H), 7.38, 7.85 (2d, J = 8 Hz, 4 H); ¹³C NMR (CDCl₃) δ 166.4 (s, N=C), 145.9 (s), 129.8 (d), 129.5 (d), and 129.2 (s, C-arom), 73.5 (s, 1-C), 31.0 (t, 2-C), 21.1 (q, ArCH₃), 13.9 (t, 3-C). Anal. Calcd. for C₁₂H₁₃NO₂S (235.3): C, 61.25; H, 5.57; N, 5.97; S, 13.62. Found: C, 61.27; H, 5.64; N, 6.05; S, 13.63.

n-C₈H₁₇I $\xrightarrow[\text{NaH, DMSO, Et}_2\text{O}]{\text{TosCH}_2\text{N=C}}$ n-C₈H₁₇–C(Tos)(N=C)–C₈H₁₇-n $\xrightarrow[\text{EtOH}]{\text{Li, NH}_3\text{ (liq)}}$ n-C₁₇H₃₆

9-Isocyano-9-tosylheptadecane and n-Heptadecane (Dialkylation of TosMIC and Reduction of the Product; TosMIC as CH₂ Connector).[105] To a suspension of prewashed sodium hydride (0.528 g, 22 mmol) in a mixture of DMSO (3 mL) and Et₂O (15 mL) was added TosMIC (1.95 g, 10 mmol) in Et₂O (15 mL) at room temperature. After 10 minutes, n-octyl iodide (4.8 g, 20 mmol) in Et₂O (15 mL) was added dropwise, the mixture was stirred for 3 hours, and poured into water. The organic layer was separated and the aqueous layer was extracted with Et₂O. The combined extracts were washed with water and brine, and dried (Na₂SO₄). After removal of the solvent, 3.77g (90%) of 9-isocyano-9-tosylheptadecane was obtained as a gummy substance; IR (neat) 2130 cm⁻¹ (N=C); ¹H NMR (CCl₄) δ 0.90 (distorted t, 6 H), 1.3 (m, 28 H), 2.45 (s, 3 H), 7.23 (d, J = 8 Hz, 2 H), 7.68 (d, J = 8 Hz, 2 H).

To freshly distilled liquid ammonia (50 mL) at −33° was added lithium (0.050 g, 7.0 mmol) in one portion, followed by 9-isocyano-9-tosylheptadecane

(0.294 g, 0.70 mmol) in a mixture of Et$_2$O (3 mL) and ethanol (0.12 mL). After 2 hours, the ammonia was allowed to evaporate. Then, water was added and the mixture was extracted with diethyl ether (5 × 10 mL). The combined extracts were washed with water (20 mL), with brine (20 mL), dried (Na$_2$SO$_4$), and concentrated. Distillation of the residue (pot temperature 125–135°, 1 mm Hg) afforded 0.156 g (93%) of heptadecane as a colorless liquid; ^{1}H NMR (CCl$_4$) δ 0.90 (distorted t, 6 H), 1.27 (m, 30 H).

5-(2-Hydroxyphenyl)-1-neopentylimidazole (Reaction of TosMIC with an Aldimine).[121] A mixture of salicylaldehyde (0.65 g, 5.3 mmol), neopentylamine (1.0 g, 11.5 mmol), and anhydrous magnesium sulfate (3.6 g, 30 mmol) was heated at reflux for 2 hours. The magnesium sulfate was removed by filtration, and the filtrate was concentrated under reduced pressure. The residue was dissolved in 20 mL of methanol, and to the mixture was added potassium carbonate (3.5 g, 25 mmol) and TosMIC (1.2 g, 6.2 mmol). After heating at reflux for 2 hours, the solvent was evaporated and water (50 mL) was added to the residue. The precipitate was collected by filtration and recrystallized from ethanol and water to afford 1.0 g (82%) of the title product, mp 247–248°; ^{1}H NMR (CDCl$_3$) δ 0.72 (s, 9 H), 3.76 (s, 2 H), 6.7–7.6 (m, 7 H).

3-Cyano-4-phenylpyrrole (Reaction of a Michael Acceptor with TosMIC).[136] To a stirred suspension of potassium *tert*-butoxide (1.05 g, ~13 mmol) in THF (10 mL) at −30° was added in 2 minutes a solution of TosMIC (1.17 g, 6.0 mmol) in THF (10 mL). The temperature rose to −25°. The mixture was stirred for 4 minutes at −30° before a solution of cinnamonitrile (0.77g, 6.0 mmol) in THF (10 mL) was added over 4 minutes. Stirring was continued for 15 minutes at −10°, and then the mixture was poured on 50 g of ice. Most of the THF was removed while the temperature was kept below 35°. The off-white precipitate was collected, washed with water, and dried in vacuum to give 0.89 g (88%) of product, mp 124–125°; IR (Nujol) 3360 (NH), 2290 cm^{-1} (CN); ^{1}H NMR (CDCl$_3$) δ 6.87 (apparent t, J = 2 Hz, 1 H), 7.1–7.8 (m, ca. 6 H), 9.0 (br s, 1 H).

3-Methyl-4-phenylpyrrole (Reaction of a Michael Donor with a 1-Isocyano-1-tosyl-1-alkene).[136] A solution of (E)-1-isocyano-2-phenyl-1-tosylethene[86] (0.566 g, 2.0 mmol) and propiophenone (0.536 g, 4.0 mmol) in DME (10 mL) was added over 10 minutes to a mixture of potassium *tert*-butoxide (0.70 g, 6.2 mmol) and DME (5 mL) at 0° and stirred for 1 hour at 20°. Methanol (0.5 mL) was added, and the mixture was stirred for another 5 minutes and then poured into ice-water (80 mL). This mixture was extracted with CH_2Cl_2 (30, 20 and 10 mL). The combined extracts were washed with brine, dried ($MgSO_4$), and concentrated. Column chromatography on alumina [activity grade II-III, Et_2O/petroleum ether (bp 40–60°) 1:5] gave 0.23 g (73%) of the title compound as a colorless oil; IR (neat) 3500 cm^{-1} (NH); ^1H NMR ($CDCl_3$) δ 2.13 (s, 3 H), 6.19–6.38 (m, 1 H), 6.49 (apparent t, J = 2 Hz, 1 H), 6.7–8.0 (m, ca. 6 H); HRMS, m/z 157.092 (M^+) (calculated 157.089).

(E)-4-Benzoyl-2-(cyclopent-1-enyl)-1-methyl-3-(2-phenylethenyl)pyrrole and 3-Benzoyl-7,8-dihydro-1-methyl-5-phenyl-6H-cyclopent[g]indole (Formation of a 2,3-Dialkenylpyrrole from a TosMIC Derivative and a Michael Acceptor, Followed by Ring Closure and Dehydrogenation to an Indole).[128] To a suspension of potassium *tert*-butoxide (1.12 g, 10 mmol) in THF (15 mL), precooled to −60°, was added dropwise a solution of (isocyanotosylmethylidene)cyclopentane[181a] (2.61 g, 10 mmol) in 20 mL of THF while maintaining the temperature below −50°. The reaction mixture turned into a transparent orange solution, which was stirred for an additional 10 minutes at −50°. A solution of 1,5-diphenylpenta-2,4-dien-1-one (2.34 g, 10 mmol) in 20 mL of THF was added. The temperature was increased to −20°. Stirring was continued for 1 hour without further cooling. Then the reaction mixture was concentrated in vacuum (at 20°), and the solid residue was suspended in 30 mL of benzene. Methyl iodide (2.82 g, 20 mmol) was added, and the mixture was treated with 30 mL of 50% aqueous KOH containing 0.25 g (1.1 mmol) of benzyltriethylammonium chloride. The reaction mixture was vigorously stirred for 1 hour at room temperature. The benzene layer was separated, washed with 10% aqueous NH_4Cl (50 mL) and then twice with water (50 mL each), and dried ($MgSO_4$), to give after concentration and crystallization from ethanol, 2.96 g (84%) of (*E*)-4-benzoyl-2-(cyclopent-1-enyl)-1-methyl-3-(2-phenylethenyl)pyrrole as pale yellow crystals, mp 141–142°; IR (KBr) 2905, 2838, 1620 (C=O, C=C), 1500, 1420, 1345, 1221 cm^{-1}; ^1H NMR ($CDCl_3$) δ 1.8–2.3 (m, 2 H), 2.4–2.8 (m, 4 H), 3.48 (s, 3 H), 5.9–6.0 (m, 1 H), 6.89 (s, 1 H), 7.0–7.4 (m, 10 H), 7.7–7.9 (m, 2 H). Anal. Calcd. for $C_{25}H_{23}NO$ (353.18): C, 84.99; H, 6.52; N, 3.97. Found: C, 85.00; H, 6.57; N, 4.00.

A solution of the foregoing product (1.76 g, 5.0 mmol) in 20 mL of dry triglyme was gently refluxed for 1 hour. The yellow-orange solution was cooled to approximately 80°, and 2,3-dichloro-5,6-dicyano-1,4-benzoquinone (DDQ, 1.18 g, 5.2 mmol) was added. The dark red mixture was stirred at 80° for 30 minutes and then poured into 200 mL of cold water. The mixture was stirred for 30 minutes. The precipitate (2 g) was collected, washed several times with cold water, and dried in air. Purification was performed over alumina (Brockmann 90, II/III, 200 g; 8:1 CH_2Cl_2/*n*-pentane) to give 1.46 g (83%) of colorless prisms, which were recrystallized from ethanol: mp 172–173°; IR (KBr) 3050, 2940, 2840, 1610 (C=O), 1520, 1450, 1440, 1362, 1225, 1140, 960 cm^{-1}; ^1H NMR ($CDCl_3$) δ 2.01 (quintet, $J = 4.2$ Hz, 2 H), 2.89 (t, $J = 4.2$ Hz, 2 H), 3.21 (t, $J = 4.2$ Hz, 2 H), 3.76 (s, 3 H), 7.0–7.3 (m, 9 H), 7.4–7.6 (m, 2 H), 8.16 (s, 1 H); Anal. Calcd. for $C_{25}H_{21}NO$ (351.16): C, 85.43; H, 6.03; N, 3.99. Found: C, 84.86; H, 5.91; N, 3.99.

(*Z*)-19-Isocyano-19-tosyltricos-9-ene and (*Z*)-Tricos-9-en-19-one (Alkylation of a Monosubstituted TosMIC Derivative and Hydrolysis of a Disubstituted TosMIC Derivative. TosMIC as CO Connector).[107] Potassium *tert*-butoxide (2.7 g, 24 mmol) was added over 5 minutes to a solution of 1-tosyl-

n-pentyl isocyanide[90] (5.02 g, 20 mmol) and (Z)-octadec-9-enyl iodide [7.90 g, ca. 20 mmol, prepared from oleyl alcohol (Merck-Schuchhardt, technical grade 88%)] in a mixture of DMSO (20 mL) and Et$_2$O (50 mL) at 10° under nitrogen. After stirring for 20 minutes at 10°, the reaction mixture was poured into a 1% solution of NH$_4$Cl in water (500 mL). The mixture was extracted three times with Et$_2$O (100, 75, and 25 mL). The combined extracts were washed with brine, dried (Na$_2$SO$_4$), and concentrated to give 10.0 g of (Z)-19-isocyano-19-tosyltricos-9-ene as a gummy material.

A solution of crude (Z)-19-isocyano-19-tosyltricos-9-ene (5.0 g, ca. 10 mmol) in Et$_2$O (50 mL) was stirred with concentrated HCl (5 mL) at 10° for 15 minutes. The organic layer was washed with a 20% aqueous solution of sodium carbonate (25 mL), and the aqueous phase was extracted with Et$_2$O (2 × 25 mL). The organic layer and extracts were combined, dried (Na$_2$SO$_4$), and concentrated to give 3.3 g of crude product, which was purified by column chromatography on silica (elution with CH$_2$Cl$_2$-pentane, 1:2). Yield 3.0 g (ca. 90%) as a viscous oil; IR (neat) 1715 cm^{-1} (C=O); ^{1}H NMR (300 MHz, CDCl$_3$) δ 0.85 (distorted t, J = 4 Hz, 6 H), 1.25 (m, 24 H), 1.52 (m, 4 H), 1.97 (m, 4 H), 2.35 (t, 4 H), 5.3 (t, J = 4 Hz, 2 H).

1,3-Diphenyl-1-hydroxypropan-2-one (Synthesis of a Tosyloxazoline, Followed by Hydrolysis to a Hydroxyketone).[73]

Powdered potassium carbonate (0.10 g, 0.7 mmol) was added to a stirred solution of 2-phenyl-1-tosylethyl isocyanide[73] ("Benzyl-TosMIC", 2.85 g, 10.0 mmol) and benzaldehyde (1.06 g, 10.0 mmol) in 20 mL of methanol at 30°. After 5 minutes, the mixture was cooled to 0° at which temperature the tosyloxazoline started to precipitate. Addition of concentrated HCl (1 mL) gave a clear solution, which was stirred for 10 minutes. Water (5 mL) was added and the mixture was neutralized with a 30% aqueous NaOH solution. The mixture was cooled to −10° for 1 hour and the separated solid was collected, dried, and crystallized from Et$_2$O/pentane (5:1) to give 1.54 g (68%) of the title compound, mp 113–114.5°; ^{1}H NMR (CDCl$_3$) δ 3.64 (s, 2 H), 4.20 (br s, 1 H), 5.18 (s, 1 H), 7.0–7.4 (m, 10 H).

N-Methyl-2-(3,4-dimethoxyphenyl)-2-hydroxyethylamine (Reaction of TosMIC with an Aldehyde, Followed by Reduction with Lithium Aluminum Hydride).[73]

A solution of TosMIC (1.95 g, 10.0 mmol) and 3,4-dimethoxybenzaldehyde (1.66 g, 10.0 mmol) in 20 mL of methanol was stirred with 0.1 g powdered potassium carbonate (0.10 g, 0.7 mmol) for 10 minutes at room temperature and for 30 minutes at 0°, then kept for 2 hours at −20°. The solid oxazoline was collected and dried (40°, 0.1 mm Hg, 30 minutes). To a solution of the crude oxazoline in 50 mL of dry THF at 0° was added lithium aluminum hydride (0.80 g, 23.5 mmol) in small portions. The mixture was stirred for 30 minutes at 25°. After cooling to 0°, 1.5 mL of water was added. THF (50 mL) was added and the mixture was filtered. The aluminum salts were washed with 50 mL of THF and the combined filtrates were concentrated. Crystallization of the residue from hexane/acetone (2:1) gave 1.49 g (71%) of the title compound; mp 102–103.5°; ^{1}H NMR (CDCl$_3$) δ 2.40 and 2.46 (br s + s, 5 H, NH, OH and CH$_3$), 2.75 (m, 2 H), 3.86 (s, 3 H), 3.88 (s, 3 H), 4.68 (m, 1 H), 6.9 (m, 3 H).

TABULAR SURVEY

The material presented in the tables is based on a *Chemical Abstracts* search covering the period 1967 (the year that TosMIC was discovered) to January 1996, and on further information available to the authors. Reference is made to all pertinent full papers that have been uncovered. Preliminary papers, as a rule, are listed only when they provide data not given in the corresponding full papers. The rather extensive patent literature is treated selectively. Only patents that provide material other than "more of the same" are cited.

The *Chemical Abstracts* search has been performed with the use of both the CAS Registry Numbers of TosMIC and its isotopically labeled counterparts, as well as the essential structure element: SO$_2$—C—N=C. The outcome of the computer search sometimes was misleading, and needed a critical evaluation. Not only have isocyanides occasionally been erroneously named isocyanates by authors, reviewers, or abstractors, they have also been formulated mistakenly as —C≡N or -N=C=O. Further, the sulfonyl group (SO$_2$) has sometimes been confused with the sulfonate group (SO$_3$).

The integral results of the "Synthetic Uses of TosMIC" are collected in eighteen tables, which follow the sequence of the various sections. The discussion of the synthesis of indoles is not accompanied by a table on indoles, since indoles are secondary products of the use of TosMIC. When closely related products are derived from different groups of substrate molecules, a subdivision of Tables is employed, as for example in Tables I-A, I-B, and I-C. The sequence within the tables is based on increasing carbon count of the substrate molecules.

Standard abbreviations used in the Tables are the following:

AIBN	azobis(isobutyronitrile)
—	data not provided
aq	aqueous
BTEAC	benzyltriethylammonium chloride
Dabco	1,4-diazabicyclo[2.2.2]octane
de	diastereomeric excess
DBU	1,8-diazabicyclo[5.4.0]undec-7-ene
DDQ	2,3-dichloro-5,6-dicyano-1,4-benzoquinone
DME	1,2-dimethoxyethane
DMF	N,N-dimethylformamide
DMSO	dimethyl sulfoxide
DPA	diisopropylamine
ee	enantiomeric excess
Ether	diethyl ether
HMPA	hexamethylphosphoric triamide
LAH	lithium aluminum hydride
LDA	lithium diisopropylamide
NMM	N-methylmorpholine
sl. add.	slow addition
TBAB	tetrabutylammonium bromide
TBAI	tetrabutylammonium iodide
TEA	triethylamine
THAI	tetrahexylammonium iodide
THF	tetrahydrofuran
THP	tetrahydropyranyl
Tos	4-tolylsulfonyl
Triton B	benzyltrimethylammonium hydroxide, 40% solution in methanol

TABLE IA. CYANIDES FROM KETONES BY REDUCTIVE CYANATION

Substrate	TosMIC (equiv.)	Conditions[a]	Product(s) and Yield(s) (%)	Refs.
C_6				
cyclohexanone	(1.0)	t-BuOK (2), DME, rt, 1.5 h	cyclohexyl-CN (80)	32,53
4-hydroxycyclohexanone	(1.5)	t-BuOK (6), DME, t-BuOH, rt, 1 h	4-hydroxycyclohexyl-CN (75)	182a
pinacolone (t-Bu methyl ketone)	(1.3)	t-BuOK (3.5), DMSO, rt, 17 h	t-Bu-CH(CH₃)-CN (70)	32
"	(1.0)	t-BuOK (2), DME, rt, 1.5 h	" (36)	32,53
"	(1.0)	1. t-BuOLi, THF, −50°, 1 h 2. t-BuOLi, THF, rt 10 min	" (85)	55
norbornadienone	(1.05)	t-BuOK (2), THF, t-BuOH, 0°, 45 min	norbornadienyl-CN (10)	182
norbornanone	(1.3)	t-BuOK (3.5), DMSO, rt, 17 h	norbornyl-CN (73) endo:exo = 4:3	32
"	(1.3)	EtONa (1.3), DME, rt, 2.5 h	" (62) endo:exo = 1:1	32
C_7				
cycloheptanone	(1.1)	t-BuOK (2), DME, rt, 1.5 h	cycloheptyl-CN (80)	32,53
i-Pr CO Pr-i (diisopropyl ketone)	(3)	t-BuOK (7), HMPA, 45°, 70 h	i-Pr-CH(Pr-i)-CN (65)	32

C8	(1.0)	t-BuOK (2), DME, rt, 1.5 h	" (<5)	32
	(1.2)	t-BuOK (2.5), DME, rt, 1.5 h	n-Pr-CH(CN)-Pr-n (74)	32, 53
	(1.0)	t-BuOK (2), DME, rt, 1.5 h	4-Br-C6H4-CH(CN)-CH3 (79)	32, 53
	(1.0)	t-BuOK (2), DME, rt, 1.5 h	Ph-CH(CN)-CH3 (68)	32
	(1.0)	t-BuOK (2), t-BuOH, DME, 0°, 45 min; rt, 1 h	" (80)	53
	(2)	t-BuOK (4), DMSO, rt, 24 h	cyclohexyl-spiro-cyclopropane-CN (70)	183
	(1.5)	—	dioxolane-spiro-cyclohexyl-CN (>60)	67
C9	(—)	t-BuOK, 0°, 1 h; rt, 2 h	indane-CN (75)	184
	(—)	EtONa (1.5), DME, EtOH, 0°, 1 h; rt, 2.5 h	indanone-CN (58)	66
	(1.5)	t-BuOK (7), t-BuOH, DME, rt, 1 h	bicyclic-CN (>54)	185

TABLE IA. CYANIDES FROM KETONES BY REDUCTIVE CYANATION (*Continued*)

Substrate	TosMIC (equiv.)	Conditions[a]	Product(s) and Yield(s) (%)	Refs.
adamantanone	(1.3)	*t*-BuOK (2.5), EtOH, DME, rt, 1 h; 40°, 1 h	2-cyanoadamantane (68)	186
2-allylcyclohexanone	(1.3)	*t*-BuOK (2.7), DMSO, rt	1-cyano-2-allylcyclohexane (63)	187
1-(ethoxycarbonyl)hexahydro-4H-azepin-4-one	(—)	—	4-cyano-1-(ethoxycarbonyl)hexahydroazepine (—)	188
2,2,6,6-tetramethyl-4-oxopiperidin-1-oxyl	(1)	*t*-BuOK (2), *t*-BuOH, DME, 0°, 45 min; rt, 1 h	4-cyano-2,2,6,6-tetramethylpiperidin-1-oxyl (90)	57
2,2,6,6-tetramethyl-4-piperidone	(1)	*t*-BuOK (2), *t*-BuOH, DME, 0°, 45 min; rt, 1 h	4-cyano-2,2,6,6-tetramethylpiperidine (76)	189
2,2,6,6-tetramethyl-3,5-heptanedione	(3)	*t*-BuOK (7), HMPA, 45°, 170 h	No reaction	32

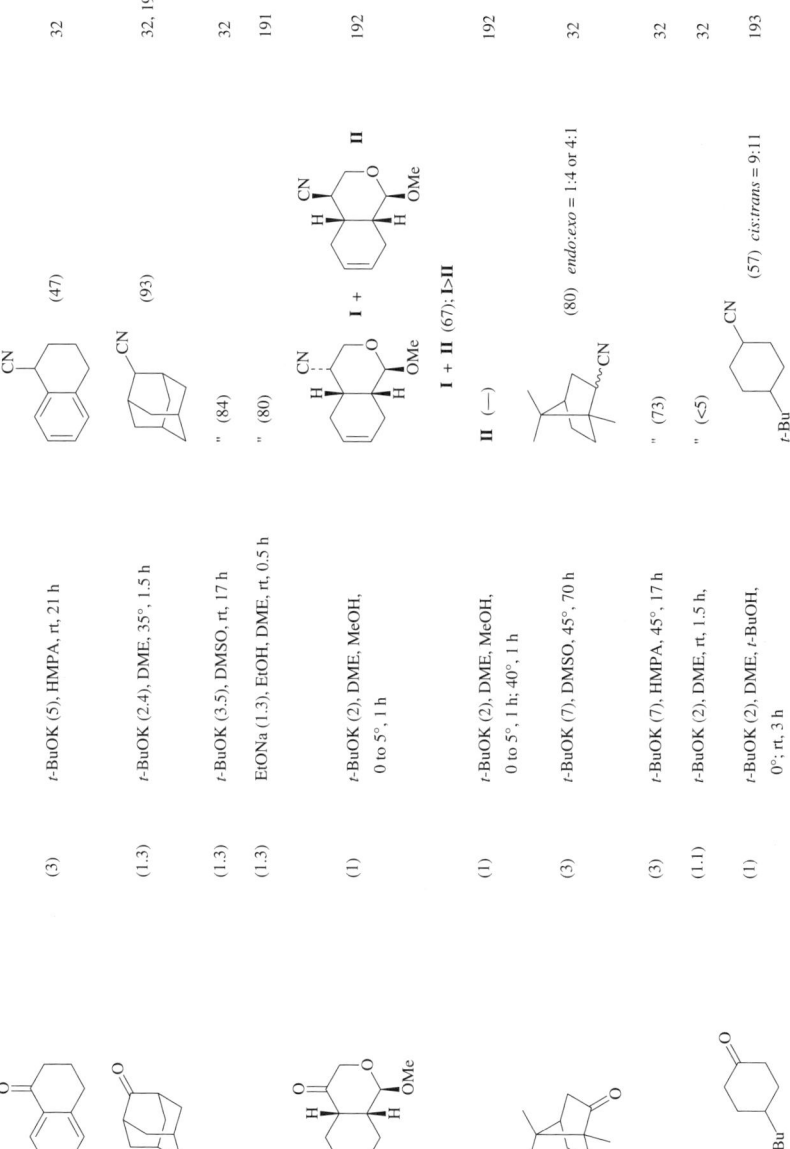

TABLE IA. CYANIDES FROM KETONES BY REDUCTIVE CYANATION (*Continued*)

Substrate	TosMIC (equiv.)	Conditions[a]	Product(s) and Yield(s) (%)	Refs.
C[11]				
(indanone with CO₂Me)	(1.5)	EtONa (1.5), EtOH, DME, 0°, 1 h; rt, 5.5 h	(37)	66
(tetralone with OMe)	(2)	*t*-BuOK (10), DME, *t*-BuOH, rt, 20 h	(77)	194
(dimethoxyindanone)	(1.3)	*t*-BuOK (7), THF, 0°, 1.5 h; rt, 72 h	(63)	195
(dimethoxyindanone)	(2)	*t*-BuOK (10), DME, *t*-BuOH, rt, 48 h	(84)	195a
(adamantanone)	(—)	—	(37)	196
C[12]				
(3-phenylcyclohexanone)	(1)	*t*-BuOK (2), DME, 0°, 45 min; 20°, 1 h	(74)	197

494

	(3)	t-BuOK (7), HMPA, 45°, 40 h	(76) 32
	(—)	t-BuOK	(80) ratio 3:2 198
C13	(1.4)	t-BuOK (5), DMSO, rt, 17 h	(84) 199
	(1.3)	t-BuOK (3.5), DMSO, rt, 17 h	(69) 32
	(1.5)	EtONa (1.5), EtOH, DME, 0°, 1 h; rt, 4 h	No reaction 66
	(1)	t-BuOK (2), t-BuOH, DME, −20°, 15 min; rt, 3.5 h	(89) 200
C14	(2)	t-BuOK (6), THF, rt, 15 h	(55) 201
	(1.3)	t-BuOK (3), DMSO, rt, 5.5 h	(0)[b] 202

TABLE IA. CYANIDES FROM KETONES BY REDUCTIVE CYANATION (*Continued*)

Substrate	TosMIC (equiv.)	Conditions	Product(s) and Yield(s) (%)	Refs.
	(—)	t-BuOK (—), DME, 0°, 20 min; rt, 1 h	(52)	203
	(1)	t-BuOK (1), THF, 0°, 1 h	(53)	205
C$_{15}$	(3)	t-BuOK (5), DME, t-BuOH, 0°, 0.5 h; rt, 8 h	(68) α:β = 22:78	206
	(3)	t-BuOK (5), DME, t-BuOH, 0°, 0.5 h; rt, 12 h	(48) α:β = 1:1	206
	(1.3)	t-BuOK (2.5), DME, EtOH, rt, 2 h	(22) + (33)	204

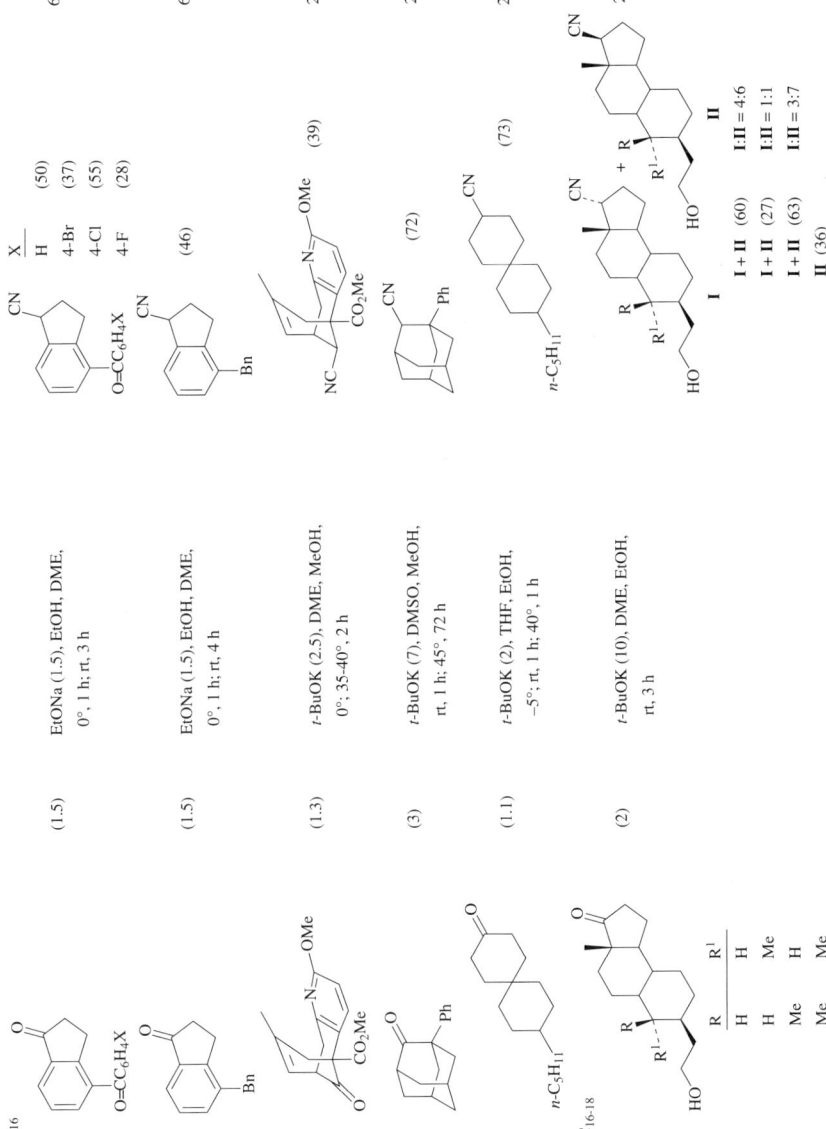

TABLE IA. CYANIDES FROM KETONES BY REDUCTIVE CYANATION (*Continued*)

Substrate	TosMIC (equ v.)	Conditions[a]	Product(s) and Yield(s) (%)	Refs.
C$_{17}$	(1.5)	EtONa (1.5), EtOH, DME, 0°, 1 h; rt, 3 h	(53)	66
C$_{17-20}$	(1.5)	EtONa (1.5), EtOH, DME, 0°, 1 h; rt, 3 h	R, R^1, R^2, R^3: Me H H H (45); Et H H H (41); Me Me H H (29); Me H Me Me (50); Bu-*t* H H H (46)	66
C$_{17}$	(10)	*t*-BuOK (10), DME, *t*-BuOH, rt, 1 h	(10)	211
	(3)	*t*-BuOK (7), DMSO, MeOH, rt, 1 h; 45°, 72 h	(59)	208
	(1.3)	*t*-BuOK (2), DME, EtOH, rt, 14 h; 60°, 5 h	No reaction	212

498

C$_{18-19}$	(1.2)	t-BuOK (10), DME, n-BuOH, rt, 1 h	I + II (52), I:II = 5:6	213
	(2)	t-BuOK (10), DME, n-BuOH, rt, 3 h	I + II (59), I:II = 2:3	210, 210, 214
C$_{19}$	(1.3)	EtONa (1.2), DME, rt, 2 h	I + II (47), I:II = 1:1	32
C$_{19-20}$	(1.3)	t-BuOK (3.5), DMSO, rt, 17 h	I (69)	32
	(2)	t-BuOK (10), DME, t-BuOH, rt, 1 h	I + II (91), I:II = 3:2	56
	(—)	t-BuOK (xs), t-BuOH, DME	I + II (85), I:II = 2:1	215
	(2)	t-BuOK (10), DME, t-BuOH, rt, 1 h	I + II (86), I:II = 7:3	56

TABLE IA. CYANIDES FROM KETONES BY REDUCTIVE CYANATION (Continued)

Substrate	TosMIC (equiv.)	Conditions[a]	Product(s) and Yield(s) (%)	Refs.
C_{21}	(2)	t-BuOK (10), DME, t-BuOH, rt, 1 h	I + II (90), I:II = 7:3; I + II (88), I:II = 7:3	56
C_{22}	(—)	t-BuOK (—), DME	I + II (60), I:II = 3:1	216, 217
C_{27}	(1.3)	t-BuOK (2.5), DME, rt, 5 h	I + II (85), I:II = 10:7	32
C_{27}	(1.5)	t-BuOK (3), DME, rt, 5 h	I + II (53), I:II = 10:9	32

[a] Equivalents of base relative to TosMIC are given in brackets after the base.
[b] 9,10-Anthraquinone was formed in 31% yield.

TABLE IB. CYANIDES FROM ALDEHYDES BY REDUCTIVE CYANATION

Substrate	TosMIC (equiv.)	Conditions[a]	Product(s) and Yield(s) (%)	Refs.
C_4 ~~~CHO	(1.1)	1. *t*-BuOK (2), DME, −50°, 50 min 2. MeOH, reflux, 15 min	~~~CN (38)	62
C_5 furan-2-CHO	(1.1)	1. *t*-BuOK (2), DME, −50°, 50 min 2. MeOH, reflux, 15 min	furan-2-CH$_2$CN (55)	62
thiophene-3-CHO	(1)	1. *t*-BuOK (2), THF, −20°, 30 min 2. MeOH, reflux, 15 min	thiophene-3-CH$_2$CN (81)	218
C_6 N-Me-pyrrole-2-CHO	(1.1)	1. *t*-BuOK (2), THF, −20°, 30 min 2. MeOH, reflux, 15 min	N-Me-pyrrole-2-CH$_2$CN (—)	65
C_{6-7} OHC-pyrrole-CHO (2,5-diCHO, N-R)	(1.1)	1. *t*-BuOK (2), DME, −45°, 30 min 2. MeOH, reflux, 15 min	" (76)	200
	(2.2)	1. *t*-BuOK (2), THF, −20°, 30 min 2. MeOH, reflux, 15 min	NC-pyrrole-CH$_2$CN (N-R) R = H (61) R = Me (73)	65
C_{7-8} 4-R-C$_6$H$_4$-CHO	(1.1)	1. *t*-BuOK (2), DME, −50°, X min 2. MeOH, reflux, 15 min	4-R-C$_6$H$_4$-CH$_2$CN R X Cl 10 (70) O$_2$N 45 (10) H 30 (67) MeO 20 (70)	62
C_9 *t*-Bu-thiophene-CHO	(1)	1. *t*-BuOK, THF, −20°, 30 min 2. MeOH, reflux, 15 min	*t*-Bu-thiophene-CH$_2$CN (89)	64

TABLE IB. CYANIDES FROM ALDEHYDES BY REDUCTIVE CYANATION (*Continued*)

Substrate	TosMIC (equiv.)	Conditions[a]	Product(s) and Yield(s) (%)	Refs.
C_{10} 3-formyl-1-methylindole	(1.1)	1. *t*-BuOK (2), DME, −50° 2. MeOH, reflux	3-(cyanomethyl)-1-methylindole (35)	219
2-formyl-1-methylindole	(1.1)	1. *t*-BuOK (2), DME, −50° 2. MeOH, reflux	2-(cyanomethyl)-1-methylindole (48)	219
C_{11} citronellal-type dienal	(1.1)	1. *t*-BuOK (2), DME, −50°, 1 h 2. MeOH, reflux, 15 min	homologated nitrile (58)	62
2-naphthaldehyde	(1.1)	1. *t*-BuOK (2), DME, −50°, 1 h 2. MeOH, reflux, 15 min	2-(cyanomethyl)naphthalene (62)	62
C_{14} 2-(p-toluoyl)-1-methyl-5-formylpyrrole	(1.1)	1. *t*-BuOK (2), DME, −45°, 30 min 2. MeOH, reflux, 15 min	5-(cyanomethyl)-1-methyl-2-(p-toluoyl)pyrrole (50)	200
C_{16-17} 1-benzyl-3-formylindole derivatives	(1.1)	1. *t*-BuOK (2), DME, −50° 2. MeOH, reflux	R^1 R^2 H H (61) Br H (54) H Me (69)	219
C_{17} 1-(benzyloxycarbonyl)-3-formylindole	(1.1)	1. *t*-BuOK (2), DME, −50° 2. MeOH, reflux	3-(cyanomethyl)-1H-indole (38)	219

[a] Equivalents of base relative to TosMIC are given in brackets after the base.

TABLE IC. CARBOXYLIC ACIDS AND ESTERS FROM KETONES OR ALDEHYDES VIA REDUCTIVE CYANATION

Substrate	Conditions[a]	Product(s) and Yield(s) (%)	Refs.
C₃ (acetone)	1. TosMIC, t-BuOK (2), THF, −10°, 5 min 2. HCl (2 N), reflux, 10 h	iPr-CO₂H (62)	85
C₅ t-BuCHO	1. TosMIC, t-BuOK (2), THF, −10°, 5 min 2. HCl (2 N), reflux, 10 h	t-BuCH₂CO₂H (60)	85
C₆ (cyclohexanone)	1. TosMIC, t-BuOK (2), THF, −10°, 5 min 2. HCl (2 N), reflux, 10 h	cyclohexyl-CO₂H (55)	85
C₇ (PhCHO)	1. PhSO₂CH₂N=C, t-BuOK (2), THF, −10°, 5 min 2. HCl (2 N), reflux, 10 h	PhCH₂CO₂H (65)	85
C₉ (cinnamaldehyde)	1. TosMIC, t-BuOK (2), THF, −10°, 5 min 2. HCl (2 N), reflux, 1 h	PhCH=CHCH₂CO₂H (67)	85
C₁₁ (bicyclic ketone)	1. TosMIC, t-BuOK (7), THF, t-BuOH, rt, 1 h 2. NaOH (2 N), reflux, 12 h	(54)	185
(sugar-CHO)	1. TosMIC, t-BuOK (2.2), THF, −10°, 8 min 2. HCl, MeOH, −10°, 2 h	(35)	220

[a] Equivalents of base relative to TosMIC are given in brackets after the base.

TABLE II. OXAZOLINES FROM KETONES OR ALDEHYDES

Substrate	Reagent	Conditions	Product(s) and Yield(s) (%)	Refs.
C_1				
CH_2O	EtCH(Tos)N=C	t-BuOK, THF, rt, 2 h	(81)	221
(steroid, MeO-aromatic, C=N-Tos)		KOH, THF, MeOH, rt, 1 h	(90)	70
(steroid, dienone, C=N-Tos)		KOH, THF, MeOH, rt, 1 h	(81)	70
(steroid, diene, MeO, C=N-Tos)		KOH, THF, MeOH, rt, 1 h	(91)	70
(steroid, enol ether MeO, C=N-Tos)		KOH, THF, MeOH, rt, 1 h	(96)	70

504

		NaOH (50%), BTEAC, C₆H₆, rt, 2 h	(87)	70
		KOH, THF, rt, 3 min	" (95)	70
		NaOH (50%), BTEAC, C₆H₆, MeOH, rt, 2 h	(90)	71, 70
		KOH, THF, MeOH, rt, 1 h	" (94)	70
		KOH, THF, MeOH, rt, 1 h	(84)	70
		KOH, THF, MeOH, rt, 1 h	(72)	70
C₂				
MeCHO	TosCH₂N=C	K₂CO₃, MeOH, rt, 10 min	(77)	26

TABLE II. OXAZOLINES FROM KETONES OR ALDEHYDES (Continued)

Substrate	Reagent	Conditions	Product(s) and Yield(s) (%)	Refs.
C2,9	EtCH(Tos)N=C	t-BuOK, THF, rt, 2 h	(>75)	221
	(steroid C=N-Tos)	KOH, THF, MeOH, rt, 1 h	mixture of diastereomers	70
RCHO	TosCH₂N=C	TfOAg (cat.), CH₂Cl₂, rt, X h	(4R,5R)-(+)-	75

R	X	n = 1	n = 2
Me	2	(94) 83% ee	(93) 75% ee
(E)-MeCH=CH	9	(96) 85% ee	(95) 83% ee
i-Pr	2	(94) 86% ee	(91) 79% ee
t-Bu	7	(93) 80% ee	(97) 85% ee
4-ClC₆H₄	1	(94) 73% ee	(94) 77% ee
Ph	2	(92) 77% ee	(96) 83% ee
4-MeOC₆H₄	2	(96) 74% ee	(95) 77% ee
3,4-(MeO)₂C₆H₃	2	(91) 80% ee	(97) 73% ee

C₃ CH₂=CHCHO	TosCH₂N=C	K₂CO₃, MeOH, rt, 2.5 h	(53)	26

TABLE II. OXAZOLINES FROM KETONES OR ALDEHYDES (*Continued*)

Substrate	Reagent	Conditions	Product(s) and Yield(s) (%)	Refs.
n-BuCHO	EtCH(Tos)N=C	*t*-BuOK, THF, rt, 2 h	(Tos, Et, *n*-Bu oxazoline) (>75)	221
C₆ MeO₂C(CH₂)₃CHO	TosCH₂N=C	NaCN, EtOH, rt, 25 min	(Tos, MeO₂C-chain oxazoline) (90)	221
cyclohexanone	TosCH₂N=C	EtOTl, DME, EtOH, rt	(EtO spiro-cyclohexane oxazoline) (72)	222
cyclohexanone	TosCH₂N=C	KOH, DME, rt, 2 h	(Tos spiro-cyclohexane oxazoline) (60)	77
t-Bu ketone	TosCH₂N=C	*t*-BuOLi, THF, −50°, 1 h	(Tos, *t*-Bu, Me oxazoline) (99)	55
t-Bu ketone	TosCH₂N=C	*n*-BuLi, THF, −50°, 1 h	(Tos, *t*-Bu, OH-C(Bu-*t*)) (99)	55
t-Bu ketone	TosCH₂N=C	EtOTl, DME, EtOH, rt	**I** (EtO, *t*-Bu oxazoline) + **II** (EtO, *t*-Bu oxazoline); **I** + **II** (35); **I:II** = 4:1	222

C₇				
4-O₂NC₆H₄CHO	TosCH₂N=C	K₂CO₃, MeOH, rt, 0.5 h	[Tos, 4-O₂NC₆H₄ oxazoline] (77)	223, 26
PhCHO	TosCH₂N=C	NaCN, EtOH, THF, 15°, 45 min	[Tos, Ph oxazoline] **I** (94)	36
	TosCH₂N=C	NaCN, EtOH, rt, 25 min	**I** (92)	221
	TosCH₂N=C	[Ferrocenyl phosphine ligand], TfOAg (cat.), CH₂Cl₂, rt, 2 h	(4S,5S)-(−)-**I** (96), 44% ee	75
[norbornanone]	TosCH₂N=C	EtOTl, DME, EtOH, rt	**I** + **II** (62); **I**:**II** = 4:1 [spiro oxazoline EtO products]	222
n-C₆H₁₃CHO	TosCH₂N=C	NaOH, MeOH, rt, 2 h	**I** + **II** (88); **I**:**II** = 7:1 [MeO, n-C₆H₁₃ oxazoline]	31
n-Pr(C=O)Pr-n	TosCH₂N=C	EtOTl, DME, EtOH, rt	[EtO, n-Pr, n-Pr oxazoline] (75)	222

TABLE II. OXAZOLINES FROM KETONES OR ALDEHYDES (*Continued*)

Substrate	Reagent	Conditions	Product(s) and Yield(s) (%)	Refs.
C₇₋₈				
4-RC₆H₄CHO	BnCH₂CH(Tos)N=C	K₂CO₃, MeOH, rt, 5 min	Tos,Bn-oxazoline with 4-RC₆H₄: R = Cl (74); H (83); MeO (36)	69
C₈				
n-C₇H₁₅CHO	TosCH₂N=C	K₂CO₃, MeOH, rt, 10 min	Tos-oxazoline with n-C₇H₁₅ (78)	26
2-RC₆H₄CHO	TosCH₂N=C	K₂CO₃, MeOH	I (Tos, 2-RC₆H₄) + II (Tos, 2-RC₆H₄)	223
R = Me		0°, 6.5 h	I + II (42); I:II = 1:2	
Me		20°, 2 h	I + II (73); I:II = 6:94	
MeO		0°, 6 h	I + II (73); I:II = 7:3	
MeO		20°, 2 h	I + II (69); I:II = 9:11	
4-MeOC₆H₄CHO	EtCH(Tos)N=C	t-BuOK, THF, rt, 2 h	Et,Tos-oxazoline with 4-MeOC₆H₄ (82)	221
	TosCH₂N=C	NaCN, EtOH, rt, 25 min	Tos-oxazoline with 4-MeOC₆H₄ (93)	221
PhC(O)CH₃	(+)-menthyl SO₂CH₂N=C, Pr-i, 90% ee	NaOH (50%), BTEAC, C₆H₆, 15°, 3 h	Ph,menthylsulfonyl-oxazoline with Pr-i (60) 18% de	40

Substrate	Conditions	Product(s) and Yield(s) (%)	Refs.
(menthyl-O-CH₂CH₂-Pr-i)SO₂CH₂N=C (−), 100% ee	NaOH (50%), BTEAC, C₆H₆, 15°, 3 h	(menthyl-oxazoline-Ph) (74) 33% de	40
(2-(sec-BuO)C₆H₄)SO₂CH₂N=C (S)-(−), 50% ee	1. n-BuLi, THF, −60°, 2 h 2. −60 to 0°, 1 h	I (76) 40% de	40
(±)-(2-(sec-BuO)C₆H₄)SO₂CH₂N=C	NaOH (50%), BTEAC, C₆H₆, 15°, 3 h	I (98) 40% de	40
(±)-(2-(1-methyl-2-methoxyethoxy)C₆H₄)SO₂CH₂N=C	NaOH (50%), BTEAC, PhMe, 15°, 3 h	I (94) 20% de	40
"	1. n-BuLi, THF, −60°, 2 h 2. −60 to 0°, 1 h	I (57) 33% de	40
(±)-(2-(tetrahydrofuran-3-yloxy)C₆H₄)SO₂CH₂N=C	NaOH (50%), BTEAC, C₆H₆, 15°, 3 h	I (100) 7% de	40
"	NaOH (50%), BTEAC, PhMe, 15°, 3 h	I (66) 20% de	40
"	1. n-BuLi, THF, −60°, 2 h 2. −60 to 0°, 1 h	I (89) 40% de	40

TABLE II. OXAZOLINES FROM KETONES OR ALDEHYDES (*Continued*)

Substrate	Reagent	Conditions	Product(s) and Yield(s) (%)	Refs.
Ph-C(=O)-CF$_3$	TosCH$_2$N=C	(EtO)$_4$Ti, CH$_2$Cl$_2$, 0°, 1 h	**I + II** (100); **I:II** = 53:47	40
	(SO$_2$CH$_2$N=C, menthyl-Pr-i) (+) 90% ee	(EtO)$_4$Ti, CH$_2$Cl$_2$, 0°, 1 h	**I + II** (96), 18% de; **I:II** = 53:47	40
	"	1. Triton B, THF, 20°, 4 h 2. CH$_2$Cl$_2$, 0°, 1 h	**I + II** (90), 18% de; **I:II** = 66:34	40
	(tetrahydrofuranyloxy-phenyl-SO$_2$-CH$_2$N=C) (R)-(−) 47% ee	(EtO)$_4$Ti, CH$_2$Cl$_2$, 0°, 1 h	**I + II** (98), 41% de; **I:II** = 60:40	40
	(±)-Ph-S(=O)-CH$_2$N=C, NTcs	(EtO)$_4$Ti, CH$_2$Cl$_2$, 0°, 1 h	**I + II** (100), 80% de; **I:II** = 94:6	40

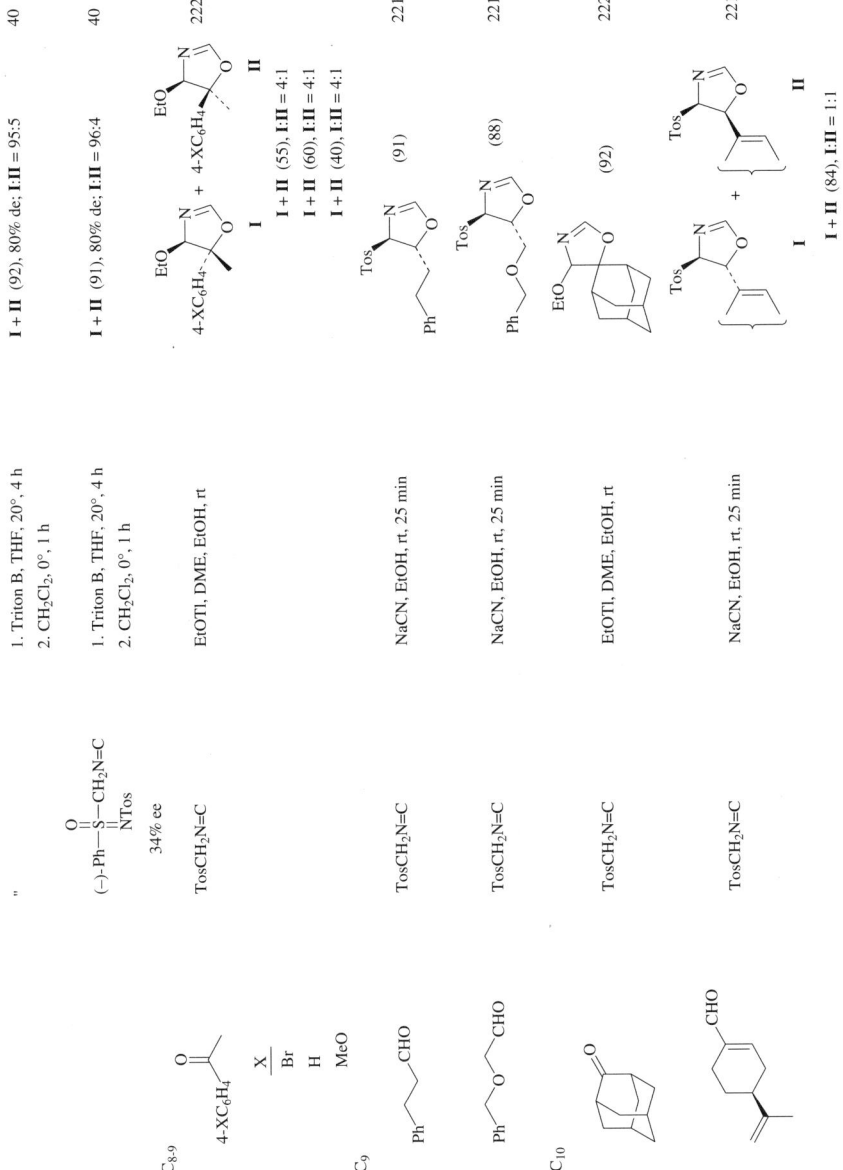

TABLE II. OXAZOLINES FROM KETONES OR ALDEHYDES (*Continued*)

Substrate	Reagent	Conditions	Product(s) and Yield(s) (%)	Refs.
(S)- citronellal-type CHO	TosCH$_2$N=C	NaCN, EtOH, rt, 25 min	**I** (Tos, oxazoline with isoprenyl chain) + **II** (epimer); **I + II** (83); **I:II** = 1:1	221
C$_{11}$ arene-Cr(CO)$_3$ CHO with R	TosCH$_2$N=C	K$_2$CO$_3$, MeOH, 0.5 h	**I** (Tos-oxazoline, Cr(CO)$_3$-arene) + **II**	223
R = Me		0°	**II** (90)	223
Me		20°	**I + II** (90); **I:II** = 2:1	223
Me, (R)-(−)		0°	(a,R,4S,5R)-(−)-**I** (95)	74
OMe		0°	**I + II** (90); **I:II** = 99:1	223
OMe		20°	**I + II** (90); **I:II** = 17:3	223
C$_{20}$ steroid (MeO, =O)	TosCH$_2$N=C	t-BuOLi, THF, −45°, 3 h	spiro-oxazoline steroid (98) ratio 1:1	54

514

TABLE III-A. OXAZOLES FROM ALDEHYDES

Substrate	Reagent	Conditions	Product(s) and Yield(s) (%)	Refs.
C_1 (CH$_2$O)$_n$	[steroid with C=N–Tos, MeO-substituted]	Na$_2$CO$_3$, MeOH, reflux, 5 h	[steroid–oxazole] **I** (96)	70
	[steroid with C=N–Tos, MeO-substituted]	Na$_2$CO$_3$, MeOH, reflux, 7 h	**I** (94)	181
	[aromatic steroid with C=N–Tos, MeO-substituted]	Na$_2$CO$_3$, MeOH, reflux, 4 h	[aromatic steroid–oxazole] (71)	181
C_2 [triazole-CHO]	TosCH$_2$N=C	K$_2$CO$_3$, MeOH, reflux, 2 h	[triazole–oxazole] (87)	81
C_3 [imidazole-CHO]	TosCH$_2$N=C	K$_2$CO$_3$, MeOH, reflux, 2 h	[imidazole–oxazole] (87)	81
C_4 [pyrazole-CHO]	TosCH$_2$N=C	K$_2$CO$_3$, MeOH, reflux, 2 h	[pyrazole–oxazole] (44)	81

TABLE III-A. OXAZOLES FROM ALDEHYDES (Continued)

Substrate	Reagent	Conditions	Product(s) and Yield(s) (%)	Refs.
imidazole-2-CHO	TosCH$_2$N=C	K$_2$CO$_3$, MeOH, reflux, 2 h	(0)[a]	81
Me$_2$N-N=CH-CHO	cyclohexylidene(Tos)(N=C)	t-BuOK, THF, −50° to rt, 1.5 h	NMe$_2$-substituted oxazole (75)	63
	Ph-C(Et)=C(Tos)-N=C	1. t-BuOK, THF, 0° 2. rt, 4 h	(80)	63
C$_5$ thiophene-3-CHO	TosCH$_2$N=C	K$_2$CO$_3$, MeOH, reflux, 24 h	(56)	218
3-methylisoxazole-5-CHO	TosCH$_2$N=C	K$_2$CO$_3$, MeOH, reflux, 1 h	(65)	224
MeO$_2$C-CH=CH-CHO	cyclohexylidene(Tos)(N=C)	t-BuOK, THF, −50° to rt, 1.5 h	CO$_2$Me-substituted oxazole (71)	63
3-methylpyrazole-5-CHO	TosCH$_2$N=C	K$_2$CO$_3$, MeOH, reflux, 2 h	(46)	81

X	R¹	Reagent	Conditions	R²	(%)	Ref.
NH	H	TosCH₂N=C	K₂CO₃, MeOH, reflux, 2 h	H	(0)[b]	78
O	H	TosCH₂N=C	K₂CO₃, MeOH, reflux, 2 h	H	(81)	78
O	H	MeCH(Tos)N=C	t-BuOK, MeOH, 40°, 20 min	Me	(64)	69
O	NO₂	TosCH₂N=C	K₂CO₃, MeOH, reflux, 30 min	H	(83)	78
O	NO₂	BnCH(Tos)N=C	K₂CO₃, MeOH, reflux, 30 min	Bn	(62)	69
S	H	TosCH₂N=C	K₂CO₃, MeOH, reflux, 2 h	H	(79)	78
S	H	TosCH₂N=C	K₂CO₃, MeOH, reflux, 24 h	H	(77)	218
S	H	BnCH(Tos)N=C	K₂CO₃, MeOH, reflux, 1 h	Bn	(71)	69
S	NO₂	TosCH₂N=C	K₂CO₃, MeOH, reflux, 1 h	H	(68)	78

C₆

				X		Ref.
		TosCH(Me)N=C	EtONa, EtOH, reflux, 2 h	O	(48)	225
				S	(76)	226
			EtONa, EtOH, reflux, 2 h	O	(15)	225
				S	(27)	226
				O	(16)	225
				S	(4)	226

TABLE III-A. OXAZOLES FROM ALDEHYDES (*Continued*)

Substrate	Reagent	Conditions	Product(s) and Yield(s) (%)	Refs.
OHC–(thiophene)–CHO	TosCH$_2$N=C	K$_2$CO$_3$, MeOH, reflux, 24 h	oxazole–thiophene–oxazole (56)	218
(pyridine)–CHO	TosCH$_2$N=C	K$_2$CO$_3$, MeOH, reflux, X h	oxazole–pyridine	78
		$\dfrac{X}{2}$	(82)	
		3	(80)	
		3	(67)	
N-Me pyrrole–CHO	TosCH$_2$N=C	K$_2$CO$_3$, MeOH, reflux, 3 h	oxazole–(N-Me pyrrole) (47)	78
EtO–CH(OEt)–CHO	cyclohexylidene(Tos)N=C	t-BuOK, THF, –80° to rt, 2 h	oxazole with CH(OEt)$_2$ and cyclohexenyl (91)	63
R^{1}CHO	TosCH(R^2)N=C		$\begin{array}{c}R^2\\ \diagup\!\!\diagdown \\ R^1\end{array}$ oxazole	
R^1	R^2			
4-O$_2$NC$_6$H$_4$	H	K$_2$CO$_3$, MeOH, reflux, 2 h	(91)	26
4-ClC$_6$H$_4$	H	K$_2$CO$_3$, MeOH, reflux, 2 h	(57)	26
	Me	K$_2$CO$_3$, MeOH, reflux, 2 h	(74)	69
	Et	t-BuOK, MeOH, 40°	(73)	69
	Bn	K$_2$CO$_3$, MeOH, reflux, 2 h	(81)	69

C$_7$

R¹	R²		Conditions	Product (%)	Refs.
Ph	H		K₂CO₃, MeOH, reflux, 2 h	(91)	26
	Me		t-BuOK, MeOH, 40°, 20 min	(75)	69
n-C₆H₁₃	Me		K₂CO₃, MeOH, reflux, 1 h	(78)	73
	Et		K₂CO₃, MeOH, reflux, 1 h	(82)	69
	H		KOH, MeOH, reflux, 4 days	(13)	31
	Me		KOH, Et₂O, rt, 3 h	(72)	31

TABLE III-A. OXAZOLES FROM ALDEHYDES (*Continued*)

Substrate	Reagent	Conditions	Product(s) and Yield(s) (%)	Refs.
		EtONa, EtOH, reflux, 2 h	(2) + (0.8)	84
		EtONa, EtOH, reflux, 2 h	(74)	84
		EtONa, EtOH, reflux, 2 h	(84)	84

Aldehyde	Isocyanide	Conditions	Product (Yield %)
1,3-benzenedicarbaldehyde	TosCH(Me)N=C	EtONa, EtOH, reflux, 2 h	(45), 84
	1,2-bis(TosCH=N-CH₂)benzene	EtONa, EtOH, reflux, 2 h	(84), 84
	1,3-bis(TosCH=N-CH₂)benzene	EtONa, EtOH, reflux, 2 h	(77), 84
	1,4-bis(TosCH=N-CH₂)benzene	EtONa, EtOH, reflux, 2 h	(52) + (15), 84

TABLE III-A. OXAZOLES FROM ALDEHYDES (Continued)

Substrate	Reagent	Conditions	Product(s) and Yield(s) (%)	Refs.
OHC–C₆H₄–CHO	TosCH(Me)N=C	EtONa, EtOH, reflux, 2 h	(33)	84
	bis(TosCH₂–N=C) m-xylylene	EtONa, EtOH, reflux, 2 h	(85)	84
C₉ indole-2-CHO	TosCH₂N=C	K₂CO₃, MeOH, reflux, 2 h	(74)	81
Ph-CH=CH-CHO	Tos-C(=cycloalkylidene)-N=C	t-BuOK, THF, −78 to 20°, 2 h	n=1 (90); n=2 (90)	80
3,4-(MeO)₂C₆H₃CHO	TosCH₂N=C	K₂CO₃, MeOH, reflux, 2 h	(90)	227
1-(t-Bu)pyrrole-3-CHO	TosCH₂N=C	Na₂CO₃, MeOH, reflux, 24 h	(62)	65

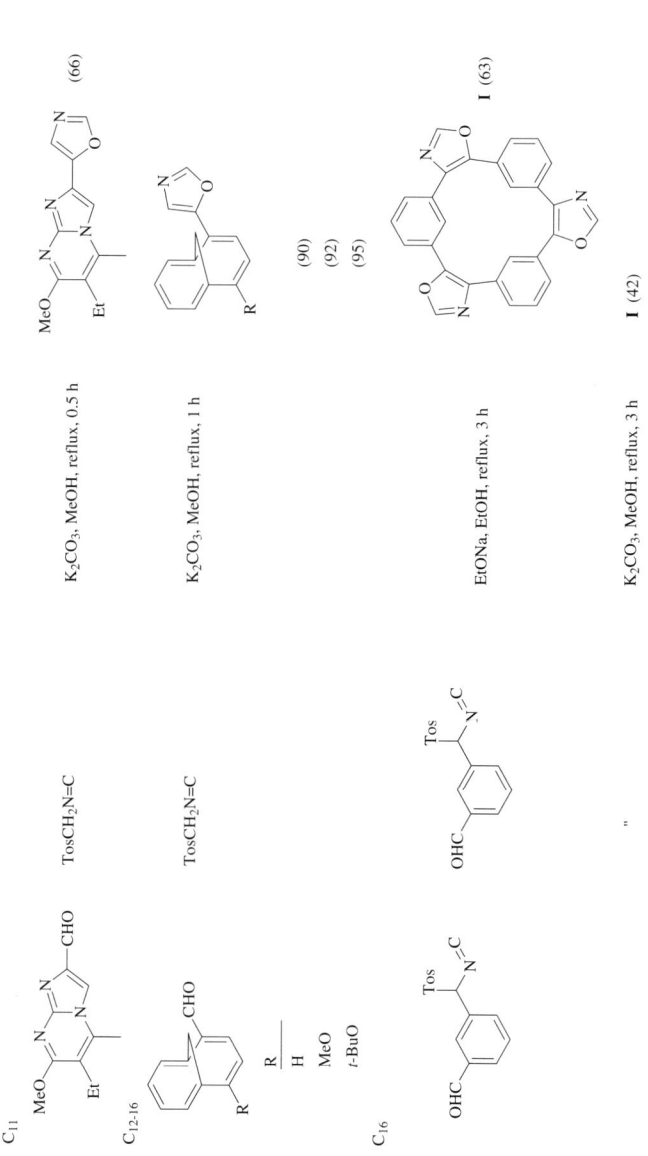

[a] See p. 440 (Eq. 46).
[b] 3-Tosylpyrrolo[1,2-c]pyrimidine was formed in ~20% yield instead of an oxazole.

TABLE III-B. TOSYLOXAZOLES FROM CARBOXYLIC ACID DERIVATIVES

Substrate	Reagent	Conditions	Product(s) and Yield(s) (%)		Refs.
			Tos─⟨N═⟩─O with R		
			R		
C₄ (Ac₂O)	TosCH₂N=C	KOH, DME, rt, 3 h	Me	(66)	26
	TosCH₂N=C	n-BuLi, THF, −60 to 20°, 2 h	Me	(73)	26
C₇ (CH₃CH=CHCH₂CO₂Me-type)	TosCH₂N=C	n-BuLi (2 eq), THF, −70 to 0°, 2 h	CH=CH-CH₂	(25)	79
PhCOCl	TosCH₂N=C	KOH, DME, rt, 3 h	Ph	(57)	26
	TosCH₂N=C	n-BuLi, THF, −60 to 20°, 2 h	Ph	(65)	26
C₈ (phthalic anhydride)	TosCH₂N=C	KOH, DME, rt, 3 h	2-HO₂CC₆H₄	(47)	26
PhCON=C=S	TosCH₂N=C	n-BuLi (1 eq), THF, −65°, 0.5 h; 0°, 2 h	Ph	(70)	120
		DBU, Cu₂O, rt, 20 h			
n-C₆H₁₃C(O)SeMe	TosCH₂N=C	n-BuLi (2 eq), THF, −70 to 0°, 20 min	n-C₆H₁₃	(40)	230
C₉ PhCO₂Et	TosCH₂N=C	n-BuLi (2 eq), THF, −70 to 0°, 20 min	Ph	(70)	79
	TosCH₂N=C	n-BuLi (2 eq), THF, −70 to 20°, 1.5 h	Ph	(0)	79
C₁₀ (Ph-CH=CH-CH₂-CO₂Me)	TosCH₂N=C	n-BuLi (2 eq), THF, −70 to 0°, 20 min	Ph-CH=CH-CH₂	(53)	79
C₁₂ (Ph-CH=CH-CH=CH-CH₂-CO₂Me)	TosCH₂N=C	n-BuLi (2 eq), THF, −70 to 0°, 20 min	Ph-CH=CH-CH=CH-CH₂	(54)	79

TABLE IV. 1-FORMAMIDO-1-TOSYLALKENES FROM ALDEHYDES OR KETONES

Substrate	Conditions	Product(s) and Yield(s) (%)	Refs.
C_1 CH$_2$O	1. TosCH$_2$N=C, t-BuOK, DME, –60°, 1 h 2. AcOH	Tos\\C=CH$_2$/NHCHO (60)	86
C_2 MeCHO	TosCH$_2$N=C, t-BuOK, THF, –70°, 15 min	NHCHO / Tos / Me (41)	63
C_3 acetone	1. TosCH$_2$N=C, t-BuOK, THF, –10°, 5 min 2. AcOH	Tos / NHCHO / Me$_2$C= (83)	36, 85
C_4 imidazole-2-CHO	TosCH$_2$N=C, K$_2$CO$_3$, MeOH, rt, 2 h	imidazolyl-CH=C(NHCHO)Tos (26)	81
i-PrCHO	1. TosCH$_2$N=C, t-BuOK, THF, –10°, 5 min 2. AcOH	i-Pr-CH=C(NHCHO)Tos (10)	36, 85
C_5 t-BuCHO	TosCH$_2$N=C, t-BuOK, DME, –30°, 30 min	t-Bu-CH=C(NHCHO)Tos **I** (65)	86
t-BuCHO	1. TosCH$_2$N=C, t-BuOK, THF, –10°, 5 min 2. AcOH	**I** (63)	36, 85
n-BuCHO	TosCH$_2$N=C, t-BuOK, DME, –40°, 10 min	n-Bu-CH=C(NHCHO)Tos (67)	136
C_6 cyclohexanone	1. TosCH$_2$N=C, t-BuOK, THF, –10°, 5 min 2. AcOH	cyclohexylidene=C(NHCHO)Tos **I** (61)	36, 85
cyclohexanone	TosCH$_2$N=C, t-BuOK, THF, –70 to –40°, 30 min	**I** (92)	63

TABLE IV. 1-FORMAMIDO-1-TOSYLALKENES FROM ALDEHYDES OR KETONES (*Continued*)

Substrate	Conditions	Product(s) and Yield(s) (%)	Refs.
C_7 *t*-Bu-CO-CH$_3$	TosCH$_2$N=C, *t*-BuOK, DME, −30°, 30 min	**I** (80)	86
	1. TosCH$_2$N=C, *t*-BuOLi, THF, −50°, 1 h 2. *t*-BuOK, THF, −50°, 10 min	*t*-Bu−C(Tos)=C=N (71)	55
2,3-Cl$_2$C$_6$H$_3$CHO	TosCH$_2$N=C, *t*-BuOK, DME, −40°, 10 min	2,3-Cl$_2$C$_6$H$_3$−CH=C(Tos)(NHCHO) (85)	136
4-R-C$_6$H$_4$CHO	TosCH$_2$N=C, *t*-BuOK, DME	4-R-C$_6$H$_4$−CH=C(Tos)(NHCHO)	
R = Br	−30°, 30 min	(79)	114
R = NO$_2$	−50°, 1 h	(77)	86
R = H	−35°, 30 min	(87)	86
PhCHO	1. Ph−S−CH$_2$−N=C, *t*-BuOK, THF, −10°, 5 min, O$_2$ 2. AcOH	Ph−CH=C(SO$_2$Ph)(NHCHO)^a (73)	36, 85
C_8 Ph-CO-CH$_3$	1. TosCH$_2$N=C, *t*-BuOK, THF, −10°, 5 min 2. AcOH	Ph−C(CH$_3$)=C(Tos)(NHCHO) (42)	36, 85
C_9 Ph-CH=CH-CHO	1. TosCH$_2$N=C, *t*-BuOK, THF, −10°, 5 min 2. AcOH	Ph−CH=CH−CH=C(Tos)(NHCHO) (73)	36, 85
Ph-CO-Et	TosCH$_2$N=C, *t*-BuOK, THF, −70 to −40°, 30 min	Ph−C(Et)=C(Tos)(NHCHO) (69)	63

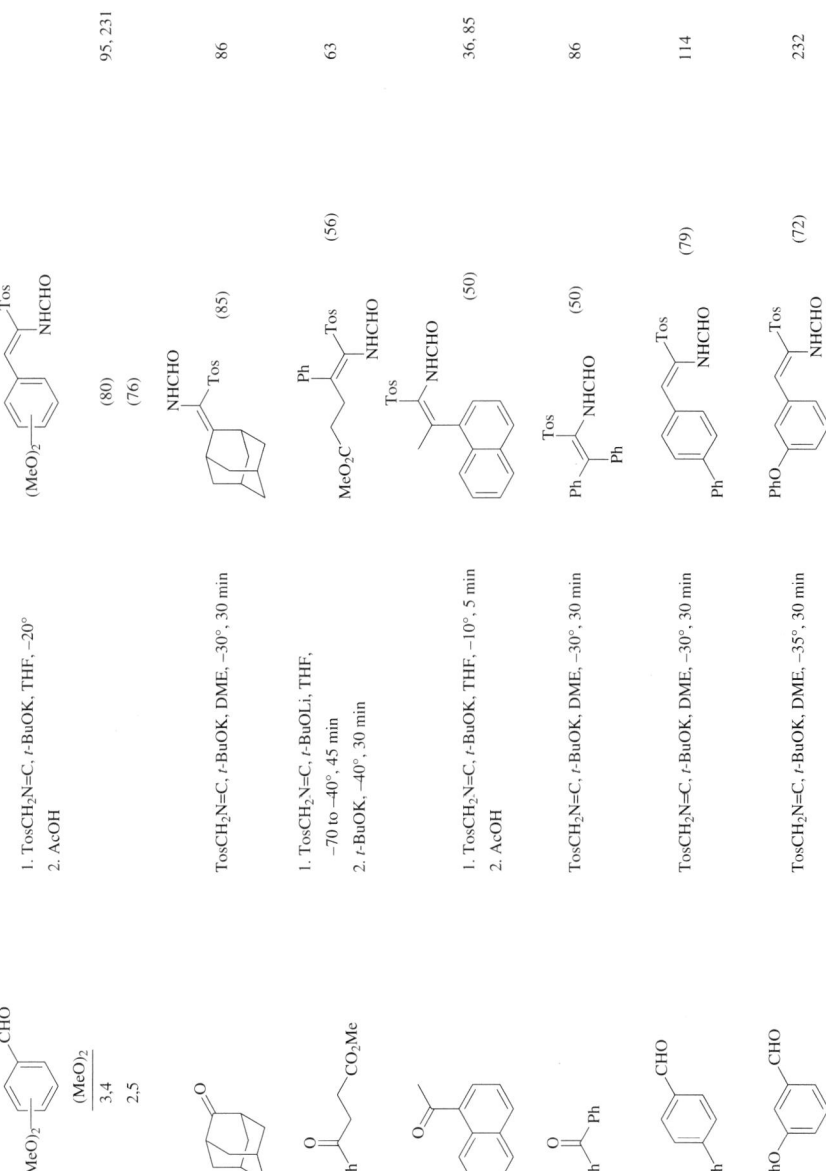

TABLE IV. 1-FORMAMIDO-1-TOSYLALKENES FROM ALDEHYDES OR KETONES (Continued)

Substrate	Conditions	Product(s) and Yield(s) (%)	Refs.
C₁₄	TosCH₂N=C, t-BuOK, THF, −60°, 20 min	(90)	63
C₁₅	1. TosCH₂N=C, t-BuOK, THF, −20° 2. AcOH	(90)	96
	TosCH₂N=C, t-BuOK, DME, −10°, 30 min	(—)	233
C₁₉	TosCH₂N=C, t-BuOK, DME	(77)	234
C₂₀	1. TosCH₂N=C, t-BuOLi, THF, −40°, 1.5 h 2. t-BuOK, −40°, 30 min	(>94)	54
	TosCH₂N=C, t-BuOLi, THF, −75°, 5 h	(29)	54

528

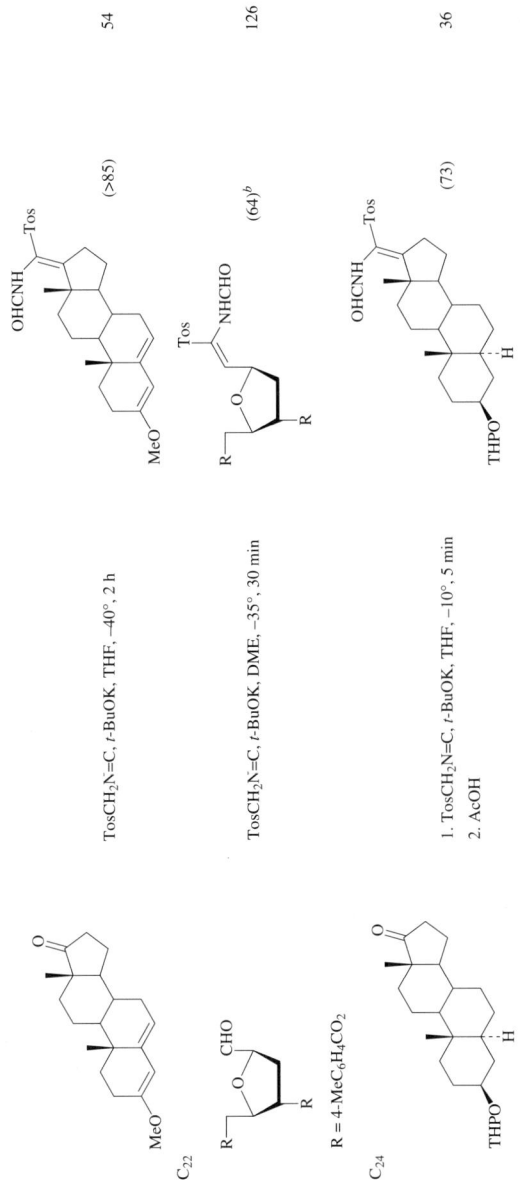

[a] Phenylsulfonylmethyl isocyanide was used instead of TosMIC.
[b] The tosyl group in Scheme 2 of ref. 128 is mistakenly depicted as TsO.

TABLE V.A. 1-ISOCYANO-1-TOSYLALKENES BY DEHYDRATION OF 1-FORMAMIDO-1-TOSYLALKENES

Substrate	Conditions	Product(s) and Yield(s) (%)	Refs.
C_{10} Tos-C(=CH$_2$)-NHCHO	POCl$_3$, TEA, DME, −10°, 1.5 h	Tos-C(=CH$_2$)-N=C (—)[a]	86
C_{11} NHCHO / Tos (cis-propenyl)	POCl$_3$, DPA, THF, −35 to −20°, 30 min	N=C / Tos (—)[b]	63
C_{12} Tos, NHCHO on isobutenyl	POCl$_3$, DPA, THF, −35 to −20°, 30 min	Tos, N=C on isobutenyl (89)	63, 86
C_{14} NHCHO, Tos, R (R = t-Bu; n-Bu)	POCl$_3$, TEA, DME, 0°, 1 h; THF, −30°, 10 min	N=C, Tos, R (77) (50)	86; 136
C_{15} Tos, NHCHO cyclohexylidene	POCl$_3$, DPA, THF, −35 to −20°, 30 min	Tos, N=C cyclohexylidene (90)	63
Tos / NHCHO, t-Bu methyl alkene	POCl$_3$, TEA, THF, 0°, 45 min	Tos / N=C, t-Bu methyl alkene (70)	55
Tos / NHCHO, t-Bu vinyl	POCl$_3$, TEA, DME, 0°, 1 h, chromatography, Al$_2$O$_3$	Tos / N=C, t-Bu vinyl (64)	55

Substrate	Conditions	Product(s) and Yield(s) (%)	Refs.

C16 aryl-substituted vinyl formamide (Tos, NHCHO on vinyl; R¹, R² on aryl)

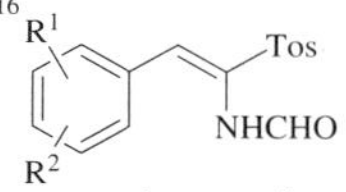

POCl₃, TEA, DME

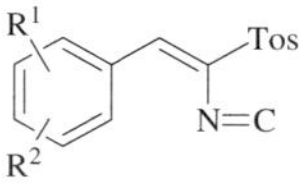

R¹	R²			
2-Cl	3-Cl	−20°, 1 h	(52)	136
4-Br	H	0°, 1 h	(—)	114
4-O₂N	H	0°, 0.5 h	(55)	86
H	H	0°, 1 h	(54)	86

C17-18 (R, Tos, NHCHO vinyl on phenyl)

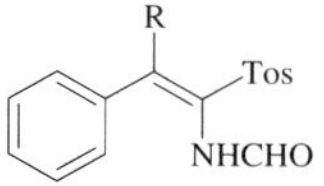

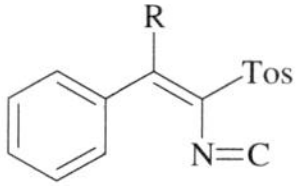

POCl₃, DPA, THF, −35 to −20°, 30 min

R		
Me	(49)	
Et	(89)	63

C18-24 (Tos, NHCHO vinyl on aryl with R, OMe)

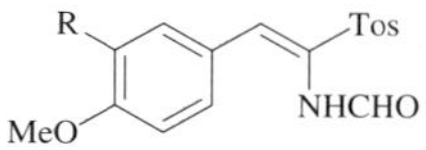

POCl₃, TEA, CH₂Cl₂, rt, 17 h

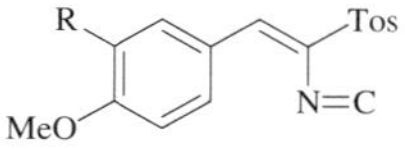

R		
MeO	(80)	95
BnO	(90)	96

C20 MeO₂C-substituted alkene (Ph, Tos, NHCHO)

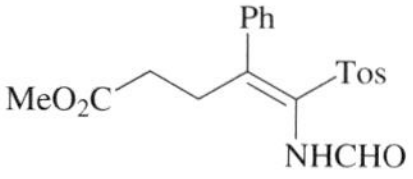

POCl₃, DPA, THF, −35 to −20°, 30 min

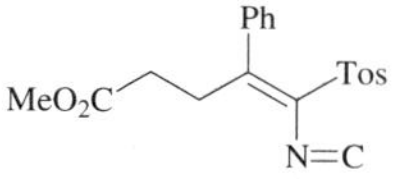

(—)[a,b] 63

C22 Ph, Ph, NHCHO, Tos vinyl

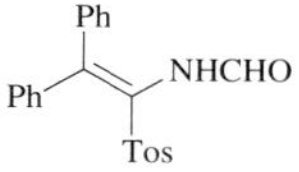

POCl₃, TEA, DME, 0°, 1 h

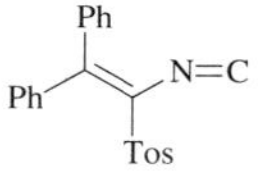

(68) 86

TABLE VA. 1-ISOCYANO-1-TOSYLALKENES BY DEHYDRATION OF 1-FORMAMIDO-1-TOSYLALKENES (*Continued*)

Substrate	Conditions	Product(s) and Yield(s) (%)	Refs.
C_{28}	$POCl_3$, TEA	(>55)[a]	234
C_{29}	$POCl_3$, DPA, THF, 0°, 1 h	(94)	54
C_{29}	$POCl_3$, DPA, THF, 0°, 1 h	(85)	54
C_{31}	$POCl_3$, TEA, DME, −5°, 1 h	(42)[c]	126

$R = 4\text{-MeC}_6H_4CO_2$

[a] The product, an unstable oil, was converted directly to the imidazole.
[b] The isocyanide was not isolated but was used in solution, owing to its instability.
[c] The tosyl group in Scheme 2 of ref 128 is mistakenly depicted as TsO.

TABLE VB. 1-ISOCYANO-1-TOSYLALKENES BY IN SITU DEHYDRATION OF 1-FORMAMIDO-1-TOSYLALKENES

Substrate	Reagent	Conditions	Product(s) and Yield(s) (%)	Refs.
C_6 cyclohexanone	$TosCH_2N=C$	1. t-BuOK, DME, −30°, 0.5 h 2. AcOH 3. $POCl_3$, TEA, 0°, 1 h	cyclohexylidene(Tos)C=N (58)	86
C_{10} adamantanone	$TosCH_2N=C$	1. t-BuOK, DME, −30°, 0.5 h 2. AcOH 3. $POCl_3$, TEA, 0°, 1 h	adamantylidene(Tos)C=N (64)	86
C_{11} octalinone	$TosCH_2N=C$	1. t-BuOK, THF, −45°, 2 h 2. H_3PO_3 3. $POCl_3$, TEA, 0°, 1.5 h	Tos,N=C octalinylidene (86)	235
C_{11-14} decalinone with R^1, R^2: R^1 / R^2 H / H OH / H OH / Me Me / OH OTMS / H	$TosCH_2N=C$	1. t-BuOK, THF, −45°, 2 h 2. H_3PO_3 3. $POCl_3$, TEA, 0°, 1.5 h	R^2, R^1 decalinylidene(Tos)C=N (91) E:Z = 1:1 (99) E:Z = 2.5:1 (98) E:Z = 1:1 (97) E:Z = 1:1 (92) E:Z = 2:1	235
C_{19} estrone methyl ether	$TosCH_2N=C$	1. t-BuOK, THF, −45°, 2 h 2. H_3PO_3 3. $POCl_3$, TEA, 0°, 1.5 h	C=N–Tos estrone methyl ether derivative (70)	54

TABLE VB. 1-ISOCYANO-1-TOSYLALKENES BY IN SITU DEHYDRATION OF 1-FORMAMIDO-1-TOSYLALKENES (*Continued*)

Substrate	Reagent	Conditions	Product(s) and Yield(s) (%)	Refs.
C₂₀ [steroid with dienone and ketone]	TosCH₂N=C and O₂S(MeO-C₆H₄)CH₂N=C	1. *t*-BuOK, THF, −45°, 2 h 2. H₃PO₃ 3. POCl₃, TEA, 0°, 1.5 h	[steroid C=N-R] (59) Tos; (84) with R = S(O₂)-C₆H₄-OMe	54
[steroid with MeO and ketone]	TosCH₂N=C	1. *t*-BuOK, THF, −45°, 2 h 2. H₃PO₃ 3. POCl₃, TEA, 0°, 1.5 h	[steroid C=N-Tos] (77)	54
[steroid with MeO, R¹, R², R³, R⁴ and ketone] R¹ R² R³ R⁴ H H H H H H H H H H H H	R⁵SO₂CH₂N=C R⁵: Me, *t*-Bu, Ph	1. *t*-BuOK, THF, −45°, 2 h 2. H₃PO₃ 3. POCl₃, TEA	[steroid C=N-SO₂R⁵] POCl₃, TEA 0°, 1.5 h (84) 0°, 1.5 h (4) 0°, 1.5 h (67)	54

R¹	R²	R³	R⁴	R⁵			
H	H	H	H	4-ClC₆H₄	(30)	0°, 1.5 h	
H	H	H	H	4-MeC₆H₄	(72)	0°, 1.5 h	
H	H	H	H	4-MeOC₆H₄	(89)	0°, 1.5 h	
H	H	H	H	n-C₁₀H₂₁	(74)	0°, 1.5 h	
H	H	H	Cl	4-MeC₆H₄	(56)	0°, 1.5 h	
H	H	OH	H	4-MeC₆H₄	(52)	0°, 1.5 h	
H	OH	H	H	4-MeOC₆H₄	(76)	−30°, 2.5 h	
H	OH	H	H	4-MeC₆H₄	(85)	0°, 1 h	
H	=O		H	4-MeC₆H₄	(71)	0°, 1 h	
OH	H	H	H	4-MeC₆H₄	(54)	0°, 1 h	
F	OH	H	H	4-MeC₆H₄	(64)	0°, 1 h	

TosCH₂N=C, 1. t-BuOK, THF, −40°, 40 min; 2. AcOH; 3. POCl₃, TEA, 0°, 50 min (70) 236

TosCH₂N=C, 1. t-BuOK, THF, −40°, 2 h; 2. H₃PO₃; 3. POCl₃, TEA, 0°, 1 h (81) 54

TosCH₂N=C, 1. t-BuOK, THF, −40°, 2 h; 2. H₃PO₃; 3. POCl₃, TEA, 0°, 1 h (70) 54

C₂₁

TABLE VB. 1-ISOCYANO-1-TOSYLALKENES BY IN SITU DEHYDRATION OF 1-FORMAMIDO-1-TOSYLALKENES (*Continued*)

Substrate	Reagent	Conditions	Product(s) and Yield(s) (%)	Refs.
C$_{23}$ (steroid ketone)	TosCH$_2$N=C	1. *t*-BuOK, THF, −40°, 2 h 2. H$_3$PO$_3$ 3. POCl$_3$, TEA, 0°, 1 h	(steroid with C=N–Tos) (63)	54

TABLE VC. 1-ISOCYANO-1-TOSYLALKENES FROM ALDEHYDES OR KETONES BY PETERSON OLEFINATION

Substrate	Reagent	Conditions	Product(s) and Yield(s) (%)	Refs.
C$_4$ cyclobutanone	TosC(Li)(TMS)N=C	THF, –60°, 10 min	Tos-C(=cyclobutylidene)N=C (64)	87
C$_5$ 5-R-2-thiophene/furan-CHO (R=NO$_2$, X=S; R=H, X=O; R=H, X=S)	TosC(Li)(TMS)N=C	THF, –80°, 10 min; –80 to –30°, 1 h; –70°, 10 min	vinyl-Tos/N=C on furan/thiophene (10), (88), (87)	87
C$_{7,9}$ 4-R-C$_6$H$_4$CHO (R = Cl, NO$_2$, H, MeO, Me$_2$N)	TosC(Li)(TMS)N=C	THF, –80 to –30°, 1 h	4-R-C$_6$H$_4$-CH=C(Tos)(N=C) (85), (55), (82), (93), (76)	87
C$_9$ cinnamaldehyde (PhCH=CHCHO)	TosC(Li)(TMS)N=C	THF, –95°, 0.5 h	PhCH=CH-CH=C(Tos)(N=C) **I** (83)	87
	TosC(Li)(TMS)N=C	THF, –80 to –30°, 30 min	**I** (>73)	34
C$_{10}$ citral (CHO)	TosC(Li)(TMS)N=C	THF, –95°, 0.5 h	geranyl/neryl-CH=C(Tos)(N=C) (90)	87

TABLE VI. MONOALKYLATION OF TOSMIC

Substrate (equiv)	Conditions	Product(s) and Yield(s) (%)	Refs.
C₁			
MeI (2)	NaOH (30%), BTEAC, CH₂Cl₂, 0°, 3 h	MeCH(Tos)N=C (95)[a]	90, 73
C₂			
EtBr (2)	NaOH (30%), TBAI, CH₂Cl₂, 0°, 2 h	EtCH(Tos)N=C (80)[a]	90
EtI (2)	NaOH (30%), TBAI, CH₂Cl₂, 0°, 2 h	EtCH(Tos)N=C (90)[a]	90
C₃			
⌒⌒Cl (2)	NaOH (30%), TBAI, CH₂Cl₂, 0°, 1.5 h	Tos\\N=C **I** (75)	90
(1.9)	NaOH (30%), BTEAC, CH₂Cl₂, 0°, 4 h	**I** (91)	91
i-PrBr (1.2)	NaH, DMSO, Et₂O, rt, 1 h	i-PrCH(Tos)N=C **I** (71)	73
i-PrCl (10)	NaOH (30%), TBAI, CH₂Cl₂, 5°, 80 h	**I** (40)	90
i-PrI (10)	NaOH (40%), TBAI, CH₂Cl₂, 6°, 80 h	**I** (42)	73
n-PrI (2)	NaOH (30%), TBAI, CH₂Cl₂, 20°, 4 h	n-PrCH(Tos)N=C (85)	90
C₄			
s-BuBr (1.2)	NaH, DMSO, Et₂O, rt, 1 h	s-BuCH(Tos)N=C (65)	73
t-BuBr (3)	NaH, DMSO, Et₂O, rt, 1 h	t-BuCH(Tos)N=C (0)[b]	73
t-BuI (3)	NaH, DMSO, Et₂O, rt, 1 h	t-BuCH(Tos)N=C (0)[b]	73
n-BuI (2)	NaOH (30%), TBAI, CH₂Cl₂, 25°, 4 h	n-BuCH(Tos)N=C (75)	90
TMSCH₂I (1.2)	NaOH (50%), TBAI, THF, rt, 1.5 h	TMSCH₂CH(Tos)N=C **I** (85)	93

C₅			
cyclopentyl-CH₂Br (1.05) (1.9)	NaOH (30%), BTEAC, CH₂Cl₂, 0°, 4 h	cyclopentyl-CH(N=C)(Tos) **I** (51) (58)	91
pentenyl-I (1.1)	NaH, DMSO, Et₂O, rt, 2 h	Tos-CH(N=C)-butenyl (78)	73
MeO₂C-pentyl-Br (1)	NaOH (40%), TBAI, CH₂Cl₂, rt, 5 h	MeO₂C-pentyl-CH(N=C)(Tos) (50)	63
C₆			
hexenyl-Br (—)	NaOH (25%), THAI, CH₂Cl₂, rt, 18 h	hexenyl-CH(N=C)(Tos) **I** (73)	160
hexenyl-I (1.1)	LDA, THF		163
t-BuO₂C-CH₂-Br (1.9)	NaOH (40%), TBAI, CH₂Cl₂, rt, 14 h	t-BuO₂C-CH₂-CH(N=C)(Tos) **I** (83) (51)	63
C₆H₁₁I (3)	NaOH (30%), BTEAC, CH₂Cl₂, 0°, 4 h	C₆H₁₁-CH(N=C)(Tos) (30)	91
Et-CH=CH-CH₂-I (1.1)	1. NaOH (30%), TBAB, CH₂Cl₂, 0°, 3 h 2. rt, 6 h	Et-CH=CH-CH₂-CH(Tos)(N=C) (80)	73
n-C₆H₁₃I (1.1)	1. NaOH (30%), TBAB, CH₂Cl₂, 0°, 3 h 2. rt, 6 h	n-C₆H₁₃-CH(Tos)(N=C) (80)	152
C₇			
BnBr (2)	NaOH (30%), TBAI, CH₂Cl₂, rt, 1.5 h	Bn-CH(Tos)(N=C) (80)	73, 90

TABLE VI. MONOALKYLATION OF TosMIC (Continued)

Substrate (equiv)	Conditions	Product(s) and Yield(s) (%)	Refs.
Pyridine-2,6-bis(CH2Br) (0.8)	NaOH (20%), TBAI, CH2Cl2, rt, 2 h, 4-RC6H4SO2CH2N=C instead of TosMIC	4-RC6H4O2S–CH(N=C)–CH2–(pyridine)–CH2–CH(N=C)–SO2C6H4R-4 (70); R = Et (—), EtO (—)	92
(E)-1-iodo-2-hexene (1)	NaOH (30%), n-Bu4NOH, CH2Cl2, 0°, 4 h	Et–CH=CH–CH2CH2–CH(Tos)(N=C) (70)	97
C8 1,2-bis(bromomethyl)benzene (0.5)	NaOH (7.5 N), TBAI, CH2Cl2, 5°, 4 h	1,2-C6H4[CH2CH(Tos)(N=C)]2 (48) + 2,2-disubstituted indane with Tos and N=C (26)	153
1,3-bis(bromomethyl)benzene (0.5)	NaOH (7.5 N), TBAI, CH2Cl2, 5°, 4 h	1,3-C6H4[CH2CH(Tos)(N=C)]2 **I** (70)	153
(same, 0.5)	NaOH (20%), TBAI, CH2Cl2, rt, 2 h	**I** (76)	92
1,4-bis(bromomethyl)benzene (0.5)	NaOH (7.5 N), TBAI, CH2Cl2, 5°, 4 h	1,4-C6H4[CH2CH(Tos)(N=C)]2 (67)	153
(same, 0.5)	NaOH (20%), TBAI, CH2Cl2, rt, 2 h, 4-EtC6H4SO2CH2N=C instead of TosMIC	1,4-C6H4[CH2CH(SO2C6H4Et-4)(N=C)]2 (—)	92
n-C8H17Br (1.2)	NaH, DMSO, Et2O, rt, 2 h	n-C8H17–CH(Tos)(N=C) **I** (71)	73

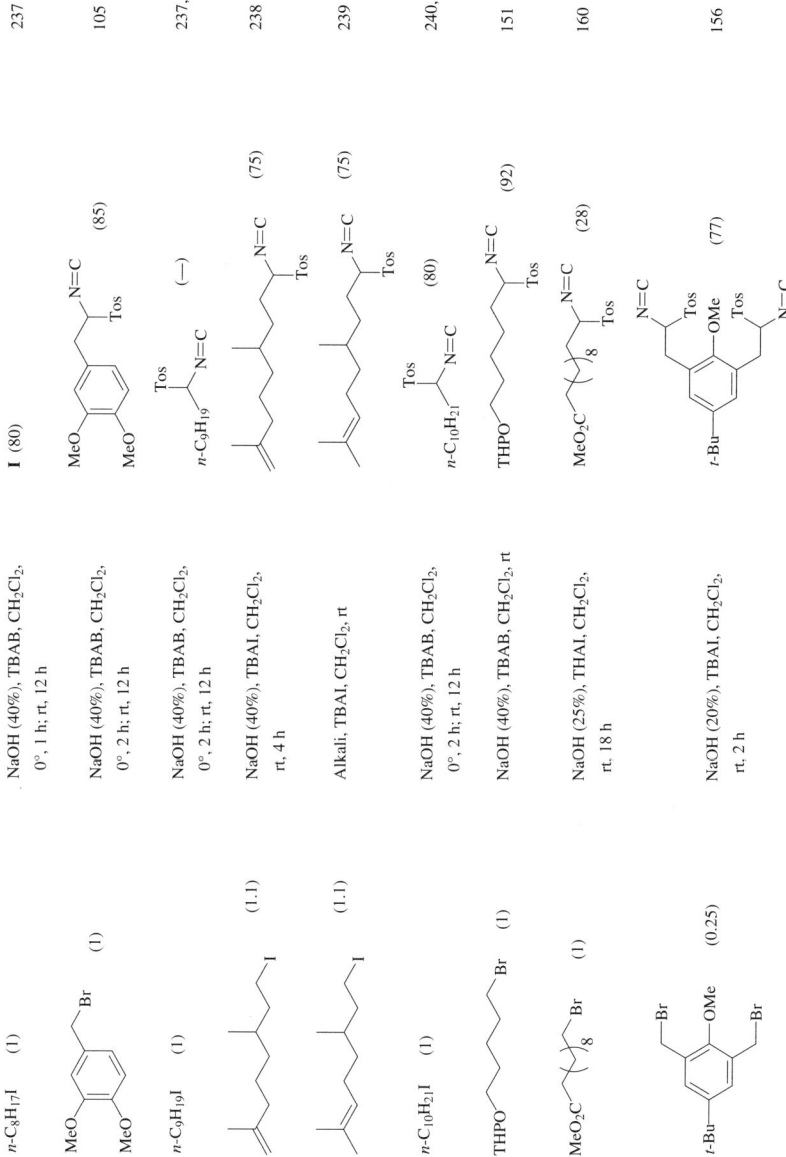

TABLE VI. MONOALKYLATION OF TOSMIC (*Continued*)

Substrate (equiv)	Conditions	Product(s) and Yield(s) (%)	Refs.
THPO-(CH₂)₆-I (1)	NaOH (40%), TBAB, CH₂Cl₂, 0°, 3 h; rt, 6 h	THPO-(CH₂)₆-CH(Tos)-N=C (80)	152
n-C₈H₁₇-C≡C-(CH₂)-I (—)	NaOH (40%), TBAB, CH₂Cl₂, 0°, 2 h; rt, 12 h	n-C₈H₁₇-C≡C-(CH₂)-CH(Tos)-N=C (78)	105
C₁₄: acetonide-OBn-CH₂-I (1)	NaOH (40%), TBAB, CH₂Cl₂, 0°, 2 h; rt, 12 h	acetonide-OBn-CH₂-CH(Tos)-N=C (85)	105
C₁₇: cyclopropylidene=N⁺(Bn)₂ TiCl₄OH⁻ (—)	LDA, THF, CH₂Cl₂, −78°, 1 h	cyclopropyl[N(Bn)₂][Tos]-N=C (25)	241
MeO₂C-(CH₂)₁₃-Br (1)	NaOH (25%), THAI, CH₂Cl₂, rt, 18 h	MeO₂C-(CH₂)₁₃-CH(Tos)-N=C (20)	160
C₁₈: n-C₁₈H₃₇X (1)	NaOH (40%), n-Bu₄NOH, CH₂Cl₂, 0°, 12 h	n-C₁₈H₃₇-CH(Tos)-N=C; X=Br (65), X=I (85)	242
C₁₉: n-C₁₉H₃₉I (1)	NaOH (40%), n-Bu₄NOH, CH₂Cl₂, rt, ~16 h	n-C₁₉H₃₉-CH(Tos)-N=C (19)	243
C₂₀: n-C₅H₁₁-CH=CH-(CH₂)₄-Br (1)	NaOH (25%), THAI, CH₂Cl₂, rt, 18 h	n-C₅H₁₁-CH=CH-(CH₂)₄-CH(Tos)-N=C (34)	160
n-C₈H₁₇-CH=CH-(CH₂)₇-Br (1)	NaOH (25%), THAI, CH₂Cl₂, rt, 18 h	n-C₈H₁₇-CH=CH-(CH₂)₇-CH(Tos)-N=C (33)	160

TABLE VI. MONOALKYLATION OF TOSMIC (Continued)

Substrate (equiv)	Conditions	Product(s) and Yield(s) (%)	Refs.
C$_{22}$			
n-C$_8$H$_{17}$—⧸═⧹—(CH$_2$)$_9$—Br	NaOH (25%), THAI, CH$_2$Cl$_2$, rt, 18 h	n-C$_8$H$_{17}$—⧸═⧹—(CH$_2$)$_9$—CH(Tos)—N=C (33)	160
n-C$_8$H$_{17}$—⧸═⧹—(CH$_2$)$_9$—I	NaOH (40%), TBAB, CH$_2$Cl$_2$, 0°, 2 h; rt, 12 h	n-C$_8$H$_{17}$—⧸═⧹—(CH$_2$)$_9$—CH(Tos)—N=C (75)	105

[a] This compound has also been prepared by the Mannich approach (Eq. 16).
[b] This compound has been obtained by a different method; see Eqs. 59 and 60.

TABLE VIIA. DISUBSTITUTED DERIVATIVES OF TOSMIC

Substrate (equiv)	Isocyanide	Conditions	Product(s) and Yield(s) (%)	Refs.
C_1				
MeI (2)	TosCH$_2$N=C	n-BuLi, THF, −70 to 0°, 0.5 h	Tos–C(Me)$_2$–N=C **I** (90)	79
(2)	TosCH$_2$N=C	NaH, DMSO, Et$_2$O, rt, 0.5 h	**I** (90)	244
(13)	TMS–CH(Tos)–N=C	t-BuOK, THF, −5° to rt, 1 h	TMS–C(Tos)(Me)–N=C (92)	93
(1.1)	Bn–CH(Tos)–N=C	NaOH (40%), TBAI, CH$_2$Cl$_2$, rt, 2 h	Bn–C(Tos)(Me)–N=C (71)	73
(1.5)	Et–CH=CH–CH$_2$CH$_2$–CH(Tos)–N=C	NaH, DMSO, Et$_2$O	Et–CH=CH–CH$_2$CH$_2$–C(Tos)(Me)–N=C (95)	97
(2)	(o-C$_6$H$_4$)(CH$_2$CH(Tos)N=C)$_2$	NaH, DMSO, rt, 3 h	(o-C$_6$H$_4$)(CH$_2$C(Me)(Tos)N=C)$_2$ (26)	83
(8)	MeO-steroid–CH(Tos)N=C	NaOH (50%), BTEAC, PhMe, 80°, 30 min	MeO-steroid–C(Tos)(Me)N=C (100)	45
C_2				
Br–CH$_2$CH$_2$–Br (1.2)	TosCH$_2$N=C	NaH, DMSO, Et$_2$O, rt, 1.5 h	cyclopropane-C(Tos)(N=C) **I** (81)	73
(—)	TosCH$_2$N=C	NaH, HMPA, THF, 0°	**I** (80)	99

	(—)	Tos—CH(N=C)— (iPr style, Tos/N=C)	NaH, DMSO, Et₂O, 35°, 1 h	Tos—C(CH₃)₂—N=C / N=C—C(CH₃)₂—Tos (18)	100
	Br—CH₂CH₂—Cl (4) (1.4)	Tos—CH₂—N=C	NaH, DMSO, Et₂O, 35°, 1 h	Tos—C(N=C)(CH₂CH₂Br) (42)	100
		Tos—CH₂—N=C	NaH, DMSO, Et₂O, 35°, 1 h	Tos—C(N=C)(CH₂CH₂Cl) (81)	100
	EtI (2)	TosCH₂N=C	NaH, DMSO, Et₂O, 20 to 35°, 2 h	Et₂C(Tos)(N=C) (80)	35
	(1.2)	Bn—CH(Tos)—N=C	NaH, DMSO, Et₂O, rt, 1 h	Et—C(Bn)(Tos)(N=C) (84)	73
	(6.5)	TMS—CH₂—CH(Tos)—N=C	t-BuOK, THF, −5° to rt, 3 h	Et—C(CH₂TMS)(Tos)(N=C) (81)	93
C₃	CH₂=CH—CH₂—Br (10)	TMS—CH₂—CH(Tos)—N=C	t-BuOK, THF, 0°, 2 h	TMS—CH₂—C(allyl)(Tos)(N=C) (79) / cyclobutane-Tos-N=C (71)	93 / 148
	Br—CH₂CH₂CH₂—Br (1.2)	TosCH₂N=C	NaH, DMSO, Et₂O, rt, 1 h		
	(0.5)	Tos—CH₂—N=C	NaH, DMSO, Et₂O, 35°, 1 h	Tos—C(N=C)(CH₃)(CH₂CH₂CH₂—) (91)	100
	(5)	Tos—CH₂—N=C	NaH, DMSO, Et₂O, 35°, 1 h	Tos—C(N=C)(CH₂CH₂CH₂Br)₂ type (53)	100
	(0.5)	Bn—CH(Tos)—N=C	NaH, DMSO, Et₂O, 35°, 1 h	Bn—C(N=C)—C(Bn)(Tos)(N=C) (70)	100

545

TABLE VIIA. DISUBSTITUTED DERIVATIVES OF TosMIC (*Continued*)

Substrate (equiv)	Isocyanide	Conditions	Product(s) and Yield(s) (%)	Refs.
Br~~~Cl (1.5)	Tos-CH(Me)-N=C	NaH, DMSO, Et₂O, 35°, 1 h	Tos-C(Me)(N=C)-~~~Cl (87)	100
Cl~~~Cl (0.5)	Tos-CH(Me)-N=C	NaH, DMSO, Et₂O, 35°, 1 h	Tos-C(Me)(N=C)-~~~-C(Me)(Tos)(N=C) (35)	100
i-PrI (2)	TosCH₂N=C	NaH, DMSO, Et₂O, rt, 1 h	*i*-Pr-C(Tos)(*i*-Pr)-N=C (>47)	98
(1.1) Bn-substrate	TosCH₂N=C	NaH, DMSO, Et₂O, rt, 1 h	Bn-C(Tos)(Bn)-N=C (>65)	98
C₄ pyridine N-oxide (—)	TosCH₂N=C	1. *n*-BuLi (2 eq), THF, −70 to 0°, 2 h 2. BnBr, 0°, 1 h	Bn-C(Tos)(propargyl)-N=C (52)	79
Br-CH(Me)-CHD-Br (—)	TosCH₂N=C	NaH, DMSO, Et₂O, rt	cyclobutane Tos, N=C, D (52)	245
Br~~~Br (1)	TosCH₂N=C	NaH, DMSO, Et₂O, rt, 1.5 h	cyclopentane-C(Tos)(N=C) (70)	73
X~~~X (0.5)	R-CH(Tos)-N=C	NaH, DMSO, Et₂O,	Tos-C(R)(N=C)-~~~-C(R)(Tos)(N=C)	100

X	R			
Br	Me	35°, 1 h	(89)	
I	Me	35°, 1 h	(94)	
Br	Ph	20-35°, 1 h	(17)	
I	Ph	20-35°, 1 h	(87)	
Br	Bn	20-35°, 1 h	(76)	

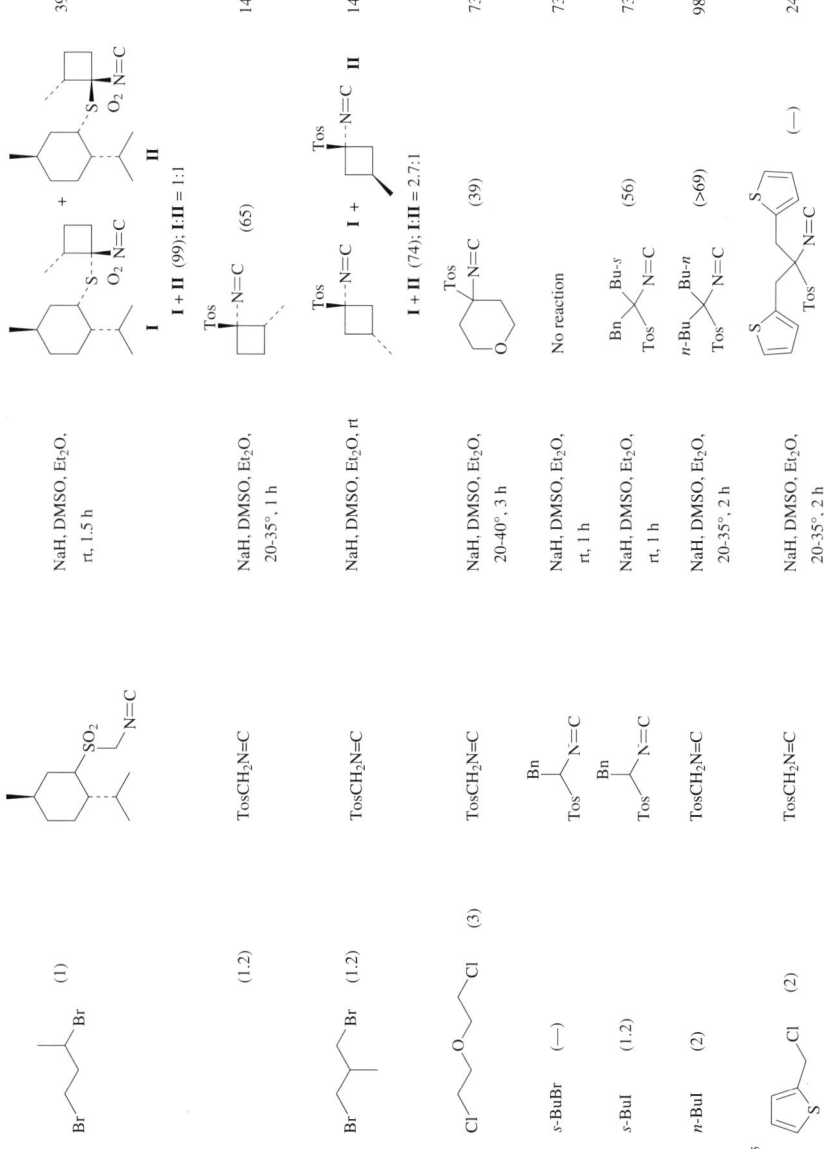

TABLE VIIA. DISUBSTITUTED DERIVATIVES OF TOSMIC (*Continued*)

Substrate (equiv)	Isocyanide	Conditions	Product(s) and Yield(s) (%)	Refs.
(—)	TosCH$_2$N=C	1. *n*-BuLi (2 eq), THF, −70 to 0°, 2 h 2. BnBr, 0°, 1 h	(64)	79
Br (1)		NaH, DMSO, Et$_2$O, 15°, 3 h	(>75)	152
Br Br (1.2)	TosCH$_2$N=C	NaH, DMSO, Et$_2$O, rt	(>20)	247
C$_6$ Cl X Cl (1.1)		sl. add. to NaOH (Y%), TBAI, CH$_2$Cl$_2$, reflux, t h Y t O 13 8 S 22 8.5	(>39) (>41)	92
Br Br (1.2)	TosCH$_2$N=C	NaH, DMSO, Et$_2$O, rt, 45 min	(83)	247
Br Br (1)	TosCH$_2$N=C	NaH, DMSO, Et$_2$O, rt, 1 h	(>51)	98
Br Br (1.2)	TosCH$_2$N=C	NaH, DMSO, Et$_2$O, rt, 1 h	(38)	247

C_7	Br-CH2CH2-C(CH3)(OCH2CH2O) (1)	TosCH(C6H13-n)N=C	NaH, DMSO, Et2O, 15°, 3 h	Tos-C(C6H13-n)(CH2CH2-C(CH3)(OCH2CH2O))-N=C (95)	152
		Et-CH=CH-CH2-CH(Tos)-N=C (1)	NaH, DMSO, Et2O, 15°, 3 h	Tos-C(CH2CH2CH=CH-Et)(C(CH3)(OCH2CH2O))-N=C (100)	152
	4-Cl-C6H4-CH2Cl (2)	TosCH2N=C	NaOH (40%), TBAI, CH2Cl2, rt, 2 h	(4-Cl-C6H4-CH2)2C(Tos)N=C (79), I (73, 98)	
	4-Cl-C6H4-CH2Br (—)	TosCH2N=C	NaOH (30%), TBAI, CH2Cl2	I (85)	161
		Bn-CH(Tos)N=C	NaOH (30%), TBAI, CH2Cl2	Bn-C(CH2-C6H4-Cl-4)(Tos)N=C (82), I	161
	4-Cl-C6H4-CH2Cl (2)	Bn-CH(Tos)N=C	NaH, DMSO, Et2O, rt, 1 h	I (66)	73
	BnBr (2)	TosCH2N=C	NaOH (40%), TBAI, CH2Cl2, rt, 40 min	Bn2C(Tos)N=C (78), I	73
	(2)	TosCH2N=C	NaH, DMSO, Et2O	I (80)	35
	TMS-CH(Tos)N=C (1)		t-BuOK, THF, −40° to rt, 2 h	TMS-C(Bn)(Tos)N=C (70)	93
	C8H17-n-CH(Tos)N=C (1)		NaH, DMSO, Et2O, 15°, 3 h	n-C8H17-C(Bn)(Tos)N=C (90)	105

TABLE VIIA. DISUBSTITUTED DERIVATIVES OF TOSMIC (Continued)

Substrate (equiv)	Isocyanide	Conditions	Product(s) and Yield(s) (%)	Refs.
2,6-bis(bromomethyl)pyridine (1)	C≡N–CH(Tos)–CH₂–C₆H₄–CH₂–CH(Tos)–N≡C	sl. add. to NaOH (16%), TBAI, CH₂Cl₂, reflux, 8.5 h	macrocycle with pyridine, Tos, N=C groups (>43)	92
(1)	bis(4-EtC₆H₄SO₂-CH(N≡C)-CH₂) pyridine	sl. add. to NaOH (16%), TBAI, CH₂Cl₂, reflux, 9 h	bis-pyridine macrocycle, 4-EtC₆H₄SO₂, N=C (>17)	92
(1)	bis(4-EtC₆H₄SO₂-CH(N≡C)-CH₂)-1,4-C₆H₄	sl. add. to NaOH (16%), TBAI, CH₂Cl₂, reflux, 13 h	pyridine-phenylene macrocycle, 4-EtC₆H₄SO₂, N=C (>57)	92
2,5-bis(bromomethyl)pyridine (1)	bis(4-EtOC₆H₄SO₂-CH(N≡C)-CH₂) pyridine	sl. add. to NaOH (10%), TBAI, CH₂Cl₂, reflux, 8.5 h	bis-pyridine macrocycle, 4-EtOC₆H₄SO₂, N=C (>10)	92

550

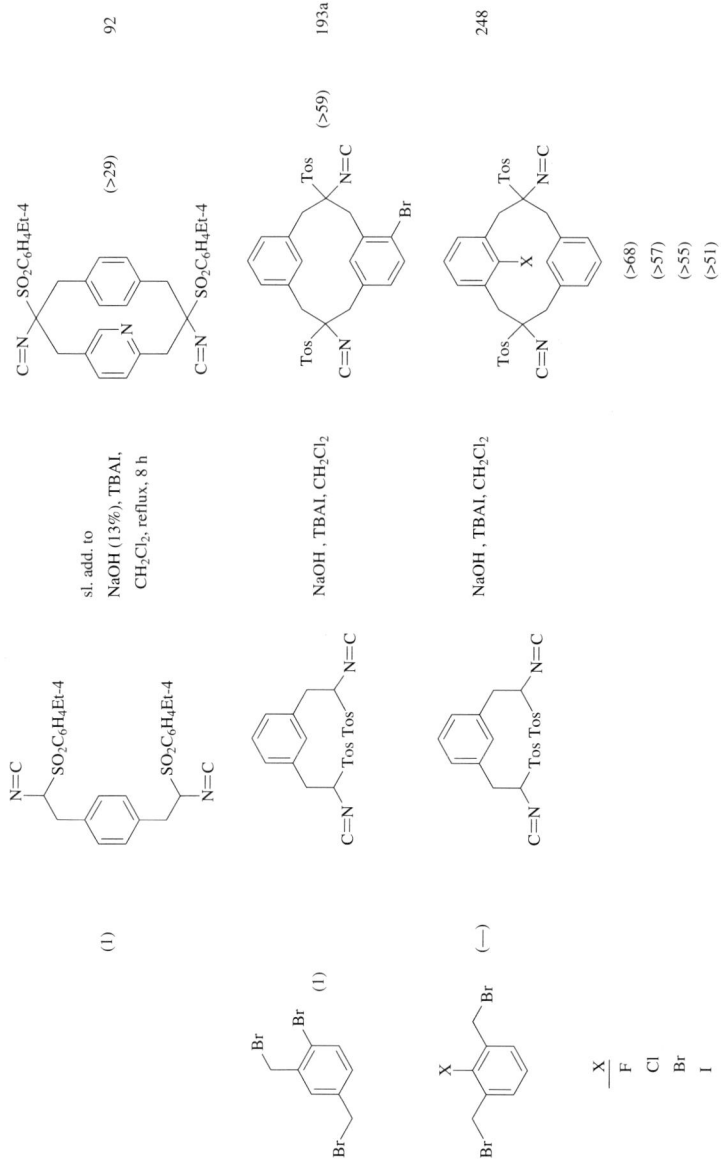

TABLE VIIA. DISUBSTITUTED DERIVATIVES OF TOSMIC (*Continued*)

Substrate (equiv)	Isocyanide	Conditions	Product(s) and Yield(s) (%)	Refs.
1,2-bis(bromomethyl)benzene (1)	TosCH$_2$N=C	NaOH (30%), TBAI, CH$_2$Cl$_2$, rt, 24 h	**I** (7) + (39)	153
(1)	1,2-bis(Tos(C≡N)CH)benzene	NaOH (30%), TBAI, CH$_2$Cl$_2$, rt, 24 h	**I** (47)	153
1,3-bis(bromomethyl)benzene (1)	TosCH$_2$N=C	NaOH (30%), TBAI, CH$_2$Cl$_2$, reflux, 2 h	**I** (>48)	155
(1)	TosCH$_2$N=C	NaOH (30%), TBAI, CH$_2$Cl$_2$, rt, 24 h	**I** (14) + **II** (—)	153

Reactants	Conditions	Products (Yield %)	Ref.
1,4-bis(bromomethyl)benzene (1.35) + TosCH₂N=C type diisocyanide (1)	NaOH (30%), TBAI, CH₂Cl₂, reflux, 7.5 h	**I** (55)	155
"	NaOH (30%), TBAI, CH₂Cl₂, reflux, 2.4 h	**I** (63) + **II** (—)	153
TosCH₂N=C + (para-xylylene diisocyanide)	NaOH (30%), TBAI, CH₂Cl₂, reflux	**I** (>14) + **II** (>23) + **III** (>12)	
TosCH₂N=C (1) + para-diisocyanide (1)	NaOH (30%), TBAI, CH₂Cl₂, rt, 24 h	**I** (0.6) + **II** (12)	153
(1)	NaOH (30%), TBAI, CH₂Cl₂, rt, 24 h	**I** (7) + **III** (—)	153

553

TABLE VIIA. DISUBSTITUTED DERIVATIVES OF TOSMIC (*Continued*)

Substrate (equiv)	Isocyanide	Conditions	Product(s) and Yield(s) (%)	Refs.
4-methylbenzyl bromide (1.35)	*m*-xylylene bis(TosCH-N=C)	NaOH (30%), TBAI, CH$_2$Cl$_2$, reflux, 7 h	cyclophane bis-isocyanide (>52)	155
(1)	*p*-xylylene bis(TosCH-N=C)	NaOH (30%), TBAI, CH$_2$Cl$_2$, rt, 6 h	cyclophane bis-isocyanide (>28)	249
4-methylbenzyl bromide (2)	TosCH$_2$N=C	NaH, DMSO, Et$_2$O, rt, 1.5 h	**I** (71)	73
(2)	TosCH$_2$N=C	NaOH (30%), TBAI, CH$_2$Cl$_2$, rt, 6 h	**I** (78)	161
2-iodo-6-methylheptane (1.1)	Tos-CH(N=C)-(CH$_2$)$_3$-CH=CH$_2$ type	NaH, DMSO, Et$_2$O, 35°, 30 min	(80)	238
(1.1)	Tos-CH(N=C)-(CH$_2$)$_2$-CH=CMe$_2$	NaH, DMSO, Et$_2$O	(80)	239

554

n-C$_8$H$_{17}$I	(2)	TosCH$_2$N=C	NaH, DMSO, Et$_2$O, rt, 3 h	n-C$_8$H$_{17}$〉〈C$_8$H$_{17}$-n Tos N=C (90)	105
	(1)	Tos〉—N=C	NaH, DMSO, Et$_2$O, rt, 1 h	n-C$_8$H$_{17}$〉 Tos N=C **I** (91)	73
	(1.1)	"	NaOH (40%), TBAI, CH$_2$Cl$_2$, rt, 1 h	**I** (>80)	98
C$_9$ 〔3,5-bis(bromomethyl)benzene〕	(1)	TosCH$_2$N=C	1. NaOH (38%), CH$_2$Cl$_2$, reflux, sl. add., 8 h 2. reflux, 2 h	(cyclophane with Tos, C≡N groups) **I** (>25)	108
	(1)	TosCH$_2$N=C	NaH, DMF, rt, 5 h	(macrocycle with OMe, Tos, C≡N) **I** (>7)	154
〔2-methoxy-1,3-bis(bromomethyl)benzene〕	(1.3)	TosCH$_2$N=C	NaH, DMF, rt, 5 h	**I** (>11) + (>14) R = OMe	109

555

TABLE VIIA. DISUBSTITUTED DERIVATIVES OF TosMIC (*Continued*)

Substrate (equiv)	Isocyanide	Conditions	Product(s) and Yield(s) (%)	Refs.
(1.2)	TosCH$_2$N=C	1. NaOH (30%), CH$_2$Cl$_2$, TBAI, reflux, sl. add., 8.5 h 2. reflux, 15 h	**I** (>19)	110
MeO–[benzyl]–Br, MeO– (1)	[m-xylylene bis-TosCH-N=C]	1. NaOH (25%), CH$_2$Cl$_2$, TBAI, reflux, sl. add., 6 h 2. reflux, 5 h	[cyclophane, R = OMe] (>57)	110
Br–[CH=CH(CH$_2$)$_3$]–Br (1)	C$_8$H$_{17-n}$–CH(Tos)–N=C	NaH, DMSO, Et$_2$O, rt, 1 h	n-C$_8$H$_{17}$–C(Tos)(N=C)–CH$_2$–[3,4-(MeO)$_2$C$_6$H$_3$] (90)	105
	TosCH$_2$N=C	KH, THF, rt, 6 h	[macrocyclic diene with Tos, N=C] (59)	149
THPO–(CH$_2$)$_n$–Br (2)	TosCH$_2$N=C	NaOH (40%), TBAB, CH$_2$Cl$_2$, rt	THPO–(CH$_2$)$_n$–C(Tos)(N=C)–(CH$_2$)$_n$–OTHP (85)	151
C$_{10}$ (2S)-OTHP–Br (2)	TosCH$_2$N=C	NaOH (40%), TBAB, CH$_2$Cl$_2$, rt	THPO–CH(Me)–CH$_2$–C(Tos)(N=C)–CH$_2$–CH(Me)–OTHP (82)	151
(1)	THPO–(CH$_2$)$_3$–CH(OTHP)–... –Tos–N=C	NaH, DMSO, Et$_2$O	THPO–(CH$_2$)$_3$–C(Tos)(N=C)–(CH$_2$)$_3$–OTHP (83)	151

556

TABLE VIIA. DISUBSTITUTED DERIVATIVES OF TOSMIC (*Continued*)

Substrate (equiv)	Isocyanide	Conditions	Product(s) and Yield(s) (%)	Refs.
	TosCH₂N=C (1)	NaOH (30%), TBAI, CH₂Cl₂, reflux, 2 h	[anthracene cyclophane bis(Tos, N=C)] (>14) + [second cyclophane] (>7)	155
		NaH, DMSO, Et₂O, 40°, 2 h	[binaphthyl tris(Tos, N=C) macrocycle]	242
MeO₂C–(CH₂)₉–Br	$C_{18}H_{37}$-n, Tos, N=C (1)		n-C₁₈H₃₇–C(Tos)(N=C)–(CH₂)₉–CO₂Me (>70)	
	[C₈H₁₇-n alkenyl]–(CH₂)₁₀–C(Tos)(N=C) (1)	NaH, DMSO, Et₂O, rt, 18 h	n-C₈H₁₇–CH=CH–(CH₂)₉–C(Tos)(N=C)–(CH₂)₉–CO₂Me (52)	160
C₁₃ Tos–C(N=C)(CH₃)–(CH₂)₃–I (1.1)	Ph–CH(Tos)(N=C)	NaH, DMSO, Et₂O, reflux, 1 h	Ph–C(Tos)(N=C)–(CH₂)₃–C(CH₃)(Tos)(C≡N) (73)	100

558

559

TABLE VIIA. DISUBSTITUTED DERIVATIVES OF TOSMIC (*Continued*)

Substrate (equiv)	Isocyanide	Conditions	Product(s) and Yield(s) (%)	Refs.
C₁₄ [bis(bromomethyl)biphenylene structure]	TosC⁻I₂N=C	NaH, DMSO, Et₂O	**I** (>17) + **II** (>17)	111
[bis(bromomethyl)biphenylene] + [bis(bromomethyl)biphenylene], 1:1	TosCH₂N=C	NaH, DMSO, Et₂O, rt, 12 h	**I** (>17) + **II** (>34)	251
[dioxolane-OBn-I substrate]	Pr-n, Tos, N=C	NaH, DMSO, Et₂O, 0°, 2 h	[dioxolane-OBn-Tos-Pr-n product] (50)	105

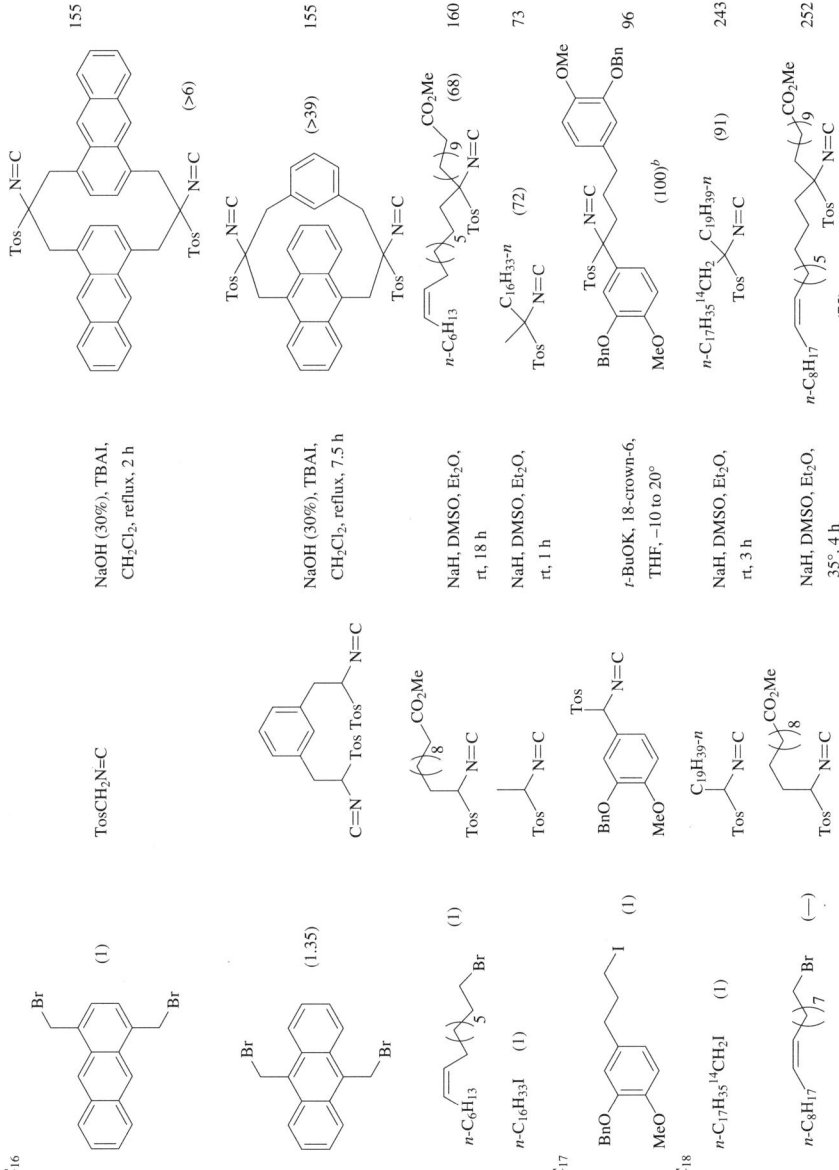

TABLE VIIA. DISUBSTITUTED DERIVATIVES OF TOSMIC (Continued)

Substrate (equiv)	Isocyanide	Conditions	Product(s) and Yield(s) (%)	Refs.
C_{20}				
(biphenyl bis-benzyl bromide) (1)	TosCH$_2$N=C	NaH, DMSO, Et$_2$O, rt, 6 h	(macrocycle with Tos, N=C) (>5)	253
n-C$_5$H$_{11}$–(CH)$_4$–Br (1)	Tos–CH(CO$_2$Me)(CH$_2$)$_8$–N=C	NaH, DMSO, Et$_2$O, rt, 18 h	n-C$_5$H$_{11}$–(CH)$_4$–C(Tos)(CO$_2$Me)(CH$_2$)$_9$–N=C (71)	160
C_{22}				
(tribromomethyl arene) (0.5)	TosCH$_2$N=C	NaOH (30%), TBAI, CH$_2$Cl$_2$, rt, 5 h	(macrocycle) (>24)	254
Et–(CH)$_6$–(CH$_2$)$_2$–Br (1)	Tos–CH(CO$_2$Me)(CH$_2$)$_8$–N=C	NaH, DMSO, Et$_2$O, rt, 18 h	Et–(CH)$_6$–C(Tos)(CO$_2$Me)(CH$_2$)$_9$–N=C (60)	160
n-C$_8$H$_{17}$–(CH)$_2$–Br (1)	Tos–CH(CO$_2$Me)(CH$_2$)$_8$–N=C	NaH, DMSO, Et$_2$O, rt, 18 h	n-C$_8$H$_{17}$–(CH)$_2$–C(Tos)(CO$_2$Me)(CH$_2$)$_9$–N=C (46)	160
n-C$_8$H$_{17}$–(CH)$_{11}$–Br (1)	Tos–CH(CO$_2$Me)(CH$_2$)$_8$–N=C	NaH, DMSO, Et$_2$O, rt, 18 h	n-C$_8$H$_{17}$–(CH)$_9$–C(Tos)(CO$_2$Me)(CH$_2$)$_9$–N=C (56)	160

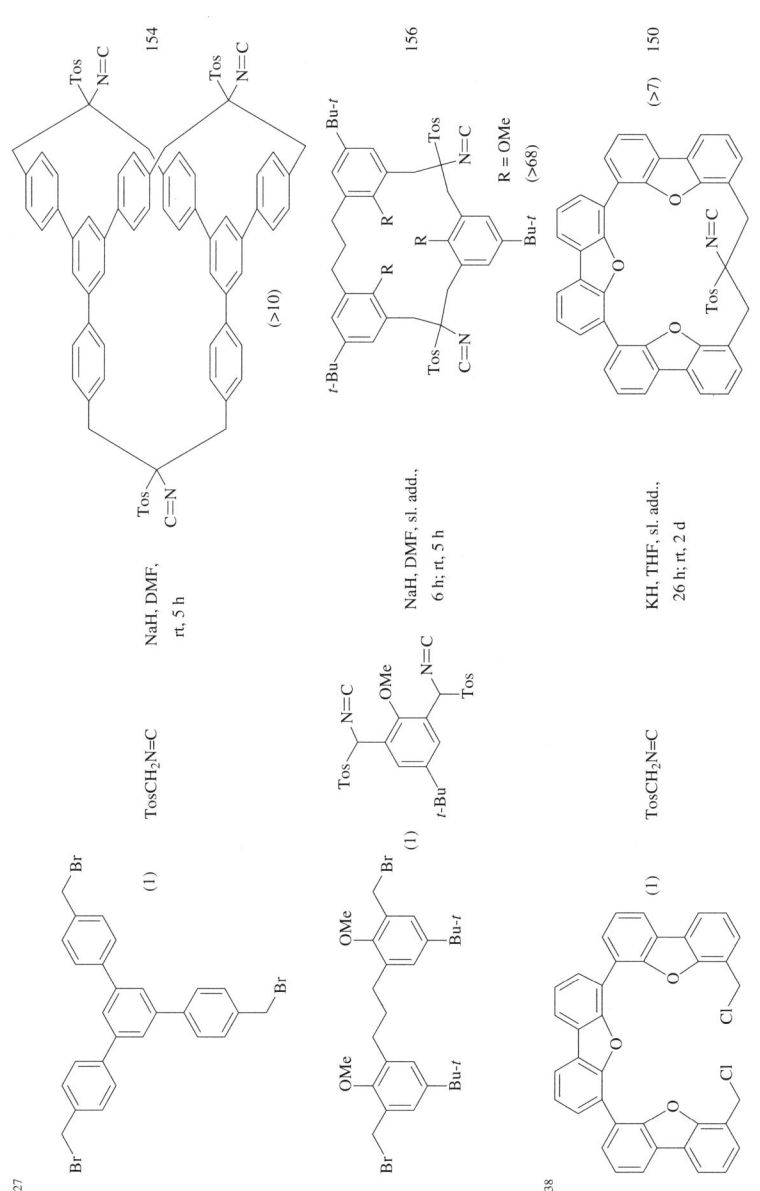

TABLE VIIB. DISUBSTITUTED DERIVATIVES OF TOSMIC FROM 1-ISOCYANO-1-SULFONYLALKENES

Substrate	Isocyanide	Conditions	Product(s) and Yield(s) (%)	Refs.
C₁				
CH₂Cl₂	(estrone-derived C=N-Tos alkene)	NaOH (50%), BTEAC, CH₂Cl₂, rt, 2 h	(Cl, Tos, N=C substituted product) (95)	101
CH₂X₂	(steroid-derived C=N-Tos alkene)	NaOH (50%), BTEAC, CH₂Cl₂, rt, 2 h	(X, Tos, N=C substituted product) X / Cl (89) / Br (76)	101
MeI	(Tos, N=C isopropenyl)	t-BuOK, DME, −30 to 20°, 1 h	(Tos, N=C gem-dimethyl alkene) (69)	103
	(Tos, N=C cycloalkylidene, n)	t-BuOK, DME, −30 to 20°, 1 h	(Tos, N=C gem-dimethylcycloalkane) n / 1 (82) / 2 (79) / 3 (81)	103
	(decalin-Tos, N=C exocyclic)	t-BuOK, DME or THF, −45 to 0°, 2 h	(decalin with gem-dimethyl C(Tos)(N=C)) (>55)	235
	(decalin R¹,R² Tos, N=C exocyclic)	t-BuOK, DME or THF, −45 to 0°, 2 h	(decalin R¹,R² with gem-dimethyl C(Tos)(N=C))	235

R¹	R²
H	H
OH	H
OH	Me
Me	OH
OTMS	H

(>58)
(>51)
(>61)
(>62)
(>81)

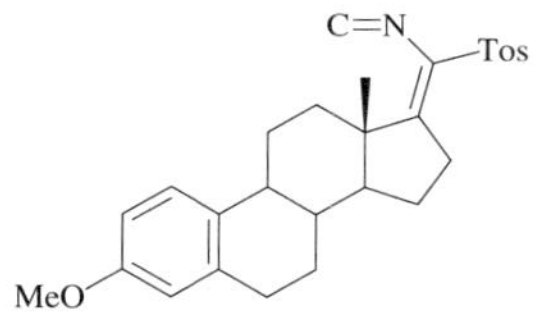

t-BuOK, DME or THF, −45 to 0°, 2 h 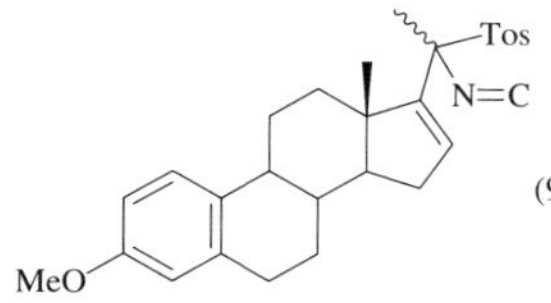 — (97) 101

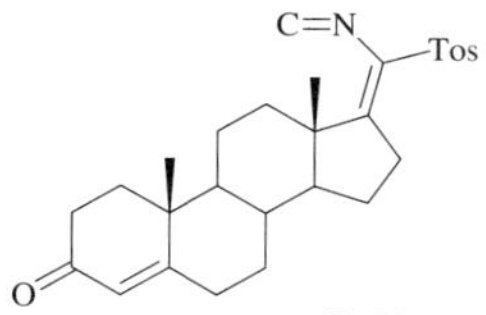

t-BuOK, DME or THF, −45 to 0°, 2 h 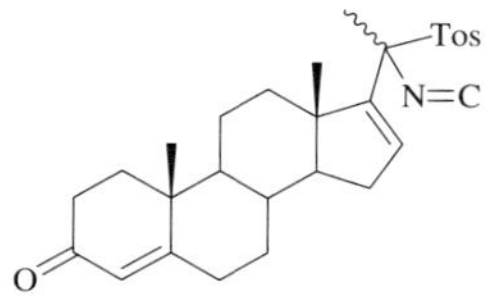 — (97) 101

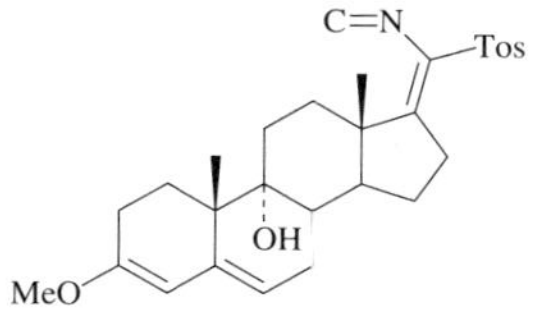

t-BuOK, DME or THF, −45 to 0°, 2 h 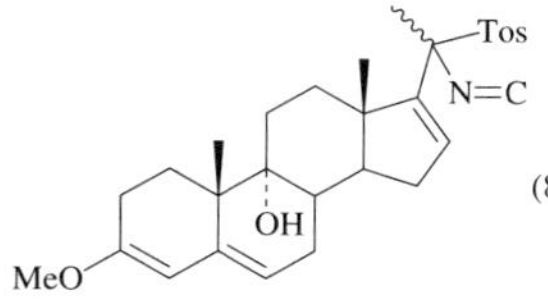 — (88) 101

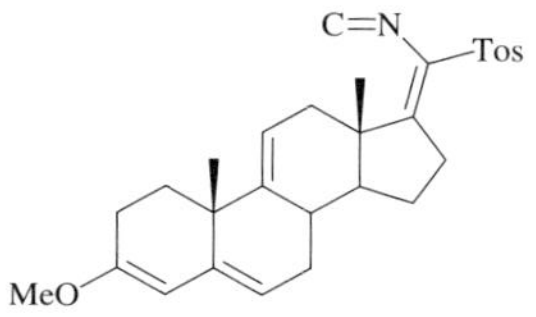

NaOH (50%), BTEAC, C₆H₆, 80°, 1 h 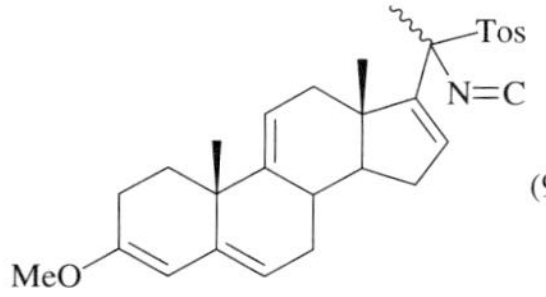 — (94) 101

TABLE VIIB. DISUBSTITUTED DERIVATIVES OF TosMIC FROM 1-ISOCYANO-1-SULFONYLALKENES (*Continued*)

Substrate	Isocyanide	Conditions	Product(s) and Yield(s) (%)	Refs.
C₂ Br~~Br	(steroid dienone with C=N, Tos)	NaOH (50%), BTEAC, C₆H₆, 80°, 1 h	(steroid with Tos, N=C) (95)	101
R–X	(methoxy steroid with C=N, Tos)	t-BuOK, DME, −30 to 20°, 1 h	(methoxy steroid with Tos, N=C) (98)	101
	Tos–C(=cyclopentylidene)–N=C	t-BuOK, DME, −30 to 20°, 1 h	Tos, N=C, cyclopentenyl-CH₂CH₂Br (69)	103
	(methoxy steroid with C=N, Tos)	NaOH (50%), BTEAC, C₆H₆, 80°, 1 h	(methoxy steroid with R, Tos, N=C) X / R I / Me (89) Cl / MeO (74)	101
C₃ ⤳Br	Tos–C(=cyclohexylidene)–N=C	t-BuOK, DME, −30 to 20°, 1 h	Tos, N=C, cyclohexenyl, allyl (76)	103

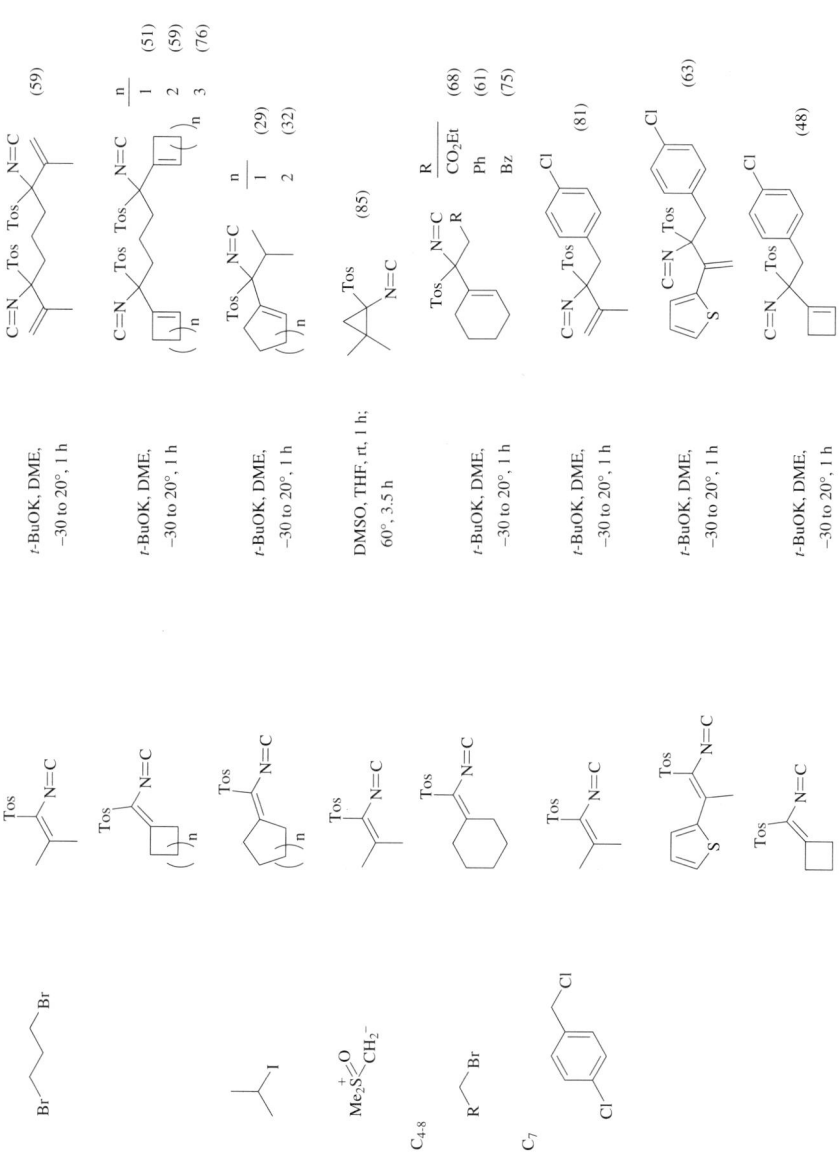

TABLE VIIB. DISUBSTITUTED DERIVATIVES OF TOSMIC FROM 1-ISOCYANO-1-SULFONYLALKENES (*Continued*)

Substrate	Isocyanide	Conditions	Product(s) and Yield(s) (%)	Refs.
C7-8 MeO2C—\—Br	Tos—⟨cyclohexenyl⟩—N=C	*t*-BuOK, DME, −30 to 20°, 1 h	Tos, N=C, cyclohexenyl, CO2Me (62)	103
MeO2C—\\—Br	Tos—⟨cyclopentenyl⟩—N=C	*t*-BuOK, DME, −30 to 20°, 1 h	Tos, N=C, cyclopentenyl, CO2Me (59)	103
⟨thiophene⟩—CH2Br, CO2R; R = Me, Et	Tos, N=C, ⟨thienyl⟩, X = O, S	*t*-BuOK, DME, −30 to 20°, 1 h	C=N Tos, ⟨thienyl-CO2R⟩ (83), (84)	103
C8 BnO—\—Cl	C=N—Tos, steroid (MeO)	NaOH (50%), BTEAC, C6H6, 80°, 1 h	BnO, Tos, N=C, steroid (MeO) (50)	101
	C=N—Tos, steroid dienyl (MeO)	NaOH (50%), BTEAC, C6H6, 80°, 1 h	BnO, Tos, N=C, steroid dienyl (MeO) (54)	101

TABLE VIIC. α-ISOCYANO-α-TOSYL KETONES FROM TOSMIC HOMOLOGS AND CARBOXYLIC CHLORIDES

Substrate	Isocyanide	Conditions	Product(s) and Yield(s) (%)	Refs.
C₁₀ (steroid with Br, CO₂Et)	(steroid C=N-Tos)	NaOH (50%), BTEAC, C₆H₆, 80°, 1 h	(steroid BnO-CH with N=C-Tos) (80)	101
(cyclohexylidene N=C-Tos)		t-BuOK, DME, −30 to 20°, 1 h	(cyclohexene with C=N-Tos, CH₂-C₆H₄-CO₂Et) (83)	103
C₂₋₇ R¹-C(O)-Cl	R²-CH(Tos)-N=C	n-BuLi, THF, −80° to rt	R¹-C(O)-C(Tos)(R²)-N=C	112
R¹	R²			
Me	Ph		(72)	
t-Bu	Bn		(54)	
Ph	Ph		(72)	
Ph	Bn		(58)	
4-O₂NC₆H₄	Ph		(57)	
4-O₂NC₆H₄	Bn		(77)	

TABLE VIIIA. IMIDAZOLES FROM ALDIMINES

Substrate $R^1\underset{N}{\diagup}\diagdown R^2$		Isocyanide $\underset{Tos}{R^3}\diagdown\underset{}{}N{\equiv}C$	Conditions	Product(s) and Yield(s) (%) $R^3\diagdown\underset{R^1}{\diagup}\underset{N}{\diagdown}\underset{R^2}{N}$	Refs.
R^1	R^2	R^3			
C_{4-23}					
C_4 Me	Et	H	t-BuNH$_2$, DME, rt, 20 h	(70)	37
Me	Et	H	K$_2$CO$_3$, MeOH, rt, 20 h	(62)	37
C_6 Me	t-Bu	H	t-BuNH$_2$, MeOH, rt, 20 h	(94)	37
Me	t-Bu	H	t-BuNH$_2$, DME, rt, 20 h	(90)	37
Me	t-Bu	H	K$_2$CO$_3$, MeOH, rt, 20 h	(54)	37
Me	t-Bu	Ph	t-BuNH$_2$, DME, rt, 0.5 h	(89)	37
t-Bu	Me	H	t-BuNH$_2$, MeOH, rt, 72 h	(96)	37
t-Bu	Me	H	K$_2$CO$_3$, MeOH, rt, 20 h	(5)	37
C_7 i-Pr	Me	H	i-PrNH$_2$, MeOH, rt, 16 h	(75)	37
C_8 4-O$_2$NC$_6$H$_4$	Me	H	K$_2$CO$_3$, MeOH, rt, 16 h	(14)	37
4-O$_2$NC$_6$H$_4$	Me	H	i-PrNH$_2$, MeOH, rt, 100 h	(0)	37
Ph	Me	H	K$_2$CO$_3$, MeOH, DME, 0°, 60 h	(10)	37
Ph	Me	H	K$_2$CO$_3$, MeOH, rt, 16 h	(90)	37
Me	C$_6$H$_{11}$	Ph	C$_6$H$_{11}$NH$_2$, MeOH, rt, 17 h	(96)	37
Me$_2$C=CH	n-Pr	H	K$_2$CO$_3$, MeOH, reflux, 3 h	(40)	115
C_9 4-MeC$_6$H$_4$	Me	H	K$_2$CO$_3$, MeOH, rt, 20 h	(37)	37
2-HOC$_6$H$_4$	Et	H	K$_2$CO$_3$, MeOH, reflux, 3 h	(50)	255
4-HOC$_6$H$_4$	Et	H	K$_2$CO$_3$, MeOH, reflux, 2 h	(30)	256
4-HOC$_6$H$_4$	n-Pr	H	K$_2$CO$_3$, MeOH, reflux, 3 h	(14)	115
Me$_2$C=CH⁀	i-Bu	H	K$_2$CO$_3$, MeOH, reflux, 3 h	(67)	114
C_{10} 2-ClC$_6$H$_4$	n-Pr	H	K$_2$CO$_3$, MeOH, reflux, 3 h	(0.8)	115
3-ClC$_6$H$_4$	n-Pr	H	K$_2$CO$_3$, MeOH, reflux, 3 h	(2.5)	115
4-FC$_6$H$_4$	n-Pr	H	K$_2$CO$_3$, MeOH, reflux, 3 h	(1)	115
3-HOC$_6$H$_4$	i-Pr	H	K$_2$CO$_3$, MeOH, reflux, 3 h	(59)	232

Bn	Et	H	K₂CO₃, MeOH, reflux, 3 h	(9)	255
C₁₁ 4-CF₃C₆H₄	n-Pr	H	K₂CO₃, MeOH, reflux, 3 h	(0.3)	115
3-HOC₆H₄	i-Bu	H	K₂CO₃, MeOH, reflux, 3 h	(43)	257
3-HOC₆H₄	i-Bu	H	K₂CO₃, MeOH, reflux, 3 h	(24)	114
2-HOC₆H₄	i-Bu	H	K₂CO₃, MeOH, reflux, 2 h	(78)	121
2-HOC₆H₄	n-Bu	H	K₂CO₃, MeOH, reflux, 2 h	(78)	121
4-HOC₆H₄	n-Bu	H	K₂CO₃, MeOH, reflux, 3 h	(75)	256
4-MeC₆H₄	n-Pr	H	K₂CO₃, MeOH, reflux, 3 h	(7)	115
C₁₂ Me₂C=CH	Bn	H	K₂CO₃, MeOH, reflux, 3 h	(20)	114
3,4-Cl₂C₆H₃	t-BuCH₂	H	K₂CO₃, MeOH, reflux, 2 h	(1)	258
4-BrC₆H₄	t-BuCH₂	H	K₂CO₃, MeOH, reflux, 2 h	(1)	258
4-O₂NC₆H₄	t-BuCH₂	H	K₂CO₃, MeOH, reflux, 2 h	(1)	258
2-HOC₆H₄	t-BuCH₂	H	K₂CO₃, MeOH, reflux, 2 h	(82)	121
3-HOC₆H₄	t-BuCH₂	H	K₂CO₃, MeOH, reflux, 2 h	(33)	258
2-HOC₆H₄	n-C₅H₁₁	H	K₂CO₃, MeOH, reflux, 2 h	(73)	121
n-C₆H₁₃⁓	n-Pr	H	K₂CO₃, MeOH, reflux, 3 h	(33)	115
C₁₃ 4-ClC₆H₄	4-ClC₆H₄	H	K₂CO₃, MeOH, DME, rt, 16 h	(43)	37
4-ClC₆H₄	4-ClC₆H₄	H	1. NaH, DME, −20°, 3 h[a]	(40)	37
4-O₂NC₆H₄	4-O₂NC₆H₄	H	K₂CO₃, MeOH, DME, rt, 16 h	(87)	37
4-O₂NC₆H₄	4-O₂NC₆H₄	H	t-BuNH₂, DME, rt, 24 h	(0)	37
Ph	4-O₂NC₆H₄	H	K₂CO₃, MeOH, DME, reflux, 1 h	(70)	37
Ph	4-O₂NC₆H₄	H	K₂CO₃, MeOH, DME, 20°, 3 h	(34)	37
Ph	4-O₂NC₆H₄	H	1. NaH, DME, −20°, 1 h[a] 2. K₂CO₃, MeOH, reflux, 0.5 h	(65)	37
Ph	4-O₂NC₆H₄	Me	NaH, DME, 0°, 1 h	(75)	37, 69
4-O₂NC₆H₄	Ph	H	K₂CO₃, MeOH, DME, rt, 16 h	(82)	37
4-O₂NC₆H₄	Ph	Me	t-BuOK, DMSO	(78)	69
4-O₂NC₆H₄	Ph	Bn	t-BuOK, DMSO	(62)	69
4-ClC₆H₄	Ph	Bn	NaH, DMSO	(68)	69
Ph	Ph	Ph	t-BuNH₂, DME, rt, 16 h	(0)	37

TABLE VIIIA. IMIDAZOLES FROM ALDIMINES (*Continued*)

Substrate			Isocyanide	Conditions	Product(s) and Yield(s) (%)	Refs.
R^1	R^2		R^3			
Ph	Ph		Ph	NaH, DME, 25°, 1 h	(0)	37
Ph	Ph		H	1. NaH, DME, –20°, 3 h^a 2. K_2CO_3, MeOH, reflux, 0.5 h	(56)	37
Ph	Ph		H	K_2CO_3, MeOH, reflux, 2.5 h	(10)	256
Ph	Ph		H	1. K_2CO_3, MeOH, rt, 70 h 2. reflux, 0.5 h	(0)	37
3,4-(OCH$_2$O)C$_6$H$_3$	t-BuCH$_2$		H	K_2CO_3, MeOH, reflux, 2 h	(4)	258
2-MeC$_6$H$_4$	t-BuCH$_2$		H	K_2CO_3, MeOH, reflux, 2 h	(6)	258
3-MeC$_6$H$_4$	t-BuCH$_2$		H	K_2CO_3, MeOH, reflux, 2 h	(4)	258
4-MeC$_6$H$_4$	t-BuCH$_2$		H	K_2CO_3, MeOH, reflux, 2 h	(4)	258
4-MeOC$_6$H$_4$	t-BuCH$_2$		H	K_2CO_3, MeOH, reflux, 2 h	(41)	258
4-HOC$_6$H$_4$	n-C$_6$H$_{13}$		H	K_2CO_3, MeOH, reflux, 2 h	(77)	256
2-HOC$_6$H$_4$	t-BuCH$_2$		H	K_2CO_3, MeOH, reflux, 2 h	(61)	121
	n-Pr		H	K_2CO_3, MeOH, reflux, 3 h	(25)	115
C$_{14}$ Ph	Bn		H	K_2CO_3, MeOH, reflux, 3 h	(4)	259
2-HOC$_6$H$_4$	Bn		H	K_2CO_3, MeOH, reflux, 2 h	(56)	121
4-MeC$_6$H$_4$	Ph		H	1. NaH, DME, –20°, 3 h^a 2. K_2CO_3, MeOH, reflux, 0.5 h	(19)	37
4-HOC$_6$H$_4$	C$_6$H$_{11}$CH$_2$		H	K_2CO_3, MeOH, reflux, 2 h	(50)	256
4-EtC$_6$H$_4$	t-BuCH$_2$		H	K_2CO_3, MeOH, reflux, 2 h	(7)	258
	s-Bu		H	K_2CO_3, MeOH, reflux, 3 h	(—)	257
	i-Bu		H	K_2CO_3, MeOH, reflux, 3 h	(23)	114
C$_{15}$ 2-HOC$_6$H$_4$	BnCH$_2$		H	K_2CO_3, MeOH, reflux, 2 h	(72)	121
4-PhOC$_6$H$_4$	Et		H	K_2CO_3, MeOH, reflux, 3 h	(47)	255
2-C$_4$H$_3$O			H	K_2CO_3, MeOH, reflux, 2 h	(5)	260

	2-C4H3S	[geranyl]	H	K2CO3, MeOH, reflux, 2 h	(2)	260
	3-C4H3S	[geranyl]	H	K2CO3, MeOH, reflux, 2 h	(11)	260
C16	4-HOC6H4	n-C8H17	H	K2CO3, MeOH, reflux, 2 h	(77)	256
	1-C10H7	n-C5H11	H	K2CO3, MeOH, reflux, 3 h	(—)	257
	2-C10H7	t-BuCH2	H	K2CO3, MeOH, reflux, 2 h	(3)	258
	2-C5H4N	t-BuCH2	H	K2CO3, MeOH, reflux, 2 h	(2)	258
		[geranyl]	H	K2CO3, MeOH, reflux, 2 h	(2)	260
	4-C5H4N	[geranyl]	H	K2CO3, MeOH, reflux, 2 h	(3)	260
C17	4-(PhCH=CH)C6H4	C6H11	H	K2CO3, MeOH, reflux, 3 h	(—)	257
	4-PhOC6H4	Et	H	K2CO3, MeOH, reflux, 3 h	(39)	255
	4-HOC6H4	i-Bu	H	K2CO3, MeOH, reflux, 3 h	(7)	257
		Bn	H	K2CO3, MeOH, reflux, 2 h	(89)	256
	$E:Z = 7:3$	[geranyl]	H	K2CO3, MeOH, reflux, 3 h	(E)-(25), (Z)-(10)	259
	$E:Z = 7:3$	4-ClC6H4CH2	H	K2CO3, MeOH, reflux, 3 h	(E)-(13)	259
	2,4-Cl2C6H3	[geranyl]	H	K2CO3, MeOH, reflux, 2 h	(2)	260
	2-ClC6H4	[geranyl]	H	K2CO3, MeOH, reflux, 2 h	(2)	260
	3-ClC6H4	[geranyl]	H	K2CO3, MeOH, reflux, 2 h	(3)	260
	4-O2NC6H4	[geranyl]	H	K2CO3, MeOH, reflux, 2 h	(4)	260
	Ph	[geranyl]	H	K2CO3, MeOH, reflux, 2 h	(12)	260
C18	4-HOC6H4	n-C10H21	H	K2CO3, MeOH, reflux, 2 h	(72)	256
	4-HOC6H4	1-C10H7CH2	H	K2CO3, MeOH, reflux, 2 h	(2)	256
	$E:Z = 7:3$	BnCH2	H	K2CO3, MeOH, reflux, 3 h	(E)-(23)	259

TABLE VIIIA. IMIDAZOLES FROM ALDIMINES (*Continued*)

Substrate			Isocyanide	Conditions	Product(s) and Yield(s) (%)	Refs.
R^1	R^2		R^3			
E:Z = 7:3	2-MeC$_6$H$_4$CH$_2$		H	K$_2$CO$_3$, MeOH, reflux, 3 h	(*E*)- (31)	259
E:Z = 7:3	4-MeC$_6$H$_4$CH$_2$		H	K$_2$CO$_3$, MeOH, reflux, 3 h	(*E*)- (24)	259
E:Z = 7:3	4-MeOC$_6$H$_4$CH$_2$		H	K$_2$CO$_3$, MeOH, reflux, 3 h	(*E*)- (31)	259
E:Z = 7:3	(*S*)-PhCH(Me)		H	K$_2$CO$_3$, MeOH, reflux, 3 h	(*E*)-(*S*)- (19)	259
E:Z = 7:3	(*R*)-PhCH(Me)		H	K$_2$CO$_3$, MeOH, reflux, 3 h	(*E*)-(*R*)- (17)	259
E:Z = 7:3	2-MeC$_6$H$_4$		H	K$_2$CO$_3$, MeOH, reflux, 2 h	(7)	260
	2-MeOC$_6$H$_4$		H	K$_2$CO$_3$, MeOH, reflux, 2 h	(31)	260
	4-MeOC$_6$H$_4$		H	K$_2$CO$_3$, MeOH, reflux, 2 h	(22)	260
	3,4-(OCH$_2$O)C$_6$H$_3$		H	K$_2$CO$_3$, MeOH, reflux, 2 h	(13)	260
C$_{19}$	2-EtOC$_6$H$_4$		H	K$_2$CO$_3$, MeOH, reflux, 2 h	(27)	260
	3,4-(MeO)$_2$C$_6$H$_3$		H	K$_2$CO$_3$, MeOH, reflux, 2 h	(14)	260
	2,4-(MeO)$_2$C$_6$H$_3$		H	K$_2$CO$_3$, MeOH, reflux, 2 h	(36)	260
C$_{20}$ *E:Z* = 7:3			H	K$_2$CO$_3$, MeOH, reflux, 3 h	(*E*)- (12)	259
C$_{21}$ 1-C$_{10}$H$_7$			H	K$_2$CO$_3$, MeOH, reflux, 2 h	(4)	260

2-C₁₀H₇	[citronellyl group]	H	K₂CO₃, MeOH, reflux, 2 h	(6)	260
BnO(CH₂)₃	BnO(CH₂)₃	H	—	(—)	261
C₂₃ BnO(CH₂)₄	BnO(CH₂)₄	H	—	(—)	261
C₅	[CH=N–TMS]	TosCH₂N=C	LiN(TMS)₂, THF, −78°, 2 h; rt, 16 h	[4-methylimidazole] (55)	113

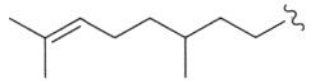

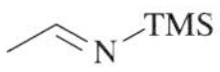

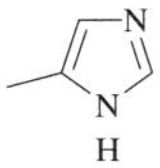

C₈

R	
H	(51)
Me	(51)
Bn	(66)

[imidazole with n-Bu and R substituents], LiN(TMS)₂, THF, −78°, 2 h; rt, 16 h, 113

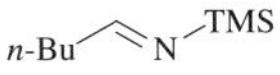

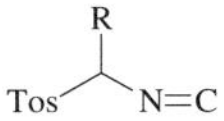

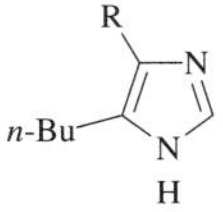

[quinoxaline] TosCH₂N=C, n-BuLi (2 eq), [CuI•(n-Bu)₃P], −70 to 0°, 4 h, [imidazo-quinoxaline] (30), 79

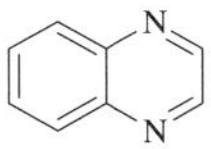

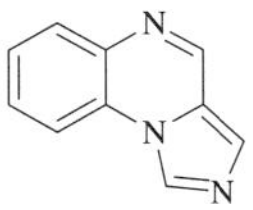

C₈₋₁₄ [R¹CH=N–R²] TosCH₂N=C
1. K₂CO₃, MeOH, reflux, 3 h
2. NaH, DMF, rt, 1 h
3. BnCl, rt, 10 h

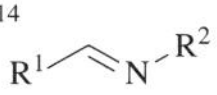

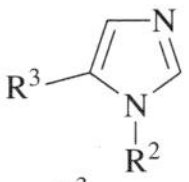

[imidazole with R³ and R² substituents]

	R¹	R²	R³	
C₈	3-HOC₆H₄	Me	3-BnOC₆H₄	(36)
C₉	3-HOC₆H₄	Et	3-BnOC₆H₄	(13)
C₁₀	2-HOC₆H₄	i-Pr	2-BnOC₆H₄	(66)
	4-HOC₆H₄	i-Pr	4-BnOC₆H₄	(19)
	3-HOC₆H₄	n-Pr	3-BnOC₆H₄	(8)
C₁₁	3-HOC₆H₄	n-Bu	3-BnOC₆H₄	(10)
C₁₂	3-HOC₆H₄	i-Bu	3-BnOC₆H₄	(5)
	3-HOC₆H₄	t-BuCH₂	3-BnOC₆H₄	(33)
C₁₄	3-HOC₆H₄	Bn	3-BnOC₆H₄	(15)

232

TABLE VIIIA. IMIDAZOLES FROM ALDIMINES (*Continued*)

Substrate	Isocyanide	Conditions	Product(s) and Yield(s) (%)	Refs.
C₉				
quinoline	TosCH₂N=C	*n*-BuLi (2 eq), LiBr, THF, −70 to 0°, 4 h	(25)	79
isoquinoline	TosCH₂N=C	*n*-BuLi (2 eq), LiBr, THF, −70 to 0°, 4 h	(44)	79
C₁₀				
Me₂N–N=⟨⟩–N–C₆H₄NO₂-4	N≡C–C(=cyclohexyl)Tos	Triton B, THF, 20–30°, 1.5 h	N–C₆H₄NO₂-4 imidazole with cyclohexenyl and CH=NMe₂ (85)	63
	Ph/Et–C=C(N≡C)Tos	Triton B, THF, 20–30°, 1.5 h	N–C₆H₄NO₂-4 imidazole with propenyl-Ph and CH=NMe₂ (99)	63
	TosCH₂N=C	LiN(TMS)₂, THF, −78°, 2 h; rt, 16 h	Ph-imidazole NH (23)	113
C₁₁				
Ph–CH=N–TMS				
pyrrolo-benzothiazepine SO₂	TosCH₂N=C	*n*-BuLi, THF, −72°, 30 min; rt, 60 h	imidazo-pyrrolo-benzothiazepine SO₂ (39)	262

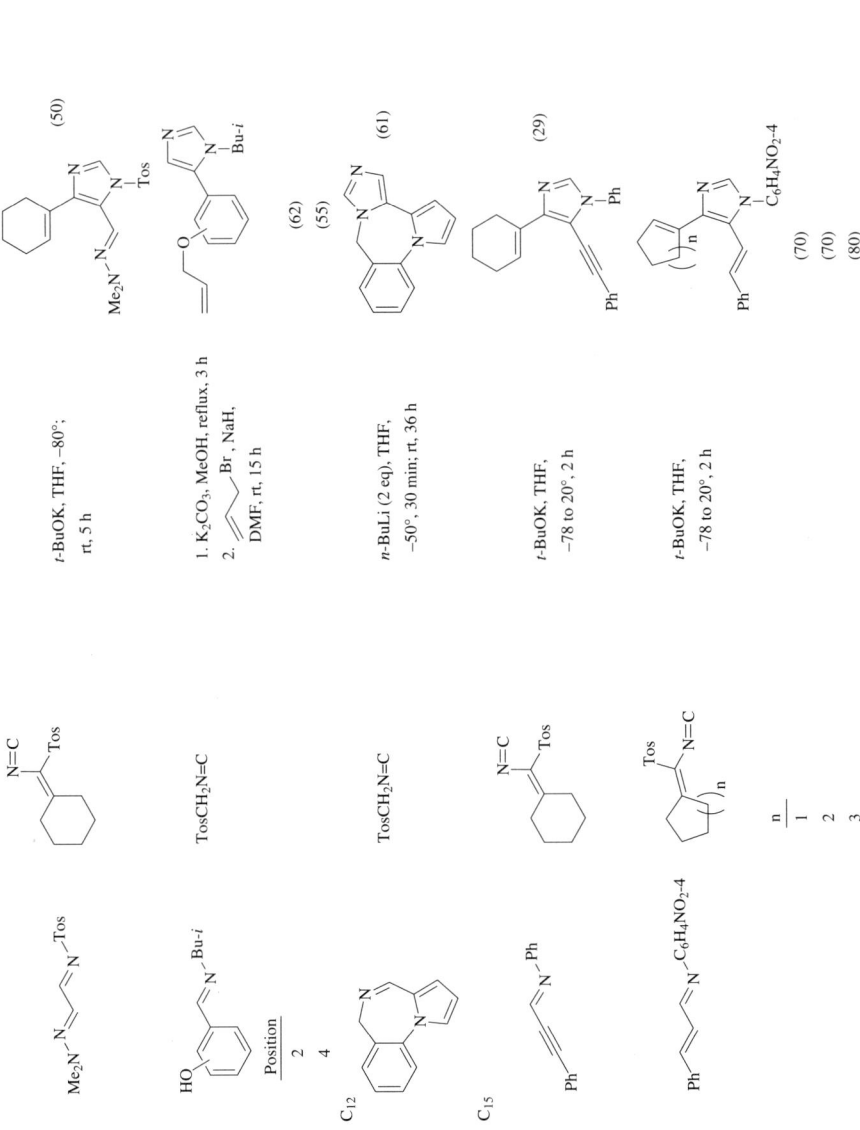

TABLE VIIIA. IMIDAZOLES FROM ALDIMINES (*Continued*)

Substrate	Isocyanide	Conditions	Product(s) and Yield(s) (%)	Refs.
C₁₆				
Ph-CH=CH-CH=N-Bn	TosCH₂N=C	*t*-BuOK, MeOH, reflux, 4 h	(imidazole with Bn, CH=CHPh) (75)	34
Ph-CH=CH-CH=N-Tos	cyclohexylidene(Tos)C=N=C	*t*-BuOK, THF, −78 to 20°, 2 h	pyrrole product (34) + imidazole product (34)	80
C₂₀				
bis(4-BrC₆H₄-N=CH)-m-C₆H₄	bis(isocyanide with Tos groups on m-xylene)	NaH, DMSO, 80°, 4 h	macrocycle (16) + larger macrocycle (6)	264

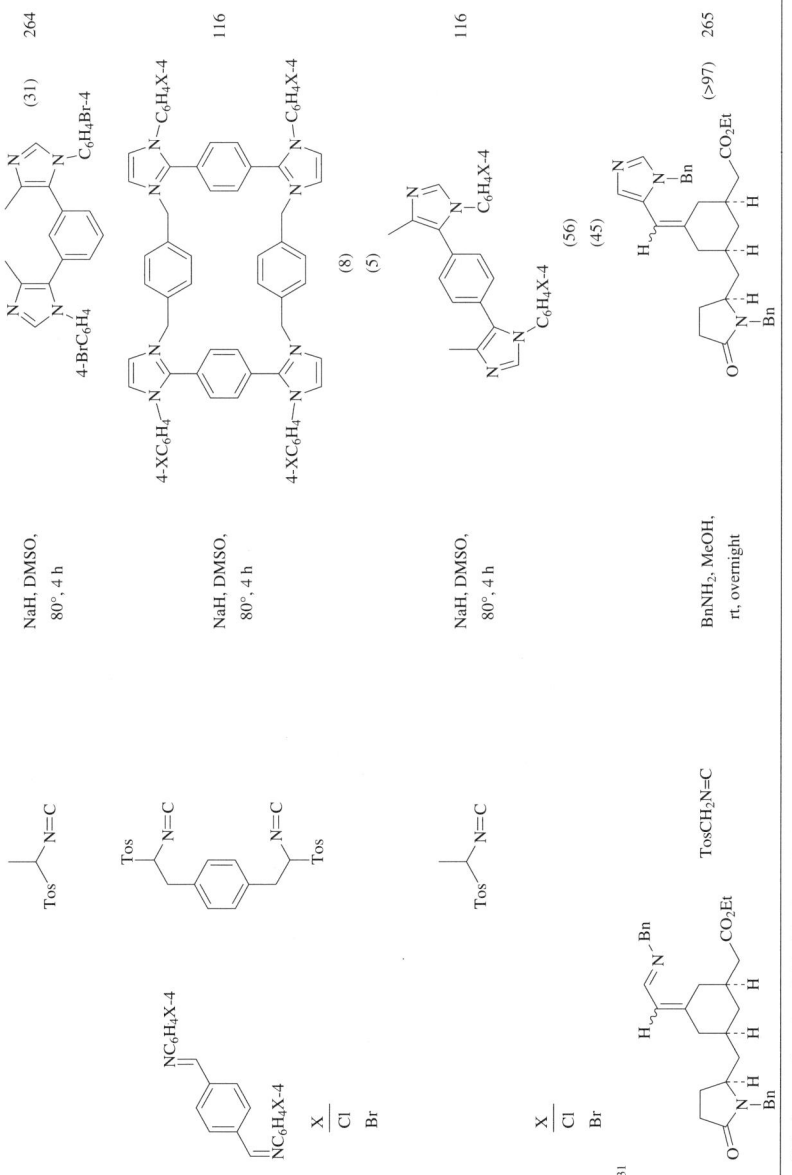

[a] The intermediate 4-tosyl-2-imidazole was isolated and used as such in a second reaction.

TABLE VIIIB. 4-TOSYLIMIDAZOLES FROM IMIDOYL CHLORIDES OR ISOTHIOCYANATES

Substrate	Isocyanide	Conditions	Product(s) and Yield(s) (%)	Refs.
C₂				
Me−N=C=S	TosCH$_2$N=C	t-BuLi (2 eq), THF, −70 to −10°, 10 min	**I** (54)	120
Me−N=C=S	TosCH$_2$N=C	NaH, DME, rt, 0.5 h	**II** (72)	120
C₃				
pyrazine[a]	TosCH$_2$N=C	EtONa, EtOH, DMF, sl. add., rt, 8 h; rt, 16 h	(28)	118
C₃₋₉				
R−N=C=S	TosCH$_2$N=C		**I** + **II**	120
R = Et		t-BuLi (2 eq), THF, −70 to −10°, 10 min	**I** (20) + **II** (50)	
4-O$_2$NC$_6$H$_4$		NaOH (4%), TBAB, C$_6$H$_6$, rt, 1 h	**I** (65) + **II** (10)	
Ph		NaH, DMSO, rt, 1 h	**I** (83)	
Ph		NaOH (4%), TBAB, CH$_2$Cl$_2$, rt, 1 h	**II** (67)	
4-Me$_2$NC$_6$H$_4$		NaH, DMSO, rt, 1 h	**I** (79) + **II** (9)	
C$_6$H$_{11}$		t-BuLi (2 eq), THF, −70 to −10°, 10 min	**I** (40) + **II** (35)	

Products:

I: Tos-substituted imidazole with HS and N−R (or N−Me for first entries)

II: Tos-substituted thiazole with RNH (or MeNH) substituent

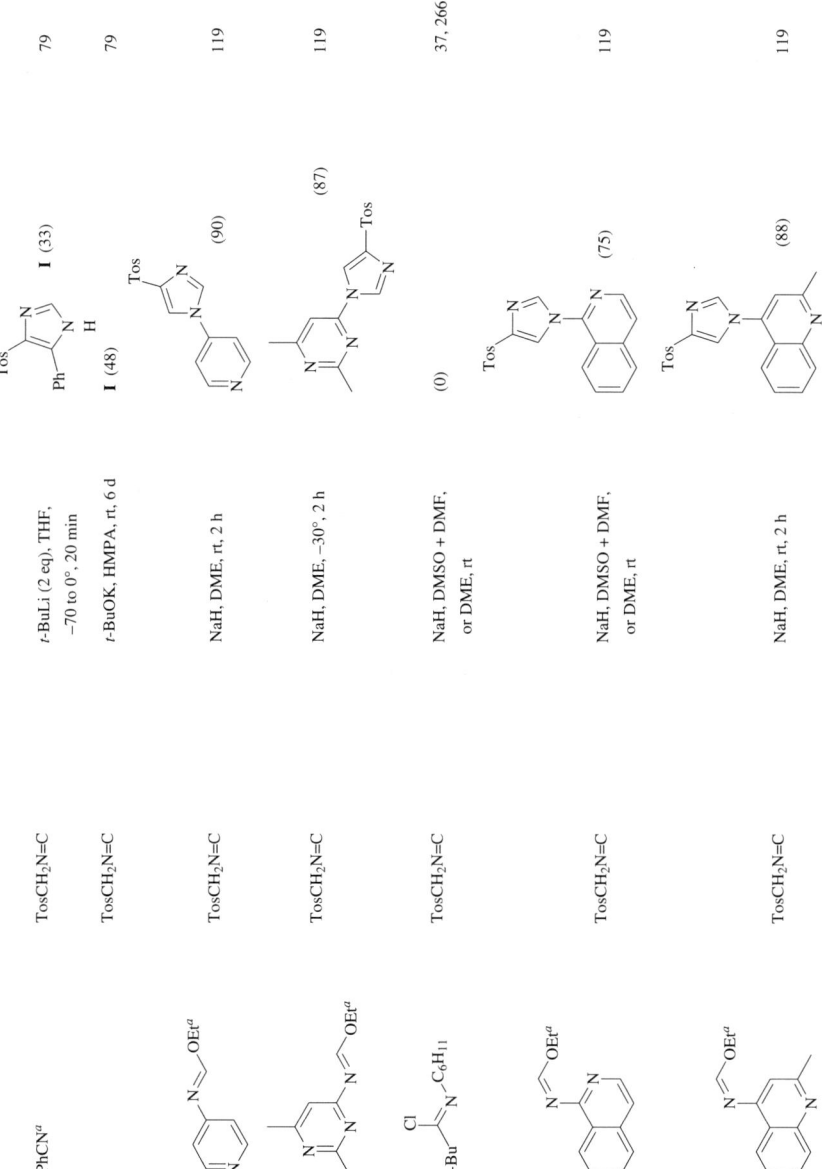

TABLE VIIIB. 4-TOSYLIMIDAZOLES FROM IMIDOYL CHLORIDES OR ISOTHIOCYANATES (*Continued*)

Substrate	Isocyanide	Conditions	Product(s) and Yield(s) (%)	Refs.
$R^1\underset{N\sim R^2}{\overset{Cl}{\diagup}}$	TosCH$_2$N=C	NaH, DME, rt, 2 h	Tos-imidazole with R^1, R^2 substituents	37, 266
R^1 / R^2				
Ph / 4-O$_2$NC$_6$H$_4$		NaH, DME, DMSO, rt, 1 h	(81)	
4-O$_2$NC$_6$H$_4$ / Ph		NaH, DME, DMSO, rt, 1 h	(85)	
Ph / Ph		NaH, DME, DMSO, rt, 1 h	(60)	
4-O$_2$NC$_6$H$_4$ / C$_6$H$_{11}$		NaH, DME, THF, rt, 1 h	(75)	
Ph / C$_6$H$_{11}$		NaH, DME, THF, rt, 1 h	(80)	
C$_{14}$ benzodiazepinone with OPO(OEt)$_2$ [a]	TosCH$_2$N=C	LDA, THF, –78°, 2 h	benzodiazepine-imidazole with Tos, N-Me (46)	117 40, 281

[a] This reaction involves a leaving group other than the Cl of an imidoyl chloride.

TABLE VIIIC. IMIDAZOLES FROM 1-ISOCYANO-1-TOSYLALKENES

Substrate	Isocyanide	Conditions	Product(s) and Yield(s) (%)	Refs.
C₁				
MeNH₂	[Tos/Bn cyclopentanone with N=C]	MeOH	[imidazole with Bn, Bn, Me] (55)	234
C₀₋₈				
R¹NH₂	[Tos furan with N=C, R² = 4-MeC₆H₄CO₂−]		[imidazole-furan with R¹, R²]	126
R¹				
H		MeOH, rt, 5 h	(86)	
Bn		MeOH, rt, 30 min	(56)	
4-O₂NC₆H₄(CH₂)₂ [a]		1. POCl₃, TEA, −5°, 1 h [b] 2. substrate, NaOMe, −5°, 30 min	(44)	
R¹NH₂	[Tos alkene with N=C, R²]	MeOH, rt	[imidazole with R¹, R²]	
R¹	R²			
C₀ H	Ph	3 h	(65)	86
C₁ Me	H	15 min	(5)	86
Me	t-Bu	10 min	(46)	86
Me	Ph	5 min	(87)	86
Me	4-O₂NC₆H₄	3 min	(88)	86
C₂ Et	Ph	12 h	(38)[c]	256
Et	4-PhC₆H₄	2 d	(18)[c]	255
C₃ CH₂CO₂Me	Ph	TEA, 7 d	(53)	121
i-Pr	3-PhOC₆H₄	24 h	(31)[c]	232

TABLE VIIIC. IMIDAZOLES FROM 1-ISOCYANO-1-TOSYLALKENES (Continued)

Substrate	Isocyanide	Conditions	Product(s) and Yield(s) (%)	Refs.
R^1NH_2	R^2	MeOH, rt	R^1 / R^2	
R^1	R^2			
C₄				
n-Pr	4-BrC₆H₄	24 h	(18)	115
n-Pr	4-ClC₆H₄	24 h	(52)	115
n-Pr	Ph	12 h	(53)^c	256, 267
i-Bu	4-BrC₆H₄	24 h	(47)^c	114
i-Bu	Ph	3 h	(93)	121
i-Bu	4-PhC₆H₄	24 h	(39)^c	114
i-Bu	3-PhOC₆H₄	24 h	(23)^c	232
t-Bu	Ph	24 h	(82)	86
n-Bu	Ph	3 h	(86)	121
n-Bu	3-ClC₆H₄	3 h	(29)	121
C₅				
t-BuCH₂	Ph	3 h	(79)	121
t-BuCH₂	4-BrC₆H₄	24 h	(41)^c	114
t-BuCH₂	2-ClC₆H₄	3 h	(59)	121
t-BuCH₂	3-ClC₆H₄	3 h	(25)	121
t-BuCH₂	4-ClC₆H₄	3 h	(60)	121
n-C₅H₁₁	3-ClC₆H₄	3 h	(25)	121
C₆				
Ph	Ph	—	(0)	86
C₆H₁₁	Ph	30 min	(97)	86
C₆H₁₁	Ph	24 h	(53)^c	256
t-Bu(CH₂)₂	Ph	3 h	(86)	121
C₇				
Bn	Ph	24 h	(—)	267
Bn	PhCH=CH	3 d	(73)	34
C₈				
BnCH₂	Ph	3 h	(78)	121
C₉				
(CH₂)₆CO₂Et	n-C₅H₁₁CH(OTBDMS)(CH₂)₂	Et₃N, 17 h	(35)^d	233

C_{10}	[structure]	Ph	24 h	(—)	267
C_{11}	1-$C_{10}H_7CH_2$	Ph	24 h	(32)[c]	256
C_{12}	n-$C_{12}H_{25}$	Ph	24 h	(—)	267

[a] The substrate was the amine hydrochloride.
[b] A substituted formamide was used instead of an isocyanide; this first step was required to convert it into the isocyanide.
[c] The yield was based on the formamide.
[d] The yield was based on TosMIC.

TABLE IX. 1,2,4-TRIAZOLES FROM DIAZONIUM SALTS

Substrate	Reagent	Conditions	Product(s) and Yield(s) (%)	Refs.
C_{5-10}				
$\overset{+}{R}N\equiv\overset{-}{N}X$	$TosCH_2N=C$	K_2CO_3, DMSO, MeOH, H_2O, −10°	**I** + **II**	122

R	X		
3-C_5H_4N	Cl	30 min	**I** (15) + **II** (3)
4-$O_2NC_6H_4$	BF_4	1 h	**I** (0) + **II** (0)[a]
Ph	BF_4	1 h	**I** (40) + **II** (18)
4-$MeOC_6H_4$	Cl	30 min	**I** (80) + **II** (12)
4-$Me_2NC_6H_4$	BF_4	1 h	**I** (94)
1-$C_{10}H_7$	BF_4	1 h	**I** (38) + **II** (3)

[a] N-Tosylmethyl-4-nitrobenzamide was obtained in 39% yield.

TABLE X. THIAZOLES FROM THIOCARBONYL COMPOUNDS

	Substrate	Isocyanide	Conditions	Product(s) and Yield(s) (%)	Refs.
C_0					
	H_2S^a	Tos-C(=CH-O-CH₂R)-NHCHO, R = 4-MeC₆H₄CO₂–	1. POCl₃, TEA, DME, –5°, 1 h 2. Excess substrate, 2 min	thiazole-substituted tetrahydrofuran (60)	126
C_1					
	CS_2	TosCH₂N=C	NaOH (10%), CHCl₃, TBAB, rt, 1.5 h	Tos-thiazole-S⁻ $(n\text{-Bu})_4N^+$ (94)	125
		TosCH₂N=C	1. NaOH (10%), CHCl₃, TBAB, rt, 1.5 h 2. RX, CHCl₃, rt, 3 h	Tos-thiazole-SR R X Me I (90) Et Br (93) CH₂=CHCH₂ Br (94) n-Bu Br (91) Bn Br (94)	125
		TosCH₂N=C	1. NaOH (10%), CHCl₃, TBAB, rt, 1.5 h 2. RCOCl, CHCl₃, 0°, 1 h	Tos-thiazole-SCOR R Me (86) Ph (91)	125

TABLE X. THIAZOLES FROM THIOCARBONYL COMPOUNDS (*Continued*)

Substrate	Isocyanide	Conditions	Product(s) and Yield(s) (%)	Refs.
C_{2-20}				
RN=C=S	TosCH$_2$N=C		Tos—[thiazole]—NHR **I** + Tos—[thiazole]—SH (NR) **II**	120
R				
Me		NaH, DME, rt, 0.5 h	**I** (72)	
Me		*n*-BuLi (2 eq), THF, −70 to −10°, 10 min	**II** (54)	
Et		NaH, DMSO, rt, 1 h	**I** (93)	
Et		*n*-BuLi (2 eq), THF, −70 to −10°, 10 min	**I** (50) + **II** (20)	
CH$_2$=CHCH$_2$		NaH, DME, rt, 0.5 h	**I** (62)	
EtO$_2$C		*n*-BuLi, THF, −65°, 0.5 h; 0°, 2 h	**I** (62)	
n-Bu		NaH, DME, rt, 0.5 h	**I** (77)	
Ph		NaOH (4%), TBAB, CH$_2$Cl$_2$, rt, 1 h	**I** (67)	
4-O$_2$NC$_6$H$_4$		*n*-BuLi, THF, −65°, 0.5 h; 0°, 2 h	**I** (56) + **II** (16)	
C$_6$H$_{11}$		NaH, DME, rt, 0.5 h	**I** (76)	
Bn		NaH, DME, rt, 0.5 h	**I** (74)	
4-Me$_2$NC$_6$H$_4$		NaH, DME, −50°, 0.5 h	**I** (68) + **II** (2)	
1-C$_{10}$H$_7$		*n*-BuLi, THF, −65°, 0.5 h; 0°, 2 h	**I** (38)	
Ph$_3$C		NaH, DME, rt, 0.5 h	**I** (39)	
C_6 [thiopyranone]	TosCH$_2$N=C	*n*-BuLi, THF, −78°, 1 h	[Tos-thiazole with CH(Me)CH$_2$CH$_2$OH] (67)	123

C$_7$ (furan-CS-S-CH$_2$-CO$_2$H)	TosCH$_2$N=C (2 eq)	KOH (3 eq), t-BuOH, rt, 6-12 h	(furan-thiazole-Tos) (58)	124
C$_{9-10}$ (4-RC$_6$H$_4$-CS-S-CH$_2$-CO$_2$H) R: Cl, H, Me, MeO	TosCH$_2$N=C (2 eq)	KOH (3 eq), t-BuOH, rt, 6-12 h	(4-RC$_6$H$_4$-thiazole-Tos) (48), (53), (62), (79)	124
C$_{11}$ (bicyclic thionolactone)	TosCH$_2$N=C	n-BuLi, THF, −20°, 2.5 h	(cyclohexane with thiazole-Tos and CH$_2$OH) (79)	268, 123

^{a} The thiazole was formed via an α,β-unsaturated isocyanide and hydrogen sulfide.

TABLE XI-A1. PYRROLES FROM TOSMIC OR TOSMIC HOMOLOGS TosCHRN=C AND MICHAEL ACCEPTORS

Substrate	Isocyanide	Conditions	Product(s) and Yield(s) (%)	Refs.
C₃				
CH₂=CH–CN	TosCH₂N=C	NaH, DMSO, Et₂O, 20-35°, 15 min	3-CN-pyrrole (10)	127
	Tos-CH(Ph)-N=C	NaH, DMSO, Et₂O, rt, 20 min	4-CN-2-Ph-pyrrole (60)	268a
C₄				
HC≡C–CO₂Me	TosCH₂N=C	DBU, DMF, −10°, 1 h	**I** (4-CO₂Me-pyrrole) + **II** (1-vinyl-CO₂Me-2-Tos-4-CO₂Me-pyrrole) (12) (3)	140
	TosCH₂N=C (0.5 eq)	DBU, DMF, rt, 3 h	**II** (30)	140
CH₃–CH=CH–CN	TosCH₂N=C	NaH, DMSO, Et₂O, 20-35°, 15 min	4-CN-3-Me-pyrrole (50) + N-(1-methyl-2-cyanoethyl)-4-CN-3-Me-pyrrole (9)	127
R¹–CH=CR²(R³)	Tos-CH(R³)-N=C		pyrrole with R¹, R², R³	

R¹	R²	R³		Conditions	Yield	Ref.
CN	CN	H		NaH, DMF, 0-5°	(79)	130
H	COMe	H		NaH, DMF, 0-5°, 15 min	(15)	127
H	COMe	Ph		NaH, DMF, 0-5°, 40 min	(64)	268a
H	CO₂Me	H		NaH, DMF, 0-5°, 15 min	(33)	127

	R^1	R^2	R^3			
C_5	CF_3	COMe	H	NaH, DMSO, Et_2O, rt, 30 min	(44)	269
	–$(CH_2)_2$CO–		H	NaH, DME, –20° to rt, 1 h	(30–50)	270
	H	CO_2Et	Ph	NaH, DMSO, Et_2O, rt, 30 min	(80)	268a
	Me	COMe	H	NaH, DMSO, Et_2O, 20-35°, 15 min	(45)	127
	Me	COMe	H	NaH, DMSO, Et_2O, 20-35°, 15 min	(84)	271
	Me	COMe	H	NaH, DMSO, Et_2O, 20-35°, 15 min	(64)	127
	Me	CO_2Me	Me	NaH, DMSO, Et_2O, rt	(71)	69
	Me	CO_2Me	Bn	NaH, DMSO, Et_2O, rt	(81)	69
C_6	2-C_4H_3S	NO_2	H	NaH, DMSO, Et_2O, 0-5°, 2 h	(43)	272
	CN	CO_2Et	H	NaH, DMSO, Et_2O, rt, 3 h	(54)	134
	CO_2Me	CO_2Me	H	NaH, DMSO, Et_2O, 20-35°	(60)	127
	CO_2Me	CO_2Me	CH_2TMS	t-BuOK, THF, rt, 17 h	(74)	93
	–$(CH_2)_3$CO–		H	NaH, DME, –20° to rt, 1 h	(30–50)	270
	Me	CO_2Et	H	NaH, DMSO, Et_2O, rt, 15 min	(70)	271, 273
	Et	COMe	H	NaH, DMSO, Et_2O, rt, 15 min	(91)	271

MeO_2C—≡—CO_2Me $TosCH_2N=C$ DBU, DMF, –10°, 1 h   **I** (21) + **II** (10)   140

Structure **I**: pyrrole with MeO_2C, CO_2Me, Tos, NH
Structure **II**: N-substituted pyrrole with CO_2Me, MeO_2C, Tos, N-CH=CH-CO_2Me

R–C(Tos)=N=C (0.5 eq)   DBU, DMF, rt, 3 h   (pyrrole with MeO_2C, OHC, R, NH) **(7)**

R			
Me	1. NaH, THF, 15°, 30 min; 2. HCl (0.75 N)	(84)	63
$CH_2CH_2CH=CH_2$ (allylic chain)	1. NaH, THF, rt, 1.5 h; 2. HCl, pH 1, 1 h	(78)	
$CH_2CH_2CH_2CH=CH_2$	1. NaH, THF, rt, 2.5 h; 2. HCl, pH 1, 1 h	(76)	

C_7 OMe / MeO–CH=CH–CO_2Me

TABLE XI-A1. PYRROLES FROM TOSMIC OR TOSMIC HOMOLOGS TosCHRN=C AND MICHAEL ACCEPTORS (Continued)

Substrate		Isocyanide	Conditions	Product(s) and Yield(s) (%)	Refs.
R^1	R^2	R^3		R^1 R^2 / R^3-N-H (pyrrole)	
n-C_3F_7	COMe	H	NaH, DMSO, Et_2O, rt, 30 min	(39)	269
Me	CO_2Me (=CH-)	H	NaH, DMSO, Et_2O, rt, 1 h	(56)	79
-(CH-CO_2Me)		H	n-BuLi, THF, −70 to 0°, 2 h	(34)	79
−(CH$_2$)$_4$CO−		H	NaH, DME, −20° to rt, 12 h	(30-50)	270
C_8 2,4-$Cl_2C_6H_3$	NO_2	H	NaH, DMSO, Et_2O, 0-5°, 2 h	(38)	272
Ph	NO_2	H	NaH, DMSO, Et_2O, 35°, 1 h	(27)	142
Ph	NO_2	H	NaH, DMSO, Et_2O, 0-5°, 2 h	(55)	272
Ph	NO_2	H	t-BuOK (2 eq), THF, −80 to −40°, 1 h	(87)	131
Ph	NO_2	Ph	n-BuLi, THF, −45°, 25 min	(100)	268a
2-C_4H_3O	COMe	H	NaH, DMSO, Et_2O, 20-35°, 15 min	(69)	127
CF_3	CO_2Bu-t	H	NaH, DMSO, Et_2O, rt, 30 min	(65)	274
Me	(=CH-CO_2Et)	H	NaH, DMSO, Et_2O, rt, 3 h	(80)	91, 129
(allyl)	CO_2Et	H	LiN(TMS)$_2$, THF	(61)	91
(E)-EtO_2C	CO_2Et	H	t-BuOK (2 eq), THF, rt, 2 h	(44)	274a
n-Pr	CO_2Et	H	NaH, DMSO, Et_2O, rt, 30 min	(74)	275
n-Pr	COEt	H	NaH, DMSO, Et_2O, rt, 2.5 h	(40)	276
n-Pr	COEt	H	NaH, DMSO, Et_2O, rt, 15 min	(74)	277
MeO-CH=CH-OMe	CO_2Me	H	t-BuOK (2 eq), THF, 0°, 3.5 h	(70)	63
Et-C(NO$_2$)=CH-CH=C(NO$_2$)-Et	TosCH$_2$N=C		DBU, THF, rt, 30 min	(Tos-pyrrole-Et with C≡C-Et) (11)	141

$C_{8,9}$

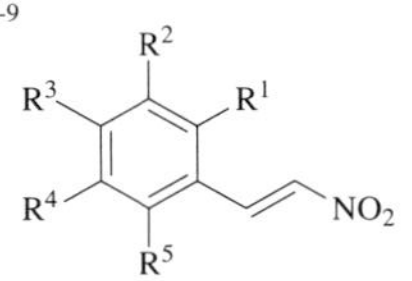

[aryl nitroalkene] + TosCH₂N=C → **277a** (aryl-nitropyrrole), NaH, DME, rt, 5 min

R¹	R²	R³	R⁴	R⁵	
Cl	H	H	H	H	(70)
H	Cl	H	H	H	(15)
H	H	Cl	H	H	(17)
Cl	H	Cl	H	H	(78)
Cl	Cl	H	H	Cl	(28)
H	Cl	Cl	Cl	H	(100)
H	H	H	Cl	H	(37)
Cl	Cl	OH	H	H	(96)
Cl	Cl	OMe	H	H	(89)

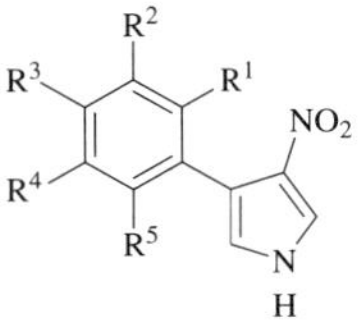

$C_{8,14}$

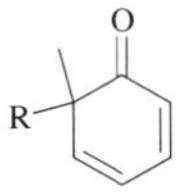

[methyl-cyclohexadienone] + TosCH₂N=C → **277b** (isoindolone NH), t-BuOK, THF, rt, 1 h

R	
Me	(75)
Ph	(60)
Bn	(56)

C_9

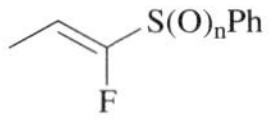

[fluoro-S(O)ₙPh alkene] + TosCH₂N=C → **139**, NaH (2 eq), THF, rt, 3 h

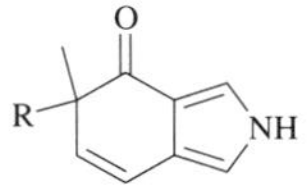

Products: 3-methyl-4-S(O)ₙPh-5-Tos-pyrrole + 3-fluoro-4-methyl-5-Tos-pyrrole

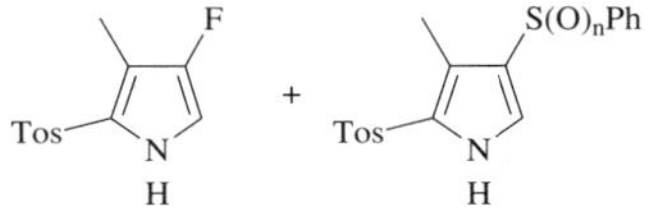

n		
1	(24)	(24)
2	(23)	(8)

TABLE XI-A1. PYRROLES FROM TOSMIC OR TOSMIC HOMOLOGS TOSCHRN=C AND MICHAEL ACCEPTORS (*Continued*)

Substrate	Isocyanide	Conditions	Product(s) and Yield(s) (%)	Refs.
Cl–(CH₂)–COPh	Ph–C(Tos)(N≡C)	NaH (3 eq), DMSO, Et₂O, rt, 5 min	Ph–pyrrole–COPh (85)	268a
R¹–CH=CH–R²	R³–C(Tos)(N≡C)		R¹,R²,R³-pyrrole-H	

R¹	R²	R³	Conditions	Yield	Refs.
Ph	CN	H	NaH, DMSO, Et₂O, 20–35°, 15 min	(35)	127
Ph	CN	H	t-BuOK (2 eq), THF, –10°, 10 min	(88)	136, 63
Ph	CN	Ph	n-BuLi, THF, rt, 5 d	(24)	268a
Ph	CN	Bn	NaH, DMSO, Et₂O, rt	(80)	69
4-MeC₆H₄	NO₂	H	NaH, DMSO, Et₂O, 0–5°, 2 h	(44)	272
4-MeOC₆H₄	NO₂	H	NaH, DMSO, Et₂O, 0–5°, 2 h	(58)	272
H	CO₂Me	Me	NaH, THF, rt, 15 h	(75)	132, 278

Substrate	Isocyanide	Conditions	Product(s) and Yield(s) (%)	Refs.
furyl-CO-C≡C-CO₂Bu-t	TosCH₂N=C (0.5 eq)	DBU, DMF, rt, 3 h	MeO₂C/furoyl/Tos pyrrole (68)ᵃ	140
C₈₋₁₃ diazepine (R¹, R²)	TosCH₂N=C	NaH, DME, –35°	pyrrolo-diazepine (R¹, R²)	279

594

R^1	R^2				
$CO_2CH_2CCl_3$	H			(20)	
COMe	H			(53)	
CO_2Et	H			(52)	
COMe	Me			(48)	
COPh	H			(50)	
Tos	H	$TosCH_2N=C$	t-BuOK, THF, rt, 1 h	(49)	277b

[structure: bicyclic cyclohexadienone fused indole-type NH with methyl and propargyl substituents]

R^1	R^2	R^3			
PhCH=CH	NO_2	H	t-BuOK (2 eq), THF, −80 to −40°, 1 h	(73)	131
H	$PhSO_2CH=CH$	Me	NaH, DMSO, THF, rt, 15 h	(64)	132, 133
Ph	COMe	H	NaH, DMSO, Et_2O, 20-35°, 15 min	(70)	127
Ph	COMe	Ph	n-BuLi, THF, 45°, 1 h	(65)	268a
Ph	CO_2Me	H	NaH, DMSO, Et_2O, 20-35°, 15 min	(70)	127
Ph	CO_2Me	Ph	NaH, DMSO, Et_2O, 20-35°, 15 min	(23)	127
Ph	CO_2Me	Ph	n-BuLi, THF, rt, 8 d	(45)	268a
2,4-$MeOC_6H_3$	NO_2	H	NaH, DMSO, Et_2O, 0-5°, 2 h	(40)	272
n-Bu	CO_2Pr-n	H	NaH, DMSO, Et_2O, rt, 30 min	(74)	275
n-Bu	CO_2Pr-n	H	NaH, DMSO, Et_2O, rt, 15 min	(69)	277
i-Bu	COPr-i	H	NaH, DMSO, Et_2O, 45°, 35 min	(52)	280
n-C_5H_{11}	CO_2Et	H	NaH, DMSO, Et_2O, rt, 15 min	(95)	281

TABLE XI-A1. PYRROLES FROM TOSMIC OR TOSMIC HOMOLOGS TosCHRN=C AND MICHAEL ACCEPTORS (*Continued*)

Substrate	Isocyanide	Conditions	Product(s) and Yield(s) (%)	Refs.
C₁₁ aryl-CH=C(CN)CO₂H; R¹=Cl, R²=Cl, R³=H	TosCH₂N=C	KOH, MeOH, CH₂Cl₂, 0-5°, 20 min	4-aryl-3-CN-pyrrole (99)	135
R¹=F, R²=Cl, R³=H	TosCH₂N=C	EtONa, CH₂Cl₂, rt, 3 h	(94)	137
R¹=H, R²=H, R³=H	PhSO₂CH₂N=C	KOH, MeOH, CH₂Cl₂, 0°, 3 h	(75)	135
R¹=H, R²=H, R³=H	4-ClC₆H₄SO₂CH₂N=C	KOH, MeOH, CH₂Cl₂, 0°, 2 h	(89)	135
R¹=H, R²=H, R³=CN	TosCH₂N=C	K₂CO₃, MeOH, CH₂Cl₂, 40°, 3 h	(81)	135
C₁₀ 4-MeO-C₆H₄-C(NO₂)=CHMe	TosCH₂N=C	DBU, THF, *i*-PrOH, rt, 17 h	2-Tos, 3-(4-MeO-C₆H₄), 4-Me pyrrole (52)	138
C₁₁ PhCO-C≡C-CO₂Me	TosCH₂N=C (0.5 eq)	DBU, DMF, rt, 3 h	pyrrole product (25)	140
MeO₂C-CH=CH-COPh	TMS-CH(Tos)-N=C	1. Triton B, THF, 0°, 5 min 2. *t*-BuOK, MeI, THF, rt, 15 min	N-Me, 2-TMS, 3-MeO₂C, 4-COPh pyrrole (82)	93

[a]

R^1	R^2	$R^3\text{-CH(Tos)-N=C}$ R^3		Product (R¹,R²,R³-pyrrole)	Conditions	Yield (%)	Ref.
4-ClC₆H₄	CO₂Et	TosCH₂N=C	H		NaH, DMSO, Et₂O, rt, 2 h	(83)	282
Ph	CO₂Et	TosCH₂N=C	H		NaH, DMSO, Et₂O, rt, 15 min	(60)	271
Ph	CO₂Et	TosCH₂N=C	H		NaH, DMSO, Et₂O, rt, 2 h	(71)	282
Ph	CO₂Et	TosCH₂N=C	Ph		n-BuLi, THF, rt, 8 d	(43)	268a
C₁₂Ph	NO₂	TosCH₂N=C	H		t-BuOK (2 eq), THF, −80 to −40°, 1 h	(75)	131
Me₂NCO	COPh	TosCH₂N=C	TMSCH₂		Triton B, THF, rt, 15 min	(90)	93
4-MeOC₆H₄	CO₂Et	TosCH₂N=C	H		NaH, DMSO, Et₂O, rt, 2 h	(72)	282
BnOCH₂	COMe	TosCH₂N=C	H		NaH, DMSO, Et₂O, 20-35°, 15 min	(83)	33
—(CH₂)₃CO—		TosCH₂N=C	H		NaH, DME, −20° to rt, 1 h	(30–50)	270
Me	CO₂C₈H₁₇-n	TosCH₂N=C	H		NaH, DMSO, Et₂O, rt, 30 min	(91)	275, 271
n-C₅H₁₁	CO₂Bu-n	TosCH₂N=C	H		NaH, DMSO, Et₂O, rt, 30 min	(34)	275
n-C₅H₁₁	CO₂Bu-n	TosCH₂N=C	H		NaH, DMSO, Et₂O, rt, 15 min	(34)	277
n-C₇H₁₅	CO₂Et	TosCH₂N=C	H		NaH, DMSO, Et₂O, rt, 30 min	(80)	281

(thiophene-CH=C(NO₂)— structure with two thienyl groups) TosCH₂N=C — DBU, THF, 30 min (40) 141

(arylidene with CN, CO₂R⁴ structure) — product: pyrrole with aryl and CN substituents — 135

R¹	R²	R³	R⁴	TosCH₂N=C reagent	Conditions	Yield (%)
Cl	H	H	Et	TosCH₂N=C	EtONa, EtOH, CH₂Cl₂, 0-3°, 1 h	(98)
H	H	H	Et	TosCH₂N=C	NaH, DME, rt, 1 h	(89)
H	H	H	Et	4-ClC₆H₄SO₂CH₂N=C	NaOMe, MeOH, CH₂Cl₂, 0°, 3 h	(89)
H	MeO	MeO	H	TosCH₂N=C	KOH, DME, rt, 2 h	(92)

TABLE XI-A1. PYRROLES FROM TOSMIC OR TOSMIC HOMOLOGS TOSCHRN=C AND MICHAEL ACCEPTORS (*Continued*)

Substrate	Isocyanide	Conditions	Product(s) and Yield(s) (%)	Refs.
C$_{13}$ (Tos-alkenyl-NO$_2$)	TosCH$_2$N=C	DBU, MeCN, 0°, 30 min	(pyrrole product) (80)	283
BnO-CO$_2$Et alkene	TosCH$_2$N=C	NaH, DMSO, Et$_2$O, 20-35°, 15 min	(—)	284
C$_{13-14}$ (COMe cyclohexenyl)	TosCH$_2$N=C	NaH, DMSO, Et$_2$O, 20-35°, 15 min	(7)	127
(phthalimide CO$_2$R alkene)	TosCH$_2$N=C	*t*-BuOK, THF, rt, 30 min NaH, DMSO, Et$_2$O, rt, 4 h	(23) (6)	134
C$_{13-19}$ (ferrocenyl-R alkene) R: CN, CO$_2$Me, COMe, COPh	TosCH$_2$N=C	NaH, DMSO, Et$_2$O, 0°, 1.5 h	(35) (56) (42) (52)	285

C₁₄				
[structure: tricyclic ketone with methyl benzofuran]	TosCH₂N=C	NaH, DMSO, Et₂O, rt, 30 min	[pyrrole-fused tricyclic product] (25)	286
Ph-CH=C(CO₂Et)(CO₂Et)	TosCH₂N=C	NaOH, EtOH, 0°, 1 h	Ph-(3-CO₂Et)pyrrole (60)	136
Et₂NCO-CH=CH-COPh	TMS-CH(Tos)-N=C	1. Triton B, THF, rt, 30 min 2. t-BuOK, MeI, THF, rt, 30 min	Et₂NCO-(COPh)(TMS-CH₂)(Me)pyrrole (77)	93

R¹	R²	R³-CH(Tos)-N=C		R¹,R²,R³-pyrrole	
R¹	R²	R³			
n-C₆H₁₃	CO₂C₅H₁₁-n	H	NaH, DMSO, Et₂O, rt, 15–30 min	(28)	277, 275
C₁₅ Ph	COPh	H	NaH, DMSO, Et₂O, 20-35°, 15 min	(70)	127
Ph	COPh	H	Triton B (2 eq), THF, rt, 15 min	(88)	136
Ph	COPh	Ph	n-BuLi, THF, rt, 3 h	(80)	268a
Ph	COPh	CH₂TMS	t-BuOK, THF, rt, 15 min	(79)	93
Ph	COPh	Me	NaH, DMSO, Et₂O, rt	(83)	69
Ph	COPh	CH₂=CHCH₂	NaH, DMSO, Et₂O, rt	(78)	69
Ph	COPh	Bn	NaH, DMSO, Et₂O, rt	(83)	69

TABLE XI-A1. PYRROLES FROM TOSMIC OR TOSMIC HOMOLOGS TOSCHRN=C AND MICHAEL ACCEPTORS (Continued)

Substrate R1	R2	R3	R4	R5	R6	Isocyanide	Conditions	Product(s) and Yield(s) (%)	Refs.
H	H	H	H	H	Cl	TosCH$_2$N=C	NaH, DMSO, Et$_2$O, rt, 5 min	(39)	287
H	H	H	H	H	F		5 min	(44)	
H	H	H	H	H	NO$_2$		30 min	(28)	
H	H	H	Cl	H	Cl		5 min	(65)	
H	H	H	H	Cl	Cl		5 min	(12)	
H	H	Cl	H	H	H		5 min	(53)	
H	H	Cl	H	H	Cl		5 min	(35)	
H	H	Cl	H	H	F		5 min	(50)	
H	H	Cl	H	H	NO$_2$		30 min	(29)	
H	H	Cl	Cl	H	Cl		5 min	(48)	
Cl	H	H	H	H	Cl		5 min	(100)	
Cl	H	H	H	H	H		5 min	(50)	
Cl	H	H	Cl	H	Cl		30 min	(40)	
Cl	H	H	H	H	F		5 min	(34)	
Cl	H	H	H	H	NO$_2$		5 min	(65)	
Cl	H	H	H	Cl	Cl		5 min	(41)	
Cl	H	Cl	H	H	H		5 min	(48)	
Cl	H	Cl	H	H	Cl		30 min	(53)	
Cl	H	Cl	Cl	H	Cl		5 min	(99)	
Cl	H	Cl	H	H	F		10 min	(48)	
Cl	H	Cl	H	H	NO$_2$		45 min	(65)	
Cl	H	Cl	H	H	Cl		10 min	(23)	

	R^1	R^2	R^3	R^4	R^5	R^6			
	Cl	H	Cl	H	Cl	Cl	20 min	(58)	
	Cl	Cl	H	H	H	H	5 min	(37)	
	Cl	Cl	H	H	H	Cl	5 min	(56)	
	Cl	Cl	H	H	H	F	5 min	(48)	
	Cl	Cl	H	H	H	NO$_2$	30 min	(31)	
	Cl	Cl	H	H	H	Cl	5 min	(45)	
	Cl	Cl	H	Cl	H	Cl	5 min	(38)	
	Cl	Cl	H	H	Cl	Cl	5 min	(47)	
C$_{16}$	H	H	H	H	H	Me	15 min	(33)	
	Cl	Cl	H	H	H	Me	5 min	(41)	
	Cl	H	Cl	H	H	Me	20 min	(49)	
	Cl	Cl	H	H	H	Me	5 min	(52)	

$R^1\text{–CH=CH–}R^2$ + TosCH$_2$N=C → pyrrole (R^1, R^2 on 3,4-positions, NH)

R^1	R^2	Conditions	Yield	Ref
PhCO	PhCO	NaH, DMSO, Et$_2$O, 20–35°, 15 min	(50)	127
n-C$_7$H$_{15}$	n-C$_6$H$_{13}$CO$_2$	NaH, DMSO, Et$_2$O, rt, 30 min	(31)	275
BnO/MeO-aryl	NO$_2$	DBU, THF, i-PrOH, −78 to −20°; −20°, 2 h	(14)	138

Ph(NO$_2$)C=C(NO$_2$)Ph + TosCH$_2$N=C → 3-Ph, 4-(phenylethynyl), 2-Tos, NH-pyrrole

DBU, THF, rt, 30 min   (72)   141

TABLE XI-A1. PYRROLES FROM TOSMIC OR TOSMIC HOMOLOGS TosCHRN=C AND MICHAEL ACCEPTORS (*Continued*)

Substrate	Isocyanide	Conditions	Product(s) and Yield(s) (%)	Refs.
(structure: CO₂Et-pyrrole with SO₂Ph and Me)	Tos–CHR–N≡C R: H, Me, CH₂=CHCH₂, TMSCH₂	NaH, (TMS)₂NH	(structure with EtO₂C, NH, R, Me, PhSO₂) H (85) Me (70) CH₂=CHCH₂ (96) TMSCH₂ (40)	91, 129 91 91, 129 91
(structure: O₂N, C₆H₁₁, NO₂, C₆H₁₁ dienedinitro)	TosCH₂N=C	DBU, THF, rt, 30 min	(pyrrole product with C₆H₁₁, alkyne-C₆H₁₁, Tos, HN) (58) + (pyrrole with O₂N, C₆H₁₁, Tos) (12) + (bipyrrole with C₆H₁₁, Tos) (9)	141

602

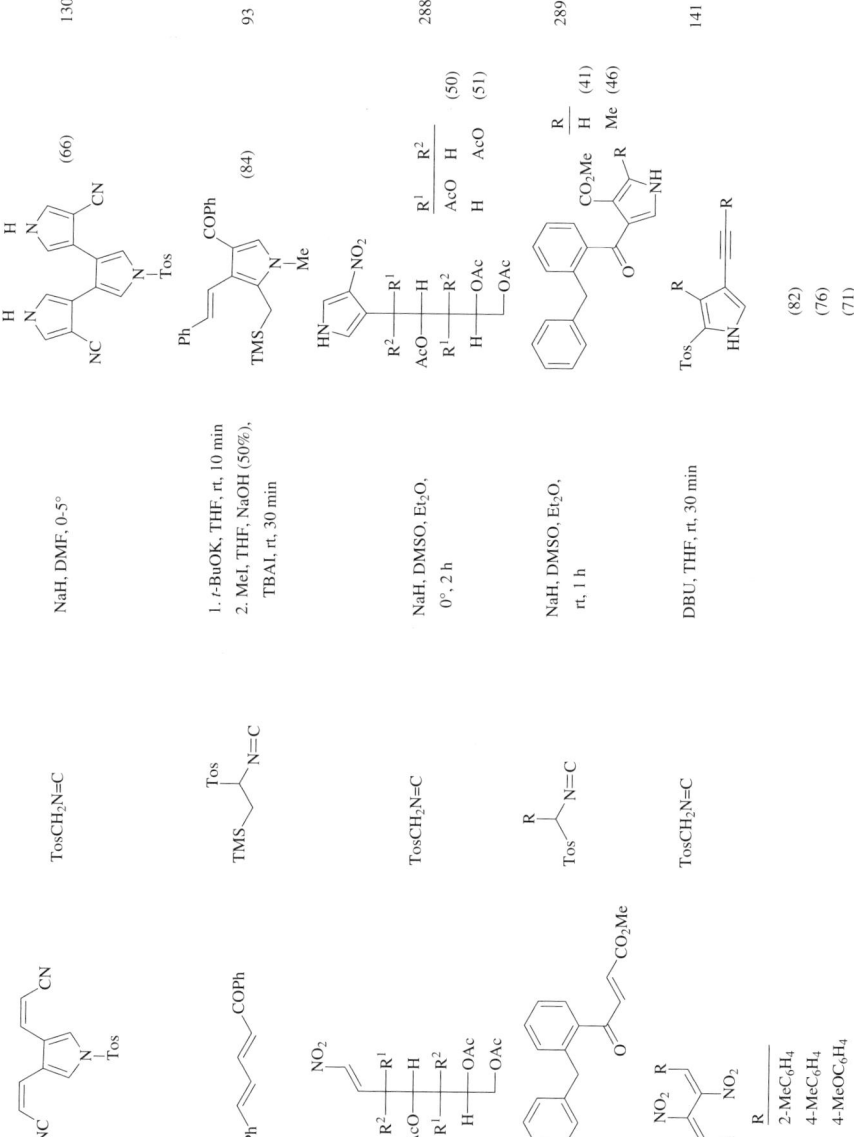

TABLE XI-A1. PYRROLES FROM TOSMIC OR TOSMIC HOMOLOGS TOSCHRN=C AND MICHAEL ACCEPTORS (Continued)

Substrate	Isocyanide	Conditions	Product(s) and Yield(s) (%)	Refs.

C_{18-19}

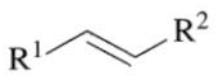

Substrate: R¹–CH=CH–R²

R¹	R²			
n-C_8H_{17}	n-$C_7H_{15}CO_2$			
4-ClC_6H_4	1-$C_{10}H_7CO$			

Isocyanide: $TosCH_2N=C$

Conditions: NaH, DMSO, Et_2O, rt; 30 min / 5 min

Product: pyrrole with R¹, R² substituents; (42), (52); Refs. 275, 287

Substrate: vinyl pyrrole with R¹ and SO₂Ph, methyl

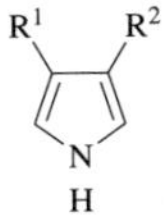

Isocyanide: R^2–CH(Tos)–N=C

R²	
H	
H	
Et	

R¹		
CO_2Bu-t		
SO_2Ph		
SO_2Ph		

Conditions: NaH, THF, rt, 1.5 h; NaH, THF, DMSO, rt; NaH, DMSO, $(TMS)_2NH$, rt

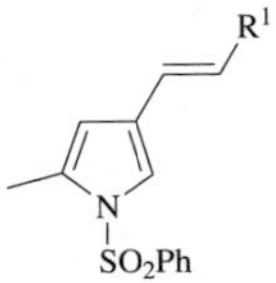

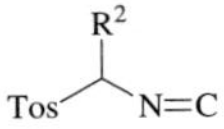

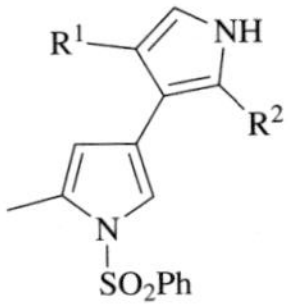

Product: 3-substituted pyrrole with NH, R², and SO_2Ph/methyl pyrrole

Yields: (74), (72), (95); Refs. 133, 278; 133, 278; 133

C_{19}

Substrate: chalcone with N-pyrrolyl phenyl ketone and R¹, R², R³ substituted phenyl

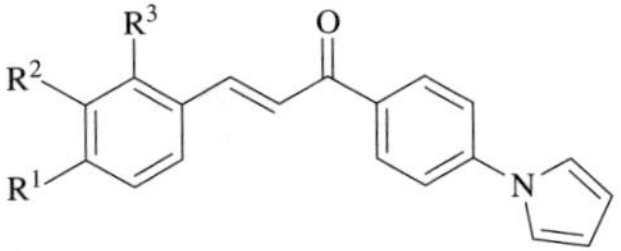

Isocyanide: $TosCH_2N=C$

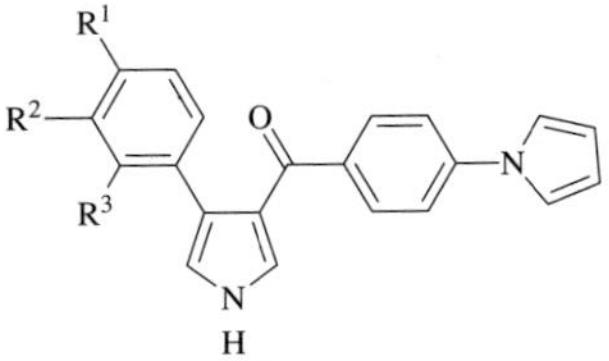

Conditions: NaH, DMSO, Et_2O, rt; 30 min / 5 min / 5 min / 20 min / 25 min

R¹	R²	R³
H	H	H
H	H	Cl
Cl	H	H
Cl	H	Cl
Cl	Cl	H

Product yields: (23), (54), (93), (33), (57); Ref. 287

C_{20}	TosCH$_2$N=C	NaH, DME, 0°		290
			(52) (36) (19) (62)	
			R^1 / R^2: H/H, Me/H, Cl/H, Cl/Cl	
C_{21}	TosCH$_2$N=C	NaH, DMSO, Et$_2$O	(25)	291
C_{23}	TosCH$_2$N=C	NaH, DMSO, Et$_2$O, rt, 5 min	(54)	287
	TosCH$_2$N=C	LiN(TMS)$_2$, THF, rt, 15 min	(72)	132

TABLE XI-A1. PYRROLES FROM TosMIC OR TosMIC HOMOLOGS TosCHRN=C AND MICHAEL ACCEPTORS (Continued)

Substrate	Isocyanide	Conditions	Product(s) and Yield(s) (%)	Refs.
C_{24} (dinaphthyl dinitro substrate)	TosCH$_2$N=C	DBU, THF, rt, 30 min	(59)	141
C_{39} (tris-pyrrole divinyl dinitrile substrate)	TosCH$_2$N=C	NaH, DMF, 0-5°	(83)	130
C_{61} (pentakis-pyrrole substrate)	TosCH$_2$N=C	NaH, DMF, 0-5°	(73)	130

[a] Since aroyl groups are more strongly directing than ester groups, it seems likely that the positions of the aroyl and ester groups at C-3 and C-4 should be exchanged. The reported spectral data are consistent with either structure.

TABLE XI-A2. PYRROLES FROM TOSMIC HOMOLOGS TosC(=CR$_2$)N=C AND MICHAEL ACCEPTORS

	Substrate				Isocyanide				Conditions	Product(s) and Yield(s) (%)	Refs.
	R^1	R^2			R^3	R^4					
C_4	CN	CN			(CH$_2$)$_4$				t-BuOK (2 eq), THF, −60° to rt, 3 h	(46)	63
	CN	CN			Ph	Me			t-BuOK (2 eq), THF, −90°, stand 45 min	(81)	63
	H	COMe			Me	H			t-BuOK, THF, −60 to 20°, 1 h	(88)	128
	H	CO$_2$Me			Me	H			t-BuOK, THF, −60 to 20°, 1 h	(90)	128
	H	COMe			(CH$_2$)$_4$				t-BuOK, THF, −60 to 20°, 1 h	(96)	128
	H	CO$_2$Me			(CH$_2$)$_4$				t-BuOK, THF, −60 to 20°, 1 h	(83)	128
C_5	CN	COMe			(CH$_2$)$_4$				Triton B (1.9 eq), THF, rt, 1.5 h	(98)	63
	CN	CO$_2$Me			(CH$_2$)$_4$				t-BuOK (2 eq), THF, −70°, stand 30 min	(79)	63
									1. t-BuOK (3 eq), THF, −70 to −55°, 1.5 h 2. HCl (0.1 N), −60 to −10°	(66)	63
									1. t-BuOK (2 eq), THF, −90 to −40°, 0.5 h 2. HCl (0.1 N), −40°	(55)	63
C_6	R^1	R^2			R^3	R^4					
	Me	CO$_2$Me			(CH$_2$)$_3$				t-BuOK, THF, −60 to 20°, 1 h	(84)	128
	CO$_2$Me	COMe			(CH$_2$)$_4$				Triton B, THF, 20-30°, 45 min	(83)	63
	CO$_2$Me	CO$_2$Me			(CH$_2$)$_4$				t-BuOK (2 eq), THF, −70°, stand 2 h	(92)[a]	63

TABLE XI-A2. PYRROLES FROM TOSMIC HOMOLOGS TosC(=CR$_2$)N=C AND MICHAEL ACCEPTORS (*Continued*)

Substrate	Isocyanide	Conditions	Product(s) and Yield(s) (%)	Refs.
C$_7$				
R^1=CH$_2$=C(OMe), R^2=CO$_2$Me	(CH$_2$)$_4$	*t*-BuOK (2 eq), THF, −30°, stand 3 h	(28)	63
R^1=MeC(NOMe), R^2=CO$_2$Me, *E/Z* = 5:1	(CH$_2$)$_4$	*t*-BuOK (2 eq), THF, −45° to rt, 1 h	(92) *E/Z* = 5:1	63
R^1=MeC(NOMe), R^2=CO$_2$Me, *E/Z* = 5:1	Ph, Me	*t*-BuOK (2 eq), THF, −15°; rt, 4.5 h	(87) *E/Z* = 5:1	63
OMe–CH=CH–CH(OMe)–CO$_2$Me	cyclohexylidene, Tos	1. NaH, DME, −20° to rt, 2 h; 2. MeI, PTC; 3. HCl (1.5 N)	OHC, CO$_2$Me pyrrole N-Me, cyclohexenyl (90)	63
	Et, Ph, Tos, N=C	1. NaH, DME, −20°, 3 h; 2. MeI, PTC; 3. HCl (1.5 N)	OHC, CO$_2$Me pyrrole N-Me, CH=CHMe-Ph (95)	63
C$_8$	cyclohexylidene, Tos, N=C		CO$_2$Et pyrrole NH, propenyl, cyclohexenyl (20) + CO$_2$Et pyrrole NH, Me, cyclohexenyl (55)	128
CH=CH–CH=CH–CO$_2$Et		*t*-BuOK, THF, −60 to 20°		

R^1	R^2	R^3	R^4		Conditions	Product (%)	Ref.
3-C$_5$H$_4$N	CN	(CH$_2$)$_4$			t-BuOK (2 eq), THF, −55° to rt, 1.5 h	(86)	63
Ph	NO$_2$	Me	H		t-BuOK (2 eq), THF, −80 to −40°, 1 h	(84)	131
Ph	NO$_2$	(CH$_2$)$_4$			t-BuOK (2 eq), THF, −80 to −40°, 1 h	(93)	131
CO$_2$Et	CO$_2$Et	Ph	Me		t-BuOK (2 eq), THF, −50°; stand 30 min	(81)	63
(dioxolane)	CO$_2$Me	(CH$_2$)$_4$			NaH, t-BuOK or BuLi, −70° to rt	(0)	63
(MeO,MeO)	CO$_2$Me	(CH$_2$)$_4$			NaH, t-BuOK or BuLi, −70° to rt	(0)	63

C$_9$
| Ph | CN | (CH$_2$)$_4$ | | t-BuOK (2 eq), THF, −45°; rt, 3 h | (74) | 63 |
| Ph | CO$_2$Me | (CH$_2$)$_4$ | | t-BuOK, THF, −60 to 20°, 1 h | (95) | 128 |

C$_{10}$
| (E)-PhCH=CH | NO$_2$ | Me | H | t-BuOK (2 eq), THF, −80 to −40°, 1 h | (78) | 131 |
| (E)-PhCH=CH | NO$_2$ | (CH$_2$)$_4$ | | t-BuOK (2 eq), THF, −80 to −40°, 1 h | (71) | 131 |

C$_{11}$
HO$_2$C–...–COPh Triton B, THF, 20-30°, 20 min (76) 63

R^1	R^2	R^3	R^4	R^5	Conditions	(%)	Ref.
2-C$_4$H$_3$O	2-C$_4$H$_3$OCO	(CH$_2$)$_3$		H	t-BuOK, THF, −60 to 20°, 1 h	(93)	128
2-C$_4$H$_3$S	2-C$_4$H$_3$SCO	(CH$_2$)$_4$		H	t-BuOK, THF, −60 to 20°, 1 h	(85)	128
CO$_2$Me	PhCO	Me	H	H	Triton B, THF, rt, 30 min	(92)	63

TABLE XI-A2. PYRROLES FROM TOSMIC HOMOLOGS TosC(=CR$_2$)N=C AND MICHAEL ACCEPTORS (Continued)

	Substrate		Isocyanide		Conditions	Product(s) and Yield(s) (%)		Refs.
	R^1	R^2	R^3	R^4		R^5		
C$_{12}$	CO$_2$Me	PhCO	(CH$_2$)$_4$		Triton B, THF, rt, 30 min	H	(83)	63
	CO$_2$Me	PhCO	Ph	Me	Triton B, THF, 0° to rt, 40 min	H	(85)	63
	Ph	CO$_2$Me	(CH$_2$)$_4$		t-BuOK (1.2 eq), THF, −40 to 20°, 1 h	H	(60)	143
	Ph	NO$_2$	Me	H	t-BuOK (2 eq), THF, −80 to −40°, 1 h	H	(89)	131
	Ph	NO$_2$	(CH$_2$)$_4$		t-BuOK (2 eq), THF, −80 to −40°, 1 h	H	(85)	131
	Ph	NO$_2$	(CH$_2$)$_4$		1. t-BuOK, THF; 2. MeI, PTC	Me	(67)	146
	Ph	NO$_2$	Ph	H	t-BuOK (2 eq), THF, −80 to −40°, 1 h	H	(80)	131
	Ph	COMe	Ph	H	t-BuOK, THF, −60 to 20°, 1 h	H	(92)	128
	Ph	COMe	(CH$_2$)$_5$		1. t-BuOK, THF, −60 to 20°, 1 h 2. MeI, KOH (50%), C$_6$H$_6$, rt, 1 h	Me	(87)	128
	Ph	CO$_2$Me	(CH$_2$)$_4$		1. t-BuOK, THF, −60 to 20°, 1 h 2. MeI, KOH (50%), C$_6$H$_6$, rt, 1 h	Me	(90)	128, 80
C$_{13}$	2-C$_4$H$_3$O	PhCO	(CH$_2$)$_5$		t-BuOK, THF, −60 to 20°, 1 h	H	(89)	128
C$_{15}$	Ph	PhCO	Me	H	t-BuOK, THF, −60 to 20°, 1 h	H	(95)	128
	Ph	PhCO	(CH$_2$)$_3$		t-BuOK, THF, −60 to 20°, 1 h	H	(95)	128
	Ph	PhN=CH	(CH$_2$)$_5$		t-BuOK, THF, −78 to 20°, 2 h	H	(70)	80
C$_{16}$	PhCO	PhCO	(CH$_2$)$_4$		Triton B, THF, 35°, 1 h	H	(84)	63
	PhCO	PhCO	Ph	H	t-BuOK, THF, −60° to rt, 40 min	H	(70)	63
	PhCO	PhCO	Ph	Me	t-BuOK (2 eq), THF, −50° to rt, 40 min	H	(75)	63
	Ph	TsN=CH	(CH$_2$)$_4$		t-BuOK, THF, −78° to rt, 2 h	H	(34)^b	80
C$_{17}$	Ph	Ph—CH=CH—CO	(CH$_2$)$_4$		t-BuOK, THF, −60 to 20°, 1 h	H	(96)	128

R¹	R²	R³	R⁴		R⁵		
Ph⤳	PhCO	Me	H	t-BuOK, THF, −60 to 20°, 1 h	H	(96)	128
Ph⤳	PhCO	(CH₂)₃		1. t-BuOK, THF, −60 to 20°, 1 h 2. MeI, KOH (50%), C₆H₆, rt, 1 h	Me	(84)	128
Ph⤳	PhCO	(CH₂)₄		1. t-BuOK, THF, −60 to 20°, 1 h	H	(91)	128, 80
Ph⤳	PhCO	(CH₂)₅		1. t-BuOK, THF, −60 to 20°, 1 h 2. MeI, KOH (50%), C₆H₆, rt, 1 h	Me	(87)	128
Ph⤳	PhCO	(CH₂)₆		1. t-BuOK, THF, −60 to 20°, 1 h 2. MeI, KOH (50%), C₆H₆, rt, 1 h	Me	(83)	128
Ph⤳	PhCO	(CH₂)₁₀		2. MeI, KOH (50%), C₆H₆, rt, 1 h	Me	(82)	128

[a] The same pyrrole was obtained in 90% yield from dimethyl maleate instead of dimethyl fumarate.
[b] An imidazole (34%) was also formed by reaction with the carbon-nitrogen double bond.

TABLE XI-B. PYRROLES FROM 1-ISOCYANO-1-TOSYLALKENES AND MICHAEL DONORS

Substrate	Michael Donor		Conditions	Product(s) and Yield(s) (%)	Refs.
$R^1\!\!-\!\!C(H)\!\!=\!\!C(Tos)\!\!-\!\!N\!\!\equiv\!\!C$	$R^2\text{-}R^3$			R^1, R^3-pyrrole (N-H)	
R^1	R^2	R^3			
C₁₄ t-Bu	H	NO₂	t-BuOK, DME, rt, 15 min	(91)	142
n-Bu	MeCO	CO₂Et	t-BuOK, EtOH, rt, 1 h	(62)	136
n-Bu	PhCO	Me	1. t-BuOK, DME, rt, 1 h 2. MeOH, rt, 5 min	(33)	136
C₁₆ 2,3-Cl₂C₆H₃	EtO₂C	CN	NaOH, EtOH, rt, 1 h	(93)	136
4-ClC₆H₄	H	NO₂	t-BuOK, DME, rt, 15 min	(86)	142
Ph	H	NO₂	t-BuOK, DME, rt, 15 min	(94)	142
Ph	H	CN	t-BuOK, DME, rt, 15 min	(0)	136
Ph	H	CO₂Et	t-BuOK, DME, rt, 15 min	(0)	136
Ph	EtO₂C	CN	Na, EtOH, rt, 30 min	(99)	136
Ph	MeCO	COMe	NaOH, MeOH, rt, 2.5 h	(86)	136
Ph	MeCO	CO₂Et	Na, EtOH, rt, 30 min	(92)	136
Ph	EtO₂C	CO₂Et	Na, EtOH, rt, 30 min	(70)	136
Ph	H	COPh	t-BuOK, DME, rt, 1 h	(57)	136
Ph	PhCO	Me	1. t-BuOK, DME, rt, 1 h 2. MeOH, rt, 5 min	(73)	136
Ph	EtO₂C	Ph	t-BuOK, DME, rt, 1 h	(57)	136
Ph	PhCO	COPh	1. NaOH, MeOH, reflux, 5 min 2. rt, 3 h	(61)	136
C₁₇ 4-MeOC₆H₄	H	NO₂	t-BuOK, DME, rt, 15 min	(88)	142

TABLE XII. KETONES BY ACID HYDROLYSIS OF DISUBSTITUTED TOSMIC DERIVATIVES

Substrate	Conditions	Product(s) and Yield(s) (%)	Refs.
C_{11} Tos-C(=N=C) (cyclopropane)	HCl (37%), THF, −10°, 5 min	cyclopropanone =O (0), Tos-C(HN—CHO) cyclopropane (79)[a]	99
C_{12} Tos-C(N=C) cyclobutane	H$_2$SO$_4$ (50%), sulfolane, 20 to 100°, 2 h	cyclobutanone (84)	148
C_{13} Tos-C(N=C) cyclobutane-D (−)	H$_2$SO$_4$ (50%), sulfolane, 20 to 100°, 2 h	cyclobutanone-D (2R,3R)-(+) (97)	245
Tos-C(N=C) gem-dimethylcyclopropane	HCl (37%), THF, −10°, 5 min	Tos-C(HN—CHO) gem-dimethylcyclopropane (84)[a]	102
	1. HCl (37%), MeOH, −10°, 1 h 2. rt, 20 h	Tos-C(NH$_3^+$ Cl$^-$) gem-dimethylcyclopropane (92)[a]	102
Tos-C(N=C) methylcyclobutane	H$_2$SO$_4$ (50%), sulfolane, 20 to 100°, 2 h	methylcyclobutanone (88)	148
Tos-C(N=C) methylcyclobutane	H$_2$SO$_4$ (50%), sulfolane, 20 to 100°, 2 h	methylcyclobutanone (88)	148
Tos-C(N=C) tetrahydropyran	HCl (37%), CH$_2$Cl$_2$, Et$_2$O rt, 5 min	tetrahydropyranone (16)[b]	98

TABLE XII. KETONES BY ACID HYDROLYSIS OF DISUBSTITUTED TOSMIC DERIVATIVES (Continued)

Substrate	Conditions	Product(s) and Yield(s) (%)	Refs.
C₁₄ (Tos, N=C, isopropenyl-methyl)	HCl (37%), Et₂O rt, <5 min	methyl isopropenyl ketone (51)	103
(Tos, N=C, cyclobutenyl-methyl)	HCl (37%), Et₂O rt, <5 min	acetyl cyclobutene (69)	103
(Tos, N=C, dimethylcyclobutyl)	H₂SO₄ (50%), sulfolane	dimethyl cyclobutanone (20)[b] $Z:E = 3:1$	247
C₁₅ (Tos, N=C, cyclopentenyl-methyl)	HCl (37%), Et₂O rt, <5 min	acetyl cyclopentene (91)	103
(Tos, N=C, cyclopropyl-cyclobutyl)	H₂SO₄ (50%), sulfolane, rt, 10 min	cyclopropyl cyclobutanone (30)	247
(Tos, N=C, isopropyl-cyclobutyl)	H₂SO₄ (50%), sulfolane, 120°, 3 h	isopropyl cyclobutanone (75)	247
(Tos, N=C, cycloheptyl)	HCl (37%), CH₂Cl₂, Et₂O rt, 5 min	cycloheptanone (51)[b]	98

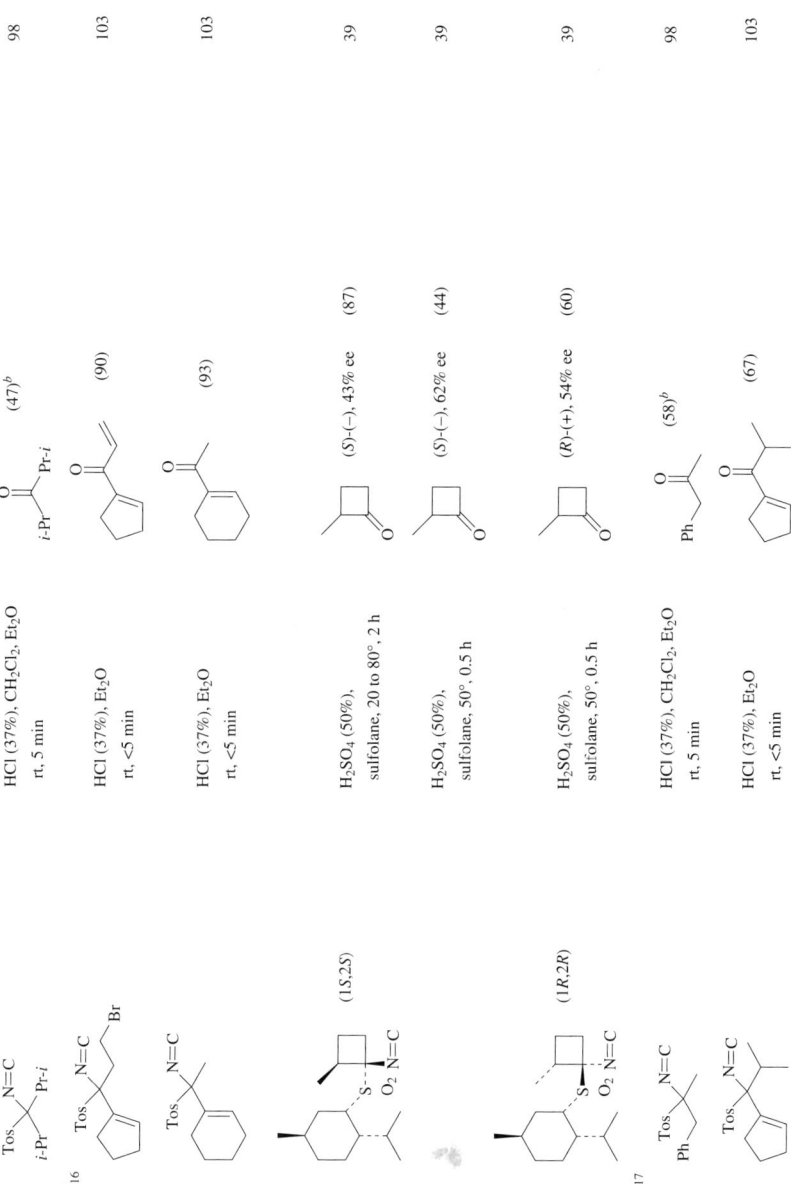

TABLE XII. KETONES BY ACID HYDROLYSIS OF DISUBSTITUTED TOSMIC DERIVATIVES (*Continued*)

Substrate	Conditions	Product(s) and Yield(s) (%)	Refs.
Tos–C(N=C)(CH₃)–CH₂CH₂CH=CHCH₂CH₃	HCl (37%), CH₂Cl₂, 0°, 10 h	methyl ketone with pentenyl chain (90)	97
Tos–C(N=C)(n-Bu)–Bu-n	HCl (37%), CH₂Cl₂, Et₂O, rt, 5 min	n-Bu–CO–Bu-n (69)[b]	98
C₁₈ Tos–C(N=C)–(cyclodecenyl)	HCl (37%), CH₂Cl₂, Et₂O, rt, 5 min	cyclodecenone (54)	149
Tos–C(N=C)(allyl)(cyclohexenyl)	HCl (37%), Et₂O, rt, <5 min	1-(cyclohex-1-enyl)but-3-en-1-one (87)	103
Tos–C(N=C)(i-Pr)(cyclohexenyl)	HCl (37%), Et₂O, rt, <5 min	1-(cyclohex-1-enyl)-2-methylpropan-1-one (90)	103
Tos–C(N=C)(n-C₈H₁₇)–CH₃	HCl (37%), CH₂Cl₂, Et₂O, rt, 5 min	n-C₈H₁₇–CO–CH₃ (80)[c]	98
C₁₉ Tos–C(N=C)(CH₂-S-thienyl)₂	HCl (37%), CH₂Cl₂, Et₂O, rt, 5 min	bis(thienylthiomethyl) ketone (—)	246
Tos–C(N=C)(CH₂C₆H₄Cl-4)(C(CH₃)=CH₂)	HCl (37%), Et₂O, rt, <5 min	1-(4-chlorophenyl)-3-methylbut-3-en-2-one (87)	103

Substrate	Conditions	Product(s) and Yield(s) (%)	Refs.
Tos−N=C−CH(iPr)−CH₂Ph	HCl (37%), CH₂Cl₂, Et₂O rt, 5 min	iPr-CH(C=O)-CH₂Ph (65)[c]	98
Tos−N=C−CH₂−C₆H₄Cl-4 (C₂₀)	HCl (37%), Et₂O rt, <5 min	cyclobutyl-C(=O)-CH₂-C₆H₄Cl-4 (86)	103
Tos−N=C−CH(sBu)−CH₂Ph	HCl (37%), CH₂Cl₂, Et₂O rt, 5 min	Ph-CH₂-C(=O)-CH(Me)Et (41)[c]	98
Tos−N=C-decalinyl (C₂₁)	HCl (37%), Et₂O rt, 10 min	2-acetyl-octahydronaphthalene (44)	235
Tos−N=C-cyclopentenyl-(CH₂)₅-CO₂Me	HCl (37%), Et₂O rt, <5 min	cyclopentenyl-C(=O)-(CH₂)₅-CO₂Me (67)	103
Tos−N=C-C(Me)(dioxolane)-(CH₂)₃-CH=CH-Et	HCl (37%), CH₂Cl₂, Et₂O rt, 15 min	MeC(=O)-(CH₂)₃-C(=O)-(CH₂)₂-CH=CH-Et (100)	152
Tos−N=C-C(Me)(dioxolane)-(CH₂)₅-CH₃	HCl (37%), CH₂Cl₂, Et₂O rt, 15 min	MeC(=O)-(CH₂)₃-C(=O)-(CH₂)₄-CH₃ (100)	152

TABLE XII. KETONES BY ACID HYDROLYSIS OF DISUBSTITUTED TOSMIC DERIVATIVES (*Continued*)

Substrate	Conditions	Product(s) and Yield(s) (%)	Refs.
C$_{21-24}$			
(bicyclic structure with Tos, N=C, R^1, R^2)		**I** + (bicyclic ketone) + **II** + (bicyclic ketone) **III** (decalin ketone with R^2)	
R^1 = H, R^2 = H	HCl (37%), Et$_2$O, rt, 10 min	(**I + II**) (58), **I:II** = 10:1	235
R^1 = OH, R^2 = H	HClO$_4$ (40%), Et$_2$O, reflux 45 min	**I** (5) + **II** (46) + **III** (3)	
R^1 = OH, R^2 = Me	"	**I** (6) + **II** (55) + **III** (6)	
R^1 = Me, R^2 = OH	"	**II** (62) + **III** R^2 = Me (3)	
R^1 = OTMS, R^2 = H	"	**I** R^1 = OH (81)	
C$_{22}$ (Tos, N=C, Ph, cyclohexenyl)	HCl (37%), Et$_2$O, rt, <5 min	(cyclohexenyl-C(O)-CH$_2$-Ph) (86)	103
(Tos, N=C, CO$_2$Me, cyclohexenyl)	HCl (37%), Et$_2$O, rt, <5 min	(cyclohexenyl-C(NHCHO)=CH-CH=CH-CO$_2$Me) (87)^d	103
(Tos-N=C...C=N-Tos bis)	HCl (37%), CH$_2$Cl$_2$, Et$_2$O, rt, 15 min	(diketone) (66)	100

C_{23}	Tos–N=C– (dibenzocyclobutene-CH₂ ketone structure)	HCl (37%), CH₂Cl₂, Et₂O, rt, 5 min	(17)[b] 111
	4-ClC₆H₄–C(Tos)(N=C)–CH₂–C(=O)–CH₃	HCl (37%), CH₂Cl₂, Et₂O, rt, 5 min	(56)[b] 4-ClC₆H₄ 98
	Tos, N=C / Ph–CH₂–... / Ph	HCl (37%), CH₂Cl₂, Et₂O, rt, 5 min	(72)[b] 98
C_{24}	Tos–C(N=C)–...–C=N–Tos	HCl (37%), CH₂Cl₂, Et₂O, rt, 15 min	(82) 100
C_{25}	Tos–C(=N)–...–C≡N	HCl (37%), CH₂Cl₂, Et₂O, rt, 15 min	(82) 100
C_{26}	C≡N–C(Tos)–cyclohexenyl–CO₂Et(aryl)	HCl (37%), Et₂O, rt, <5 min	(86) 103
	Tos, N=C, n-C₁₆H₃₃	HCl (37%), CH₂Cl₂, Et₂O, rt, 5 min	(63)[c] 98
C_{27}	Tos–C(isopropenyl)–C≡N...C=N–Tos	HCl (37%), CH₂Cl₂, Et₂O, rt, 5 min	(76) 103

TABLE XII. KETONES BY ACID HYDROLYSIS OF DISUBSTITUTED TOSMIC DERIVATIVES (Continued)

Substrate	Conditions	Product(s) and Yield(s) (%)	Refs.
	HCl (37%), CH$_2$Cl$_2$, Et$_2$O rt, 15 min	(75)[c]	152
	H$_2$SO$_4$ (2 N), MeOH	(75)[e]	151
	HCl (37%), CH$_2$Cl$_2$, Et$_2$O rt, 2 h	(67)[c]	237
	HCl (37%), CH$_2$Cl$_2$, Et$_2$O rt, 5 min	(85)	238
	HCl (37%), CH$_2$Cl$_2$, Et$_2$O rt, 5 min	(85)	239
C$_{28}$	HCl (37%), CH$_2$Cl$_2$, Et$_2$O rt, 15 min	(95)	100
	HCl (37%), CH$_2$Cl$_2$, Et$_2$O rt, 2 h	(—)	237
C$_{29}$	HCl (37%), Et$_2$O. rt, 5 min	(71)	103
	HCl (37%), CH$_2$Cl$_2$, Et$_2$O rt, 15 min	(93)	100

620

Starting Material	Conditions	Product	Yield (%)	Ref.
(steroid with Tos, N=C, Cl, MeO)	HCl (37%), CH$_2$Cl$_2$, Et$_2$O, rt, 10 min	(29) chloroketone steroid		101
(steroid with Tos, N=C, MeO)	HCl (37%), CH$_2$Cl$_2$, Et$_2$O, rt, 2 min	(97) acetyl steroid		101
(dienone steroid with Tos, N=C)	HCl (37%), CH$_2$Cl$_2$, Et$_2$O, rt, 2 min	(68)		101
(OTHP, Tos, N=C, OTHP chain)	H$_2$SO$_4$ (2 N), MeOH	(2R,6R,8S) (67)^e		151
(OTHP, Tos, N=C, THPO chain)	H$_2$SO$_4$ (2 N), MeOH	(2S,6R) (55)^{b,e}		151
(n-C$_5$H$_{11}$, Tos, N=C, C$_{10}$H$_{21}$-n)	HCl (37%), CH$_2$Cl$_2$, rt, 2 h	(n-C$_5$H$_{11}$, C$_{10}$H$_{21}$-n ketone) (70)^c		240

621

TABLE XII. KETONES BY ACID HYDROLYSIS OF DISUBSTITUTED TosMIC DERIVATIVES (*Continued*)

Substrate	Conditions	Product(s) and Yield(s) (%)	Refs.
C_{30}			
	HCl (6 N), CH_2Cl_2, Et_2O, 0°, 1.5 h	(60)	101
	$HClO_4$ (40%), Et_2O, 30°, 10 min	X: Br (80), Cl (50)	101
	$HClO_4$ (40%), Et_2O, CH_2Cl_2, rt, 3 min	(90)	101
	HCl (5 N), CH_2Cl_2, Et_2O, rt, 4 h	(90)	101
	HCl (8 N), CH_2Cl_2, Et_2O, rt, 10 min	(50)	101

C$_{31}$ (structure with Tos, N=C, MeO steroid)	HCl (6 N), CH$_2$Cl$_2$, rt, 45 min	(structure, acetyl steroid) (90)	45
(Tos, N=C bis-cyclopentenyl structure)	HCl (37%), Et$_2$O, rt, 5 min	(bis-cyclopentenyl diketone) (89)	103
(Et, Tos, N=C steroid, MeO)	HCl (6 N), CH$_2$Cl$_2$, Et$_2$O, 0°, 1.5 h	(Et ester steroid) (60)	101
(MeO, Tos, N=C steroid)	HCl (8 N), C$_6$H$_6$, rt, 3 h	(MeO-acetyl steroid) (38)	101
C$_{32}$ (macrocycle with Tos, N=C, X, pyrrole/furan)	HCl (37%), CH$_2$Cl$_2$, rt, 30 min	(macrocyclic diketone) X: O — 5 min (39)^c; S — 30 min (41)^c	92

TABLE XII. KETONES BY ACID HYDROLYSIS OF DISUBSTITUTED TOSMIC DERIVATIVES (Continued)

Substrate	Conditions	Product(s) and Yield(s) (%)	Refs.
C₃₃ [macrocycle with Tos, N=C, pyridine]	HCl (37%), CH$_2$Cl$_2$, rt, 0.5 h	[diketone macrocycle with pyridine] (43)[c]	92
[bis-cyclohexenyl TosMIC]	HCl (37%), Et$_2$O, rt, 5 min	[bis-cyclohexenyl diketone] (90)	103
C₃₄ [Br-substituted bis-Tos macrocycle]	HCl (37%), CH$_2$Cl$_2$, Et$_2$O	[Br-substituted diketone macrocycle] (59)[b]	193
[X-substituted bis-Tos macrocycle]	HCl (37%), CH$_2$Cl$_2$, Et$_2$O	[X-substituted diketone macrocycle] $\begin{array}{c|c} X & \\ \hline F & (68)^b \\ Cl & (57)^b \\ Br & (55)^b \\ I & (51)^b \end{array}$	248

Substrate	Conditions	Product (%)	Refs.
(structure with Tos, N=C groups on dibenzosuberane)	HCl (35%), CH₂Cl₂, Et₂O, rt, 1 h	(45) ketone	153
(structure with Tos, N=C on [3.3]metacyclophane)	HCl (37%), CH₂Cl₂, rt, 10 min	(55)^c diketone, " (81)	155
(structure with Tos, N=C on [2.2]paracyclophane-like)	HCl (35%), CH₂Cl₂, rt, 1 h	(52)^c	153
	HCl (37%), CH₂Cl₂, rt, 10 min		155
(structure with Tos, N=C on larger paracyclophane)	HCl (37%), CH₂Cl₂, rt, 10 min	(14)^b	155

TABLE XII. KETONES BY ACID HYDROLYSIS OF DISUBSTITUTED TOSMIC DERIVATIVES (Continued)

Substrate	Conditions	Product(s) and Yield(s) (%)	Refs.
[cyclophane with 4-EtC6H4SO2 groups, N=C groups, and two pyridine-like N rings]	HCl (35%), CH2Cl2, rt, 1 h	[cyclophane diketone with two para-phenylene bridges] (83)	153
[macrocycle with 4-EtOC6H4SO2 groups, N=C groups, and two pyridine rings]	HCl (37%), CH2Cl2, rt, 30 min	[macrocyclic diketone with two pyridine rings] (17)^f	92
[macrocycle with 4-EtOC6H4SO2 groups, N=C groups, and two pyridine rings]	HCl (37%), CH2Cl2, rt, 30 min	[macrocyclic diketone with two pyridine rings] (10)^g	92
Tos-C(Ph)(N=C)-(CH2)n-C(Ph)(Tos)(C=N)	HCl (37%), CH2Cl2, Et2O, rt, 1 h	Ph-C(=O)-(CH2)4-C(=O)-Ph (90)	100

C$_{35}$			
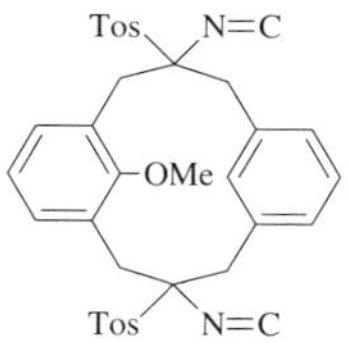 (structure: Tos-N=C, OMe bridged bicyclic with two benzene rings, Tos-N=C)	HCl (37%), CH$_2$Cl$_2$, rt, 1 h	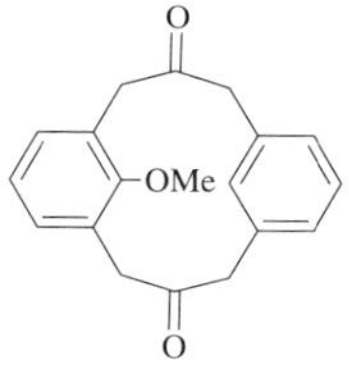 (57)^c (macrocycle with OMe, two C=O, two benzene rings)	110
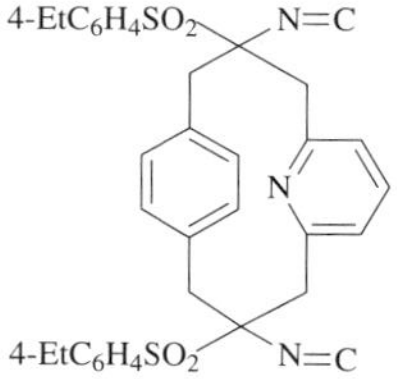 (structure: 4-EtC$_6$H$_4$SO$_2$, N=C, pyridine and benzene rings, N=C, 4-EtC$_6$H$_4$SO$_2$)	HCl (37%), CH$_2$Cl$_2$, rt, 30 min	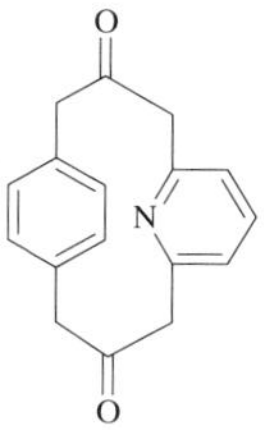 (57)^h (macrocycle with pyridine, two C=O, benzene)	92
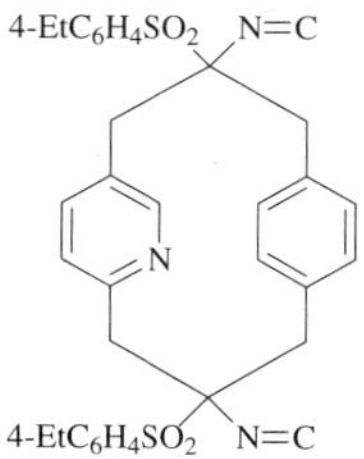 (structure: 4-EtC$_6$H$_4$SO$_2$, N=C, benzene and pyridine rings, N=C, 4-EtC$_6$H$_4$SO$_2$)	HCl (37%), CH$_2$Cl$_2$, rt, 30 min	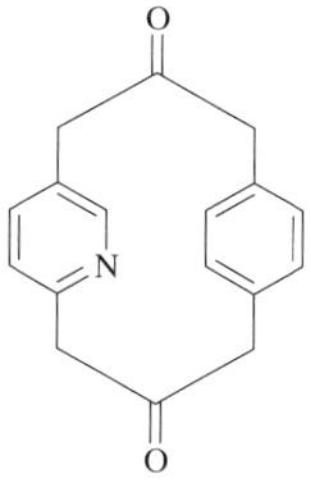 (29)^h (macrocycle with benzene, two C=O, pyridine)	92
(structure: C=N-Ph, Tos, Tos, N=C-Ph acyclic)	HCl (37%), CH$_2$Cl$_2$, Et$_2$O, rt, 15 min	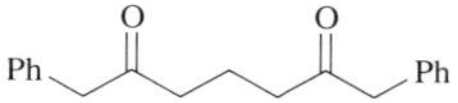 (91) Ph-CH$_2$-C(=O)-(CH$_2$)$_3$-C(=O)-CH$_2$-Ph	100

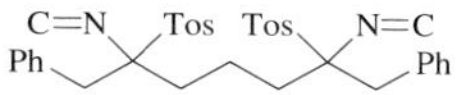

TABLE XII. KETONES BY ACID HYDROLYSIS OF DISUBSTITUTED TOSMIC DERIVATIVES (Continued)

Substrate	Conditions	Product(s) and Yield(s) (%)	Refs.
C_{36}	HCl (37%), CH_2Cl_2, rt, 1 h	(28)[b]	249
	HCl (37%), CH_2Cl_2, Et_2O, rt, 15 min	(11)[b]	109
	HCl (37%), CH_2Cl_2, Et_2O, rt, 15 min	(93)	100
	HCl (37%), CH_2Cl_2, rt, 2 h	" (19)[b]	110
	HCl (37%), CH_2Cl_2, Et_2O, rt, 1.5 min	(47)	101

C_{37}	HCl (8 N), CH_2Cl_2, Et_2O, 0°, 30 min	(50)	101
C_{38}	HCl (37%), CH_2Cl_2, rt, 10 min	(44)[c]	155
C_{39}	HCl (37%), Et_2O, rt, 2 h	(—)	252
C_{40}	HCl (37%), Et_2O, rt, 2 h	(70)[c]	242
C_{41}	AcOH, THF, reflux, 20 h	(85)[c]	96, 254
C_{42}	$Hg(NO_3)_2$, THF, rt, 7 h[i]	(68)[c]	96

TABLE XII. KETONES BY ACID HYDROLYSIS OF DISUBSTITUTED TOSMIC DERIVATIVES (*Continued*)

Substrate	Conditions	Product(s) and Yield(s) (%)	Refs.
(C43)	HCl (37%), CH$_2$Cl$_2$, rt, 10 min	(14)[b]	155
	HCl (37%), CH$_2$Cl$_2$, rt, 10 min	(39)[c]	155
(C44)	HCl (37%), CH$_2$Cl$_2$, rt, 2 h	(94)	160
	HCl (37%), CH$_2$Cl$_2$, Et$_2$O, rt, 15 min	(10)[b]	156

630

C45	HCl (37%), CH₂Cl₂, rt, 2 h	(25)[b]	108, 154
C46	HCl (37%), CH₂Cl₂, Et₂O, rt, 5 min	(34)[b]	251
C47	HCl (37%), CH₂Cl₂, rt, 2 h	(85)	243
	HCl (2 N), C₆H₆, rt, 6 h	(7)[b]	150

TABLE XII. KETONES BY ACID HYDROLYSIS OF DISUBSTITUTED TOSMIC DERIVATIVES (*Continued*)

Substrate	Conditions	Product(s) and Yield(s) (%)	Refs.
C$_{50}$	HCl (37%), CH$_2$Cl$_2$, rt, 10 min	(7)[b]	155
C$_{51}$	HCl (35%), CH$_2$Cl$_2$, rt, 1 h	(72)	153, 155
C$_{54}$	HCl (37%), CH$_2$Cl$_2$, Et$_2$O, rt, 15 min	(14)[b]	109

C$_{57}$		HCl	(11)[b] 250
C$_{58}$		HCl (37%), CH$_2$Cl$_2$, rt, 10 min	(5)[b] 253
		HCl (37%), CH$_2$Cl$_2$, DMF, rt, 15 min	(68)[b] 156

TABLE XII. KETONES BY ACID HYDROLYSIS OF DISUBSTITUTED TOSMIC DERIVATIVES (Continued)

Substrate	Conditions	Product(s) and Yield(s) (%)	Refs.
C_{63}	HCl (37%), CH$_2$Cl$_2$, rt, 10 min	(7)[b]	155
C_{66}	HCl (37%), CH$_2$Cl$_2$, DMF, rt, 15 min	(10)[b]	156

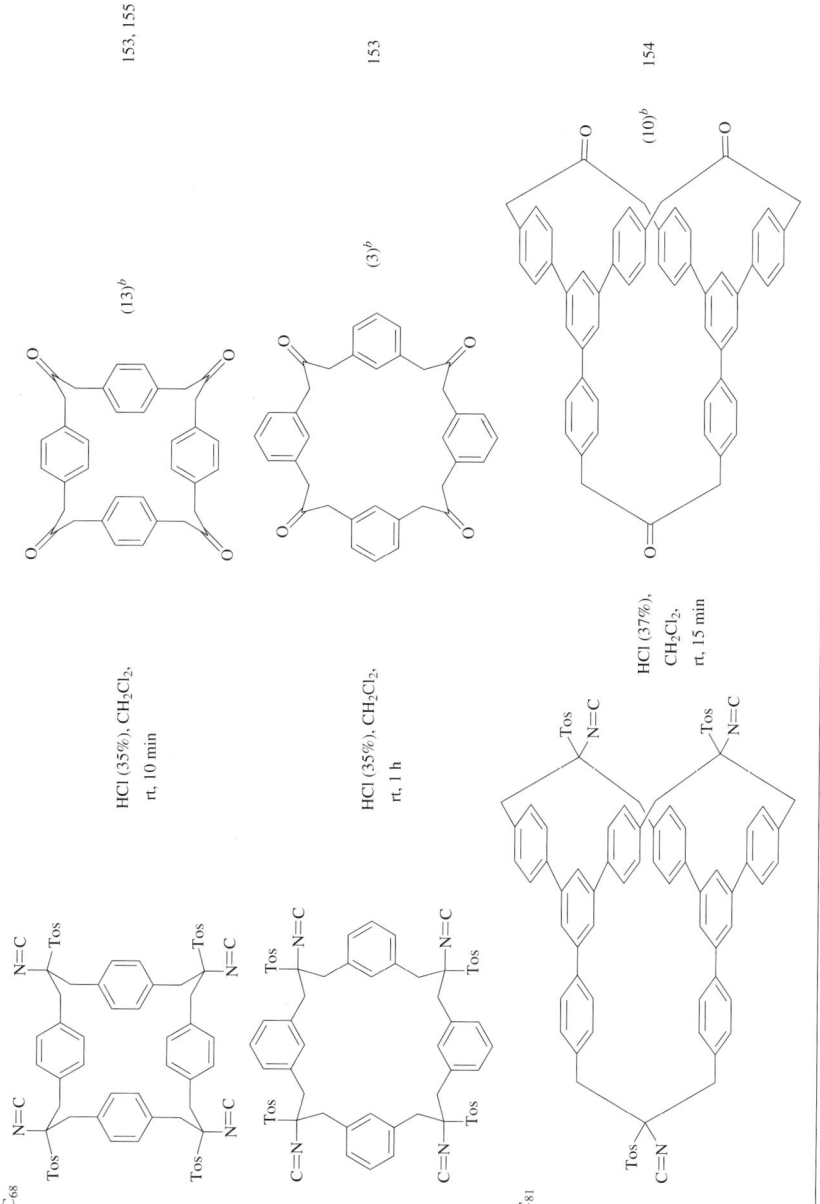

TABLE XII. KETONES BY ACID HYDROLYSIS OF DISUBSTITUTED TOSMIC DERIVATIVES (*Continued*)

Substrate	Conditions	Product(s) and Yield(s) (%)	Refs.

[a] Acid hydrolysis does not lead to a cyclopropanone.
[b] The yield was based on TosMIC.
[c] The yield was based on the monosubstituted TosMIC.
[d] The enamide was formed instead of a ketone.
[e] The ketone was converted to an acetal in situ.
[f] The yield was based on the substituted TosMIC.
[g] The yield was based on the isocyanide.
[h] The yield was based on the monosubstituted isocyanide.
[i] This product was obtained after chromatography over alumina, which effected hydrolysis via the isocyanate.

TABLE XIII. 1,2-DIKETONES FROM ACID CHLORIDES

Substrate	Isocyanide	Conditions	Product(s) and Yield(s) (%)	Refs.
(—)	Tos-C(Me)(Ph)-N=C	HCl (38%), THF, rt, 2.5 h	MeC(O)C(O)Ph (87)	112
(—)	Tos-C(Bn)(t-BuC(O))-N=C	HCl (38%), EtOH, heat, 5 min	t-BuC(O)C(O)Bn (73)	112
(—)	Tos-C(Ph)(PhC(O))-N=C	HCl (38%), THF, rt, 2.5 h	PhC(O)C(O)Ph (89)	112
(—)	Tos-C(Ph)(4-O₂NC₆H₄C(O))-N=C	HCl (38%), THF, rt, 2.5 h	4-O₂NC₆H₄C(O)C(O)Ph (63)	112
(—)	Tos-C(Bn)(PhC(O))-N=C	HCl (38%), THF, rt, 2.5 h	PhC(O)C(O)Bn (50)	112
C₂ MeC(O)Cl	Tos-CH(Ph)-N=C	1. n-BuLi, THF, –80° to rt 2. HCl (38%), THF, rt, 2.5 h	MeC(O)C(O)Ph (73)	112
C₅ 2-thienyl-COCl	Tos-CH(Ph)-N=C	1. n-BuLi, THF, –80° to rt 2. HCl (38%), THF, rt, 2.5 h	2-thienyl-C(O)C(O)Ph (71)	112
	Tos-CH(Bn)-N=C	1. n-BuLi, THF, –80° to rt 2. HCl (38%), THF, rt, 2.5 h	2-thienyl-C(O)C(O)Bn (67)	112, 100

637

TABLE XIII. 1,2-DIKETONES FROM ACID CHLORIDES (*Continued*)

Substrate	Isocyanide	Conditions	Product(s) and Yield(s) (%)	Refs.
t-Bu-C(O)Cl	Tos-CH(Bn)-N=C	1. n-BuLi, THF, −80° to rt 2. HCl (38%), THF, rt, 2.5 h	t-Bu-C(O)-C(O)-Bn (52)	112
4-O₂NC₆H₄-C(O)Cl	Tos-CH(Ph)-N=C	1. n-BuLi, THF, −80° to rt 2. HCl (38%), THF, rt, 2.5 h	4-O₂NC₆H₄-C(O)-C(O)-Ph (51)	112
4-O₂NC₆H₄-C(O)Cl	Tos-CH(Bn)-N=C	1. n-BuLi, THF, −80° to rt 2. HCl (38%), THF, rt, 2.5 h	4-O₂NC₆H₄-C(O)-C(O)-Bn (65)	112
Ph-C(O)Cl	Tos-CH(iPr)-N=C	1. n-BuLi, THF, −80° to rt 2. HCl (38%), THF, rt, 2.5 h	Ph-C(O)-C(O)-iPr (56)	112
Ph-C(O)Cl	Tos-CH(Ph)-N=C	1. n-BuLi, THF, −80° to rt 2. HCl (38%), THF, rt, 2.5 h	Ph-C(O)-C(O)-Ph (68)	112
Ph-C(O)Cl	Tos-CH(Bn)-N=C	1. n-BuLi, THF, −80° to rt 2. HCl (38%), THF, rt, 2.5 h	Ph-C(O)-C(O)-Bn (51)	112
Bn-C(O)Cl	Tos-CH(Ph)-N=C	1. n-BuLi, THF, −80° to rt 2. HCl (38%), THF, rt, 2.5 h	Bn-C(O)-C(O)-Ph (54)	112
4-MeOC₆H₄-C(O)Cl	Tos-CH(Me)-N=C	1. n-BuLi, THF, −80° to rt 2. HCl (38%), THF, rt, 2.5 h	4-MeOC₆H₄-C(O)-C(O)-Me (57)	112

C₇ group: rows 1–6
C₈ group: rows 7–8

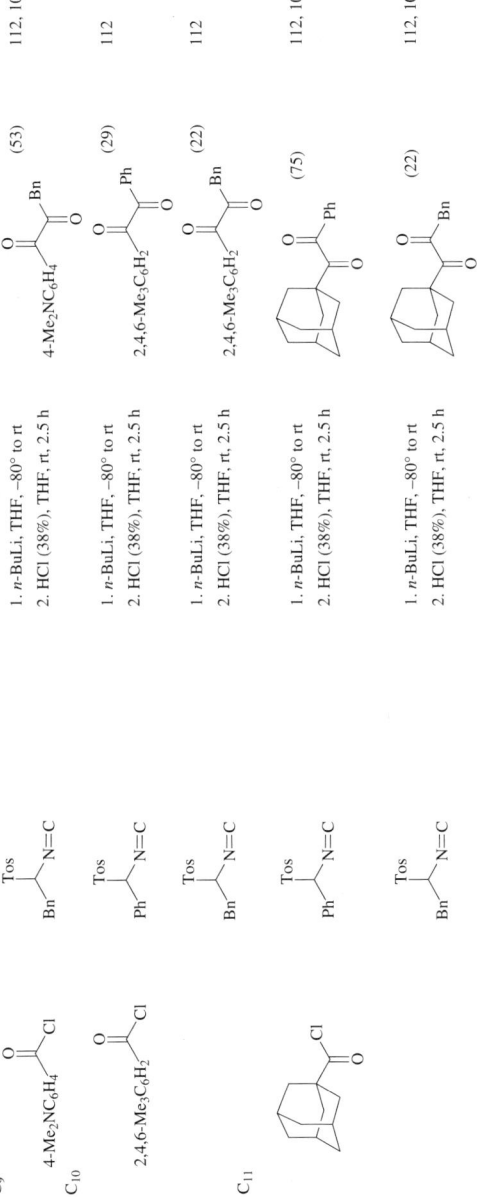

TABLE XIV. α-HYDROXY KETONES AND α-HYDROXY ALDEHYDES BY HYDROLYSIS OF OXAZOLINES

Substrate	Isocyanide	Conditions	Product(s) and Yield(s) (%)	Refs.
C_1				
CH_2O				
		1. MeOH (10 eq), NaOH (50%), BTEAC, PhMe, rt, 1 h 2. HCO$_2$H (60%), rt, 45 min	OHCO... (69)	70
		1. MeOH (10 eq), NaOH (50%), BTEAC, PhMe, rt, 1 h 2. H$_2$SO$_4$ (4 N), rt, 17 h	HO... **I** (92)	158
		1. KOH, THF, rt, 3 min 2. H$_2$SO$_4$ (4 N), rt, 17 h	HO...Tos (60) + **I** (30)	70
		1. MeOH (5 eq), KOH, THF, rt, 1 h 2. HCO$_2$H (60%), rt, 45 min	OHCO... OH (75)	70

	1. MeOH (5 eq), KOH, THF, rt, 1 h 2. H$_2$SO$_4$ (2 N), rt, 22 h	(43)	70
	1. MeOH (10 eq), NaOH (50%), BTEAC, PhMe, rt, 1 h 2. HCO$_2$H (60%), rt, 45 min	(57)	70
	1. MeOH (10 eq), NaOH (50%), BTEAC, PhMe, rt, 1 h 2. H$_2$SO$_4$ (4 N), rt, 17 h	(93)	158
	1. MeOH (10 eq), NaOH (50%), BTEAC, PhMe, rt, 1 h 2. H$_2$SO$_4$ (4 N), rt, 17 h	(68)	158
	1. MeOH (10 eq), NaOH (50%), BTEAC, PhMe, rt, 1 h 2. H$_2$SO$_4$ (4 N), rt, 17 h	(71)	158

TABLE XIV. α–HYDROXY KETONES AND α–HYDROXY ALDEHYDES BY HYDROLYSIS OF OXAZOLINES (*Continued*)

Substrate	Isocyanide	Conditions	Product(s) and Yield(s) (%)	Refs.
		1. MeOH (10 eq), NaOH (50%), BTEAC, PhMe, rt, 1 h 2. H$_2$SO$_4$ (4 N), rt, 17 h	(73)	158
		1. MeOH (10 eq), NaOH (50%), BTEAC, PhMe, rt, 1 h 2. H$_2$SO$_4$ (4 N), rt, 17 h	(73)	158
		1. MeOH (10 eq), NaOH (50%), BTEAC, PhMe, rt, 1 h 2. H$_2$SO$_4$ (4 N), rt, 17 h	(92)	158
		1. MeOH, Triton B, NaOH (50%), C$_6$H$_6$, rt, 25 min 2. H$_2$SO$_4$ (4 N), rt, 24 h	(58)	236

Carbonyl	Isocyanide	Conditions	Product (Yield %)	Ref.
C_2 CH$_3$CHO	Tos–CH(Bn)–N=C	1. K$_2$CO$_3$, MeOH, 30°, 5 min; 2. HCl (conc), rt, 10 min	PhCH$_2$–C(O)–CH(OH)–CH$_3$ (65)	73
C_3 C$_2$H$_5$CHO	Tos–CH(Bn)–N=C	1. K$_2$CO$_3$, MeOH, 30°, 5 min; 2. HCl (conc), rt, 10 min	Et–CH(OH)–C(O)–CH$_2$Bn (52)	73
C_6 cyclohexanone	Tos–CH$_2$–N=C	1. TlOEt, DME, EtOH, rt[a]; 2. HCl (dil.), THF, rt, 17 h	1-hydroxycyclohexyl-CHO (38)[b]	72
t-Bu–C(O)–CH$_3$	Tos–CH$_2$–N=C	1. TlOEt, DME, EtOH, rt[a]; 2. HCl (dil.), THF, rt, 17 h	t-Bu–C(OH)(Me)–CHO (70)[b]	72
C_7 norbornanone	Tos–CH$_2$–N=C	1. TlOEt, DME, EtOH, rt[a]; 2. HCl (dil.), THF, rt, 17 h	2-hydroxy-2-norbornyl-CHO (78)[b]	72
n-Pr–C(O)–Pr-n	Tos–CH$_2$–N=C	1. TlOEt, DME, EtOH, rt[a]; 2. HCl (dil.), THF, rt, 17 h	n-Pr–C(OH)(Pr-n)–CHO (52)[b]	72
RC$_6$H$_4$CHO	Tos–CH(Bn)–N=C	1. K$_2$CO$_3$, MeOH, 30°, 5 min; 2. HCl (conc), rt, 10 min	RC$_6$H$_4$–CH(OH)–C(O)–CH$_2$Bn	73

R	
H	(68)
4-Cl	(67)
4-OMe	(48)
4-Pr-i	(60)
3-Me	(52)

TABLE XIV. α–HYDROXY KETONES AND α–HYDROXY ALDEHYDES BY HYDROLYSIS OF OXAZOLINES (*Continued*)

Substrate	Isocyanide	Conditions	Product(s) and Yield(s) (%)	Refs.
C$_8$ Ph-CO-CH$_3$	Tos-CH$_2$-N=C	1. K$_2$CO$_3$, MeOH, 30°, 5 min 2. HCl (conc), rt, 10 min	Ph-C(OH)(CH$_3$)-CHO (70)[b]	72
4-BrC$_6$H$_4$-CO-CH$_3$	Tos-CH$_2$-N=C	1. K$_2$CO$_3$, MeOH, 30°, 5 min 2. HCl (conc), rt, 10 min	4-BrC$_6$H$_4$-C(OH)(CH$_3$)-CHO (71)[b]	72
C$_27$ (cholestanone)	Tos-CH$_2$-N=C	1. TlOEt, DME, EtOH, rt 2. HCl (dil.), THF, rt, 17 h	(cholestane-2-OH-3-CHO) (70)[b]	72

[a] The 4-ethoxyoxazoline was purified.
[b] The product was a mixture of monomer and dimer.

TABLE XV. ALKANES, *N*-METHYLAMINES, AND ISOCYANIDES BY REDUCTION OF TOSMIC DERIVATIVES

Substrate	Conditions	Product(s) and Yield(s) (%)	Refs.
C_{15} Ph–C(Tos)(H)–N=C	LAH, THF, 40°, 30 min	Ph–CH2–NHMe (48)	73
C_{16} Bn–C(Tos)(H)–N=C	Electroreduction	Bn–CH2–N=C (46)	161
	LAH, THF, 0° to reflux	Bn–CH2–NHMe (72)	73
C_{17} Bn–C(Tos)(Me)–N=C	Electroreduction	Bn–C(Me)=N=C (51)	161
n-Bu–C(Tos)(*n*-Bu)–N=C	LAH, THF, 0° to reflux	*n*-Bu–CH(*n*-Bu)–NHMe (72)	73
C_{18} *n*-C8H17–C(Tos)(H)–N=C	LAH, THF, 40°, 30 min	*n*-C8H17–CH2–NHMe (69)	73
(MeO)2C6H3–CH2–C(Tos)(H)–N=C	Li, NH3 (liq), EtOH, –33°, 2 h	(MeO)2C6H3–CH2–CH2–CH3 (90)	105
C_{19} *n*-C10H21–C(Tos)(H)–N=C	Li, NH3 (liq), EtOH, –33°, 2 h	*n*-C11H24 (92)	105
C_{20} *n*-Bu–C(Tos)(Bn)–N=C	LAH, THF, 0° to reflux	*n*-Bu–CH(Bn)–NHMe (58)	73

TABLE XV. ALKANES, *N*-METHYLAMINES, AND ISOCYANIDES BY REDUCTION OF TOSMIC DERIVATIVES (*Continued*)

Substrate	Conditions	Product(s) and Yield(s) (%)	Refs.
C$_{22}$			
n-C$_8$H$_{17}$-CH(Tos)(N=C)	Li, NH$_3$ (liq), EtOH, −33°, 2 h	n-C$_8$H$_{17}$–C≡C– (92)	105
C$_{23}$			
4-ClC$_6$H$_4$CH$_2$–C(Tos)(N=C)–CH$_2$C$_6$H$_4$Cl-4	Electroreduction	4-ClC$_6$H$_4$CH$_2$–CH(N=C)–CH$_2$C$_6$H$_4$Cl-4 (75)	161
4-ClC$_6$H$_4$CH$_2$–C(Tos)(N=C)–Bn	Electroreduction	4-ClC$_6$H$_4$CH$_2$–CH(N=C)–Bn (80)	161
Bn–C(Tos)(N=C)–Bn	Electroreduction	Bn–CH(N=C)–Bn (83)	161
	LAH, THF, 0° to reflux	Bn–CH(NHMe)–Bn (81)	73
(OBn sugar-Tos)	Li, NH$_3$ (liq), EtOH, −33°, 1.5 h	(OH sugar) (70)	105
C$_{24}$			
n-C$_8$H$_{17}$–C(Tos)(N=C)–Bn	Li, NH$_3$ (liq), EtOH, −33°, 2 h	n-C$_8$H$_{17}$–CH(Bn)– (95)	105
C$_{25}$			
4-MeC$_6$H$_4$CH$_2$–C(Tos)(N=C)–CH$_2$C$_6$H$_4$Me-4	Electroreduction	4-MeC$_6$H$_4$CH$_2$–CH(N=C)–CH$_2$C$_6$H$_4$Me-4 (74)	161

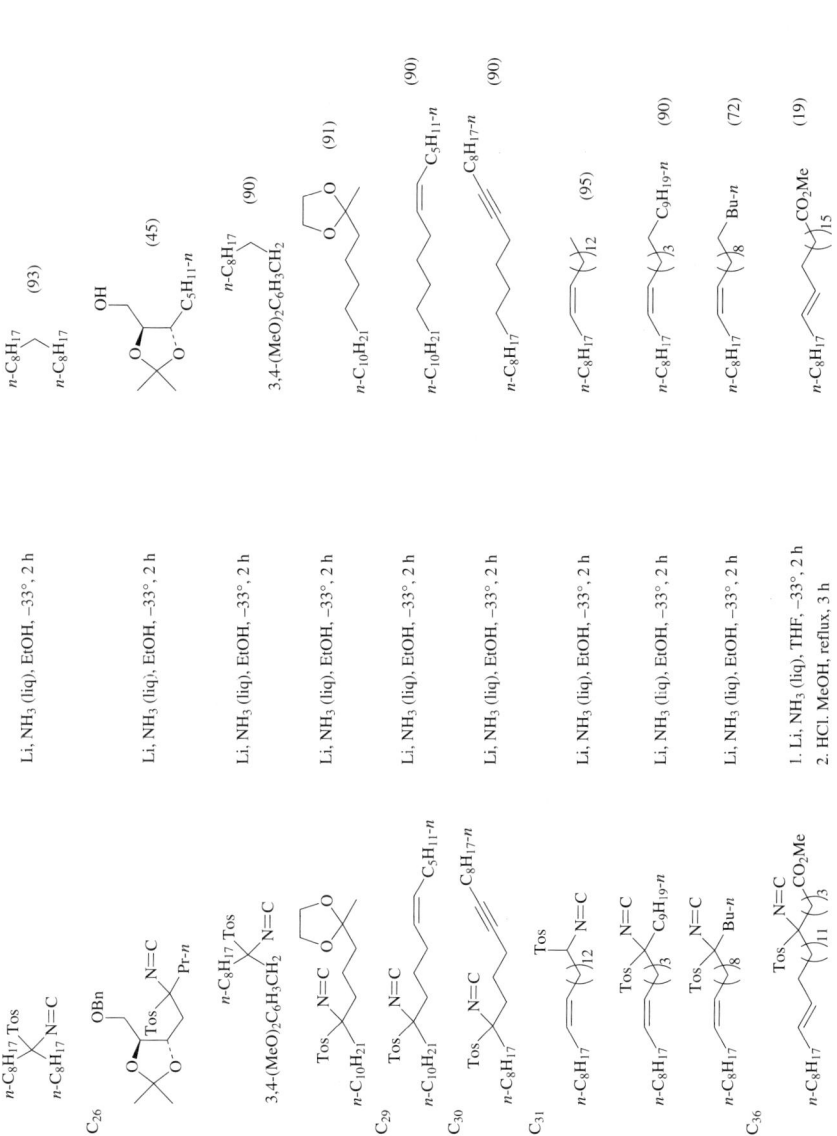

TABLE XV. ALKANES, N-METHYLAMINES, AND ISOCYANIDES BY REDUCTION OF TosMIC DERIVATIVES (*Continued*)

Substrate	Conditions	Product(s) and Yield(s) (%)	Refs.
C_{37} n-C_6H_{13}–CH=CH–$(CH_2)_7$–C(Tos)(N=C)–$(CH_2)_{10}$–CO_2Me	1. Li, NH_3 (liq), THF, $-33°$, 2 h 2. HCl. MeOH, reflux, 3 h	n-C_5H_{11}–CH=CH–$(CH_2)_{19}$–CO_2Me (11)	160
C_{41} n-C_5H_{11}–CH=CH–$(CH_2)_4$–C(Tos)(N=C)–$(CH_2)_{10}$–CO_2Me	1. Li, NH_3 (liq), THF, $-33°$, 2 h 2. HCl. MeOH, reflux, 3 h	n-C_5H_{11}–CH=CH–$(CH_2)_{4}$–$(CH_2)_{12}$–CO_2Me (12)	160
C_{43} C_2H_5–CH=CH–$(CH_2)_6$–C(Tos)(N=C)–$(CH_2)_9$–CO_2Me	1. Li, NH_3 (liq), THF, $-33°$, 2 h 2. HCl. MeOH, reflux, 3 h	C_2H_5–CH=CH–$(CH_2)_6$–$(CH_2)_{11}$–CO_2Me (8)[a]	160
n-C_8H_{17}–CH=CH–$(CH_2)_{11}$–C(Tos)(N=C)–$(CH_2)_{10}$–CO_2Me	1. Li, NH_3 (liq), THF, $-33°$, 2 h 2. HCl. MeOH, reflux, 3 h	n-C_8H_{17}–CH=CH–$(CH_2)_{22}$–CO_2Me (15)	160

[a] The product was an impure mixture owing to overreduction of double bonds.

TABLE XVI. β–HYDROXY N-METHYLAMINES BY REDUCTION OF 4-TOSYLOXAZOLINES

Substrate	Isocyanide	Conditions	Product(s) and Yield(s) (%)	Refs.
C₂ CH₃CHO	Tos–N=C	1. [Fe ferrocene ligand with Me, PPh₂, piperidine] CF₃SO₃Ag (cat.), CH₂Cl₂, rt, 2 h 2. LAH, rt	(−)-(R)- [OH, NHMe] (67)	75
C₃ [acetone]	Tos–N=C	1. K₂CO₃, MeOH, rt, 10 min 2. LAH, THF, 0° 3. 35°, 30 min	[OH, t-Bu, NHMe] (54)	73
C₄ i-PrCHO	Tos–N=C	1. K₂CO₃, MeOH, rt, 10 min 2. LAH, THF, 0° 3. 35°, 30 min	[OH, i-Pr, NHMe] (58)	73
	Tos–N=C	1. [Fe ferrocene ligand with Me, PPh₂, piperidine] CF₃SO₃Ag (cat.), CH₂Cl₂, rt, 2 h 2. LAH, rt	(−)-(R)-i-Pr [OH, NHMe] (67)	75
[2-butanone]	Tos–N=C	1. K₂CO₃, MeOH, 40°, 25 min 2. LAH, THF, 0° 3. 35°, 30 min	[OH, Et, Me, NHMe] (55)	73

TABLE XVI. β–HYDROXY N-METHYLAMINES BY REDUCTION OF 4-TOSYLOXAZOLINES (*Continued*)

Substrate	Isocyanide	Conditions	Product(s) and Yield(s) (%)	Refs.
C5 ∕∕∕CHO	Tos—N=C	1. [Fe ferrocene ligand with Me, PPh2, PPh2, N-piperidinyl]; CF3SO3Ag (cat.), CH2Cl2, rt, 2 h; 2. LAH, rt	(−)-(R)- ∕=∕—CH(OH)—CH2—NH(Me) (63)	75
t-BuCHO	Tos—N=C	1. [same Fe ferrocene ligand]; CF3SO3Ag (cat.), CH2Cl2, rt, 2 h; 2. LAH, rt	(−)-(R)-t-Bu—CH(OH)—CH2—NH(Me) (68)	75
C6 cyclohexanone	Tos—N=C	1. K2CO3, MeOH, 40°, 25 min; 2. LAH, THF, 0°; 3. 35°, 30 min	cyclohexyl(OH)—CH2—NH(Me) (52)	73
C7 4-Cl-C6H4-CHO	Tos—N=C	1. K2CO3, MeOH, rt, 10 min; 2. LAH, THF, 0°; 3. 35°, 30 min	4-Cl-C6H4—CH(OH)—CH2—NH(Me) (66)	73

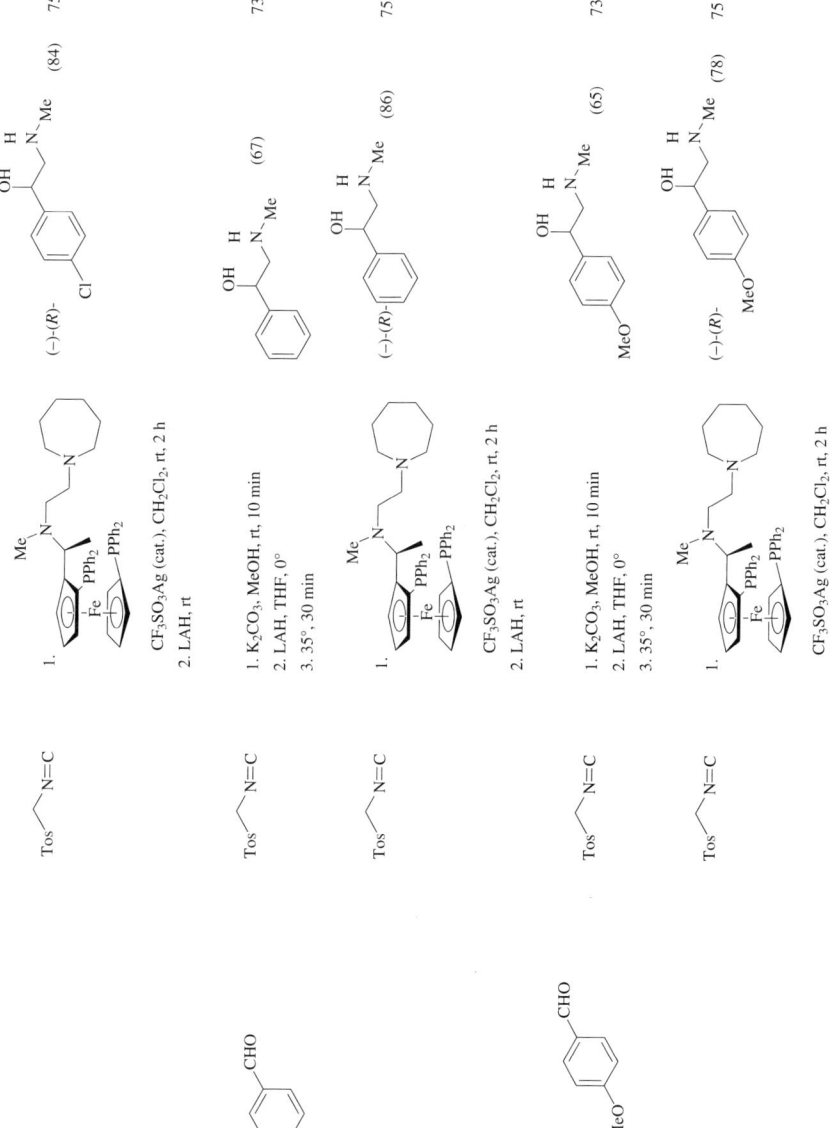

TABLE XVI. β–HYDROXY N-METHYLAMINES BY REDUCTION OF 4-TOSYLOXAZOLINES (*Continued*)

Substrate	Isocyanide	Conditions	Product(s) and Yield(s) (%)	Refs.
C₉				
MeO–C₆H₃(OMe)–CHO	Tos–CH₂–N=C	1. K₂CO₃, MeOH, rt, 10 min 2. LAH, THF, 0° 3. 35°, 30 min	MeO–C₆H₃(OMe)–CH(OH)–CH₂–NHMe (71)	73
	Tos–CH₂–N=C	1. [Ferrocenyl ligand with Me, PPh₂, PPh₂, and N-CH₂CH₂-piperidine] CF₃SO₃Ag (cat.), CH₂Cl₂, rt, 2 h 2. LAH, rt	(−)-(R)- MeO–C₆H₃(OMe)–CH(OH)–CH₂–NHMe (83)	75
C₁₁				
(−)-(R) 2-Me-C₆H₄–CHO·Cr(CO)₃	Tos–CH₂–N=C	1. K₂CO₃, MeOH, 0°, 30 min 2. CH₂Cl₂, daylight, 3 d 3. LAH, THF, rt, 3 h	(+)-(S)- 2-Me-C₆H₄–CH(OH)–CH₂–NHMe (57)	74

TABLE XVII. α–ADDITIONS TO THE ISOCYANO CARBON OF TOSMIC

Substrate	Isocyanide	Conditions	Product(s) and Yield(s) (%)	Refs.
C₀ Cl₂	Tos–N=C	CH₂Cl₂, −5°	Tos–N=CCl₂ (68)	47, 51
C₁ MeSCl	Tos–N=C	(—)	Tos–N=CClSMe (—)	162
C₄ (CF₃CO)₂O	Tos–N=C	1. CH₂Cl₂, rt, 10 h 2. H₂O	Tos–N(C(O))–C(CF₃)(OH)(H)(OH) (100)	165
C₅ t-BuCOCl	Tos–N=C	60°, 48 h	Tos–C(Cl)=N–C(O)Bu-t (100)	163
	Tos–CH(–(CH₂)₄–CH=CH₂)–N=C	—	Tos–C(Cl)(–(CH₂)₄–CH=CH₂)–N=C(Bu-t)(O) (100)	163
C₆ ClOC–C(CH₃)₂–COCl	Tos–N=C	C₆H₆, reflux, 18 h	Tos–C(Cl)=N–C(C(CH₃)₂)(COCl)(O) (67)	164
	Tos–N=C	—	Tos–C(Cl)=N–S–C₆H₄–NO₂ (—)	162
O₂N–C₆H₄–SCl				
C₆H₅–SCl	Tos–N=C	—	Tos–C(Cl)=N–S–C₆H₅ (—)	162
C₈ 1,2-(ClOC)₂C₆H₄	Tos–N=C	C₆H₆, reflux, 18 h	2-(COCl)-C₆H₄–C(O)–C(Cl)=N–CH₂–Tos (70)	164

653

TABLE XVII. α–ADDITIONS TO THE ISOCYANO CARBON OF TOSMIC (*Continued*)

$C_{10\text{-}18}$

Substrate: PhCH$_2$O-C(O)-NH-CH(R^1)-CO$_2$H (S), R^2 + H$_2$N-CH(CO$_2$Me)(R^2) (S) + i-PrCHO

Isocyanide: Tos-CH$_2$-N=C

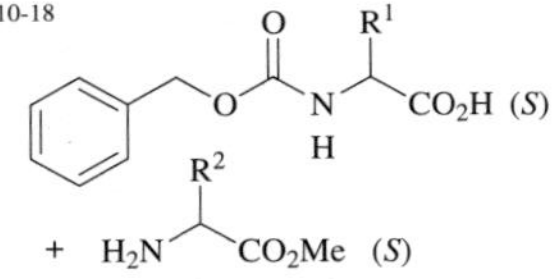

Conditions: 1. MeOH, 0°, 1 h; 2. rt, 7 d

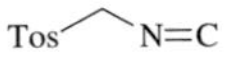

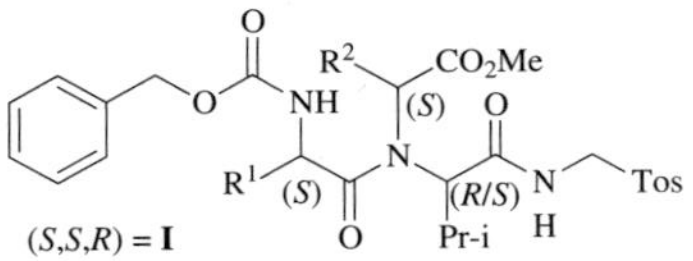

Product(s): (S,S,R) = **I**; (S,S,S) = **II**

Refs.: 168

R^1	R^2	I + II	I:II
H	H	(84)	—
H	Me	(72)	58:42
Me	H	(73)	50:50
Me	Me	(80)	60:40
H	Pr-*i*	(84)	81:19
Pr-*i*	H	(39)	—
H	Bu-*i*	(68)	78:22
H	Bu-*t*	(59)	70:30
Me	Pr-*i*	(80)	81:19
Pr-*i*	Me	(73)	59:41
H	CH$_2$Bu-*t*	(65)	81:19
Me	Bu-*i*	(66)	79:21
Me	Bu-*i*	(37)	69:31
Bu-*t*	Me	(75)	59:41
H	Ph	(77)	54:46
Me	CH$_2$Bu-*t*	(87)	81:19
Pr-*i*	Pr-*i*	(76)	81:19
H	Bn	(78)	77:23

Me	Ph		(60)	55:45
Pr-i	Bu-i		(77)	78:22
Me	Bn		(69)	79:21
Pr-i	CH₂Bu-t		(59)	82:18
Pr-i	Ph		(78)	55:45
Pr-i	Bn		(78)	78:22

C₁₁

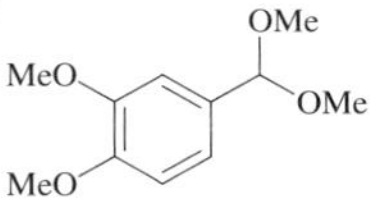

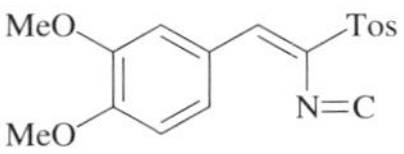

(Ar)CH(OMe)(OMe) [3,4-dimethoxyphenyl methoxymethyl] + (E)-Tos-CH=CH-Ar'(N=C) [where Ar' = 3,4-dimethoxyphenyl], CF₃CO₂H, C₆H₆, rt, 4 h → Tos-CH=CH-Ar'-NH-C(=O)-CH(OMe)-Ar (70), 96

(Ar)CH(OMe)(OMe) + Tos-CH(Ar')-N=C, CF₃CO₂H, C₆H₆, rt, 2 h → Tos-CH(CH₂Ar')-NH-C(=O)-CH(OMe)-Ar (88), 96

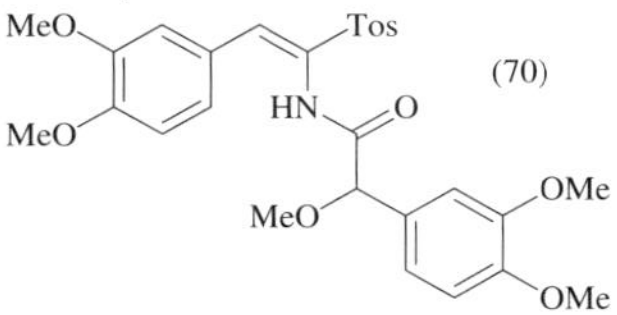

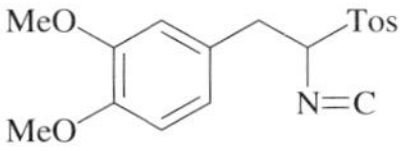

[thione-pyridine ester with Ph(CH₂)₂ chain], Tos-CH₂-N=C, 1. C₆H₆, hν, 45°, 1 h; 2. 45°, 15 min → [pyridyl-S-C(=N-CH₂Tos)-CH₂CH₂Ph] (45), 166

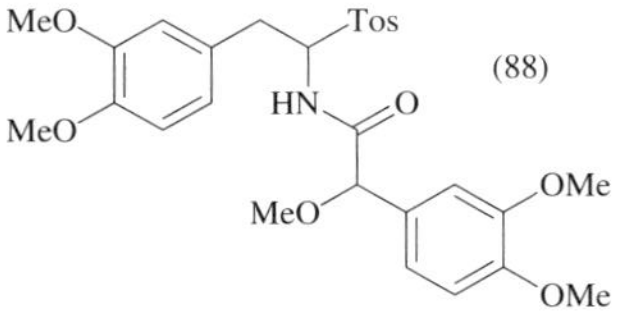

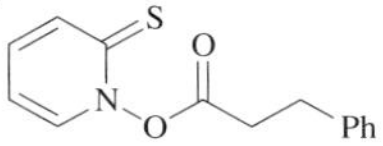

C₁₄

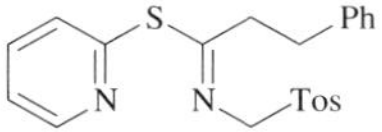

[acetylated amino sugar] + i-PrCHO + HCO₂H, Tos-CH₂-N=C, ZnCl₂, Et₂O, THF, -40°, 10 d → [acetylated sugar-N(CH(Pr-i)C(=O)NH-CH₂-Tos)(CHO)] (10), 167

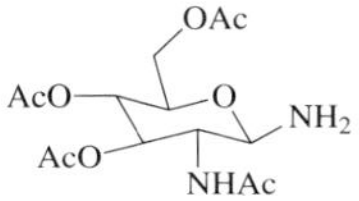

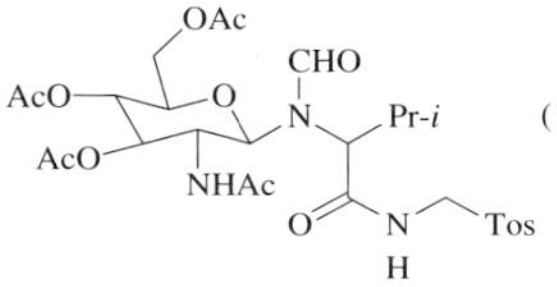

655

TABLE XVIII. CYCLOADDITIONS TO THE ISOCYANO CARBON OF TosMIC

Substrate	Isocyanide	Conditions	Product(s) and Yield(s) (%)	Refs.
C$_6$				
[Cr(CN)(CO)$_5$]$^-$ + HBF$_4$ + Me$_2$CO	Tos−N=C	rt, overnight	(OC)$_5$Cr−[oxazoline with HN, Tos] (53)	292
C$_7$ t-Bu−C(=O)−NC	Tos−N=C	C$_6$H$_6$, rt, 1 h	[dioxolane with Bu-t, CN, NC, Tos] (89)	169
C$_8$ [azaboretidine: CF$_3$, TMS, Me, B−N−Me, CF$_3$]	Tos−N=C	CHCl$_3$, rt, 5 h	[TMS, Me−N−Me, B−CF$_3$, N−Tos ring] (96)	175
C$_9$ [imidazolidine with P(O)(OEt)$_2$, N=C, H, H]	Tos−N=C	PhMe, rt, 4 d	[bicyclic with P(O)(OEt)$_2$, N−Tos] (47)	171
C$_{12}$ [dioxane with two allyl groups]	Tos−N=C	1. Cp$_2$Zr(Bu-n)$_2$, THF, −70 to 20° 2. TosMIC, 67°, 15 h 3. MeOH, H$_2$O	[bicyclic ketone with spiro-dioxane, H, H] (46)	178
C$_{14}$ [pyrimidinedione with Me, OH, piperidine, cyclopropyl]	Tos−N=C	1. MeONa, MeOH, rt, 5 min 2. TosMIC, CH$_2$Cl$_2$, 0° 3. AcCl, 0°, ~30 min 4. rt, 1.5 h	[fused ring with N−Tos, Me, spiro-cyclopropyl, Me] (43)	173

656

C₁₅ PhCO−C(N₂)−COPh	Tos−CH₂−N=C	Xylene, 100°, 1 h	(74) structure with Ph, Ph, O, N-Tos	170
C₁₆ PhCO−C≡C−COPh	Tos−CH₂−N=C	PhMe, 60°, 1 h	(34) bicyclic furan structure with Ph, Ph, N-Tos, N-Tos	172
C₁₇ 4-MeOC₆H₄CO−C(N₂)−COC₆H₄OMe-4	Tos−CH₂−N=C	Xylene, 100°, 1 h	(83) furanone with 4-MeOC₆H₄, 4-MeOC₆H₄, N-Tos	170
C₁₈ cyclohexenone with OAc and N-piperidyl-cyclopropyl	Tos−CH₂−N=C	CH₂Cl₂, rt, 24 h	(64) spirocyclic structure with N-Tos	173
C₁₉ (CO)₅Cr=C(OMe)(2-Ph-cyclohexenyl)	Tos−CH₂−N=C	1. THF, ZnCl₂, rt, 4 h 2. reflux, 5 h	(84) phenanthrene with OMe, H, N-Tos	177
C₂₁ cyclobutadiene with CO₂Bu-t, Bu-t, Bu-t, Bu-t	Tos−CH₂−N=C	C₅H₁₂/CHCl₃ 4:1, rt, 2 weeks	(49) cyclopentadiene with CO₂Bu-t, Bu-t, Bu-t, t-Bu, N-Tos	293

657

TABLE XVIII. CYCLOADDITIONS TO THE ISOCYANO CARBON OF TOSMIC (*Continued*)

Substrate	Isocyanide	Conditions	Product(s) and Yield(s) (%)	Refs.
C$_{26}$ (Tos–CH–N=C and Tos–CH–N=C on benzene ring)		KOH, ROH, reflux, 3 h	benzo-fused 8-membered ring with OR, N, N=C R / Me (69) / Et (74) / i-Pr (71)	82a
		t-BuOK, *t*-BuOH, reflux, 3 h	benzo-fused 8-membered ring with O, NH, N=C (69)	82a
C$_{30}$ PhCO, COPh, Ph, Ph (cyclopropene)	Tos–CH$_2$–N=C	MeCN, reflux, 5 h	cyclobutene with COPh, PhCO, N-Tos, Ph, Ph (69)	174

REFERENCES

[1] Matsumoto, K.; Suzuki, M. *Encyclopedia of Reagents for Organic Synthesis*; Paquette, L. A., Ed.; Vol. 4, Wiley: New York, 1995, p. 2474 [($C_2H_5O_2CCH_2N=C$)].

[1a] Fieser and Fieser's *Reagents for Organic Synthesis*; Vol. 16, Wiley: New York, 1992, p. 164 [($C_2H_5O_2CCH_2N=C$, leading references)].

[2] van Leusen, A. M.; van Leusen, D. *Encyclopedia of Reagents for Organic Synthesis*; Paquette, L. A., Ed.; Vol. 3, Wiley: New York, 1995, p. 1820 [($C_2H_5O)_2P(O)CH_2N=C$].

[2a] Fieser and Fieser's *Reagents for Organic Synthesis*; Vol. 14, Wiley: New York, 1989, p. 135, and Coll. Index Vol. 1–12 [($C_2H_5O)_2P(O)CH_2N=C$].

[3] van Leusen, A. M.; van Leusen, D. *Encyclopedia of Reagents for Organic Synthesis*; Paquette, L. A., Ed.; Vol. 7, Wiley: New York, 1995, p. 4979 [($CH_3C_6H_4SCH_2N=C$)].

[3a] Fieser and Fieser's *Reagents for Organic Synthesis*; Vol. 4, Wiley: New York, 1974, p. 515 [(leading references) ($CH_3C_6H_4SCH_2N=C$)].

[4] Katritzky, A. R.; Xie, L.; Fan, W. Q. *Synthesis* **1993**, 45 [(leading reference) ($R_2NCHR(N=C)$)].

[5] Versleijen, J. P. G.; Faber, P. M.; Bodewes, H. H.; Braker, A. H.; van Leusen, D.; van Leusen, A. M. *Tetrahedron Lett.* **1995**, *36*, 2109 [($(RO)_2BCH_2N=C$)].

[6] van Leusen, A. M. *Lect. Heterocyclic Chem.* **1980**, *5*, S111.

[7] van Leusen, A. M. In *Perspectives in the Organic Chemistry of Sulfur*; Zwanenburg, B., Klunder, A. J., Eds.; Elsevier: Amsterdam, 1987, p. 119.

[8] Di Santo, R.; Massa, S.; Artico, M. *Farmaco* **1993**, *48*, 209; *Chem. Abstr.* **1993**, *118*, 254774r.

[9] van Leusen, A. M.; van Leusen, D. *Encyclopedia of Reagents for Organic Synthesis*; Paquette, L. A., Ed.; Vol. 7, Wiley: New York, 1995, p. 4973 [($CH_3C_6H_4SO_2CH_2N=C$)].

[10] Malatesta, L.; Bonati, F. *Isocyanide Complexes of Metals*; Wiley: New York, 1969.

[11] Schöllkopf, U. *Angew. Chem., Int. Ed. Engl.* **1970**, *9*, 763.

[12] *Isonitrile Chemistry*, Ugi, I., Ed.; Academic: New York, 1971.

[13] Millich, F. *Chem. Rev.* **1972**, *72*, 101.

[14] Hoppe, D. *Angew. Chem., Int. Ed. Engl.* **1974**, *13*, 789.

[15] Schöllkopf, U. *Angew. Chem., Int. Ed. Engl.* **1977**, *16*, 339.

[16] Periasamy, M. P.; Walborsky, H. M. *Org. Prep. Proc. Int.* **1979**, *11*, 293.

[17] Schöllkopf, U. *Pure Appl. Chem.* **1979**, *51*, 1347.

[18] Drenth, W.; Nolte, R. J. M. *Acc. Chem. Res.* **1979**, *12*, 30.

[19] Singleton, E.; Oosthuizen, H. E. *Adv. Organomet. Chem.* **1983**, *22*, 209.

[20] Grundmann, C. in *Methoden der organischen Chemie* (Houben Weyl) Band E5, Teil 2; Falbe, J., Ed.; Thieme: Stuttgart, 1985, p. 1611.

[21] Edenborough, M. S.; Herbert, R. B. *Nat. Prod. Rep.* **1988**, *5*, 229.

[22] Aumann, R. *Angew. Chem., Int. Ed. Engl.* **1988**, *27*, 1456.

[23] Togni, A.; Pastor, S. D. *Chirality* **1991**, *3*, 331.

[24] Scheuer, P. J. *Acc. Chem. Res.* **1992**, *25*, 432.

[25] Marcaccini, S.; Torroba, T. *Org. Prep. Proced. Int.* **1993**, *25*, 141.

[26] van Leusen, A. M.; Hoogenboom, B. E.; Siderius, H. *Tetrahedron Lett.* **1972**, 2369.

[27] van Leusen, A. M.; Strating, J. *Quart. Rep. Sulfur Chem.* **1970**, *5*, 67.

[28] Hoogenboom, B. E.; Oldenziel, O. H.; van Leusen, A. M. *Org. Synth.* **1977**, *57*, 102; *Coll. Vol. VI* **1988**, 987.

[29] Tezaki, K.; Nakayama, S.; Miyazaki, Y.; Sugita, Y. JP 61,186,359; *Chem. Abstr.* **1987**, *106*, 18138t.

[30] Barendse, N. C. M. E. EP 242,001; *Chem. Abstr.* **1988**, *109*, 24508s.

[31] van Leusen, D., unpublished results.

[32] Oldenziel, O. H.; van Leusen, D.; van Leusen, A. M. *J. Org. Chem.* **1977**, *42*, 3114.

[33] Gossauer, A.; Suhl, K. *Helv. Chim. Acta* **1976**, *59*, 1698.

[34] Cappon, J. J.; Witters, K. D.; Verdegem, P. J. E.; Hoek, A. C.; Luiten, R. J. H.; Raap, J.; Lugtenburg, J. *Recl. Trav. Chim. Pays Bas* **1994**, *113*, 318.

[35] van Leusen, A. M.; Boerma, G. J. M.; Helmholdt, R. B.; Siderius, H.; Strating, J. *Tetrahedron Lett.* **1972**, 2367.

[36] Schöllkopf, U.; Schröder, R.; Blume, E. *Justus Liebigs Ann. Chem.* **1972**, *766*, 130.
[37] van Leusen, A. M.; Wildeman, J.; Oldenziel, O. H. *J. Org. Chem.* **1977**, *42*, 1153.
[38] Olijnsma, T.; Engberts, J. B. F. N.; Strating, J. *Recl. Trav. Chim. Pays Bas* **1972**, *91*, 209.
[39] van Leusen, D.; Rouwette, P. H. F. M.; van Leusen, A. M. *J. Org. Chem.* **1981**, *46*, 5159.
[40] Hundscheid, F. J. A.; Tandon, V. K.; Rouwette, P. H. F. M.; van Leusen, A. M. *Tetrahedron* **1987**, *43*, 5073.
[41] Bringmann, G.; Schneider, S. *Synthesis* **1983**, 139.
[42] Obrecht, R.; Herrmann, R.; Ugi, I. *Synthesis* **1985**, 400.
[43] van Leusen, D.; van Leusen, A. M. *Recl. Trav. Chim. Pays Bas* **1984**, *103*, 41.
[44] Kamogawa, H.; Maeda, K. *J. Polym. Sci., Polym. Chem. Ed.* **1984**, *22*, 1393.
[45] van Leusen, D.; van Leusen, A. M. *Synthesis* **1991**, 531.
[46] van Leusen, A. M.; Jeuring, H. J.; Wildeman, J.; van Nispen, S. P. J. M. *J. Org. Chem.* **1981**, *46*, 2069.
[47] Houwing, H. A.; Wildeman, J.; van Leusen, A. M. *Tetrahedron Lett.* **1976**, 143.
[48] Holland, G. F. US 4,282,242 (1981); *Chem. Abstr.* **1981**, *95*, 187068e.
[49] Kuhn, D. G.; Kamhi, V. M.; Furch, J. A.; Diehl, R. E.; Lowen, G. T.; Kameswaran, V. *Pestic. Sci.* **1994**, *41*, 279.
[50] van Leusen, A. M.; van Leusen, D. *Encyclopedia of Reagents for Organic Synthesis*; Paquette, L. A., Ed.; Vol. 5, Wiley: New York, 1995, p. 3605 [($CH_3C_6H_4SO_2CH_2N=C(SCH_3)C_6H_5$)].
[51] Houwing, H. A.; van Leusen, A. M. *J. Heterocycl. Chem.* **1981**, *18*, 1127.
[52] Houwing, H. A.; Wildeman, J.; van Leusen, A. M. *J. Heterocycl. Chem.* **1981**, *18*, 1133.
[53] Oldenziel, O. H.; van Leusen, A. M. *Tetrahedron Lett.* **1973**, 1357.
[54] van Leusen, D.; van Leusen, A. M. *Recl. Trav. Chim. Pays Bas* **1991**, *110*, 393.
[55] van Leusen, D.; van Leusen, A. M. *Recl. Trav. Chim. Pays Bas* **1991**, *110*, 402.
[56] Bull, J. R.; Tuinman, A. *Tetrahedron* **1975**, *31*, 2151.
[57] Rauckman, E. J.; Rosen, G. M.; Abou-Donia, M. B. *J. Org. Chem.* **1976**, *41*, 564.
[58] Ziegler, F. E.; Wender, P. A. *J. Am. Chem. Soc.* **1971**, *93*, 4318.
[59] Wender, P. A.; Eissenstat, M. A.; Sapuppo, N.; Ziegler, F. E. *Org. Synth.* **1978**, *58*, 101; *Coll. Vol. VI* **1988**, 334.
[60] Yoneda, R.; Harusawa, S.; Kurihara, T. *J. Org. Chem.* **1991**, *56*, 1827.
[61] Konieczny, M. T.; Toma, P. H.; Cushman, M. *J. Org. Chem.* **1993**, *58*, 4619.
[62] van Leusen, A. M.; Oomkes, P. G. *Synth. Commun.* **1980**, *10*, 399.
[63] Leusink, F. R., Ph.D. Thesis, Groningen University, 1993.
[64] Garrigues, B.; Oussaid, B.; Hubert, C. *Bull. Soc. Chim. Fr.* **1993**, *130*, 58.
[65] Oussaid, B.; Hubert, C.; Fayet, J. P.; Garrigues, B. *Bull. Soc. Chim. Fr.* **1993**, *130*, 86.
[66] Aono, T.; Kishimoto, S.; Araki, Y.; Noguchi, S. *Chem. Pharm. Bull.* **1978**, *26*, 1776.
[67] Flynn, D. L.; Becker, D. P.; Spangler, D. P.; Nosal, R.; Gullikson, G. W.; Moummi, C.; Yang, D. C. *Bioorg. Med. Chem. Lett.* **1992**, *2*, 1613.
[68] Donetti, A.; Boniardi, O.; Ezhaya, A. *Synthesis* **1980**, 1009.
[69] Possel, O.; van Leusen, A. M. *Heterocycles* **1977**, *7*, 77.
[70] van Leusen, D.; Batist, J. N. M.; Lei, J.; van Echten, E.; Brouwer, A. C.; van Leusen, A. M. *J. Org. Chem.* **1994**, *59*, 5650.
[71] van Leusen, A. M.; van Leusen, D. US 4,548,749 (1985); *Chem. Abstr.* **1985**, *102*, 149625q.
[72] Oldenziel, O. H.; van Leusen, A. M. *Tetrahedron Lett.* **1974**, 167.
[73] Possel, O., Ph.D. Thesis, Groningen University, 1978.
[74] Solladié-Cavallo, A.; Quazzotti, S.; Colonna, S.; Manfredi, A.; Fischer, J.; DeCian, A. *Tetrahedron: Asymmetry* **1992**, *3*, 287.
[75] Sawamura, M.; Hamashima, H.; Ito, Y. *J. Org. Chem.* **1990**, *55*, 5935.
[76] Ito, Y.; Sawamura, M.; Hayashi, T. *J. Am. Chem. Soc.* **1986**, *108*, 6405.
[77] Oldenziel, O. H., Ph.D. Thesis, Groningen University, 1975.
[78] Saikachi, H.; Kitagawa, T.; Sasaki, H.; van Leusen, A. M. *Chem. Pharm. Bull.* **1979**, *27*, 793.
[79] van Nispen, S. P. J. M.; Mensink, C.; van Leusen, A. M. *Tetrahedron Lett.* **1980**, *21*, 3723.
[80] Moskal, J.; van Stralen, R.; Postma, D.; van Leusen, A. M. *Tetrahedron Lett.* **1986**, *27*, 2173.

[81] Saikachi, H.; Kitagawa, T.; Sasaki, H.; van Leusen, A. M. *Chem. Pharm. Bull.* **1982**, *30*, 4199.
[82] Sasaki, H.; Nakagawa, H.; Khuhara, M.; Kitagawa, T. *Chem. Lett.* **1988**, 1531.
[83] Sasaki, H.; Kitagawa, T. *Chem. Pharm. Bull.* **1983**, *31*, 756.
[84] Sasaki, H.; Kitagawa, T. *Chem. Pharm. Bull.* **1987**, *35*, 4747.
[85] Schöllkopf, U.; Schröder, R. *Angew. Chem., Int. Ed. Engl.* **1972**, *11*, 311.
[86] van Leusen, A. M.; Schaart, F. J.; van Leusen, D. *Recl. Trav. Chim. Pays Bas* **1979**, *98*, 258.
[87] van Leusen, A. M.; Wildeman, J. *Recl. Trav. Chim. Pays Bas* **1982**, *110*, 202.
[88] Spek, A. L. *Cryst. Struct. Commun.* **1979**, *8*, 123.
[89] Herdeis, C.; Beck, W. *Chem. Ber.* **1983**, *116*, 3205.
[90] van Leusen, A. M.; Bouma, R. J.; Possel, O. *Tetrahedron Lett.* **1975**, 3487.
[91] Magnus, P.; Gallagher, T.; Schultz, J.; Or, Y. S.; Ananthanarayan, T. P. *J. Am. Chem. Soc.* **1987**, *109*, 2706.
[92] Shinmyozu, T.; Hirai, Y.; Inazu, T. *J. Org. Chem.* **1986**, *51*, 1551.
[93] van den Berg, K. J., Ph.D. Thesis, Groningen University, 1990.
[94] van Echten, E., unpublished results.
[95] Barrett, A. G. M.; Barton, D. H. R.; Falck, J. R.; Papaioannou, D.; Widdowson, D. A. *J. Chem. Soc., Perkin Trans. 1* **1979**, 652.
[96] Barrett, A. G. M.; Barton, D. H. R.; Franckowiak, G.; Papaioannou, D.; Widdowson, D. A. *J. Chem. Soc., Perkin Trans. 1* **1979**, 662.
[97] Reddy, P. S.; Vidyasagar, V.; Yadav, J. S. *Synth. Commun.* **1984**, *14*, 197.
[98] Possel, O.; van Leusen, A. M. *Tetrahedron Lett.* **1977**, 4229.
[99] Matthies, D.; Büchling, U. *Arch. Pharm. (Weinheim, Ger.)* **1983**, *316*, 598.
[100] van Leusen, A. M.; Oosterwijk, R.; van Echten, E.; van Leusen, D. *Recl. Trav. Chim. Pays Bas* **1985**, *104*, 50.
[101] van Leusen, D.; van Echten, E.; van Leusen, A. M. *Recl. Trav. Chim. Pays Bas* **1992**, *111*, 469.
[102] Matthies, D.; Büchling, U. *Chem. Ztg.* **1982**, *106*, 440.
[103] Moskal, J.; van Leusen, A. M. *Tetrahedron Lett.* **1984**, *25*, 2585.
[104] Trost, B. M. *Acc. Chem. Res.* **1978**, *11*, 453.
[105] Yadav, J. S.; Reddy, P. S.; Joshi, B. V. *Tetrahedron* **1988**, *44*, 7243.
[106] Yadav, J. S.; Reddy, P. S. *Tetrahedron Lett.* **1984**, *25*, 4025.
[107] Moskal, J.; Wildeman, J.; van Leusen, A. M., unpublished results.
[108] Meno, T.; Sako, K.; Suenaga, M.; Mouri, M.; Shinmyozu, T.; Inazu, T.; Takemura, H. *Can. J. Chem.* **1990**, *68*, 440.
[109] Breitenbach, J.; Vögtle, F. *Synthesis* **1992**, 41.
[110] Osada, S.; Suenaga, M.; Miyahara, Y.; Shinmyozu, T.; Inazu, T. *Mem. Fac. Sci., Kyushu Univ., Ser. C* **1993**, *19*, 33; *Chem. Abstr.* **1994**, *120*, 8320w.
[111] Vögtle, F.; Schulz, J. E.; Rissanen, K. *J. Chem. Soc., Chem. Commun.* **1992**, 120.
[112] van Leusen, D.; van Leusen, A. M. *Tetrahedron Lett.* **1977**, 4233.
[113] Shih, N. Y. *Tetrahedron Lett.* **1993**, *34*, 595.
[114] Kuwano, E.; Hisano, T.; Eto, M.; Suzuki, K.; Unnithan, G. C.; Bowers, W. S. *Pestic. Sci.* **1992**, *34*, 263; *Chem. Abstr.* **1992**, *116*, 230183e.
[115] Yamada, N.; Kuwano, E.; Eto, M. *Z. Naturforsch., Teil C* **1993**, *48*, 301.
[116] Sasaki, H.; Kitagawa, T. *Chem. Pharm. Bull.* **1988**, *36*, 3646.
[117] Ananthan, S.; Clayton, S. D.; Ealick, S. E.; Wong, G.; Evoniuk, G. E.; Skolnick, P. *J. Med. Chem.* **1993**, *36*, 479.
[118] Kreutzberger, A.; Kolter, K. *Chem. Ztg.* **1986**, *110*, 256.
[119] Taylor, E. C.; LaMattina, J. L.; Tseng, C. P. *J. Org. Chem* **1982**, *47*, 2043.
[120] van Nispen, S. P. J. M.; Bregman, J. H.; van Engen, D. G.; van Leusen, A. M.; Saikachi, H.; Kitagawa, T.; Sasaki, H. *Recl. Trav. Chim. Pays Bas* **1982**, *101*, 28.
[121] Kikuchi, M.; Kuwano, E.; Eto, M. *J. Fac. Agric., Kyushu Univ.* **1990**, *34*, 397; *Chem. Abstr.* **1990**, *113*, 167296q.
[122] van Leusen, A. M.; Hoogenboom, B. E.; Houwing, H. A. *J. Org. Chem.* **1976**, *41*, 711.

[123] Jacobi, P. A.; Egbertson, M.; Frechette, R. F.; Miao, C. K.; Weiss, K. T. *Tetrahedron* **1988**, *44*, 3327.
[124] Oldenziel, O. H.; van Leusen, A. M. *Tetrahedron Lett.* **1972**, 2777.
[125] van Leusen, A. M.; Wildeman, J. *Synthesis* **1977**, 501.
[126] Bergstrom, D. E.; Zhang, P.; Zhou, J. *J. Chem. Soc., Perkin Trans. 1* **1994**, 3029.
[127] van Leusen, A. M.; Siderius, H.; Hoogenboom, B. E.; van Leusen, D. *Tetrahedron Lett.* **1972**, 5337.
[128] Moskal, J.; van Leusen, A. M. *J. Org. Chem.* **1986**, *51*, 4131.
[129] Magnus, P.; Or, Y. S. *J. Chem. Soc., Chem. Commun.* **1983**, 26.
[130] Magnus, P.; Danikiewicz, W.; Katoh, T.; Huffman, J. C.; Folting, K. *J. Am. Chem. Soc.* **1990**, *112*, 2465.
[131] ten Have, R.; Leusink, F. R.; van Leusen, A. M. *Synthesis* **1996**, 871.
[132] Carter, P.; Fitzjohn, S.; Halazy, S.; Magnus, P. *J. Am. Chem. Soc.* **1987**, *109*, 2711.
[133] Halazy, S.; Magnus, P. *Tetrahedron Lett.* **1984**, *25*, 1421.
[134] Arnold, D. P.; Brown, R. F. C.; Nitschinsk, L. J.; Perlmutter, P.; Tope, H. K. *Aust. J. Chem.* **1994**, *47*, 975.
[135] Genda, Y.; Muro, H.; Nakayama, K.; Miyazaki, Y.; Sugita, Y. DE 3,601,285 (1987); *Chem. Abstr.* **1987**, *107*, 198076y.
[136] van Leusen, D.; van Echten, E.; van Leusen, A. M. *J. Org. Chem.* **1992**, *57*, 2245.
[137] Wollweber, D. DE 3,800,387 (1989); *Chem. Abstr.* **1990**, *112*, 55586g.
[138] Barton, D. H. R.; Kervagoret, J.; Zard, S. Z. *Tetrahedron* **1990**, *46*, 7587.
[139] Uno, H.; Sakamoto, K.; Tominaga, T.; Ono, N. *Bull. Chem. Soc. Jpn.* **1994**, *67*, 1441.
[140] Saikachi, H.; Kitagawa, T.; Sasaki, H. *Chem. Pharm. Bull.* **1979**, *27*, 2857.
[141] Dell'Erba, C.; Giglio, A.; Mugnoli, A.; Novi, M.; Petrillo, G.; Stagnaro, P. *Tetrahedron* **1995**, *51*, 5181.
[142] van Leusen, D.; Flentge, E.; van Leusen, A. M. *Tetrahedron* **1991**, *47*, 4639.
[143] Postma, D.; van Leusen, A. M., unpublished results.
[144] Brown, R. K. In *The Chemistry of Heterocyclic Compounds*; Weissberger, A., Taylor, E. C., Eds.; Vol 25 (Indoles, Part One), Wiley: New York, 1972, p. 277.
[145] Döpp, H.; Döpp, D.; Langer, U.; Gerding, B. `Indole' in *Methoden der organischen Chemie* (Houben-Weyl), Band E6 b1 and Band E6 1994, b2 (Hexarenes I, Teil 2a und Teil 2b); Thieme: Stuttgart, pp. 546–1336.
[146] Leusink, F. R.; ten Have, R.; van den Berg, K. J.; van Leusen, A. M. *J. Chem. Soc., Chem. Commun.* **1992**, 1401.
[146a] Kolb, M. In *Encyclopedia of Reagents for Organic Synthesis*; Paquette, L. A., Ed.; Vol. 4, Wiley: New York, 1995, p. 2307.
[147] Ogura, K. In *Encyclopedia of Reagents for Organic Synthesis*; Paquette, L. A., Ed.; Vol. 5, Wiley: New York, 1995, p. 3589.
[148] van Leusen, D.; van Leusen, A. M. *Synthesis* **1980**, 325.
[149] Frazza, M. S.; Roberts, B. W. *Tetrahedron Lett.* **1981**, *22*, 4193.
[150] Schwartz, E. B.; Knobler, C. B.; Cram, D. J. *J. Am. Chem. Soc.* **1992**, *114*, 10775.
[151] Yadav, J. S.; Gadgil, V. R. *Tetrahedron Lett.* **1990**, *31*, 6217.
[152] Rao, A. V. R.; Deshpande, V. H.; Reddy, S. P. *Synth. Commun.* **1984**, *14*, 469.
[153] Sasaki, H.; Kitagawa, T. *Chem. Pharm. Bull.* **1983**, *31*, 2868.
[154] Breitenbach, J.; Ott, F.; Vögtle, F. *Angew. Chem., Int. Ed. Engl.* **1992**, *31*, 307.
[155] Kurosawa, K.; Suenaga, M.; Inazu, T.; Yoshino, T. *Tetrahedron Lett.* **1982**, *23*, 5335.
[156] Yamato, T.; Doamekpor, L. K.; Koizumi, K.; Kishi, K.; Haraguchi, M.; Tashiro, M. *Justus Liebigs Ann. Chem.* **1995**, 1259.
[157] Kobayashi, M.; Miura, A. *Phosphorus and Sulfur* **1987**, *32*, 169.
[158] van Leusen, D.; van Leusen, A. M. *Tetrahedron Lett.* **1984**, *25*, 2581.
[159] van Leusen, D.; van Leusen, A. M. *J. Org. Chem.* **1994**, *59*, 7534.
[159a] Grossert, J. S. In *The Chemistry of Sulphones and Sulphoxides*; Patai, S., Rappoport, Z., Stirling, C., Eds; Wiley: New York, 1988, p. 925.
[160] Johnson, D. W. *Chem. Phys. Lipids* **1990**, *56*, 65.
[161] Heβ, U.; Brosig, H.; Fehlhammer, W. P. *Tetrahedron Lett.* **1991**, *32*, 5539.

[162] Berrée, F.; Marchand, E.; Morel, G. *Tetrahedron Lett.* **1992**, *33*, 6155.
[163] Tian, W. S.; Livinghouse, T. *J. Chem. Soc., Chem. Commun.* **1989**, 819.
[164] Capuano, L.; Hell, W.; Wamprecht, C. *Justus Liebigs Ann. Chem.* **1986**, 132.
[165] El Kaim, L. *Tetrahedron Lett.* **1994**, *35*, 6669.
[166] Barton, D. H. R.; Ozbalik, N.; Vacher, B. *Tetrahedron* **1988**, *44*, 3501.
[167] Lehnhoff, S.; Goebel, M.; Karl, R. M.; Klösel, R.; Ugi, I. *Angew. Chem., Int. Ed. Engl.* **1995**, *34*, 1104.
[168] Yamada, T.; Motoyama, N.; Taniguchi, T.; Kazuta, Y.; Miyazawa, T.; Kuwata, S.; Matsumoto, K.; Sugiura, M. *Chem. Lett.* **1987**, 723.
[169] Moore, H. W.; Yu, C. C. *J. Org. Chem.* **1981**, *46*, 4935.
[170] Capuano, L.; Tammer, T. *Chem. Ber.* **1981**, *114*, 456.
[171] Schnell, M.; Ramm, M.; Köckritz, A. *J. Prakt. Chem.* **1994**, *336*, 29.
[172] Ott, W.; Kollenz, G.; Peters, K.; Peters, E. M.; von Schnering, H. G.; Quast, H. *Justus Liebigs Ann. Chem.* **1983**, 635.
[173] Vilsmaier, E.; Baumheier, R.; Lemmert, M. *Synthesis* **1990**, 995.
[174] Eicher, T.; Stapperfenne, U. *Synthesis* **1987**, 619.
[175] Brauer, D. J.; Bürger, H.; Hagen, T.; Pawelke, G. *J. Organomet. Chem.* **1994**, *484*, 107.
[176] Kojima, H.; Yamamoto, K.; Kinoshita, Y.; Inoue, H. *J. Heterocycl. Chem.* **1993**, *30*, 1691.
[177] Merlic, C. A.; Burns, E. E. *Tetrahedron Lett.* **1993**, 34, 5401.
[178] Davis, J. M.; Whitby, R. J.; Jaxa-Chamiec, A. *Tetrahedron Lett.* **1994**, *35*, 1445.
[179] Bouquillon, S.; Fabre, P. L.; Dartiguenave, M. *Inorg. Chim. Acta* **1993**, *210*, 135.
[180] Rouwette, P. H. F. M., Ph.D. Thesis, Groningen University, 1979.
[181] van Leusen, D., Ph.D. Thesis, Groningen University, 1990.
[181a] van Leusen, A. M.; Akkerboom, P. J. Eur. Pat. Appl. 7672 (1980); *Chem. Abstr.* **1980**, *93*, 114167h.
[182] Sustmann, R.; Brandes, D.; Lange, F.; Nüchter, U. *Chem. Ber.* **1985**, *118*, 3500.
[182a] Praefcke, K.; Schmidt, C. *Z. Naturforsch., Teil B* **1980**, *35*, 1451.
[183] Darmon, M. J.; Schuster, G. B. *J. Org. Chem.* **1982**, *47*, 4658.
[184] Becker, D. P.; Flynn, D. L. *Synthesis* **1992**, 1080.
[185] Kirmse, W.; Feldmann, G. *Chem. Ber.* **1989**, *122*, 1531.
[186] Nakazaki, M.; Naemura, K.; Hashimoto, M. *J. Org. Chem.* **1983**, *48*, 2289.
[187] Wolff, S.; Agosta, W. C. *J. Am. Chem. Soc.* **1983**, *105*, 1292.
[188] Lowe, J. A., III; Drozda, S. E.; Snider, R. M.; Longo, K. P.; Rizzi, J. P. *Bioorg. Med. Chem. Lett.* **1993**, *3*, 921.
[189] Houssin, R.; Bernier, J.-L.; Henichart, J.-P. *J. Heterocycl. Chem.* **1984**, *21*, 465.
[190] Oldenziel, O. H.; Wildeman, J.; van Leusen, A. M. *Org. Synth.* **1977**, *57*, 8.
[191] Oldenziel, O. H.; van Leusen, A. M. *Synth. Commun.* **1972**, *2*, 281.
[192] Sisk, S. A.; Hutchinson, C. R. *J. Org. Chem.* **1979**, *44*, 3500.
[193] Gerlach, U.; Haubenreich, T.; Hünig, S.; Keita, Y. *Chem. Ber.* **1993**, *126*, 1205.
[193a] Sako, K.; Shinmyozu, T.; Takemura, H.; Suenaga, M.; Inazu, T. *J. Org. Chem.* **1992**, *57*, 6536.
[194] Beetz, T.; Meuleman, D. G.; Wieringa, J. H. *J. Med. Chem.* **1982**, *25*, 714.
[195] Tombari, D. G.; Moglioni, A. G.; Dominici, F. P.; Moltrasio de Iglesias, G. Y. *Org. Prep. Proced. Int.* **1992**, *24*, 45.
[195a] Crombie, L.; Tuchinda, P.; Powell, M. J. *J. Chem. Soc., Perkin Trans. 1* **1982**, 1477.
[196] Sasaki, T.; Eguchi, S.; Mizutani, M. *Tetrahedron Lett.* **1975**, 2685.
[197] Freerksen, R. W.; Watt, D. S. *Synth. Commun.* **1976**, *6*, 447.
[198] Buchi, G. H. *Perfum. Flavor.* **1978**, *3*, 1, 3, 4, 6, 8,10.
[199] Mbela, T. K. N.; Poupaert, J. H.; Cumps, J.; Moussebois, C.; Haemers, A.; Borloo, M.; Dumont, P. *J. Pharm. Pharmacol.* **1995**, *47*, 237.
[200] Di Santo, R.; Costi, R.; Massa, S.; Artico, M. *Synth. Commun.* **1995**, *25*, 787.
[201] Romanelli, M. N.; Teodori, E.; Scapecchi, S.; Dei, S.; Budriesi, R.; Chiarini, A. *Farmaco* **1991**, *46*, 1121; *Chem. Abstr.* **1992**, *116*, 227651a.
[202] Sindelar, K.; Holubek, J.; Ryska, M.; Svátek, E.; Urban, J.; Protiva, M. *Collect. Czech. Chem. Commun.* **1983**, *48*, 1898; *Chem. Abstr.* **1984**, *100*, 6305h.
[203] Corey, E. J.; Behforouz, M.; Ishiguro, M. *J. Am. Chem. Soc.* **1979**, *101*, 1608.

[204] Wünsch, B.; Zott, M.; Höfner, G. *Arch. Pharm. (Weinheim, Ger.)* **1993**, *326*, 823.
[205] Sugita, S. I.; Toda, S.; Yoshiyasu, T.; Teraji, T. *Mol. Cryst. Liq. Cryst.* **1994**, *239*, 113.
[206] Roush, W. R. *J. Am. Chem. Soc.* **1980**, *102*, 1390.
[207] Campiani, G.; Sun, L. Q.; Kozikowski, A. P.; Aagaard, P.; McKinney, M. *J. Org. Chem.* **1993**, *58*, 7660.
[208] Tseng, C. C.; Handa, I.; Abdel-Sayed, A. N.; Bauer, L. *Tetrahedron* **1988**, *44*, 1893.
[209] Schmidt, W.; Vögtle, F.; Poetsch, E. *Justus Liebigs Ann. Chem.* **1995**, 1319.
[210] Hu, Y.; Zorumski, C. F.; Covey, D. F. *J. Med. Chem.* **1993**, *36*, 3956.
[211] Kaminsky, J. J.; Bristol, J. A.; Puchalski, C.; Lovey, R. G.; Elliott, A. J.; Guzik, H.; Solomon, D. M.; Conn, D. J.; Domalski, M. S.; Wong, S. C.; Gold, E. H.; Long, J. F.; Chiu, P. J. S.; Steinberg, M.; McPhail, A. T. *J. Med. Chem.* **1985**, *28*, 876.
[212] Armarego, W. L. F.; Tucker, P. G. *Aust. J. Chem.* **1979**, *32*, 1805.
[213] Popp, F. D.; Watts, R. F. *J. Pharm. Sci.* **1978**, *67*, 871.
[214] Purdy, R. H.; Morrow, A. L.; Blinn, J. R.; Paul, S. M. *J. Med. Chem.* **1990**, *33*, 1572.
[215] Bull, J. R.; Thomson, R. I. *S. Afr. J. Chem.* **1982**, *35*, 101.
[216] Ottow, E.; Rohde, R.; Schwede, W.; Wiechert, R. *Tetrahedron Lett.* **1993**, *34*, 5253.
[217] Schwede, W.; Cleve, A.; Ottow, E.; Wiechert, R. *Tetrahedron Lett.* **1993**, *34*, 5257.
[218] Oussaid, B.; Moeini, L.; Garrigues, B.; Villemin, D. *Phosphorus, Sulfur, and Silicon* **1993**, *85*, 23.
[219] Merour, J. Y.; Buzas, A. *Synth. Commun.* **1988**, *18*, 2331.
[220] Procter, G.; Genin, D.; Challenger, S. *Carbohydr. Res.* **1990**, *202*, 81.
[221] Horne, D. A.; Yakushijin, K.; Büchi, G. *Heterocycles* **1994**, *39*, 139.
[222] Oldenziel, O. H.; van Leusen, A. M. *Tetrahedron Lett.* **1974**, 163.
[223] Solladié-Cavallo, A.; Quazzotti, S.; Colonna, S.; Manfredi, A. *Tetrahedron Lett.* **1989**, *30*, 2933.
[224] Crowe, E.; Hossner, F.; Hughes, M. J. *Tetrahedron* **1995**, *51*, 8889.
[225] Sasaki, H.; Egi, R.; Kawanishi, K.; Kitagawa, T.; Shingu, T. *Chem. Pharm. Bull.* **1989**, *37*, 1176.
[226] Sasaki, H.; Kawanishi, K.; Kitagawa, T.; Shingu, T. *Chem. Pharm. Bull.* **1989**, *37*, 2303.
[227] Kozikowski, A. P.; Ames, A. *J. Org. Chem.* **1980**, *45*, 2548.
[228] Tully, W. R.; Gardner, C. R.; Gillespie, R. J.; Westwood, R. *J. Med. Chem.* **1991**, *34*, 2060.
[229] Suschitzky, H.; Kramer, W.; Neidlein, R.; Rosyk, P.; Bohn, T. *J. Chem. Soc., Perkin Trans. 1* **1991**, 923.
[230] Kozikowski, A. P.; Ames, A. *J. Am. Chem. Soc.* **1980**, *102*, 860.
[231] Kleinschroth, J.; Hartenstein, J. *Synthesis* **1988**, 970.
[232] Kuwano, E.; Hisano, T.; Eto, M. *Agric. Biol. Chem.* **1991**, *55*, 2999.
[233] Amino, Y.; Eto, H.; Eguchi, C. *Chem. Pharm. Bull.* **1989**, *37*, 1481.
[234] Naito, T.; Honda, Y.; Miyata, O.; Ninomiya, I. *Chem. Pharm. Bull.* **1993**, *41*, 217.
[235] Blay, G.; Schrijvers, R.; Wijnberg, J. B. P. A.; de Groot, A. *J. Org. Chem.* **1995**, *60*, 2188
[236] Kochanny, M. J.; VanBrocklin, H. F.; Kym, P. R.; Carlson, K. E.; O'Neil, J. P.; Bonasera, T. A.; Welch, M. J.; Katzenellenbogen, J. A. *J. Med. Chem.* **1993**, *36*, 1120.
[237] Yadagiri, P.; Yadav, J. S. *Synth. Commun.* **1983**, *13*, 1067.
[238] Kamath, S. V.; Rangaishenvi, M. V.; Bapat, B. V.; Hiremath, S. V.; Kulkarni, S. N. *Indian J. Chem., Sect. B* **1983**, *22B*, 921; *Chem. Abstr.* **1984**, *100*, 210190.
[239] Kamath, S. V.; Rangaishenvi, M. V.; Kulkarni, S. N. *Indian J. Chem., Sec. B* **1983**, *22B*, 1264; *Chem. Abstr.* **1984**, *101*, 38692.
[240] Reddy, P. S.; Sahasrabudhe, A. B.; Yadav, J. S. *Synth. Commun.* **1983**, *13*, 379.
[241] Mertin, A.; Thiemann, T.; Hanss, I.; de Meijere, A. *Synlett* **1991**, 87.
[242] Rao, A. V. R.; Yadav, J. S.; Annapurna, G. S. *Synth. Commun.* **1983**, *13*, 331.
[243] Hassam, S. B. *J. Labelled Compd. Radiopharm.* **1987**, *24*, 107.
[244] van Nispen, S. P. J. M., Ph.D. Thesis, Groningen University, 1980.
[245] Harris, R. N.; Sundararaman, P.; Djerassi, C. *J. Am. Chem. Soc.* **1983**, *105*, 2408.

[246] Kawase, T.; Ohsawa, T.; Enomoto, T.; Oda, M. *Chem. Lett.* **1994**, 1333.
[247] Hanack, M.; Auchter, G. *J. Am. Chem. Soc.* **1985**, *107*, 5238.
[248] Osada, S.; Miyahara, Y.; Shinmyozu, T.; Inazu, T. *Mem. Fac. Sci., Kyushu Univ., Ser. C* **1993**, *19*, 39; *Chem. Abstr.* **1994**, *120*, 8321.
[249] Sako, K.; Meno, T.; Takemura, H.; Shinmyozu, T.; Inazu, T. *Chem. Ber.* **1990**, *123*, 639
[250] Kobiro, K.; Takahashi, M.; Nishikawa, N.; Kakiuchi, K.; Tobe, Y.; Odaira, Y. *Tetrahedron Lett.* **1987**, *28*, 3825.
[251] Schulz, J. E.; Rissanen, K.; Vögtle, F. *Chem. Ber.* **1992**, *125*, 2239
[252] Johnson, D. W.; Poulos, A. *Tetrahedron Lett.* **1992**, *33*, 2045.
[253] Schmohel, E.; Ott, F.; Breitenbach, J.; Nieger, M.; Vögtle, F. *Chem. Ber.* **1993**, *126*, 2477.
[254] Shinmyozu, T.; Kusumoto, S.; Nomura, S.; Kawase, H.; Inazu, T. *Chem. Ber.* **1993**, *126*, 1815.
[255] Yamada, N.; Kuwano, E.; Kikuchi, M.; Eto, M. *Biosci., Biotech., Biochem.* **1992**, *56*, 1943; *Chem. Abstr.* **1993**, *118*, 185719.
[256] Kikuchi, M.; Kuwano, E.; Nakashima, Y.; Eto, M. *J. Fac. Agric., Kyushu Univ.* **1991**, *36*, 83; *Chem. Abstr.* **1992**, *116*, 189538.
[257] Pratt, G. E.; Kuwano, E.; Farnsworth, D. E.; Feyereisen, R. *Pestic. Biochem. Physiol.* **1990**, *38*, 223; *Chem. Abstr.* **1991**, *114*, 98632.
[258] Kuwano, E.; Kikuchi, M.; Eto, M. *J. Fac. Agric., Kyushu Univ.* **1990**, *35*, 35; *Chem. Abstr.* **1992**, *115*, 87436.
[259] Kuwano, E.; Takeya, R. Eto, M. *Agric. Biol. Chem.* **1985**, *49*, 483.
[260] Kuwano, E.; Takeya, R.; Eto, M. *Agric. Biol. Chem.* **1984**, *48*, 3115.
[261] Battersby, A. R.; Bartholomew, S. A. J.; Nitta, T. *J. Chem. Soc., Chem. Commun.* **1983**, 1291.
[262] Silvestri, R.; Artico, M.; Pagnozzi, E.; Stefancich, G. *J. Heterocycl. Chem.* **1994**, *31*, 1033.
[263] Massa, S.; Di Santo, R.; Costi, R.; Artico, M. *J. Heterocycl. Chem.* **1993**, *30*, 749.
[264] Sasaki, H.; Ogawa, K.; Iijima, Y.; Kitagawa, T.; Shingu, T. *Chem. Pharm. Bull.* **1988**, *36*, 1990.
[265] Olson, G. L.; Cheung, H. C.; Chiang, E.; Madison, V. S.; Sepinwall, J.; Vincent, G. P.; Winokur, A.; Gary, K. A. *J. Med. Chem.* **1995**, *38*, 2866.
[266] van Leusen, A. M.; Oldenziel, O. H. *Tetrahedron Lett.* **1972**, 2373.
[267] Kikuchi, M.; Kuwano, E.; Nakashima, Y.; Eto, M. *Biosci. Biotech. Biochem.* **1992**, *56*, 161; *Chem. Abstr.* **1992**, *116*, 198544.
[268] Jacobi, P. A.; Frechette, R. F. *Tetrahedron Lett.* **1987**, *28*, 2937.
[268a] Di Santo, R.; Costi, R.; Massa, S.; Artico, M. *Synth. Commun.* **1995**, *25*, 795.
[269] Aoyagi, K.; Toi, H.; Aoyama, Y.; Ogoshi, H. *Chem. Lett.* **1988**, 1891.
[270] Rühe, J.; Kröhnke, C.; Ezquerra, T. A.; Kremer, F.; Wegner, G. *Ber. Bunsenges. Phys. Chem.* **1987**, *91*, 885; *Chem. Abstr.* **1988**, *109*, 30651.
[271] Cheng, D. O.; LeGoff, E. *Tetrahedron Lett.* **1977**, 1469.
[272] Ono, N.; Muratani, E.; Ogawa. T. *J. Heterocycl. Chem.* **1991**, *28*, 2053.
[273] Cheng, D. O.; Bowman, T. L.; LeGoff, E. *J. Heterocycl. Chem.* **1976**, *13*, 1145.
[274] Leroy, J. *J. Fluorine Chem.* **1991**, *53*, 61.
[274a] Arnold, D. P.; Nitschinsk, L. J.; Kennard, C. H. L.; Smith, G. *Aust. J. Chem.* **1991**, *44*, 323.
[275] Chamberlin, K. S.; LeGoff, E. *Heterocycles* **1979**, *12*, 1567.
[276] Teo, K. E.; Barnett, G. H.; Anderson, H. J.; Loader, C. E. *Can. J. Chem.* **1978**, *56*, 221.
[277] Chamberlin, K. S.; LeGoff, E. *Synth. Commun.* **1978**, *8*, 579.
[277a] Massa, S.; Di Santo, R.; Costi, R.; Mai, A.; Artico, M. *Med. Chem. Res.* **1993**, *3*, 192.
[277b] Spreitzer, H.; Mustafa, S. *Chem. Ber.* **1990**, *123*, 413.
[278] Magnus, P.; Halazy, S. *Tetrahedron Lett.* **1985**, *26*, 2985.
[279] Harris, D.; Syren, S.; Streith, J. *Tetrahedron Lett.* **1978**, 4093.
[280] Franck, B.; Nonn, A.; Fuchs, K.; Gosmann, M. *Justus Liebigs Ann. Chem.* **1994**, 503.
[281] Ahmed, F. R.; Cheung, K. M.; Toube, T. P.; Utley, J. H. P. *J. Mater. Chem.* **1995**, *5*, 837.
[282] Massa, S.; Di Santo, R.; Mai, A.; Botta, M.; Artico, M.; Panico, S.; Simonetti, G. *Farmaco* **1990**, *45*, 833; *Chem. Abstr.* **1991**, *114*, 74753.

[283] Kohori, K.; Hashimoto, M.; Kinoshita, H.; Inomata, K. *Bull. Chem. Soc. Jpn.* **1994**, *67*, 3088.
[284] Bohlmann, F.; Klose, W.; Nickisch, K. *Tetrahedron Lett.* **1979**, 3699.
[285] Nemeroff, N. H.; McDonnell, M. E.; Axten, J. M.; Buckley, L. J. *Synth. Commun.* **1992**, *22*, 3271.
[286] Bird, C. W.; Chauhan, Y. P. S.; Turton, D. R. *Tetrahedron* **1981**, *37*, 1277.
[287] Artico, M.; Di Santo, R.; Costi, R.; Massa, S.; Retico, A.; Artico, M.; Apuzzo, G.; Simonetti, G; Strippoli, V. *J. Med. Chem.* **1995**, *38*, 4223.
[288] Del Valle, J. L.; Polo, C.; Torroba, T.; Marcaccini, S. *J. Heterocycl. Chem.* **1995**, *32*, 899.
[289] Baxter, A. J. G.; Dixon, J.; Ince, F.; Manners, C. N.; Teague, S. J. *J. Med. Chem.* **1993**, *36*, 2739.
[290] Di Santo, R.; Massa, S.; Costi, R.; Simonetti, G. *Farmaco* **1994**, *49*, 229; *Chem. Abstr.* **1994**, *121*, 57393.
[291] Kroszczynski, W. *Rocz. Chem.* **1975**, *49*, 813; *Chem. Abstr.* **1975**, *83*, 114724.
[292] Rieger, D.; Lotz, S. D.; Kernbach, U.; André, C.; Bertran-Nadal, J.; Fehlhammer, W. P. *J. Organomet. Chem.* **1995**, *491*, 135.
[293] Fink, J.; Regitz, M. *Chem. Ber.* **1986**, *119*, 2159.

CUMULATIVE CHAPTER TITLES
BY VOLUME

Volume 1 (1942)

1. **The Reformatsky Reaction**: Ralph L. Shriner

2. **The Arndt-Eistert Reaction**: W. E. Bachmann and W. S. Struve

3. **Chloromethylation of Aromatic Compounds**: Reynold C. Fuson and C. H. McKeever

4. **The Amination of Heterocyclic Bases by Alkali Amides**: Marlin T. Leffler

5. **The Bucherer Reaction**: Nathan L. Drake

6. **The Elbs Reaction**: Louis F. Fieser

7. **The Clemmensen Reduction**: Elmore L. Martin

8. **The Perkin Reaction and Related Reactions**: John R. Johnson

9. **The Acetoacetic Ester Condensation and Certain Related Reactions**: Charles R. Hauser and Boyd E. Hudson, Jr.

10. **The Mannich Reaction**: F. F. Blicke

11. **The Fries Reaction**: A. H. Blatt

12. **The Jacobson Reaction**: Lee Irvin Smith

Volume 2 (1944)

1. **The Claisen Rearrangement**: D. Stanley Tarbell

2. **The Preparation of Aliphatic Fluorine Compounds**: Albert L. Henne

3. **The Cannizzaro Reaction**: T. A. Geissman

4. **The Formation of Cyclic Ketones by Intramolecular Acylation**: William S. Johnson

5. **Reduction with Aluminum Alkoxides (The Meerwein-Ponndorf-Verley Reduction)**: A. L. Wilds

6. **The Preparation of Unsymmetrical Biaryls by the Diazo Reaction and the Nitrosoacetylamine Reaction**: Werner E. Bachmann and Roger A. Hoffman

7. **Replacement of the Aromatic Primary Amino Group by Hydrogen**: Nathan Kornblum

8. **Periodic Acid Oxidation**: Ernest L. Jackson

9. **The Resolution of Alcohols**: A. W. Ingersoll

10. **The Preparation of Aromatic Arsonic and Arsinic Acids by the Bart, Béchamp, and Rosenmund Reactions**: Cliff S. Hamilton and Jack F. Morgan

Volume 3 (1946)

1. **The Alkylation of Aromatic Compounds by the Friedel-Crafts Method**: Charles C. Price

2. **The Willgerodt Reaction**: Marvin Carmack and M. A. Spielman

3. **Preparation of Ketenes and Ketene Dimers**: W. E. Hanford and John C. Sauer

4. **Direct Sulfonation of Aromatic Hydrocarbons and Their Halogen Derivatives**: C. M. Suter and Arthur W. Weston

5. **Azlactones**: H. E. Carter

6. **Substitution and Addition Reactions of Thiocyanogen**: John L. Wood

7. **The Hofmann Reaction**: Everett L. Wallis and John F. Lane

8. **The Schmidt Reaction**: Hans Wolff

9. **The Curtius Reaction**: Peter A. S. Smith

Volume 4 (1948)

1. **The Diels-Alder Reaction with Maleic Anhydride**: Milton C. Kloetzel

2. **The Diels-Alder Reaction: Ethylenic and Acetylenic Dienophiles**: H. L. Holmes

3. **The Preparation of Amines by Reductive Alkylation**: William S. Emerson

4. **The Acyloins**: S. M. McElvain

5. **The Synthesis of Benzoins**: Walter S. Ide and Johannes S. Buck

6. **Synthesis of Benzoquinones by Oxidation**: James Cason

7. **The Rosenmund Reduction of Acid Chlorides to Aldehydes**: Erich Mosettig and Ralph Mozingo

8. **The Wolff-Kishner Reduction**: David Todd

CUMULATIVE CHAPTER TITLES BY VOLUME

Volume 5 (1949)

1. **The Synthesis of Acetylenes**: Thomas L. Jacobs

2. **Cyanoethylation**: Herman L. Bruson

3. **The Diels-Alder Reaction: Quinones and Other Cyclenones**: Lewis L. Butz and Anton W. Rytina

4. **Preparation of Aromatic Fluorine Compounds from Diazonium Fluoborates: The Schiemann Reaction**: Arthur Roe

5. **The Friedel and Crafts Reaction with Aliphatic Dibasic Acid Anhydrides**: Ernst Berliner

6. **The Gattermann-Koch Reaction**: Nathan N. Crounse

7. **The Leuckart Reaction**: Maurice L. Moore

8. **Selenium Dioxide Oxidation**: Norman Rabjohn

9. **The Hoesch Synthesis**: Paul E. Spoerri and Adrien S. DuBois

10. **The Darzens Glycidic Ester Condensation**: Melvin S. Newman and Barney J. Magerlein

Volume 6 (1951)

1. **The Stobbe Condensation**: William S. Johnson and Guido H. Daub

2. **The Preparation of 3,4-Dihydroisoquinolines and Related Compounds by the Bischler-Napieralski Reaction**: Wilson M. Whaley and Tutucorin R. Govindachari

3. **The Pictet-Spengler Synthesis of Tetrahydroisoquinolines and Related Compounds**: Wilson M. Whaley and Tutucorin R. Govindachari

4. **The Synthesis of Isoquinolines by the Pomeranz-Fritsch Reaction**: Walter J. Gensler

5. **The Oppenauer Oxidation**: Carl Djerassi

6. **The Synthesis of Phosphonic and Phosphinic Acids**: Gennady M. Kosolapoff

7. **The Halogen-Metal Interconversion Reaction with Organolithium Compounds**: Reuben G. Jones and Henry Gilman

8. **The Preparation of Thiazoles**: Richard H. Wiley, D. C. England, and Lyell C. Behr

9. **The Preparation of Thiophenes and Tetrahydrothiophenes**: Donald E. Wolf and Karl Folkers

10. **Reductions by Lithium Aluminum Hydride**: Weldon G. Brown

Volume 7 (1953)

1. **The Pechmann Reaction**: Suresh Sethna and Ragini Phadke

2. **The Skraup Synthesis of Quinolines**: R. H. F. Manske and Marshall Kulka

3. **Carbon-Carbon Alkylations with Amines and Ammonium Salts**: James H. Brewster and Ernest L. Eliel

4. **The von Braun Cyanogen Bromide Reaction**: Howard A. Hageman

5. **Hydrogenolysis of Benzyl Groups Attached to Oxygen, Nitrogen, or Sulfur**: Walter H. Hartung and Robert Simonoff

6. **The Nitrosation of Aliphatic Carbon Atoms**: Oscar Touster

7. **Epoxidation and Hydroxylation of Ethylenic Compounds with Organic Peracids**: Daniel Swern

Volume 8 (1954)

1. **Catalytic Hydrogenation of Esters to Alcohols**: Homer Adkins

2. **The Synthesis of Ketones from Acid Halides and Organometallic Compounds of Magnesium, Zinc, and Cadmium**: David A. Shirley

3. **The Acylation of Ketones to Form β-Diketones or β-Keto Aldehydes**: Charles R. Hauser, Frederic W. Swamer, and Joe T. Adams

4. **The Sommelet Reaction**: S. J. Angyal

5. **The Synthesis of Aldehydes from Carboxylic Acids**: Erich Mosettig

6. **The Metalation Reaction with Organolithium Compounds**: Henry Gilman and John W. Morton, Jr.

7. **β-Lactones**: Harold E. Zaugg

8. **The Reaction of Diazomethane and Its Derivatives with Aldehydes and Ketones**: C. David Gutsche

Volume 9 (1957)

1. **The Cleavage of Non-enolizable Ketones with Sodium Amide**: K. E. Hamlin and Arthur W. Weston

2. **The Gattermann Synthesis of Aldehydes**: William E. Truce

3. **The Baeyer-Villiger Oxidation of Aldehydes and Ketones**: C. H. Hassall

4. **The Alkylation of Esters and Nitriles**: Arthur C. Cope, H. L. Holmes, and Herbert O. House

5. **The Reaction of Halogens with Silver Salts of Carboxylic Acids**:   C. V. Wilson

6. **The Synthesis of β-Lactams**:   John C. Sheehan and Elias J. Corey

7. **The Pschorr Synthesis and Related Diazonium Ring Closure Reactions**: DeLos F. DeTar

Volume 10 (1959)

1. **The Coupling of Diazonium Salts with Aliphatic Carbon Atoms**: Stanley J. Parmerter

2. **The Japp-Klingemann Reaction**:   Robert R. Phillips

3. **The Michael Reaction**:   Ernst D. Bergmann, David Ginsburg, and Raphael Pappo

Volume 11 (1960)

1. **The Beckmann Rearrangement**:   L. Guy Donaruma and Walter Z. Heldt

2. **The Demjanov and Tiffeneau-Demjanov Ring Expansions**:   Peter A. S. Smith and Donald R. Baer

3. **Arylation of Unsaturated Compounds by Diazonium Salts**: Christian S. Rondestvedt, Jr.

4. **The Favorskii Rearrangement of Haloketones**:   Andrew S. Kende

5. **Olefins from Amines: The Hofmann Elimination Reaction and Amine Oxide Pyrolysis**:   Arthur C. Cope and Elmer R. Trumbull

Volume 12 (1962)

1. **Cyclobutane Derivatives from Thermal Cycloaddition Reactions**:   John D. Roberts and Clay M. Sharts

2. **The Preparation of Olefins by the Pyrolysis of Xanthates. The Chugaev Reaction**:   Harold R. Nace

3. **The Synthesis of Aliphatic and Alicyclic Nitro Compounds**:   Nathan Kornblum

4. **Synthesis of Peptides with Mixed Anhydrides**:   Noel F. Albertson

5. **Desulfurization with Raney Nickel**:   George R. Pettit and Eugene E. van Tamelen

Volume 13 (1963)

1. **Hydration of Olefins, Dienes, and Acetylenes via Hydroboration**:   George Zweifel and Herbert C. Brown

2. **Halocyclopropanes from Halocarbenes**: William E. Parham and Edward E. Schweizer

3. **Free Radical Addition to Olefins to Form Carbon-Carbon Bonds**: Cheves Walling and Earl S. Huyser

4. **Formation of Carbon-Heteroatom Bonds by Free Radical Chain Additions to Carbon-Carbon Multiple Bonds**: F. W. Stacey and J. F. Harris, Jr.

Volume 14 (1965)

1. **The Chapman Rearrangement**: J. W. Schulenberg and S. Archer

2. **α-Amidoalkylations at Carbon**: Harold E. Zaugg and William B. Martin

3. **The Wittig Reaction**: Adalbert Maercker

Volume 15 (1967)

1. **The Dieckmann Condensation**: John P. Schaefer and Jordan J. Bloomfield

2. **The Knoevenagel Condensation**: G. Jones

Volume 16 (1968)

1. **The Aldol Condensation**: Arnold T. Nielsen and William J. Houlihan

Volume 17 (1969)

1. **The Synthesis of Substituted Ferrocenes and Other π-Cyclopentadienyl-Transition Metal Compounds**: Donald E. Bublitz and Kenneth L. Rinehart, Jr.

2. **The γ-Alkylation and γ-Arylation of Dianions of β-Dicarbonyl Compounds**: Thomas M. Harris and Constance M. Harris

3. **The Ritter Reaction**: L. I. Krimen and Donald J. Cota

Volume 18 (1970)

1. **Preparation of Ketones from the Reaction of Organolithium Reagents with Carboxylic Acids**: Margaret J. Jorgenson

2. **The Smiles and Related Rearrangements of Aromatic Systems**: W. E. Truce, Eunice M. Kreider, and William W. Brand

3. **The Reactions of Diazoacetic Esters with Alkenes, Alkynes, Heterocyclic, and Aromatic Compounds**: Vinod Dave and E. W. Warnhoff

4. **The Base-Promoted Rearrangements of Quaternary Ammonium Salts**: Stanley H. Pine

Volume 19 (1972)

1. **Conjugate Addition Reactions of Organocopper Reagents**: Gary H. Posner

2. **Formation of Carbon-Carbon Bonds via π-Allylnickel Compounds**: Martin F. Semmelhack

3. **The Thiele-Winter Acetoxylation of Quinones**: J. F. W. McOmie and J. M. Blatchly

4. **Oxidative Decarboxylation of Acids by Lead Tetraacetate**: Roger A. Sheldon and Jay K. Kochi

Volume 20 (1973)

1. **Cyclopropanes from Unsaturated Compounds, Methylene Iodide, and Zinc-Copper Couple**: H. E. Simmons, T. L. Cairns, Susan A. Vladuchick, and Connie M. Hoiness

2. **Sensitized Photooxygenation of Olefins**: R. W. Denny and A. Nickon

3. **The Synthesis of 5-Hydroxyindoles by the Nenitzescu Reaction**: George R. Allen, Jr.

4. **The Zinin Reaction of Nitroarenes**: H. K. Porter

Volume 21 (1974)

1. **Fluorination with Sulfur Tetrafluoride**: G. A. Boswell, Jr., W. C. Ripka, R. M. Scribner, and C. W. Tullock

2. **Modern Methods to Prepare Monofluoroaliphatic Compounds**: Clay M. Sharts and William Λ. Sheppard

Volume 22 (1975)

1. **The Claisen and Cope Rearrangements**: Sara Jane Rhoads and N. Rebecca Raulins

2. **Substitution Reactions Using Organocopper Reagents**: Gary H. Posner

3. **Clemmensen Reduction of Ketones in Anhydrous Organic Solvents**: E. Vedejs

4. **The Reformatsky Reaction**: Michael W. Rathke

Volume 23 (1976)

1. **Reduction and Related Reactions of α,β-Unsaturated Compounds with Metals in Liquid Ammonia**: Drury Caine

2. **The Acyloin Condensation**: Jordan J. Bloomfield, Dennis C. Owsley, and Janice M. Nelke

3. **Alkenes from Tosylhydrazones**: Robert H. Shapiro

Volume 24 (1976)

1. **Homogeneous Hydrogenation Catalysts in Organic Solvents**: Arthur J. Birch and David H. Williamson

2. **Ester Cleavages via S_N2-Type Dealkylation**: John E. McMurry

3. **Arylation of Unsaturated Compounds by Diazonium Salts (The Meerwein Arylation Reaction)**: Christian S. Rondestvedt, Jr.

4. **Selenium Dioxide Oxidation**: Norman Rabjohn

Volume 25 (1977)

1. **The Ramberg-Bäcklund Rearrangement**: Leo A. Paquette

2. **Synthetic Applications of Phosphoryl-Stabilized Anions**: William S. Wadsworth, Jr.

3. **Hydrocyanation of Conjugated Carbonyl Compounds**: Wataru Nagata and Mitsuru Yoshioka

Volume 26 (1979)

1. **Heteroatom-Facilitated Lithiations**: Heinz W. Gschwend and Herman R. Rodriguez

2. **Intramolecular Reactions of Diazocarbonyl Compounds**: Steven D. Burke and Paul A. Grieco

Volume 27 (1982)

1. **Allylic and Benzylic Carbanions Substituted by Heteroatoms**: Jean-François Biellmann and Jean-Bernard Ducep

2. **Palladium-Catalyzed Vinylation of Organic Halides**: Richard F. Heck

Volume 28 (1982)

1. **The Reimer-Tiemann Reaction**: Hans Wynberg and Egbert W. Meijer

2. **The Friedländer Synthesis of Quinolines**: Chia-Chung Cheng and Shou-Jen Yan

3. **The Directed Aldol Reaction**: Teruaki Mukaiyama

Volume 29 (1983)

1. **Replacement of Alcoholic Hydroxy Groups by Halogens and Other Nucleophiles via Oxyphosphonium Intermediates**: Bertrand R. Castro

2. **Reductive Dehalogenation of Polyhalo Ketones with Low-Valent Metals and Related Reducing Agents**: Ryoji Noyori and Yoshihiro Hayakawa

3. **Base-Promoted Isomerizations of Epoxides**: Jack K. Crandall and Marcel Apparu

Volume 30 (1984)

1. **Photocyclization of Stilbenes and Related Molecules**: Frank B. Mallory and Clelia W. Mallory

2. **Olefin Synthesis via Deoxygenation of Vicinal Diols**: Eric Block

Volume 31 (1984)

1. **Addition and Substitution Reactions of Nitrile-Stabilized Carbanions**: Siméon Arseniyadis, Keith S. Kyler, and David S. Watt

Volume 32 (1984)

1. **The Intramolecular Diels-Alder Reaction**: Engelbert Ciganek

2. **Synthesis Using Alkyne-Derived Alkenyl- and Alkynylaluminum Compounds**: George Zweifel and Joseph A. Miller

Volume 33 (1985)

1. **Formation of Carbon-Carbon and Carbon-Heteroatom Bonds via Organoboranes and Organoborates**: Ei-Ichi Negishi and Michael J. Idacavage

2. **The Vinylcyclopropane-Cyclopentene Rearrangement**: Tomáš Hudlický, Toni M. Kutchan, and Saiyid M. Naqvi

Volume 34 (1985)

1. **Reductions by Metal Alkoxyaluminum Hydrides**: Jaroslav Málek

2. **Fluorination by Sulfur Tetrafluoride**: Chia-Lin J. Wang

Volume 35 (1988)

1. **The Beckmann Reactions: Rearrangements, Elimination-Additions, Fragmentations, and Rearrangement-Cyclizations**: Robert E. Gawley

2. **The Persulfate Oxidation of Phenols and Arylamines (The Elbs and the Boyland-Sims Oxidations)**: E. J. Behrman

3. **Fluorination with Diethylaminosulfur Trifluoride and Related Aminofluorosulfuranes**: Miloš Hudlický

Volume 36 (1988)

1. **The [3 + 2] Nitrone-Olefin Cycloaddition Reaction**: Pat N. Confalone and Edward M. Huie

2. **Phosphorus Addition at sp^2 Carbon**: Robert Engel

3. **Reduction by Metal Alkoxyaluminum Hydrides. Part II. Carboxylic Acids and Derivatives, Nitrogen Compounds, and Sulfur Compounds**: Jaroslav Málek

Volume 37 (1989)

1. **Chiral Synthons by Ester Hydrolysis Catalyzed by Pig Liver Esterase**: Masaji Ohno and Masami Otsuka

2. **The Electrophilic Substitution of Allylsilanes and Vinylsilanes**: Ian Fleming, Jacques Dunoguès, and Roger Smithers

Volume 38 (1990)

1. **The Peterson Olefination Reaction**: David J. Ager

2. **Tandem Vicinal Difunctionalization: β-Addition to α,β-Unsaturated Carbonyl Substrates Followed by α-Functionalization:** Marc J. Chapdelaine and Martin Hulce

3. **The Nef Reaction**: Harold W. Pinnick

Volume 39 (1990)

1. **Lithioalkenes from Arenesulfonylhydrazones**: A. Richard Chamberlin and Steven H. Bloom

2. **The Polonovski Reaction**: David Grierson

3. **Oxidation of Alcohols to Carbonyl Compounds via Alkoxysulfonium Ylides: The Moffatt, Swern, and Related Oxidations**: Thomas T. Tidwell

Volume 40 (1991)

1. **The Pauson-Khand Cycloaddition Reaction for Synthesis of Cyclopentenones**: Neil E. Schore

2. **Reduction with Diimide**: Daniel J. Pasto and Richard T. Taylor

3. **The Pummerer Reaction of Sulfinyl Compounds**: Ottorino DeLucchi, Umberto Miotti, and Giorgio Modena

4. **The Catalyzed Nucleophilic Addition of Aldehydes to Electrophilic Double Bonds**: Hermann Stetter and Heinrich Kuhlmann

Volume 41 (1992)

1. **Divinylcyclopropane-Cycloheptadiene Rearrangement**: Tomáš Hudlický, Rulin Fan, Josephine W. Reed, and Kumar G. Gadamasetti

2. **Organocopper Reagents: Substitution, Conjugate Addition, Carbo/Metallocupration, and Other Reactions**: Bruce H. Lipshutz and Saumitra Sengupta

Volume 42 (1992)

1. **The Birch Reduction of Aromatic Compounds**: Peter W. Rabideau and Zbigniew Marcinow

2. **The Mitsunobu Reaction**: David L. Hughes

Volume 43 (1993)

1. **Carbonyl Methylenation and Alkylidenation Using Titanium-Based Reagents**: Stanley H. Pine

2. **Anion-Assisted Sigmatropic Rearrangements**: Stephen R. Wilson

3. **The Baeyer-Villiger Oxidation of Ketones and Aldehydes**: Grant R. Krow

Volume 44 (1993)

1. **Preparation of α,β-Unsaturated Carbonyl Compounds and Nitriles by Selenoxide Elimination**: Hans J. Reich and Susan Wollowitz

2. **Enone Olefin [2 + 2] Photochemical Cyclizations**: Michael T. Crimmins and Tracy L. Reinhold

Volume 45 (1994)

1. **The Nazarov Cyclization**: Karl L. Habermas, Scott E. Denmark, and Todd K. Jones

2. **Ketene Cycloadditions**: John A. Hyatt and Peter W. Raynolds

Volume 46 (1994)

1. **Tin(II) Enolates in the Aldol, Michael, and Related Reactions**: Teruaki Mukaiyama and Shū Kobayashi

2. **The [2,3]-Wittig Reaction**: Takeshi Nakai and Koichi Mikami

3. **Reductions with Samarium(II) Iodide**: Gary A. Molander

Volume 47 (1995)

1. **Lateral Lithiation Reactions Promoted by Heteroatomic Substituents**: Robin D. Clark and Alam Jahangir

2. **The Intramolecular Michael Reaction**: R. Daniel Little, Mohammad R. Masjedizadeh, Olof Wallquist (in part), and Jim I. McLoughlin (in part)

Volume 48 (1996)

1. **Asymmetric Epoxidation of Allylic Alcohols: The Katsuki–Sharpless Epoxidation Reaction**: Tsutomu Katsuki and Victor S. Martin

2. **Radical Cyclization Reactions**: B. Giese, B. Kopping, T. Göbel, J. Dickhaut, G. Thoma, K. J. Kulicke, and F. Trach

Volume 49 (1997)

1. **The Vilsmeier Reaction of Fully Conjugated Carbocycles and Heterocycles**: Gurnos Jones and Stephen P. Stanforth

2. **[6 + 4] Cycloaddition Reactions**: James H. Rigby

3. **Carbon–Carbon Bond-Forming Reactions Promoted by Trivalent Manganese**: Gagik G. Melikyan

Volume 50 (1997)

1. **The Stille Reaction**: Vittorio Farina, Venkat Krishnamurthy, and William J. Scott

Volume 51 (1997)

1. **Asymmetric Aldol Reactions Using Boron Enolates**: Cameron J. Cowden and Ian Paterson

2. **The Catalyzed α-Hydroxylation and α-Aminoalkylation of Activated Olefins (The Morita–Baylis–Hillman Reaction)**: Engelbert Ciganek

3. **[4 + 3] Cycloaddition Reactions**: James H. Rigby and F. Christopher Pigge

Volume 52 (1998)

1. **The Retro-Diels–Alder Reaction. Part I. C—C Dienophiles**: Bruce Rickborn

2. **Enantioselective Reduction of Ketones**: Shinichi Itsuno

Volume 53 (1998)

1. **The Oxidation of Alcohols by Modified Oxochromium(VI)-Amine Reagents**: Frederick A. Luzzio

2. **The Retro–Diels–Alder Reaction. Part II. Dienophiles with One or More Heteroatoms**: Bruce Rickborn

Volume 54 (1999)

1. **Aromatic Substitution by the $S_{RN}1$ Reaction**: Roberto Rossi, Adriana B. Pierini, and Ana N. Santiago

2. **Oxidation of Carbonyl Compounds with Organohypervalent Iodine Reagents**: Robert M. Moriarty and Om Prakash

Volume 55 (1999)

1. **Synthesis of Nucleosides**: Helmut Vorbrüggen and Carmen Ruh-Pohlenz

Volume 56 (2000)

1. **The Hydroformylation Reaction**: Iwao Ojima, Chung-Ying Tsai, Maria Tzamarioudaki, and Dominique Bonafoux

2. **The Vilsmeier Reaction. 2. Reactions with Compounds Other Than Fully Conjugated Carbocycles and Heterocycles**: Gurnos Jones and Stephen P. Stanforth

AUTHOR INDEX, VOLUMES 1–57

Volume number only is designated in this index.

Adams, Joe T., 8
Adkins, Homer, 8
Ager, David J., 38
Albertson, Noel F., 12
Allen, George R., Jr., 20
Angyal, S. J., 8
Antoulinkis, Evan G., 57
Apparu, Marcel, 29
Archer, S., 14
Arseniyadis, Siméon, 31

Bachmann, W. E., 1, 2
Baer, Donald R., 11
Behr, Lyell C., 6
Behrman, E. J., 35
Bergmann, Ernst D., 10
Berliner, Ernst, 5
Biellmann, Jean-François, 27
Birch, Arthur J., 24
Blatchly, J. M., 19
Blatt, A. H., 1
Blicke, F. F., 1
Block, Eric, 30
Bloom, Steven H., 39
Bloomfield, Jordan J., 15, 23
Bonafoux, Dominique, 56
Boswell, G. A., Jr., 21
Brand, William W., 18
Brewster, James H., 7
Brown, Herbert C., 13
Brown, Weldon G., 6
Bruson, Herman Alexander, 5
Bublitz, Donald E., 17
Buck, Johannes S., 4
Burke, Steven D., 26
Butz, Lewis W., 5

Caine, Drury, 23
Cairns, Theodore L., 20
Carmack, Marvin, 3
Carter, H. E., 3
Cason, James, 4
Castro, Bertrand R., 29

Chamberlin, A. Richard, 39
Chapdelaine, Marc J., 38
Cheng, Chia-Chung, 28
Ciganek, Engelbert, 32, 51
Clark, Robin D., 47
Confalone, Pat N., 36
Cope, Arthur C., 9, 11
Corey, Elias J., 9
Cota, Donald J., 17
Cowden, Cameron J., 51
Crandall, Jack K., 29
Crimmins, Michael T., 44
Crounse, Nathan N., 5

Daub, Guido H., 6
Dave, Vinod, 18
Davies, Huw M. L., 57
Denmark, Scott E., 45
Denny, R. W., 20
DeLucchi, Ottorino, 40
DeTar, DeLos F., 9
Dickhaut, J., 48
Djerassi, Carl, 6
Donaruma, L. Guy, 11
Drake, Nathan L., 1
DuBois, Adrien S., 5
Ducep, Jean-Bernard, 27
Dunoguès, Jacques, 37

Eliel, Ernest L., 7
Emerson, William S., 4
Engel, Robert, 36
England, D. C. 6

Fan, Rulin, 41
Farina, Vittorio, 50
Fieser, Louis F., 1
Fleming, Ian, 37
Folkers, Karl, 6
Fuson, Reynold C., 1

Gadamasetti, Kumar G., 41
Gawley, Robert E., 35

Geissman, T. A., 2
Gensler, Walter J., 6
Giese, B., 48
Gilman, Henry, 6, 8
Ginsburg, David, 10
Göbel, T., 48
Govindachari, Tuticorin R., 6
Grieco, Paul A., 26
Grierson, David, 39
Gschwend, Heinz W., 26
Gutsche, C. David, 8

Habermas, Karl L., 45
Hageman, Howard A., 7
Hamilton, Cliff S., 2
Hamlin, K. E., 9
Hanford, W. E., 3
Harris, Constance M., 17
Harris, J. F., Jr., 13
Harris, Thomas M., 17
Hartung, Walter H., 7
Hassall, C. H., 9
Hauser, Charles R., 1, 8
Hayakawa, Yoshihiro, 29
Heck, Richard F., 27
Heldt, Walter Z., 11
Henne, Albert L., 2
Hoffman, Roger A., 2
Hoiness, Connie M., 20
Holmes, H. L., 4, 9
Houlihan, William J., 16
House, Herbert O., 9
Hudlický, Miloš, 35
Hudlický, Tomáš, 33, 41
Hudson, Boyd E., Jr., 1
Hughes, David L., 42
Huie, E. M., 36
Hulce, Martin, 38
Huyser, Earl S., 13
Hyatt, John A., 45

Idacavage, Michael J., 33
Ide, Walter S., 4
Ingersoll, A. W., 2
Itsuno, Shinichi, 52

Jackson, Ernest L., 2
Jacobs, Thomas L., 5
Jahangir, Alam, 47
Johnson, John R., 1
Johnson, William S., 2, 6
Jones, Gurnos, 15, 49, 56
Jones, Reuben G., 6
Jones, Todd K., 45
Jorgenson, Margaret J., 18

Katsuki, Tsutomu, 48
Kende, Andrew S., 11
Kloetzel, Milton C., 4
Kobayashi, Shū, 46
Kochi, Jay K., 19
Kopping, B., 48
Kornblum, Nathan, 2, 12
Kosolapoff, Gennady M., 6
Kreider, Eunice M., 18
Krimen, L. I., 17
Krishnamurthy, Venkat, 50
Krow, Grant R., 43
Kuhlmann, Heinrich, 40
Kulicke, K. J., 48
Kulka, Marshall, 7
Kutchan, Toni M., 33
Kyler, Keith S., 31

Lane, John F., 3
Leffler, Marlin T., 1
Little, R. Daniel, 47
Lipshutz, Bruce H., 41
Luzzio, Frederick A., 53

McElvain, S. M., 4
McKeever, C. H., 1
McLoughlin, Jim I., 47
McMurry, John E., 24
McOmie, J. F. W., 19
Maercker, Adalbert, 14
Magerlein, Barney J., 5
Málek, Jaroslav, 34, 36
Mallory, Clelia W., 30
Mallory, Frank B., 30
Manske, Richard H. F., 7
Marcinow, Zbigniew, 42
Martin, Elmore L., 1
Martin, Victor S., 48
Martin, William B., 14
Masjedizadeh, Mohammad R., 47
Meijer, Egbert W., 28
Melikyan, Gagik G., 49
Mikami, Koichi, 46
Miller, Joseph A., 32
Miotti, Umberto, 40
Modena, Giorgio, 40
Molander, Gary, 46
Moore, Maurice L., 5
Morgan, Jack F., 2
Moriarty, Robert M., 54, 57
Morton, John W., Jr., 8
Mosettig, Erich, 4, 8
Mozingo, Ralph, 4
Mukaiyama, Teruaki, 28, 46

Nace, Harold R., 12
Nagata, Wataru, 25
Nakai, Takeshi, 46
Naqvi, Saiyid M., 33
Negishi, Ei-Ichi, 33
Nelke, Janice M., 23
Newman, Melvin S., 5
Nickon, A., 20
Nielsen, Arnold T., 16
Noyori, Ryoji, 29

Ohno, Masaji, 37
Ojima, Iwao, 56
Otsuka, Masami, 37
Owsley, Dennis C., 23

Pappo, Raphael, 10
Paquette, Leo A., 25
Parham, William E., 13
Parmerter, Stanley M., 10
Pasto, Daniel J., 40
Paterson, Ian, 51
Pettit, George R., 12
Phadke, Ragini, 7
Phillips, Robert R., 10
Pierini, Adriana B., 54
Pigge, F. Christopher, 51
Pine, Stanley H., 18, 43
Pinnick, Harold W., 38
Porter, H. K., 20
Posner, Gary H., 19, 22
Prakash, Om, 54, 57
Price, Charles C., 3

Rabideau, Peter W., 42
Rabjohn, Norman, 5, 24
Rathke, Michael W., 22
Raulins, N. Rebecca, 22
Raynolds, Peter W., 45
Reed, Josephine W., 41
Reich, Hans J., 44
Reinhold, Tracy L., 44
Rhoads, Sara Jane, 22
Rickborn, Bruce, 52, 53
Rigby, James H., 49, 51
Rinehart, Kenneth L., Jr., 17
Ripka, W. C., 21
Roberts, John D., 12
Rodriguez, Herman R., 26
Roe, Arthur, 5
Rondestvedt, Christian S., Jr., 11, 24
Rossi, Roberto A., 54

Ruh-Pohlenz, Carmen, 55
Rytina, Anton W., 5

Santiago, Ana N., 54
Sauer, John C., 3
Schaefer, John P., 15
Schore, Neil E., 40
Schulenberg, J. W., 14
Schweizer, Edward E., 13
Scott, William J., 50
Scribner, R. M., 21
Semmelhack, Martin F., 19
Sengupta, Saumitra, 41
Sethna, Suresh, 7
Shapiro, Robert H., 23
Sharts, Clay M., 12, 21
Sheehan, John C., 9
Sheldon, Roger A., 19
Sheppard, W. A., 21
Shirley, David A., 8
Shriner, Ralph L., 1
Simmons, Howard E., 20
Simonoff, Robert, 7
Smith, Lee Irvin, 1
Smith, Peter A. S., 3, 11
Smithers, Roger, 37
Spielman, M. A., 3
Spoerri, Paul E., 5
Stacey, F. W., 13
Stanforth, Stephen P., 49, 56
Stetter, Hermann, 40
Struve, W. S., 1
Suter, C. M., 3
Swamer, Frederic W., 8
Swern, Daniel, 7

Tarbell, D. Stanley, 2
Taylor, Richard T., 40
Thoma, G., 48
Tidwell, Thomas T., 39
Todd, David, 4
Touster, Oscar, 7
Trach, F., 48
Truce, William E., 9, 18
Trumbull, Elmer R., 11
Tsai, Chung-Ying, 56
Tullock, C. W., 21
Tzamarioudaki, Maria, 56

van Leusen, Albert M., 57
van Leusen, Daan, 57
van Tamelen, Eugene E., 12
Vedejs, E., 22

Vladuchick, Susan A., 20
Vorbrüggen, Helmut, 55
Wadsworth, William S., Jr., 25
Walling, Cheves, 13
Wallis, Everett S., 3
Wallquist, Olof, 47
Wang, Chia-Lin L., 34
Warnhoff, E. W., 18
Watt, David S., 31
Weston, Arthur W., 3, 9
Whaley, Wilson M., 6
Wilds, A. L., 2
Wiley, Richard H., 6
Williamson, David H., 24

Wilson, C. V., 9
Wilson, Stephen R., 43
Wolf, Donald E., 6
Wolff, Hans, 3
Wollowitz, Susan, 44
Wood, John L., 3
Wynberg, Hans, 28

Yan, Shou-Jen, 28
Yoshioka, Mitsuru, 25

Zaugg, Harold E., 8, 14
Zweifel, George, 13, 32

CHAPTER AND TOPIC INDEX, VOLUMES 1–57

Many chapters contain brief discussions of reactions and comparisons of alternative synthetic methods related to the reaction that is the subject of the chapter. These related reactions and alternative methods are not usually listed in this index. In this index, the volume number is in **boldface**, the chapter number is in ordinary type.

Acetoacetic ester condensation, **1**, 9
Acetylenes, synthesis of, **5**, 1; **23**, 3; **32**, 2
Acid halides:
   reactions with esters, **1**, 9
   reactions with organometallic compounds, **8**, 2
α-Acylamino acid mixed anhydrides, **12**, 4
α-Acylamino acids, azlactonization of, **3**, 5
Acylation:
   of esters with acid chlorides, **1**, 9
   intramolecular, to form cyclic ketones, **2**, 4; **23**, 2
   of ketones to form diketones, **8**, 3
Acyl fluorides, synthesis of, **21**, 1; **34**, 2; **35**, 3
Acyl hypohalites, reactions of, **9**, 5
Acyloins, **4**, 4; **15**, 1; **23**, 2
Alcohols:
   conversion to fluorides, **21**, 1, 2; **34**, 2; **35**, 3
   conversion to olefins, **12**, 2
   oxidation of, **6**, 5; **39**, 3; **53**, 1
   replacement of hydroxy group by nucleophiles, **29**, 1; **42**, 2
   resolution of, **2**, 9
Alcohols, synthesis:
   by base-promoted isomerization of epoxides, **29**, 3
   by hydroboration, **13**, 1
   by hydroxylation of ethylenic compounds, **7**, 7
   from organoboranes, **33**, 1
   by reduction, **6**, 10; **8**, 1
Aldehydes, catalyzed addition to double bonds, **40**, 4
Aldehydes, synthesis of, **4**, 7; **5**, 10; **8**, 4, 5; **9**, 2; **33**, 1
Aldol condensation, **16**
   directed, **28**, 3
   with boron enolates, **51**, 1
Aliphatic fluorides, **2**, 2; **21**, 1, 2; **34**, 2; **35**, 3
Alkenes:
   arylation of, **11**, 3; **24**, 3; **27**, 2
   cyclopropanes from, **20**, 1
   epoxidation and hydroxylation of, **7**, 7
   free-radical additions to, **13**, 3, 4
   hydroboration of, **13**, 1
   hydrogenation with homogeneous catalysts, **24**, 1
   reactions with diazoacetic esters, **18**, 3
   reactions with nitrones, **36**, 1
   reduction by alkoxyaluminum hydrides, **34**, 1
Alkenes, synthesis:
   from amines, **11**, 5
   from aryl and vinyl halides, **27**, 2
   by Bamford–Stevens reaction, **23**, 3
   by Claisen and Cope rearrangements, **22**, 1
   by dehydrocyanation of nitriles, **31**
   by deoxygenation of vicinal diols, **30**, 2
   from α-halosulfones, **25**, 1
   by palladium-catalyzed vinylation, **27**, 2
   from phosphoryl-stabilized anions, **25**, 2
   by pyrolysis of xanthates, **12**, 2
   from silicon-stabilized anions, **38**, 1
   from tosylhydrazones, **23**, 3; **39**, 1
   by Wittig reaction, **14**, 3
Alkene reduction by diimide, **40**, 2
Alkenyl- and alkynylaluminum reagents, **32**, 2
Alkenyllithiums, formation of **39**, 1
Alkoxyaluminum hydride reductions, **34**, 1; **36**, 3
Alkoxyphosphonium cations, nucleophilic displacements on, **29**, 1
Alkylation:
   of allylic and benzylic carbanions, **27**, 1
   with amines and ammonium salts, **7**, 3
   of aromatic compounds, **3**, 1
   of esters and nitriles, **9**, 4
   γ-, of dianions of β-dicarbonyl compounds, **17**, 2
   of metallic acetylides, **5**, 1
   of nitrile-stabilized carbanions, **31**
   with organopalladium complexes, **27**, 2

Alkylidenation by titanium-based reagents, **43**, 1
Alkylidenesuccinic acids, synthesis and reactions of, **6**, 1
Alkylidene triphenylphosphoranes, synthesis and reactions of, **14**, 3
Allenylsilanes, electrophilic substitution reactions of, **37**, 2
Allylic alcohols, synthesis:
 from epoxides, **29**, 3
 by Wittig rearrangement, **46**, 2
Allylic and benzylic carbanions, heteroatom-substituted, **27**, 1
Allylic hydroperoxides, in photooxygenations, **20**, 2
π-Allylnickel complexes, **19**, 2
Allylphenols, synthesis by Claisen rearrangement, **2**, 1; **22**, 1
Allylsilanes, electrophilic substitution reactions of, **37**, 2
Aluminum alkoxides:
 in Meerwein–Ponndorf–Verley reduction, **2**, 5
 in Oppenauer oxidation, **6**, 5
Amide formation by oxime rearrangement, **35**, 1
α-Amidoalkylations at carbon, **14**, 2
Amination:
 of heterocyclic bases by alkali amides, **1**, 4
 of hydroxy compounds by Bucherer reaction, **1**, 5
Amine oxides:
 Polonovski reaction of, **39**, 2
 pyrolysis of, **11**, 5
Amines:
 synthesis from organoboranes, **33**, 1
 synthesis by reductive alkylation, **4**, 3; **5**, 7
 synthesis by Zinin reaction, **20**, 4
 reactions with cyanogen bromide, **7**, 4
α-Aminoalkylation of activated olefins, **51**, 2
Aminophenols from anilines, **35**, 2
Anhydrides of aliphatic dibasic acids, Friedel–Crafts reaction with, **5**, 5
Anion-assisted sigmatropic rearrangements, **43**, 2
Anthracene homologs, synthesis of, **1**, 6
Anti-Markownikoff hydration of alkenes, **13**, 1
π-Arenechromium tricarbonyls, reaction with nitrile-stabilized carbanions, **31**
Arndt–Eistert reaction, **1**, 2
Aromatic aldehydes, synthesis of, **5**, 6; **28**, 1
Aromatic fluorides, synthesis of, **5**, 4

Aromatic hydrocarbons, synthesis of, **1**, 6; **30**, 1
Aromatic substitution by the $S_{RN}1$ reaction, **54**, 1
Arsinic acids, **2**, 10
Arsonic acids, **2**, 10
Arylacetic acids, synthesis of, **1**, 2; **22**, 4
β-Arylacrylic acids, synthesis of, **1**, 8
Arylamines, synthesis and reactions of, **1**, 5
Arylation:
 by aryl halides, **27**, 2
 by diazonium salts, **11**, 3; **24**, 3
 γ-, of dianions of β-dicarbonyl compounds, **17**, 2
 of nitrile-stabilized carbanions, **31**
 of alkenes, **11**, 3; **24**, 3; **27**, 2
Arylglyoxals, condensation with aromatic hydrocarbons, **4**, 5
Arylsulfonic acids, synthesis of, **3**, 4
Aryl thiocyanates, **3**, 6
Asymmetric aldol reactions using boron enolates, **51**, 1
Asymmetric cyclopropanation, **57**, 1
Asymmetric epoxidation, **48**, 1
Atom transfer preparation of radicals, **48**, 2
Azaphenanthrenes, synthesis by photocyclization, **30**, 1
Azides, synthesis and rearrangement of, **3**, 9
Azlactones, **3**, 5

Baeyer–Villiger reaction, **9**, 3; **43**, 3
Bamford–Stevens reaction, **23**, 3
Bart reaction, **2**, 10
Barton fragmentation reaction, **48**, 2
Béchamp reaction, **2**, 10
Beckmann rearrangement, **11**, 1; **35**, 1
Benzils, reduction of, **4**, 5
Benzoin condensation, **4**, 5
Benzoquinones:
 acetoxylation of, **19**, 3
 in Nenitzescu reaction, **20**, 3
 synthesis of, **4**, 6
Benzylic carbanions, **27**, 1
Biaryls, synthesis of, **2**, 6
Bicyclobutanes, from cyclopropenes, **18**, 3
Birch reaction, **23**, 1; **42**, 1
Bischler–Napieralski reaction, **6**, 2
Bis(chloromethyl) ether, **1**, 3; **19**, *warning*
Borane reduction, chiral, **52**, 2
Borohydride reduction, chiral, **52**, 2
Boron enolates, **51**, 1
Boyland–Sims oxidation, **35**, 2
Bucherer reaction, **1**, 5

Cannizzaro reaction, **2**, 3
Carbenes, **13**, 2; **26**, 2; **28**, 1
Carbenoid cyclopropanation reactions, **57**, 1
Carbohydrates, deoxy, synthesis of, **30**, 2
Carbo/metallocupration, **41**, 2
Carbon–carbon bond formation:
 by acetoacetic ester condensation, **1**, 9
 by acyloin condensation, **23**, 2
 by aldol condensation, **16**; **28**, 3; **46**, 1
 by alkylation with amines and ammonium salts, **7**, 3
 by γ-alkylation and arylation, **17**, 2
 by allylic and benzylic carbanions, **27**, 1
 by amidoalkylation, **14**, 2
 by Cannizzaro reaction, **2**, 3
 by Claisen rearrangement, **2**, 1; **22**, 1
 by Cope rearrangement, **22**, 1
 by cyclopropanation reaction, **13**, 2; **20**, 1
 by Darzens condensation, **5**, 10
 by diazonium salt coupling, **10**, 1; **11**, 3; **24**, 3
 by Dieckmann condensation, **15**, 1
 by Diels–Alder reaction, **4**, 1, 2; **5**, 3; **32**, 1
 by free-radical additions to alkenes, **13**, 3
 by Friedel–Crafts reaction, **3**, 1; **5**, 5
 by Knoevenagel condensation, **15**, 2
 by Mannich reaction, **1**, 10; **7**, 3
 by Michael addition, **10**, 3
 by nitrile-stabilized carbanions, **31**
 by organoboranes and organoborates, **33**, 1
 by organocopper reagents, **19**, 1; **38**, 2; **41**, 2
 by organopalladium complexes, **27**, 2
 by organozinc reagents, **20**, 1
 by rearrangement of α-halosulfones, **25**, 1
 by Reformatsky reaction, **1**, 1; **28**, 3
 by trivalent manganese, **49**, 3
 by Vilsmeier reaction, **49**, 1; **56**, 2
 by vinylcyclopropane-cyclopentene rearrangement, **33**, 2
Carbon–halogen bond formation, by replacement of hydroxy groups, **29**, 1
Carbon–heteroatom bond formation:
 by free-radical chain additions to carbon–carbon multiple bonds, **13**, 4
 by organoboranes and organoborates, **33**, 1
Carbon–phosphorus bond formation, **36**, 2
Carbonyl compounds, α,β-unsaturated:
 formation by selenoxide elimination, **44**, 1
 vicinal difunctionalization of, **38**, 2
Carbonyl compounds, from nitro compounds, **38**, 3
 oxidation with hypervalent iodine reagents, **54**, 2
Carboxylic acid derivatives, conversion to fluorides, **21**, 1, 2; **34**, 2; **35**, 3

Carboxylic acids:
 reaction with organolithium reagents, **18**, 1
 synthesis from organoboranes, **33**, 1
Chapman rearrangement, **14**, 1; **18**, 2
Chloromethylation of aromatic compounds, **2**, 3; **9**, *warning*
Cholanthrenes, synthesis of, **1**, 6
Chugaev reaction, **12**, 2
Claisen condensation, **1**, 8
Claisen rearrangement, **2**, 1; **22**, 1
Cleavage:
 of benzyl–oxygen, benzyl–nitrogen, and benzyl–sulfur bonds, **7**, 5
 of carbon–carbon bonds by periodic acid, **2**, 8
 of esters via S_N2-type dealkylation, **24**, 2
 of non-enolizable ketones with sodium amide, **9**, 1
 in sensitized photooxidation, **20**, 2
Clemmensen reduction, **1**, 7; **22**, 3
Collins Reagent, **53**, 1
Condensation:
 acetoacetic ester, **1**, 9
 acyloin, **4**, 4; **23**, 2
 aldol, **16**
 benzoin, **4**, 5
 Claisen, **1**, 8
 Darzens, **5**, 10; **31**
 Dieckmann, **1**, 9; **6**, 9; **15**, 1
 directed aldol, **28**, 3
 Knoevenagel, **1**, 8; **15**, 2
 Stobbe, **6**, 1
 Thorpe–Ziegler, **15**, 1; **31**
Conjugate addition:
 of hydrogen cyanide, **25**, 3
 of organocopper reagents, **19**, 1; **41**, 2
Cope rearrangement, **22**, 1; **41**, 1; **43**, 2
Copper–Grignard complexes, conjugate additions of, **19**, 1; **41**, 2
Corey–Winter reaction, **30**, 2
Coumarins, synthesis of, **7**, 1; **20**, 3
Coupling reaction of organostannanes, **50**, 1
Cuprate reagents, **19**, 1; **38**, 2; **41**, 2
Curtius rearrangement, **3**, 7, 9
Cyanoethylation, **5**, 2
Cyanogen bromide, reactions with tertiary amines, **7**, 4
Cyclic ketones, formation by intramolecular acylation, **2**, 4; **23**, 2
Cyclization:
 of alkyl dihalides, **19**, 2
 of aryl-substituted aliphatic acids, acid chlorides, and anhydrides, **2**, 4; **23**, 2
 of α-carbonyl carbenes and carbenoids, **26**, 2
 cycloheptenones from α-bromoketones, **29**, 2

Cyclization (*Continued*)
of diesters and dinitriles, **15**, 1
Fischer indole, **10**, 2
intramolecular by acylation, **2**, 4
intramolecular by acyloin condensation, **4**, 4
intramolecular by Diels–Alder reaction, **32**, 1
intramolecular by Michael reaction, **47**, 2
Nazarov, **45**, 1
by radical reactions, **48**, 2
of stilbenes, **30**, 1
Cycloaddition reactions:
of cyclenones and quinones, **5**, 3
cyclobutanes, synthesis of, **12**, 1; **44**, 2
Diels–Alder, acetylenes and alkenes, **4**, 2
Diels–Alder, maleic anhydride, **4**, 1
[4 + 3], **51**, 3
of enones, **44**, 2
of ketenes, **45**, 2
of nitrones and alkenes, **36**, 1
Pauson–Khand, **40**, 1
photochemical, **44**, 2
retro Diels–Alder reaction, **52**, 1; **53**, 2
[6 + 4], **49**, 2
Cyclobutanes, synthesis:
from nitrile-stabilized carbanions, **31**
by thermal cycloaddition reactions, **12**, 1
Cycloheptadienes, from:
divinylcyclopropanes, **41**, 1
polyhalo ketones, **29**, 2
π-Cyclopentadienyl transition metal
carbonyls, **17**, 1
Cyclopentenones:
annulation, **45**, 1
synthesis, **40**, 1; **45**, 1
Cyclopropane carboxylates, from diazoacetic
esters, **18**, 3
Cyclopropanes:
from α-diazocarbonyl compounds, **26**, 2
from metal-catalyzed decomposition of
diazo compounds, **57**, 1
from nitrile-stabilized carbanions, **31**
from tosylhydrazones, **23**, 3
from unsaturated compounds, methylene
iodide, and zinc-copper couple, **20**, 1
Cyclopropenes, synthesis of, **18**, 3

Darzens glycidic ester condensation, **5**, 10; **31**
DAST, **34**, 2; **35**, 3
Deamination of aromatic primary amines, **2**, 7
Debenzylation, **7**, 5; **18**, 4
Decarboxylation of acids, **9**, 5; **19**, 4
Dehalogenation of α-haloacyl halides, **3**, 3

Dehydrogenation:
in synthesis of ketenes, **3**, 3
in synthesis of acetylenes, **5**, 1
Demjanov reaction, **11**, 2
Deoxygenation of vicinal diols, **30**, 2
Desoxybenzoins, conversion to benzoins, **4**, 5
Dess-Martin oxidation, **53**, 1
Desulfurization:
of α-(alkylthio)nitriles, **31**
in alkene synthesis, **30**, 2
with Raney nickel, **12**, 5
Diazo compounds, carbenoids derived from,
57, 1
Diazoacetic esters, reactions with alkenes,
alkynes, heterocyclic and aromatic
compounds, **18**, 3; **26**, 2
α-Diazocarbonyl compounds, insertion and
addition reactions, **26**, 2
Diazomethane:
in Arndt–Eistert reaction, **1**, 2
reactions with aldehydes and ketones, **8**, 8
Diazonium fluoroborates, synthesis and
decomposition, **5**, 4
Diazonium salts:
coupling with aliphatic compounds, **10**, 1, 2
in deamination of aromatic primary
amines, **2**, 7
in Meerwein arylation reaction, **11**, 3; **24**, 3
in ring closure reactions, **9**, 7
in synthesis of biaryls and aryl quinones,
2, 6
Dieckmann condensation, **1**, 9; **15**, 1
for synthesis of tetrahydrothiophenes, **6**, 9
Diels–Alder reaction:
intramolecular, **32**, 1
retro–Diels–Alder reaction, **52**, 1; **53**, 2
with alkynyl and alkenyl dienophiles, **4**, 2
with cyclenones and quinones, **5**, 3
with maleic anhydride, **4**, 1
Diimide, **40**, 2
Diketones:
pyrolysis of diaryl, **1**, 6
reduction by acid in organic solvents, **22**, 3
synthesis by acylation of ketones, **8**, 3
synthesis by alkylation of β-diketone
anions, **17**, 2
Dimethyl sulfide, in oxidation reactions, **39**, 3
Dimethyl sulfoxide, in oxidation reactions,
39, 3
Diols:
deoxygenation of, **30**, 2
oxidation of, **2**, 8
Dioxetanes, **20**, 2

Divinyl-aziridines, -cyclopropanes, -oxiranes, and -thiiranes, rearrangements of, **41**, 1
Doebner reaction, **1**, 8

Eastwood reaction, **30**, 2
Elbs reaction, **1**, 6; **35**, 2
Enamines, reaction with quinones, **20**, 3
Ene reaction, in photosensitized oxygenation, **20**, 2
Enolates, in directed aldol reactions, **28**, 3; **46**, 1; **51**, 1
Ene reaction, in photosensitized oxygenation, **20**, 2
Enolates, in directed aldol reactions, **28**, 3; **46**, 1; **51**, 1
Enone cycloadditions, **44**, 2
Enzymatic reduction, **52**, 2
Enzymatic resolution, **37**, 1
Epoxidation:
  of allylic alcohols, **48**, 1
  with organic peracids, **7**, 7
Epoxide isomerizations, **29**, 3
Esters:
  acylation with acid chlorides, **1**, 9
  alkylation of, **9**, 4
  alkylidenation of, **43**, 1
  cleavage via S_N2-type dealkylation, **24**, 2
  dimerization, **23**, 2
  glycidic, synthesis of, **5**, 10
  hydrolysis, catalyzed by pig liver esterase, **37**, 1
  β-hydroxy, synthesis of, **1**, 1; **22**, 4
  β-keto, synthesis of, **15**, 1
  reaction with organolithium reagents, **18**, 1
  reduction of, **8**, 1
  synthesis from diazoacetic esters, **18**, 3
  synthesis by Mitsunobu reaction, **42**, 2
Ethers, synthesis by Mitsunobu reaction, **42**, 2
Exhaustive methylation, Hofmann, **11**, 5

Favorskii rearrangement, **11**, 4
Ferrocenes, **17**, 1
Fischer indole cyclization, **10**, 2
Fluorination of aliphatic compounds, **2**, 2; **21**, 1, 2; **34**, 2; **35**, 3
Fluorination by DAST, **35**, 3
Fluorination by sulfur tetrafluoride, **21**, 1; **34**, 2
Formylation:
  by hydroformulation, **56**, 1
  of alkylphenols, **28**, 1
  of aromatic hydrocarbons, **5**, 6
  of aromatic compounds, **49**, 1
  of nonaromatic compounds, **56**, 2
Free radical additions:
  to alkenes and alkynes to form carbon–heteroatom bonds, **13**, 4
  to alkenes to form carbon-carbon bonds, **13**, 3
Friedel-Crafts catalysts, in nucleoside synthesis, **55**, 1
Friedel–Crafts reaction, **2**, 4; **3**, 1; **5**, 5; **18**, 1
Friedländer synthesis of quinolines, **28**, 2
Fries reaction, **1**, 11

Gattermann aldehyde synthesis, **9**, 2
Gattermann–Koch reaction, **5**, 6
Germanes, addition to alkenes and alkynes, **13**, 4
Glycidic esters, synthesis and reactions of, **5**, 10
Gomberg–Bachmann reaction, **2**, 6; **9**, 7
Grundmann synthesis of aldehydes, **8**, 5

Halides, displacement reactions of, **22**, 2; **27**, 2
Halides, synthesis:
  from alcohols, **34**, 2
  by chloromethylation, **1**, 3
  from organoboranes, **33**, 1
  from primary and secondary alcohols, **29**, 1
Haller–Bauer reaction, **9**, 1
Halocarbenes, synthesis and reactions of, **13**, 2
Halocyclopropanes, reactions of, **13**, 2
Halogen-metal interconversion reactions, **6**, 7
α-Haloketones, rearrangement of, **11**, 4
α-Halosulfones, synthesis and reactions of, **25**, 1
Helicenes, synthesis by photocyclization, **30**, 1
Heterocyclic aromatic systems, lithiation of, **26**, 1
Heterocyclic bases, amination of, **1**, 4
  in nucleosides, **55**, 1
Heterodienophiles, **53**, 2
Hilbert-Johnson method, **55**, 1
Hoesch reaction, **5**, 9
Hofmann elimination reaction, **11**, 5; **18**, 4
Hofmann reaction of amides, **3**, 7, 9
Homogeneous hydrogenation catalysts, **24**, 1
Hunsdiecker reaction, **9**, 5; **19**, 4
Hydration of alkenes, dienes, and alkynes, **13**, 1
Hydrazoic acid, reactions and generation of, **3**, 8

Hydroboration, **13**, 1
Hydrocyanation of conjugated carbonyl compounds, **25**, 3
Hydroformylation, **56**, 1
Hydrogenation catalysts, homogeneous, **24**, 1
Hydrogenation of esters, with copper chromite and Raney nickel, **8**, 1
Hydrohalogenation, **13**, 4
Hydroxyaldehydes, aromatic, **28**, 1
α-Hydroxyalkylation of activated olefins, **51**, 2
α-Hydroxyketones, synthesis of, **23**, 2
Hydroxylation of ethylenic compounds with organic peracids, **7**, 7
Hypervalent iodine reagents, **54**, 2; **57**, 2

Imidates, rearrangement of, **14**, 1
Iminium ions, **39**, 2
Indoles, by Nenitzescu reaction, **20**, 3
  via reaction with TosMIC, **57**, 3
Isocyanides, sulfonylmethyl, reactions of, **57**, 3
Isoquinolines, synthesis of, **6**, 2, 3, 4; **20**, 3

Jacobsen reaction, **1**, 12
Japp–Klingemann reaction, **10**, 2

Katsuki–Sharpless epoxidation, **48**, 1
Ketene cycloadditions, **45**, 2
Ketenes and ketene dimers, synthesis of, **3**, 3; **45**, 2
Ketones:
  acylation of, **8**, 3
  alkylidenation of, **43**, 1
  Baeyer–Villiger oxidation of, **9**, 3; **43**, 3
  cleavage of non-enolizable, **9**, 1
  comparison of synthetic methods, **18**, 1
  conversion to amides, **3**, 8; **11**, 1
  conversion to fluorides, **34**, 2; **35**, 3
  cyclic, synthesis of, **2**, 4; **23**, 2
  cyclization of divinyl ketones, **45**, 1
  synthesis from acid chlorides and organometallic compounds, **8**, 2; **18**, 1
  synthesis from organoboranes, **33**, 1
  synthesis from α,β-unsaturated carbonyl compounds and metals in liquid ammonia, **23**, 1
  reaction with diazomethane, **8**, 8
  reduction to aliphatic compounds, **4**, 8
  reduction by alkoxyaluminum hydrides, **34**, 1
  reduction in anhydrous organic solvents, **22**, 3
  synthesis from organolithium reagents and carboxylic acids, **18**, 1

synthesis by oxidation of alcohols, **6**, 5; **39**, 3
Kindler modification of Willgerodt reaction, **3**, 2
Knoevenagel condensation, **1**, 8; **15**, 2; **57**, 3
Koch–Haaf reaction, **17**, 3
Kornblum oxidation, **39**, 3
Kostaneki synthesis of chromanes, flavones, and isoflavones, **8**, 3

β-Lactams, synthesis of, **9**, 6; **26**, 2
β-Lactones, synthesis and reactions of, **8**, 7
Leuckart reaction, **5**, 7
Lithiation:
  of allylic and benzylic systems, **27**, 1
  by halogen-metal exchange, **6**, 7
  heteroatom facilitated, **26**, 1; **47**, 1
  of heterocyclic and olefinic compounds, **26**, 1
Lithioorganocuprates, **19**, 1; **22**, 2; **41**, 2
Lithium aluminum hydride reductions, **6**, 2
  chirally modified, **52**, 2
Lossen rearrangement, **3**, 7, 9

Mannich reaction, **1**, 10; **7**, 3
Meerwein arylation reaction, **11**, 3; **24**, 3
Meerwein–Ponndorf–Verley reduction, **2**, 5
Mercury hydride method to prepare radicals, **48**, 2
Metalations with organolithium compounds, **8**, 6; **26**, 1; **27**, 1
Methylenation of carbonyl groups, **43**, 1
Methylene-transfer reactions, **18**, 3; **20**, 1
Michael reaction, **10**, 3; **15**, 1, 2; **19**, 1; **20**, 3; **46**, 1; **47**, 2
Mitsunobu reaction, **42**, 2
Moffatt oxidation, **39**, 3; **53**, 1
Morita–Baylis–Hillman reaction, **51**, 2

Nazarov cyclization, **45**, 1
Nef reaction, **38**, 3
Nenitzescu reaction, **20**, 3
Nitriles:
  formation from oximes, **35**, 2
  synthesis from organoboranes, **33**, 1
  α,β-unsaturated:
    by elimination of selenoxides, **44**, 1
Nitrile-stabilized carbanions:
  alkylation and arylation of, **31**
Nitroamines, **20**, 4
Nitro compounds, conversion to carbonyl compounds, **38**, 3
Nitro compounds, synthesis of, **12**, 3
Nitrone-olefin cycloadditions, **36**, 1
Nitrosation, **2**, 6; **7**, 6
Nucleosides, synthesis of, **55**, 1

Olefins, hydroformylation of, **56**, 1
Oligomerization of 1,3-dienes, **19**, 2
Oppenauer oxidation, **6**, 5
Organoboranes:
  formation of carbon–carbon and carbon–heteroatom bonds from, **33**, 1
  isomerization and oxidation of, **13**, 1
  reaction with anions of α-chloronitriles, **31**
Organohypervalent iodine reagents, **54**, 2; **57**, 2
Organometallic compounds:
  of aluminum, **25**, 3
  of copper, **19**, 1; **22**, 2; **38**, 2; **41**, 2
  of lithium, **6**, 7; **8**, 6; **18**, 1; **27**, 1
  of magnesium, zinc, and cadmium, **8**, 2;
  of palladium, **27**, 2
  of tin, **50**, 1
  of zinc, **1**, 1; **20**, 1; **22**, 4
Oxidation:
  of alcohols and polyhydroxy compounds, **6**, 5; **39**, 3; **53**, 1
  of aldehydes and ketones, Baeyer–Villiger reaction, **9**, 3; **43**, 3
  with hypervalent iodine reagents, **54**, 2
  of amines, phenols, aminophenols, diamines, hydroquinones, and halophenols, **4**, 6; **35**, 2
  of α-glycols, α-amino alcohols, and polyhydroxy compounds by periodic acid, **2**, 8
  of organoboranes, **13**, 1
  of phenolic compounds, **57**, 2
  with peracids, **7**, 7
  by photooxygenation, **20**, 2
  with selenium dioxide, **5**, 8; **24**, 4
Oxidative decarboxylation, **19**, 4
Oximes, formation by nitrosation, **7**, 6
Oxochromium(VI)-amine complexes, **53**, 1
Oxo process, **56**, 1

Palladium-catalyzed vinylic substitution, **27**, 2
Palladium-catalyzed coupling of organostannane, **50**, 1
Pauson–Khand reaction to prepare cyclopentenones, **40**, 1
Pechmann reaction, **7**, 1
Peptides, synthesis of, **3**, 5; **12**, 4
Peracids, epoxidation and hydroxylation with, **7**, 7
  in Baeyer–Villiger oxidation, **9**, 3; **43**, 3
Periodic acid oxidation, **2**, 8
Perkin reaction, **1**, 8
Persulfate oxidation, **35**, 2
Peterson olefination, **38**, 1

Phenanthrenes, synthesis by photocyclization, **30**, 1
Phenols, dihydric from phenols, **35**, 2
  oxidation of **57**, 2
Phosphinic acids, synthesis of, **6**, 6
Phosphonic acids, synthesis of, **6**, 6
Phosphonium salts:
  halide synthesis, use in, **29**, 1
  synthesis and reactions of, **14**, 3
Phosphorus compounds, addition to carbonyl group, **6**, 6; **14**, 3; **25**, 2; **36**, 2
  addition reactions at imine carbon, **36**, 2
Phosphoryl-stabilized anions, **25**, 2
Photochemical cycloadditions, **44**, 2
Photocyclization of stilbenes, **30**, 1
Photooxygenation of olefins, **20**, 2
Photosensitizers, **20**, 2
Pictet–Spengler reaction, **6**, 3
Pig liver esterase, **37**, 1
Polonovski reaction, **39**, 2
Polyalkylbenzenes, in Jacobsen reaction, **1**, 12
Polycyclic aromatic compounds, synthesis by photocyclization of stilbenes, **30**, 1
Polyhalo ketones, reductive dehalogenation of, **29**, 2
Pomeranz–Fritsch reaction, **6**, 4
Prévost reaction, **9**, 5
Pschorr synthesis, **2**, 6; **9**, 7
Pummerer reaction, **40**, 3
Pyrazolines, intermediates in diazoacetic ester reactions, **18**, 3
Pyridinium chlorochromate, **53**, 1
Pyrolysis:
  of amine oxides, phosphates, and acyl derivatives, **11**, 5
  of ketones and diketones, **1**, 6
  for synthesis of ketenes, **3**, 3
  of xanthates, **12**, 2

Quaternary ammonium salts, rearrangements of, **18**, 4
Quinolines, synthesis of:
  by Friedländer synthesis, **28**, 2
  by Skraup synthesis, **7**, 2
Quinones:
  acetoxylation of, **19**, 3
  diene additions to, **5**, 3
  synthesis of, **4**, 6
  in synthesis of 5-hydroxyindoles, **20**, 3

Radical formation and cyclization, **48**, 2
Ramberg–Bäcklund rearrangement, **25**, 1

Rearrangements:
  anion-assisted sigmatropic, **43**, 2
  Beckmann, **11**, 1; **35**, 1
  Chapman, **14**, 1; **18**, 2
  Claisen, **2**, 1; **22**, 1
  Cope, **22**, 1; **41**, 1, **43**, 2
  Curtius, **3**, 7, 9
  divinylcyclopropane, **41**, 1
  Favorskii, **11**, 4
  Lossen, **3**, 7, 9
  Ramberg–Bäcklund, **25**, 1
  Smiles, **18**, 2
  Sommelet–Hauser, **18**, 4
  Stevens, 18, 4
  [2,3] Wittig, **46**, 2
  vinylcyclopropane-cyclopentene, **33**, 2
Reduction:
  of acid chlorides to aldehydes, **4**, 7; **8**, 5
  of aromatic compounds, **42**, 1
  of benzils, **4**, 5
  of ketones, enantioselective, **52**, 2
  Clemmensen, **1**, 7; **22**, 3
  desulfurization, **12**, 5
  with diimide, **40**, 2
  by dissolving metal, **42**, 1
  by homogeneous hydrogenation catalysts, **24**, 1
  by hydrogenation of esters with copper chromite and Raney nickel, **8**, 1
  hydrogenolysis of benzyl groups, **7**, 5
  by lithium aluminum hydride, **6**, 10
  by Meerwein–Ponndorf–Verley reaction, **2**, 5
  chiral, **52**, 2
  by metal alkoxyaluminum hydrides, **34**, 1; **36**, 3
  of mono- and polynitroarenes, **20**, 4
  of olefins by diimide, **40**, 2
  of α,β-unsaturated carbonyl compounds, **23**, 1
  by samarium(II) iodide, **46**, 3
  by Wolff–Kishner reaction, **4**, 8
Reductive alkylation, synthesis of amines, **4**, 3; **5**, 7
Reductive cyanation, **57**, 3
Reductive desulfurization of thiol esters, **8**, 5
Reformatsky reaction, **1**, 1; **22**, 4
Reimer–Tiemann reaction, **13**, 2; **28**, 1
Resolution of alcohols, **2**, 9
Retro Diels–Alder reactions, **52**, 1; **53**, 2
Ritter reaction, **17**, 3
Rosenmund reaction for synthesis of arsonic acids, **2**, 10
Rosenmund reduction, **4**, 7

Samarium(II) iodide, **46**, 3
Sandmeyer reaction, **2**, 7
Schiemann reaction, **5**, 4
Schmidt reaction, **3**, 8, 9
Selenium dioxide oxidation, **5**, 8; **24**, 4
Seleno–Pummerer reaction, **40**, 3
Selenoxide elimination, **44**, 1
Shapiro reaction, **23**, 3; **39**, 1
Silanes:
  addition to olefins and acetylenes, **13**, 4
  electrophilic substitution reactions, **37**, 2
Sila–Pummerer reaction, **40**, 3
Silyl carbanions, **38**, 1
Simmons–Smith reaction, **20**, 1
Simonini reaction, **9**, 5
Singlet oxygen, **20**, 2
Skraup synthesis, **7**, 2; **28**, 2
Smiles rearrangement, **18**, 2
Sommelet–Hauser rearrangement, **18**, 4
Sommelet reaction, **8**, 4
$S_{RN}1$ reactions of aromatic systems, **54**, 1
Stevens rearrangement, **18**, 4
Stetter reaction of aldehydes with olefins, **40**, 4
Stilbenes, photocyclization of, **30**, 1
Stille reaction, **50**, 1
Stobbe condensation, **6**, 1
Substitution reactions using organocopper reagents, **22**, 2; **41**, 2
Sulfide reduction of nitroarenes, **20**, 4
Sulfonation of aromatic hydrocarbons and aryl halides, **3**, 4
Swern oxidation, **39**, 3; **53**, 1

Tetrahydroisoquinolines, synthesis of, **6**, 3
Tetrahydrothiophenes, synthesis of, **6**, 9
Thiazoles, synthesis of, **6**, 8
Thiele–Winter acetoxylation of quinones, **19**, 3
Thiocarbonates, synthesis of, **17**, 3
Thiocyanation of aromatic amines, phenols, and polynuclear hydrocarbons, **3**, 6
Thiophenes, synthesis of, **6**, 9
Thorpe–Ziegler condensation, **15**, 1; **31**
Tiemann reaction, **3**, 9
Tiffeneau–Demjanov reaction, **11**, 2
Tin(II) enolates, **46**, 1
Tin hydride method to prepare radicals, **48**, 2
Tipson–Cohen reaction, **30**, 2
Tosylhydrazones, **23**, 3; **39**, 1
Tosylmethyl isocyanide (TosMIC), **57**, 3

Ullman reaction:
    in synthesis of diphenylamines, **14**, 1
    in synthesis of unsymmetrical biaryls, **2**, 6
Unsaturated compounds, synthesis
    with alkenyl- and alkynylaluminum
    reagents, **32**, 2

Vilsmeier reaction, **49**, 1; **56**, 2
Vinylcyclopropanes, rearrangement to
    cyclopentenes, **33**, 2
Vinyllithiums, from sulfonylhydrazones, **39**, 1
Vinylsilanes, electrophilic substitution
    reactions of, **37**, 2
Vinyl substitution, catalyzed by palladium
    complexes, **27**, 2
von Braun cyanogen bromide reaction, **7**, 4
Vorbrüggen reaction, **55**, 1

Willgerodt reaction, **3**, 2
Wittig reaction, **14**, 3; **31**
[2,3]-Wittig rearrangement, **46**, 2
Wolff–Kishner reaction, **4**, 8

Xanthates, synthesis and pyrolysis of, **12**, 2

Ylides:
    in Stevens rearrangement, **18**, 4
    in Wittig reaction, structure and properties,
    14, 3

Zinc–copper couple, **20**, 1
Zinin reduction of nitroarenes, **20**, 4